REINFORCED MASONRY ENGINEERING HANDBOOK

CLAY and CONCRETE MASONRY

JAMES E. AMRHEIN
Civil & Structural Engineer

Fifth Edition

This handbook was prepared in keeping with the most current information and practice for the present state of the art of masonry design, engineering and construction. This handbook expresses the opinion of the author, and care has been taken to ensure that all data and information furnished are as accurate as possible. Although he has endeavored to supplement this data by conferences with experts and obtained qualified advice, the author and the publisher cannot assume or accept any responsibility or liability, including liability for negligence, for errors or oversights in this data and information and in the use of such information or in the preparation of engineering plans.

Published by
Masonry Institute of America
2550 Beverly Boulevard
Los Angeles, California 90057

Printed in the United States of America

PREFACE

It was recognized several years ago that a comprehensive reinforced masonry engineering design handbook was needed which would encompass the coefficients, tables, charts and design data required for the design of reinforced masonry structures. The author, recognizing this need, tried to fulfill these requirements with the first edition of this publication. Since then, the subsequent editions have been improved and expanded to comply with applicable editions of the Uniform Building Code, and keep pace with the growth of reinforced masonry engineering.

It is the desire of the author that this book be as useful as possible to designers of reinforced masonry in eliminating repetitious and routine calculations. This publication will reduce the time required for masonry design.

The allowable stresses plus the detail and design requirements included in this book are based upon the 1991 edition of the Uniform Building Code published by the International Conference of Building Officials. Also included in this edition is information and design tables based on the ACI 530-88/ASCE 5-88 Masonry Code and Specifications.

In addition to the code requirements, sound engineering practice has been included in this publication to serve as a guide to the engineer and designer using it.

It is realized that there may be several design and analysis methods and the results for the design can be somewhat different. The techniques included in this publication have been reviewed by competent engineers who have found the results to be satisfactory and safe. The author welcomes recommendations for the extension and improvement of the material and any new design techniques for future editions.

THE MASONRY INSTITUTE OF AMERICA

The Masonry Institute of America, founded in 1957 under the name Masonry Research, is a professional promotion, technical and research organization established to improve and extend the uses of masonry. It is supported by the mason contractors through a labor management contract between the unions and contractors. The Masonry Institute of America is active in Los Angeles County, California in promoting new ideas and masonry work, improving building codes, conducting research, presenting design, construction and inspection seminars, and writing technical and non-technical papers, all for the purpose of improving the masonry industry. The Masonry Institute of America does not compete with architects or engineers in design or in bidding, contracting or supervision of masonry construction.

BIOGRAPHY OF THE AUTHOR

JAMES E. AMRHEIN, Executive Director of the Masonry Institute of America, has more than 40 years experience in construction, engineering, technical promotion, teaching, structural design and earthquake engineering. He was a project engineer with Stone & Webster Engineering Corporation in Boston, Massachusetts, Supervising Structural Engineer for the Portland Cement Association in Los Angeles, and has been active in seismic design and research, including the investigation of the structural performance of buildings subjected to earthquakes throughout the world. His B.C.E. was earned at Manhattan College followed by a M.S.C.E. from Columbia University in New York City. He was elected to Tau Beta Pi and Chi Epsilon honorary engineering societies.

In 1983 Mr. Amrhein received the Outstanding Engineer Merit Award from the Institute for the Advancement of Engineering and the Steven B. Barnes Award from the Structural Engineers Association of Southern California for his contributions in the field of masonry research and education. He also received the Distinguished Service Award from the Western States Clay Products Association.

Mr. Amrhein is a Registered Civil, Structural and Quality Engineer in California and a Licensed Professional Engineer in New York. He is a Fellow in the American Society of Civil Engineers and the American Concrete Institute, and a member of numerous other professional organizations including the International Conference of Building Officials, the Structural Engineers Association of Southern California, and the Earthquake Engineering Research Institute. He is a founding member and past president of The Masonry Society.

Mr. Amrhein is a Navy veteran who served overseas during World War II and the Korean Incident with the Seabees. From 1961 to 1980 he was on the evening Civil Engineering faculty of California State University at Long Beach, as an adjunct (full) professor. He has presented masonry design seminars for the American Society of Civil Engineers in their continuing education program and has lectured at many universities throughout the United States and around the world.

ACKNOWLEDGMENTS

I gratefully acknowledge the recommendations and suggestions of the professionals who helped improve and prepare this publication.

I am particularly appreciative to Chester Schultz and Ralph McLean of the office McLean & Schultz for their assistance with the first edition. John Arias and Phil Kim of that office were very helpful with their contributions.

My thanks to Edward M. McDermott, who spent many hours assisting me in make up and expediting the publication of the first several editions. His constant support helped with this book.

Other professionals who assisted were Joseph Oddo, Juan Giron, Steve Tanikawa, and Rulon Fronk.

Engineers, Michael Yevtovich and Amir Aryan, were cooperative in preparing preliminary designs on some of the projects which were used.

For the Fifth edition, which is a complete re-write and update from the 4th Edition, Phillip J. Samblanet, Professional Engineer and Staff Engineer of MIA was very thorough in reviewing and preparing much of the material. His continued constructive comments, suggestions, and input have helped to improve this publication.

Drawings were made by Thomas Escobar, a dedicated draftsman and future architect. He is attending California State Polytechnic University Pomona.

Arnold Bookbinder S.E. was helpful in providing some connection details.

Mark Amrhein, Civil and Geotech Engineer provided information on earth loads. Don Lee S.E. plotted many of the design curves.

I am pleased to acknowledge the work of Petra Stadtfeld, who retained a cheerful and very cooperative attitude no matter how many changes were made. Amy K. Hall proof read and aided in editing.

David Willis set the type and was cooperative in meeting our changes and gave continuous assistance as we improved our texts, tables, and diagrams.

I appreciate the support of the Board of Trustees of the Masonry Institute of America, of Rennie Tejeda, Chairman, George Angevine, John Chrysler, Robert Hatch, Douglas Williams and R.E. Williams, who have given full cooperation to see that this publication has always been a success.

SPECIAL RECOGNITION

The author appreciates all those who contributed to this book, especially those authors whose publications served as a resource for this document.

Portions of the 1991 Uniform Building Code, copyright © 1991, and the 1991 Building Code Standards, copyright © 1991, are reproduced, herein, with permission of the publisher, the International Conference of Building Officials.

Likewise, portions of the text and tables were reproduced from the following publications with their organization's permission:

ACI 530-88/ASCE 5-88 Building Code Requirement for Masonry Structures.

ACI 530.1-88/ASCE 6-88 Specifications for Masonry Structures.

Recommended Lateral Force Requirements and Commentary, 1990 Edition, by the Structural Engineer's Association of California.

Permission was also received from the American Society for Testing and Materials, the Portland Cement Association, the Brick Institute of America and the National Concrete Masonry Association to use information and data from their many excellent publications.

SYMBOLS AND NOTATIONS

a = depth of equivalent rectangular stress block for strength design.

= ground acceleration.

a_b = depth of stress block of member for strength design as determined by UBC Chapter 24, Equation 12-10.

A = area of floor or roof supported by a member.

= cross sectional area of a member.

A_l = loaded area.

= bearing area, square inches.

A_b = cross-sectional area of anchor bolt, square inches.

A_e = effective area of masonry, square inches.

A_f = area of flange of intersecting wall.

A_g = gross cross-sectional area of masonry section, square inches.

A_{jh} = total area of special horizontal shear reinforcement in a masonry frame equal to $0.5V_{jh}/f_{yh}$.

A_{mv} = net area of masonry section bounded by wall thickness and length of section in the direction of shear force considered, square inches.

A_n = net area of masonry, square inches.

A_p = area of tension (pullout) cone of an embedded anchor bolt projected onto the surface of masonry, square inches.

A_s = effective cross-sectional area of reinforcement in a column or flexural member, square inches.

A_{se} = effective area of steel for slender wall design, square inches.

A_{st} = total area of longitudinal reinforcement in columns.

A'_s = effective cross-sectional area of compression reinforcement in a flexural member, square inches.

A_{tr} = total cross-sectional area of transverse reinforcement (stirrup or tie) within a spacing s and perpendicular to plane of bars being spliced or developed, square inches.

$a_u = \phi f_y(1 - 0.59q)$. Coefficient for computing steel area A_s.

A_v = area of steel required for shear reinforcement perpendicular to the longitudinal reinforcement, square inches.

A_x = the torsional amplification factor at Level x.

ACI = American Concrete Institute.

ANSI = American National Standards Institute.

ASCE = American Society of Civil Engineers.

ASTM = American Society for Testing and Materials.

Avg. = average.

b = effective width of rectangular member or width of flange for T and I sections, inches.

= column dimension, inches.

b' = width of web in T and I members.

b_t = computed tension force on anchor bolts, pounds.

b_v = computed shear force on anchor bolts, pounds.

B_a = allowable tension force on anchor bolts for the ACI/ASCE code, pounds.

B_t = allowable tension force on anchor bolts, pounds.

B_v = allowable shear force on anchor bolts, pounds.

BTU = British Thermal Units.

c = coefficient that determines the distance to the neutral axis in a beam in strength design.

= distance from the neutral axis to extreme fiber.

= total compression force, pounds, kips.

= numerical coefficient specified in UBC Section 2334(b)1.

C_d = masonry shear strength coefficient as obtained from UBC Table No. 24-0.

C_e = combined height, exposure and gust factor as given in Table 3-6 (UBC Table No. 23-G).

= snow exposure factor as given in UBC Table No. A-23-S.

C_f = compression on the flange.

Ch. = Chapter.

cm = Centimetre.

Comp. = compressive.

C_p = numerical coefficient specified in UBC Section 2336 and given in UBC Table No. 23-P (shown partially in Table 3-15).

C_q = pressure coefficient for the structure or portion of the structure under consideration as given in Table 3-7 (UBC Table No. 23-H).

CM = center of mass.

CMU = Concrete Masonry Unit.

CR = center of rigidity.

C_s = slope reduction factor given in UBC Appendix Chapter 23, Division I, Sec. 2341.

C_t = numerical coefficient given in UBC Section 2334(b)2.

C_w = compression on the web.

cu. = cubic.

d = distance from the compression face of a flexural member to the centroid of longitudinal tensile reinforcement, inches.

d_b = diameter of the reinforcing bar, inches.

d_{bb} = diameter of largest beam longitudinal reinforcing bar passing through or anchored in, the joint inches.

d_{bp} = diameter of largest pier longitudinal reinforcing bar passing through the joint, inches.

d_1 or d' = distance from compression face of a flexural member to the centroid of longitudinal compressive reinforcement.

d_x = distance in x direction from center of rigidity to shear wall.

d_y = distance in y direction from center of rigidity to shear wall.

D = nominal diameter of reinforcing bar, inches.

= dimension of a building in direction parallel to the applied force.

= dead loads, or related internal moments and forces.

D_i = inside diameter, inches.

D_o = outside diameter, inches.

DL = dead load.

D_s = the plan dimension of the building of the vertical lateral force resisting system.

e = eccentricity measured from the vertical axis of a section to the load.

= eccentricity of P_{uf}.

e' = eccentricity measured from tensile steel axis to the load.

e_k = eccentricity to kern point.

e_m = strain in masonry.

e_{mu} = maximum useable compressive strain of masonry.

e_s = strain in steel.

e_x = eccentricity in x direction of center of mass to center of rigidity.

e_y = eccentricity in y direction of center of mass to center of rigidity.

E = load effects of earthquake, or related internal moments and forces.

= East.

E' = eccentricity measured from tensile steel axis to the load, ft.

EST = Equivalent Solid Thickness.

E_g = modulus of elasticity of grout in compression.

E_c = modulus of elasticity of concrete in compression, $33w^{1.5}f_c'$ psi.

E_m = modulus of elasticity of masonry in compression, psi. In the UBC, $E_m = 750 f_m'$.

Eq = equation.

E_s = modulus of elasticity of steel = 29,000,000 psi.

$ET_{1.1}$ $ET_{1.2}$ $ET_{1.4}$ = specified equivalent thickness for items 5-1.1, 5-1.2, and 5-1.4 of UBC Table No. 43-B.

E_v = modulus of rigidity (shear modulus) of masonry in shear, $0.4E_m$, psi.

E.F.P. = equivalent fluid pressure of lateral earth loads.

f_a = computed axial compressive stress due to design axial load, psi.

f_b = computed flexural stress in the extreme fiber due to design bending loads only, psi.

f_c = concrete compressive stress in extreme fiber in flexure, psi.

f_{ct} = average splitting tensile strength of lightweight aggregate concrete, psi.

f_c' = specified compressive strength of concrete, psi.

f_g = compressive strength of grout, psi.

f_g' = specified compressive strength of grout, psi.

f_i = lateral force at Level i for use in Formula (34-5).

f_m = actual compressive masonry stress from combined flexural and axial loading, $f_m = f_a + f_b$, psi.

f'_m = specified compressive strength of masonry at the age of 28 days, psi.

f'_{mu} = ultimate compressive strength of the masonry, psi.

f_{md} = computed compressive strength in masonry due to dead load only.

f_r = modulus of rupture, psi.

f_s = computed stress in reinforcement due to design loads, psi.

f'_s = stress in compressive reinforcement in flexural members, psi.

f_{sb} = soil bearing pressure, psf.

f_y = specified yield strength of reinforcement, psi, ksi.

f_t = flexural tensile stress in masonry, psi.

ft = feet.

ft kips = foot kips, moment.

ft lbs = foot pounds, moment.

f_v = computed shear stress due to design load, psi.

f_y = tensile yield stress of reinforcement, psi.

f_{yh} = tensile yield stress of horizontal reinforcment, psi.

F = dimensional coefficient equal to M/K or $bd^2/1200$ and used in the determination of resisting moment of masonry sections.

= loads due to weight and pressure of fluids or related moments and forces.

F_a = allowable average axial compressive stress for centroidally applied axial load only, psi.

F_b = allowable flexural compressive stress if members were carrying bending load only, psi.

F_{br} = allowable bearing stress, psi.

F_i, F_n, F_x = lateral force applied to level i, n or x respectively.

F_p = lateral forces on the part of the structure.

F.R. = frictional sliding resistance.

F_s = allowable stress in reinforcement, psi.

F_{sc} = allowable compressive stress in column reinforcement, psi.

FST = face shell thickness of hollow masonry units, inches.

F_{su} = ultimate tensile stress of steel, psi.

F_t = that portion of the base shear, V, considered concentrated at the top of the structure in addition to F_n.

= allowable flexural tensile stress in masonry, psi.

F_v = allowable shear stress in masonry, psi.

g = acceleration due to gravity.

= gram.

gal = gallons.

G = shear modulus (modulus of rigidity) of the masonry, $0.4E_m$, psi.

h = height of a wall or column between points of support, feet, inches.

= hour.

h_b = beam depth in a masonry frame equal to $1800 d_{bp}/f'^{0.5}_g$.

h'_c = pier depth in the plane of the frame, inches.

= beam depth, inches.

h' = effective height or length of column or wall, feet, inches.

h_i, h_n, h_x = height in feet above the base to Level i, n or x respectively.

h_p = pier depth in a masonry frame equal to $4800 d_{bb}/f'^{0.5}_g$.

H = loads due to weight and pressure of soil, water in soil or related internal moments and forces.

= height of block or brick using specified dimensions as defined in UBC Chapter 24, in inches.

Hz = Hertz, cycles per second.

i.e. = for example.

in. = inches.

in. lbs = inch pounds, moment.

I = moment of inertia about the neutral axis of the cross-sectional area, in.^{4}

= importance factor as given in Table 3-5 (UBC Table No. 23-L) or, for snow loads, as given in UBC Table No. A-23-T).

= impact loads or related internal moments and forces.

ICBO = International Conference of Building Officials.

I_g, I_{cr} = gross, cracked moment of inertia of the wall cross section, in.^{4}

i = interval.

I = importance factor.

IRA = initial rate of absorption.

j = ratio or distance between centroid of flexural compressive forces and centroid of tensile forces to depth, d.

jd = moment arm.

j_w = moment arm coefficient for web.

k = the ratio of depth of the compressive stress in a flexural member to the depth, d.

= kip, 1000 pounds.

= kilo, 1000.

k_h = coefficients for lateral earth pressure of backfill against a cantilever retaining wall.

kip = 1000 pounds.

K_{hr} = coefficient for lateral earth pressure of backfill against a retaining wall supported at top.

km = kilometers.

kN = kilonewtons.

kPa = kilopascals.

k_v = coefficient for vertical earth pressure of backfill against a cantilever retaining wall.

K = $^1/_2 f_b jk$ for flexural computations, psi.

= $f_s p j$ for flexural computations, psi.

K_a = active (Rankine) earth pressure coefficient.

K_b = flexural coefficient for balanced design conditions.

K_u = flexural coefficient for strength design equal to M_u / bd^2.

kg = kilogram.

l, L = length of a wall or segment, feet, inches.

= lap splice length;

= embedment length.

= length of block or brick using specified dimensions as defined in UBC Chapter 24.

l' = length of the compression area.

l_b = embedment depth of anchor bolts, inch.

l_{be} = anchor bolt edge distance, the least length measured from the edge of masonry to the surface of the anchor bolt, inches.

lbs = pounds.

l_d = required development length of the reinforcement.

l_{db} = basic development length, inches.

L = live loads, or related internal moments and forces.

LL = live load.

level i = level of structure referred to by the subscript i. "$i = 1$" designates the first level above the base.

level n = that level which is uppermost in the main portion of the structure.

level x = that level which is under design consideration. "$x = 1$" designates the first level above the base.

lin. = linear.

M = design moment.

= mass of a structure.

= mega, 1,000,000.

max. = maximum.

M_B = overturning moment at the base of the building or structure.

M_c = moment capacity of compression steel in a flexural member about the centroid of the tensile force.

M_{cr} = cracking moment strength of the masonry wall.

MG = Megagram.

min. = minimum.

M_m = the moment of the compressive force in the masonry about the centroid of the tensile force in the reinforcement.

M_n = nominal moment strength of the masonry wall.

M_{OT} = overturning moment.

MPa = Megapascals.

M_R = resisting moment.

M_s = the moment of the tensile force in the reinforcement about the centroid of the compressive force in masonry.

M_{ser} = service moment at the midheight of the panel, including $P\Delta$ effects.

M_T = torsional moment.

M_u = factored moment.

M_x = the overturning moment at level x.

m = metre.

= milli, one thousandth, 0.001.

mm = millimetre.

mph = miles per hour.

M.M. = Modified Mercali Intensity Scale.

n = ratio of modulus of elasticity of steel (E_s) to that of masonry (E_m) or concrete (E_c). For masonry the modular ratio, n, is equal to E_s/E_m.

N = Newton, force.

= North.

= number of bars in a layer being spliced or developed at a critical section.

NA = neutral axis.

No. = number.

OTM = overturning moment.

o.c. = on center.

p = ratio of the area of flexural tensile reinforcement, A_s, to the area (bd).

p' = ratio of area of compressive reinforcement to the effective area of masonry (bd).

p_b = reinforcement ratio producing balanced design conditions.

p_g = ratio of the area of vertical reinforcement to the gross area, A_g.

p_n = ratio of area of shear reinforcement to masonry area, A_{mv}.

= ratio of distributed shear reinforcement on a plane perpendicular to plane or A_{mv}.

P = design axial load, pounds, kips.

= design wind pressure, pounds per square foot.

P_a = allowable centroidal axial load for reinforced masonry columns.

= force from the active soil pressure.

Pa = Pascals.

P_b = nominal balanced design axial strength.

P_{br} = bearing load.

pcf = pounds per cubic foot, unit weight.

P_f = minimum roof snow load, pounds per square foot.

= load from tributary floor or roof area.

P_g = basic ground snow load, pounds per square foot.

plf = pounds per linear foot.

P_m = compressive capacity of the masonry only in a tied column, pounds.

P_o = nominal axial load strength without bending, pounds.

P_p = passive soil pressure.

P_s = Compressive capacity of the reinforcing steel only in a tied masonry column, pounds.

psf = pounds per square foot.

psi = pounds per square inch.

P_u = factored axial load, pounds.

P_{uf} = factored load from tributary floor or roof loads, pounds.

P_{uw} = factored weight of the wall tributary to the section under consideration, pounds.

P_w = unfactored weight of the wall tributary to section under consideration.

q = ratio coefficient for strength design = $p\left(f_y/f'_m\right)$.

q_s = surcharge load.

= wind stagnation pressure, psf.

= wind stagnation pressure at the standard height of 33 feet as set forth in Table 3-3 (UBC Table No. 23-F), pounds per square foot.

r = radius of gyration.

= rate of reduction equal to 0.08 percent for floors. See Table 3-2 (UBC Table No. 23-C) for roofs.

R = h'/t reduction factor for walls and columns.

= reduction in percent.

= support reaction, pounds, kips.

= the resultant force from the weight of soil and the frictional resistance.

r_b = ratio of the are of bars cut off to the total area of bars at the section.

R_C = coefficient or rigidity for cantilever piers or walls.

R_{cx} = rigidity of cantilever wall in x direction.

R_{cy} = rigidity of cantilever wall in y direction.

R_F = coefficient of rigidity for fixed piers or walls.

R_s = snow load in pounds per square foot per degree of pitch over 20 degrees.

R_w = structural framing coefficient for seismic design numerical coefficient given in UBC Tables Nos. 23-0 and 23-Q (the portion of UBC Table No. 23-O relating to masonry structures is shown in Table 3-14 of this book).

R_x = rigidity of wall in x direction.

R_y = rigidity of wall in y direction.

s = spacing of stirrups or bent bars in the direction parallel to that of the main reinforcement.

= section modulus, in.^{3}

= site coefficient for soil characteristics given in Table 3-13 (UBC Table No. 23-J).

= total snow load, pounds per square foot.

= South.

S = snow load, psf.

= site coefficient, soils characteristics and site geology.

S_a = acceleration spectra.

Sec. = Section.

SI = International Systems of Measurements as adopted by the General Conference of Weights and Measures.

sq. in. = square inches.

sq. ft = square feet.

STC = sound transmission coefficient.

STR. DES.= strength design.

t = thickness of wall or column, inches.

t' = effective thickness of a wythe, wall or column, inches.

t_p = least actual lateral dimension of a prism, inches.

T = tension force, pounds.

= effects of temperature, creep, shrinkage and differential settlement.

= fundamental period of vibration, in seconds, of the structure in the direction under consideration.

T_E = equivalent thickness, inches.

T_{eq} = equivalent tension force.

TL = total load.

UBC = Uniform Building Code.

u = bond stress per unit of surface area of bar.

U = required strength to resist factored loads, or related internal moments and forces.

v = shear stress, psi.

v' = shear stress taken by shear reinforcement, psi.

v_c = allowable shear stress for concrete, psi.

v_m = allowable shear stress for masonry, psi.

V = total design shear force, pounds.

= the total design lateral load or shear at the base.

= basic wind speed, miles per hour.

$V_{1.1}$
$V_{1.2}$ = volume of aggregates expressed as a percentage of the total aggregate volume for Items 5-1.1, 5-1.2, and 5-1.4 of UBC Table
$V_{1.4}$ 43-B.

V_c = nominal shear strength provided by the masonry.

V_{jh} = total horizontal joint shear in a masonry frame.

V_{jv} = vertical force acting on joint core.

V_m = nominal shear strength provided by masonry, pounds.

V_n = nominal shear strength, pounds.

= net volume (gross volume less volume of voids), cubic inches.

V_s = nominal shear strength provided by shear reinforcement, pounds.

V_u = factored shear force at section.

V_x = the design story shear in Story x.

w = uniformly distributed load.

= width of beam, wall, or column, inches.

w_b = width of beam in a masonry frame, inches.

w_i, w_x = that portion of W which is located at or is assigned to level i or x respectively.

w_s = unit weight of the soil, pounds per cubic foot.

w_u = factored distributed lateral load.

W = wind load, or related internal moments in forces.

= the total seismic dead load defined in UBC Section 2334(a).

= weight of soil wedge.

= West.

W_a = Actual width of masonry unit, inches.

W_p = the weight of an element or component.

= the weight of a part or a portion of a structure.

w_{px} = the weight of the diaphragm and the elements tributary thereto at Level x, including applicable portions of other loads defined in Section 2334(a).

WSD = working stress design.

WT = equivalent web thickness of hollow masonry units, inches.

Wt = weight, pounds, kips.

x_{CR} = distance from y axis to center of rigidity.

y_{CR} = distance from x axis to center of rigidity.

y = distance from centroidal axis of the section to centroid of area considered.

z = ratio of distance (zkd) between extreme fiber and resultant of compressive forces to distance kd.

Z = seismic zone factor given in Table 3-11 (UBC Table No. 23-I).

β = angle of the backfill slope from a horizontal level plane.

γ_i = horizontal displacement at Level i.

γ_s = unit weight of soil, pounds per cubic foot.

δ = angle of the wall friction to a horizontal level plane.

δ_i, δ_n = deflections at levels i and n respectively, relative to the base, due to applied lateral forces.

Δ = deflection of element.

Δ_C = coefficient of deflection for cantilever piers or walls.

Δ_F = coefficient of deflection for fixed piers or walls.

ΔL = unrestrained expansion, inches.

= change in length.

Δ_m = deflection due to moment.

Δ_s = the midheight deflection limitation for slender walls under service lateral and vertical loads, inches.

ΔT = change in temperature.

Δ_v = deflection due to shear.

Δ_u = horizontal deflection at midheight under factored load; $P\Delta$ effect must be included in the deflection calculation.

μ = coefficient of sliding friction.

Σ_o = sum of perimeters of all the longitudinal reinforcement.

ϕ = angle of internal friction; angle of shearing resistance in Coulomb's equation, degrees.

= strength reduction factor.

$°C$ = degrees Celcius.

$°F$ = degrees Fahrenheit.

$\%$ = percent.

$\#$ = number.

Table of Contents

SECTION 6, DESIGN OF STRUCTURAL MEMBERS — STRENGTH DESIGN

SECTION 10, FORMULAS FOR REINFORCED MASONRY DESIGN

SECTION 11, DESIGN OF ONE-STORY INDUSTRIAL BUILDING

SECTION 12, DESIGN OF SEVEN STORY MASONRY LOAD BEARING APARTMENT BUILDING

SECTION 13, RETAINING WALLS

SECTION 14, EXPLANATION OF TABLES AND DIAGRAMS

WORKING STRESS DESIGN TABLES AND DIAGRAMS

xxi

STRENGTH DESIGN TABLES AND DIAGRAMS

ACI/ASCE TABLES

DEDICATION

This book is dedicated to my wife of many years, Laurette Amrhein. Laurette has always been very supportive of my work and the time required to accomplish these tasks.

REINFORCED MASONRY ENGINEERING HANDBOOK

CLAY and CONCRETE MASONRY

JAMES E. AMRHEIN
Civil & Structural Engineer

FROM THE CODE OF HAMMURABI (2200 B.C.)

If a builder builds a house for a man and does not make its construction firm and the house collapses and causes the death of the owner of the house — that builder shall be put to death. If it causes the death of a son of that owner — they shall put to death the son of that builder. If it causes the death of a slave of the owner — he shall give to the owner a slave of equal value.

If it destroys property — he shall restore whatever it destroyed and because he did not make the house firm he shall rebuild the house which collapsed at his own expense. If a builder builds a house and does not make its construction meet the requirements and a wall falls in — that builder shall strengthen the wall at his own expense.

REINFORCED MASONRY ENGINEERING HANDBOOK

> "...They said to one another, 'Come, let us make bricks and bake them.' They used bricks for stone and bitumen for mortar. Then they said, 'Let us build ourselves a city and a tower with its top in the heavens.'"
>
> *from the Old Testament of the Holy Bible, Book of Genesis, Chapter XI, Verses 3 and 4.*

INTRODUCTION

Masonry structures have been constructed since the earliest days of mankind, not only for homes but also for works of beauty and grandeur. Stone was the first masonry unit and was used for primitive but breathtaking structures such as the 4000 year old Stonehenge ring on England's Salisbury Plains. Stone was also used around 2500 B.C. to build the Egyptian pyramids in Giza. Limestone veneer which once clad the pyramids can now only be seen at the top of the great pyramid, Cheops, since much of the limestone facing was later removed and re-used.

The 1500 mile Great Wall of China was constructed of brick and stone between 200 B.C. to 220 A.D.

Egyptian Pyramids located in Giza were constructed around 2500 B.C. Note limestone veneer at the top of the great pyramid, Cheops.

As with the Egyptian Pyramids, numerous other structures such as the 1500 mile long Great Wall of China testify to the durability of masonry.

Additionally, structures such as the 1500 year old stone pyramids of Yucatan, Mexico, demonstrate the skill of ancient masons. In fact, the stone walls at Machu Picchu in Peru have masonry unit joints so tight that it is difficult to insert a knife blade between masonry units.

The stone walls at Machu Picchu in Peru were built between 1200 and 1400 A.D.

Masonry has been used worldwide to construct impressive structures, such as St. Basil's Cathedral in Moscow, the Taj Mahal in Agra, India, as well as homes, churches, bridges, and roads.

Built between 1631 and 1653, the Taj Mahal's grandeur is in its symmetry, and skillful marble work.

The outer walls of St. Basil's Cathedral in Moscow, were built in 1492, while the remainder of this impressive cathedral was constructed in the 17th century.

In the United States, masonry was used from Boston to Los Angeles and has been the primary material for building construction from the 18th to the 20th centuries.

Built in 1891, the 16 story brick Monadnock Building in Chicago is still in use today.

In the early 1900's concrete block masonry units were introduced to the construction industry. Later, between 1930 and 1940 reinforcing steel began to be embedded in masonry construction to provide increased resistance to lateral dynamic forces from earthquakes.

Prior to the development of reinforced masonry, most masonry structures were designed only to support gravity loads, and the forces from wind and earthquakes were ignored. Massive dead loads from the thick and heavy walls stabilized the unreinforced structures against lateral forces.

The introduction of reinforced masonry allowed wall thicknesses to be decreased dramatically and provided a rational method to design walls to resist dynamic lateral loads from winds and earthquakes.

An excellent example of the benefits of reinforced masonry is the 13-story Pasadena Hilton Hotel in California, completed in 1971. The load bearing, high strength concrete block walls are 12" CMU for the bottom three floors and 8" CMU for the upper 10 floors.

13 Story Pasadena Hilton Hotel. Completed in 1971.

The Pasadena Hilton, like the newer 16 story Queens Surf in Long Beach, California and the 19 story Holiday Inn in Burbank, California is located in one of the severest seismic areas in the world.

Currently, the tallest reinforced masonry structure is the 28 story Excalibur Hotel in Las Vegas, Nevada. This large high-rise complex consists of four buildings each containing 1008 sleeping rooms. The load bearing walls for the complex required masonry with a specified compressive strength of 4,000 psi.

Tallest concrete masonry building in the world 28-story Excalibur Hotel, Las Vegas, Nevada.

Although taller masonry buildings may someday be constructed, it is of more importance that the benefits of reinforced masonry are appropriate not only for multi-story buildings, but for buildings of every size and type, even single story dwellings.

BASIS OF DESIGN

The basis of design for masonry structures described in this publication are the requirements found in the *Uniform Building Code,* (UBC) published by the International Conference of Building Officials, and, to a lesser extent, the requirements of the *Building Code Requirements for Masonry Structures* (ACI 530-88/ASCE 5-88) and the *Specifications for Masonry Structures* (ACI 530.1-88/ASCE 6-88). The allowable stresses for masonry and reinforcing steel, dead loads, live loads and lateral forces as prescribed by the UBC are used primarily herein, although ACI/ASCE allowable stresses equations are given as well, in Section 10. However, there are many regions of the United States that utilize other codes or modification, or opt for different editions of the UBC. The use of the arbitrary requirements and the limitations of allowable stresses and other parameters may be waived in certain jurisdictions.

Similar to past editions, numerous tables and diagrams have been provided at the end of this book to facilitate the design of masonry structures. Additional tables have been included to simplify strength design procedures and the ACI/ASCE design methods, while some of the seldom used old tables were deleted. Note, however, to avoid confusion, the table and diagram numbers were kept the same to be consistent with past editions — thus some gaps exist in the table numbering.

Section 14 has been included to provide explanation for the tables and diagrams. Additionally, numerous example problems are provided throughout the book, which demonstrate these tables and diagrams. Cross references have also been included at the top of most tables and diagrams to direct the reader to appropriate examples.

Included in this publication is information, tables and some design charts that conform to the requirements of the *Building Code Requirements for Masonry Structures* (ACI 530-88/ASCE 5-88) and the *Specifications for Masonry Structures* (ACI 530.1-88/ASCE 6-88).

As an engineer and designer, one should not get lost in the precision of the numbers listed in the design tables of this handbook, and lose sight of the fact that loads for which the structures are designed are arbitrary and in many cases significantly different than the actual loads.

Judgment in design and detailing which insures both safety and economy is the mark of a professional engineer.

SECTION 1

Materials

The four principal materials used in reinforced masonry are the masonry units, mortar, grout and reinforcing steel. These materials are assembled into a homogeneous structural system.

1-1 MASONRY UNITS

The masonry units considered in this publication are clay brick, concrete brick, hollow clay bricks and hollow concrete blocks. However, the structural principles given in this publication apply to all types of masonry by using the appropriate allowable stress values.

Masonry units are available in a variety of sizes, shapes, colors, and textures. Always check with the local manufacturer or supplier for the properties and availability of the desired units.

1-1.A Clay Masonry

Clay masonry is manufactured to comply with the ASTM C 62; *Specification for Building Brick (Solid Masonry Units Made From Clay or Shale)*, C 216; *Specification for Facing Brick (Solid Masonry Units Made From Clay or Shale)* and C 652; *Specification for Hollow Brick (Hollow Masonry Units Made From Clay or Shale)* or UBC Standard No. 24-1; *Building Brick, Facing Brick and Hollow Brick (Made From Clay or Shale)*. It is made by firing clay in a kiln for 40 to 150 hours depending upon the type of kiln, size and volume of the units and other variables. The clay is fired at a fusing temperature between 1600°F to 2700°F, depending on the type of clay. For building brick and face brick the temperature is controlled between 1600°F and 2200°F, while the temperature ranges between 2400°F and 2700°F for fire brick.

Clays, unlike metals, soften slowly and fuse gradually when subjected to elevated temperatures. This softening property of clay allows it to harden into a solid and durable unit when properly fired.

Fusing takes place in three stages:

1) Incipient fusion— occurs when the clay particles become sufficiently soft causing the mass to stick together

2) Vitrification— characterized by extensive fluxing as the mass densifies and solidifies

Independence Hall in Philadelphia, Pennsylvania, constructed in 1730 of fired brick.

3) Viscous fusion— the point at which the clay mass begins to break down and becomes molten.

The key to the firing process is to control the temperature in the kiln so that incipient fusion is complete, and partial vitrification occurs but viscous fusion is avoided.

After the temperature reaches the maximum and is maintained for a prescribed time, the cooling process begins. Usually 48 to 72 hours are required for proper cooling in periodic kilns, but in tunnel kilns the cooling period seldom exceeds 48 hours. The rate of cooling has a direct effect on color and the finished quality. Additionally, excessively rapid cooling causes cracking and checking of the units, and therefore must be controlled closely.

Clays shrink during both drying and firing; therefore, allowances must be made in the size of the finished product. Both drying shrinkage and firing shrinkage vary for different clays, usually falling within the following ranges:

Drying Shrinkage 2 to 8 percent

Firing Shrinkage   2.5 to 10 percent

Firing shrinkage increases with higher temperatures which, in turn, produce darker shades. Consequently, when a wide range of colors is desired, some variation between the final sizes of the dark and light units is inevitable.

To obtain products of uniform size, manufacturers attempt to control factors contributing to shrinkage. However, because of variations in the raw materials and because of temperature variations within kilns, absolute uniformity is unattainable. Specifications for brick include permissible size variations.

Clay units are manufactured in accordance with the prescribed standards of the American Society for Testing and Materials (ASTM). They are classified into solid units and hollow units.

1-1.A.1 Solid Clay Units

A solid clay masonry unit, as specified in ASTM C 62 and C 216, is a unit whose net cross-sectional area, in every plane parallel to the bearing surface, is 75% or more of its gross cross-sectional area measured in the same plane. A solid brick may have a maximum coring of 25%.

No void   *Voids 25% or less*
                   of cross-sectional area

Figure 1-1 Solid clay brick.

Solid clay units are specified in ASTM C 62, ASTM C 216 and in UBC Standard 24-1.

Building bricks are classified as solid masonry units used where appearance is not a consideration. ASTM C 62 includes three grades of building brick which relate the physical requirements to the durability of a brick unit.

Facing bricks are solid masonry units used where the appearance of the units is a consideration. Limits on chippage and cracks, as well as tolerances on the dimensions and distortions of facing brick are included in ASTM C 216. This standard covers two grades of facing brick based on their resistance to weathering.

The recommended uses, physical requirements and grade requirements are the same as for Grades SW and MW under ASTM C 62.

1-1.A.1(a) Grades of Building and Facing Bricks

Bricks are graded according to their weathering resistance.

The effect of weathering on a brick is related to the weathering index which, for any locality, is the product of the average annual number of freezing cycle days and the average annual winter rainfall in inches. Grade requirements for face exposures are listed in Table 1-1 and are described below. The physical requirements for each grade included in ASTM C 62 and C 216 are given in Table 1-3 except that facing brick is classified only into Grades SW and MW.

TABLE 1-1   Grade Requirements for Face Exposures[1]

Exposure	Weathering Index		
	Less than 50	50 to 500	500 and greater
In vertical surfaces:			
In contact with earth	MW	SW	SW
Not in contact with earth	MW	SW	SW
In other than vertical surfaces:			
In contact with earth	SW	SW	SW
Not in contact with earth	MW	SW	SW

1. Table 1 of ASTM C 62 and C 216.

GRADE SW (Severe Weathering) Bricks are intended for use where a high and uniform degree of resistance to frost action and disintegration by weathering is desired and the exposure is such that the brick may freeze when permeated with water.

GRADE MW (Moderate Weathering) bricks are used where they will be exposed to temperatures below freezing, but unlikely to be permeated with water, and where a moderate and somewhat non-uniform degree of resistance to frost action is permissible.

GRADE NW (Negligible Weathering) applies to building brick only and is intended for use in backup or interior masonry.

1-1.A.1(b) Types of Facing Bricks

Included in ASTM C 216 are three types of face brick based upon factors affecting the appearance of the finished wall. These types of face bricks are described as follows:

TYPE FBS (Face Brick Standard) brick is for general use in exposed masonry construction. Most bricks are manufactured to meet the requirement of Type FBS.

TYPE FBX (Face Brick Extra) brick is for general use in exposed masonry construction where a higher degree of precision and a lower permissible variation in size than that permitted for Type FBS brick is required.

TYPE FBA (Face Brick Architectural) brick is manufactured and selected to produce characteristic architectural effects resulting from non-uniformity in size and texture of the individual units.

1-1.A.1(c) Solid Clay Brick Sizes

There are no standard solid clay brick sizes and therefore it is always necessary to check with the brick manufacturer or supplier for the actual brick dimensions. As a guide some typical brick sizes are shown below:

	Height	Width	Length
STANDARD BUILDING BRICK:	$2\frac{1}{2}''\times$	$3\frac{1}{2}''\times$	$7\frac{1}{2}''$
OVERSIZE BUILDING BRICK:	$3\frac{1}{2}''\times$	$3''$	$\times 9\frac{1}{2}''$
MODULAR BUILDING BRICK:	$3\frac{3}{8}''\times$	$3''$	$\times 11\frac{1}{2}''$
COMMON BUILDING BRICK:	$2\frac{1}{2}''\times$	$3\frac{1}{8}''\times$	$8\frac{1}{4}''$
STANDARD FACE BRICK:	$2\frac{1}{4}''\times$	$3\frac{3}{4}''\times$	$8''$
MODULAR FACE BRICK:	$3\frac{1}{2}''\times$	$3''$	$\times 11\frac{1}{2}''$
OVERSIZE FACE BRICK:	$3\frac{1}{2}''\times$	$3''$	$\times 9\frac{1}{2}''$

1-1.A.2 Hollow Clay Units

A hollow clay masonry unit as specified in UBC Standard 24-1 and ASTM C 652, is a unit whose net cross-sectional area in every plane parallel to the bearing surface is less than 75% of its gross cross-sectional area measured in the same plane. Hollow clay units are classified by Grade, Type and Class as outlined below from the above Standards.

Solid shell	Double shell	Cored shell
hollow	hollow	hollow
brick units	brick units	brick units

Figure 1-2 Hollow clay brick.

1-1.A.2(a) Grades of Hollow Brick

Two grades of hollow brick are covered: Grade SW and Grade MW. These grades are similar to the grades for solid brick.

1-1.A.2(b) Types of Hollow Brick

Four types of hollow brick are covered in ASTM C 652.

TYPE HBS (Hollow Brick Standard) is for general use in exposed exterior and interior masonry walls and partitions where a wider color range and a greater variation in size than is permitted for Type HBX hollow brick.

TYPE HBX (Hollow Brick Extra) is for general use in exposed exterior and interior masonry walls and partitions where a high degree of mechanical perfection, a narrow color range, and a minimal variation in size is required.

TYPE HBA (Hollow Brick Architectural) is manufactured and selected to produce characteristic architectural effects resulting from nonuniformity in size, color and texture of the individual units.

TYPE HBB (Hollow Brick Basic) is for general use in masonry walls and partitions where color and texture are not a consideration, and where a greater variation in size is permitted than is required by Type HBX hollow brick.

1-1.A.2(c) Classes of Hollow Brick

Two classes of hollow brick are covered in ASTM C 652:

Class H4OV — Hollow brick intended for use where void areas or hollow spaces are between 25% to 40% of the gross cross-sectional area of the unit measured in any plane parallel to the bearing surface are desired.

Class H60V — Hollow brick intended for use where larger void areas are desired than allowed for class H40V brick. The sum of the void areas for class H60V must be greater than 40%, but not greater than 60%, of the gross cross-sectional area of the unit measured in any plane parallel to the bearing surface. The void spaces, the web thicknesses, and the shell thicknesses must comply with the minimum requirements contained in Table 1-2.

TABLE 1-2 Class H60V - Hollow Brick Minimum Thickness of Face Shells and Webs[1]

Nominal Width of Units (in.)	Face Shell Thicknesses (in.) Solid	Cored or Double Shell	End Shells or End Webs (in.)
3 and 4	$^3/_4$	...	$^3/_4$
6	1	$1^1/_2$	1
8	$1^1/_4$	$1^1/_2$	1
10	$1^3/_8$	$1^5/_8$	$1^1/_8$
12	$1^1/_2$	2	$1^1/_8$

1. ASTM C 652, Table 1.

1-1.A.2(d) Sizes of Hollow Brick

Hollow clay brick, like solid brick, are available in a variety of sizes but are customarily manufactured in nominal 4, 6 or 8 inch thicknesses. Actual thicknesses, however, are about $1/_2$ inch less than the nominal thicknesses (i.e., a 6″ nominal hollow brick is actually about $5^1/_2$″ thick.)

1-1.A.3 Physical Requirements of Clay Masonry Units

1-1.A.3(a) General

The physical requirements for each grade of solid and hollow brick are compressive strength, water absorption and the saturation coefficient as shown in Table 1-3.

TABLE 1-3 Physical Requirements, Solid and Hollow Bricks[1]

Desig-nation	Minimum Compressive Strength (Brick Flatwise),psi Gross Area		Maximum Water Absorption by 5 Hour Boiling Per Cent		Maximum Saturation Coefficient[2]	
	Aver-age of 5 Bricks	Indi-vidual	Aver-age of 5 Bricks	Indi-vidual	Aver-age of 5 Bricks	Indi-vidual
Grade SW	3000	2500	17.0	20.0	0.78	0.80
Grade MW	2500	2200	22.0	25.0	0.88	0.90
Grade NW[3]	1500	1250	no limit	no limit	no limit	no limit

1. Based on ASTM C 62, C 216 or C 652.

2. The saturation coefficient or C/B Ratio, is the ratio of absorption by 24-hour submersion in cold water to that after 5-hour submersion in boiling water.

3. Does not apply for C 216 and C 652.

1-1.A.3(b) Water Absorption and Saturation Coefficient

The water absorption rate and saturation coefficient (known as the C/B ratio) are indications of the freeze-thaw resistance of a brick. Their values for Grade SW brick and Grade MW brick, indicate that there are more voids or pores in Grade SW units which allows water to expand as it changes state into ice.

1-1.A.3(c) Tolerances

Table 1-4 shows the allowable tolerances for face brick per ASTM C 216 and for hollow clay brick as per ASTM C 652. Dimensional tolerances for building brick conforming to ASTM C 62 are the same as for Type FBS. For tolerances on distortion see ASTM C 216 and C 652.

TABLE 1-4 Dimensional Tolerances[1]

Specified Dimension (inches)	Maximum Permissible Variation From Specific Dimensions, Plus or Minus (inches)	
	Type FBX; HBX	Type FBS; HBS & HBB
3 and under	$^1/_{16}$	$^3/_{32}$
Over 3 to 4, incl.	$^3/_{32}$	$^1/_8$
Over 4 to 6, incl.	$^1/_8$	$^3/_{16}$
Over 6 to 8, incl.	$^5/_{32}$	$^4/_{16}$
Over 8 to 12, incl.	$^7/_{32}$	$^5/_{16}$
Over 12 to 16, incl.	$^9/_{32}$	$^3/_8$

1. Based on ASTM C 216 and C 652. Building brick conforming to ASTM C 62 are required to meet the dimensional toloerance of Type FBS.

ASTM C 67 *Test Methods of Sampling and Testing Brick and Structural Clay Tile* outlines the methods for measuring water absorption and the saturation coefficient.

The saturation coefficient, commonly called the C/B (Cold/Boiling) ratio, is the percent absorption of the twenty-four hour cold water test divided by the percent absorption of the five-hour boiling test.

The C/B ratio is based on the concept that only a portion of the pores will be filled during the cold water test, and that all the pores which can possibly be filled, will be filled during the boiling test.

1-1.A.3(d) Initial Rate of Absorption, I.R.A.

The initial rate of absorption (suction) of a brick has an important effect on the bond between the brick and the mortar. It is defined as the amount of water in grams per minute absorbed by 30 square inches of brick in one minute. Maximum bond strength occurs when the suction of the brick at the time of laying is between 5 and 20 grams of water per 30 sq.in. of brick when the surface area is immersed in $1/8''$ of water for one minute.

Note that there is no consistent relationship between total absorption and suction or I.R.A. Some bricks with high absorption have low suction (I.R.A.) and vice versa. Since the suction of the brick while being laid is of primary importance and since suction can be controlled at the job site by wetting, it is not controlled by product specifications.

Dry bricks and bricks with high suction rates tend to absorb large quantities of water from the mortar which often results in poor bond adhesion. Therefore, it is sometimes advisable to wet the dry bricks a few hours prior to laying them so their cores are moist while their surface is dry. Bricks in this condition, with a dry surface and wet core, are preferred since they tend to bond well with the mortar. Note that very wet or saturated bricks should be avoided since they may not bond well to the mortar. These saturated bricks tend to move easily and not stay in position (float), thus making bricklaying extremely difficult and slow.

To check the internal moisture condition of a brick, the bricklayer or inspector should occasionally break a brick and observe its interior dampness condition.

Brick products and properties often vary significantly depending on the clay type and the manufacturer. It is always advisable to consult with the local brick manufacturer for specific information on the intended brick for a project.

1-1.B Concrete Masonry

Concrete masonry units for load bearing systems may be either concrete brick as specified by ASTM C 55, *Specification for Concrete Building Brick* or UBC Standard No. 24-3, *Concrete Building Brick* or hollow load bearing concrete masonry units as specified by ASTM C 90, *Specification for Hollow Load-Bearing Concrete Masonry Units* or UBC Standard 24-4, *Hollow and Solid Load-Bearing Concrete Masonry Units.*

Concrete brick and hollow units are primarily made from portland cement, water and suitable aggregates with or without the inclusion of other materials.

Concrete brick and hollow units may be made from lightweight or normal weight aggregates or both.

1-1.B.1 Concrete Brick

Concrete brick is classified by grade and type.

1-1.B.1(a) Grade of Concrete Brick

Grade N concrete bricks are for use as architec-

tural veneer and facing units in exterior walls and for use where high strength and resistance to moisture penetration and severe frost action is desired.

Grade S concrete bricks are for general use where moderate strength and resistance to frost action and moisture penetration is required.

1-1.B.1(b) Types of Concrete Brick

Type I, Moisture-controlled concrete brick must conform to the requirements of Table 1-5.

Type II, Non-moisture-controlled units need not meet the water absorption requirements of Table 1-5.

1-1.B.1(c) Physical Properties Requirements

The strength and absorption requirements for concrete brick are given in Table 1-5.

TABLE 1-5 Strength and Absorption Requirements[1]

Compressive Strength, Min., psi (Concrete Brick Tested Flatwise)		Water Absorption, Max., (Avg of 3 brick) With Oven dry weight of concrete Lb/Ft^3		
Average Gross Area		Weight Classification		
Grade	Avg. of 3 Concrete Brick	Individual Concrete Brick	Lightweight Less than 105	Medium Weight Less than 125 to 105	Normal Weight 125 or More
N-I	3500	3000	15	13	10
N-II	3500	3000	15	13	10
S-I	2500	2000	18	15	13
S-II	2500	2000	18	15	13

1. UBC Standard Table No. 24-3-B or ASTM C 55, Table 2.

1-1.B.2 Hollow Load Bearing Concrete Masonry Units.

UBC Standard 24-4 and its reference standard, ASTM C 90-86, classify concrete masonry units according to grade and type as hereafter described.

ASTM C 90-90 *Specification for Load Bearing Concrete Masonry Units* combines the requirements for hollow load bearing units (C 90) and for solid load bearing units (C 145). It also deletes grade classifications and requires all load bearing concrete masonry units to meet the requirements of the old Grade N designation.

1-1.B.2(a) Grades of Hollow Concrete Units

Grade N are units having a weight classification

of 85 pcf or greater. These units are suitable for use in exterior walls both below and above grade which may or may not be exposed to moisture penetration or the weather. They may also be used for interior walls and backup wythes.

Grade S are units having a weight classification of less than 85 pcf and are limited to above grade installation in either exterior walls with weather-protective coatings or in walls not exposed to the weather (see footnote 1, Table 1-6).

1-1.B.2(b) Types of Hollow Concrete Units

Similar to concrete brick, hollow concrete units are classified into Types I and II as shown in Table 1-6.

1-1.B.2(c) Physical Properties Requirements

UBC Standard No. 24-4 requires concrete masonry units to meet the strength and moisture absorption requirements listed in Table 1-6.

The water absorption requirements are based on three weight classifications for hollow concrete masonry units:

(1) normal weight units over 125 pcf when dry.

(2) medium weight units ranging between 105 and 125 pcf when dry.

(3) lightweight units weigh between 85 to 105 pcf.

TABLE 1-6 Strength and Absorption Requirements[2]

Compressive Strength, Min., psi		Water absorption, Max., (Avg of 3 Units) With Oven-Dry Weight of Concrete Lb/Cu. Ft			
Average Net Area		Weight Classification			
Grade	Avg. of 3 Units	Individual unit	Light-weight		Medium weight less than 125 to 105	Normal weight 125 or more
			Less than 85	Less than 105		
N-I	1900	1500	—	18	15	13
N-II	1900	1500	—	18	15	13
S-I[1]	1300	1100	20	—	—	—
S-II[1]	1300	1100	20	—	—	—

1. Limited to use above grade in exterior walls with weather-protective coatings and in walls not exposed to the weather.

NOTE: To prevent water penetration, protective coating should be applied on the exterior face of the basement walls and when required on the face of exterior walls above grade.

2. UBC Standard Table No. 24-4-B.

1-1.B.2(d) Categories of Hollow Concrete Units

There are two categories of hollow concrete masonry units:

Precision Units require that no overall dimension (width, height and length) differ by more than $\frac{1}{8}$ inch from the specified standard dimensions.

Particular Feature Units have dimensions specified in accordance with the following (local suppliers should be consulted to determine achievable dimensional tolerances):

(1) For molded face units, no overall dimension may vary by more than $\frac{1}{8}$ inch from the specified standard dimension. Dimensions of molded features (ribs, scores, hex-shapes, pattern, etc.) must be within $\frac{1}{16}$ inch of the specified standard dimensions and must be within $\frac{1}{16}$ inch of the specified placement on the unit.

(2) For split-faced units, all non-split overall dimensions may differ by no more than $\frac{1}{8}$ inch from the specified standard dimensions. On faces that are split, overall dimensions will vary.

(3) For slumped units, no overall height dimension may differ by more than $\frac{1}{8}$ inch from the specified standard dimension. On faces that are slumped, overall dimensions will vary.

1-1.B.2(e) Sizes of Hollow Concrete Masonry Units

Concrete blocks have customarily been manufactured in modular nominal dimensions which are multiples of 8″ (i.e., standard block are nominally 8″ high by 16″ long). The actual block dimensions, however, are typically $\frac{3}{8}″$ less than the nominal dimensions to account for a standard thickness mortar joint. Accordingly, a precision $8 \times 8 \times 16$ nominal block is actually $7\frac{5}{8}″ \times 7\frac{5}{8}″ \times 15\frac{5}{8}″$ (See Figure 1-3).

Slumped block units are equal to the standard manufacturer's dimensions plus $\frac{1}{2}$ inch to account for the thicker mortar joints used with these irregular units. Note also that the nominal dimensions of non-modular size units usually exceed the standard dimensions by $\frac{1}{8}$ to $\frac{1}{4}$ inch.

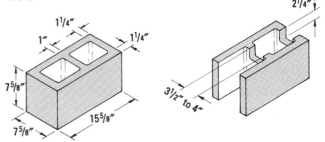

2 cell CMU   Open end bond beam CMU

Figure 1-3 Typical nominal 8″ hollow concrete masonry units.

Face-shell thicknesses (FST) and web thicknesses (WT) concrete masonry units are required to conform to the values listed in Table No. 1-7.

Special unit designs (often called face shell units or expandable units) involving corrosion-resistant metal ties between face shells may be approved.

TABLE 1-7 Minimum Thickness of Face-Shells and Webs[5]

Nominal Width, In.	Actual Width, W_A In.	Face-Shell Thickness (FST) Min. In.[1,4]	Webs[1] Min, In.	Web Thickness (WT) Equivalent Web Thickness, Min in./Lin Ft[2]
4	$3^5/_8$	$^3/_4$	$^3/_4$	$1^5/_8$
6	$5^5/_8$	1	1	$2^1/_4$
8	$7^5/_8$	$1^1/_4$	1	$2^1/_4$
10	$9^5/_8$	$1^3/_8$ / $1^1/_4$ [3]	$1^1/_8$	$2^1/_2$
12	$11^5/_8$	$1^1/_2$ / $1^1/_4$ [3]	$1^1/_8$	$2^1/_2$

1. Average of measurements on three units taken at the thinnest point. UBC Standard No. 24-7.

2. Sum of the measured thickness of all webs in the unit, multiplied by 12, and divided by the length of the unit. In the case of open-ended units where the open-ended portion is solid grouted, the length of that open-ended portion shall be deducted from the overall length of the unit.

3. This face-shell thickness (FST) is applicable where allowable design load is reduced in proportion to the reduction in thicknesses shown, except that allowable design load on solid-grouted units shall not be reduced.

4. For split-faced units, a maximum of 10 percent of a shipment may have face-shell thicknesses less than those shown, but in no case less than $^3/_4$ inch.

5. Based on UBC Standard Table No. 24-4-C and ASTM C 90, Table 2.

1-1.B.3 Moisture Content for Concrete Brick and Hollow Masonry Units

The primary concept of moisture-controlled units is to limit the shrinkage of the concrete block due to moisture loss. In arid areas, more moisture, if available, will leave the unit, and thus more shrinkage will occur. To limit this shrinkage in arid areas, a lower allowable moisture content is required than in humid areas. An example of this would be the use of concrete block in the dry desert area of Las Vegas, Nevada, compared to the rainy, moist climate of Seattle, Washington. In Las Vegas, concrete block will readily dry out, while in Seattle it may never be completely dry.

The selection of a Type I, moisture-controlled unit, or Type II, non-moisture-controlled unit, is dependent on the location, use, manufacturing conditions, construction methods and design details of the concrete masonry wall.

For moisture controlled units, Type I, the relative percentage moisture in the unit when delivered to the job site should conform to Table 1-8.

TABLE 1-8 Moisture Content Requirements for Type I Concrete Units[4]

Moisture Content, Max. Percent of Total Absorption (Average of 3 Units)			
Linear Shrinkage, Percent	Humid[1]	Intermediate[2]	Arid[3]
0.03 or less	45	40	35
From 0.03 to 0.045	40	35	30
0.045 to 0.065 max.	35	30	25

1. Average annual relative humidity above 75 percent.

2. Average annual relative humidity 50 to 75 percent.

3. Average annual relative humidity less than 50 percent.

4. Based on UBC Standard Table No. 24-4-A and ASTM C 90, Table 1.

The use of the product may also determine whether Type I or Type II should be used. For fences, enclosures, and retaining walls the importance of moisture-controlled units is certainly downgraded and the extra cost, if there is any, need not be incurred.

The determination of a Type I unit is based on its moisture content as delivered to the job site. This implies that it might have to be protected from the weather after its manufacture and during storage. If it is manufactured in a moist, rainy area, the units would have to be stored under cover after they have been sufficiently dehydrated. If they are manufactured in a dry area, they could be stored outside and the dry weather may continue the dehydration process.

Concrete block, if stored for a period of time, as stated above, can achieve climatic balance and perform satisfactorily with a minimum of shrinkage. Thus, concrete block units should be protected from the weather even during storage at the job site.

However, if moisture controlled units are not covered and are exposed to rain or snow at the job site, they will no longer meet Type I moisture requirements.

Type II units, non-moisture-controlled, should not be so moist or green as to cause excessive shrinkage cracks. They should be aged a sufficient period of time to achieve a climatic moisture balance condition. This period of time is dependent on the materials, the moisture content, the density or permeability of the block and the humidity of the area.

It is recommended that all blocks, whether Type I or II, be cured.

Construction methods have a significant influence on the performance of concrete masonry units. As the wall is constructed, the units are partially restrained by the mortar head joint and the adjacent units. When fluid, high slump grout is pumped into the cells, the excess water is absorbed into the block, increasing its moisture content probably more than that specified for a Type I unit. The block will expand and, upon drying out, shrink even more than its original size prior to grouting. This condition is difficult to avoid since it is necessary to use fluid high slump grout in reinforced masonry walls.

EXAMPLE 1-A   Application of Table 1-8 Moisture Content Requirements for Type I, Concrete Block Units.

Determination of moisture controlled units.

Concrete block as received

Total moist weight =	29.4	lbs
Oven dry weight =	−28.5	lbs
Weight of water =	0.9	lbs

Block fully saturated weight =	31.6	lbs
Oven dry weight =	−28.5	lbs
Total absorbed water =	3.1	lbs

Assume project is in

a) Humid location; Relative humidity > 75%

b) Arid location; Relative humidity < 50%

Total linear shrinkage of block from 100% saturated to 100% oven dry = 0.052%

The maximum allowable moisture content as a percentage of the total absorption from Table 1-8. The linear shrinkage 0.052% is between 0.045% and 0.065%, therefore the allowable maximum moisture content as a percentage of the total absorption is

1) Humid location 35%

2) Arid location 25%

The absorption of the block as received is

$$\frac{\text{actual moisture content}}{\text{total absorbed moisture}} = \frac{0.9}{3.1} \times 100 = 29\%$$

This is less than the maximum allowable of 35% for the humid location and more than the allowable moisture content, 25% for the arid area.

Therefore, this unit will qualify as a Type I moisture controlled unit for the humid location, but does not qualify for the arid location. The block would be classified as Type II, non-moisture controlled unit.

EXAMPLE 1-B   How Much Moisture Can a Concrete Masonry Unit Have and Still Qualify as a Type I Moisture Controlled Unit?

Project located in intermediate humidity conditions, relative humidity: 50% to 75%

Oven dry weight of block = 28.5 lbs

Weight of block 100% saturated = 32.0 lbs

Weight of absorbed water:

32.0 − 28.5 lbs = 3.5 lbs

Total shrinkage of block from 100%

Saturated to oven dry = 0.043%

From Table 1-8 the maximum moisture content for the intermediate location with the shrinkage between 0.03% to 0.045% is 35%.

The maximum moisture in the masonry unit for it to qualify as Type I is 3.5 lb × 0.35 = 1.23 lbs water

Therefore, a concrete masonry unit can easily be checked by its weight:

Oven dry =	28.5	lbs
Allowable moisture =	+1.23	lbs
Total =	29.73	lbs

Thus the unit should weigh $29\,^3/_4$ lbs or less.

Constructed primarily of concrete masonry units, the Queen's Surf in Long Beach, California (Seismic Zone No. 4), rises 16 stories.

1-2 MORTAR

1-2.A General

Mortar is a plastic mixture of materials used to bind masonry units into a structural mass. It is used for the following purposes:

1. It serves as a bedding or seating material for the masonry units.

2. It allows the units to be leveled and properly placed.

3. It bonds the units together.

4. It provides compressive strength.

5. It provides shear strength, particularly parallel to the wall.

6. It allows some movement and elasticity between units.

7. It seals irregularities of the masonry units.

8. It can provide color to the wall by using color additives.

9. It can provide an architectural appearance by using various types of joints, as shown in Figure 1.9.

Historically, mortar has been made from a variety of materials. Plain mud, clay, earth with ashes, and sand with lime mortars have all been used. Modern mortar consists of cementitious materials and well graded sand.

1-2.B Types of Mortar

The requirements for mortar are provided in ASTM C 270, *Mortar for Unit Masonry* and in UBC Standard No. 24-20, *Mortar for Unit Masonry and Reinforced Masonry Other Than Gypsum.*

There were originally five types of mortar which were designated as M, S, N, O, and K. The types are identified by every other letter of the word MaSoNwOrK. Type K is no longer referred to in the Uniform Building Code or in ASTM C 270.

1-2.B.1 Selection of Mortar Types

The performance of masonry is influenced by various mortar properties such as workability, water retentivity, bond strength, durability, extensibility, and compressive strength. Since these properties vary with mortar type, it is important to select the proper mortar type for each particular application. Tables 1-9 and 1-10 are general guides for the selection of mortar type. Selection of mortar type should also consider all applicable building codes and engineering practice standards.

TABLE 1-9 Mortar Types for Classes of Construction

ASTM Mortar Type Designation	Construction Suitability
M	Masonry subjected to high compressive loads, severe frost action, or high lateral loads from earth pressures, hurricane winds, or earthquakes. Structures below or against grade such as retaining walls, etc.
S	Structures requiring high flexural bond strength, and subject to compressive and lateral loads.
N	General use in above grade masonry. Residential basement construction, interior walls and partitions. Masonry veneer and non-structural masonry partitions.
O	Non-load-bearing walls and partitions. Solid load bearing masonry with an actual compressive strength not exceeding 100 psi and not subject to weathering.

TABLE 1-10 Guide for the Selection of Masonry Mortars[1,4]

Location	Building Segment	Mortar Type	
		Rec.	Alt.
Exterior, above grade	Load-bearing wall	N	S or M
	Non-load bearing wall	O[2]	N or S
	Parapet wall	N	S
Exterior, at or below grade	Foundation wall, retaining wall, manholes, sewers, pavements, walks and patios	S[3]	M or N[3]
Interior	Load-bearing wall	N	S or M
	Non-bearing partitions	O[2]	N

1. This table does not provide for many specialized mortar uses, such as chimney, reinforced masonry, and acid-resistant mortars.

2. Type O mortar is recommended for use where the masonry is unlikely to be frozen when saturated or unlikely to be subjected to high winds or other significant lateral loads. Type N or S mortar should be used in other cases.

3. Masonry exposed to weather in a nominally horizontal surface is extremely vulnerable to weathering. Mortar for such masonry should be selected with due caution.

4. Based on ASTM C 270, Table X1.1. Rec. = Recommended, Alt. = Alternative.

In Seismic Zone Nos. 3 and 4, both the Uniform Building Code and ACI 530/ASCE 5 require that mortar for structural uses be type S or M. This requirement provides additional strength and bond in structures located in high seismic risk areas.

1-2.B.2 Specifying Mortar

Mortar may be specified by either property or proportion specifications.

1-2.B.2(a) Property Specifications

Property specifications are those in which the acceptability of the mortar is based on the properties of the ingredients (materials) and the properties (water retention, air content, and compressive strength) of samples of the mortar mixed and tested in the laboratory.

Property specifications are used for research so that the physical characteristics of a mortar can be determined and reproduced in subsequent tests.

The property requirements for mortar are given in Table 1-11.

TABLE 1-11   Property Specifications for Mortar[1,4]

Mortar	Type	Avg. Comp. Strength at 28 day Min. psi	Water Retention, Min. %	Air Content Max. %	Aggregate Ratio (measured in damp, loose conditions)
Cement-Lime	M	2500	75	12	
	S	1800	75	12	
	N	750	75	14 [2]	
	O	350	75	14 [2]	
Masonry Cement	M	2500	75	... [3]	Not less than 2¼ and not more than 3½ times the sum of the separate volume of cementitious materials
	S	1800	75	... [3]	
	N	750	75	... [3]	
	O	350	75	...	

1. Laboratory-prepared mortar only.

2. When structural reinforcement is incorporated in cement-lime mortar, the maximum air content shall be 12 percent.

3. When structural reinforcement is incorporated in masonry cement mortar, the maximum air content shall be 18 percent.

4. Based on ASTM C 270, Table 2.

Compressive strength is usually the only property or characteristic which a specifier who is not a researcher would require. Two methods are used to determine the compressive strength of mortar. The first method tests 2" cubes of mortar in compression after having cured for 28 days. The second method, based on UBC Standard 24-22 *Field Tests Specimens for Mortar*, uses mortar specimens 2" in diameter by 4" high. These cylinders must have a minimum compressive strength of 1500 psi. Although no qualification is made for the age of the compression test cylinders, it may be assumed as 28 days.

Table 1-12 is a comparison of the equivalent strength between cylinders and cube specimens for three types of mortar.

TABLE 1-12   Compressive Strength of Mortar[1] (psi)

Mortar Type	2" dia x 4" high Cylinder Specimen	2" Cube Specimen
M	2100	2500
S	1500	1800
N	625	750

1. Lesser periods of time for testing may be used provided the relation between early tested strength and the 28-day strength of the mortar is established.

The field strength of mortar should be used only as a quality control test, rather than a quantification evaluation. The in-place mortar strength can be much higher than the test values. The higher in-place strength is attributed to the inherent difficulty in failing the thin and wide (¼" to ⅝" high by 1¼" to 4" wide) mortar joints in compression. Additionally, the masonry units above and below the mortar joint, as well as the grout, confine the mortar so that the in-place mortar strength is much higher than the strengths of the test specimens. NCMA TEK 107 *Laboratory and Field Testing of Mortar and Grout* and Figure 1-4 dramatically show that a ⅜" to ⅝" mortar specimen has a strength far exceeding the strength of the 2" test specimens.

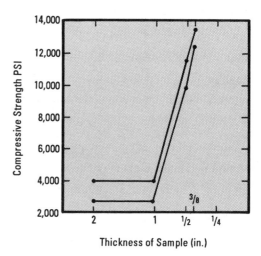

Figure 1-4 Effect of specimen thickness on compressive strength.

Because the in-place mortar strength exceeds the cube and cylinder test strengths, mortar will perform well even when tests on mortar are less than the specified strength of the mortar specimens. Additionally, because the in-place strength is quite high, mortar performs well even when the compressive strength of the entire masonry assemblage, f'_m, is higher than the cylinder and cube strengths. In fact, the National Concrete Masonry Association conducted a research project in which the apparent strength of mortar ranged from 500 psi to 3500 psi and the apparent strength of the wall increased only 4%, (NCMA TEK NOTE 15 *Compressive Strength of Concrete Masonry*).

NCMA TEK NOTE 70 *Concrete Masonry Prism Strength* indicates that the compressive strength of masonry structures built with type S or M mortar may be as much as 40% higher than masonry prisms built with type N Mortar. Additionally, tests by the Brick Institute of America indicate that the effect on the prism strength using type M mortar instead of type N may be as high as 34%. Conservatively, the UBC and ACI/ASCE have required that type M or S mortars be used for all masonry structures located in seismic zones 3 and 4.

Besides compressive strength requirements, it is suggested that the bond shear strength be investigated, particularly where lateral forces from winds or earthquakes must be considered.

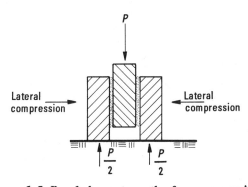

Figure 1-5 Bond shear strength of masonry unit and mortar.

The Office of the State Architect of California specifies that the bond shear strength between brick and mortar should be a minimum of 20 psi, after 14 days of curing. The actual bond strength in the wall is usually many times higher than the required 20 psi, since the vertical load on the wall and the development of shear friction also counteract the shear forces.

A lack of bond at the interface of mortar and masonry unit may permit moisture penetration through the unbonded areas. The use of lime in the mortar increases bond between mortar and the masonry unit. Workmanship

can also greatly affect bond strength, and the time lapse between spreading mortar and placing the masonry unit should be kept to a minimum since the bond of the mortar will be reduced by a long delay in placing the units.

1-2.B.2(b) Proportion Specifications

Proportion specifications limit the amount of the constituent parts by volume. Water content, however, may be adjusted by the mason to provide proper workability under various field conditions. When the proportions of ingredients are not specified, the proportions by mortar type must be used as given in Table 1-13 (UBC Table No. 24-A).

Mortars other than those approved in Table 1-13 may be used when laboratory or field tests demonstrate that the mortar, when combined with the masonry units, will achieve the specified compressive strength, f'_m.

The most common cement-lime mortar proportions by volume are:

Type M mortar; 1 Portland Cement: $^1/_4$ Lime: $3^1/_2$ Sand

Type S mortar; 1 Portland Cement: $^1/_2$ Lime: $4^1/_2$ Sand

Type N mortar; 1 Portland Cement: 1 Lime: 6 Sand

Type O mortar; 1 Portland Cement: 2 Lime: 9 Sand

1-2.C Mortar Materials

The principal mortar constituents are cement, lime, sand and water — each making a unique contribution to a mortar's performance. Cement contributes to mortar durability, high early strength and high compressive strength. Lime contributes to workability, water retentivity and elasticity. Both contribute to bond strength. Sand acts as a filler and contributes to the strength. Water is the ingredient which creates a plastic, workable mortar and is required for the hydration of the cement.

1-2.C.1 Cements

Three types of cement are now permitted to be used in mortar by the Uniform Building Code: portland cement, masonry cement and mortar cement.

1-2.C.1(a) Portland Cement

The basic cementitious ingredient in most mortar is portland cement. This material must meet the requirements of ASTM C 150 *Portland Cement* or UBC Standard No. 26-1, *Portland Cement and Blended Hydraulic Cements*. In mortar, the type of portland cement is limited to type I, II or III. The use of air-entraining portland cement (Type IA, IIA or IIIA) is not recommended for masonry mortar because air entrainment can reduce the bond between mortar and the masonry units.

TABLE 1-13 Mortar Proportions for Unit Masonry[4]

Mortar	Type	Portland Cement or Blended Cement[1]	Masonry Cement[2] M	S	N	Mortar Cement[3] M	S	N	Hydrated Lime or Lime Putty[1]	Measured in a Damp Loose Condition
Cement—lime	M	1	—	—	—	—	—	—	$1/4$	Not less than $2\,1/4$ and not more than 3 times the sum of the separate volumes of cementitious materials.
	S	1	—	—	—	—	—	—	over $1/4$ to $1/2$	
	N	1	—	—	—	—	—	—	over $1/2$ to $1\,1/4$	
	O	1	—	—	—	—	—	—	over $1\,1/4$ to $2\,1/2$	
Mortar cement	M	1	—	—	—	—	—	1	—	
	M	—	—	—	—	1	—	—	—	
	S	$1/2$	—	—	—	—	—	1	—	
	S	—	—	—	—	—	1	—	—	
	N	—	—	—	—	—	—	1	—	
Masonry cement	M	1	—	—	1	—	—	—	—	
	M	—	1	—	—	—	—	—	—	
	S	$1/2$	—	—	1	—	—	—	—	
	S	—	—	1	—	—	—	—	—	
	N	—	—	—	1	—	—	—	—	
	O	—	—	—	1	—	—	—	—	

1. When plastic cement is used in lieu of portland cement, hydrated lime or putty may be added, but not in excess of one tenth of the volume of cement. [UBC Sec. 2402(b)2.D. states: "Plastic cement shall meet the requirements for portland cement as set forth in ASTM C 150 except in respect to limitations on insoluble residue, air entrainment, and additions subsequent to calcination. Approved types of plasticizing agents shall be added to portland cement Type I or II in the manufacturing process, but not in excess of 12 percent of the total volume." (Plastic cement is not permitted for structural masonry in Seismic Zone Nos. 2, 3 and 4.) Author]

2. Masonry cement conforming to the requirements of U.B.C. Standard No. 24-16.

3. Mortar cement conforming to the requirements of U.B.C. Standard No. 24-19.

4. UBC Table 24-A.

Portland cement is the primary adhesive material, and based on the water cement ratio can produce high strength mortars. Hydrated lime is used in conjunction with the portland cement to provide the desired strength, workability and board life (board life is defined as the period of time during which mortar is still plastic and workable).

1-2.C.1(b) Masonry Cement

Masonry cement is a proprietary blend of portland cement and plasticizers such as ground inert fillers and other additives for workability. Masonry cement must meet the requirements of ASTM C 91 *Masonry Cement* or the UBC Standard No. 24-16, *Cement, Masonry* and is available for types N, S and M mortar.

The use of masonry cement for mortar is prohibited in Seismic Zone Nos. 2, 3 and 4.

1-2.C.1(c) Mortar Cement.

Mortar cement is also a portland cement based material which meets the requirements of UBC Standard No. 24-19, *Mortar Cement*. Mortar cement may be used for mortar in all seismic zones.

There are three types of mortar cement:

(1) **Type N.** Contains the cementitious materials used in the preparation of UBC Standard No. 24-20, Type N or Type O mortars. Type N mortar cement may also be used in combination with portland or blended hydraulic cement to prepare Type S or Type M mortars.

(2) **Type S.** Contains the cementitious materials used in the preparation of UBC Standard No. 24-20, Type S mortar.

(3) **Type M.** Contains the cementitious materials used in the preparation of UBC Standard No. 24-20, Type M mortar.

As mentioned previously, mortar cement may be used in all seismic zones while masonry cement is prohibited in Seismic Zone Nos. 2, 3 and 4. Reasons included are:

(a) Mortar cement limits or restricts the quantity of certain materials (see Table 1-14). In masonry cement, the ingredients are seldom known.

TABLE 1-14 Restricted Materials in Mortar Cements[1]

Materials	Maximum Limit
Chloride Salts	0.06%
Caboxylic Acids	0.25%
Sugars	1.00%
Glycols	1.00%
Lignin and derivatives	0.50%
Stearates	0.50%
Fly Ash	No limit
Clay (except fireclay)	5.00%

1. UBC Standard Table No. 24-19B.

(b) A minimum 28-day flexural bond strength for mortar cement is determined in accordance with UBC Standard No. 24-30, *Standard Test Method for Flexural Bond Strength.* No such test is required for masonry cement.

(c) The air content in mortar cement is restricted. Often times the air content in masonry cement is unknown.

Flexural Bond Strength of Mortar

The flexural bond strength of mortar cement is based on a laboratory evaluation of a standardized mortar and a standardized masonry unit. Minimum flexural bond strength values from Table 1-15 are 71 psi for Type N Mortar, 104 psi for Type S Mortar and 116 psi for Type M Mortar.

The test apparatus consists of a metal frame used to support a specimen as shown in Figure 1-6. The support system must be adjustable to support prisms of various height (See UBC Standard No. 24-30 for additional information on this test).

TABLE 1-15 Physical Requirements[3]

Mortar Cement Type	N	S	M
Fineness, residue on a No. 325 sieve Maximum, percent	24	24	24
Autoclave expansion Maximum, percent	1.0	1.0	1.0
Time of setting,Gillmore Method Initial set, Minimum, hour Final set, Maximum, hour	2 24	1½ 24	1½ 24
Compressive strength[1] 7 days, minimum, psi 28 days, minimum, psi	500 900	1300 2100	1800 2900
Flexural bond strength[2] 28 days, minimum, psi	71	104	116
Air content of mortar Minimum percent by volume Maximum percent by volume	8 16	8 14	8 14
Water retention Minimum, percent	70	70	70

1. Compressive strength shall be based on the average of 3 mortar cubes composed of 1 part mortar cement and 3 parts blended sand (half graded Ottawa sand, and half standard 20-30 Ottawa sand) by volume and tested in accordance with this standard.

2. Flexural bond strength shall be determined in accordance with UBC Standard No. 24-30.

3. UBC Standard Table No. 24-19-A.

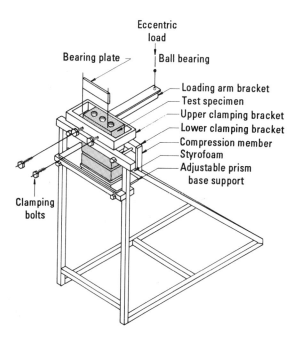

Figure 1-6 Bond wrench test apparatus.

1-2.C.2 Hydrated Lime

Hydrated lime is manufactured from calcining limestone (calcium carbonate with the water of crystallization, $CaCO_3H_2O$). The high heat generated in the kiln drives off the water of crystallization, H_2O, and the carbon dioxide, CO_2, resulting in quicklime, CaO.

The quicklime can then be slaked by placing it in water thus making hydrated lime, lime putty or slaked lime $Ca(OH)_2$. The hydrated lime can then be dried and ground, producing a white pulverized hydrated lime which is sacked and used in mortar.

Hydrated lime can be used without delay making it more convenient to use than quicklime.

The Uniform Building Code Standard 24-18 and ASTM C 207, both titled *Hydrated Lime for Masonry Purposes*, is available in the following four types, S, SA, N and NA. Type S and N hydrated limes contain no air entraining admixtures. However, types NA and SA limes provide more entrained air in the mortar than allowed by UBC or ASTM requirements and therefore may not be used. Additionally, unhydrated oxides are not controlled in Type N or NA limes thus making only type S hydrated lime suitable for masonry mortar.

Lime in mortar provides cementitious properties to the mortar and is not considered to be an admixture. Used in mortar it:

a) Improves the plasticity or workability of the mortar.

b) Improves the water tightness of the wall.

c) Improves the water retentivity or board life of the mortar.

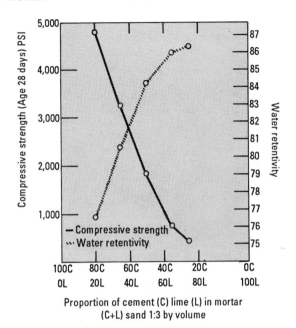

Figure 1-7 Relation between mortar composition, compressive strength, and water retentivity.

Figure 1-7 shows the relationship between various proportions of cement and lime versus mortar strength or water retentivity.

1-2.C.3 Mortar Sand

For masonry mortar, sand aggregate is required to meet ASTM C 144, *Aggregates for Masonry Mortar* per UBC Section 2402(b)1A.

Sand used in preparing mortar can be either natural or manufactured. Manufactured sand is obtained by crushing stone, gravel or air-cooled blast-furnace slag. It is characterized by sharp and angular particles producing mortars with workability properties different than mortars made with natural sand which generally have round, smooth particles.

Sand gradation is most often specified or defined by referring to a standard sieve analysis. For mortar, sand is graded within the limits given in Table 1-16.

Sand should be free of injurious amounts of deleterious substances and organic impurities. ASTM C 144 gives guidelines on determining if an aggregate has excessive impurities.

TABLE 1-16 Sand for Masonry Mortar[1]

Sieve Size	Percent Passing	
	Natural Sand	Manufactured Sand
No. 4	100	100
No. 8	95 to 100	95 to 100
No. 16	70 to 100	70 to 100
No. 30	40 to 75	40 to 75
No. 50	10 to 35	20 to 40
No. 100	2 to 15	10 to 25
No. 200	—	0 to 10

1. Based on ASTM C 144, Section 4.1

Concrete sand should not be used in mortar because the maximum grain size is too large. Additionally the fine particles which are needed in masonry sand have often been washed out of concrete sand thus creating a harsh, coarse sand unsuitable for mortar. Mortar sand needs at least 5% fines which pass the 200 sieve to aid plasticity, workability and water retention of mortar.

Mortar sand, like all mortar ingredients, should be stored in a level, dry, clean place. Ideally, it should be located near the mixer so it can be measured and added with minimum handling and can be kept from contamination by harmful substances.

1-2.C.4 Water

Water must be clean and free of deleterious amounts of acids, alkalies or organic materials. Water containing soluble salts such as potassium and sodium sulfates should be avoided since these salts can contribute to efflorescence.

1-2.C.5 Admixtures

There are numerous admixtures which may be added to mortar to affect its properties. One of these, called a retarding set admixture, delays the set and stiffening of mortar. In fact, the set may be delayed for 36 hours or more if desired.

There are also admixtures that are used to replace lime. These may be an air entraining chemical or a pulverized fire clay or bentonite clay to provide workability. Care should be taken with these admixtures since the bond between the mortar and the masonry units may be reduced.

The use of any admixtures must be approved by the architect or engineer and should be acceptable to the building official.

1-2.C.6 Color

Mortar colors are generally mineral oxides or carbon black. Iron oxide is used for red, yellow, and brown colors; chromium oxide for green, and cobalt oxide for blue colors. Commercially prepared colors for mortars also offer a wide variety of colors and shades.

The amount of color additive depends on the color and intensity desired. Generally the amount of color additive ranges from 0.5% to 7.0% for the mineral oxides with a maximum of 3% for carbon black. These percentages are based on the weight of cement content and the maximum percentages are far greater than the normal amounts of color additives generally required.

Mixing time of the mortar should be long enough for a uniform, even color to be obtained and should be the same length of time for every mortar batch. Additionally the mixing sequence should be the same for each batch.

Retempering of colored mortar must be kept to a minimum to reduce the variations in color of the mortar. For best results, mortar should not be retempered at all.

Finally, the source, manufacturer and amount of each ingredient should remain the same for all colored mortar on a project so as to obtain the same color throughout. Prepackaged mineral color additives that can be added to the mix based on full sacks of portland cement generally provide a consistent mortar color.

1-2.D Mixing

1-2.D.1 Measurement of Mortar Materials

The method of measuring materials for mortar must be such that the specified proportions of the mortar materials can be controlled and accurately maintained. A reasonable method to control the mortar proportions is to use full sacks of cement per batch and to use measuring boxes for the proper amounts of lime and sand. Dry preblended mixes are also available.

1-2.D.2 Job Site Mixed Mortar

Mortar mixing is best accomplished in a paddle type mixer. About one-half of the water and one quarter of the sand are put into the operating mixer first, then the cement, lime, color (if any), and the remaining water and sand. All materials should mix for three to ten minutes in a mechanical mixer with the amount of water required to provide the desired workability. Small amounts of mortar can be hand mixed. Dry mixes for mortar which are blended in a factory should be mixed at the job site in a mechanical mixer until workable, but not more than 10 minutes.

Figure 1-8 Plaster or paddle mortar mixer.

Figure 1-8, shows a paddle mixer with a stationary drum. The blades rotate through the mortar materials for thorough mixing.

A drum or barrel mixer, shown in Figure 1-9 rotates the drum in which the materials are placed. The materials are carried to the top of the rotation and then drops down to achieve mixing.

Figure 1-9 Drum or barrel concrete mixer.

1-2.D.3 Ready Mixed Mortar

ASTM C 1142-90, *Standard Specification for Ready-Mixed Mortar for Unit Masonry* covers the requirements for this material. Ready-mixed mortar consists of cementitious materials, aggregate, water and an admixture for set-control which are measured and mixed at a central location, using weight-or-volume-control equipment. This mortar is delivered to a construction site and is usable for a period in excess of 2½ hours.

There are four types of ready mix mortar, RM, RS, RN, and RO. These types of mortar can be manufactured with one of the four mortar formulations: portland cement, portland cement-lime, masonry cement, or masonry cement with portland cements.

TABLE 1-17 Property Specification Requirements[3]

Mortar Type	Avg[1] Compressive Strength at 28 days, Min psi, Cubes	Water Retention Min %	Air Content,[2] Max %
RM	2500	75	18
RS	1800	75	18
RN	750	75	18
RO	350	75	18

1. Twenty-eight days old from date of casting. The strength values as shown are the standard values. Intermediate values may be specified in accordance with project requirements.

2. When structural reinforcement is incorporated in mortar, the maximum air content shall be 12%, or bond strength. Test data shall be provided to justify higher air content.

3. Based on ASTM C 1142, Table 1.

Ready-mixed mortar is selected by type and the length of workable time required. The consistency based on the mason's use should be specified. Otherwise the ready-mixed mortar is required to have a cone penetration consistency of 55 ± 5 mm as measured by ASTM C 780, *Test Methods for Preconstruction and Construction Evaluation of Mortars for Plain and Reinforced Unit Masonry.*

1-2.D.4 Retempering

Mortar may be retempered one time with water when needed to maintain workability. This should be done on wet mortar boards by forming a basin or hollow in the mortar, adding water, and then reworking the mortar into the water. Splashing water over the top of the mortar is not permissible.

Harsh mortar that has begun to stiffen or harden due to hydration, should be thrown out. Generally, mortar should be used within two-and-one-half hours after the initial water has been added to the dry ingredients at the job site. Retempering color mortar should be avoided to limit color variations.

1-2.E Types of Mortar Joints.

Shown in Figure 1-10 are nine examples of commonly used mortar joints. Each joint provides a different architectural appearance to the wall. However, because some joints provide poor weather resistance, care must be taken in the selection of the proper type of mortar joint. Joints with ledges such as weather, squeezed, raked and struck joints tend to perform poorly in exterior applications and allow moisture penetration. Tooled joints are recommended for exterior applications since the tooling compacts the mortar tightly preventing moisture penetration.

Figure 1-10 Mortar joint types.
(Continued on next page)

a) Concave Joint
Most common joint used, tooling works the mortar tight into the joint to produce a good weather joint. Pattern is emphasized and small irregularities in laying are concealed.

b) "V" Joint
Tooling works the mortar tight and provides a good weather joint. Used to emphasize joints and conceal small irregularities in laying and provide a line in center of mortar joint.

c) Weather Joint
Use to emphasize horizontal joints. Acceptable weather joint with proper tooling.

d) Flush Joint
Use where wall is to be plastered or where it is desired to hide joints under paint. Special care is required to make joint weatherproof. Mortar joints must be compressed to assure intimate contact with the block.

e) Squeezed Joint
Provides a rustic, high texture look. Satisfactory indoors and exterior fences. Not recommended for exterior building walls.

f) Beaded Joint
Special effect, poor exterior weather joint because of exposed ledge — not recommended.

g) Raked Joint
Strongly emphasizes joints. Poor weather joint — Not recommended if exposed to weather unless tooled at bottom of mortar joint.

h) Struck Joint
Use to emphasize horizontal joints. Poor weather joint — Not recommended as water will penetrate on lower ledge.

i) Grapevine Joint
Shows a horizontal indentation.

Figure 1-10 (Continued from previous page).

1-3 GROUT

1-3.A General

Grout is a mixture of portland cement, sand, pea gravel and water mixed to fluid consistency so that it will have a slump of 8 to 10 inches. Grout is placed in the cores of hollow masonry units or between the wythes of solid units to bind the reinforcing steel and the masonry into a structural system. Additionally, grout provides:

a) More cross-sectional area allowing a grouted wall to support greater vertical and lateral shear forces than a non-grouted wall.

b) Added sound transmission resistance thus reducing the sound passing through the wall.

c) Increased fire resistance and an improved fire rating of the wall.

d) Improved energy storage capabilities of a wall.

e) Greater weight thus improving the overturning resistance of retaining walls.

Requirements for grout are given in ASTM C 476, *Grout for Masonry* and UBC Standard No. 24-29, *Grout for Masonry*.

Grouting of two wythe brick wall.

1-3.B   Types of Grout

The Uniform Building Code identifies two types of grout for masonry construction: fine grout and coarse grout. As their names imply, these two types of grouts differ primarily in the maximum allowable size of aggregates. The fineness or coarseness of the grout is selected based on the size of grout space and the height of the grout pour. Table 1-18, Grouting Limitations (UBC Table No. 24-G) covers the requirements for the selection of the grout type.

1-3.B.1   Fine Grout

Fine grout is used where the grout space is small, narrow, or too congested with reinforcing steel. When fine grout is used, there must be a clearance of $1/4''$ or more between the reinforcing steel and the masonry unit.

The normal proportions by volume for fine grout are as follows:

1 part portland cement

$2\frac{1}{2}$ to 3 parts sand

Water for a slump of 8 to 10 inches

1-3.B.2   Coarse Grout

Coarse grout may be used where the grout space for 2 wythe masonry is at least $1\frac{1}{2}$ inches in width horizontally, or where the minimum block cell dimension is $1\frac{1}{2} \times 3$ inches.

Although approved aggregates for grout (sand and pea gravel) are limited to a maximum size of $3/8$ inch, a coarse grout using $3/4$ inch aggregate may be used if the grout space is especially wide, (8 inches or more horizontally). Larger size aggregates take up more volume, thus requiring less cement for an equivalent strength mix that used smaller aggregates. Larger aggregates also tend to reduce the shrinkage of the grout and allow the slump of grout to be reduced to 7 or 8 inches for easier placement. Note, to place grout with $3/4$ inch aggregate, a concrete pump is usually required.

When coarse grout is made with pea gravel, there must be a minimum clearance of $1/2''$ between the reinforcing steel and the masonry unit. Accordingly, if coarse grout is made using larger sized aggregates, this clearance between the reinforcing and the masonry unit must be increased to approximately $1/4''$ more than the largest size aggregate.

The typical proportions by volume for coarse grout are as follows:

1 part portland cement

$2\frac{1}{4}$ to 3 parts sand

1 to 2 parts pea gravel

Water for a slump of 8 to 10 inches

1-3.C   Slump of Grout

The water content of grout is adjusted to provide fluidity (slump) allowing proper grout placement for various job conditions. The high slump allows the grout to flow into openings and around steel reinforcement. Excess water in the grout is absorbed by the masonry units, reducing the apparently high water/cement ratio. Additionally the moist masonry acts in curing the grout.

Fluidity is measured by a slump cone test. Both types of grout, fine and coarse, must contain enough water to provide a slump of 8 to 10 inches.

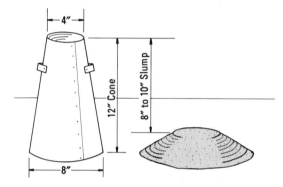

Figure 1-11 Slump cone and slump of grout.

TABLE 1-18   Grouting Limitations[4]

Grout Type	Grout Pour Maximum Height (feet)[1]	Minimum Dimension of the Total Clear Areas within Grout Spaces and Cells[2,3]	
		Multiwythe Masonry	Hollow-Unit Masonry
Fine	1	$3/4$	$1\frac{1}{2} \times 2$
Fine	5	$1\frac{1}{2}$	$1\frac{1}{2} \times 2$
Fine	8	$1\frac{1}{2}$	$1\frac{1}{2} \times 3$
Fine	12	$1\frac{1}{2}$	$1\frac{3}{4} \times 3$
Fine	24	2	3×3
Coarse	1	$1\frac{1}{2}$	$1\frac{1}{2} \times 3$
Coarse	5	2	$2\frac{1}{2} \times 3$
Coarse	8	2	3×3
Coarse	12	$2\frac{1}{2}$	3×3
Coarse	24	3	3×4

1. See also UBC Section 2404(f).

2. The actual grout space or grout cell dimensions must be larger than the sum of the following items: (1) The required minimum dimensions of total clear areas in Table No. 24-G; (2) The width of any mortar projections within the space; and (3) The horizontal projections of the diameters of the horizontal reinforcing bars within a cross-section of the grout space or cells.

3. The minimum dimensions of the total clear areas shall be made up of one or more open areas, with at least one area being $3/4$ inch or greater in width.

4. UBC Table No. 24-G

1-3.D Proportions

Grout ingredient proportions are commonly selected from Table 1-19, Grout Proportions By Volume (UBC Table No. 24-B). Proportions of the grout ingredients may also be determined by laboratory testing or field experience, if a satisfactory history of the grout's performance is available. Note that any grout performance history must be based on grout, mortar, and masonry units, which are similar to those intended for use on the new project. Additionally the history results should have been determined in accordance with UBC Standard No. 24-26, *Test Method for Compressive Strength of Masonry Prisms* or UBC Standard No. 24-29, *Grout for Masonry*.

The use of 70% sand and 30% pea gravel requires six sacks of portland cement per cubic yard and results in a pumpable grout that provides the minimum strength of 2000 psi required by UBC Standard No. 24-29. Grout must have adequate strength for satisfactory f'_m values and for sufficient bonding to the reinforcing steel and the masonry units. Without adequate bonding, stresses cannot be properly transferred between the various materials. Adequate strength is also needed to assure the embedded anchor bolts will be anchored securely.

Experience has shown that grout proportions based on Table 1-19 (UBC Table No. 24-B and Sec. 2403(d)2) are successful for normal load-bearing concrete masonry construction.

1-3.D.1 Aggregates for Grout

Aggregates for grout should meet the requirements of ASTM C 404, *Aggregates for Grout* per UBC Section 2402(b)1.B. Grading of the aggregate should be in accordance with Table 1-20, Grading Requirements (ASTM C 404, Table 1).

TABLE 1-20 Grading Requirements[1]

Sieve Size	Amounts Finer than Each Laboratory Sieve (Square openings), Percent by Weight				
	Fine Aggregate			Coarse Aggregate	
	Size No. 1	Size No. 2		Size No. 8	Size No. 89
		Natural	Manufactured		
1/2-inch	—	—	—	100	100
3/8-inch	100	—	—	85 to 100	90 to 100
No. 4	95 to 100	100	100	10 to 30	20 to 55
No. 8	80 to 100	95 to 100	95 to 100	0 to 10	5 to 30
No.16	50 to 85	60 to 100	60 to 100	0 to 5	0 to 10
No. 30	25 to 60	35 to 70	35 to 70	—	0 to 5
No. 50	10 to 30	15 to 35	20 to 40	—	—
No. 100	2 to 10	2 to 15	10 to 25	—	—
No. 200	—	—	0 to 10	—	—

1. Based on ASTM C 404, Table 1.

1-3.E Mixing

Grout prepared at the job site should be mixed for 3 to 10 minutes, in order to assure thorough blending of all ingredients. Enough water must be used in the mixing process to achieve a high slump of 8 to 10 inches. Dry grout mixes which are blended at a factory should be mixed at the job site in a mechanical mixer until workable, but not more than 10 minutes.

1-3.F Grout Admixtures

Admixtures are any materials other than water, cement and aggregate which are added to the grout, either before or during mixing, in order to improve the properties of the fresh or hardened grout or to decrease its cost.

The four most common types of grout admixtures are:

1. Shrinkage Compensating Admixtures — Used to counteract the loss of water and the shrinkage of the cement by creating expansive gases in the grout.

2. Plasticizer Admixtures — Used to obtain the high slump required for grout without the use of excess water. By adding a plasticizer to a 4″ slump grout mix, an 8 to 10″ slump can be achieved.

TABLE 1-19 Grout Proportions by Volume[1,2]

Type	Parts by Volume of Portland Cement or Blended Cement	Parts by Volume of Hydrated Lime or Lime Putty	Aggregate Measured in a Damp, Loose Condition	
			Fine	Coarse
Fine grout	1	0 to 1/10	2 1/4 to 3 times the sum of the volumes of the cementitious materials	
Coarse grout	1	0 to 1/10	2 1/4 to 3 times the sum of the volumes of the cementitious materials	1 to 2 times the sum of the volumes of the cementitious materials

1. Grout shall attain a minimum compressive strength at 28 days of 2,000 psi. The building official may require a compressive field strength test of grout made in accordance with the UBC Standard No. 24-28.

2. UBC Table No. 24-B.

3. Cement Replacement Admixtures — Used to decrease the amount of cement in the grout without adversely affecting the compressive and bond strengths of the grout. Types C and F fly ash are by far the most common cement replacement admixtures. Current practice allows 15 to 20% of the portland cement by weight to be replaced with fly ash as long as the strength characteristics are maintained.

4. Accelerator admixtures — Used in cold weather construction to reduce the time that the wall must be protected from freezing. Accelerators decrease the setting time of the grout and speed up its strength gain. Accelerators also increase the heat of hydration preventing the grout from freezing under most circumstances.

Careful consideration must be given prior to the use of all admixtures since an admixture may adversely affect certain grout properties while improving the intended properties. Admixtures containing chloride and antifreeze liquids may not be used per UBC Section 2403(e) despite their benefits, since chlorides cause corrosion of the reinforcing steel. Admixtures can significantly reduce the compressive and bond strengths of the grout.

Similarly, care should be taken when using two or more admixtures in a grout batch since the combination of admixtures often produces unexpected results. Under all circumstances, information regarding laboratory and field performance of an admixture should be obtained from the manufacturer prior to its use in a grout. Additionally, it should be noted that UBC Section 2403(e) requires that the building official approves the use of all grout admixtures prior to their use.

UBC Sec 2403(e)

(e) Additives and Admixtures 1. General. Additives and admixtures to mortar or grout shall not be used unless approved by the building official.

2. Antifreeze compounds. Antifreeze liquids, chloride salts or other such substances shall not be used in mortar or grout.

3. Air entrainment. Air-entraining substances shall not be used in mortar or grout unless tests are conducted to determine compliance with the requirements of this code.

4. Colors. Only pure mineral oxide, carbon black or synthetic colors may be used. Carbon black shall be limited to a maximum of 3 percent of the weight of the cement.

1-3.G Grout Strength Requirements

Section 24.2904 of the Uniform Building Code Standard No. 24-29 states: "grout shall have a minimum compressive strength when tested in accordance with UBC Standard No. 24-28 equal to its specified strength, but no less than 2000 psi."

The required minimum compressive strength of 2000 psi is needed in order to achieve adequate bond between the grout, and the reinforcing steel, and the masonry unit. This minimum value is satisfactory for masonry construction in which the specified design strength, f'_m equals 1500 psi, and the masonry unit has a compressive strength of 1900 psi. However, footnote 4 to UBC Table No. 24-C (Table 2-3) requires that the grout strength equal or exceed the compressive strength of higher strength concrete masonry units. It is also recommended that the compressive strength of the grout in concrete masonry construction be 1.25 to 1.40 times the design strength of the masonry assemblage, f'_m. An example of this requirement is that 2000 psi grout is required for a masonry assemblage specified strength, f'_m, of 1500 psi.

For clay masonry construction, it is recommended that the grout be proportioned in accordance with Table 1-19 Grout Proportions by Volume (UBC Table No. 24-B and ASTM C 476, Table 1).

If grout tests are required, the following schedule is suggested.

1. At the start of grouting operations, take one test per day for the first three days. The tests should consist of three specimens which are made as outlined in Section 1-3.I of this book and in accordance with UBC Standard No. 24-28 *Method of Sampling and Testing Grout* or ASTM C 1019, *Standard Method of Sampling and Testing Grout.*

2. After the initial three tests, specimens for continuing quality control should be taken at least once each week. Additionally, specimens should be taken more frequently for every 25 cubic yards of grout, or for every 2500 square feet of wall, whichever comes first.

1-3.H Testing Grout Strength

In order to determine the compressive strength of grout, specimens are made that will represent the hardened grout in the wall. The specimen is made in a mold consisting of masonry units identical to those being used in construction and at the same moisture condition as those units being laid. The units are arranged to form a space approximately 3 to 4 inches square and twice as high as it is wide (Figures 1-12 and 1-13).

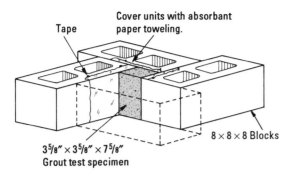

Figure 1-12 Typical arrangement for making a grout specimen for block.

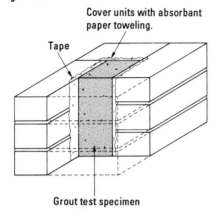

Figure 1-13 Typical arrangement for making a grout specimen for brick.

To prevent the grout from bonding to the masonry units, the space is lined with a permeable paper or porous separator, which still allows any excess water to be absorbed into the units.

The representative samples of the grout are placed in the molds, puddled and kept damp, and undisturbed for 48 hours. Afterwards, the grout specimens are taken to a laboratory where they are placed in a fog room until tested.

1-3.1 Methods of Grouting Masonry Walls

There are several methods of constructing and grouting masonry walls that will result in strong, homogeneous and satisfactory walls. The method selected is influenced by the type of masonry, the area and length of wall, the equipment available, and the experience of the contractor.

1-3.l.1 Grout Pour and Lift

The total height of masonry to be grouted prior to the erection of additional masonry is called a grout pour. Grout is placed in increments called lifts. A grout lift is the height of grout placed in a single continuous operation prior to consolidation.

Though lifts may not exceed 5 or 6 feet in height, a grout pour may consist of several lifts. For example, if the wall is built 18 feet high, the total grout pour could be the entire 18 feet. For this situation, the contractor could place the grout in 3 lifts of 6 feet each.

Currently UBC Table No. 24-G (See Table 1-18) limits a grout pour to a maximum height of 24 feet.

1-3.l.2 Low Lift and High Lift Grouting

Although the terms low lift and high lift grouting were deleted from the Uniform Building Code in recent years, they are still commonly used when referring to grouting methods.

In general, low lift grouting may be used when the height of the grout pour is 5 feet or less. High lift grouting may be used only when cleanout holes are provided, and the height of the masonry wall prior to grouting may exceed 5 feet as stated in UBC Table No. 24-G Grouting Limitations. (See Table 1-18)

1-3.l.3 Low Lift Grouting Procedure

When the low lift grouting procedure is used, masonry walls may be built to a height of 5 feet. Because of this limited pour height which allows for easy inspection of the walls, cleanout openings are not required.

For two wythe masonry walls, it is necessary to tie the wythes together with wire ties or joint reinforcing whenever the grout pour height is more than 12 inches to prevent the wythes from bulging or blowing out. (Figure 1-14) These ties should be spaced no more than 24 inches on centers horizontally and 16 inches apart vertically for running bond. For stacked bond construction these ties must be spaced no more than 12 inches on centers vertically.

Hollow unit masonry does not require ties since the cross-webs and end shells support the face shells and resist bulging and blowouts.

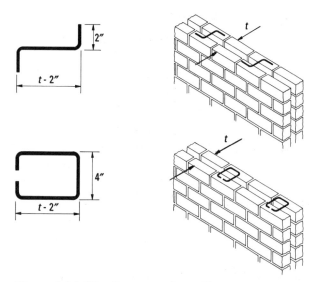

Figure 1-14 Ties for two wythe walls.

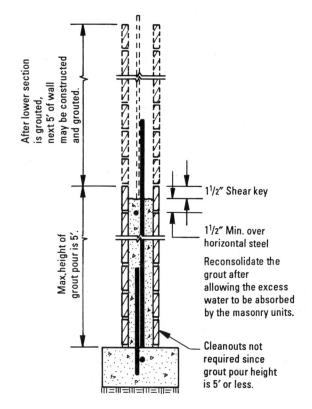

1¹/₂″ Shear key

1¹/₂″ Min. over horizontal steel

Reconsolidate the grout after allowing the excess water to be absorbed by the masonry units.

Cleanouts not required since grout pour height is 5′ or less.

Figure 1-15 Low lift grouting, cleanouts not required.

Grout may not be placed until all the masonry units, ties, reinforcing steel and embedded anchor bolts are in place up to the top of the grout pour. Once these are in place the wall may be fully grouted. For grout pours 12 inches high or less, the grout may be consolidated by puddling with a puddling stick such as a 1″ × 2″ piece of wood. However, grout pours in excess of 12 inches in height must be consolidated by means of a mechanical vibrator. The grout must also be reconsolidated after the excess water is absorbed by the units (usually after 3 to 5 minutes) to close any voids due to the water lost.

Masonry units, ties, reinforcing steel, and anchor bolts for the next pour may be placed, once the grout has been thoroughly reconsolidated.

Horizontal construction joints should be formed between grout pours by stopping the grout pour 1¹/₂ inches below the top of the masonry. Where bond beams occur, these joints may be reduced to ¹/₂ inch deep to allow sufficient grout above the horizontal reinforcing steel.

At the top of the wall, the grout should be placed flush with the masonry units.

1-3.I.4 High Lift Grouting Procedure

Grouting after a wall is constructed to its full height is often quite economical. This method allows the mason to continually lay masonry units without waiting for the walls to be grouted. High lift grouting procedures must be used when grout pours exceed 5 feet. Currently the maximum pour height the Uniform Building Code allows is 24 feet.

Cleanout openings must be provided in walls which are to be high lift grouted. UBC Section 2404(f) requires cleanouts in the bottom course at every vertical bar. However, in solid grouted walls, cleanouts must be provided at no more than 32 inches on center, even if the reinforcing steel is spaced at a greater spacing (Figure 1-16). For partially grouted walls it is recommended that the maximum spacing of cleanouts be no more than 48 inches on center.

Two wythe masonry walls must be tied together with wire ties or joint reinforcing, as outlined in the low lift grouting section to prevent blowouts and bulging (Figure 1-18).

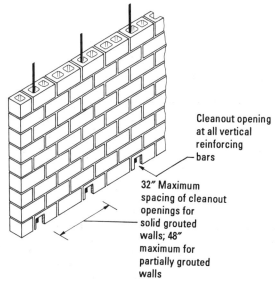

Cleanout opening at all vertical reinforcing bars

32" Maximum spacing of cleanout openings for solid grouted walls; 48" maximum for partially grouted walls

Figure 1-16 Maximum spacing of cleanout holes.

Grout lifts may be up to 6 feet high and must be mechanically consolidated. After a delay of 3 to 5 minutes, the grout should be reconsolidated to close any voids due to water loss.

Because of the fluidity of the grout and the tendency of the aggregate to segregate, control barriers must be placed in walls to confine the flow of grout. These barriers, which are constructed with masonry units laid in the grout space, must extend the full height of the grout pour. The Uniform Building Code limits the spacing of these barriers to no more than 30 feet on center. It is advisable that the full height of the wall between control barriers be grouted in one day.

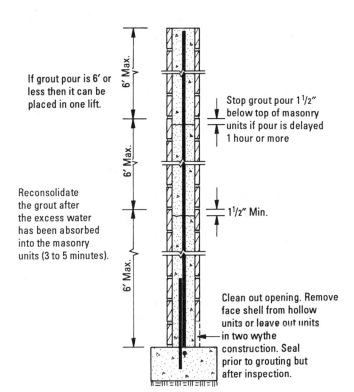

If grout pour is 6' or less then it can be placed in one lift.

6' Max.

6' Max.

Reconsolidate the grout after the excess water has been absorbed into the masonry units (3 to 5 minutes).

6' Max.

Stop grout pour 1¹/₂" below top of masonry units if pour is delayed 1 hour or more

1¹/₂" Min.

Clean out opening. Remove face shell from hollow units or leave out units in two wythe construction. Seal prior to grouting but after inspection.

Figure 1-17 High lift grouting block wall.

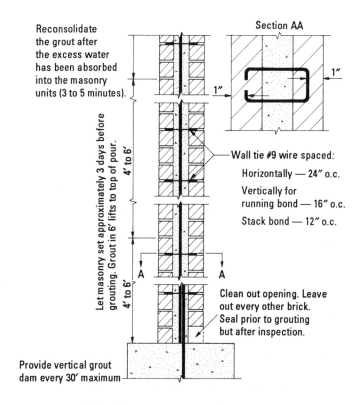

Reconsolidate the grout after the excess water has been absorbed into the masonry units (3 to 5 minutes).

Section AA

1"

1"

Let masonry set approximately 3 days before grouting. Grout in 6' lifts to top of pour.

4' to 6'

4' to 6'

Wall tie #9 wire spaced:

Horizontally — 24" o.c.

Vertically for running bond — 16" o.c.

Stack bond — 12" o.c.

A   A

Clean out opening. Leave out every other brick. Seal prior to grouting but after inspection.

Provide vertical grout dam every 30' maximum

Figure 1-18 High lift method of grouting 2 wythe walls, with cleanout openings.

At the bottom of the wall the grout space may be covered with a layer of loose sand to prevent mortar droppings from sticking to the foundations. The mortar droppings and sand are then removed from the grout space by blowing it out, washing it out, or cleaning it out by hand.

Once the cleanout holes have been cleaned and inspected they may be sealed with a masonry unit, a face shell, or a form board which is then braced to resist the pressure of the poured grout.

1-3.J Consolidation

Grout must be consolidated just like concrete. Consolidation eliminates voids and causes grout to flow around the reinforcement and into small openings or voids.

Consolidation may be performed using a puddle stick if the lifts are not higher than 12 inches. Lifts greater than 12 inches high however, must be consolidated by mechanical vibrators. As there is generally only a small volume of grout to be consolidated in a cell or grout space, the mechanical vibrator need only be used for a few seconds in any location. It is important not to over vibrate so as to avoid the possibility of blowing out face shells or dislodging masonry units.

1-4 REINFORCING STEEL

1-4.A General

Reinforcing steel in masonry has been used extensively in the West since the 1930's, revitalizing the masonry industry in earthquake prone areas. Reinforcing steel extends the characteristics of ductility, toughness and energy absorption that is so necessary in structures subjected to the dynamic forces of earthquakes.

Reinforced masonry performs well because the materials, steel, masonry, grout, and mortar, work together as a single structural unit. The temperature coefficient for steel, mortar, grout, and the masonry units are very similar. This similarity of thermal coefficients allows the different component materials to act together through normal temperature ranges. Disruptive stresses are not created at the interface between the steel and the grout which would destroy the bond between these materials and prevent force transfer.

Structures subjected to severe lateral dynamic loads such as earthquakes must be capable of providing the necessary strength or energy absorbing capacity and ductility to withstand these forces. Reinforcing steel serves to resist the shear and tensile forces generated by the dynamic loads. It can also provide sufficient ductility to the masonry structure so that the structure can sustain load reversals beyond the capability of plain, unreinforced masonry.

In order for the reinforcing steel to provide adequate ductility and strength, it is of prime importance that the reinforcing steel is placed properly so as to provide a continuous load path throughout the structure. The engineer must pay special attention to reinforcing steel details to ensure continuity. The following items must be provided:

(1) The proper size and amount of reinforcement which complies with the limited minimum and maximum percentages of reinforcement and other code requirements.

(2) The minimum required rebar protection.

(3) The proper spacing of longitudinal and transversal reinforcement.

(4) Sufficient anchorage of flexural and shear reinforcing bars.

(5) Adequate lapping of the reinforcing bars.

(6) Sufficient stirrups, ties, metal plates, spirals, etc., in order to provide confinement.

1-4.B Types of Reinforcement

1-4.B.1 Reinforcing Bars

For reinforced masonry construction, deformed bars range in size from #3 ($^3/_8''$ diameter) to a maximum of #11 ($1^3/_8''$ diameter) per UBC Section 2402(b)10.B. This reinforcing steel must conform to ASTM A 615, A 616, A 617, A 706, A 767 or A 775 which specify the physical characteristics of the reinforcing steel.

ASTM A 615, A 616 and A 617 cover reinforcing steel manufactured from billet, rail and axle steel respectively. ASTM A 707, A 767 and A 775 are generally not applicable since they cover low alloy, zinc-coated and epoxy-coated reinforcing steel which are currently seldom used in masonry construction.

Reinforcing steel may be either Grade 40, with a minimum yield strength of 40,000 psi or Grade 60 minimum yield strength of 60,000 psi. Grade 60 steel is furnished in all sizes, while Grade 40 steel bars are normally only available in #3, #4, #5 and #6 sizes. If Grade 40 steel is required, special note must be made to ensure delivery. It is always good practice to determine the grade of steel and sizes available in the area where the project is to be built.

The identification marks are shown in the following order:

1st— Producing Mill (usually an initial)

2nd—Bar Size Number (#3 through #18)

3rd— Type of reinforcing (Type N for New Billet, A for Axle, I for Rail, W for Low Allow.)

4th— Grade of reinforcing for Grade 60 steel (grade is shown as a marked 60 or One (1) grade mark line. The grade mark line is smaller and between the two main longitudinal ribs which are on opposite sides of all U.S. made bars.)

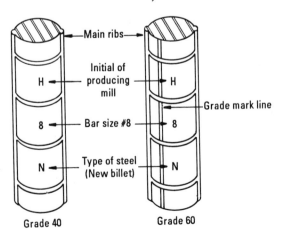

Figure 1-19 Identification marks. Line system of grade marks.

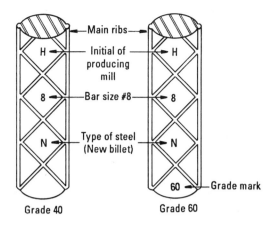

Figure 1-20 Identification marks. Number system of grade marks.

1-4.B.2   Joint Reinforcing

When high strength steel wire fabricated in ladder or truss type configurations is placed in the bed joints to reinforce the wall in the horizontal directions, it is called joint reinforcing.

The most common uses of joint reinforcing are:

(1) to control shrinkage cracking in concrete masonry walls.

(2) to provide part or all of the minimum steel required.

(3) to function as designed reinforcement that resists forces in the masonry, such as tension and shear.

(4) to act as a continuous tie system for veneer and cavity walls.

Joint reinforcement must meet the requirements of UBC Standard No. 24-15, Part 1, *Joint Reinforcement for Masonry.* See Section 7 of this book for additional information on joint reinforcing.

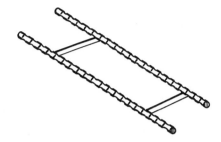

Figure 1-21 Ladder type joint reinforcing.

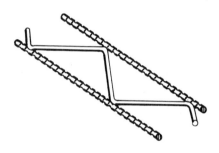

Figure 1-22 Truss type joint reinforcing.

1-5   QUESTIONS AND PROBLEMS

1-1   What three ASTM specifications give the requirements for unit clay masonry?

1-2   What is the range of firing temperatures for building brick and for face brick?

1-3   State the three stages of fusing clay and describe each stage.

1-4   What is the approximate time required for the firing of brick in a kiln?

1-5   What is the difference between a solid clay unit and a hollow clay unit? Can solid units have voids? If so, what is the maximum percentage of voids that is permissible? What are the minimum and maximum percentages of voids in hollow units?

1-6   State the three grades of building brick and describe each grade.

1-7   Describe each type of face brick.

1-8   What are the grades of hollow brick and how are they classified?

1-9   Describe each type of hollow brick.

1-10   What are the three basic physical requirements for clay brick?

1-11   What is the significance of the water absorption rate and the saturation coefficient?

1-12   What is the initial rate of absorption and how does it compare to water absorption?

1-13   Why should clay brick have the proper moisture content before laying? Explain the consequences if it is too wet or too dry.

1-14   Describe each grade of concrete brick. What are the minimum strength requirements for each grade?

1-15   What are the types of concrete brick and what is the difference between them?

1-16   What are the weight ranges for light weight, medium weight and normal weight concrete masonry units?

1-17   Describe each grade of hollow concrete masonry units?

1-18   What are the types of hollow concrete units and what is the significance of each type?

1-19   Define Grade N and Type I and Type II concrete block? What are their net area strength requirements?

1-20   A Type I concrete block is made for use in an area where the relative humidity is 65%. If the linear shrinkage of the unit is 0.058%, what is the maximum moisture content as a percentage of the total absorption that the block may have? If the block weighs 30 pounds per unit dry and 34 pounds per unit saturated, how much moisture can the unit have to qualify for a Type I moisture control unit.

1-21   A wall is constructed with normal weight hollow concrete masonry units. What is the weight of the wall if it is made of nominal 8″ units and is grouted at 48″ o.c.? Compare this to a 12″ solid grouted wall.

1-22   What is meant by the equivalent solid thickness of a hollow unit?

1-23   A Grade N, Type I concrete block unit is made from material which weighs 110 pounds per cubic foot. What is its weight classification? If it is made from material which weighs 127 pound per cubic foot, what is its weight classification?

1-24   What is the significance of a Type I moisture control concrete masonry unit? What are its advantages and disadvantages?

1-25   What is the purpose of mortar? Give six reasons for using mortar.

1-26   Give a classification and description for each type of mortar based upon strength properties.

1-27   What types of mortar are required in Seismic Zone 3 or 4 for structural masonry.

1-28   What are standard proportions for Type M, S, N mortar using portland cement and lime?

1-29   What types of cement may be used in mortar?

1-30   What are the benefits of using hydrated lime in a mortar mix? What are the disadvantages?

1-31   What is the significance of proper grading of sand for masonry mortar? What ASTM specification gives the requirements for mortar sand.

1-32   Are coloring agents for a mortar considered admixtures?

1-33   How long should mortar generally be mixed? What is the effect of over-mixing mortar? What is retempering and how often may mortar be retempered?

1-34   Name and describe four different mortar joints types.

1-35   What is grout? What are its ingredients?

1-36   Give five reasons for using grout.

1-37   What is fine grout and coarse grout?

1-38   What are the normal proportions for fine grout? For coarse grout?

1-39   What is the average slump for grout to be used in a 6″ CMU masonry wall? What should its minimum strength be for fine grout or coarse grout?

1-40   What should the range of slumps be for grout? Why is it allowed to be so fluid?

1-41   Name three admixtures for grout and the reasons to use them.

1-42   Describe the method of making a grout test specimen.

1-43   Describe low-lift grouting.

1-44   Describe high-lift grouting.

1-45   Why must grout be consolidated?

1-46   Sketch a reinforcing bar and show its identification marks.

1-47   What are the advantages of using joint reinforcing?

SECTION 2

Strength of Masonry Assemblies

2-1 GENERAL

Masonry assemblies are comprised of the masonry unit, mortar and grout. Grouted masonry excels in compressive strength and this is the characteristic by which masonry is generally structurally rated.

The ultimate compressive strength of the masonry assembly is given the symbol, f'_{mu}, to distinguish it from the specified compressive strength, f'_m.

To obtain the ultimate compressive strength value, f'_{mu}, prisms are made and tested in accordance with UBC Section 2405(c)2, UBC Standard No. 24-26 *Test Methods for Compressive Strength of Masonry Prisms* and ASTM E 447 *Standard Test Methods for Compressive Strength of Masonry Prisms*. A prism is a test specimen made up of masonry units, mortar and sometimes grout. The masonry units are laid up in stack bond and tested in compression. From the results of the prism test, a value for f'_m can be confidently stated.

2-2 VERIFICATION OF THE SPECIFIED STRENGTH, f'_m.

The required or specified value, f'_m is used as the basis for structural engineering design and must be obtained or verified in accordance with prescribe code requirements.

The 1991 Uniform Building Code has provided the following three methods to verify the specified strength of the masonry assembly, f'_m:

a) *Masonry Prism Testing* - UBC Section 2405(c)1

b) *Masonry Prism Test Records* - UBC Section 2405(c)2

c) *Unit Strength Method* - UBC Section 2405(c)3; Table No. 24-C.

2-3 VERIFICATION BY PRISM TESTS

2-3.A Prism Testing

Prism testing is primarily used when the specified strength, f'_m, is required to be higher than 1500 psi for concrete masonry, or 2600 psi for clay masonry. It is important that prior to construction, adequate lead time be allowed to prepare prisms since retesting could be

Masonry Prism Test

required. The strength developed depends on many factors, including workmanship and materials.

Masonry prisms are built one unit or less in length and in a stack bond arrangement. The construction of a prism with running bond introduces head joints in the specimen forming a vertical plane of weakness, allowing splitting to occur at a much lower value than the actual strength of the wall. In a wall laid up in running bond, the masonry units are confined by the total wall and the effect of the head joints is significantly diminished.

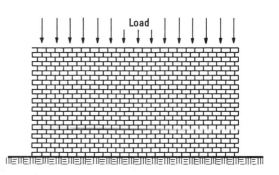

Figure 2-1 Masonry units are confined in the wall and cannot move laterally in plane of wall.

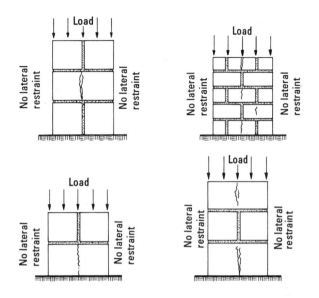

Figure 2-2 Running bond prisms result in low strength not representative of the strength of the wall. Half units are not restrained.

When large masonry prisms are tested in compression, the bearing area of the spherical bearing head block of the testing machine may not be large enough to cover the full area of the specimen. In this case, a solid steel plate should be placed between the bearing block and the specimen so that the entire area of the specimen is covered. The solid plate should have a thickness at least equal to the distance from the edge of the spherical bearing to the most distant corner of the specimen. It is also recommended that the top plate be a minimum of 3½ inches thick.

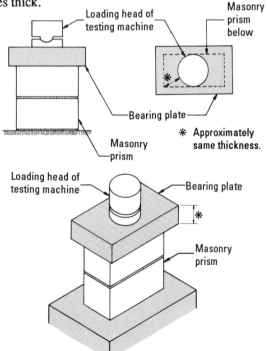

Figure 2-3 Bearing plate thickness.

2-3.B Construction of Prisms

Prisms are to be made using the actual materials that will be used in the construction of the wall. The brick or hollow units, sand and cement that will be used in the structure, should be used in the construction of the prisms. In the prisms, the mortar bedding, the thickness and tooling of joints, the grouting and the condition of the units should be, insofar as possible, the same as will be used in the structure, except that no reinforcement should be included. The prisms should be built in stack bond in accordance with ASTM E 447, *Section 5.4.1, Method A.*

Prisms are to be constructed on a level base and in an opened plastic moisture-tight bag, large enough to enclose the completed prism. Where the cross sections of units vary due to architectural surfaces or taper of the cells, the same placement should be used as specified in the project construction. Prisms should be laid up in stack bond as discussed previously.

The length of masonry prisms can be reduced by saw cutting. Prisms composed of regular shaped hollow units should have at least one complete cell with one full-width cross web on each end. Irregular-shaped units for prisms can be cut to obtain as symmetrical a cross section as possible. The minimum allowable length of saw-cut prisms is 4 inches, but it is recommended that it be at least equal to its width.

Prisms should be a minimum of two units in height, but not less than 1.3 times the least thickness nor more than 5.0 times the least thickness.

When the project construction is to be solid grouted, the prisms are to be solid grouted. The grout should be placed between one to two days following the laying up of the prism. Consolidation of the grout should be the same as that used in the construction. After reconsolidation and settlement due to water loss additional grout is be placed in the prism to level off the top. When open-ended units are used, masonry units may be used to confine the grout during placement.

When the project construction is to be partially grouted, two sets of prisms are to be constructed; one set is grouted solid and the other set remains ungrouted.

Where the walls are multiwythe composite masonry, prisms using each type of unit should be built and tested separately.

Prisms should be left undisturbed in the plastic bags for at least two days following construction.

2-3.C Standard Prism Tests

Uniform Building Code Standard No. 24-26 is based on ASTM Standard E 447-80, which requires a prism two-units high with one mortar joint, as shown in Figure 2-4.

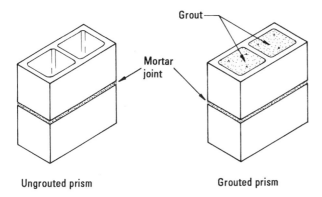

Ungrouted prism   Grouted prism

Figure 2-4 Masonry prism construction for UBC Standard No. 24-26.

However, ASTM E 447-84 requires that prisms have at least two mortar joints and can be made up of a half height unit, a full height unit, and a half height unit.

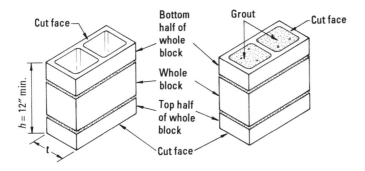

Figure 2-5 Masonry prism construction for ASTM E 447-84.

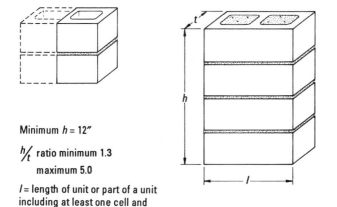

Minimum *h* = 12″

h/t ratio minimum 1.3
          maximum 5.0

l = length of unit or part of a unit including at least one cell and adjacent web but not less than 4″.

Figure 2-6 Size of prism specimen.

In accordance with UBC Sec. 2405(c), a set of five masonry prisms should be made and tested prior to the start of construction of the actual wall so that the required f_m' can be verified for the actual materials. The prisms are tested at 28 days after construction. It is suggested that more than five prisms be made so that a seven day strength can be obtained and the five highest test results can be used for the average valve.

If the full allowable stresses are used in the design, a continuing field check is made with three prisms for each 5000 sq. ft of wall area, but not less than one set of three prisms for the project.

If half stresses are used in the design, additional prism construction and testing is not required after the five initial prisms. The supplier of the material used to initially verify the f_m' can supply a letter of certification that the materials are the same as those used for the initial verification prism tests.

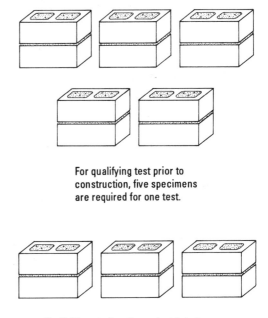

For qualifying test prior to construction, five specimens are required for one test.

For field control as the project is being constructed, three specimens are required for one test when full stresses are used.

Figure 2-7 Number of specimens for prism testing.

It may be difficult to cap and test a full size 8″ × 8″ × 16″ masonry unit prism, particularly if it is high strength clay or concrete masonry. It is recommended that approximately half length units be made into a prism and tested. The half length unit should include the full thickness of the middle cross web.

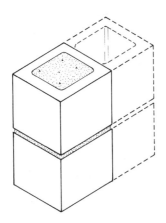

Figure 2-8 Prism of half hollow masonry unit.

The specimen would be approximately 8″ wide by 9″ long. It can be made, transported, capped and tested much easier than a full unit. The results are more consistent in that there is significantly less chance of eccentric loading and uneven capping.

Additionally, smaller prisms do not require special testing machines while full size high strength masonry unit prisms often require testing equipment with a capacity in excess of 750,000 pounds.

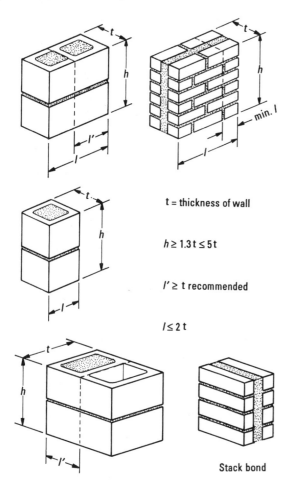

t = thickness of wall

$h \geq 1.3t \leq 5t$

$l' \geq t$ recommended

$l \leq 2t$

Stack bond

Figure 2-9 Sizes of masonry prisms.

For example, brick with a unit strength of 14,000 psi may have an assumed $f'_m = 5300$ psi, however, properly constructed prisms should result in greater strengths. A grouted two-wythe prism 9 inches thick, 18 inches high and $11\frac{1}{2}$ inches long (one unit) would require a testing machine with a capacity of 750,000 pounds. However, if the prism were only 9 inches in length, it would require approximately a 430,000 pound capacity testing machine.

Seven-day tests may also be made to obtain a relationship between the seven day and the 28-day strength. When seven-day tests are made, extrapolation could determine whether projected 28-day tests results will be satisfactory and meet the 28-day strength requirement.

2-3.D   Test Results

The compressive strength of the masonry prisms determined in accordance with UBC Standard No. 24-26 or ASTM E 447 is the ultimate compressive strength, f'_{mu}, and the average for each set of prisms must equal or exceed the specified compressive strength, f'_m.

If the compressive strength of the masonry is to be determined by the result of the prism test, it is to be the lesser of the average strength of the prisms in the set, or 1.25 times the least prism strength multiplied by the prism height to thickness correction factor. Test results are to be multiplied by the correction factors given in Table 2-1 and applies to both concrete and clay masonry prisms.

TABLE 2-1   Prism Correction Factor[2]

Prisms h/t_p[1]	1.30	1.50	2.00	2.50	3.00	4.00	5.00
Correction factor	0.75	0.86	1.00	1.04	1.07	1.15	1.22

1. h/t_p ratio of prism height to least actual lateral dimension of prism.

2. UBC Standard Table No. 24-26-A.

In the ACI 530.1/ASCE 6-88 *Specifications for Masonry Structures* the same correction factors as shown in Table 2-1 (UBC Standard Table No. 24-26-A) apply only to concrete masonry prisms. Another table is provided for the correction factors for clay masonry prisms as shown in Table 2.2.

TABLE 2-2 Correction Factors for Clay Masonry Strength[1]

Prism height to thickness ratio	2.0	2.5	3.0	3.5	4.0	4.5	5.0
Correction Factor	0.82	0.85	0.88	0.91	0.94	0.97	1.0

1. ACI 530.1 Table 1.6.3.3.(b).

The author suggests that Table 2-2 should be applicable to both concrete and clay hollow unit prisms and two wythe grouted prisms while Table 2-3 could be used to determined the strength of single wythe concrete brick and clay brick specimens.

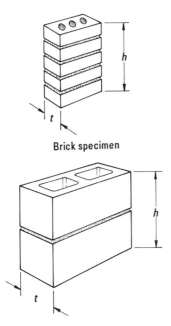

Brick specimen

Two wythe and hollow unit specimens

Figure 2-10 Typical test specimens.

2-3.E Strength of Component Materials

When the compression strength of the masonry assemblage, f'_m, is specified, the component materials, grout and masonry units must be stronger than the specified strength.

Higher strength materials must be specified in order to obtain a satisfactory strength of the wall because there are differences in the Modulus of Elasticity and the Poisson's Ratio between the masonry units and the grout. These differences cause a reduction in the strength of the total masonry assembly and must be compensated for by starting out with higher strength grout and masonry units. In addition, the workmanship in the construction of the prisms and the masonry walls has a significant influence on the strength of the masonry system.

2-3.E.1 Hollow Concrete Masonry

The specified strength is the minimum strength that must be obtained in the wall. It is recommended that for concrete block systems the strength of the masonry unit and grout be 25 to 40 percent more than the specified strength. This may be adjusted if the strength relationship has been established between the materials used and the prism strength. Accordingly, for a specified f'_m of 3000 psi, the concrete masonry units and grout should have a strength of 3700 to 4200 psi. When the masonry unit and the grout are combined and tested, the strength obtained for the prism, f'_m, should be at least 3000 psi.

When specifying masonry units, specify a minimum strength only, not a range of strengths. This minimum strength would be the average of three units tested with the strength of no single unit being less that 20% below the specified minimum average.

2-3.E.2 Clay Brick and Hollow Brick Masonry

Clay brick and hollow brick are generally of high strength in that the clays are fired and fused together to create a strong body or masonry unit. The strength of the units depends on the clays or shale used, the firing temperature and the duration of firing. Table 2-3 (UBC Table No. 24-C) provides a conservative guide for the required strength of clay units to obtain a specified f'_m.

The strength of clay units are normally at least one third more than the specified f'_m. Grout should be mixed to the proportions provided in Section 1-3.B (UBC Table No. 24-B) or prisms may be made to determine the required strength of grout to obtain the f'_m strength. It is suggested that the grout have a strength at least equal to the specified f'_m.

2-3.E.3 Mortar

As specified in the Uniform Building Code, Section 2407(h)4.A *Special Provisions for Seismic Zones Nos. 3 and 4*, only Type S, mortar with a minimum laboratory cube strength of 1800 psi, or Type M mortar, with a minimum laboratory cube strength of 2500 psi, should be used in reinforced grouted masonry. Because of the relatively thin mortar joints, Type S or M mortar used in masonry may have an in-place strength of f'_m = 3000 psi or more. The h/t ratio of the mortar in the joint is very small, enabling the mortar to exhibit strengths far higher than the strengths obtained from cube tests of mortar.

When the compressive strength of mortar is desired, the mortar should be tested in accordance with UBC Standard No. 24-22 (see Section 1-2.B.2(a)).

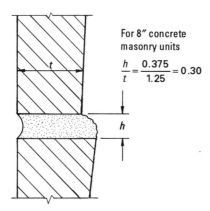

For 8″ concrete masonry units

$$\frac{h}{t} = \frac{0.375}{1.25} = 0.30$$

Figure 2-11 h/t for confined mortar bed joint.

2-3.E.4 Grout

When required, grout strengths are found by making grout specimens (see Figures 1-12 and 1-13) in accordance with UBC Standard No. 24-28 or ASTM C 1019. The minimum strength should not be less than 2000 psi nor should it be less than the strength of the units for concrete masonry construction. For additional information on grout testing, see Section 1-3.H and the above referenced standards.

2-4 VERIFICATION BASED ON PRISM TEST RECORDS

2-4.A Selection of f'_m Based on Experience

Prism test records have demonstrated that f'_m values can be reliably obtained when the same materials are used in their construction. Therefore, UBC Section 2405(c)2 allows f'_m to be based on previous testing history.

The code states that when there is a representative masonry prism test record of at least 30 masonry prism tests conducted in accordance with UBC Standard No. 24-26, the specified compressive strength of masonry, f'_m, may be selected on the basis of this record. The value of f'_m may not exceed 75 percent of the average value of the prism test record (the average compressive strength of the 30 test record prisms must equal or exceed 1.33 f'_m).

During construction, sets of three prisms must still be made for each 5000 square feet of wall area to assure the resulting strength of the prisms is adequate. Not less than one set of three prisms must be made per project when full stresses are used in design.

When half stresses are used in design, no field testing is necessary but a letter of certification is required from the supplier of material stating the materials are representative to those of the prism test record.

2-5 SELECTION BY UNIT STRENGTH

2-5.A Selection of f'_m from Code Tables

The specified compressive strength of masonry, f'_m, may be selected from tables that are based on the strength of the masonry unit and mortar used. These tables are conservative and higher values may be obtained by conducting prism tests.

In order to use full allowable stresses, the masonry units and grout must be tested prior to construction and for each 5000 square feet of wall area during construction to assure compliance with UBC Table No. 24-C (Table 2-3).

When half stresses are used in design, unit testing is not required but a letter of certification from the masonry unit manufacturer must be supplied stating that the units conform to the required compressive strength.

Similarly, for design using full stresses, grout must be tested for each 5000 square feet of wall for concrete masonry, with not less than one test per project, to demonstrate compliance to UBC Table No. 24-C (Table 2-3), Footnote 4. For half stress design, a letter of certification from the grout supplier must be obtained to assure the grout strength is adequate.

For clay masonry using full stresses, grout proportions are to be in accordance with UBC Table No. 24-B and are to be verified by the engineer, special inspector or approved agency. If half stresses are used, a letter of certification of compliance to the proportions of UBC Table No. 24-B (Table 1-19) can be furnished by the grout supplier.

Table 2-3 (UBC Table No. 24-C) shows the allowable UBC f'_m values based on the strength of the concrete or clay unit and the type of mortar used.

TABLE 2-3 Specified Compressive Strength of Masonry, f'_m (psi)[2], Based on Specifying the Compressive Strength of Masonry Units[5]

Compressive Strength of Clay Masonry Units[1] (psi)	Specified Compressive Strength of Masonry, f'_m	
	Type M or S Mortar[3] (psi)	Type N Mortar[3] (psi)
14,000 or more	5,300	4,400
12,000	4,700	3,800
10,000	4,000	3,300
8,000	3,350	2,700
6,000	2,700	2,200
4,000	2,000	1,600

Compressive Strength of Concrete Masonry Units[4] (psi)	Specified Compressive Strength of Masonry, f'_m	
	Type M or S Mortar[3] (psi)	Type N Mortar[3] (psi)
4,800 or more	3,000	2,800
3,750	2,500	2,350
2,800	2,000	1,850
1,900	1,500	1,350

1. Compressive strength of solid clay masonry units is based on gross area. Compressive strength of hollow clay masonry units is based on minimum net area. Values may be interpolated. When hollow clay masonry units are grouted, the grout shall conform to the proportions in Table No. 24-B.

2. Assumed assemblage. The specified compressive strength of masonry f'_m is based on gross area strength when using solid units or solid grouted masonry and net area strength when using ungrouted hollow units.

3. Mortar for unit masonry, proportion specification, as specified in Table No. 24-A. These values apply to portland cement-lime mortars without added air-entraining materials.

4. Values may be interpolated. In grouted concrete masonry the compressive strength of grout shall be equal to or greater than the compressive strength of the concrete masonry units.

5. UBC Table No. 24-C.

ACI 530.1-88/ASCE 6-88 also provides tables for the selection of f'_m based on the strength of the masonry unit and type of mortar used as shown in Tables 2-4 and 2-5 (ACI/ASCE Tables 1.6.2.1 and 1.6.2.2).

TABLE 2-4 Compressive Strength of Masonry Based on the Compressive Strength of the Clay Masonry Units and Type of Mortar Used in Construction[1]

Net Area Compressive Strength of Clay Masonry Units, psi		Net Area Compressive Strength of Masonry, psi, f'_m
Type M or S Mortar	Type N Mortar	
2400	3000	1000
4400	5500	1500
6400	8000	2000
8400	10500	2500
10400	13000	3000
12400	-	3500
14400	-	4000

1. ACI/ASCE Table 1.6.2.1.

TABLE 2-5 Compressive Strength of Masonry Based on the Compressive Strength of Concrete Masonry Units and Type of Mortar Used in Construction[2]

Net Area Compressive Strength of Concrete Masonry Units, psi		Net Area Compressive Strength of Masonry, psi[1], f'_m
Type M or S Mortar	Type N Mortar	
1250	1300	1000
1900	2150	1500
2800	3050	2000
3750	4050	2500
4800	5250	3000

1. For units of less than 4 in. height, 85 percent of values listed.

2. ACI/ASCE Table 1.6.2.2.

It is recommended that the selection or verification of f'_m based on the compressive strength of the masonry units be limited to 2500 psi for concrete masonry and 3500 psi for clay masonry. For greater values of f'_m, it is advisable to conduct prism tests.

2-5.B f'_m for Solid Grouted vs Partially Grouted Masonry

In the 1982 and previous editions of the Uniform Building Code there was a differential of 10% between solid grouted hollow masonry, f'_m = 1500 psi, and partially grouted masonry (grouted only at the location of the reinforcing steel) f'_m = 1350 psi. The differential between solid grouted and partially grouted masonry has been deleted and the values given in Table 2-3 can be used for either solid or partially grouted.

In the majority of cases for walls 4 to 8 inches thick, there is no concern or difference when out-of-plane forces are considered. However, for vertical load and forces parallel to the wall, the actual bedded and grouted area should be used in design or analysis.

For 10 or 12 inches thick partially grouted walls which are subjected to forces or moments perpendicular to the plane of the wall and where the principle steel is placed for maximum d distance, the development of a tee section is much more possible and should be investigated. It would not be unreasonable to reduce allowable masonry stresses 10% and design as a solid fully grouted section as was provided for by the 1982 edition of the Uniform Building Code.

2-5.B.1 Solid Grouted Walls

The use of solid grouted walls has many advantages including:

a. Increased cross-sectional area provides greater capacity for shear and vertical loads.

b. Increased fire rating. An 8″ CMU wall not solid grouted has a fire rating of one hour while a solid grouted wall has a four hour fire rating.

c. In retaining walls, the increased weight improves the stability of the wall.

d. Improved Sound Transmission Coefficient, STC. Solid grouted walls do not easily transmit sound.

e. Design is easier for a solid section.

Some disadvantages are:

a. It requires more material (grout).

b. Wall is heavier and foundation may have to be bigger.

c. Seismic load on wall is greater because it weighs more.

2-5.B.2 Partial Grouted Walls

The advantages to partially grouted walls are as follows:

a. Less material (grout) is needed.

b. Wall is lighter and seismic forces are decreased.

Disadvantages to partially grouted walls are:

a. Decreases cross-sectional area and provides less capacity for shear and vertical loads.

b. Decreased fire rating.

c. In retaining walls, the decreased weight lessens the stability of the wall.

d. Sound transmits more easily through partially grouted walls.

e. Design may be slightly more difficult for a hollow section.

2-6 STRESS DISTRIBUTION IN A WALL

Brick masonry generally has high unit compressive strength and therefore for out of plane forces, the outside brick shells resist the maximum stresses. This offers a great advantage in reinforced brick masonry construction and it may be assumed that f'_m for brick may be 2600 psi as selected from UBC Table No. 24-C.

The net area strength of concrete masonry units must be specified for a project if the specified strength is greater than $f'_m = 1500$ psi. The value of $f'_m = 1500$ psi is based on the standard net area strength of 1900 psi for ASTM C 90 *Standard Specifications for Load-Bearing Concrete Masonry Units*.

If masonry walls or columns are not subjected to flexural stresses and support vertical load only, a deficiency in the strength of the masonry unit may be compensated for by an increase in the strength of the grout. However, this is not a satisfactory solution for stresses perpendicular to the plane of the wall.

Figure 2-12 shows the flexural stress distribution on a cross-section of a wall with maximum flexural compressive stresses on the outside of the wall. The masonry is subjected to compression and the grout may not stressed due to flexural moment. The strength of grout would not contribute greatly to the flexural strength of the wall and the strength of the masonry unit is the governing factor that controls the moment capacity of the wall when moment is perpendicular to the plane of the wall.

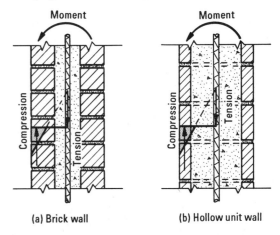

Figure 2-12 Stress distribution in a wall subjected to bending perpendicular to plane of wall.

If the masonry wall is subjected to an overturning moment parallel to the wall, as is the case of a shear wall resisting lateral wind and seismic forces (Figure 2-13), the use of high strength grout to compensate for lower strength masonry may be reasonable. It is, however, recommended that the strength of the component materials be as specified in Section 2-3.E.

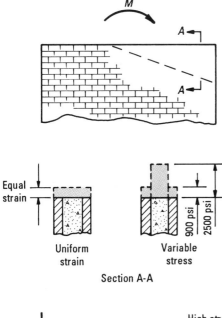

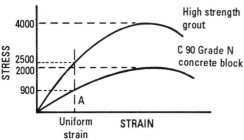

Figure 2-13 Moment parallel to wall, stress and strain distribution.

2-7 WALLS OF COMPOSITE MASONRY MATERIALS

Many times walls are constructed with a combination of masonry materials of different characteristics and strength. If the individual masonry elements of such a composite wall are not bonded together, they would be considered to act structurally independent. In most cases, one masonry element is considered to be the structural wall and the other to be a veneer.

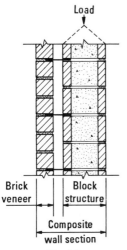

Figure 2-14 Masonry wall with masonry veneer.

If the masonry materials are bonded together it may be assumed that they act as a total structural system, distributing stresses between the wythes. The thickness would be the total thickness of the wall, and the ultimate strength, f'_m, for axial compression, would be limited to the strength of the weakest masonry unit.

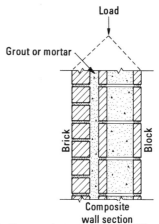

Figure 2-15 Composite masonry wall in which all materials act structurally.

When the wall is subjected to flexure, the ultimate compression strength should be governed by the strength of the masonry unit that is resisting the flexural compression stress. The bond between units would be achieved by grout or mortar as the units are laid.

To obtain a higher strength than the limiting value of the weakest masonry unit, prisms should be constructed and tested in accordance with UBC Standard No. 24-26 or ASTM E 447, as outlined previously. Additional consideration may be given to the relative strengths of masonry materials making up the wall.

Eccentric loads and moments on a wall cause higher stresses on one side than on the other. Higher strength masonry could advantageously be used on the side of higher stress.

An example of this would be in a cantilever retaining wall to use high strength brick on the outside of the wall and lower strength concrete brick on the inside.

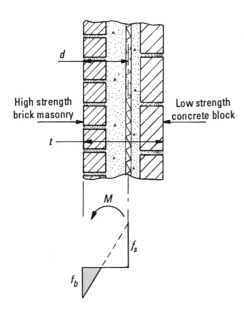

Figure 2-16 Cantilever retaining wall with masonry of different strengths.

2-8   HIGH STRENGTH MASONRY SYSTEMS

With the advancement in construction techniques, the limitations of masonry systems are now pushed to greater extents. Higher strength masonry such as the specified f'_m of 4000 psi used for the 28 story Excalibur Hotel in Las Vegas, Nevada, is more frequently specified and used.

Excalibur Hotel Casino under construction, Las Vegas, Nevada.

Originally, the maximum strength of masonry was low but this value has slowly been increasing. Concrete masonry conforming to ASTM C 90 requirements is designed for f'_m of 1500 psi which is now considered a low design value.

High strength systems for concrete masonry are regularly specified from f'_m = 3000 to 4000 psi and, for clay masonry, the f'_m may be as high as 5000 psi.

2-9   CORES AND PRISMS

Occasionally it is necessary to determine the strength of an existing wall in order to structurally analyze it for possible additional loads, or to insure that the strength of the wall is as prescribed by plans or specifications.

To determine the strength of an existing masonry wall or one just constructed, it is recommended that prisms be sawed from the wall rather that coring with a drill and obtaining a horizontal cylinder. The size of the prism should be in accordance with the principles outlined in Section 2-3 and should be taken at a location that will avoid vertical reinforcing steel. The rectangular prisms would then be loaded in the same direction as they are oriented in the wall. The full load would be shared by the masonry units and the grout and the resulting strength would be reasonably near the true in-place strength of the masonry in the wall. (Note that the saw cutting may reduce the strength, so the wall strength may actually be 10 to 20 percent greater.)

Cores taken 90º to the plane of the wall are tested perpendicular to its normal position and loading in the wall. The full load would be carried directly through the masonry unit, grout and masonry unit. The results of a load test on such a core are thus subject to question as to its indication of the true strength of the wall.

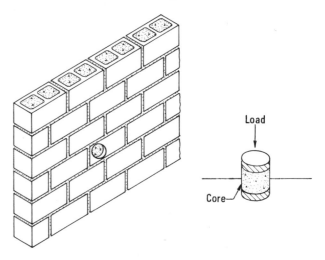

Figure 2-17 Test of core from wall. Full load through all components of specimen.

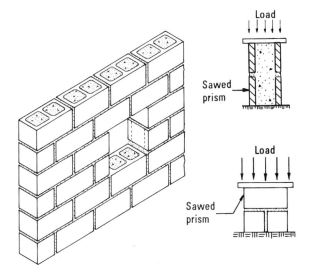

Figure 2-18 Test of prism sawed from wall. Load on specimen causes uniform strain, load is shared by all components of specimen.

2-10 MODULUS OF ELASTICITY, E_m

2-10.A General

The physical measure of a material to deform under load is called the modulus of elasticity, E_m. It is the ratio of the stress to the strain of a material or combination of materials as is the case for grouted masonry.

By definition, the modulus of elasticity, E_m, is to be determined by the secant method in which the slope of the line is taken from $0.05\ f'_m$ to a point on the curve at $0.33\ f'_m$.

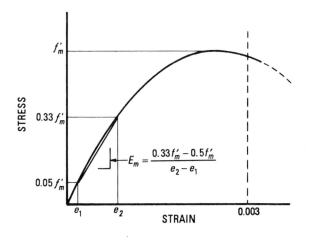

Figure 2-19 Stress-strain curve for grouted masonry prism and slope of line for modulus of elasticity

Originally, E_m for masonry was the same as for concrete, namely $1000\ f'_c$ or for masonry, $1000\ f'_m$. This value changed for concrete in the 1967 UBC to $33w^{1.5}(f'_c)^{0.5}$ to reflect the influence of the unit weight of concrete and the curvature of the stress strain curve.

The value for masonry assemblies was maintained as $E_m = 1000\ f'_m$ until 1988 when it was changed to $750\ f'_m$. This change recognized that masonry is not as stiff as concrete and has a lower modulus.

However, no accommodation was made to further define the E_m based on weight, strength or volume of component materials. Thomas Holme, of the Solite Corporation, has suggested the equation, $E_m = 22w^{1.5}(f'_m)^{0.5}$, to reflect the influence of light weight masonry and the strength of the assembly. Similarly, the Colorado building code has recognized that clay masonry has a lower E_m and thus uses $500\ f'_m$ as the modulus of elasticity of clay masonry.

When using the ACI/ASCE Masonry Code, the modulus of elasticity is given in this book in the A/A Tables A-1 and C-1 (ACI/ASCE Specification Tables 1.6.2.1 and 1.6.2.2.)

2-10.B Proposed Evaluation of Modulus of Elasticity

The modulus of elasticity (E_m) is made up of multiple parameters including the strength of the masonry unit, mortar and grout; the unit weight of the unit, mortar and grout; the volume of each of the components and the material of the masonry unit (clay or concrete).

It can be recognized that the influence of the grout will be greater on a 10 inch concrete masonry unit (CMU) wall than a 6 inch CMU wall. Also if light weight units are used versus normal weight units the modulus will be different. Even varying the type of mortar or the height of the units can affect the modulus of elasticity.

All the above can change the modulus of elasticity but sensitivity evaluations can be made to determine the influence of each parameter. The wide variation in materials, workmanship and quality control may make the detailed determination of the E_m unnecessary or even unrealistic.

As we develop more sophisticated design methods such as strength design and limit state design it would be advantageous to narrow down the modulus of masonry assemblies.

2-11 INSPECTION OF MASONRY DURING CONSTRUCTION

2-11.A Special Inspection

Reinforced masonry is normally built in place at the job site. Accordingly, it is imperative that there be some assurance that the masonry units, mortar, grout, reinforcing steel conform to the specifications and that the construction, steel placement and grouting be as shown in the plans and specifications. This assurance takes the form of observation by a qualified masonry construction inspector as outlined in UBC Sections 305 and 306.

Special Inspection Section 306

(a) General. In addition to the inspections required by Section 305, the owner or the engineer or architect of record acting as the owner's agent shall employ one or more special inspectors who shall provide inspections during construction on the following types of work:

7. Structural Masonry: During preparation of masonry wall prisms, sampling and placing of all masonry units, placement of reinforcement, inspection of grout space, immediately prior to closing of cleanouts, and during all grouting operations.

Exceptions: 1. Special inspection need not be provided when design stresses have been adjusted to permit noncontinuous inspection.

2. For closed-end hollow-unit masonry where f'_m is no more than 1,500 psi for concrete units or 2,600 psi for clay units and cleanouts are provided at the bottom course of every grout pour at each vertical bar, special inspection for placing of units may be performed on a periodic basis in accordance with Section 306(e).

3. For open-ended hollow-unit masonry where cleanouts are provided at the bottom course of every grout pour at each vertical bar, special inspection for placing units may be performed on a periodic basis in accordance with Section 306(e).

When masonry systems are constructed with continuous special inspection as defined above, full allowable stresses may be used to design all structural elements.

2-11.B   Advantages of Inspection

Special inspection has a great advantage in providing concerned parties such as the owner, architect, engineer, building official and masonry contractors assurance that all facets of the masonry construction are in accordance with the plans and specifications.

When using special inspection, full allowable stresses may be used to design the masonry. This can result in smaller members, higher, thinner walls and less reinforcing steel. These beneficial factors can often offset the cost of the inspection.

Special inspection is required when masonry is designed using the strength design methods of UBC Section 2411 and 2412.

2-11.C   Periodic Inspection

UBC Section 306(e) permits the use of full allowable stresses when special inspection is done only on a periodic basis. Inspections must be made to assure that the steel is proper size, in the correct location and that the grout is placed and consolidated correctly.

Periodic special inspection must be performed as specified in the plans and specifications and as approved by the building official.

2-12   REDUCED ALLOWABLE STRESSES

For many reinforced masonry projects, stresses are low or there are no critical conditions in the project. For these cases, the final design may be governed by minimum code requirements. Thus the cost of special inspection may not be warranted. UBC Section 2406(c) states that special inspection need not be provided if the allowable masonry stresses are reduced by one half. This qualification results in designs in which the masonry element may be bigger and/or require more reinforcing steel.

2-13   SUMMARY OF QUALITY CONTROL REQUIREMENTS

To summarize all of the conditions for verification of the strength of masonry systems along with the testing and inspection requirements, the Concrete Masonry Association of California and Nevada developed Table 2-6 Quality Control Requirements, and Table 2-7, Special Inspections Requirements.

TABLE 2-6 Quality Control Requirements

Certification of f'_m UBC Sec. 2405(c)	Design	Testing or Verification Required		
		Prior to Construction All Seismic Zones	During Construction	
			f'_m Zones 0,1, 2 all values, Zones 3 & 4 CMU ≤ 1500 psi[1] CLAY ≤ 2600 psi[1]	f'_m Zones 3, 4 CMU > 1500 psi[1] CLAY > 2600 psi[1]
METHOD 1[2] Masonry prism testing, UBC Sec. 2504(c)1	Full Stress	5 Prisms[3]	3 Prisms ea. 5,000 sq. ft of Wall[4]	3 Prisms ea. 5,000 sq. ft of Wall[4]
	Half Stress	5 Prisms[3]	Letter of Certification[5]	3 Prisms ea. 5,000 sq. ft of Wall[4]
METHOD 2[2] Masonry prism test record (30 prisms), UBC Sec. 2405(c)2	Full Stress	Approved 30 Prism Record[6]	3 Prisms ea. 5,000 sq. ft of Wall[4]	3 Prisms ea. 5,000 sq. ft of Wall[4]
	Half Stress	Approved 30 Prism Record[6]	Letter of Certification[5]	3 Prisms ea. 5,000 sq. ft of Wall[4]
METHOD 3[2] Unit Strength Method, (UBC Table No. 24-C), UBC Sec. 2405(c)3	Full Stress	Units & Grout or 5 Prisms[7]	Units & Grout or 3 Prisms ea. 5,000 sq. ft of Wall[7]	Units & Grout or 3 Prisms ea. 5,000 sq. ft of Wall[7]
	Half Stress	Letter of Certification[8] Except Zones 3, 4, f'_m CMU > 1500 psi CLAY > 2600 psi Units & Grouts or 5 Prisms[7]	None	Units & Grout or 3 Prisms ea. 5,000 sq. ft of Wall[7]

1. UBC Section 2407(h)4 limits f'_m in Seismic Zone Nos. 3 and 4 to a maximum of 1500 psi for concrete masonry and 2600 psi for clay masonry unless f'_m is verified by testing.

2. Grout must achieve a minimum compressive strength at 28 days of 2000 psi. The Building Official may require a field compressive strength test of the grout be made in accordance with UBC Standard No. 24-28.

3. Prism testing in accordance with UBC Section 2405(c)1A.

4. Prism testing in accordance with UBC Section 2405(c)1B (or 2405(c)2D for masonry prism test record).

5. UBC Section 2405(c)1C requires a letter of certification from the responsible party who verified the f'_m in accordance with UBC Section 2405(c)1A (or 2405(c)2A for masonry prism test record) to assure the materials used in the construction of the project are representative of the materials used to construct the prisms prior to construction.

6. Prism testing in accordance with UBC Section 2405(c)2A.

7. Unit testing in accordance with UBC Section 2405(c)3A and grout testing or verification in accordance with UBC 2405(c)3D or F. The exception in UBC Section 2405(c)3A allows prism testing to be performed in lieu of unit testing and grout testing or verification.

8. A letter of certification is required in accordance with UBC Section 2405(c)3B for units and UBC Section 2405(c)3E or C for grout.

TABLE 2-7 Special Inspection Requirements

Design Stress	Method of Design in Accordance with UBC Sections 2407, 2408, 2409, 2410	Method of Design in accordance with UBC Section 2411 or 2412
Full Allowable	Yes	Yes
Half Allowable	No	Not applicable

2-14 QUESTIONS AND PROBLEMS

2-1   What three methods are described in the Uniform Building Code for verifying the specified strength in masonry?

2-2   When must prisms be made? How many prisms are required prior to construction? How many prisms for full stress design should be made during construction?

2-3   Is it necessary to make and test prisms for concrete masonry when $f'_m = 1500$ psi and half stresses are used?

2-4   Are prisms required before and during construction for inspected or uninspected work if $f'_m = 2700$ psi for clay masonry?

2-5   What can the assumed f'_m be for a wall if you use solid clay units for a structure that has a gross strength of 6000 psi? What should be the strength of the grout? Are prisms tests required?

2-6   What are the correction factors based on the UBC for concrete masonry prisms, (a) 12″ thick, 18″ high and 24″ long, (b) 6″ thick, 24″ high and 16″ long? What are the correction factors based upon ACI requirements for (a) hollow clay units 6″ wide, 12″ high and 12″ long, (b) for solid clay units that are 4″ wide, 20″ high and 12″ long?

2-7   What are the strengths of concrete block and grout required to provide a f'_m of at least 3000 psi? What type of mortar should be used and what should be its strength?

2-8   What is the maximum f'_m allowed if the results of five compression tests are as follows: 3250 psi, 2700 psi, 2600 psi, 3400 psi, 3160 psi? If the test results are 4308 psi, 4410 psi, 3560 psi, 3010 psi, 3900 psi, what is the f'_m?

2-9   Why must the strength of the masonry unit be greater than the desired f'_m?

2-10   What are the UBC and ACI equations for the modulus of elasticity?

2-11   What is the influence of the strength of grout and mortar on the modulus of elasticity?

2-12   Explain what is meant by inspected masonry? What are the advantages and disadvantages of inspection during construction?

2-13   Describe the benefits of prisms testing?

2-14   What is meant by periodic inspection and does it qualify for the use of full allowable stresses?

2-15   Why is the compressive strength of grouted masonry systems not governed by the water-to-cement ratio of the mortar or grout as is concrete? State in words why it is better to let a mason use judgment when adding water to a mortar mix rather than specifying a certain amount that must be used.

SECTION 3

Loads

3-1 GENERAL

All structures must be designed to support their own weight along with any superimposed forces, such as the dead loads from other materials, live loads, wind pressures, seismic forces and earth pressures. These vertical and lateral loads may be of short duration such as those from earthquakes, or they may be of longer duration, such as the dead loads of machinery and equipment.

Proper design must consider all possible applied forces along with the interaction of these forces on the structure.

3-2 LOAD COMBINATIONS

Because various loads may act on a structure simultaneously, load combinations should be evaluated to determine the most severe conditions for the design of the structure.

These combinations are given in UBC Section 2303(f) as follows[1]:

1. Dead plus floor live[4] plus roof live (or snow).[2]

2. Dead plus floor live[4] plus wind[2] (or seismic).

3. Dead plus floor live[4] plus wind plus one half snow.[2]

4. Dead plus floor live[4] plus snow plus one half wind.[2]

5. Dead plus floor live[4] plus snow[3] plus seismic.

   Footnotes:

   1. Include lateral earth pressures in the design where they result in a more critical combination.

   2. The Uniform Building Code does not require crane hook loads to be combined with roof live loads nor with more than three fourths of the snow load or one-half of the wind load.

   3. Snow loads exceeding 30 psf may be reduced 75 percent upon the approval of the building official, and snow loads 30 psf or less need not be combined with seismic loads.

   4. The floor live load should not be included if its inclusion would result in lower stresses for the structure or member being designed.

3-3 DEAD LOADS

Dead loads are long term stationary forces which include the self weight of the structure and the weights of permanent equipment and machinery. Additionally UBC Section 2304(d) requires a uniformly distributed dead load of 20 psf when partitions are used.

3-4 LIVE LOADS

Live loads are short duration forces which are variable in magnitude and location. Examples of live load items include people, furniture and snow.

Building codes provide live loads based on the use of the structure. For instance, office areas must be designed for 50 psf live loads (LL), residences for 40 psf LL and corridors for 100 psf LL. Table 3-1 provides a more complete list of design live loads based on use.

3-4.A Floor Loads

Floor live loads are based on the use of a structure and can be found in Table 3-1. If expected floor loads exceed the values in Table 3-1, the actual loads should be used in the design.

Floor loads found in Table 3-1 and roof live loads determined in Table 3-2, Method 2 may sometimes be reduced in accordance with UBC Section 2306 and the following formula:

$$R = r(A\text{-}150) \qquad \text{(UBC Chapter 23, Equation 6-1)}$$

Where:

R = Reduction in percent.

r = Rate of reduction equal to 0.08 percent for floors or as given in Table 3-2 for roofs.

A = Area of floor or roof supported by the member being designed.

Note that this reduction may only be applied to members supporting more than 150 square feet of tributary area and that it may not be applied to floors in areas of public access. Additionally, the reduction may not be applied when the live load is in excess of 100 pounds per square foot except that columns supporting storage loads in excess of 100 psf may receive a 20 percent reduction.

The reduction, R, is limited to 40 percent for members supporting only one floor level and 60 percent for other members. Additionally, the R value found by Equation 6-1 may not exceed:

$$R = 23.1(1 + D/L) \qquad \text{(UBC Chapter 23, Equation 6-2)}$$

Where the terms in UBC Eq. 6-2 are defined as:

D = Dead load per square foot of area supported by the member.

L = Unit live load per square foot of area supported by the member.

Roof live loads are further limited by a maximum reduction, R, and rate of reduction, r, by Table 3-2.

UBC Section 2306 places a final limitation of a maximum 40 percent live load reduction for garages supporting private pleasure cars having capacities of no more than nine passengers per vehicle.

3-4.B Concentrated Loads

Concentrated loads are considered occupying a space 2.5 feet square and are applied to a floor which carries no other live loads. For further details on concentrated load refer to UBC Section 2304(c).

TABLE 3-1   Uniform and Concentrated Loads[9]

Use or Occupancy		Uniform Load[1]	Concentrated Load
Category	Description		
1. Access floor systems	Office use	50	2,000[2]
	Computer use	100	2,000[2]
2. Armories		150	0
3. Assembly areas[3] and auditoriums and balconies therewith	Fixed seating areas	50	0
	Movable seating and other areas	100	0
	Stage areas and enclosed platforms	125	0
4. Cornices, marquees, and residential balconies		60	0
5. Exit facilities[4]		100	0[5]
6. Garages	General storage and/or repair	100	See Footnote No. 6
	Private or pleasure-type motor vehicle storage	50	See Footnote No. 6
7. Hospitals	Wards and rooms	40	1,000[2]
8. Libraries	Reading rooms	60	1,000[2]
	Stack rooms	125	1,500[2]
9. Manufacturing	Light	75	2,000[2]
	Heavy	125	3,000[2]
10. Offices		50	2,000[2]
11. Printing plants	Press rooms	150	2,500[2]
	Composing and Linotype rooms	100	2,000[2]
12. Residential[7]		40	0[5]
13. Rest rooms[8]			
14. Reviewing stands, grandstands, bleachers, and folding and telescoping seating		100	0

Use or Occupancy		Uniform Load[1]	Concentrated Load
Category	Description		
15. Roof decks	Same as area served or for the type of occupancy accommodated		
16. Schools	Classrooms	40	1,000[2]
17. Sidewalks and driveways	Public access	250	See Footnote No. 6
18. Storage	Light	125	
	Heavy	250	
19. Stores	Retail	75	2,000[2]
	Wholesale	100	3,000[2]

1. See UBC Section 2306 for live load reductions.

2. See UBC Section 2304(c), first paragraph, for area of load application.

3. Assembly areas include such occupancies as dance halls, drill rooms, gymnasiums, playgrounds, plazas, terraces and similar occupancies which are generally accessible to the public.

4. Exit facilities shall include such uses as corridors serving an occupant load of 10 or more persons, exterior exit balconies, stairways, fire escapes and similar uses.

5. Individual stair treads shall be designed to support a 300-pound concentrated load placed in a position which would cause maximum stress. Stair stringers may be designed for the uniform load set forth in the table.

6. See UBC Section 2304(c), second paragraph, for concentrated loads.

7. Residential occupancies include private dwellings, apartments and hotel guest rooms.

8. Rest room loads shall be not less than the load for the occupancy with which they are associated, but need not exceed 50 pounds per square foot.

9. UBC Table No. 23-A.

3-4.C   Roof loads

Building codes recognize that roofs carry lower loads than floors since roofs are not occupied or subjected to other high live loads. However, if the roof is used for personnel occupancy, the live load for occupancy must be used in design.

TABLE 3-2   Minimum Roof Live Loads[1,5]

Roof Slope	Method 1			Method 2		
	Tributary Loaded Area in Square Feet for any Structural Member			Uniform Load[2]	Rate of Reduction r (percent)	Maximum Reduction R (percent)
	0 to 200	201 to 600	Over 600			
1. Flat or rise less than 4″ per foot. Arch or dome with rise less than 1/8 of span	20	16	12	20	.08	40
2. Rise 4″ per foot to less than 12″ per foot. Arch or dome with rise 1/8 of span to less than 3/8 of span	16	14	12	16	.06	25
3. Rise 12″ per foot and greater. Arch or dome with rise 3/8 of span or greater	12	12	12	12	No reductions permitted	
4. Awnings except cloth covered[3]	5	5	5	5		
5. Greenhouses, lath houses and agricultural buildings[4]	10	10	10	10		

1. Where snow loads occur, the roof structure shall be designed for such loads as determined by the building official. See UBC Section 2305(d). For special-purpose roofs, see UBC Section 2305(e).

2. See UBC Section 2306 for live load reductions. The rate of reduction r in UBC Section 2306 Formula (6-1) shall be as indicated in the table. The maximum reduction R shall not exceed the value indicated in the table.

3. As defined in UBC Section 4506.

4. See UBC Section 2305(e) for concentrated load requirements for greenhouse roof members.

5. UBC Table No. 23-C.

3-4.C.1   Snow loads

Snow loads are generally established by the local building official. The weight of snow, depth of snow and depth of snow drifts should be obtained from the local jurisdiction where the structure is to be built. Snow loads should be considered in place of the roof live loads from Table 3-2, when their effect will result in larger members.

UBC Section 2305(d) allows snow loads to be reduced if they are in excess of 20 psf for roofs with a pitch over 20 degrees. The reduction is based on:

$$R_s = \frac{S}{40} - \frac{1}{2}$$

Where:

R_s = Snow load reduction in pounds per square foot per degree of pitch over 20 degrees.

S = Total snow load in pounds per square foot.

The Appendix to UBC Chapter 23, Division 1 provides an alternate method of determining snow loads. It is based on the ground snow load, P_g, as shown on UBC Appendix Figures Nos. A-5a, A-5b, and A-5c. The building official should determine the basic ground snow loads.

The roof snow load is calculated as:

$$P_f = C_e I P_g \qquad \text{(UBC Appendix Equation 43-1A)}$$

Where:

C_e is the snow exposure coefficient and is:

0.6 for roofs of buildings which are surrounded by one half mile or more of open terrain,

0.9 for structures in sheltered or densely forested areas, or

0.7 for all other buildings.

I is the importance factor based on occupancy and is:

1.15 for essential facilities and assembly buildings for more than 300 people,

0.9 for agricultural and miscellaneous buildings, or

1.0 for all other buildings.

Roof snow loads are assumed to act vertically on the horizontal projected roof area. When snow loads are more than 20 pounds per square foot and the slope of the roof is more than 30 degrees, the roof load, P_f, may be multiplied by a slope reduction factor, C_s.

For more detailed information on snow loads, drifts and unbalanced snow loads, see UBC Appendix Chapter 23, Snow Load Design.

3-4.C.2 Special Roof Loads

Water can quickly pond on roofs which are not sufficiently sloped or drained. Thus designers must consider the possibility of ponding water which can create substantial additional roof loads and leakage. Likewise special purpose roofs require extra attention and detailing. See UBC Section 2305(e) and (f) for code design requirements if these conditions apply.

3-5 WIND LOADS

Part II of UBC Chapter 23 and UBC Section 2303 cover the principle wind design parameters and requirements. New detailed design and construction requirements have also been added to the 1991 UBC Appendix to Chapter 24 for masonry structures located in high-wind areas.

The design wind pressure, P, is determined by the formula:

$$P = C_e C_q q_s I \qquad \text{(UBC Chapter 23, Equation 16-1)}$$

Where each term is defined and found as follows:

P = Design Wind Pressure in pounds per square foot.

q_s = Wind Stagnation Pressure (psf) at a height of 33 feet as given in Table 3-3.

The wind stagnation pressure is directly related to the wind velocity which the code refers to as the basic wind speed, V. Figure 3-1 of the text can be used to find basic wind velocities for various areas in the United States. Use higher wind speed values if 50-year wind speeds for an area at the standard height of 33 feet above grade are larger than those found in Figure 3-1.

Once the basic wind speed is determined, the stagnation pressure is found in Table 3-3 which is based on the equation $q_s = 0.00256 V^2$.

TABLE 3-3 Wind Stagnation Pressure (q_s) At Standard Height of 33 feet[2]

Basic wind speed (mph)[1]	70	80	90	100	110	120	130
Pressure q_s (psf)	12.6	16.4	20.8	25.6	31.0	36.9	43.3

1. Wind speed from UBC Section 2314.

2. UBC Table No. 23-F

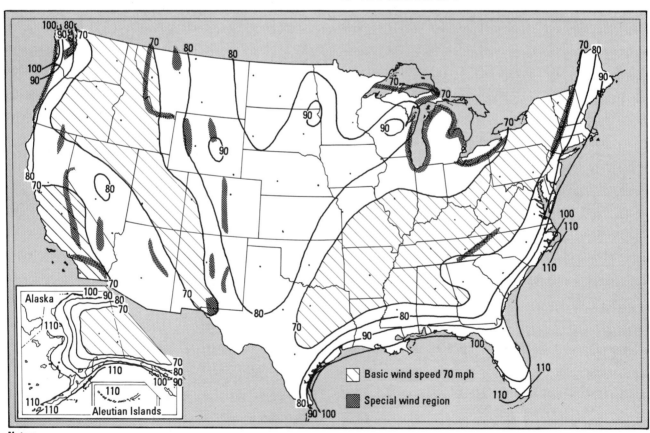

Notes:
1. Linear interpolation between wind speed contours is acceptable.
2. Caution in use of wind speed contours in mountainous regions of Alaska is advised.
3. Wind speed for Hawaii is 80, Puerto Rico is 95 and the Virgin Islands is 110.
4. Wind speed may be assumed to be constant between the coastline and the nearest inland contour.

Figure 3-1 Minimum basic wind speeds in miles per hours (UBC Figure No. 23-1).

For miscellaneous structures such as fences less than 12 feet in height, the stagnation pressure may be reduced to three fourths of q_s but not less than 10 psf per UBC Section 2320.

I = Importance Factor

The code incorporates importance factors in both the wind and seismic design equations to increase the design loads on certain types of facilities. (See Tables 3-4 and 3-5.)

TABLE 3-4   Occupancy Categories[2]

Occupancy Categories	Occupancy Type or Functions of Structure
I. Essential Facilities[1]	Hospitals and other medical facilities having surgery and emergency treatment areas.
	Fire and police stations.
	Tanks or other structures containing, housing or supporting water or other fire-suppression materials or equipment required for the protection of essential or hazardous facilities, or special occupancy structures.
	Emergency vehicle shelters and garages.
	Structures and equipment in emergency-preparedness centers.
	Standby power-generating equipment for essential facilities.
	Structures and equipment in government communication centers and other facilities required for emergency response.
II. Hazardous Facilities	Structures housing, supporting or containing sufficient quantities of toxic or explosive substances to be dangerous to the safety of general public if released.
III. Special Occupancy Structure	Covered structures whose primary occupancy is public assembly—capacity > 300 persons.
	Buildings for schools through secondary or day-care centers—capacity > 250 students.
	Buildings for colleges or adult education schools—Capacity > 500 students.
	Medical facilities with 50 or more residents incapacitated patients, but not included above.
	Jails and detention facilities.
	All structures with occupancy > 5,000 persons.
	Structures and equipment in power generating stations and other public facilities not included above, and required for continued operation.
IV. Standard Occupancy Structures	All structures having occupancies or functions not listed above.

1. Essential facilities are those structures which are necessary for emergency operations subsequent to a natural disaster.

2. UBC Table No. 23-K.

TABLE 3-5   Occupancy Requirement[3]

Occupancy Category[1]	Importance Factor, I	
	Earthquake[2]	Wind
I. Essential Facilities	1.25	1.15
II. Hazardous Facilities	1.25	1.15
III. Special Occupancy Structures	1.00	1.00
IV. Standard Occupancy Structures	1.00	1.00

1. Occupancy types or functions of structures within each category are listed in UBC Table No. 23-K and structural observation requirements are given in UBC Section 305, 306 and 307.

2. For life-safety-related equipment, see UBC Section 2336(a).

3. UBC Table No. 23-L.

C_e = Combined Height, Exposure and Gust Factor Coefficient

This coefficient incorporates the influences of a number of effects and is found in Table 3-6 based on the height of the structure above the ground and the exposure condition of the site.

TABLE 3-6   Combined Height, Exposure and Gust Factor Coefficient(C_e)[1,2]

Height Above Average Level of Adjoining Ground (feet)	Exposure D	Exposure C	Exposure B
0-15	1.39	1.06	0.62
20	1.45	1.13	0.67
25	1.50	1.19	0.72
30	1.54	1.23	0.76
40	1.62	1.31	0.84
60	1.73	1.43	0.95
80	1.81	1.53	1.04
100	1.88	1.61	1.13
120	1.93	1.67	1.20
160	2.02	1.79	1.31
200	2.10	1.87	1.42
300	2.23	2.05	1.63
400	2.34	2.19	1.80

1. Values for intermediate heights above 15 feet may be interpolated.

2. UBC Table No. 23-G.

The Uniform Building Code recognizes only three of the four exposure classifications that were originally outlined in the American National Standard Institute's (ANSI) publication A 58.1. Exposure A which is characterized by building sites in large city centers, has not been adopted.

The UBC classification, Exposure B, is characterized by a site located in a terrain which has buildings, forests or surface irregularities over 20 feet in height covering over 20 percent of the area within one mile of the site. Although a site may be located in Exposure B, often times local building codes require buildings to be designed as if they were located in the more severe Exposure C. This category is characterized by generally flat and open terrain which extends around a building for at least one-half mile. Exposure D represents the most severe exposure and is applicable to structures in flat and unobstructed areas near large bodies of water. Areas subjected to the severe wind forces of hurricanes such as Florida may fall into this category.

C_q = Pressure Coefficient

Pressure coefficients must be determined from Table 3-7 or UBC Table No. 23-H for both the primary load resisting system and the components of the structure. The methods to find the appropriate pressure coefficients for each are listed briefly in Sections 3-5.A and 3-5.B.

TABLE 3-7   Pressure Coefficient (C_q)[9]

Structure or part thereof	Description	C_q Factor
1. Primary frames and systems	Method 1 (Normal force method) Walls:	
	Windward wall	0.8 inward
	Leeward wall	0.5 outward
	Roofs[1]:	
	Wind perpendicular to ridge	
	Leeward roof or flat roof	0.7 outward
	Windward roof	
	less than 2:12	0.7 outward
	Slope 2:12 to less than 9:12	0.9 outward or 0.3 inward
	Slope 9:12 to 12:12	0.4 inward
	Slope >12:12	0.7 inward
	Wind parallel to ridge and flat roofs	0.7 outward
	Method 2 (Projected area method) On vertical projected area	
	Structures 40 feet in height	1.3 horizontal any direction
	Structures over 40 feet in height	1.4 horizontal any direction
	On horizontal projected area[1]	0.7 upward
2. Elements and components not in areas of discontinuity[2]	Wall elements All structures	1.2 inward
	Enclosed & unenclosed structures	1.2 outward
	Open structures	1.6 outward
	Parapet walls	1.3 inward or outward
	Roof elements[3] Enclosed & unenclosed structures	
	Slope < 7:12	1.3 outward
	Slope 7:12 to 12:12	1.3 outward or inward
	Open structures	
	Slope < 2:12	1.7 outward
	Slope 2:12 to 7:12	1.6 outward or 0.8 inward
	Slope > 7:12 to 12:12	1.7 outward or inward

Structure or part thereof	Description	C_q Factor
3. Elements and components in areas of discontinuities[2,4,6]	Wall corners[7]	1.5 outward or 1.2 inward
	Roof eaves, rakes or ridges without overhangs[7]	
	Slope < 2:12	2.3 upward
	Slope 2:12 to 7:12	2.6 upward
	Slope >7:12 to 12:12	1.6 outward
	For slopes less than 2:12 Overhangs at roof eaves, rakes or ridges, and canopies	0.5 added to values above
4. Chimneys, tanks and solid towers	Square or rectangular	1.4 any direct.
	Hexagonal or octagonal	1.1 any direct.
	Round or elliptical	0.8 any direct.

1. For one story or the top story of multistory open structures, an additional value of 0.5 shall be added to the outward C_q. The most critical combination shall be used for design. For definition of open structure see UBC Section 2312.

2. C_q values listed are for 10-square-foot tributary areas. For tributary areas of 100 square feet the value of 0.3 may be subtracted from C_q, except for areas at discontinuities with slopes less than 7:12 where the value of 0.8 may be subtracted for C_q. Interpolation may be used for tributary areas between 10 and 100 square feet. For tributary areas greater than 1,000 square feet, use primary frame values.

3. For slopes greater than 12:12, use wall element values.

4. Local pressures shall apply over a distance from the discontinuity of 10 feet or 0.1 times the least width of the structure, whichever is smaller.

6. Discontinuities at wall corners or roof ridges are defined as discontinuous breaks in the surface where the included interior angle measures 170 degrees or less.

7. Load is to be applied on either side of discontinuity but not simultaneously on both sides.

9. Portion of UBC Table No. 23-H.

3-5.A Primary Load Resisting System

Two methods are outlined in UBC 2317 for the determination of wind pressures and pressure coefficients for the design of the primary load resisting system.

(1) Method 1, called the Normal Force Method, assumes that wind pressures act normal to all exterior building surfaces at one time (see Example 3-A and Figure 3-2). This method may be used to determine wind pressures on any type of structure but it must be used when the building has a gabled rigid frame.

(2) Method 2, called the Projected Area Method, simultaneously applies horizontal pressures to the entire vertical projected area of a structure and vertical pressures to the entire horizontal projected area (See Example 3-A and Figure 3-3). This method is particularly useful in determining the overall stability of structures subjected to wind pressures. It is, however, limited to buildings less than 200 feet high which are not gabled rigid frames.

Whether Method 1 or Method 2 is utilized, the overall stability of the structure must be checked. The overturning moment at the base of a structure is limited to two thirds of the resisting moment from the dead loads of the structure. It should be noted that per UBC Section 2317(a), the weight of the soil on the footings and foundations may be considered in the dead load resisting moment. If a building is no more than 60 feet high and the height-to-width ratio in the wind direction is 0.5 or less, the combined effect of uplift and overturning may be reduced by one-third.

EXAMPLE 3-A Wind Pressure Determination.

A small office building which is to be constructed near Fresno, California. The building will be situated in a relatively open area with only scattered obstructions. Calculate the design wind pressures for the Primary Load Resisting System using both the Normal Force Method and the Projected Area Method.

The structure has a gable roof which does not have a system of rigid frames.

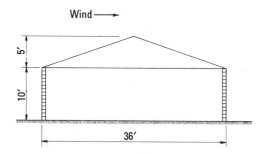

Solution 3-A:

From Figure 3-1, V = 70 mph for Fresno. Therefore, from Table 3-3, q_s = 12.6 psf.

Since this is an ordinary small building, an Importance Factor of 1.0 applies. Also, because the building is located in open terrain, use an Exposure Classification C. Therefore from Table 3-6, C_e = 1.06 for exposure classification C and h = 15'. The pressure coefficient valves, C_q, can be found for the various building elements from Table 3-7. Based on these C_q valves, the design wind pressures valves, P, are determined as shown in Table 3-8.

TABLE 3-8 Method 1 — Normal Force Method

Item	C_q	Design Wind Pressure (psf) $P = C_q C_e q_s I$
Windward wall	0.8	0.8(1.06)(12.6)(1.0) = 10.7 inward
Leeward wall	0.5	0.5(1.06)(12.6)(1.0) = 6.7 outward
Windward roof	0.9 or	0.9(1.06)(12.6)(1.0) = 12.0 outward
	0.3	0.3(1.06)(12.6)(1.0) = 4.0 inward
Leeward roof	0.7	0.7(1.06)(12.6)(1.0) = 9.3 outward

See Figure 3-2 for sketch of applied design pressures.

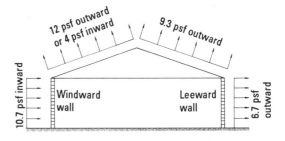

Figure 3-2 Method 1. Wind pressure distribution.

Only the C_q valves will change in the projected area method. From Table 3-7 appropriate C_q valves are found. These are listed along with the design pressure, P, in Table 3-9.

TABLE 3-9 Method 2 — Projected Area Method

Item	C_q	Design Wind Pressure (psf) $P = C_q C_e q_s I$
Vertical projected area	1.3	1.3(1.06)(12.6)(1) = 17.4 psf any direction
Horizontal proj. area	0.7	0.7(1.06)(12.6)(1) = 9.3 psf upward

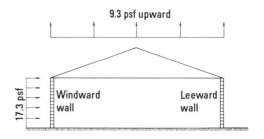

Figure 3-3 Method 2. Wind pressure distribution.

3-5.B Elements and Components of Structures

Wind pressures higher than those determined for the primary load-resisting system by UBC Section 2316, Equation 16-1 are often experienced on small areas of the overall structure, especially at areas of discontinuities such as eaves, ridges and building corners. Because these high pressures are generally distributed over only small areas at any one time, they do not threaten the overall stability of the structure. However, these high pressures can cause failure of individual elements or components of a structure if they are not properly designed and secured with adequate connections. The UBC requires components and elements to be designed according to UBC Section 2316, Equation 16-1 based on the severest of the following loadings:

1. The design pressure found using C_q values for elements and components in Table 3-7 and spread over the entire tributary area of the component.

2. The pressure found based on C_q values for local areas of discontinuities and spread over an area with a width equal to 10 feet or 0.1 times the least width of the structure, whichever is less.

The severest of these loadings should be used to design the component by applying the wind pressure perpendicular to the component's surface. Note that for outward acting forces, the C_e value found in Table 3-6 should be based on a mean roof height and the resulting outward pressure should be applied over the entire height of the structure. For inward forces, however, C_e may be found at the actual height of the element.

3-6 EARTHQUAKE LOADS

3-6.A General

One of the most important lateral force considerations are the loads imposed on a structure by seismic actions. These earthquake loads are sudden, dynamic and can be of immense intensity.

The basic premise of the seismic provisions are:

a. In minor earthquakes, structures should experience no damage.

b. In moderate earthquakes, structural elements should experience no damage, but there may be some damage to non-structural elements.

c. In major earthquakes, structural and non-structural damage may be severe, but the structure should not collapse. Designers are counting on ductility and proper detailing to prevent collapse.

The Uniform Building Code recognizes the probability of earthquake occurrence and severity, site conditions, importance (based on occupancy) of the structure, and type of structural system.

3-6.B Earthquake Zones, *Z*

There are only a few areas in the United States that are not subject to earthquakes. Figure No. 3-4 shows a seismic zone map of the United States which delineates each of the earthquake zones from 0 to 4 and Table 3-10 shows some of the seismic zones of various areas of the world. Each zone indicates the expected intensity of an earthquake that may occur within that area.

The following list briefly describes the amount of damage from an earthquake based on the Mercalli Intensity Scale:

Zone 0 — No damage

Zone 1 — Minor damage; distant earthquakes may cause damage to structures with fundamental periods greater than 1.0 second; corresponds to intensities V and VI of the M.M.* Scale.

Zone 2 — Moderate damage; corresponds to intensity VII of the M.M.* Scale.

Zone 3 — Major damage; corresponds to intensity VII and higher of the M.M.* Scale.

Zone 4 — Those areas within Zone No. 3 determined by the proximity to certain major fault systems.

✳ Modified Mercalli Intensity Scale of 1931.

TABLE 3-10 Seismic Zone Tabulation For Areas Outside the United States

Location	Seismic Zone	Location	Seismic Zone
ASIA:		PACIFIC OCEAN AREA:	
Turkey		Caroline Island	
Ankara	2	Koror, Paulau	2
Karamursel	3	Ponape	0
ATLANTIC OCEAN AREA:		Johnson Island	1
Azores	2	Mariana Islands	
Bermuda	1	Guam	
CARIBBEAN SEA:		Kwajalein	
Bahama Islands	1	Saipan	
Canal Zone	2	Tinian	1
Leeward Islands	3	Marcus Island	1
Puerto Rico	3	Okinawa	3
Trinidad Island	2	Philippine Islands	3
NORTH AMERICA:		Samoa Islands	3
Greenland	1	Wake Island	0
Iceland			
Keflavik	3		

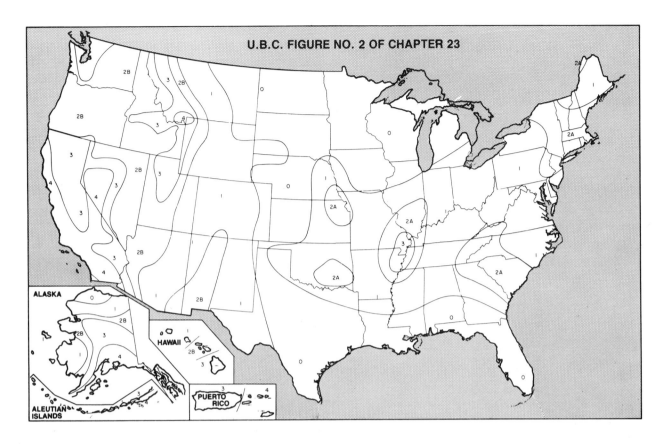

U.B.C. FIGURE NO. 2 OF CHAPTER 23

Figure 3-4 Seismic zone map of the United States (UBC Figure No. 23-2). Note: Pima County, Arizona and Oahu County, Hawaii have been designated Seismic Zone No. 2A.

3-6.C Seismic Forces

The Uniform Building Code, Sections 2330 through 2339 provide requirements for the seismic design of structures. It is based upon the 1988 Recommended Lateral Force Requirements of the Seismology Committee of the Structural Engineers Association of California.

The design procedure considers the earthquake zone, soil and site geology characteristics, the occupancy of the building, the structural systems, the configuration of the structure both vertically and horizontally and the height of the building. It includes both static equivalent load and dynamic analysis. The minimum forces that must be used in design are determined by the static lateral force analysis. The structure may require dynamic analysis depending upon configuration, structural framing and other criteria. Refer to Section 2335 of the 1991 Uniform Building Code for detailed requirements of structures that must be designed by dynamic analysis.

The dynamic motion of the earth when a crustal plate moves and ruptures along a fault line causes a structure to move with an acceleration. This motion and acceleration induces forces within the structure due to its weight and inertial force.

Basically the force on a structure is:

$$F = Ma$$

Where:

M = mass of the structure

a = acceleration of the structure

Figure 3-5 shows the response of both a tall flexible building and a short stiff building subjected to a ground acceleration, a.

The most common technique of determining the lateral seismic forces on a structure is the static equivalent force procedure. Using this procedure, the dynamic seismic force is translated into an equivalent static force on the building and is distributed throughout the height of the building to each of the resisting elements. The static seismic force is assumed to be an external base shear force, V, that is applied to the structure. Vertical effects of an earthquake are neglected.

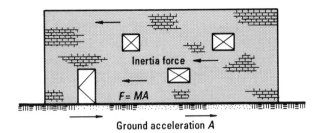

Ground acceleration A

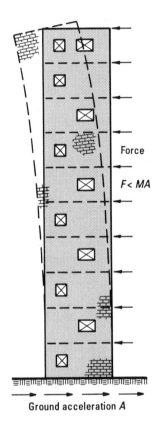

Force

$F < MA$

Ground acceleration A

Figure 3-5 Response of a short stiff building and a tall flexible building.

The use of the static force design procedure assumes that the seismic force is an external force, V, applied to the structure. This is similar to the design of wind forces on the building.

3-6.C.1 Base Shear, V

When using the static force design procedure, the seismic base shear force, V, that acts on the building in a given direction is as follows:

$$V = \frac{ZIC}{R_w} W \quad \text{(UBC Chapter 23, Equation 34-1)}$$

The nomenclature and symbols are explained as follows:

V = The Total Design Lateral Force or Shear at the Base of a Building

Z = Seismic Zone Factor

Table 3-11 correlates the seismic zone numbers shown in Figure 3-4 to zone factors, Z, which reflect the effective peak lateral ground acceleration as a percent of gravity. Thus, for Seismic Zone No. 4, Z = 0.4, which means that the design is based on a lateral ground acceleration of 40 percent of gravity. The other zones have been scaled down to reflect lower anticipated ground accelerations.

TABLE 3-11 Seismic Zone Factor, Z[1]

Zone	1	2A	2B	3	4
Z	0.075	0.15	0.20	0.30	0.40

1. UBC Table No. 23-I.

I = Importance Factor

Structures are classified for design based upon their use and importance. Essential facilities such as hospitals, fire stations, emergency centers and communication centers must remain functioning in a catastrophe and are therefore designed for greater safety factors using these I values.

Table No. 3-4 lists various occupancy categories and Table 3-5 lists the appropriate seismic importance factors.

C = Numerical Coefficient determined by the equation:

$$C = \frac{1.25\, S}{T^{2/3}} \quad \text{(UBC Chapter 23, Equation 34-2)}$$

Where S is the site coefficient and T is the period as described later in this section. Note that C need not exceed 2.75. Additionally, the value C/R_w must not be less than 0.075 except as noted in UBC Section 2334(b)1.

T = Fundamental Period of Vibration (in seconds)

The period of a structure is influential in its response to an earthquake and is dependent on the framing and the height of the building.

During an earthquake, a building will vibrate in at least one mode of vibration for a period of time (See Figure 3-6). It may vibrate only back and forth in the simple first mode of vibration or it may vibrate in higher modes depending on the acceleration and duration of the earthquake. Since earthquakes produce erratic ground motions in various directions, most buildings respond in higher modes of vibration, allowing one part of the building to move in one direction while another part of the building moves in the opposite direction. Note, however, that just after an earthquake, buildings may vibrate into lower modes which may cause even more severe stresses than those generated during the earthquake.

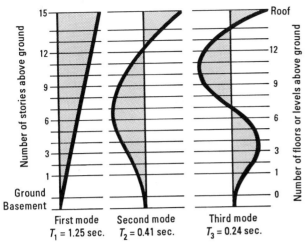

Figure 3-6 Three modes of vibration that a building may respond to in an earthquake (Blume, Newmark and Corning, 1961).

Two methods are currently outlined in the UBC to determine the fundamental period of vibration, T:

Method A: The value, T, may be approximated for all structures as:

$$T = C_t (h_n)^{3/4} \quad \text{(UBC Chapter 23, Equation 34-3)}$$

The value C_t may be taken as $0.1/\sqrt{A_c}$ for masonry shear walls as described in UBC Section 2334(b)2.A. Normally, however, C_t is approximated as 0.02 for masonry buildings and thus:

$$T = 0.020 (h_n)^{3/4}$$

Table 3-12 lists the periods of masonry structures of various heights based on UBC Eq. 34-3 and $C_t = 0.02$. Similarly, Figure 3-7 is useful in easily approximating the seismic coefficient, C, based on the period and the site coefficient.

TABLE 3-12 Period of Masonry Structures (Seconds)

Height; h_n	Period; T^1 (seconds)
40	0.32
60	0.43
80	0.53
100	0.63
120	0.73
140	0.81
160	0.90

1. Based on UBC Chapter 23, Eq 34-3 with $C_t = 0.020$

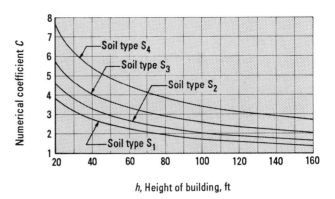

Figure 3-7 Seismic coefficient, C, based on the period, T, and the site coefficient, S.

METHOD B: Alternately, the fundamental period, T, may be determined using the structural properties and the deflection characteristics of the resisting elements by using the Rayleigh type formula:

$$T = 2\pi \sqrt{\left(\sum_{i=1}^{n} w_i \, \delta_i{}^2 \right) \div \left(g \sum_{i=1}^{n} f_i \, \delta_i \right)} \quad \begin{array}{l}\text{(UBC Chapter 23;}\\ \text{Equation 34-5)}\end{array}$$

Where:

δ_i = Deflection at level i relative to the base, due to applied lateral forces

f_i = Portion of a total distributed lateral force at level i

The values of f_i and δ_i may be determined from a lateral force distributed approximately in accordance with the principles shown in this handbook or any other rational distribution. The elastic deflections, δ_i, must be calculated using the applied lateral forces, f_i.

Note, if Method B is used to find the period, T, the resulting value of, C, is not permitted to be less than 80% of the value obtained by using, T, from Method A.

S = Site Coefficient; Soil Characteristics and Site Geology

The inter-relationship between the soil characteristics and the structure significantly affect the seismic forces imposed on a structure.

A flexible building founded on a soft soil will respond to the ground acceleration and will be subjected to high seismic forces because the building and soil will have longer periods. Conversely, a flexible building founded on a stiff, bedrock foundation will not be subjected to nearly as high forces because of the difference of periods between the foundation and the building. This phenomenon was evident in the Caracas earthquake of 1967 and the Mexico City earthquake of 1985. In Caracas, standard concrete framed, eight story apartment buildings were located throughout the city. When founded on hardpan soil or rock, these buildings performed very well, but buildings founded on soft alluvium soil were seriously damaged.

Accordingly, the items to be considered for the response of a building is the soil type and the period of the building.

Soil or site coefficients are given in Table 3-13.

TABLE 3-13   Site Coefficients,[1,2] S

Type	Description	S Factor
S_1	A soil profile with either: (a) A rock-like material characterized by a shear-wave velocity greater than 2,500 feet per second or by other suitable means of classification, or (b) Stiff or dense soil condition, where the soil depth is less than 200 feet.	1.0
S_2	A soil profile with dense or stiff soil conditions, where the soil depth exceeds 200 feet.	1.2
S_3	A soil profile 70 feet or more in depth and containing more than 20 feet of soft to medium stiff clay but not more than 40 feet of soft clay.	1.5
S_4	A soil profile containing more than 40 feet of soft clay characterized by a shear wave velocity less than 500 feet per second.	2.0

1. The site factor shall be established from properly substantiated geotechnical data. In locations where the soil properties are not known in sufficient detail to determine the soil profile, type soil profile S_3 shall be used. Soil profile S_4 need not be assumed unless the building official determines that soil profile S_4 may be present at the site, or in the event that soil profile S_4 is established by geotechnical data.

2. UBC Table No. 23-J.

Four types of soil are listed from type S_1 which is a bedrock or stiff soil with a factor of 1 to a soft soil of Type S_4 with a soil factor of 2.0. The technical data supplied by a soils or geotechnical engineer can often be used to find appropriate soil factors for a site.

R_w = Framing Factor

The framing factor R_w is based on the materials, the ductility and the type of framing system. UBC Table No. 23-O, *Structural Systems* provides values for numerous structural systems and materials. Table 3-14 shows a portion of that table which pertains to masonry shear wall systems. For each system, a height limitation is arbitrarily imposed by the UBC as shown.

Load bearing masonry wall buildings have a box system of framing in which the walls carry both vertical and horizontal loads. These rigid type of structures would be designed with a $R_w = 6$ to insure minimum architectural and structural damage in small and moderate earthquakes. In great earthquakes all types of structures will be damaged but should remain standing.

W = Total Seismic Dead load

The seismic dead load, W, is the total dead load of the structure plus:

1. 25% of floor live loads for storage and warehouse buildings.

2. 10 pounds per square foot if partitions are used.

3. Snow loads when they are greater than 30 psf. These may be reduced up to 75% as noted in UBC Section 2334(a)3.

4. Weights of all permanent equipment.

TABLE 3-14   Structural Systems,[8] R_w

Basic Structural System[1]	Lateral Load Resisting System Description	R_w[2]	H[3]
A. Bearing Wall System	2. Shear walls b. Masonry	6	160
B. Building Frame System	3. Shear walls b. Masonry	8	160
D. Dual System	1. Shear Walls d. Masonry with SMRF e. Masonry with steel OMRF[8] f. Masonry with Concrete IMRF[4,8]	8 6 7	160 160 —

1. Basic structural systems are defined in UBC Sec. 2333(f).

2. See UBC Sec. 2334(c) for combination of structural system.

3. H-Height limit applicable to Seismic Zones Nos. 3 and 4. See UBC Sec. 2333(g).

4. Prohibited in Seismic Zones Nos. 3 and 4.

8. Portion of UBC Table No. 23-O, where IMRF: Intermediate Moment Resisting Frame; where SMRF: Special Moment Resisting Frame; where OMRF: Ordinary Moment Resisting Frame.

3-6.D Vertical Distribution of Total Seismic Forces

The inertial force of a building caused by an earthquake acceleration is distributed vertically over the height of the building according to equation:

$$V = F_t + \sum_{i=1}^{n} F_i \quad \text{(UBC Chapter 23, Equation 34-6)}$$

F_t is the force at the top of the building caused by the whiplash action as the building accelerates back and forth due to the motion of the earthquake. It may be determined by the equation:

$$F_t = 0.07 T V \quad \text{(UBC Chapter 23, Equation 34-7)}$$

F_t need not exceed $0.25\,V$ and may be considered zero when the period, T, is 0.7 seconds or less. The period, T, in this equation is the same, T, used to determine the response factor, C.

The balance of the seismic force, $(V - F_t)$ is distributed over the height of the structure including Level, n, according to the equation

$$F_x = \frac{(V - F_t) w_x h_x}{\sum_{i=1}^{n} w_i h_i} \quad \text{(UBC Chapter 23, Equation 34-8)}$$

At each level, x, the lateral force, F_x, is applied over the area of the building based on the mass distribution at that level. Forces to the shear walls are calculated due to the forces F_x and F_t applied at the appropriate levels above the base (see Figure 3-8).

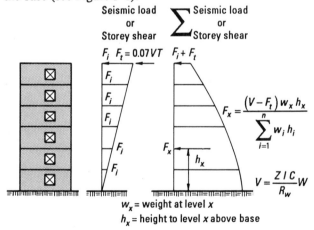

Figure 3-8 Seismic shear on building.

3-6.E Seismic Loads on Elements of a Structure

3-6.E.1 Elements

Individual elements of a building such as walls, parapets, partitions, etc. must also be designed to resist forces due to seismic motions.

The seismic forces on the elements, F_p, are calculated using the horizontal force coefficients factor, C_p, as given in Table 3-15, the importance factor, I, the seismic zone factor, Z, and the weight of the element or part, W_p, per the equation:

$$F_p = Z I C_p W_p \quad \text{(UBC Chapter 23, Equation 36-1)}$$

TABLE 3-15 Horizontal Force Factor[6], C_p

Elements of Structures & Nonstructural Components[1]	Value of C_p	Footnote No.
I. Part or Portion of Structure		
1. Walls, including the following		
a. Unbraced (cantilevered) parapets.	2.00	
b. Other interior walls above the ground floor.	0.75	2, 3
c. All interior bearing and nonbearing walls and partitions.	0.75	3
d. Masonry or concrete fences over 6 feet high.	0.75	
2. Penthouse (except where framed by an extension of the structural frame).	0.75	
3. Connections for prefabricated structural elements other than walls, with force applied at center of gravity.	0.75	4
4. Diaphragms.	—	5
II. Nonstructural Components		
1. Exterior and interior ornamentations and appendages.	2.00	
2. Chimneys, stacks, trussed towers, and tanks on legs.		
a. Supported on or projecting as an unbraced cantilever above the roof more than 1/2 its total height.	2.00	
b. All others, including those supported below the roof with unbraced projection above the roof less than 1/2 its height, or braced or guyed to the structural frame at or above their centers of mass.	0.75	

1. See UBC Sec. 2336(b) for items supported at or below grade.

2. See UBC Sec. 2337(b)4C and UBC Sec. 2336(b).

3. Where flexible diaphragms, as defined in UBC Sec. 2334(f), provide lateral support for walls and partitions, the value of C_p for anchorage shall be increased 50 percent for the center one half of the diaphragm span.

4. Applies to Seismic Zones Nos. 2, 3 and 4 only.

5. See UBC Sec. 2337(b)9.

6. Portion of UBC Table No. 23-P.

EXAMPLE 3-B   Seismic Force Factors.

Determine the seismic design force factor on a wall and parapet for earthquake zone 4.

Assume $I = 1.00$

$Z = 0.40$ for zone 4

$C_p = 0.75$ for walls

$C_p = 2.0$ for parapets

Solution 3-B:

$F_p = ZIC_pW_p$

Therefore: for the wall,

$F_p = 0.40 \times 1.00 \times 0.75W_p = 0.30W_p$

Parapet design $F_p = 0.40 \times 1.00 \times 2.00W_p = 0.80W_p$

However the maximum total seismic lateral loading on a one story masonry shear wall building would be

$V = \dfrac{ZIC}{R_w}W$ where $C \le 2.75$ and $R_w = 6$

(Table 3-14)

$= \dfrac{(0.40 \times 1.00 \times 2.75)W}{6}$

$= 0.183W$

Therefore, the design force for a wall ($0.30W_p$) is 164% greater than for the building ($0.183W$).

The lower total lateral force coefficient for the entire building, 0.183, vs 0.30 which is the lateral force coefficient for the wall, is due to evaluating the response of the building as a total structure while individual elements are designed for higher loads to prevent local failures.

Many engineers reason that if the shear resisting walls of the building constitute the major portion of the building, as in the case of low rise buildings, they need only design the walls for the lateral load on the building as a whole ($0.183W$). It is recommended that connections and individual elements still be designed for the higher seismic coefficients.

The individual elements of a building are considered rigid with a period of 0.06 seconds or less and thus designed for higher seismic forces than the building as a whole. The forces, F_p, are used to design the elements and the connections to the structure so that they may transmit the seismic forces through the system.

3-6.E.2   Anchorage of Masonry Walls

UBC Section 2310 requires that masonry walls be anchored to the structure to resist horizontal forces, F_p, or a minimum of 200 pounds per linear foot of wall, whichever is greater.

UBC Section 2310

Concrete or masonry walls shall be anchored to all floors, roofs and other structural elements which provide required lateral support for the wall. Such anchorage shall provide a positive direct connection capable of resisting the horizontal forces specified in this chapter or a minimum force of 200 pounds per lineal foot of wall, whichever is greater. Walls shall be designed to resist bending between anchors where the anchor spacing exceeds 4 feet. Required anchors in masonry walls of hollow units or cavity walls shall be embedded in a reinforced grouted structural element of the wall. See Sections 2336, 2337(b) 8 and 9.

For additional UBC requirements on the anchorage of masonry walls to diaphragms and sub-diaphragms, see Section 4-2A.

3-6.E.3   Interior Walls

UBC Section 2309 requires that all interior walls and partitions in excess of 6 feet in height be designed for a minimum lateral force perpendicular to the wall of 5 psf.

3-7   RETAINING WALL LOADS

3-7.A   Loading Conditions

Loads imposed on cantilevered retaining walls are primarily a result of the active earth pressure exerted by the soil. The weights of the wall and the soil over its heel plus the overturning lateral load exert a vertical force which must be less than the allowable bearing capacity of the foundation soils.

Retaining walls should always be designed with adequate drainage behind the wall. The following discussion and recommended earth pressures are based on a well-drained backfill with no build-up of hydrostatic pressures behind the wall. Ideally, to minimize lateral earth pressures and to prevent the build-up of hydrostatic pressure, all backfill placed behind the wall should consist of free-draining, granular material. However, as a minimum, a free-draining, granular material zone at least 24-inches wide should be placed adjacent to the wall for its full height. A lateral footing drain should be placed behind the wall, consisting of perforated pipe surrounded by at least 4 inches of pea gravel on all sides. The footing drain should be sloped so that water will flow by gravity to a suitable discharge point.

If silty soil is used for backfill behind retaining walls, far greater lateral pressures develop. It is difficult to evaluate the magnitude of these lateral earth pressures since the density and moisture content of such low permeability soils play a significant role. If much of the soil is loose, the soil will readily absorb water and become a saturated mass. This will further increase the wall's lateral pressure since the water pressure and the soil components are additive. Another major problem with using silty soils as backfill is that the fines in the soil can plug the washed rock and the footing drains causing full hydrostatic pressure to develop against the wall.

3-7.B Active Earth Pressure

The science of soil mechanics provides methods of determining the active earth pressures acting on a retaining wall. This active earth pressure depends on the type of soil, the height of retained soil, the slope of the soil above grade, the effectiveness of soil drainage, the type of restraint, any possible surcharges, etc.

To achieve an active condition, the top of the wall must be free to deflect at least 0.1 percent of the wall height during backfilling. Since such a small amount of movement is difficult to prevent, most design cases can safely assume an active soil pressure condition except for walls restrained by stiff floor diaphragms or prestressed tie-back anchors which were constructed prior to backfilling. If the wall is restrained prior to backfilling, an at-rest soil pressure condition results, increasing the lateral earth pressures significantly.

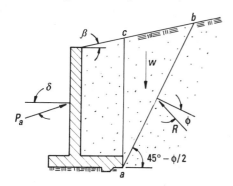

a. Soil wedge exerting pressure on wall.

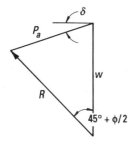

b. Force diagram.

Figure 3-9 Soil pressures acting on wall.

The lateral and vertical pressures acting on a wall in Figure 3-9 result from the weight of soil wedge *abc* straining along plane *ab*. The weight of the soil is partially supported as a force normal to *ab* and as a frictional resistance tangential to *ab*. These two resisting forces can be combined into a resultant, R, acting at an angle of ϕ from the normal of plane *ab*. The weight of the soil which is not resisted by the forces acting along *ab* must be carried by the retaining wall as shown by the force diagram, Figure 3-9b. This resulting pressure force acts on the wall at an angle δ below the horizontal.

The active force applied on a wall is often assumed to be a pure horizontal force. However, as can be seen from the force diagram in Figure 3-9b, the active force acts at an angle δ below the horizontal and therefore has a vertical component. This angle δ is sometimes referred to as the angle of wall friction due to its dependence upon the vertical plane *ac* of the active soil wedge. This angle is sometimes approximated as $2/3\phi$ for a masonry wall and $1/3\phi$ for a steel wall. However, since most retaining walls have a heel extending from the wall, this vertical plane does not act along the wall face but rather up through the edge of the heel. Therefore, this angle may approach ϕ of the soil for most retaining walls.

The active pressure applied against a retaining wall can be expressed as an equivalent fluid unit weight. The equivalent fluid unit weight models the soil as a fluid with a triangular pressure distribution. The Uniform Building Code states in Section 2308 that an equivalent fluid unit weight of not less than 30 pounds per cubic foot should be used for the depth of the retained earth. The City of Los Angeles Building Code provides a table of equivalent fluid weights for drained granular soils. Table 3-16 shows these values from a level soil condition to a 1 to 1 sloped soil condition.

TABLE 3-16 Equivalent Fluid Weight[2]

Surface Slope of Retained Material[1] — Horiz. to Vert.	Equivalent Fluid Weight lb/cu. ft
Level	30
5 to 1	32
4 to 1	35
3 to 1	38
2 to 1	43
1½ to 1	55
1 to 1	80

1. Where the surface slope of the retained earth varies, the design slope shall be obtained by connecting a line from the top of the wall to the highest point on the slope whose limits are within the horizontal distance from the stem equal to the stem height of the wall.

2. Los Angeles City Code Table No. 29-E.

To better evaluate the anticipated active earth pressures, Terzaghi and Peck in *Soil Mechanics in Engineering Practice* have described five different soil types, as shown in Table 3-17. The soil characterization should be based upon the type of soil found at the active wedge failure surface located along plane *ab*. This plane typically extends from the heel of the wall up at approximately a 45 to 60 degree angle. It should be noted that the soil type may not be granular soil unless the failure zone described above is filled with granular soil.

TABLE 3-17 Types of Backfill for Retaining Walls

Type	Description
1.	Coarse-grained soil without admixture of fine soil particles, very permeable (clean sand or gravel).
2.	Coarse-grained soil of low permeability due to admixture of particles of silt size.
3.	Residual soil with stones, fine silty sand, and granular materials with conspicuous clay content.
4.	Very soft or soft clay, organic silts, or silty clays.
5.	Medium or stiff clay, deposited in chunks and protected in such a way that a negligible amount of water enters the spaces between the chunks during floods or heavy rains. If this condition cannot be satisfied, the clay should not be used as backfill material. With increasing stiffness of the clay, danger to the wall due to infiltration of water increases rapidly.

Tables 3-18 and 3-19 summarize typical horizontal and vertical force values acting on a retaining wall based on the soil type and the back slope angle, β. These values, summarized from Terzaghi and Peck, are only estimates based on empirical data. Specific values for each site should be based on field investigations by a qualified professional.

TABLE 3-18 Lateral Earth Pressure, pcf

Soil Type	Back slope Angle, β				
	Level	5H:1V	4H:1V	3H:1V	2H:1V
1	30	31	32	33	39
2	36	37	38	40	45
3	45	47	49	50	59
4	100	100	103	105	N/A
5	120	125	130	133	155

TABLE 3-19 Vertical Earth Pressure, pcf

Soil Type	Back slope Angle, β				
	Level	5H:1V	4H:1V	3H:1V	2H:1V
1	0	5	9	10	17
2	0	5	9	12	20
3	0	6	12	15	70
4	0	0	0	0	0
5	0	17	28	40	N/A

3-7.C Surcharge Loading

In addition to the normal active soil pressures imposed on a retaining wall, surcharge pressures may exist above and behind the wall. For instance, parking lots are often placed adjacent to the top of retaining walls thus imposing a surcharge loading on the wall. Oftentimes parking lot loads from normal passenger automobiles are modeled as an additional two feet of soil above the proposed final grade. Surcharges due to construction equipment, staging areas, soil stockpiles and freezing water in the backfill, etc., must be evaluated by a geotechnical engineer and added to the forces acting on the wall.

3-7.D Resisting Forces

The resisting forces supporting a retaining wall include the passive earth pressure of the soil at the wall's toe, the friction of the soil along the base of the footing, the weight of the wall and the weight of the soil on the wall's heel. These resisting forces must prevent the wall from lateral sliding, rotating and overturning about the toe of the wall. Retaining walls must be designed to resist sliding and overturning with a factor of safety of at least 1.5.

The allowable bearing capacity of the foundation soils must be capable of supporting the maximum load exerted from the footing. This maximum load occurs at the toe of the footing and is a result of the weight of the wall and the soil over its heel, and the lateral overturning force.

The UBC provides some conservative values for the allowable bearing capacity, the passive earth pressure (as an equivalent fluid unit weight) and the coefficient of sliding resistance as shown in Table 3-20. These values should serve as guidelines only, and specific or site case values should be obtained from a geotechnical engineer.

Table 3-20 Allowable Foundation and Lateral Pressure[8]

Class of Materials[2]	Allowable Foundation Pressure lbs/sq. ft[3]	Lateral Bearing lbs/sq. ft/ft of Depth Below Natural Grade[4]	Lateral Sliding[1]	
			Coefficient[5]	Resistance lbs/sq. ft[6]
1. Massive Crystalline Bedrock	4000	1200	0.70	
2. Sedimentary and Foliated Rock	2000	400	0.35	
3. Sandy Gravel and/or Gravel (GW and GP)	2000	200	0.35	
4. Sand, Silty Sand, Clayey Sand, Silty Gravel and Clayey Gravel (SW, SP, SM, SC, GM, and GC)	1500	150	0.25	
5. Clay, Sandy Clay, Silty Clay, and Clayey Silt (CL, ML, MH, and CH)	1000[7]	100		130

1. Lateral bearing and lateral sliding resistance may be combined.

2. For soil classifications OL, OH and PT (i.e., organic clays and peat), a foundation investigation shall be required.

3. All values of allowable foundation pressure are for footings having a minimum width of 12 inches and a minimum depth of 12 inches into natural grade. Except as in Footnote 7 below, increase of 20 percent allowed for each additional foot of width or depth to a maximum value of three times the designated value.

4. May be increased the amount of the designated value for each additional foot of depth to a maximum of 15 times the designated value. Isolated poles for uses such as flagpoles or signs and poles used to support buildings which are not adversely affected by a $1/2$-inch motion at ground surface due to short-term lateral loads may be designed using lateral bearing values equal to two times the tabulated values.

5. Coefficient to be multiplied by the dead load.

6. Lateral sliding resistance value to be multiplied by the contact area. In no case shall the lateral sliding resistance exceed one half the dead load.

7. No increase for width is allowed.

8. UBC Table No. 29-B.

The lateral bearing value on a shear key will be in proportion to the soil pressure in front of the key. For instance, if the soil pressure is 900 psf, this would be equivalent to an earth fill or depth of earth of $900/110 = 8.2$ ft (110 psf is the weight of earth). The lateral bearing resistance for sandy gravel would be $200 \times 8.2 = 1640$ psf.

3-7.E Soil Pressure Due to Earthquakes

Lateral earth pressures on retaining walls will increase due to earthquake loading. Methods have been developed to calculate this increase for cantilevered walls which are free to yield a sufficient amount to mobilize the soil strength in the soil backfill. The most common method of calculating lateral seismic forces on a cantilevered retaining wall is based upon a static approach developed by Mononobe (1929) and Okabe (1926). This method assumes a static active pressure computed by the Coulomb theory with the addition of a dynamic horizontal force. This analysis is described in detail by Seed and Whitman (1970) and the Applied Technology Council (ATC) (1979).

3-8 QUESTIONS AND PROBLEMS

3-1   Define dead load and live load.

3-2   What are the design live loads for apartments, office buildings, schools and corridors?

3-3   A member supports 300 sq. ft of a floor dead load which is 80 pounds per square foot and a floor live load of 50 pounds per square foot. What is the allowable live load reduction?

3-4   What are the five load combinations to be considered in the design of a structure?

3-5   What is the area considered for a concentrated load? What are the design concentrated loads for a library and a manufacturing plant?

3-6   What is the minimum roof live load for a flat roof in which the tributary area for the structural member is over 600 sq. ft?

3-7   What is the uniform load for a roof that has a rise of 4 on 12 and an area of 425 sq. ft?

3-8   A roof in Alaska has a pitch of 5 inches per foot and a potential snow live load of 100 lbs per sq. ft. What is the design snow load for the roof if the structure is sheltered and has an importance factor is 1.15?

3-9   Figure 3-1 shows the minimum basic wind speeds for various areas in the United States. Explain the significance of these wind speeds and describe the importance of the special wind speed regions. What is the standard height where wind velocities are measured? How does this affect the wind speed at ground level?

3-10   What is the wind load to be considered in the design of a masonry building 90 ft high located in Seattle.

3-11   What are the factors to be considered in the design for wind pressure.

3-12   What are occupancy categories and the importance factors based upon these occupancy categories?

3-13   Describe wind exposure B, C, and D and explain their significance. What is the pressure coefficient C_q? Explain its use for primary frames and elements or components not in areas of discontinuity and chimneys.

3-14   What is the lateral load perpendicular to a 6″ thick solid grouted interior masonry wall which is to be built in (a) Denver, Colorado, (b) San Francisco, California, and (c) Phoenix, Arizona?

3-15   Given a two-story building shown in the Figure below, determine the wind loads on the structure and on the pier elements A, B and C to be used in the lateral force calculations based upon UBC Wind Loading. Assume exposure B with a wind speed of 70 mph and an importance factor 1.0. What are the maximum pressures windward and leeward to be considered on the wall and on the roof?

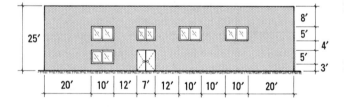

3-16   What is the factor of safety for the stabilizing moment of the dead load against an overturning moment from wind pressure?

3-17   In the design of a structure for earthquake loading, what are the three basic premises upon which the seismic provisions are based?

3-18   What are earthquake zones and how are they categorized? How do they relate to the Modified Mercalli Intensity Scale? In what earthquake zone is St. Louis, Missouri; Boston, Massachusetts; and San Francisco, California; the Island of Hawaii in the Hawaiian Islands and Bermuda?

3-19   What is the basic equation for base shear given in the 1991 Uniform Building Code? Define the factors Z, I, C, R_w and W.

3-20   What is the significance of the fundamental period of vibration of a structure? What is the equation for this period? What is the whiplash effect and when must it be considered?

3-21   What is meant by modes of vibration? What is the first mode of vibration?

3-22   What is the effect of foundation soils on the period of a building? If a stiff building is founded on soft soil as opposed to base rock? What are the consequences?

3-23   What is the significance of the framing factor, R_w, and how do shear wall buildings compare to frame buildings? What is the effect of each on drift of the structure?

3-24   What is the period in each direction for a 10 story shear wall building 120 feet high and 60 feet wide?

3-25   Why is the lateral seismic force on an element greater than the force on the building?

3-26   Give the equation for the seismic force on an element and explain each of the terms. Why is the lateral force coefficient, C_p, greater for a parapet than for a wall?

3-27   What is the minimum anchorage force that a wall must be designed for when connecting it to a floor or a roof diaphragm?

3-28   An 8 foot high cantilevered wall retains a back fill with a slope of 2 to 1. What is the lateral force and overturning moment on the wall?

3-29   A 6 foot high cantilever wall retains a level backfill of type 3 soil and has a surcharge from a parking lot of 200 lbs per square foot. What is the lateral force on the wall?

3-30   What is the minimum factor of safety to be considered for a retaining wall for sliding and overturning?

3-31   What are the allowable foundation and the lateral force resistance pressures for a sandy gravel soil and for a clay and sand clay soil?

3-32   What are the lateral sliding coefficients for bed rock, sandy gravel and sandy silty gravel? What is the sliding resistance for sand clay soil?

SECTION 4

Distribution and Analysis for Lateral Forces

4-1 GENERAL

Buildings must resist not only vertical dead and live loads but also lateral forces from winds and earthquakes. Generally, these lateral forces are resisted by shear walls and/or moment resistant space frames. This section will discuss shear walls primarily, although there is a brief explanation of the new concept of ductile masonry frames in Section 6-9.

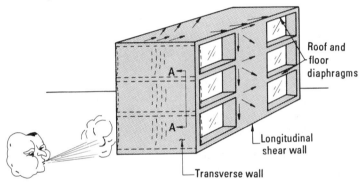

Figure 4-1 Lateral force distribution in a shear wall type building.

As shown in Figures 4-1 and 4-2, lateral forces from severe winds or earthquakes bend transverse walls between the floors. In box type buildings, the lateral loads are transmitted from these transverse walls to the side shear walls by horizontal floor and roof diaphragms.

Masonry shear wall with openings

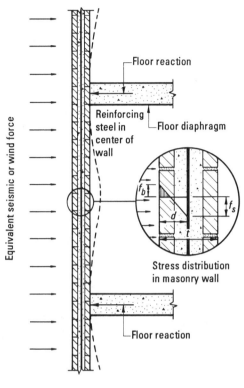

Section A-A

Figure 4-2 Load and stress distribution on wall.

4-2 HORIZONTAL DIAPHRAGMS

Diaphragms are often designed as horizontal beams where the roof or floor systems act as the webs and the bond beams or edge members act as the flange elements. Lateral forces imposed on the horizontal diaphragm cause it to deflect in beam action between the resisting shear walls and/or drag struts (Figure 4-3). As the diaphragm deflects and shear develops, the loads are transferred into the diaphragm ledger members (Figure 4-4). These chord members in turn, transmit the applied forces through anchor bolts into the masonry shear walls or the drag struts. These shear walls must be capable of resisting shear and overturning forces while the drag struts must carry both axial and flexural forces. Likewise, the masonry bond beams which act as flange elements for the diaphragm, must be adequately reinforced to resist the applied tension and compression forces.

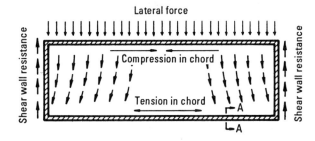

Figure 4-3 Beam action of diaphragm.

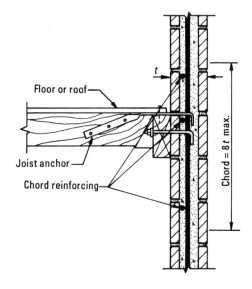

Figure 4-4 Diaphragm chord, Section AA.

Diaphragms do differ somewhat from beams in several special ways, however, as noted in the 1990 SEAOC *Recommended Lateral Force Requirements and Commentary*, Section 1 H.2.j.

"1. The span (of the diaphragm) is usually very short relative to depth; therefore, plane sections are not likely to remain plane, contrary to the usual assumption in the analysis of bending.

2. Web shear stresses and deflections due to shear are relatively more significant (in diaphragms) than stresses and deflections due to flexural action.

3. The diaphragm's components (flange, web, and connection devices) often are made of different materials. The "flanges" may be the walls normal to the direction of loading of the diaphragm, and the "flange" forces at the midspan of the diaphragm would be progressively diminished by the reduction in bending moment toward the diaphragm ends. It is intended that the chord or boundary members that resist these "flange" forces be located near the vicinity of the plane of the diaphragm.

4. Relative and absolute deflections under prescribed lateral loading are often important design limitations."

Numerous types of diaphragm systems are used, most of which are composed of reinforced concrete, metal or wood. Diaphragms may be flat, inclined or curved and may have openings although large openings should be avoided.

4-2.A Diaphragm Anchorage Requirements

The damage resulting from the 1971 San Fernando earthquake indicated that connections between walls and diaphragms were often inadequate. Accordingly, the Uniform Building Code, Sections 2310, 2336 and 2337 now outline much more stringent anchorage requirements. Along with these anchorage requirements are guidelines in using parts of the diaphragm, called sub-diaphragms, to assist in transferring anchor loads from the walls to the diaphragms. The following is a brief list of some of the major UBC anchorage and sub-diaphragm requirements:

1. Masonry walls must be positively anchored to all diaphragms with reinforcing steel, anchor bolts or joist anchors (UBC Section 2310). Connections relying on shear friction are not permitted (UBC Section 2336(a)).

2. Connections must be capable of resisting the larger of the forces determined by UBC Section 2336 (See Section 3-6.E), or, 200 pounds per linear foot of horizontal force in any direction (UBC Section 2310).

3. Anchors may be spaced no more than 4 feet on centers unless the wall is designed to resists bending between the anchors. (UBC Section 2310).

4. Anchors must be embedded in a structural, reinforced grouted element such as a bond beam (UBC Section 2310).

5. Diaphragms which support masonry walls must have continuous ties or struts between diaphragm chords to properly distribute anchorage forces. Sub-diaphragms may be used to transmit the anchorage forces into the main diaphragm (UBC Section 2337(b)9.C).

6. In Seismic Zone Nos. 2, 3 and 4 anchorage of wood diaphragms may not be accomplished with toenails or nails subject to withdrawal. Additionally, wood ledgers may not be used in cross-grain bending (UBC Section 2337(b)9.D).

7. Per UBC Section 2337(b)9.E, the one-third stress increase usually allowed by UBC Section 2303(e) for wind and seismic loads, is not permitted to be included for the design of the diaphragm connections of certain types of buildings.

EXAMPLE 4-A Lateral Load on Diaphragm.

A 40 ft by 100 ft building is subjected to lateral load at its roof line of 700 pounds per linear foot. What is the stress in the chord?

Solution 4-A

Calculate the moment and chord forces

$$M = \frac{wl^2}{8} = \frac{700 \times 100^2}{8} = 875,000 \text{ ft lbs}$$

$$\text{Tension or compression in chord} = \frac{M}{d} = \frac{875,000}{40}$$
$$= 21,875 \text{ lbs}$$

The steel required in a wall bond beam at the roof line may be found as follows:

$$A_s = \frac{T}{F_s} \text{ where } F_s = 1.33 \times 24,000 \text{ psi}$$
$$= 32,000 \text{ psi}$$

$$A_s = \frac{21,875}{32,000} = 0.68 \text{ sq. in.}$$

Conservatively use two #6 bars (A_s = 0.88 sq.in.)

Shear between the ledger and bond beam flange elements

$$= \frac{21,875}{\frac{1}{2} \times 100}$$
$$= 438 \text{ lbs / ft}$$

Use $^5/_8''$ anchor bolts, from Table A-8 (UBC Table No. 24-E)

Allowable shear in masonry = 1330 lbs

$$\text{Spacing of bolts on long wall} = \frac{1330 \times 1.33}{438} \times 12$$
$$= 48'' \text{ o.c.}$$

$$\text{Shear to end walls (shear walls)} = \frac{700 \times 50}{40}$$
$$= 875 \text{ plf}$$

$$\text{Spacing of bolts on short wall} = \frac{1330 \times 1.33 \times 12}{875}$$
$$= 24'' \text{ o.c.}$$

4-2.B Deflection of Diaphragms and Walls

Lateral loads on the walls due to wind or earthquake will cause the diaphragm to deflect and will thus allow the walls to translate relative to their horizontal support at the bottom. Since masonry walls are relatively flexible perpendicular to the plane of the wall, they can tolerate a significant amount of bending and translation without impairing their shear resisting capacity parallel to the wall. The numerous horizontal mortar joints that can crack and open up provide an articulated wall which

allows significant deflections up to 0.007h. The Slender Wall Research Project (1980-1982) conducted by an ACI - SEOASC Task Committee demonstrated this effectively. Overstressing the masonry is not critical as there is a significant safety factor built in.

The deflection of the diaphragm can be calculated by assuming that the walls are the flange elements which resist the bending and deflection. These flange elements can be considered as half the distance between floors or parapet plus half the height of the wall from the floor to the ledger member. Their flange height may also conservatively be assumed as 6 times the wall thickness.

EXAMPLE 4-B Diaphragm Deflection.

Assume that the diaphragm in Figure 4-5 is 100 ft long by 40 ft wide, the parapet is three ft. high and the wall is 14 ft, from the floor to the ledger. The grouted brick wall is 9 inches thick and the lateral load is 500 lbs/ft. Calculate the diaphragm deflection.

Solution 4-B

Use f'_m = 1500 psi and E_m = 1,125,000 psi

$$d = w/2 = 40/2 = 20 \text{ ft}$$

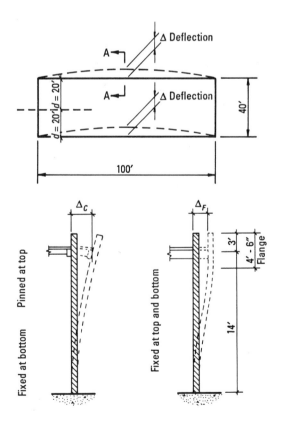

Figure 4-5 Deflection of diaphragm and walls.

Effective width of diaphragm flange
$6t = 6 \times 9 = 54$ in.

Area of flange = $9 \times 54 = 486$ sq. in.

$I = 2Ad^2 = 2 \times 486 \times (20 \times 12)^2$
$\quad = 56 \times 10^6$ in.^{4}

For a simply supported beam subjected to a uniform load:

$$= \frac{5wl^4}{384EI} = \frac{5 \times 500 \times 100^4 \times 1728}{384 \times 1.125 \times 10^6 \times 56 \times 10^6}$$
$$= 0.017 \text{ inches}$$

The moment of inertia is based only on the chords (walls) and not the type of diaphragm. Section 4.2.C describes various types of diaphragms that influence the deflection.

The deflection of walls is prescribed by UBC Section 2411 as a service limitation and is stated as $0.007h$.

For Example 4-B the deflection limitation of the wall is

$\Delta = 0.007(14)12$
$\quad = 1.176''$

This is more than the diaphragm deflection of $0.017''$ and this is a satisfactory design.

For additional information on the design and deflection control of diaphragms see *Diaphragms* by the American Plywood Institute, *Diaphragm Design Manual* by the Steel Deck Institute and other referenced publications.

4-2.C   Types of Diaphragms

As mentioned previously, diaphragms may be constructed of concrete, metal, wood or other suitable materials. They may be flat, inclined, curved, warped or folded and may have openings. The Tri-Services Technical Manual, *Seismic Design for Buildings* classifies diaphragms in five categories, i.e. very flexible, flexible, semi-flexible, semi-rigid and rigid, and are based on an F factor. The factor F is equal to the average deflection, in micro inches, of the diaphragm web per foot of span when stressed with a shear of one pound per foot. Generally, diaphragms are classified as either flexible or rigid depending on the diaphragm deflection relative to the deflections of the resisting vertical walls.

4-2.C.1   Flexible Diaphragms

Since wood and plywood sheathing floors and roofs are relatively flexible in comparison to the much stiffer masonry walls, they are considered as flexible diaphragms. Because of this flexibility, they are assumed to load the shear walls in proportion to the tributary area supported by each wall (See Example 4-C). They are also considered incapable of transmitting rotational or torsional forces.

EXAMPLE 4-C   Shear Force to Walls.

Find the shear force on walls A and B assuming, the roof is a flexible diaphragm.

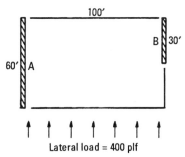

Lateral load = 400 plf

Lateral load to wall A = $400 \times 100/2 = 20,000$ lbs
Lateral load per foot to wall A = $20,000/60 = 333$ plf

Lateral load to wall B = $400 \times 100/2 = 20,000$ lbs
Lateral load per foot to wall B = $20,000/30 = 667$ plf

Table 4-1 shows UBC limitations for the span to width and the height to width ratios of flexible diaphragms.

TABLE 4-1   Maximum Diaphragm Dimension Ratios

Material	Horizontal Diaphragms Maximum Span-Width Ratios	Vertical Diaphragms Maximum Height-Width Ratios
1. Diagonal sheathing, conventional	3:1	2:1
2. Diagonal sheathing, special	4:1	3½:1
3. Plywood and particleboard, nailed all edges	4:1	3½:1
4. Plywood and particleboard, blocking omitted at intermediate joints	4:1	2:1

UBC Table No. 25-I

Flexible diaphragms that have plans in the shape of a T, L or Z can generate variable and incompatible deflections under lateral loads due to the discontinuities in the structure. Figure 4-6(a) illustrates that the deflection of diaphragm A is not compatible with the deflection of diaphragm B.

Thus substantial tearing forces can develop along the boundary between diaphragms A and B especially at point 4.

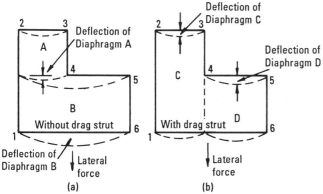

Figure 4-6 Relative deflection of diaphragms in building with irregular plan.

In order to resist these tearing forces and to resolve these incompatible deflections, members, called drag struts, are used to subdivide the irregular plans into a series of rectangular diaphragms such as C and D in Figure 4-6(b). The drag struts also provide the diaphragm with satisfactory boundary elements.

Lateral forces are transmitted from a diaphragm into a drag strut by shear while the drag strut transmits the load into the shear walls by appropriate anchorage. Depending upon the direction of the wind or earthquake forces, the drag strut may be in tension or compression and must be designed for either force.

EXAMPLE 4-D   Determination of Lateral Shear Force to Walls, Flexible Diaphragm.

Calculate the shear force in the shear walls and the drag strut.

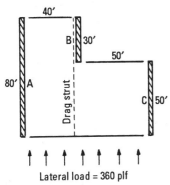

Lateral load to:

$$\text{Wall A} = 360 \times \frac{40}{2} = 7,200 \text{ lbs}$$

$$\text{Wall B} = 360 \times \frac{40+50}{2} = 16,200 \text{ lbs}$$

$$\text{Wall C} = 360 \times \frac{50}{2} = 9,000 \text{ lbs}$$

Lateral load per foot:

$$\text{Wall B and drag strut must resist} = \frac{16,200}{80}$$
$$= 202.5 \text{ plf}$$

Drag strut delivers $202.5 \times 50 = 10,125$ lbs to wall B

$$\text{Wall B must resist} = \frac{16,200}{30} = 540 \text{ plf}$$

Use $^5/_8''$ anchor bolts; Table A-8, (UBC Table No. 24-E) Allowable Shear = 1330 lbs

$$\text{Spacing of anchor bolts} = \frac{1330 \times 1.33 \times 12}{540}$$
$$= 39'' \text{ o.c. max.}$$

As shown in Example 4-D, flexible diaphragms with irregular plans such as L, T, Z, etc., are designed so that each rectangular element will transmit shear forces to their respective resisting elements. The amount of force transferred to the shear resisting elements is in proportion to the tributary areas they support since flexible diaphragms are considered incapable of distributing forces in relation to the rigidity of the shear walls.

Figures 4-7 and 4-8 show plans of irregular buildings along with the tributary areas supported by each resisting element.

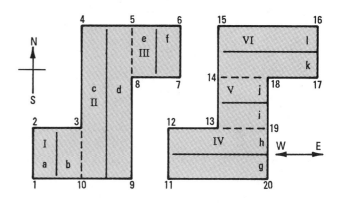

(a) Lateral force in
N-S direction

(b) Lateral force in
E-W direction

Figure 4-7 Tributary load areas to lateral force resisting shear wall in Z plan building.

Force in the N-S direction, Figure 4-7(a).

Tributary Load Areas

Shear Wall 1-2.   The tributary load area is a

Shear Wall 3-4.   The tributary load areas are b and c

Shear Wall 8-9.   The tributary load areas are d and e

Shear Wall 6-7.   The tributary load area is f

Diaphragm I is resisted by shear wall 1-2 and drag strut 3-10 which transmits the force to shear wall 3-4.

Diaphragm II is resisted by shear wall 3-4 and drag strut 3-10 which transmits the force to wall 3-4 on the west side and on the east side by shear wall 8-9 and drag strut 5-8 which transmits the force to wall 8-9.

Diaphragm III is resisted by shear wall 6-7 and drag strut 5-8 which transmits the force to wall 8-9.

Force in the E-W direction Figure 4-7(b).

Tributary Load Areas

Shear Wall 11-20.   The tributary load area is g

Shear Wall 12-13.   The tributary load areas are h an i

Shear Wall 17-18.   The tributary load areas are j and k

Shear Wall 15-16.   The tributary load area is l

Diaphragm IV is resisted by shear walls 11-20, 12-13 and drag strut 13-19 which transmits the force to shear wall 12-13.

Diaphragm V is resisted by drag strut 14-18 which transmits the force to shear walls 18-17 and by drag strut 13-19 which transmits the force to shear wall 12-13.

Diaphragm VI is resisted by shear walls 15-16, 17-18 and drag strut 14-18 which transmits the force to shear wall 18-17.

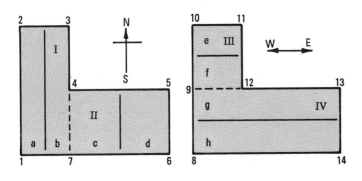

| (a) Lateral force in N-S direction | (b) Lateral force in E-W direction |

Figure 4-8   Tributary load areas to lateral force resisting shear walls in L plan building.

Force in the N-S direction, Figure 4-8(a).

Tributary Load Areas

Shear Wall 1-2.   The tributary load area is a

Shear Wall 3-4.   The tributary load areas are b and c

Shear Wall 5-6.   The tributary load area is d

Diaphragm I is resisted by shear walls 1-2 and 3-4 and drag strut 4-7 which transmits the force to shear wall 3-4.

Diaphragm II is resisted by shear wall 5-6 and drag strut 4-7 which transmits the force to shear wall 3-4.

Force in the E-W direction Figure 4-8(b).

Tributary Load Areas

Shear Wall 10-11.   The tributary load area is e

Shear Wall 12-13.   The tributary load areas are f an g

Shear Wall 8-14.   The tributary load area is h

Diaphragm III is resisted by shear wall 10-11 and drag strut 9-12 which transmits the force to shear wall 12-13.

Diaphragm IV is resisted by shear wall 8-14 and drag strut 9-12 which transmits the force to shear wall 12-13.

4-2.C.2   Rigid Diaphragms

Floors or roofs constructed of concrete and poured gypsum on steel decking are generally considered as rigid diaphragms which can transmit both shear and rotational forces into the shear walls. Because of their stiffness, rigid diaphragms are assumed to load shear wall resisting elements in proportion to the walls' relative rigidities. Thus even if a rigid diaphragm is loaded uniformly along its edge, it is assumed that it will distribute the load to the shear walls in proportion to their rigidity or stiffness. The more rigid and stiff walls will proportionately receive more force from the diaphragm.

EXAMPLE 4-E   Rigid Diaphragm, Distribution of Lateral Force to Shear Walls.

A lateral wind or seismic load of 120 kips is imposed on a building with a rigid diaphragm roof. If the end shear walls have relative rigidities of 3 and 5, how much lateral force does each wall resist? Ignore torsional effect. Distribute direct lateral force only.

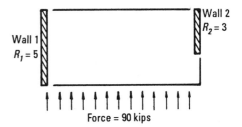

Solution 4-E

Total resistance $= \Sigma R = R_1 + R_2 = 5 + 3 = 8$

Force to wall 1

$$\text{Force} \times \frac{R_1}{\Sigma R} = 120 \times \frac{5}{8} = \ 75 \ \text{kips}$$

Force to wall 2

$$\text{Force} \times \frac{R_2}{\Sigma R} = 120 \times \frac{3}{8} = \ 45 \ \text{kips}$$

$$\text{Sum of Forces} = \overline{120 \ \text{kips}}$$

4-3 WALL RIGIDITIES

The rigidity of a wall element is dependent on its dimensions, the modulus of elasticity, E_m, the modulus of rigidity or shear modulus, E_v, or, G, and the conditions of support at the top and the bottom of the wall.

If the wall is fixed securely to the foundation but the top is free to translate and rotate, it is considered a cantilever wall. This is similar to a cantilever beam which deflects and rotates at the ends.

If the pier or wall is fixed at the top as well as the bottom, it is considered a fixed or restrained wall. This is similar to a beam fixed at both end.

The rigidity of the wall is defined as the reciprocal of the total deflection which is made up of both moment deflection and shear deflection.

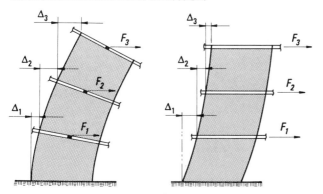

(a) Flexural deformation   (b) Shear deformation

Figure 4-9 Shear wall deformation.

4-3.A Cantilever Pier or Wall

For a pier or wall fixed at only the bottom, cantilevering from the foundation, the deflection is:

$$\Delta_c = \Delta_m + \Delta_v = \frac{Ph^3}{3E_m I} + \frac{1.2Ph}{AE_v}$$

Where

Δ_m = deflection due to flexual bending, inches

Δ_v = deflection due to shear, inches

P = lateral force on pier, lbs

h = height of pier, inches

A = cross-sectional area of pier, sq. in.

I = cross-sectional movement of inertia of pier in direction of bending, (inches^4). $I = td^3/12$.

E_m = modulus of elasticity in compression, psi

E_v = G = modulus of elasticity in shear, psi

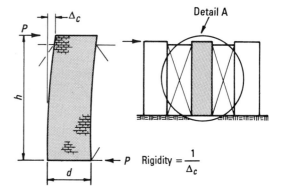

Figure 4-10 Wall pier displaced at top and cantilevering from fixed bottom.

For masonry design, assume E_m and E_v are constant, $E_v = 0.4\ E_m$, and the same strength material is throughout the wall. If it is also assumed that $E_m = 1,000,000$ psi, the wall thickness t, is 1 inch and $P = 100,000$ lbs, the deflection equations becomes

$$\Delta_c = \Delta\ \text{cantilever} = 0.4\left(\frac{h}{d}\right)^3 + 0.3\frac{h}{d}$$

$$\text{Rigidity of Cantilever Pier } R_c = \frac{1}{\Delta\ \text{cantilever}} = \frac{1}{\Delta_c}$$

4-3.B Fixed Pier or Wall

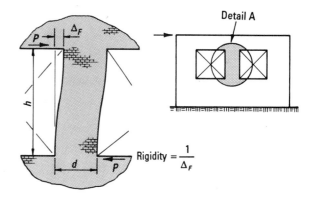

Figure 4-11 Wall pier with top displaced and fixed top and bottom.

For a pier or wall fixed at the top and the bottom, the deflection resulting from a force, P is

$$\Delta_f = \Delta_m + \Delta_v = \frac{Ph^3}{12E_m I} + \frac{1.2Ph}{AE_v},$$

Assuming $E_m = 1,000,000$ psi and the wall thickness is constant, $t = 1$ inch and $P = 100,000$ lbs, the deflection equations become

$$\Delta_f = \Delta\ \text{fixed} = 0.1\left(\frac{h}{d}\right)^3 + 0.3\left(\frac{h}{d}\right)$$

$$\text{Rigidity of Fixed Pier } R_f = \frac{1}{\Delta\ \text{fixed}} = \frac{1}{\Delta_f}$$

Tables T-1a to T-1f give the deflection coefficients and rigidities for both fixed and cantilever walls based on a wall thickness of 1 inch, a lateral force = 100 kips, a modulus of elasticity of 1,000,000 psi and modulus of rigidity of 400,000 psi.

To determine the absolute deflection of a wall, factor the table values by the actual values of modulus of elasticity, shear modulus, thickness and lateral force. Note, the effects of rotation could also be considered.

4-3.C Combinations of Walls

Wall elements can be individual walls resisting lateral forces or portions of walls that are added to increase the resisting capacity of the wall system. The wall systems may be combined and the relative rigidity calculated. High rise walls may be considered as cantilevering from the foundation, and their rigidity is determined for each floor level based on the properties of the wall element below that floor level. They may also be considered as fixed between floors, and their rigidity calculation is then based on a height between floors.

EXAMPLE 4-F   Relative Rigidity, One Story.

What is the relative rigidity of a wall 105 feet long consisting of two windows and three masonry walls cantilevering from the foundation? Assume the walls are connected to a rigid diaphragm and therefore deflect the same amount.

Solution 4-F

The resistance of each wall is additive to obtain the total resistance of the full length of the wall. Assume all walls are the same thickness and strength.

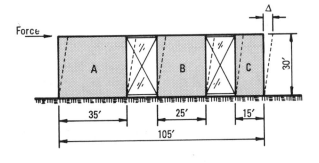

Wall	h/l or h/d	R_c^*
A	0.86	1.952
B	1.20	0.951
C	2.00	0.263

* From Table T-1 $\Sigma = 3.166$

Rigidity of wall $= \Sigma R + R_A + R_B + R_C = 3.166$

Deflection of wall $= \Delta = \dfrac{1}{\Sigma R} = \dfrac{1}{3.166} = 0.316$

If the wall was continuous in one element, 75 ft long, (35 ft + 25 ft + 15 ft) and all the glass was at one end, the $h/l = 30/75 = 0.40$, the rigidity would be 6.868 and the deflection would be equal to 0.146. This wall would thus be approximately twice as stiff as the above example.

EXAMPLE 4-G   Relative Rigidity, Multi-Story.

What is the relative rigidity of the 45 foot long three story wall shown below? Walls D, E and F are connected and the deflection of each wall adds to the deflection of the walls above.

Assume all walls are the same thickness and strength. Also assume floor-to-floor cantilever action.

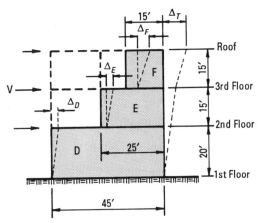

Solution 4-G

Deflection $\Delta_T = \Delta_D + \Delta_E + \Delta_F +$ rotational effects.

For simplicity, ignore rotational effects.

The deflection of walls D, E, and F are due to force V.

Wall	$\dfrac{h}{l}$	Δ_c^*
F	1.00	0.700
E	0.60	0.266
D	0.44	0.166

* From Table T-1 $\Delta_T = \overline{1.132}$

$$R_{DEF} = \frac{1}{\Delta_T} = \frac{1}{1.132} = 0.883$$

If the wall was solid 50 ft high and 45 ft long, the $h/l = 50/45 = 1.11$, the deflection, $\Delta = 0.88$ the rigidity $R_c = 1.136$.

4-3.D High Rise Walls

EXAMPLE 4-H Relative Rigidity, High Rise.

For the elevation shown below what is the relative rigidity of the wall at each floor level? Wall strengths and equivalent solid thicknesses (E.S.T.) are given.

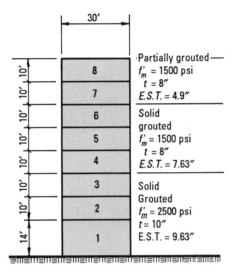

Solution 4-H

Table T-1 is based on $t = 1''$ and $E_m = 1,000,000$ psi. Corrections to the cantilever deflection value, Δ_c can be made by multiplying the value given by

$$\frac{1,000,000}{750 f'_m} \text{ or } \frac{1333}{f'_m} \text{ and } \frac{1}{t}.$$

Correction coefficient for

Walls 7, 8 $= \Delta_C = \dfrac{1333}{1500} \times \dfrac{1}{4.9} = 0.1814 \Delta_C$

Walls 4, 5, 6 $= \Delta_C = \dfrac{1333}{1500} \times \dfrac{1}{7.63} = 0.1165 \Delta_C$

Walls 1, 2, 3 $= \Delta_C = \dfrac{1333}{2500} \times \dfrac{1}{9.63} = 0.0554 \Delta_C$

Rigidity of 8 Story Wall

Floor Level	h (ft)	l (ft)	$\dfrac{h}{l}$	Δ_c from T Tables	Correction Coefficient	Actual Δ_c	$\Sigma\Delta$	Rigidity $\dfrac{1}{\Sigma\Delta}$
8	10	30	0.33	0.113	0.1814	0.0205	0.1033	9.68
7	10	30	0.33	0.113	0.1814	0.0205	0.0828	12.08
6	10	30	0.33	0.113	0.1165	0.0132	0.0623	16.05
5	10	30	0.33	0.113	0.1165	0.0132	0.0491	20.37
4	10	30	0.33	0.113	0.1165	0.0132	0.0359	27.86
3	10	30	0.33	0.113	0.0554	0.0063	0.0227	44.05
2	10	30	0.33	0.113	0.0554	0.0063	0.0164	60.98
1	14	30	0.47	0.183	0.0554	0.0101	0.0101	99.01

4-3.E Relative Stiffness of Walls

Walls with different configurations can have different stiffnesses or rigidities which in turn will change the period of the building, the response of the building and the amount of force resisted by each wall or configuration. For instance, walls with expansion joints will have much lower rigidities than solid walls of equal size.

EXAMPLE 4-I Wall Rigidities.

Using Table T-1 compute the rigidity of the walls shown, assuming they are cantilevered from the base.

a) Solid wall

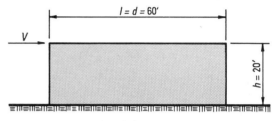

$$\frac{h}{d} = \frac{20}{60} = 0.33$$

$$R_C = 8.820$$

b) Wall with vertical slots (no head joints)

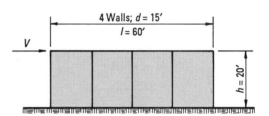

$$\frac{h}{d} = \frac{20}{15} = 1.33$$

$$R_C = 0.746 \quad \text{(Table - 1a)}$$

$$4R_C = 4 \times 0.746 = 2.984$$

c) Wall with vertical slots and wall elements are assumed to be cracked; $k = 0.50$

compression length $kd = 0.50 \times 15 = 7.5'$

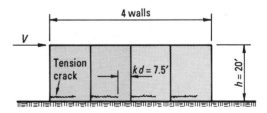

$$\frac{h}{kd} = \frac{20}{7.5} = 2.67$$

$$R_C = 0.119 \quad \text{(Table - 1c)}$$

$$4R_C = 4 \times 0.119 = 0.476$$

d) Wall contains a window opening

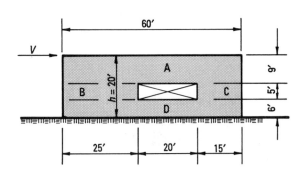

1) Deduct from solid wall the effect of the opening

Solid Wall ABCD

$$\frac{h}{d} = \frac{20}{60} = 0.33$$

$R_{solid} = 8.82$ (Table T-1a) $\Delta = 0.113$

Deduct deflection of middle strip

$$\frac{h}{d} = \frac{4}{60} = 0.067 \qquad \qquad \frac{-\Delta = 0.020}{\Delta = 0.093}$$

2) Add deflection of fixed wall piers B + C

Pier B

$$\frac{h}{d} = \frac{4}{25} = 0.16 R_B = 20.657$$

Pier C

$$\frac{h}{d} = \frac{4}{15} = 0.27 R_C = 12.053$$

$$\overline{\Sigma(R_B + R_C) = 32.710}$$

$$\frac{1}{\Sigma R_{BC}} = \frac{1}{32.710} = \Delta = 0.031$$

$$\overline{\Sigma \Delta = 0.124}$$

$$R_{ABCD} = \frac{1}{\Sigma \Delta} = \frac{1}{0.124} = 8.06 < 8.82$$

(Solid wall)

e) Wall contains window and door openings

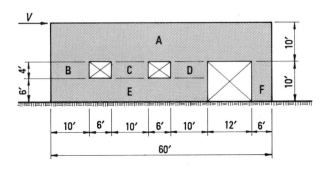

1) Solid wall ABCDEF

$$\frac{h}{d} = \frac{20}{60} = 0.33$$

$R_{solid} = 8.820$ $\Delta = 0.113$

2) Deduct bottom strip BCDEF

$$\frac{h}{d} = \frac{10}{60} = 0.17$$

From Table T-1 for a fixed pier,

$$\frac{-\Delta = 0.051}{\Delta_A = 0.062}$$

3) Add back the fixed piers B, C and D

For one pier

$$\frac{h}{d} = \frac{4}{10} = 0.40$$

$R_B = 7.911$

$$\Sigma(R_B + R_C + R_D) = 3R_B = 23.27$$

$$\Delta_{BCD} = 0.043$$

Add Pier E

$$\frac{h}{d} = \frac{6}{42} = 0.14; \qquad \frac{\Delta_E = 0.042}{\Sigma \Delta = 0.085}$$

4) $R_{BCDE} = \dfrac{1}{\Sigma \Delta} = \dfrac{1}{0.085} = 11.76$

5) Add pier F

$$\frac{h}{d} = \frac{10}{6} = 1.67; \quad R_F = 1.034$$

$$\Sigma R_F + R_{BCDE} = 11.76 + 1.034 = 12.80$$

$$\Delta_{BCDEF} = \frac{1}{R_{BCDEF}} = \frac{1}{12.80} = 0.078$$

$$\Sigma \Delta = \Delta_A + \Delta_{BCDEF} = 0.062 + 0.078 = 0.140$$

6) $R_{ABCDEF} = \dfrac{1}{\Sigma \Delta}$

$$= \frac{1}{0.140} = 7.14 < 8.82 \text{ for a solid wall}$$

EXAMPLE 4-J Shear Stresses in Walls with a Rigid Diaphragm.

Calculate the shear stresses in the walls shown below, assuming a rigid diaphragm transmits a total seismic force of 90 kips to 9″ thick reinforced brick shear walls. These walls are designed assuming $f'_m = 1500$ psi and no special inspection will be provided. Elevations of the end walls are as shown below. Do not include torsional effects.

Use Tables and Diagrams A-5 and A-6 to find the allowable shear stress.

Solution 4-J

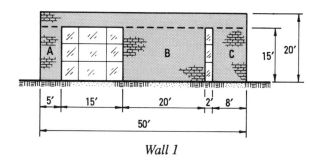

Wall 1

Relative rigidity of wall 1

Pier	h (ft)	d (ft)	h/d	R_f (From Table T-1)
A	15	5	3.00	0.278
B	15	20	0.75	3.743
C	15	8	1.88	0.814

$$\Sigma R = \overline{4.835}$$

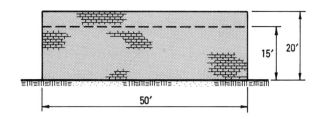

Wall 2

Elevation of end walls

**Relative rigidity of wall 2
(Rigidity of total wall)**

h (ft)	d (ft)	h/d	R_c (From Table T-1)
15	50	0.30	9.921

The force from the rigid diaphragm to the shear walls must be increased by 50% per UBC Section 2407(h)4.F. Therefore, distribute a diaphragm force of $1.5(90) = 135$ kips into the shear walls based on their rigidity. Note, this increase is for shear design only and does not apply to overturning forces.

$$\text{Wall 1} = \frac{4.835}{4.835+9.921} \times 135$$
$$= 0.328 \times 135 = 44 \text{ kips}$$

$$\text{Wall 2} = \frac{9.921}{4.835+9.921} \times 135$$
$$= 0.672 \times 135 = 91 \text{ kips}$$

Wall 1 resists 33% of the load and Wall 2 will resists 67% of the load.

Distribute the shear force into wall 1

$$V_{\text{Pier A}} = \frac{0.278}{4.835} \times 44 = 2.5 \text{ kips}$$
$$f_v = \frac{V}{td} = \frac{2500}{9 \times 60} = 4.6 \text{ psi}$$

From Table A-5a for $h/d = 3.0$; the allowable shear stress is 18 psi. Increase by one-third for wind or seismic forces:

$$F_v = 18 \times 1.33 = 24 \text{ psi} > 4.6 \text{ psi} \text{ O.K.}$$
$$V_{\text{Pier B}} = \frac{3.743}{4.837} \times 44 = 34.1 \text{ kips}$$
$$f_v = \frac{V}{td} = \frac{34,100}{9 \times 240} = 15.8 \text{ psi}$$
For $h/d = 0.75$; $F_v = 21 \times 1.33$
$$= 28 \text{ psi} > 15.8 \text{ psi} \text{ O.K.}$$

$$V_{\text{Pier C}} = \frac{1.88}{4.835} \times 44 = 7.4 \text{ kips}$$
$$f_v = \frac{V}{td} = \frac{7400}{9 \times 96} = 8.6 \text{ psi}$$
For $h/d = 1.88$; $F_v = 21 \times 1.33$
$$= 28 \text{ psi} > 8.6 \text{ psi} \text{ O.K.}$$

No shear reinforcing is required in any of the piers. Use minimum temperature steel; $A_s = 0.0007 \, bt$ min.

4-4 OVERTURNING

The lateral forces from winds and earthquakes can create severe overturning moments on buildings. If the overturning moment is great enough, it may overcome the dead weight of the structure and induce tension at the ends of shear walls. It will also cause high compression forces that may require an increase in the specified masonry strength, f'_m, an increase in the amount of compression steel in the wall, or an increase in the thickness or size of the shear wall.

In evaluating the stabilizing effect of the dead load to the overturning moment, 85 percent of the dead load should be used as per UBC Section 2337(a). This will account for any vertical acceleration of the ground as occurred in the San Fernando Earthquake of February 9, 1971.

The overturning moment at the base of a structure is found by the equation:

$$OTM = F_t h_n + \sum_{i=1}^{n} F_i h_i$$

As the equation shows, the OTM equals the force at the top, F_t, times its height above the base, h_n, plus the sum of the forces at each level, F_i, times their heights above the base, h_i. This is for all floors, n, taken at each level, $i = 1$.

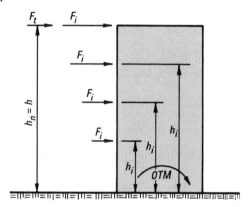

Figure 4-12   Overturning moment at base.

The overturning moment for each wall should also be determined at various floor levels to establish the amount of reinforcing required and the loads and stresses on the masonry.

$$OTM_x = F_t(h_n - h_x) + \sum_{i=1}^{n} F_i(h_i - h_x)$$

The overturning moment at level, x, above the base is equal to force at the top, F_t times the height from level x to the top $(h_n - h_x)$, plus the sum of the forces at each level F_i times the height from level i to level x $(h_i - h_x)$.

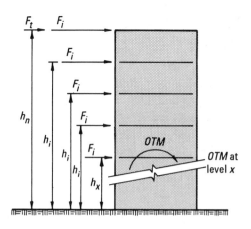

Figure 4-13   Overturning moment at any level.

Determine the base shear, story shear and overturning moment for the masonry shear wall structure shown. The structure is located in Seismic Zone 4 on a deep layer of dense, stiff soil.

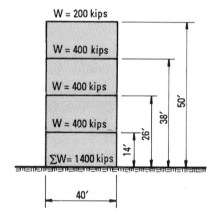

Solution 4-K

$$V = \frac{ZIC}{R_w} W$$

$Z = 0.40$ for zone 4

$I = 1.00$ importance factor

$R_w = 6.00$

$$C = \frac{1.25S}{T^{2/3}}$$

$S = $ site coeff. $= 1.2$ for dense, stiff soil with d > 200'

$T = C_t(h_n)^{3/4}; \ C_t = 0.020; \ h_n = 50'$

$T = 0.020(50)^{3/4} = 0.38$

$C = \dfrac{1.25s}{T^{2/3}} = \dfrac{1.25(1.2)}{0.38^{2/3}} = 2.86 \quad \max C = 2.75$

$V = \dfrac{ZIC}{R_w} W = \dfrac{0.40 \times 1 \times 2.75}{6} W = 0.183W$

$V = 0.183(1400) = 256$ kips

Concentrated Force at top (whiplash)

$$F_t = 0.07TV; \quad \text{Max } F_t \leq 0.25\,V$$

$$F_t = 0 \text{ when } T \leq 0.7$$

$$T = 0.38 \text{ sec.} < 0.7 \text{ sec. } F_t = 0$$

Distribution of Forces and Overturning Moments

$$F_x = \frac{(V - F_t)w_x h_x}{\sum\limits_{i=1}^{n} w_i h_i}$$

Level	w_i or w_x (kips)	h_i or h_x (ft)	$w_i h_i$	F_t	F_x (kips)	Lateral Force (kips)	Story Shear (kips)	Over- turning Moment (ft kips)
4	200	50	10,000	0	62	62	—	—
3	400	38	15,200	—	94	94	62	744
2	400	26	10,400	—	65	65	156	2616
1	400	14	5,600	—	35	35	221	5268
Base	1400		—	—	—	256	256	8852

$$\Sigma = \overline{41,200} \qquad \Sigma = \overline{256}$$

Example for Level 3

$$F_x = \frac{(256 - 0)(400)(38)}{41,200} = \frac{256(15,200)}{41,200} = 94 \text{ kips}$$

Overturning moment; OTM

$$M_x = F_t(h_t - h_x) + \sum_{i=1}^{n} F_i(h_i - h_x)$$

$$M_{Base} = F_t(h_t) + \sum_{i=1}^{n} F_i(h_i)$$

$$M_3 = 0(50 - 38) + 62(50 - 38)$$
$$= 0 + 744 = 744 \text{ ft kips}$$

$$M_2 = 62(24) + 94(12) = 2616 \text{ ft kips}$$

$$M_1 = 62(36) + 94(24) + 65(12) = 5268 \text{ ft kips}$$

$$M_B = 62(50) + 94(38) + 65(26) + 35(14)$$
$$= 8,852 \text{ ft kips}$$

4-5 TORSION

In a shear wall building with rigid floor and roof diaphragms, the seismic forces are resisted by the shear wall elements in proportion to their rigidities. If all of the lateral force resisting elements are the same size and symmetrically located, they will be loaded by the lateral forces equally.

However, if some walls are stiffer than others, or if they are unsymmetrically located, some lateral force resisting elements will resist more load than others. This condition of the center of rigidity not coinciding with the center of mass creates torsional moments. The center of mass tends to rotate about the center of rigidity.

If a building has an open front, severe torsional stresses may occur since a large eccentricity exists between the building's center of mass and the center of rigidity (see Figure 4-14). Because of the torsion, the lateral forces resisted by some shear walls will be significantly increased.

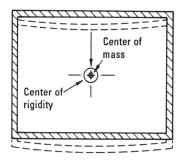

a. Equal deflection of walls

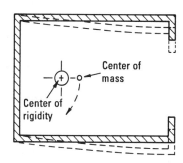

b. Unequal deflection of walls due to torsion

Figure 4-14 Lateral distortions of buildings.

For safety, all buildings having rigid diaphragms should be designed considering at least a 5 percent accidental torsional eccentricity to account for variances in the materials and the locations of walls. The UBC Section 2334(e) requires this eccentricity to be added to the calculated eccentricity (see Figure 4-15). Additionally, negative torsional effects must be ignored.

$$\text{Torsional moment} = V_x(e_y)$$
$$= V_y(e_x)$$

Note:

$$e_x = e_x \text{ (calculated)} + 0.05\,L$$
$$e_y = e_y \text{ (calculated)} + 0.05\,W$$

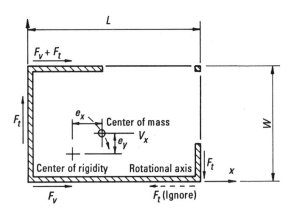

Figure 4-15 Plan of building showing location of center of mass and center of rigidity. Shear and torsional forces are shown.

EXAMPLE 4-L Center of Rigidity.

Locate the center of rigidity for the building shown below, and determine the force distribution to each 16′ high wall. Neglect accidental eccentricity in the y direction for simplicity of this problem.

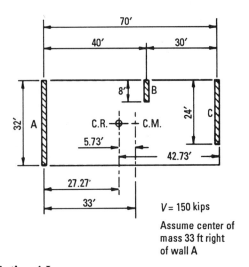

V = 150 kips

Assume center of mass 33 ft right of wall A

Solution 4-L

Locate the Center of Rigidity.

Wall	h (ft)	d (ft)	h/d	R_c	x	$R_c x$
A	16	32	0.50	5.000	0	0
B	16	8	2.00	0.263	40	10.52
C	16	24	0.67	3.112	70	217.84

$$\Sigma R_c = \overline{8.375} \qquad \Sigma R_c x = \overline{228.36}$$

Calculate the design eccentricity.

$$x_{cr} = \frac{228.36}{8.375} = 27.27 \text{ ft}$$

$$e_x = 33.0 - 27.27 = 5.73 \text{ ft}$$

$$\text{Minimum } e = (0.05 \times 70) + 5.73$$

$$= 9.23 \text{ ft}$$

$$\text{Torsional moment} = T = 150 \text{ kips} \times 9.23 \text{ ft}$$

$$= 1384.5 \text{ ft kips}$$

Calculate the total force to each wall.

Wall	R	d_x	Rd_x	Rd_x^2
A	5.000	27.27	136.35	3718
B	0.263	12.73	3.35	43
C	3.112	42.73	132.98	5682

$$\Sigma R = \overline{8.375} \qquad \Sigma Rd_x^2 = \overline{9443}$$

$$\text{Force to wall} = F_V + F_T = V \times \frac{R}{\Sigma R} + T \frac{Rd_x}{\Sigma Rd_x^2}$$

$$\qquad\qquad F_V \qquad\qquad\quad F_T$$

$$A = 150 \times \frac{5.000}{8.375} - 1384.5 \times \frac{136}{9443} = 89.6 - 19.9$$

$$= 89.6 \text{ kips}$$

$$B = 150 \times \frac{0.263}{8.375} + 1384.5 \times \frac{3.4}{9443} = 4.7 + 0.5$$

$$= 5.2 \text{ kips}$$

$$C = 150 \times \frac{3.112}{8.375} + 1384.5 \times \frac{133}{9443} = 55.7 + 19.5$$

$$= 75.2 \text{ kips}$$

EXAMPLE 4-M Forces to Walls, Rigid Diaphragm.

The figure below shows a plan view of a one-story shear wall masonry structure with a rigid diaphragm roof. The relative rigidity of each shear wall is given.

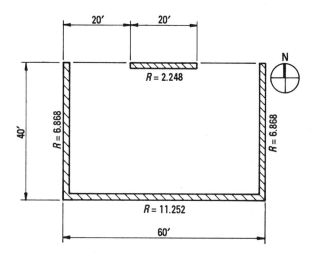

Determine a. The center of mass and the center of rigidity

   b. The minimum base shear and torsion values for both N-S and E-W lateral forces

   c. The forces in each shear wall for a N-S earthquake

Given: Building is a one story box system;
All walls are a total of 19 ft high;
16 ft between supports with a 3 ft parapet.
Earthquake zone 3; $Z = 0.30$
$R_w = 6$; $C = 2.75$
$I = 1.00$
Weights:
$\quad$ Roof $\quad\quad$ = 75 psf
$\quad$ N Wall $\quad$ = 75 psf
$\quad$ S wall $\quad$ = 100 psf
$\quad$ E, W walls = 75 psf

Solution 4-M Part a; Centers of Mass and Rigidity

Find the weights of each building component and determine the location of the center of mass.

$\quad$ Use $h = 16/2 + 3 = 11$

Thus the weights of the E and W walls are 75 (11) (40) = 33,000 lbs (see Table below). Note that since the building is symmetrical with respect to the y axis, it is expected that $x_{cm} = 60/2 = 30'$. However, to show the methodology, calculate x_{cm}.

Item	Weight	x	y	Wx	Wy
Roof	180 kips	30	20	5400	3600
W Wall	33 kips	0	20	0	660
E Wall	33 kips	60	20	1980	660
N Wall	16.5 kips	30	40	495	660
S Wall	66 kips	30	0	1980	0

$\quad \Sigma w = \overline{328.5}$ kips $\quad\quad \Sigma wx = \overline{9855}$
$\quad\quad\quad\quad\quad\quad\quad\quad\quad\quad\quad \Sigma wy = \overline{5580}$

$\bar{y}_{cm} = \dfrac{\Sigma Wy}{\Sigma W} = \dfrac{5580}{328.5}$

$\quad$ C.M. = 17.0 ft north of the south wall

$\bar{x}_{cm} = \dfrac{\Sigma W_x}{\Sigma W} = \dfrac{9855}{328.5}$

$\quad\quad\quad = 30$ ft to the east of the west wall

(This lies on the symmetrical centerline, as expected.)

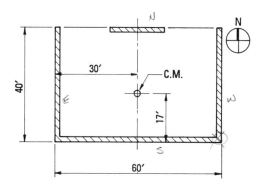

Calculate the center of rigidity, use $h = 16'-0''$ (neglect parapet)

Wall	L	h/l	R_{cy}	R_{cx}	x	y	yR_{cx}	xR_{cy}
N	20	0.80	—	2.248	—	40	89.9	—
S	60	0.27	—	11.252	—	0	0	—
E	40	0.40	6.868	—	60	—	—	412.1
W	40	0.40	6.868	—	0	—	—	0

$\quad\quad\quad\quad \Sigma R_{cx} = 13.500 \quad\quad \Sigma yR_x = 89.9$

$\quad\quad \Sigma R_{cy} = 13.736 \quad\quad\quad\quad \Sigma xR_{cy} = 412.1$

C.R. y direction $= \dfrac{yR_{cx}}{\Sigma R_{cx}} = \dfrac{89.9}{13.5} = 6.7$ ft

C.R. x direction $= \dfrac{xR_{cy}}{\Sigma R_{cy}} = \dfrac{412.1}{13.736} = 30$ ft

Calculate torsional eccentricity

Eccentricity between center of mass and center of rigidity.

$e_y = 17 - 6.7 = 10.3'$

Add minimum 5% accidental eccentricity

$0.05 \times 40 = 2.0'$

$e_y = 10.3 + 2.0 = 12.3$ ft

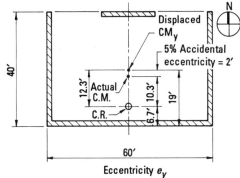

Eccentricity e_y

$e_x = 30 - 30 = 0'$

Include minimum 5% accidental eccentricity

$\quad 0.05 \times 60 = \pm 3.0'$

$\quad e_x = 0 \pm 3.0 = 3$ ft

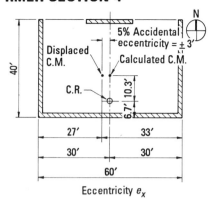

Eccentricity e_x

Solution 4-M Part b; Base Shear

Calculate the seismic base shear

$$V = \frac{ZIC}{R_w}W$$

$$= \frac{0.3 \times 1.0 \times 2.75}{6} \times W$$

$$= 0.14W = 0.14 \times 328.5 = 46 \text{ kips}$$

Determine torsional moments

The torsional moments due to a N-S seismic force rotating about C.R. is:

$$T = Ve_x = 46 \times 3 \text{ ft}$$

$$= \pm 138 \text{ ft kips}$$

Likewise the torsional moment due to an E-W seismic force

$$T = Ve_y = 46 \times 12.3 \text{ ft}$$

$$= \pm 566 \text{ ft kips}$$

Solution 4-M Part c; Forces to Shear Walls

Determine the forces on each shear wall from a N-S earthquake, $V = 46$ kips; $T = 138$ ft kips

Force due to shear $F_v = V \dfrac{R_y}{\Sigma R_y}$

Forces due to torsion $F_t = T \dfrac{Rd}{\Sigma Rd^2}$

Where $V = 46$ kips, and $T = 138$ ft kips

EXAMPLE 4-N Center of Mass and Rigidity.

Locate the center of mass, C.M., and the center of rigidity, C.R., for the industrial structure shown. This is only an example of how to combine walls of different strengths and thicknesses. It is generally recommended to maintain a consistent strength requirement and uniform thickness throughout the structure.

All walls are 18 feet high. There are no openings, windows or doors in the walls. The roof is a rigid concrete slab which is 8 inches thick and weighs 70 psf.

Walls are cantilevered from the base.

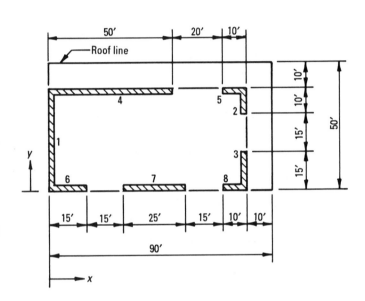

Solution 4-N

The values for rigidity, R_c, from Table T-1 are based on $t = 1''$ and $E_m = 1,000,000$ psi. It is acceptable, based on linear variation, to equate an 8" thickness to the base 1" and correct walls of other thicknesses by multiplying R_c by $1/8$. Correct R_c also for variations in the modulus of elasticity by multiplying R_c by $E_m/1,000,000$. However, because $E_m = 750f'_m$, the value R_c may be corrected by $750f'_m/1,000,000 = f'_m/1333$.

Distribution of forces for a seismic force on the N direction (Solution 4-M Part c)

Wall	R_y	R_x	d_x	d_y	Rd	Rd^2	Direct Force F_v	Torsional Force F_t	Total Force $F_v + F_t$
N	—	2.25	—	35.3	79.4	2803.7	—	-0.7	-0.7
S	—	11.25	—	4.7	52.9	248.5	—	+0.5	+0.5
E	6.87	—	33	—	226.7	7481.4	23.6	+2.0	25.0
W	6.87	—	27	—	185.5	5008.2	23.0	-1.6	21.6 *

$\Sigma R_y = 13.7$ $\Sigma Rd^2 = 15,541.8$ $\Sigma = 46.4 = V$

$\Sigma R_x = 13.5$

* Since the East and West walls are symmetrical, use $F = 25.0$ kips for both walls (Earthquake force can act in either N or S direction).

Properties of Each Wall

Wall No.	Thickness (Inches)	Thickness Correction $t/8$	f'_m (psi)	E_m Correction $f'_m/1333$	Combined Correction for R_c	Weight of Wall (psf)
1	8	1.00	1350	1.01	1.01	80
2	12	1.50	3000	2.25	3.38	120
3	12	1.50	3000	2.25	3.38	120
4	8	1.00	1500	1.13	1.13	80
5	12	1.50	1500	1.13	1.70	120
6	10	1.25	2000	1.50	1.88	100
7	10	1.25	2000	1.50	1.88	100
8	10	1.25	2000	1.50	1.88	100

Determination of Center of Rigidity $h = 18'-0''$

Wall No.	Direction	Length (feet)	h/l	R_c from T-1	Correction Coeficient	Corrected R_c	x (ft)	xR_{cy}	y (ft)	yR_{cx}
1	y	40	0.45	5.833	1.01	5.891	0.33	1.96		
2	y	10	1.80	0.348	3.38	1.176	79.5	93.51		
3	y	15	1.20	0.951	3.38	3.214	79.5	255.54		
4	x	50	0.36	7.895	1.13	8.921			39.67	353.90
5	x	10	1.80	0.348	1.70	0.592			39.50	23.38
6	x	15	1.20	0.951	1.88	1.788			0.42	0.75
7	x	25	0.72	2.738	1.88	5.147			0.52	2.68
8	x	10	1.80	0.348	1.88	0.654			0.52	0.34

$$\Sigma xR_c = 351.01 \qquad \Sigma yR_c = 381.05$$

Location of center of rigidity

ΣR_c in the y direction $= 10.28$

ΣR_c in the x direction $= 17.10$

$$x = \frac{\Sigma xR_{cy}}{\Sigma R_{cy}} = \frac{351.01}{10.28} = 34.14 \text{ ft}$$

$$y = \frac{\Sigma yR_{cx}}{\Sigma R_{cx}} = \frac{381.05}{17.10} = 22.28 \text{ ft}$$

Determination of Center Mass

Wall No.	W (psf)	Length (ft)	Area 18xL	W kips	Direction	x (ft)	xW	y (ft)	yW
1	80	40	720	57.6	y	0.33	19.0	20.00	1,152.0
2	120	10	180	21.6	y	79.50	1,717.2	35.00	756.0
3	120	15	270	32.4	y	79.50	2,575.8	7.50	243.0
4	80	50	900	72.0	x	25.00	1,800.0	39.67	2,856.2
5	120	10	180	21.6	x	75.00	1,620.0	39.50	853.2
6	100	15	270	27.0	x	7.50	202.5	0.42	11.3
7	100	25	450	45.0	x	42.50	1,912.5	0.42	18.9
8	100	10	180	18.0	x	75.0	1,350.0	0.42	7.6

$$\Sigma W = 295.2 \text{ kips} \qquad \Sigma yW = 5,898.2$$
$$\Sigma xW = 11,197.0$$

Location of center of mass of walls

$$\bar{x} = \frac{\Sigma xW}{\Sigma W} = \frac{11,197}{295.2} = 37.93 \text{ ft}$$

$$\bar{y} = \frac{\Sigma yW}{\Sigma W} = \frac{5,898.2}{295.2} = 19.98 \text{ ft}$$

Assume center of mass of roof coincides with geometric center of roof

$$\bar{x} = 45 \text{ ft} \qquad \bar{y} = 25 \text{ ft}$$

Weight of roof $= 90 \times 50 \times 0.07 \text{ ksf} = 315 \text{ kips}$

Combined center of mass

$$\bar{x} = \frac{\overset{\text{(walls)}}{295.2 \times 37.93} + \overset{\text{(roof)}}{315 \times 45}}{295.2 + 315} = 41.58 \text{ ft}$$

$$\bar{y} = \frac{\overset{\text{(walls)}}{295.2 \times 19.98} + \overset{\text{(roof)}}{315 \times 25}}{295.2 + 315} = 22.57 \text{ ft}$$

Eccentricity $= $ C.M. $-$ C.R.

x direction $= 41.58 - 34.14 = 7.44 \text{ ft}$

y direction $= 22.57 - 22.28 = 0.29 \text{ ft}$

The design eccentricity is increased by 5% of the building dimension perpendicular to the direction of the seismic force (UBC Section 2334(e)).

x direction = $7.44 + 0.05 \times 80 = 11.44$ ft

y direction = $0.29 + 0.05 \times 40 = 2.29$ ft

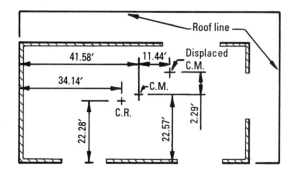

4-6 BASE ISOLATION

4-6.A General

For many years it has been known that if a structure could be floated or isolated from seismic motions it would not be subjected to high earthquake forces (see Figure 4-16). This technique of isolating the base of a structure is now an acceptable design and construction alternative and holds great promise for future structures.

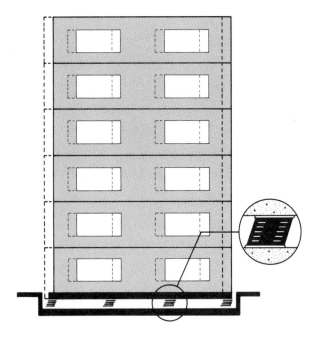

Figure 4-16 Building constructed on base isolators.

Base isolators are a horizontally flexible and vertically rigid structural element which allow large lateral deformations due to seismic loads.

Essentially, base isolation prevents the violent seismic shaking of the earth from being transmitted to the structure. In effect, it decouples the structure from the ground and changes the response of the building. This shift in response significantly reduces the buildings acceleration and interstory drift.

The isolation system may be quite varied depending on structure, availability, economy, etc. The system should provide a significant change in the period of motion between the earth and the structure to adequately decouple the building from the ground. This period differential should be at least one second or more.

A good example of the differential in period between the soil and a structure was shown dramatically in the October 17, 1985 Mexico City Earthquake. Frame buildings which had a long period of vibration founded on solid rock or alluvial soil having short periods of vibration survived the shaking well. Similar buildings founded on the deep soft soil of the Mexico City lake bed were significantly damaged since the long period of the soil was close to the period of the tall frame buildings. Thus the vibrations magnified through the soft soils and into the buildings.

Likewise, stiff buildings with very short periods of vibration founded on the soft mud of Mexico City performed very well, while rigid buildings on rock or stiff soil were damaged.

The performance of these buildings in Mexico exemplify the principle of seismic isolation in that there must be a large differential in soil/site period to the building period. Base isolators create such a differential above and below the isolation interface.

According to Dr. Ron Mayes of Dynamic Isolation System, Inc., Berkeley, California, an isolation system should have a flexible support system to lengthen the period of vibration and to reduce the response of the structure. It should absorb energy and be a damper to control deflection. Note that it must also be sufficiently rigid at low wind loads.

An excellent example of base isolators are lead-filled elastomeric bearings, Figure 4-17, which provide the required flexibility, damping and low load rigidity. They have been used successfully on many structures and have been proven by performance in actual seismic events.

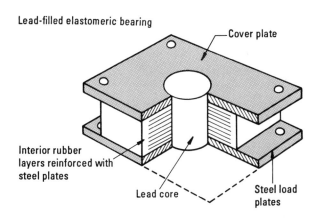

Figure 4-17 Steel, lead and rubber mechanical energy dissipating device.

4-6.B Principles of Seismic Isolation

The principles of seismic isolation are represented in Figure 4-18.

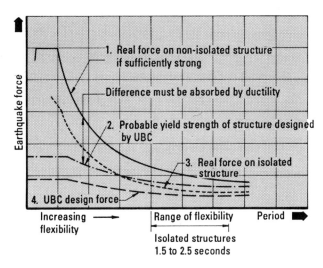

Figure 4-18 Design principles of seismic isolation

Curve 1 plots the real force on a non-isolated structure that will be imposed on the building due to seismic action. Note that as the period increases, the earthquake force is reduced.

Curve 2 is the probable yield strength that the structure may have when designed to UBC requirements. This indicates that the structure would reach plastic yielding and thus the period would be increased. However, the structure may be seriously damaged.

Curve 3 plots the real force on a structure that is isolated on energy dissipators. As the period increases, the capacity demand drops down to where it can be be-

low the yield strength of the structure. This keeps the structure into an elastic response condition.

Curve 4 indicates the seismic force used in design as required by the UBC regulation. It is considerably below the real force to which a structure may be exposed.

It should be noted that the base isolated structures the periods are from 1.5 to 2.5 seconds, thus the real force on the structure approaches the UBC design force and can be less than the yield strength of the structure.

Shear walls buildings are generally very stiff and have very short periods. Accordingly, they are subjected to high seismic forces and must be designed for high force levels. By isolating a shear wall building from the seismic acceleration of the earth and thus decoupling it, the period is lengthened and its response and force level significantly reduced.

4-7 QUESTIONS AND PROBLEMS

4-1 What is a horizontal diaphragm and how does it function to resist lateral forces on a building?

4-2 What are the requirements for diaphragm anchorage?

4-3 What are the effects of the deflection of a diaphragm on the load on a wall?

4-4 Given a building 60 feet by 180 feet with 9 inch thick brick walls ($w = 90$ psf), that are 18 feet high located in Seismic Zone No. 4. Assume the roof dead load is 15 lb per square foot and the live load is 20 lbs per square foot. What is the shear force per linear foot which the roof diaphragm delivers to the side walls? Specify the anchor bolts required for a 4×12 ledger on the side walls and on the longitudinal walls.

4-5 What are flexible and rigid diaphragms? Given the following plan, what is the force to each of the walls A, B, C, D and E if a flexible diaphragm is used? What are the forces in these walls if a rigid diaphragm is used? Assume the walls are cantilevered from the foundation and are 20 feet high. The lateral force is 750 plf.

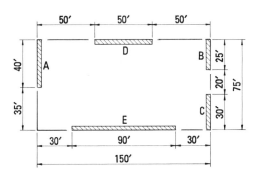

4-6   Compare the following wall, shown with openings to a similar wall without any openings? Determine the rigidity of the wall in each case. If a lateral force on the wall with openings is 50 kips, what is the shear force in each of the wall elements?

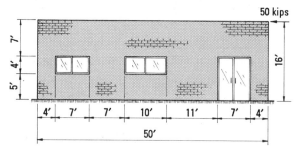

4-7   Locate the center of mass and the center of rigidity for the plan shown. Assume the roof is a rigid diaphragm that is 4 inches of concrete on a metal deck ($w = 55$ psf). What are the forces to each of the walls shown if there is a lateral force on the wall of 700 lbs per linear foot? Assume all walls are 24 ft high and cantilevered from the base.

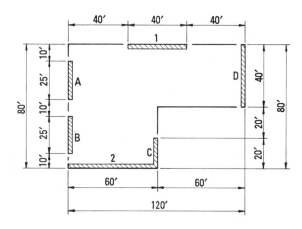

4-8   Determine the centers of mass and rigidity of the building shown. The walls cantilevered from the foundation are 24 feet high. Assume the rigid concrete roof weighs 65 lbs per square foot, the walls weigh 78 lbs per square foot and the columns weigh 270 lbs per linear foot. The base shear is 150 kips in either direction. Determine the force in each of the walls.

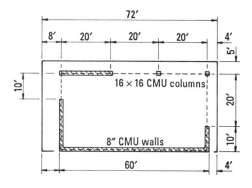

4-9   The 8 inch interior shear wall shown is solid grouted masonry with $f'_m = 1500$ psi, $f_y = 60,000$ psi and no special inspection (use half stresses). The seismic load from the flexible roof diaphragm is 30 kips applied at the top wall.

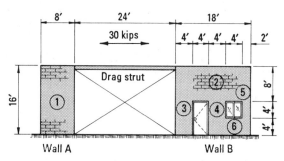

Determine the lateral load in piers 1, 2, 3, 4, 5 and 6 due to the 30 kips load, neglecting the weight of the walls for seismic effects. Also determine the maximum anchorage load from the drags struts to the walls. Assume pin ends and no axial deformation of the strut. If the load at the top of wall B is 25 kips what will be the axial load in pier Number 5?

4-10   How are torsional shear forces distributed in a building? What is the minimum eccentricity that must be used in the calculations for torsion in a building. Are negative torsional shears deducted from the direct force shear?

4-11   What is base isolation and how does it function? Is it advantageous to use base isolation in resisting wind loads? Is base isolation beneficial if (a) there is a soft soil and a flexible building? (b) if the soil is rock and the building stiff? (c) if the soil is soft and the building is rigid?

SECTION 5

Design of Structural Members—Working Stress Design

5-1 HISTORY

Prior to the 1933 Long Beach, California earthquake, masonry structures were generally designed by empirical procedures based on the past performance of similar structures.

Since reinforcing steel was not utilized, early masonry structures tended to be massive in order to effectively resist lateral as well as vertical loads. Although this empirical or rational procedure is still permitted to be used on a limited basis, the Long Beach earthquake showed engineers that a more defined and logical procedure was necessary to design structures that would effectively withstand such high seismic forces.

During this time, elastic working stress design procedures were being used to design reinforced concrete structures. Based on this elastic design approach, engineers began reinforcing masonry so that the steel could resist tensile forces while the masonry carried compressive forces.

By 1937, the Uniform Building Code included working stress design procedures for masonry which allowed engineers to size masonry members by ensuring that anticipated service loads did not exceed allowable design stresses.

With the working stress design method, engineers have designed masonry structures throughout much of the 20th century.

5-2 PRINCIPLES OF WORKING STRESS DESIGN

5-2.A General; Flexural Stress

The design and analysis of reinforced masonry structural systems have traditionally been by the straight line, elastic working stress method. This procedure assumes the masonry resists compressive forces and reinforcing steel resists tensile forces.

In Working Stress Design (WSD), the limits of allowable stress (Tables A-3 and A-4) for the materials are established based on the properties of each material. The actual or code live loads and dead loads must not cause stresses in the structural section that exceed these allowable values.

The procedure presented is based on the working stress or straight line assumptions where all stresses are in the elastic range and:

1. Plane sections before bending remain plane during and after bending.

2. Stress is proportional to strain which is proportional to distance from the neutral axis.

3. Modulus of elasticity is constant throughout the member.

4. Masonry carries no tensile stresses.

5. Span of the member is large compared to the depth.

6. Masonry elements combine to form a homogeneous and isotropic member.

7. External and internal moments and forces are in equilibrium.

8. Steel is stressed about the center of gravity of the bars equally.

9. The member is straight and of uniform cross-section.

These assumptions are in keeping with homogeneous elastic materials. For heterogeneous materials, such as reinforced masonry, these assumptions are satisfactory for normal working stress levels. For high stress levels many of the assumptions may not be applicable, particularly item 2, since stress may not be proportional to strain.

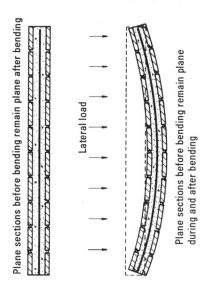

Figure 5-1 Wall in flexure, illustrates assumption 1.

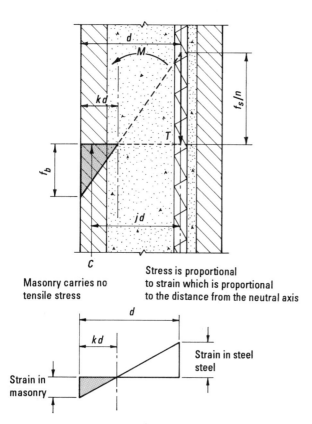

Figure 5-2 Stresses and strains in wall due to out of plane lateral loads, perpendicular to the plane of the wall.

Masonry carries no tensile stress

Stress is proportional to strain which is proportional to the distance from the neutral axis

5-3 DERIVATION OF FLEXURAL FORMULAS

The basis of the flexural equations for working stress design, WSD, techniques of heterogeneous systems in which one material resists compression and the other material with different physical properties resists tension is the concept of the modular ratio. The modular ratio, n, is the ratio of the modulus of elasticity of steel, E_s, to the modulus of elasticity of masonry, E_m.

$$n = \frac{E_s}{E_m}$$

By use of the modular ratio, n, the steel area can be transformed into an equivalent masonry area. The strain is in proportion to the distance from the neutral axis and therefore the strain of the steel can be converted to stress in the steel. In order to establish the ratio of stresses and strains between the materials, it is necessary to locate the neutral axis.

5-3.A Location of Neutral Axis

The location of the neutral axis is defined by the dimension, kd, which is dependent on the modular ratio, n, and the reinforcing steel ratio, $p = A_s/bd$. For a given modular ratio, n, the neutral axis can be raised by decreasing the amount of steel (reducing p) or it can be lowered by increasing the amount of steel (increasing p).

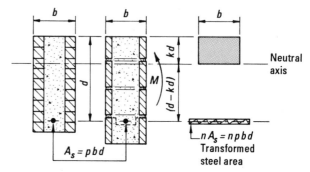

Figure 5-3 Location of neutral axis for a beam.

Take moments of the stress areas about the neutral axis.

Compressive stress area	×	Moment arm	=	Tensile stress area	×	Moment arm

$$(bkd) \times (\tfrac{1}{2}kd) = (npdb) \times (d-kd)$$

$$\tfrac{1}{2}bd^2k^2 = np(bd^2 - kbd^2)$$

$$\tfrac{1}{2}bd^2k^2 - npbd^2(1-k) = 0$$

Divide by bd^2 and multiply by 2

$$k^2 - 2np(1-k) = 0$$

Solving for k

$$k = \sqrt{(np)^2 + 2np} - np$$

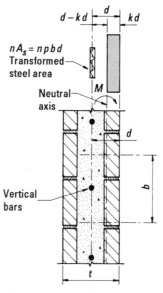

Figure 5-4 Location of neutral axis for a wall.

Reinforced masonry beam subjected to lateral forces.

5-3.B Variation of Coefficients *k*, *j* and Flexural Coefficient *K*

The coefficient *k* defines the depth of the compression area, *kd*, and is the location of the neutral axis for the section. The neutral axis is determined by the modular ratio and the steel ratio. For under-reinforced sections where the reinforcing steel is stressed to its allowable value, the coefficient *k* will increase as the amount of steel increases. Accordingly, the depth of the compression area will also increase until the stress in the masonry increases up to the allowable compressive stress. When the maximum allowable masonry stress is attained, the section is considered in a balanced stress condition, since the steel stress is already at its maximum allowable value. If the area of steel is increased, and the masonry stress is held at its maximum value, the stress in the steel decreases and the compression stress block deepens, increasing the coefficient *k*. It is determined by the equation

$$k = \sqrt{(np)^2 + 2np} - np \quad \text{or} \quad k = \frac{1}{1 + \dfrac{f_s}{nf_b}}$$

The coefficient *j* defines the distance between the centroid of the compression area and the centroid of the

tensile steel area. The lever arm, *jd*, is used to compute the internal resistance moment. This lever arm, *jd*, decreases from a maximum value to a minimum value as the depth of the compressive stress block and is determined by the equation

$$j = 1 - \frac{k}{3}$$

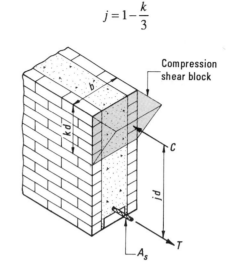

Figure 5-5 Stresses on a beam in flexure.

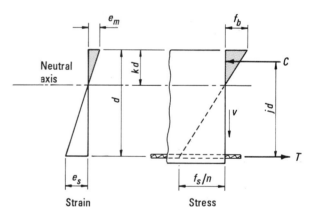

Figure 5-6 Stress and strain representation for a beam in flexure.

The flexural coefficient *K* is a combination of values, i.e., $f_s p j$ or $\frac{1}{2} f_b k j$, that defines the moment or flexural capacity of the section.

If the steel tensile stress is maintained at its maximum allowable stress, the value of *K* will vary from minimum to maximum as the masonry compressive stress f_b increases. The value of *K* also increases as the steel tensile stress is reduced while the compressive stress in the masonry is maintained at its maximum allowable stress. The E Tables may be used to find *K* values easily. Alternately, it may be determined either based on steel stress as:

$$K = f_s p j$$

or based on masonry stress as:

$$K = \frac{kj}{2} f_b$$

See the F Diagrams for variation of K vs p for different stresses in the masonry and steel

5-3.C   Moment Capacity of a Section

The moment capacity of a reinforced structural masonry wall or beam can be limited by the allowable masonry stress, (over-reinforced), allowable steel stress, (under-reinforced), or both, in which case it would be a balanced design condition.

When a member is designed for the specified loads and the masonry and reinforcing steel are stressed to their maximum allowable stresses, the design is said to be a "balanced" design. This balanced design is different than the balanced design for strength design method. (See Section 6-4.A.1.) For working stresses, balanced design occurs when the masonry is stressed to its maximum allowable compressive stress and the steel is stressed to its maximum allowable tensile stress.

However, in many cases, the "balanced" design does not satisfy the conditions for the materials available or for the predetermined member size or the economy of the project. It may be advantageous to under-stress (under-reinforce) the masonry or under-stress (over-reinforce) the steel so that the size of the member can be maintained.

The moment capability of a section based on the steel stress is defined as:

$$M_s = \text{force} \times \text{moment arm}$$

Where:

Force in the steel, $T = A_s f_s = pbdf_s$

Moment Arm $= jd$

$$M_s = T \times jd = A_s f_s \, jd$$

$$M_s = pbd \, f_s \, jd = f_s \, pjbd^2$$

Also, since $K = f_s pj$,

$$M_s = Kbd^2$$

The moment capability of a section based on the masonry stress is defined as:

$$M_m = \text{force} \times \text{moment arm}$$

Where:

Force in the masonry, $C = \frac{1}{2} f_b (kd) b = \frac{1}{2} f_b \, kbd$

Moment Arm $= jd$

$$M_m = C \times jd = \frac{1}{2} f_b \, kbdjd$$

$$M_m = \frac{1}{2} f_b \, kbd^2$$

Since $K = \frac{1}{2} f_b \, kj$,

$$M_m = Kbd^2$$

5-3.D   Summary

The above discussion shows the general derivations for moment on a section for any stress level within the elastic straight line stress range. It assumes the section only has tensile reinforcement steel.

The primary WSD formulas for design or analysis are:

$$M = Kbd^2 \text{ (in. lbs)}$$

or $M_m = \frac{1}{2} f_b \, kjbd^2$ (in. lbs)

or $M_s = f_s \, pjbd^2$ (in. lbs)

and $K = \dfrac{M}{bd^2}$

Where:

M is the moment on the member, or moment per unit width in in. lbs or in. lbs/ft

b is the width of the member in inches

d is the depth from the outer compression fiber to the centroid of the tension reinforcing steel in inches.

K is the flexural coefficient and is determined by the formulas above and is $K = f_s pj$ or $K = \frac{1}{2} f_s kj$ psi. This concept is similar to the classic stress equation:

$$f = \frac{Mc}{I} = \frac{M}{S} \quad \text{where} \quad S = \frac{I}{c}$$

Moment = stress × section modulus

$$M = f \times S$$

For a solid rectangular section:

$$S = \frac{bd^2}{6}$$

Thus, $\text{Stress} = \dfrac{M}{S} = \dfrac{M}{bd^2} \times 6$

This is similar to

$$f_b = \frac{M}{bd^2} \times \frac{2}{jk}$$

and $f_s = \dfrac{M}{bd^2} \times \dfrac{1}{jp}$

A reinforced masonry section is not symmetrical about the neutral axis. The value of c, which is the distance from the neutral axis to the extreme tension or compression fiber, is different for the stress in the masonry and the steel. Therefore, the section modulus, I/c will be different when determining the stress in the masonry or the steel.

$$S = bd^2 \div \frac{2}{jk} \text{ for masonry, and}$$

$$S = bd^2 pj \text{ for steel}$$

EXAMPLE 5-A Determination of Moment Capacity of a Wall.

A partially grouted 8″ concrete masonry wall is reinforced with #6 bars at 24″ o.c. The steel is 5.3″ from the compression face and is Grade 60. If f'_m = 2500 psi, what is the moment capacity of the wall?

Solution 5-A

For f'_m = 2500 psi

$$F_b = 0.33 f'_m = 825 \text{ psi}$$

$$E_m = 750 f'_m = 1,875,000 \text{ psi}$$

Also for f_y = 60,000 psi

$$F_s = 24,000 \text{ psi}$$

$$E_s = 29,000,000 \text{ psi.}$$

Steel ratio, $p = \dfrac{A_s}{bd}$

$$= \frac{0.44}{24 \times 5.3} = 0.0035$$

Modular ratio, $n = \dfrac{E_s}{E_m}$

$$= \frac{29,000,000}{750 \times 2500} = 15.5$$

$$k = \sqrt{(np)^2 + 2np} - np$$

$$= \sqrt{(15.5 \times 0.0035)^2 + 2(15.5 \times 0.0035)}$$
$$\quad - (15.5 \times 0.0035)$$

$$= 0.280$$

$$j = 1 - \frac{k}{3} = 1 - \frac{0.280}{3}$$
$$= 0.907$$

$$M_m = \tfrac{1}{2} f_b k j b d^2 = \tfrac{1}{2}(825)(0.28)(0.907)(12)(5.3)^2$$
$$= 35,312 \text{ in. lbs/ft}$$
$$= 2.94 \text{ ft k/ft}$$

$$M_s = f_s p j b d^2 = 24,000(0.0035)(0.907)(12)(5.3)^2$$
$$= 25,681 \text{ in. lbs/ft}$$
$$= 2.14 \text{ ft k/ft} \leftarrow \text{Controls}$$

Alternately,

From Table E-3b for p = 0.0035 find

$$K = 76.0 \qquad f_b = 600 \text{ psi} \qquad f_s = 24,000 \text{ psi}$$
$$k = 0.276 \qquad j = 0.907 \qquad 2/jk = 7.90$$

Moment capacity $= Kbd^2$
$$= 76 \times 12 \times 5.3^2$$
$$= 25,618 \text{ in. lbs/ft}$$
$$= 2135 \text{ ft lb/ft} \quad \text{(same as above)}$$

5-3.D.1 Strain Compatibility

Two of the basic assumptions of the working stress design are that plane sections before bending remain plane during and after bending and that stress is proportional to strain which is proportional to the distance from neutral axis.

The above assumptions provide the basis for straight line values for stress and strain on the cross-section of a member subjected to moment and is illustrated by Figures 5-1, 5-2 and 5-7.

The location of the neutral axis is explained in Section 5-3A and is denoted as a distance, kd, from the compression face.

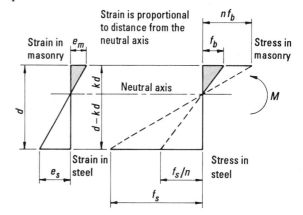

Figure 5-7 Relationship of stress and strain.

Stress in masonry: $f_b = e_m E_m$

Stress in steel: $f_s = e_s E_s$

Ratio of strains: $\dfrac{e_m}{e_s} = \dfrac{(kd)}{(d-kd)}$ straight line variation

Ratio of stresses: $\dfrac{f_b}{f_s} = \dfrac{e_m E_m}{e_s E_s} = \dfrac{e_m}{e_s} \times \dfrac{1}{n}$

$$\frac{f_b}{f_s} = \frac{(kd)}{(d-kd)} \times \frac{1}{n}$$

$$f_b = \frac{(kd)}{(d-kd)} \times \frac{f_s}{n}$$

This shows straight line variation of stresses when f_s is divided by modular ratio n.

EXAMPLE 5-B   Flexural Design — Tension Reinforcement.

Determine the tension reinforcement required for a 14 foot long, simply supported clay brick beam using both the Uniform Building Code and the ACI/ASCE Code. The beam is 9″ wide by 24″ deep with an effective depth, d, of 20″. A superimposed live load of 1400 plf is carried by the beam as well as its own weight.

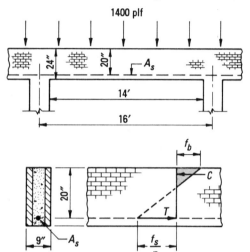

Figure 5-8 Beam in flexure.

Assume

$f_m' = 2500$ psi

$F_s = 24,000$ psi

Special inspection will be provided.

Clay masonry units will be set with type S mortar

Solution 5-B

Part (a) Design per the Uniform Building Code

Find the self weight of the beam from Table B-4 as 90 psf

$$DL = 90 \times \frac{24″}{12 \text{ in./ft}} = 180 \text{ plf}$$
$$LL = 1400 \text{ plf}$$
$$\text{Total } w = 1580 \text{ plf}$$

Calculate the simple beam moment

$$M = \frac{wl^2}{8} = \frac{1580(14)^2}{8}$$
$$= 38,710 \text{ ft lbs}$$

$$K = \frac{M}{bd^2} = \frac{38,710 \text{ ft lbs} \times 12 \text{ in./ft}}{9 \times 20^2}$$
$$= 129$$

Enter Table E-4b for $f_m' = 2500$ psi with $K = 129$:

estimate $p = 0.0062$

Therefore, the required area of steel is

$A_s = 0.0062 \times 9 \times 20 = 1.12$ sq. in.

From Table C-3, select 4 - #5 bars ($A_s = 1.24$ sq. in.) or 2 - #7 ($A_s = 1.20$ sq. in.)

Part (b) Design per ACI/ASCE Code

For $K = 129$, enter A/A Table B-3a for clay masonry with type S mortar and $f_m' = 2500$ psi to find

$p = 0.00082$

$A_s = 0.0082(9)(20) = 1.48$ sq. in.

From Table C-3, use 2 - #8 bars (1.56 sq. in.)

The additional amount of reinforcing required by the ACI/ASCE code is a result of the higher value of E_m and thus lower value of n.

EXAMPLE 5-C   Stresses in Masonry and Reinforcing Steel.

A 10″ thick reinforced brick wall (see Figure 5-9) was constructed with #7 bars at 24″ o.c. in the center of the wall. After construction it was found that the designer used a lower moment than the required design moment of 2.5 ft kips/ft. Check the masonry and steel stresses to ensure the wall is not over stressed.

Special inspection was provided during the wall's construction. Use $f_m' = 2000$ psi and $F_y = 60,000$ psi

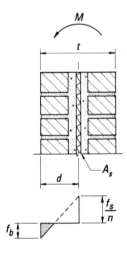

Figure 5-9 Stresses in wall.

Solution 5-C

(1) From Table A-3 and A-4 the allowable stresses are:

$F_b = 667$ psi and $F_s = 24,000$ psi

(2) Compute the flexural coefficient, K

$$K = \frac{M}{bd^2} = \frac{2.5 \times 12,000}{12 \times (5)^2}$$

$$K = 100$$

(3) Compute the reinforcement ratio, p

$p = A_s/bd$ or from Table C-8c for #7 bars @ 24″ with $d = 5″$, $p = 0.0050$

(4) By plotting $K = 100$ and $p = 0.005$ in Diagram F-2 determine the actual stresses:

$$f_b = 650 \text{ psi and } f_s = 23,000 \text{ psi}$$

Both stresses are below the allowable values and the wall will be sufficient to withstand the increased loading.

5-3.D.2 Variation in Stress Levels of the Materials

The following outlines the conditions of variable stress for the materials, masonry and reinforcing steel in which:

1) The reinforcing steel is at the maximum allowable tension stress, (the section is under-reinforced), while the masonry stress is variable from a low value up to its maximum allowable compressive stress.

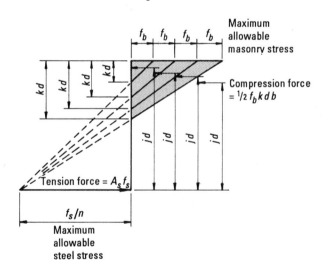

Figure 5-10 Maximum tensile stress and variable compression stress, under-reinforced.

2) The masonry is at the maximum allowable compression stress, (the section is over-reinforced), while the stress in the reinforcing steel is variable from a low value to the maximum allowable tension stress.

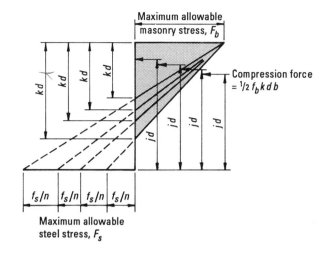

Figure 5-11 Maximum compressive stress with variable steel stress, over-reinforced.

EXAMPLE 5-D   Flexural Design; Determination of Beam Depth and Reinforcing Steel.

For balanced working stress design conditions, find the minimum lintel depth and the required area of reinforcing based on (a) the UBC with no inspection, (b) the UBC with special inspection, and (c) the ACI/ASCE code.

Design Data:

Brick masonry lintel constructed with type S mortar

$M = 45$ ft k

$b = 9″$

$f'_m = 1500$ psi

$F_s = 24,000$ psi

Neglect weight of lintel beam

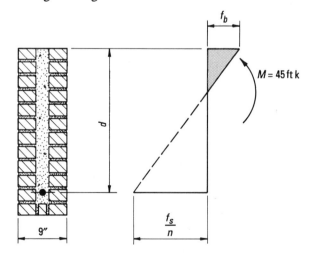

Figure 5-12 Moment on a beam in flexure.

Solution 5-D

Balanced design conditions occur when the maximum allowable masonry and steel stresses occur simultaneously.

Part (a)   Using UBC with no inspection

Refer to Table E-1a for non-inspected masonry with $f'_m = 1500$ psi and $F_s = 24,000$ psi

(1) For balanced conditions where $f_b = \frac{1}{2}(0.33 \times f'_m) = 250$ psi and $F_s = 24,000$ psi:

$$K_b = 24.6 \qquad p_b = 0.0011$$

(2) $K = \dfrac{M}{bd^2}$; Convert M into in. lbs and solve for d

$$d_{min} = \sqrt{\frac{M \times 12,000}{Kb}}$$
$$= \sqrt{\frac{45 \times 12,000}{24.6 \times 9}}$$
$$= 49.4'', \text{ use } d = 50''.$$

If the cover is assumed as approximately 6", the total depth, $D = 56''$

(3) $A_s = pbd = 0.0011(9)(50) = 0.50$ sq. in.

(4) Find reinforcing steel to provide at least $A_s = 0.50$ sq. in. from Table C-3.

Use 1 - #7 (As = 0.60 sq. in.) or 2 - #5 bars ($A_s = 0.62$ sq. in.)

Part (b)   Using UBC with special inspection

(1) In Table E-1b, find $K_{bal} = 77.2$, $p_{bal} = 0.0036$

(2) $d_{min} = \sqrt{\dfrac{45 \times 12,000}{77.2 \times 9}} = 27.9''$ Use 28"

Use total depth $D = 28 + 6 = 34''$

(3) $A_s = pbd = 0.0036(9)(28) = 0.91$ sq. in.

(4) From Table C-3 choose 1 - #9 bar ($A_s = 1.0$ sq. in.)

Part (c)   Using ACI/ASCE Code with special inspection

Refer to A/A Table B-1a for clay masonry constructed with Type S mortar and $f'_m = 1500$ psi.

(1) $K_{bal} = 64.8$, $p_{bal} = 0.0030$

(2) $d_{min} = \sqrt{\dfrac{45 \times 12,000}{64.8 \times 9}} = 30.4''$ say 32"

Thus, the total depth $D = 32'' + 6 = 38''$

(3) $A_s = pbd = 0.0030(9)(32) = 0.86$ sq. in.

(4) From Table C-4 use 2 - #6 bars ($A_s = 0.88$ sq. in.)

EXAMPLE 5-E   Moment Capacity of Beam.

Determine the moment capacity of the lintel beam shown in Figure 5-13.

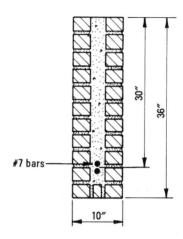

Figure 5-13 Beam cross section.

Given:

$$b = 10 \text{ inches}$$
$$\text{total depth} = 36 \text{ inches}$$
$$d \text{ to steel} = 30 \text{ inches}$$
$$A_s = 2 \text{ - \# 7 bars}$$
$$f'_m = 2000 \text{ psi} \quad \text{No special inspection}$$
$$\text{Use half stresses}$$
$$F_s = 24,000 \text{ psi}$$

Solution 5-E

(1) From Table A-3, Allowable Stresses,

$f_b = 333$ psi maximum

(2) From Table C-3, Area of Steel

for 2 - #7 bars $A_s = 1.20$ sq. in.

steel ratio $p = \dfrac{A_s}{bd} = \dfrac{1.20}{10 \times 30} = 0.0040$

(3) Enter Diagram F-2 with $p = 0.0040$

and $f_b = 333$ psi maximum
and $f_s = 24,000$ psi maximum

Proceed vertically up the $p = 0.0040$ line until either the limiting f_b line or f_s line is intersected.

The $f_b = 333$ psi is intersected first at the ordinate $K = 48$. Also read $f_s = 13,000$ psi.

(4) Moment capacity,

$$M = Kbd^2 = Kbd^2 = 48 \times 10 \times 30^2$$
$$= 432,000 \text{ in. lbs}$$
$$= 36.0 \text{ ft kips}$$

5-3.E Design Using *npj* and 2/*jk* Values

The various tables given in this handbook are based on specific E_m and n values. However, there are instances when the materials used or analyzed do not have the same properties.

Therefore, a technique of design has been developed that is applicable to any material, modulus of elasticity, E_m, modular ratio, n, or stress value. It is called the the Universal Elastic Flexural Design Technique in which values for 2/jk and npj are obtained and then values of np, j, k and p are determined. Table E-9 provides the data to determine np, 2/jk, npj, j and k.

Since the moment based on allowable flexural compressive masonry stress, F_b is:

$$M = bd^2\left(\frac{jk}{2}\right)F_b$$

A value for 2/jk can be found by rearranging the equation as follows:

$$\frac{2}{jk} = bd^2 \frac{F_b}{M}$$

Similarly, since the moment based on the allowable tensile steel stress F_s, is:

$$M = bd^2(pj)F_s$$

A value of npj can be found by multiplying both sides by n and solving for npj:

$$npj = \frac{nM}{bd^2 F_s}$$

With the values of 2/jk and npj, np values can be obtained from Table E-9 and the required steel ratio is calculated using the actual modular ratio:

$$p = np/n$$

The area of steel can then be determined:

$$A_s = pbd$$

Therefore for design, given the moment on the section, the effective depth, d, the width, b, the specified strength of the masonry, f'_m, the allowable stress of the steel, F_s, and calculating the modular ratio, n, the values, 2/jk and npj can be calculated and the required steel can be determined.

Given a moment requirement of 2150 ft lbs/ft, determine the reinforcing steel required for an 8″ nominal CMU if f'_m = 2500 psi, f_s = 24,000 psi and d = 5.3″. Use full masonry stresses.

Solution 5-F

$$F_b = \tfrac{1}{3} f'_m = \tfrac{1}{3}(2500) = 833 \text{ psi}$$

$$n = \frac{E_s}{E_m} = \frac{29{,}000{,}000}{750 f'_m} = 15.5$$

Determine 2/jk and npj to find np from Table E-9. Use the maximum value to obtain the required steel ratio.

$$\frac{2}{jk} = bd^2 \frac{F_b}{M}$$

$$= 12 \times 5.3^2 \times \frac{833}{2150 \times 12}$$

$$= 10.883$$

From Table E-9 for 2/jk = 10.883:

$$np = 0.024$$

$$npj = \frac{nM}{bd^2 F_s}$$

$$= \frac{15.5 \times 2150 \times 12}{12 \times 5.3^2 \times 24{,}000}$$

$$= 0.0494$$

From Table E-9 for npj = 0.0494:

$$np = 0.054$$

Steel stress governs since np is larger.

$$p = \frac{np}{n} = \frac{0.054}{15.5}$$

$$= 0.0035$$

$$A_s = pbd = 0.0035 \times 12 \times 5.3$$

$$= 0.22 \text{ sq. in./ft}$$

Use #6 at 24″ o.c.

For analysis, the physical properties and the moment are given or calculated and the stress in the masonry and steel can then be determined as:

$$f_b = \frac{M}{bd^2}\left(\frac{2}{jk}\right)$$

$$f_s = \frac{M}{bd^2}\left(\frac{1}{pj}\right)$$

Where $p = A_s/bd$ and $n = E_s/E_m$ the values and 2/jk and j are easily obtained from Table E-9 based on the calculated np value.

5-3.F Partially Grouted Walls

In order to reduce the weight of a wall and to minimize the amount of grout used, only cells containing reinforcing steel are grouted in partially grouted hollow unit walls. This reduces the cross-sectional area of the wall and consideration should be given to reduced vertical load capacity, reduced shear capacity parallel to the wall and flexural capacity for out of plane forces.

Walls grouted only at the reinforcing steel develop a rectangular or a tee stress block when they are subjected to lateral forces perpendicular to the wall. If the compression area or kd distance to the neutral axis is within the face shells, the wall would be analyzed as a rectangular section.

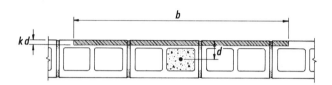

Figure 5-14 Partially grouted wall, rectangular stress block.

If the neutral axis, kd, is below the face shell the section would have a Tee section stress block.

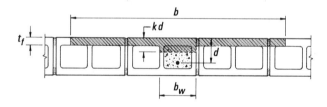

Figure 5-15 Partially grouted wall, Tee stress block.

For an 8 inch hollow unit wall the neutral axis will generally be within the face shell and the wall can be designed or analyzed as a rectangular section.

For larger units where the reinforcing steel is placed at a maximum d distance, a Tee section stress block may develop. The compression force, C, is resisted by both the face shell flange and part of the web.

Compression on flange

$$C_f = \tfrac{1}{2} f_b \left(1 + \frac{kd - t_f}{kd} \right) bt_f$$

Compression on web

$$C_w = \tfrac{1}{2} f_b \left(\frac{kd - t_f}{kd} \right) b_w \left(kd - t_f \right)$$

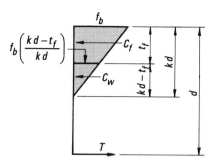

Figure 5-16 Stress diagram for tee section.

The determination of the depth of the stress block, kd, is based on the modular ratio, n, steel ratio p, thickness of the face shell, t_f, and depth to the steel, d.

$$kd = \frac{np + \tfrac{1}{2}\left(\dfrac{t_f}{d}\right)^2}{np + \left(\dfrac{t_f}{d}\right)}$$

The moment resistance for the Tee section becomes

$$M = C_f j_f d + C_w j_w d$$

The compression force on the web is small and can be ignored. The evaluation of the $j_f d$ value becomes complex and can be reasonably estimated by conservatively assuming the lever arm $jd = (d - t_f/2)$.

The value of the compression force can be determined by using the face shell area only and the average stress on it.

The compression force is

$$C = \tfrac{1}{2} f_b \left(1 + \frac{kd - t_f}{kd} \right) bt_f$$

The moment based on masonry stress is

$$M_m = \tfrac{1}{2} f_b \left(1 + \frac{kd - t_f}{kd} \right) bt_f \left(d - \frac{t_f}{2} \right)$$

The moment based on steel stress is

$$M_s = A_s f_s jd = pbd f_s \left(d - \frac{t_f}{2} \right)$$

EXAMPLE 5-G Design of a Partially Grouted Wall.

Determine the reinforcing steel required for a nominal 10 inch CMU wall, 25 feet high and subjected to a lateral wind force of 25 psf.

Assume $f'_m = 1500$ psi, $n = 25.8$, $f_s = 24,000$ psi, $d = t/2 = 4.81$ and no special inspection.

The wall is to be partially grouted at the vertical reinforcing steel bars which are spaced at 48" o.c.

Solution 5-G

$$\text{Moment} = \frac{wh^2}{8} \quad \text{(pinned each end)}$$

$$= \frac{25 \times 25^2}{8} \times 4$$

Moment per bar = 7812 ft lbs / bar

$$M = A_s f_s \left(d - \frac{t_f}{2} \right)$$

$$A_s = \frac{M}{f_s \left(d - \frac{t_f}{2} \right)}$$

$$= \frac{7812 \times 12}{1.33 \times 24,000 \left(4.81 - \frac{1.25}{2} \right)}$$

$$= 0.70 \text{ sq. in./48 in.}$$

Use #8 bars at 48 in. o.c., (A_s = 0.79 sq. in.)

Check minimum area of steel.

Use $0.007bt$ min. for horizontal (temperature and shrinkage) steel and $0.0013bt$ min. for vertical steel.

Minimum A_s = 0.0013 × 48 × 9.63
  = 0.60 < 0.79 sq. in.   O.K.

Determine masonry stress and kd distance

$$p = \frac{A_s}{bd} = \frac{0.79}{48 \times 4.81}$$

$$= 0.0034$$

$$np = 25.8 \times 0.0034$$

$$= 0.088$$

$$\frac{t_f}{d} = \frac{1.25}{4.81} = 0.26$$

$$k = \frac{np + \frac{1}{2} \left(\frac{t_f}{d} \right)^2}{np + \left(\frac{t_f}{d} \right)}$$

$$= \frac{0.088 + 0.5(0.26)^2}{0.088 + (0.26)}$$

$$= 0.350$$

$$kd = 0.35 \times 4.81$$

$$= 1.68 > t_f = 1.25 \text{ in.}$$

Therefore the stress block is a Tee section.

Allowable masonry stress

$$F_b = \frac{1}{2} \times \frac{1}{3} \times \frac{4}{3} \times 1500 = 333 \text{ psi}$$

Masonry stress

$$f_b = \frac{2M}{\left(1 + \frac{kd - t_f}{kd} \right) \left(d - \frac{t_f}{2} \right) bdt_f}$$

$$= \frac{2 \times 7812 \times 12}{\left(1 + \frac{1.68 - 1.25}{1.68} \right) \left(4.81 - \frac{1.25}{2} \right) (48 \times 4.81 \times 1.25)}$$

$$= 124 \text{ psi} < 333 \text{ psi}   \text{O.K.}$$

Allowable steel stress

$$F_s = \frac{4}{3} \times 0.4 f_y = 32,000 \text{ psi}$$

Steel stress

$$f_s = \frac{M}{A_s \left(d - \frac{t_f}{2} \right)}$$

$$= \frac{7812 \times 12}{0.79 \left(4.81 - \frac{1.25}{2} \right)}$$

$$= 28,388 \text{ psi} < 32,000 \text{ psi}   \text{O.K.}$$

Horizontal steel; use minimum A_s

$$A_s = 0.0007bt$$

$$= 0.0007 \times 48 \times 9.63$$

$$= 0.32 \text{ sq. in./48 in.}$$

Use #5 at 48 in. o.c.

5-3.G Compression Reinforcement

Masonry elements seldom require compression steel to obtain the required moment capacity since masonry sections are generally large and deep. However, in order not to overstress the masonry, in some cases it may be beneficial to use compression steel. In walls and piers subjected to overturning moments, jamb steel at each end acts in both tension and compression and increases the moment capacity of the wall or pier.

The use of compression reinforcement in masonry increases the moment capacity of the section by increasing the compression capacity of the masonry. It increases the moment arm distance, jd, producing an increase in the flexural moment capacity.

5-3.G.1 Compression Steel — Modular Ratio

The Uniform Building Code, Section 2626(f)5 states that the area of compression steel is multiplied by $2n$ to obtain the transformed area in flexural members reinforced with compression steel. The stress in the compression steel must not exceed the allowable tensile stress.

Based on the working stress, elastic design theory, it is assumed that the strain between the masonry and the steel is the same, therefore, the sharing of load between the masonry and compression steel would be in direct relation to modular ratio so that the stress in the steel would be as shown in calculations based upon using an "n" value.

As the stress strain curve for masonry is not linear and the strain increases in a non-linear fashion, strain in the steel is increased thus more load is taken by the steel than is initially calculated.

In addition, there is plastic flow and creep that takes place in the masonry. The masonry is still capable of taking its share of the load but there is an increased strain in the masonry and with this increase in strain in the masonry a similar strain is introduced into the steel subjecting the steel to a greater load or stress condition. Accordingly, the value for design and calculations of $2n$ is more in keeping with the actual stresses in the member with compression steel. This condition also utilizes, to a much more efficient degree, the use of steel by the introduction of the $2n$ value in keeping with the ACI and UBC statements.

Tables N-1 to N-8 and Diagrams P-1 to P-8 are provided for the design and analysis of walls and beams using compression reinforcement.

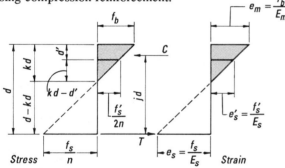

Figure 5-17 Stress and strain compatibility in flexural member with compression steel.

These Tables and Diagrams, are based on a value for the transformed area of steel in compression which is doubled, i.e., transformed compression steel area = $2nA'_s$ = $2np'bd$. In computing the location of the neutral axis, it is easier to maintain the compression area of the masonry as kdb and to account for the area displaced by steel by $(2n-1)A'_s$.

Maximum stress of compression steel at the maximum allowable masonry stress is found as follows:

$$\frac{f_b}{kd} = \frac{f'_s}{2n(kd-d')}$$

$$f'_s = 2nf_b\left(\frac{kd-d'}{kd}\right)$$

EXAMPLE 5-H   Compresion Steel Stress.

Determine the stress in the compression steel for a section with:

$$f'_m = 1500 \text{ psi}$$
$$f_b = 0.33f'_m = 500 \text{ psi}$$
$$n = 25.8$$
$$d = 40''; \ d' = 4''$$
$$k = 0.30$$

Solution 5-H

$$f'_s = 2nf_b\left(\frac{kd-d'}{kd}\right)$$
$$= 2 \times 25.8(500)\left(\frac{0.30 \times 40 - 4}{0.30 \times 40}\right)$$
$$= 17,200 \text{ psi}$$

The stress in the steel is limited by the allowable stress in the masonry and the d' distance. Although the compression steel is not stressed to its maximum allowable stress ($f'_s = 0.4f_y$ max. or 24,000 psi), it still improves the compression and moment capacity of the section.

Compression steel is effective only if d' is less than kd.

EXAMPLE 5-I   Flexural Design, Tension and Compression Reinforcement.

A masonry beam is subjected to bending moment, M. Determine the reinforcing steel required (a) with tension steel, A_s, only (b) with tension steel, A_s, and compression steel, A'_s.

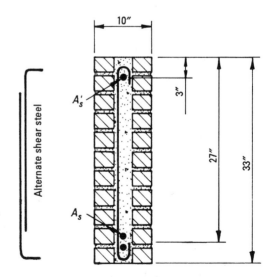

Figure 5-18 Beam with tension and compression steel.

Given:

Moment M = 55 ft kips
f'_m = 1500 psi
f_y = 60,000 psi
Special inspection will be provided

Solution 5-I

From Table A-3 and A-4:

f_b = 500 psi
n = 25.8
F_s = 24,000 psi tension steel
F_{sc} = 24,000 psi compression steel
b = 10″, d = 27″, d' = 3″

Part (a) Tension steel, A_s, only

(1) Determine the flexural coeffficient, K

$$K = \frac{M}{bd^2} = \frac{55 \times 12,000}{10 \times 27^2} = 90.5$$

This is greater than the balanced K which is K_b = 77.2 as given in Table E-1b. Either (a) over-reinforce the beam, (b) use compression steel, (c) increase the size of the member, (d) provide special inspection and use full stresses.

(2) Determine the steel area reinforced for tension only.

From Table E-1b,

For K = 89.8: p = 0.0058, and
for K = 92.3: p = 0.0063

interpolate between the two values above for K = 90.5;

Steel ratio p = 0.0059

Area of steel $A_s = pbd$
$= 0.0059 \times 10 \times 27$
$= 1.59$ sq. in.

From Table C-3, selection of size and amount of steel.

Use 2 - #8 bars (A_s = 1.58 sq. in.), or
1 - #9 and 1 - #7 (A_s = 1.60 sq. in.)

Part (b) Tension steel, A_s, and compression steel, A'_s

(3) Using Table N-1b, Coefficients for Tension and Compression Steel, or Diagram P-1b, p vs K.

for $\dfrac{d'}{d} = \dfrac{3''}{27''} = 0.11$ and K = 90.5;

read tension steel ratio p = 0.0042
compression steel ratio p' = 0.0009

Tension steel $A_s = pbd = 0.0042 \times 10 \times 27$
$= 1.13$ sq. in.

Use 2 - #7 bars (A_s = 1.20 sq. in.)

Compression steel $A'_s = p'bd$
$= 0.0009 \times 10 \times 27$
$= 0.24$ sq. in.

Use 1 - #5 bar (A'_s = 0.31 sq. in.)

Total area of steel used:
1.20 + 0.31 = 1.51 sq. in.

5-4 SHEAR

5-4.A General

Structural elements such as walls, piers and beams are subjected to shear forces as well as flexural stresses. The unit shear stress is computed based on the formula

$$f_v = \frac{V}{bjd} = \frac{V}{bd} \text{ or } \frac{V}{bl}$$

The deletion of the j coefficient is usually not significant as the actual shear stress distribution is not fully understood and the refinement of jd is unwarranted. In fact, the j coefficient is not included in the calculation of the shear stress for concrete in either UBC Chapter 26, Equation 26-1 or in ACI 318-8a, Equation A-1. Therefore j is often ignored or approximated as 0.9 for the preliminary design.

Shear design analysis and criteria have been based on tests and experience and the limiting allowable stresses are conservative.

If the unit shear stress does not exceed the allowable shear stress for masonry as given in Tables A-3 and A-5 and Diagrams A-4, A-5 no shear reinforcing is required. If the unit shear exceeds the allowable shear stress for masonry, shear reinforcing steel must be provided to resist all the shear forces. Tables and Diagrams A-6 and K-1 to K-4 can be used to size the shear reinforcing steel.

If the unit shear stress exceeds the maximum allowable shear stress for the reinforcing steel, the section must be increased in size and/or higher strength masonry must be specified.

5-4.B Beam Shear

When masonry flexural members are designed to resist shear forces without the use of reinforcing steel, the calculated shear stress may not exceed $1.0(f'_m)^{1/2}$ nor 50 psi when special inspection is provided. When there will be no special inspection, the allowable shear stress must be reduced by one half of these values.

Web reinforcement must be provided to carry the entire shear in excess of 20 pounds per square inch whenever there is required negative reinforcement and for a distance of one-sixteenth the clear span beyond the point of inflection in a continuous beam or where there is continuity at the support. (UBC Section 2406(c)6.A. Exception.)

Should the unit shear stress exceed the allowable masonry shear stress, all the shear stress must be resisted by reinforcing steel.

For flexural members with reinforcing steel resisting all the shear force, the maximum allowable shear stress is $3.0(f'_m)^{1/2}$ psi with 150 psi as a maximum. This is with special inspection during construction. Without special inspection, the maximum allowable shear stress must be reduced by 50 percent.

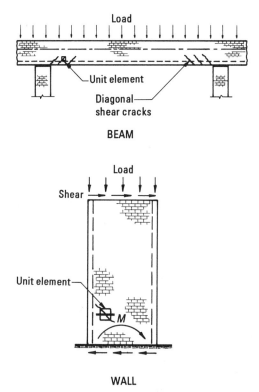

WALL

Figure 5-19 Diagonal tension cracks in a flexural member.

The principle of shear reinforcing is to provide steel to resist the diagonal tension stresses developed in an member. Figures 5-19 and 5-20 demonstrate the diagonal tension principle.

Diagonal tension stresses are due to the combined vertical and horizontal shear, and although it can be reasoned that reinforcing steel in either direction will resist the diagonal tension stresses it is generally accepted that shear reinforcing should be parallel to the direction of the external applied loads or shear forces. Therefore shear reinforcing is vertical in a beam and horizontal in a wall.

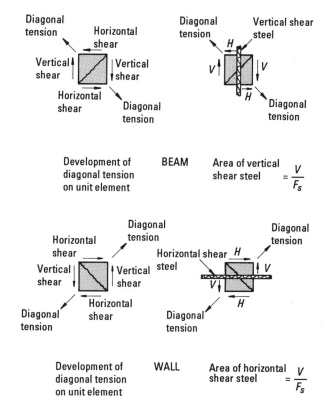

Figure 5-20 Development of shear in unit element.

The steel resists the shear by tension and it must be anchored in the compression zone of the beam or the wall.

The unit shear, f_v, is used to determine the shear steel spacing based on the formula:

$$\text{Spacing, } s = \frac{A_v F_s}{f_v b}$$

Diagrams K-1 to K-4 can be used to quickly find the required shear reinforcing size and spacing. Likewise Tables K-1 to K-4 give the allowable shear stress, F_v, can be found for a given size and spacing of steel.

$$F_v = \frac{A_v F_s}{bs}$$

For continuous or fixed beams, the shear value used to determine the shear steel spacing may be taken at a distance d from the face of the support. The shear value at the face of the support should be used to calculate the shear steel spacing in simple beams.

The maximum spacing of shear steel may not exceed $d/2$. The first shear reinforcing bar should be located at half the calculated spacing but no more than $d/4$ from the face of support.

For a more conservative analysis, one may wish to determine the shear in a continuous or fixed beam at a distance $d/2$ from the support or even at the support.

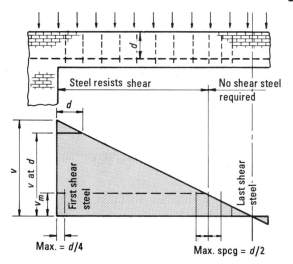

Figure 5-21 Spacing of shear reinforcing in a beam.

The thickness of a member or wall for shear calculations may be influenced by the treatment of the joints. Masonry with flush or concave tooled joints would have the total thickness effective. However, if joints are raked, consideration should be given to the reduction in the width of the wall caused by raking.

EXAMPLE 5-J Flexural Design — Unit Shear Stress.

Determine the unit shear stress for the following continuous masonry beam:

$DL = 150$ plf
$LL = 400$ plf
Span = 14 ft (continuous span)
$d = 20$ inches
$b = 9$ inches
$f'_m = 1500$ psi — no special inspection

Solution 5-J

(1) Total load = 400 + 150 = 550 lb/ft

(2) Total shear $V = \frac{1}{2} \times 550 \times 14$

$$= 3850 \text{ lbs}$$

Compute shear at a distance d from face of support

$$V = 3850 - \frac{20}{12} \times 550$$

$$= 2933 \text{ lbs}$$

(3) Calculate the shear stress (Assume $j = 0.88$)

$$f_v = \frac{V}{bjd} = \frac{2933}{9 \times 0.88 \times 20}$$

$$= 18.5 \text{ psi}$$

From Table A-1, the allowable flexural shear stress with no shear reinforcing is:

$$f_v = 20 \text{ psi} > 18.5$$

No shear reinforcing is required.

EXAMPLE 5-K Beam — Shear Reinforcing.

A concrete masonry spandrel beam is subjected to a shear force

$$V = 13 \text{ kips}$$

Design the shear reinforcing for the simply supported beam if:

Nominal $b = 8''$, Actual $b = 7.625''$, $d = 36''$
$f_v = 24,000$ psi, $f'_m = 1500$ psi
No special inspection will be provided.

Solution 5-K

(1) From Table A-3, the allowable flexural shear stress with shear reinforcing is

$F_v = 58$ psi; assume $j = 0.9$

$$\text{Shear stress, } f_v = \frac{V}{bjd} = \frac{13,000}{7.625 \times 0.9 \times 36}$$

$$= 53 < 58 \text{ psi}$$

(2) From Diagram K-2, spacing of shear steel

for $b = 7.625''$ and $f_v = 53$ psi

use #5 at 16" o.c. (Shear capacity, $F_v = 60$ psi)

EXAMPLE 5-L   Beam Shear Reinforcing Size and Spacing.

Determine the shear reinforcing required in the 8" solid grouted masonry beam shown in Figure 5-22.

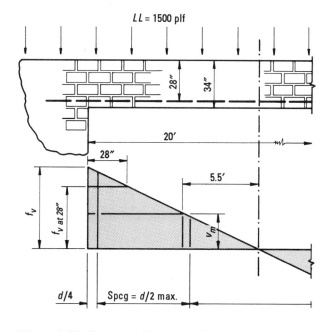

Figure 5-22 Shear reinforcing in beam.

Design data:

$f'_m = 2500$ psi; $f_y = 60,000$ psi

$d = 28''$; Inspection will be provided

Solution 5-L

(1) Conservatively, assume the beam is constructed of normal weight concrete block and grout. Thus, from Table B-3a, the weight of solid grouted hollow concrete block = 84 psf

$$DL = \frac{84 \times 34}{12} = 224 \text{ plf}$$

$$LL = 1500 \text{ plf}$$

$$TL = w = 1724 \text{ plf}$$

$$\text{Total shear } V = \frac{wl}{2} = \frac{1724 \times 20}{2}$$

$$= 17,240 \text{ lbs}$$

(2) Calculate the shear stress. For 8″ concrete masonry units, $t = 7.63″$. Also assume $j = 0.89$.

$$f_v = \frac{V}{bjd} = \frac{17,240}{7.63 \times 0.89 \times 28}$$

$$= 90.7 \text{ psi}$$

(3) Check the capacity of the masonry without shear reinforcing.

From Table A-3 for inspected, 2500 psi masonry;

$$F_v = 50 \text{ psi} < f_v$$

(4) Find where shear reinforcing is required.

$$V = F_v bjd = 50 \times 7.63 \times 0.89 \times 28$$

$$= 9510 \text{ lbs}$$

Distance from center of beam where no shear reinforcing is required.

$$\text{Distance } S = \frac{V}{w} = \frac{9510}{1724} = 5.52 \text{ ft}$$

(5) Calculate the shear at a distance d from the support and determine the size and spacing of the reinforcing steel.

$$V = 17,240 - 1724(28″/12)$$

$$= 13,217 \text{ lbs}$$

$$\text{Unit shear, } f_v = \frac{V}{bjd} = \frac{13,217}{7.63(0.89)(28)}$$

$$= 69.5 \text{ psi say } 70 \text{ psi}$$

From Diagram K-2 for $t = 7.63″$ and $f_v = 70$ psi, try either #4 @ 8″ or #6 @ 16″.

Maximum spacing of shear reinforcing is limited to $d/2 = 28/2 = 14″$

Therefore use #4 @ 8″

Place the first bar at $s/2 = 8/2 = 4″$

Continue the reinforcing past the point where shear reinforcing is no longer required.

$$\text{Number of space} = \frac{(5.5 \times 12 \text{ in./ft}) - 4″}{8}$$

Use at least 8 spaces @ 8″ = 5′ - 4″.

5-4.C Shear Parallel to Wall

Walls which resist lateral forces, particularly forces due to wind or earthquake, are called shear walls. These walls may be load bearing or non-load bearing. Shear walls may also resist lateral forces due to earth or water.

The allowable shear stress for walls, based on M/Vd and whether special inspection is required, is given in UBC Section 2406(c)7 and Tables and Diagrams A-5 and A-6 of this book.

When M/Vd is less than one, the maximum allowable shear stress in the masonry is determined by the equation:

$$F_v = \tfrac{1}{3}\left(4 - \frac{M}{Vd}\right)\sqrt{f'_m} \qquad \text{(UBC Chapter 24, Equation 6-10)}$$

with a maximum value of

$$F_{v(max)} = \left(80 - 45\frac{M}{Vd}\right) \text{ (psi)}$$

When M/Vd is one or greater, the maximum allowable masonry shear stress is:

$$F_v = 1.0\sqrt{f'_m}; \text{ 35 psi maximum} \qquad \text{(UBC Chapter 24, Equation 6-11)}$$

When the shear stress, f_v, exceeds the allowable masonry shear stress given above, reinforcing steel must be provided to resist all the shear.

The allowable shear steel for reinforced walls when M/Vd is less than one:

$$F_v = \tfrac{1}{2}\left(4 - \frac{M}{Vd}\right)\sqrt{f'_m} \qquad \text{(UBC Chapter 24, Equation 6-12)}$$

with a maximum value of:

$$F_{v(max)} = \left(120 - 45\frac{M}{Vd}\right) \text{ (psi)}$$

For $\dfrac{M}{Vd} \geq 1$,

$$F_v = 1.5\sqrt{f'_m}; \text{ 75 psi maximum} \qquad \text{(UBC Chapter 24, Equation 6-13)}$$

The reduction in allowable shear stress based on the M/Vd ratio is related to the decreased shear capability from a pure shear condition, i.e., $M/Vd = 0$; to a flexural shear condition in which the wall element is acting as a flexural beam element as well as a shear resisting wall.

All allowable stresses may be increased one third when the lateral force is due to wind or seismic action.

When calculating shear or diagonal tension stresses, shear walls in Seismic Zone Nos. 3 and 4 must be designed to resist 1.5 times the forces determined by:

$$V = \frac{(ZICW)}{R_W}$$

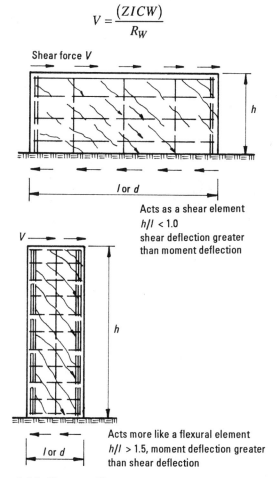

Figure 5-23 Shear walls.

The requirement that the reinforcing steel be designed to resist all the shear is a conservative approach for the masonry still has a shear capability which is ignored. In strength design for shear walls, (UBC Section 2412), credit is given to masonry as a basic resistance valve. However, under dynamic reversal forces which occur in an earthquake this conservative approach is warranted.

Design the horizontal shear reinforcing in a brick pier for a lateral seismic force, V, of 19.2 kips if:

$f'_m = 1500$ psi; $f_y = 60,000$ psi;

$w = 48$ in.; $d = 42$ in.; $t = 10$ in.

Inspection will be provided

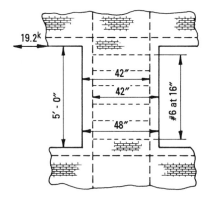

Figure 5-24 Pier with shear reinforcing.

Solution 5-M

(1) Calculate the actual shear stress (assume $j = 0.9$).

Per UBC Sec. 2407(n)4.F, increase the design shear force by 1.5 times the applied force

$$f_v = \frac{1.5V}{bjd} = \frac{1.5(19,200)}{10(0.9)(42)} = 76 \text{ psi}$$

(2) Find the allowable shear stress by calculating M/Vd. For a fixed pier subjected to a deflection, Δ:

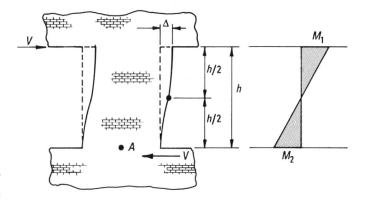

Figure 5-25 Fixed pier subjected to displacement, Δ.

$$\Sigma M_A = 0$$
$$0 = M_1 + M_2 - Vh$$
$$M = \frac{Vh}{2}$$
$$\frac{M}{Vd} = \frac{Vh/2}{Vd} = \frac{h}{2d}$$

Therefore for this pier,

$$\frac{M}{Vd} = \frac{h}{2d} = \frac{5' \times 12 \text{ in./ft}}{2(42'')}$$
$$= 0.71$$

From Table A-5b for $M/Vd = 0.71$ and $f'_m = 1500$ psi, the allowable shear stress for the masonry is:

$$F_v = 43 \text{ psi} \times 1.33 = 57 \text{ psi} < 76 \text{ psi} \quad \text{N.G.}$$

Reinforcing steel must be provided and designed to carry all the shear load. From Table A-6b,

$$F_v = 64 \text{ psi} \times 1.33 = 85 \text{ psi} > 76 \text{ psi} \quad \text{O.K.}$$

Size the shear steel from Diagram K-3 for $F_s = 32,000$ psi, $t = 10''$, $f'_m = 1500$ psi and $f_v = 76$ psi.

Choose #6 at 16" o.c. vertically.

5-4.D Shear Perpendicular to Wall

To compute the unit shear stress perpendicular to a masonry wall, the dimension d to the steel reinforcing could be used.

To determine the unit shear at the base of the wall, it would be satisfactory to determine the unit shear stress $f_v = V/bt$ as unreinforced masonry or $f_v = V/bd$ as reinforced masonry.

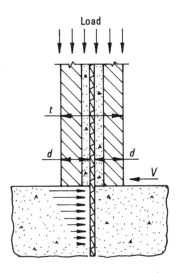

Figure 5-26 Shear resistance at floor joint.

The shear capacity of a masonry wall is influenced by vertical forces or loads on the wall. Vertical loads on a wall will increase its shear capacity by the added frictional resistance between the wall and the concrete footing or floor. The range and normal applicable coefficients for static friction are listed in Table 5-1.

TABLE 5-1 Coefficient of Static Friction

Materials	Range	Normal[1]
Masonry and masonry	0.65 - 0.75	0.70
Masonry and concrete	0.65 - 0.75	0.70
Masonry and dry earth	0.30 - 0.50	0.35
Masonry and metal	0.30 - 0.50	0.40
Concrete and dry earth	0.30 - 0.50	0.35
Masonry and wood	0.50 - 0.60	0.50

1. The normal coefficient values are reasonable to use to consider lateral frictional shear resistance.

Shear resistance of reinforcing steel at the floor joint can be conservatively assumed as the same as for anchor bolts. Values are given in Table A-8. The connection between the floor, roof diaphragms and the walls must be capable of resisting a lateral force in any direction of at least 200 plf (UBC Sec. 2310).

EXAMPLE 5-N Determination of Shear Stresses for a Partially Grouted Wall.

Calculate the shear stress for an 8" hollow unit masonry wall shown below with steel grouted at 32" o.c. and a shear force of 200 plf.

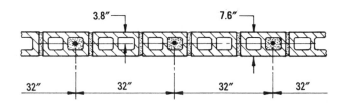

Figure 5-27 Plan section of hollow unit masonry wall.

Solution 5-N

Shear perpendicular to wall.

a. Minimum shear area; grouted cell + web + end wall

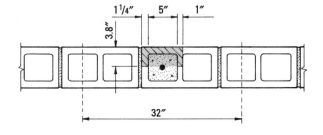

Min. Shear width per 32″

b = cell width + web + end wall
= 5 + 1 + 1.25
= 7.25″ per 32″

$$f_v = \frac{V}{bd} = \frac{200 \times 2.67}{7.25 \times 3.8} = 19.4 \text{ psi}$$

b. Shear area using grouted cell, web, end wall and one mortared face shell.

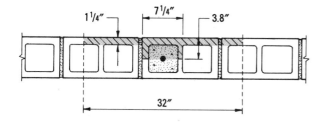

Shear area = (5 + 1 + 1.25)3.8
+ (32 − 7.25)1.25 = 58.5 sq. in.

$$f_v = \frac{V}{\text{shear area}} = \frac{200 \times 2.67}{58.5} = 9.1 \text{ psi}$$

c. Shear area for walls with compression stress use grouted cell, web, wall and both mortared face shells.

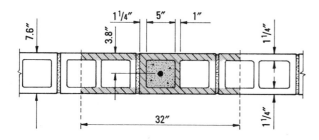

Shear area = (5 + 1 + 1.25)7.63
+ (32 − 7.25)1.25 × 2
= 55.3 + 61.9
= 117.2 sq. in.

$$f_v = \frac{V}{\text{shear area}} = \frac{200 \times 2.67}{117.2} = 4.6 \text{ psi}$$

Shear area parallel to wall (same as part c above)

A_v = 117.2 sq. in. per 32″

$$f_v = \frac{V}{bd} = \frac{200 \times 2.67}{117.2} = 4.6 \text{ psi; shear parallel to wall}$$

5-5 BOND

5-5.A Bond in Masonry

Properly designed and placed mortar and grout will develop sufficient bond strength with the masonry units which will result in a homogeneous mass for design considerations within the range of allowable stresses. High bond strength results when the clay masonry units are saturated surface dry (s.s.d.) and have a suction or initial rate of absorption of approximately 20 grams or less at time of being laid (See Section 1-1A). Mortar type M or S, which is workable and mixed with maximum amount of water produces the strongest bond strength.

The Office of the State Architect of California specifies that a minimum shear bond strength between brick masonry units and mortar shall be 20 psi and between brick masonry units and grout, 100 psi.

UBC Standard 24-19 requires a minimum flexural bond strength of

Type N mortar, 71 psi
Type S mortar, 104 psi
Type M mortar, 116 psi

Although these values are for a standardized concrete masonry unit and mortar cement, they give an indication of the magnitude of bond strength that may be achieved.

5-5.B Bond between Grout and Steel

Bond between mortar or grout and reinforcing steel is vital and necessary to insure that stresses will be transferred between the steel, the grout and the masonry. The bond strength is developed by the adhesion of the portland cement paste and the mechanical interlock with the deformation of reinforcing steel.

$$\text{Bond stress } u = \frac{V}{\Sigma_o jd} \qquad \text{(UBC Chapter 24, Equation 9-8)}$$

TABLE 5-2 Allowable Bond Stress, u psi

	No Special Inspection	Special Inspection
Plain Bars	30	60
Deformed Bars 1988 UBC	70	140
Deformed Bars 1991 UBC	100	200

In the report, *Bond and Splices in Reinforced Masonry*, by Soric and Tulin, 1987, it is noted that the allowable bond stress could be 400 psi based on an experimental minimum test result of 1000 psi, before failure, with a factor of safety of 2.5 applied.

EXAMPLE 5-O   Determination of Bond Stress.

Calculate the bond stress, u, for a masonry beam reinforced with (a) two #6 bars, and (b) one #7 bar.

Given:

Span = 14 ft;
DL = 180 plf; LL = 400 plf
d = 20 inches; b = 9 inches
F_s = 24,000 psi; f'_m = 2500 psi
no special inspection (use half stresses)

Solution 5-O

1) From footnote 9 of Table A-3, the allowable bond stress in the masonry, $u = 100$ psi

2) Total shear, $V = \frac{1}{2} \times (180 + 400) \times 14 = 4060$ lbs

3) From Table C-3 the perimeter of the steel bars are:

   a) Two #6 bars, $\Sigma_o = 4.7''$
   b) One #7 bar, $\Sigma_o = 2.7''$

4) Calculate bond stress

$$u = \frac{V}{\Sigma_o jd}$$

Assume $j = 0.88$

a) Two #6 bars

$$u = \frac{4060}{4.7 \times 0.88 \times 20}$$
$$= 49 \text{ psi} < 100 \text{ psi   O.K.}$$

b) One #7 bar

$$u = \frac{4060}{2.7 \times 0.88 \times 20}$$
$$= 85 \text{ psi} < 100 \text{ psi   O.K.}$$

5) Note: Since the reinforcing bars are embedded in at least 2000 psi concrete grout, it would also be reasonable to use the allowable bond stress for concrete.

EXAMPLE 5-P   Development Length.

What is the development length required for a No. 5 reinforcing bar grade 60, $F_s = 24,000$ psi, for tension or compression conditions?

a) Without a hook, straight bar only

b) With a hook at the end

Solution 5-P

Part (a)   Tension condition

The development length for deformed reinforcing steel in tension may be calculated as follows:

1. Without a hook:

   $l_d = 0.002 d_b f_s$ (UBC Chapter 24, Eq. 9-11)
   $= 0.002(0.625)(24,000)$
   $= 30$ inches or 48 bar diameters

2. With a hook:

   Value of hook = 7500 psi (UBC Sec. 2409(e)5.E)

   Therefore the balance of the stress is to be achieved by development length.

   $l_d = 0.002(0.625)(24,000 - 7500)$
   $= 21$ inches or 33 bar diameters

Part (b)   Compression condition

The development length for deformed reinforcing steel in compression may be calculated as follows.

1. Without a hook:

   $l_d = 0.0015 d_b f_s$ (UBC Chapter 24, Equation 9-12)
   $= 0.0015 (0.625)(24,000)$
   $= 23$ inches or 36 bar diameters

2. With a hook:

   Note: Hooks may not be considered effective for compression (UBC Sec. 2409 (e) 5.F)

   Note: Table 7-6 gives the development length for straight bars in tension or compression.

5-6   COMPRESSION IN WALLS AND COLUMNS

5-6.A   Walls

5-6.A.1   General

Load bearing reinforced masonry walls are limited to an axial load of:

$$P = F_a A_e$$

Where:

$$F_a = 0.20 f'_m \left[1 - \left(\frac{h'}{42t} \right)^3 \right]$$

(UBC Chapter 24, Equation 6-3)

A_e = effective cross-sectional area of masonry which includes grouted and mortared areas. For cavity walls consider only the loaded wythes. If mortar joints are raked, reduce the effective area accordingly.

Any vertical wall reinforcing is not considered effective in carrying vertical loads since it is not confined by ties. Thus the reinforcing steel is considered effective only for resisting lateral loads parallel and perpendicular to the wall. The allowable load bearing wall stress, F_a, is the same for both reinforced and unreinforced masonry.

5-6.A.2 Stress Reduction and Effective Height

A stress reduction factor is applied to the axial compressive capacity based on the effective height to thickness ratio, h'/t, with the equation:

$$\left[1-\left(\frac{h'}{42t}\right)^3\right]$$

Figure 5-29 shows conditions that describe the effective height of a wall. For members not supported at the top normal to the plane of the wall, the effective height, h', is considered twice the height of the member above the base.

If a wall spans horizontally, it can be considered to be continuous over vertical supports such as pilasters or intersecting walls. Such a continuous wall would have inflection points approximately at the quarter points although they are often conservatively assumed to be $0.2l$ from the supports (See Figure 5-28). The effective length (or h') of the wall is the distance between points of inflection or $0.6l$.

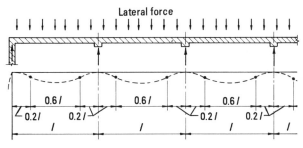

Effective $l = 0.6 l$

Figure 5-28 Longitudinal continuity of wall.

In the 1982 and prior editions of the Uniform Building Code, there was a limitation on the height or length to thickness ratio of walls. This was based on the masonry material and the use of the wall whether it was load bearing or non-load bearing.

Although there is no current code limitation on the h'/t ratio, UBC Section 2409(b)2 requires that when the h'/t is greater than 30 the design should consider duration of the loads, effects of deflections on moments and forces, influence of axial loads and the variable moment of inertia on member stiffness and fixed-end moments. The use of the slender wall design procedure in UBC Section 2411 considers many of the above requirements (see Section 6-5).

The ACI/ASCE 530 Masonry Code and Specification requires that the axial load reduction equation be based on the radius of gyration of the section.

For $\frac{h}{r} < 99$

$$F_a = 0.25 f'_m \left[1-\left(\frac{h}{140r}\right)^2\right] \quad \text{(ACI/ASCE Standard Equation 7-1)}$$

For $\frac{h}{r} \geq 99$

$$F_a = 0.25 f'_m \left(\frac{70r}{h}\right)^2 \quad \text{(ACI/ASCE Standard Equation 7-2)}$$

5-6.A.3. Effective Width

The effective width of a flexural wall member may be either horizontal or vertical depending on the way the wall spans. There should be consideration as to whether the wall is laid up in running bond or stack bond and whether the units are solid grouted, or open ended units.

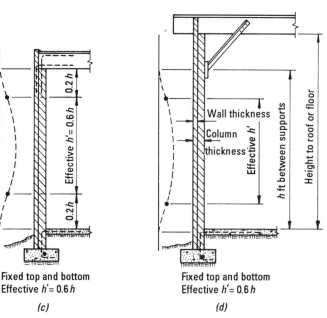

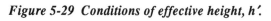

Pinned at supports Effective $h'= h'$	Fixed at base Effective $h' = 0.8h$	Fixed top and bottom Effective $h'= 0.6h$	Fixed top and bottom Effective $h'= 0.6h$
(a)	(b)	(c)	(d)

Figure 5-29 Conditions of effective height, h'.

For running bond, the effective width used in computing flexural stresses must not be greater than six times the wall thickness nor the center to center distance between the reinforcing bars (UBC Sec. 2409(c)4).

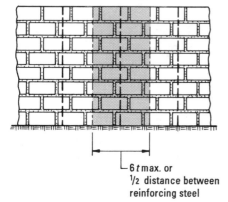

6 t max. or
¹/₂ distance between
reinforcing steel

* Maximum spacing of steel up to 8 ft. has been shown to be effective based on an Effective 'b' research program by the Masonry Institute of America.

Figure 5-30 Effective width of flexural member, running or common bond.

UBC Section 2407(b)2 states that a wall is considered to be laid in stack bond if less than 75 percent of the units in a vertical plane lap the ends of the units below a distance of less than one half the height of the unit or one fourth the length of the unit.

Where stack bond is used, the effective width considered may not exceed three times the wall thickness nor the center to center distance between the reinforcing

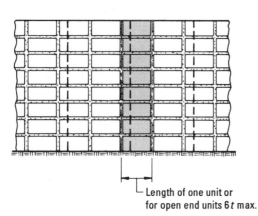

Length of one unit or
for open end units 6 t max.

Figure 5-31 Effective width of flexural member, stack bond.

bars nor the length of one unit, unless the wall is grouted solid using open-ended units (UBC Sec. 2409(c)4). Solid grouted, open end units in stack bond provide the required continuity for equivalent running bond.

Buildings in Seismic Zone No. 2, where stack bond is used, must be reinforced with a minimum A_s of $0.0007bt$ distributed uniformly with joint reinforcement or reinforcing steel spaced at a maximum of four feet on centers.

In Seismic Zone Nos. 3 and 4, however, stack bond masonry walls must be reinforced horizontally with a minimum area of steel of $0.0015bt$ unless open end units are solidly grouted. Then the horizontal steel area need only be $0.0007bt$.

EXAMPLE 5-Q   Lateral Wind Force on Wall, Flexural Design.

Determine the required flexural reinforcement for a 16'- 0" high, 8" concrete masonry wall subjected to a 20 psf lateral wind load.

Given: $d = 3.75"$, $f'_m = 2000$ psi (use half stresses), and
$F_s = 24,000$ psi

Solution 5-Q

$f_b = 333$ psi, $n = 19.3$ (Table A-3)

$1.33f_b = 444$ psi   (one third increase allowed for wind load by UBC Sec. 2303(d))

$1.33F_s = 32,000$ psi

(1)   Assume pin connections at top and bottom of wall;

$$M = \frac{wl^2}{8} = \frac{20 \times 16^2}{8} \times 12$$
$$= 7680 \text{ in. lbs / ft}$$

(2)   Flexural coefficient $K = \dfrac{M}{bd^2} = \dfrac{7680}{12 \times 3.75^2}$
$$= 45.5$$

(3)   Enter diagram F-9 with the flexural coeffcient, $K = 45.5$ and $f_b = 444$ psi. Read $np = 0.033$.

(4)   Steel ratio $p = \dfrac{np}{n} = \dfrac{0.033}{19.3}$
$$= 0.0017$$

(5)   From Table C-86 for $d = 3.75"$ and $p = 0.0017$, choose #5 at 48" o.c. ($A_s = 0.31$ sq. in./ft)

Alternate method, use Table E-2a

(6)   From Table E-2a for $1.33K = 45.7$

$f_b = 444$ psi

$f_s \approx 30,000$ psi $p \approx 0.0017$

(7)   Again choose #5 at 48".

EXAMPLE 5-R Minimum Wall Thickness.

For a solid grouted brick, non-load bearing exterior wall subjected to a lateral wind force, determine the minimum wall thickness when the steel is located in the center of the wall.

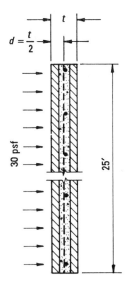

Figure 5- 32 Lateral load on wall.

Given:

Height of wall = 25 feet

Wind load   = 30 psf

f'_m = 2000 psf (special inspection will be provided
                    — use full allowable stresses.)

f_y = 60,000 psi

p = 0.0013

Solution 5-R

(1) From Tables A-3 and A-4, find the allowable stresses. These allowable stresses may be increased by one-third since load is due to wind (UBC Sec. 2303(d)).

$$f_b = 667 \times 1.33 = 887 \text{ psi}$$
$$F_s = 24,000 \times 1.33 = 32,000 \text{ psi}$$

(2) Enter Diagram F-2 and proceed vertically up the p = 0.0013 line until it intersects with f_s or f_b. Find minimum K by proceeding to the left of the lowest intersection. Read $K_{min} \approx 38$.

(3) $K = M/bd^2$. Calculate M by assuming the wall is pinned at the top and bottom.

$$M = \frac{wl^2}{8} = \frac{30(25)^2}{8} = 2344 \text{ ft lbs/ft}$$
$$= 28,125 \text{ in. lbs/ft}$$

Since $b = 12''$/ft, the above equation can be solved for d.

$$(d_{min})^2 = \frac{M}{bK}$$

$$d_{min} = \sqrt{\frac{28,125}{12(38)}}$$

$$= 7.86''$$

Since $d = t/2$, use a 16″ thick wall.

(4) Check stresses with $d = 8''$

$$K = \frac{M}{bd^2}$$

$$= \frac{28,125}{12(8)^2}$$

$$= 36.6$$

Enter Diagram F-2 with $K = 36.6$ and $p = 0.0013$ read:

$$f_b \approx 400 \text{ psi} < 890 \text{ psi} \text{ O.K.}$$

$$f_s \approx 31,000 \text{ psi} < 32,000 \text{ psi} \text{ O.K.}$$

Determine the moment capacity of a grouted masonry wall which spans vertically and is reinforced with the minimum area of steel as shown. Also, find the allowable uniform pressure in Figure 5-33 the wall can support if it spans 15 ft vertically.

Assume:

f'_m = 3000 psi (no special inspection)

f_y = 60,000 psi and F_s = 24,000 psi

t = 9″

Vertical steel, $A_s = 0.0013bt$

Horizontal steel, $A_s = 0.0007bt$

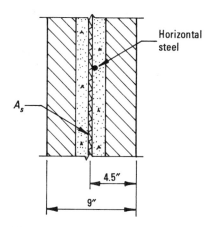

Figure 5- 33 Location of steel in wall.

Solution 5-S

Part (a) Moment Capacity

From Table J-2 with $A_s = 0.0013bt$, $d = 4.5''$, $f'_m = 3000$ psi and $F_s = 24,000$ psi:

$$M_m = 1.65 \text{ ft k/ft}$$

$$M_s = 0.64 \text{ ft k/ft}$$

Note that this table assumes special inspection is provided. As per footnote 8, reduce M_m by one half. $M_m = 1.65/2 = 0.82$ ft k/ft which is still larger than M_s.

Therefore, M_s controls the design and the

Moment capacity of wall $= 0.64$ ft k/ft

$$= 640 \text{ ft lbs/ft}$$

Part (b)   Lateral Load

Assume the wall is simply supported at the top and bottom. Thus the maximum lateral load the wall can safely support is:

$$M = \frac{wL^2}{8} \text{ or } w = \frac{8M}{L^2}$$

$$w = \frac{8(640)}{15^2} = 22.8 \text{ psf}$$

5-6.B Columns

5-6.B.1 General

A masonry column is a vertical structural member designed primarily to support vertical and axial loads. In a reinforced column the masonry and the reinforcing steel share in supporting the imposed vertical loads and any overturning moment. The reinforcing steel is secured with horizontal ties to properly locate the steel and provide some confinement in accordance with UBC Sections 2407(h).4. and 2409(b).5.

The area of vertical reinforcement in a masonry column may not be less than 0.5% or more than 4% of the effective area of the column. At least four #3 vertical reinforcing bars must be provided in all columns.

Details of reinforcing and ties are shown in Section 7-14 of this text.

The maximum allowable axial load on a reinforced masonry column is:

$$P_a = \left(0.20 f'_m A_e + 0.65 A_s F_{sc}\right)\left[1 - \left(\frac{h'}{42t}\right)^3\right] \quad \text{(UBC Ch. 24, Eq. 6-4)}$$

The maximum allowable unit axial stress is:

$$F_a = \frac{P_a}{A_e} \quad \text{(UBC Chapter 24, Equation 6-5)}$$

The reduction factor based on the h'/t ratio is the same for reinforced columns as for walls. The same consideration is made for the determination of the effective height, h'.

The effective thickness, t, is the specified thickness in the direction considered. For non-rectangular columns the effective thickness is the thickness of a square column with the same moment of inertia (UBC Sec. 2407(b)3.D).

In Seismic Zone Nos. 3 and 4, the minimum nominal column thickness is 12 inches unless the allowable stresses are reduced to one half the values given in this section and UBC Section 2406 in which case the minimum thickness may be 8 inches (UBC Section 2407(h)4.E.(ii)).

EXAMPLE 5-T Brick Column Capacity.

The brick masonry column shown in Figure 5-34 is located in Seismic Zone No. 2A and supports an axial load, P, of 222 kips. Determine the required vertical steel and ties.

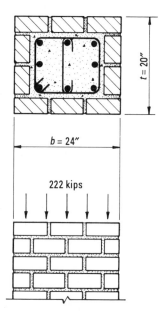

Figure 5-34 Brick column.

Given:

$P = 222$ kips

Effective height of column, $h' = 29' - 4''$

$f'_m = 2500$ psi

$f_y = 60,000$ psi

No inspection will be provided

Solution 5-T

$$P_a = 222 \text{ k} \leq \frac{\left(0.20 f'_m A_e + 0.65 f_{sc}\, p A_e\right)\left[1-\left(\dfrac{h'}{42t}\right)^3\right]}{1000}$$

$$R = 1 - \left(\frac{h'}{42t}\right)^3 = 1 - \left(\frac{29.33 \text{ ft} \times 12 \text{ in./ft}}{42 \times 20''}\right)^3$$

$$= 0.926$$

$$P_{masonry} = 0.20 f'_m A_e \quad \text{and}$$

$$P_{steel} = 0.65 F_{sc}\, p A_e = 0.65 F_{sc} A_s$$

Therefore:

$$222 \leq \frac{P_m + P_s}{1000 \ (R)}$$

$$\text{or } P_m + P_s \geq \frac{222,000}{0.926} = 239,740 \text{ pounds}$$

$$= 240 \text{ k}$$

Use Table S-1b to find P_m. Note that this table is based on full allowable stresses per footnote 2. Therefore, since the masonry will not be inspected, reduce P_m by one-half.

For $f'_m = 2500$ psi and for a $20'' \times 24''$ column

$P_m = 231.8$ k (Table S-1b for full stresses)

$P_m = \frac{1}{2}(231.8) = 116$ psi

Thus $P_s \geq 240 - P_m = 240 - 116$

$$= 124 \text{ k}$$

From Table A-4 for grade 60 reinforcing, $F_{sc} = 24,000$ psi. Use Table S-4 to find the required vertical reinforcing for $P_s \geq 124$ k.

Choose 8 - #9 vetical bars ($A_s = 8.0$ sq. in.)

Based on ease of construction and minimum tie size for bars larger than No. 7, choose #3 ties. Note, ties must be spaced at least 18″ o.c., but no less than the least dimension of the column. Since minimum $t = 20''$, space ties no more than 18″ o.c. Check Tables 7-11 and 7-12, based on #9 longitudinal bar and the #3 tie bar size. Both tables require a maximum tie spacing of 18″.

Thus use:

#3 Ties @ 18″ o.c. max.

Check whether ties are needed on the center #9 bars. The largest distance between the #9 bars is approximately:

$$\frac{24 - 2(3.5'' \text{ brick}) - 2(1'' \text{ clear})}{2} = 7.5''$$

Since the UBC requires ties on all longitudinal bars farther than 6″ from a tie anchor, tie all longitudinal bars with #3 ties @ 18″ o.c.

Column tie requirements are located in Section 7-14.D of this book, or in the UBC Section 2409(b)5.B.

5-6.B.2 Projecting Pilaster

Columns located in a wall but which project from the plane of the wall are called pilasters.

Pilasters are built integrally with the wall and in addition to supporting vertical loads can also be designed to carry lateral loads from adjacent wall sections. The magnitude of lateral load to the pilaster is dependent on the height of the pilaster and the spacing between pilasters. For tall, closely spaced pilasters with a height to spacing ratio of 2 or more, it may be assumed that the walls span horizontally.

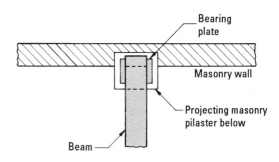

Plan of pilaster

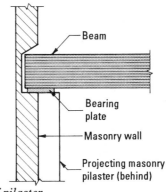

Elevation of pilaster

Figure 5-35 Masonry pilaster.

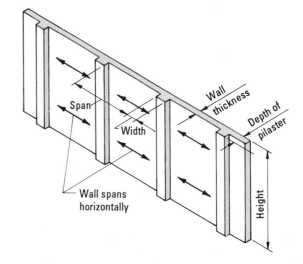

Figure 5-36 Wall loads to pilaster.

For lower walls with a wider spacing of pilaster and a height/spacing ≤ 1, it may be assumed that the walls span vertically and a triangular section of laterally loaded wall is carried by the pilasters. The triangular area is often assumed as 45 degrees to the horizontal.

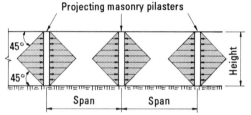

Figure 5-37 Lateral wall loads to pilaster.

5-6.B.3 Design of Pilasters

For the support of the vertical load, a projecting pilaster can be designed as a reinforced masonry column utilizing the rectangular cross-section of the element.

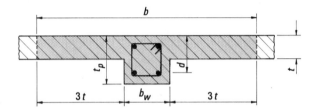

Figure 5-38 Projecting pilaster and width of effective wall section.

The lateral loads and eccentric vertical loads on a pilaster impose a moment on the wall and pilaster. Two conditions of loading may be considered.

a. Loads causing tension on the wall and compression on the projecting pilaster.

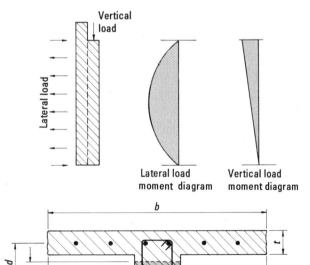

Figure 5-39 Wall and pilaster with loads causing tension on wall and compression on the projecting pilaster.

b. Loads causing compression on the wall and tension on the projection pilaster.

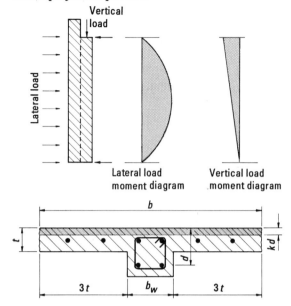

Figure 5-40 Wall and pilaster with lateral load causing compression on wall.

Generally the critical loading is the condition shown in Figure 5-39 where the projecting pilaster is in compression. The design for combined load and moment can be made using the methods outlined in Section 5-8 of this text.

The design of a pilaster with vertical load and lateral load can be easily accomplished by satisfying the revised unity equation.

$$\frac{P}{P_a} + \frac{f_b}{F_b} \leq 1.00 \leq 1.33$$

The ratio of the actual load, P, and the maximum allowable load, P_a, is determined. The limiting masonry stress is used to calculate the maximum masonry stress is used to calculate the maximum masonry bending stress, f_b, to satisfy the unity equation. See Sec. 5-8.B.1.

$$f_b = \left[(1.00 \text{ or } 1.33) - \frac{P}{P_a} \right] F_b$$

5-6.B.4 Flush Wall Pilasters

In order to simplify construction of a wall and to provide support of a beam framing into it, flush wall pilasters can be used. This permits construction of a wall without projections which speeds construction and provides more floor area.

In Seismic Zone Nos. 3 and 4, if a pilaster is between 8 to 12 inches and is designed as a column, the allowable design stresses must be reduced by one half per UBC Section 2407(h)4E(ii).

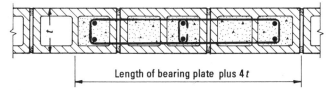

Length of bearing plate plus 4 *t*

Figure 5-41 Flush wall pilaster designed as a column.

A flush wall pilaster can be designed as a reinforced column in which case the vertical reinforcing steel supports part of the load. However, for the steel to be considered effective, it must be tied in accordance with Sec. 7-14.D. The minimum dimension, which is the thickness of the wall, governs in determining the h'/t reduction factor.

Alternately, a flush wall pilaster may be designed as a reinforced load bearing wall and the reinforcing is considered to resist only tension from lateral forces and eccentric vertical loads.

The maximum effective width of the in-the-wall columns can be considered to be the length of the bearing plate or angle plus four times the wall thickness, t, (UBC Sec. 2407(c)1.A.) The effective width may be considered less than the maximum and the area of column steel is based on the minimum steel ratio.

5-6.C Bearing

Base plates, beams, steel angles, etc. which support structural elements transfer load to the masonry support. If these bearing elements cover the masonry support fully, the masonry bearing stress is limited to:

$F_{br} = 0.26 f'_m$ (UBC Chapter 24, Equation 6-14)

If only one third or less of the area is loaded, the bearing stress is limited to:

$F_{br} = 0.38 f'_m$ (UBC Chapter 24, Equation 6-15)

For this allowable increase to apply there should be at least an edge distance of one quarter of the parallel side dimension to the loaded area. Values between the full loaded area and one third or more loaded area can be interpolated.

When special inspection is not provided these values are reduced by half.

The allowable bearing values are higher than the allowable axial compressive stress for walls since the load and stress rapidly dissipate throughout the wall. The compressive capacity of a wall ($F_a = 0.2 f'_m R$) will control over the bearing capacity of the wall (either $0.26 f'_m$ or $0.38 f'_m$). The bearing capacity of columns will occasionally control over their axial compressive capacity thus mandating larger column sizes. For instance assume in Example 5-T that the entire column area was covered by

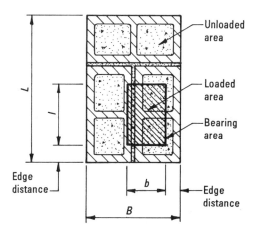

Figure 5-42 Relationship of bearing area.

a beam bearing plate. The bearing capacity of this column would only be:

$$P_{br} = f_{br} \times \text{area}$$
$$P_{br} = \frac{0.26(1500)}{2} \times (24 \times 20)$$
$$= 156,000 \text{ lbs}$$

This capacity is much less than both the applied load of 222 kips and the axial compressive capacity of about 240 kips and the resulting column size would accordingly have to be increased.

EXAMPLE 5-U Bearing Stresses.

A 16″ × 16″ nominal masonry with an 11″ × 11″ steel bearing plate is to support a beam load, $f'_m = 2000$ psi, no special inspection. Determine the maximum load that can be put on the bearing plate.

Solution 5-U

Area of column 15.63 × 15.63 = 244.3 sq. in.
Area of bearing plate 11 × 11 = 121 sq. in.

Ratio of areas $= \dfrac{121}{244.3} = 0.50$

Allowable bearing value
  Full area = $\frac{1}{2}(0.26 f'_m) = 260$ psi
  One third area = $\frac{1}{2}(0.38 f'_m) = 380$ psi

Interpolate for 0.50

$$\frac{380 - 260}{1 - 0.33} = \frac{260 - x}{1 - 0.50}$$
$$x = 350 \text{ psi}$$

Therefore bearing capacity = 121 × 350 = 42,350 lbs

The effective length over which concentrated loads are distributed from bearing plates or angles is the distance between loads or the length of bearing plate or angle plus four times the wall thickness, t, whichever is the least (UBC Sec. 2407(c)1.A.).

The masonry element under a concentrated load may be designed as a column with reinforcing steel supporting some of the load or as a load bearing wall in which the steel is neglected.

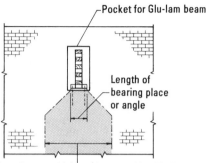

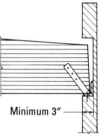

Maximum length over which concentrated load is distributed is distance between loads or length of bearing plate or angle plus four times wall thickness *t*, whichever is minimum.

Figure 5-43 Distribution of concentrated loads.

5-7 EMBEDDED ANCHOR BOLTS

Embedded anchor bolts are structural connections used to secure beams, columns, angles and other load bearing systems to masonry. The embedded bolts may be stressed in tension, shear or combined tension and shear.

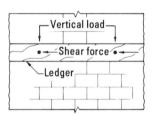

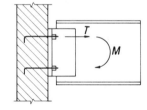

Anchor bolts in ledger subjected to vertical load and lateral shear

Anchor bolts in connecting angle subjected to vertical shear and tension

Figure 5-44 Typical loads on anchor bolts.

The maximum allowable tension on the masonry from an embedded anchor bolt is given by the equation.

$$B_t = 0.5 A_p \sqrt{f'_m} \quad \text{(UBC chapter 24, Equation 6-27)}$$

The limiting area, A_p, is the lesser of the following two equations based on depth of embedment, l_b, or the edge distance, l_{be}.

$$A_p = \pi l_b{}^2 \quad \text{(UBC Chapter 24, Equation 6-29)}$$

$$A_p = \pi l_{be}{}^2 \quad \text{(UBC Chapter 24, Equation 6-30)}$$

When the projected areas of adjacent anchor bolts overlap, the A_p of each anchor bolt is to be reduced by one half of the overlapping area. See Table A-7c for the percent capacity reduction of anchor bolts in tension based on embedment and spacing.

The maximum allowable tension on the anchor bolt is given by the equation.

$$B_t = 0.2 A_b f_y \quad \text{(UBC Chapter 24, Equation 6-28)}$$

The limiting value for B_t must be used for design.

The maximum allowable shear load is the lesser of the shear load on the masonry or on the bolt as determined by the following equations

$$B_v = 350 \sqrt[4]{f'_m A_b} \quad \text{(UBC Chapter 24, Equation 6-31)}$$

$$B_v = 0.12 A_b f_y \quad \text{(UBC Chapter 24, Equation 6-32)}$$

The anchor bolt edge distance, l_{be}, in the direction of the shear load should be 12 bolt diameter for UBC Equation 6-31 but the shear stress may be reduced linearly to zero when the l_{be} is $1^1/2$ inches (see Table A-8b).

When anchor bolts are closer than eight bolt diameters, the shear stress should be reduced by linear interpolation to 0.75 times the allowable shear value for a spacing of four bolt diameters (see Table A-8c).

For combined tension and shear on anchor bolts, the unity equation (UBC Chapter 24, Equation 6-33) must be satisfied.

$$\frac{b_t}{B_t} + \frac{b_v}{B_v} \le 1.0 \text{ or } \le 1.33 \text{ (wind or seismic)}$$

The minimum edge distance, l_{be}, is $1^1/2$ inches. The minimum embedment depth, l_b, is four bolt diameters but not less than 2 inches. The minimum center to center spacing of anchor bolts is four bolt diameters.

EXAMPLE 5-V Anchor Bolt Analysis.

Determine the adequacy of an embedded anchor connection supporting a cantilever steel beam with a load of 600 lbs as shown.

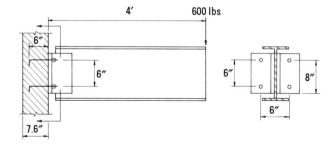

Figure 5-45 Section of cantilever beam connection.

Given: f'_m = 1500 psi; Nominal 8″ CMU solid grouted; $^3/_4$″ anchor bolts embedded 6″ into the wall.

Solution 5-V

Moment on connection

$M = Pl = 600 \times 4 = 2400$ ft lbs

Assume moment resistance on connection is as shown:

Tension pull on bolt

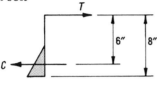

$$b_t = \frac{M}{d} = \frac{2400 \times 12}{8 \times 2}$$
$$= 1800 \text{ lbs/bolt}$$

Shear on bolts

$$b_v = \frac{P}{\text{No. of bolts}} = \frac{600}{4}$$
$$= 150 \text{ lbs/bolt}$$

Allowable tension on $^3/_4$″ dia. anchor bolts with 6″ embedment and 8″ spacing.

From Table A-7a, masonry value

B_t = 2190 lbs/bolt for a spacing of 2 lbs or more

From Table A-7c, find the percent capacity of the anchor bolts:

spacing = 8″, l_b = 6″

$\frac{8}{6} = 1.3l_b$

% capacity = 88%

Allowable masonry value = 2190 × 0.88 = 1927 lbs/bolt

From Table A-7b, steel value

B_t = 3180 lbs/bolt

Reduce for bolt spacing

Steel value = 3180 × 0.88 = 2198 lbs/bolt

Tension on masonry governs

Allowable shear on bolts

From Table A-8, B_v = 1780 lbs

Check compliance with interaction unity equation

$$\frac{b_t}{B_t} + \frac{b_v}{B_v} \le 1.00$$

$$\frac{1800}{1927} + \frac{150}{1780} = 0.93 + 0.08$$
$$= 1.01 \approx 1.00$$

Embedded anchor bolt connection is satisfactory.

5-8 COMBINED BENDING AND AXIAL LOADS

5-8.A General

Most walls and columns are subjected to both axial and bending loads. This is particularly true of bearing walls that carry the loads of floors and roofs and are subjected to a lateral wind or earthquake force. Lateral loads may also be imposed by earth pressure on the wall.

The interaction of vertical load and bending forces will also occur if the vertical load is eccentric to the axis of the wall or column. The interaction of combined stresses may also result when a moment is imposed on the wall or column in addition to the axial load.

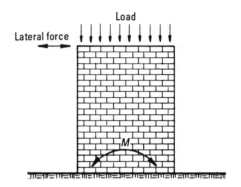

Figure 5-46 Combinations of loading causing combined stresses.

The interaction of these forces results in increased compressive stress on the masonry. Tension in the reinforcing steel may also occur if the moment is large enough to overcome the effect of the compressive stress due to vertical load.

When a masonry wall or column is subjected to both axial load and moment or eccentric vertical load, an analysis must be made considering the combined effects of the axial and bending stresses.

Such members must be designed in accordance with accepted principles of mechanics or in accordance with the unity equation.

The interaction of load and moment on a section is complex and is represented by the curves in Figure 5-47. The unity equation, Method 1, is represented by Curve 1 and considers each stress from vertical load and moment independently. Curve 2 recognize the capability of the section but limits the stress to the combination of vertical stress and flexural stress. The maximum vertical stress is limited to F_a, while the maximum flexural stress is limited to F_b. Curve 2 is based on Method 2.

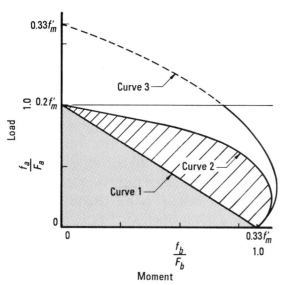

Figure 5-47 Graphic representation of interaction.

Curve 3 is similar to Curve 2 except the maximum stress is permitted to be $0.33f'_m$ with the axial load cut off based on $0.2f'_m$. This interaction method is based on ACI/ASCE equations and limitations.

5-8.B Methods of Design for Interaction of Load and Moment

There are several methods by which structural elements can be designed for interaction of loads and moments, three of which are presented. Some methods are more conservative than others and the designing engineer should evaluate the methods accordingly.

Method 1. Assumes that the vertical load and moment act independently and the stresses are determined for each condition. The unity equation is checked to determine compliance.

Modified Method 1. This modification of Method 1 assumes that the vertical load counteracts the tension stress caused by the moment up to the point where the tension stress exceeds the vertical compression stress. The limiting condition for this is when $e > t/6$ or $l/6$. The initial determination of flexural stress can be by assuming a homogeneous section and using the equation $f = M/S$ or Mc/I. When the tension stress exceeds the compression stress, or the allowable tension stress, consider each condition for vertical load and moment independently and proceed similar to Method 1.

Method 2. This method determines the axial stress, then the maximum allowable flexural compressive stress that will satisfy the unity equation. With these values and the applied loads, the statics of the section are evaluated based on; sum of vertical forces equal $0 (\Sigma F_v = 0)$, and sum of the moments equals $0 (\Sigma M = 0)$. The stress in the steel is calculated and the required area of steel determined.

These equations were developed by Ralph McLean, structural engineer, of the firm McLean and Schultz, Consulting Engineers, Architect and Planners of Fullerton, California.

Method 3. This method assumes that the section is homogeneous and uncracked. The stresses are determined by $\dfrac{P}{A} \pm \dfrac{Mc}{I}$ with the moment of inertia based on the gross section. If P/A axial compressive stress is less than the flexural stress, Mc/I, then there will be tension on the section and it must be reinforced for this tension force.

The axial and flexural stresses as determined by P/A and Mc/I must be checked against the maximum allowable stresses to insure compliance with the unity equation.

5-8.B.1 Unity Equation

The classic approach to the interaction of load and moment is the code unity equation. This approach limits the ratio of the actual axial stress to the maximum allowable axial stress, plus the actual flexural stress to the maximum allowable flexural stress, to 1.00.

The combination of stresses may not exceed the unity equation:

$$\frac{f_a}{F_a} + \frac{f_b}{F_b} \leq 1 \quad \text{(Walls)}$$

Where:
$$\frac{P}{P_a} + \frac{f_b}{F_b} \leq 1 \quad \text{(Columns)}$$

f_a = computed actual axial unit stress due to the load determined from total axial load and effective area (psi):

$$= \frac{P}{bt} \quad \text{(Walls)}$$

bt = actual cross-sectional solid area of wall (sq. in.)

F_a = maximum allowable axial stress if the member were carrying axial load only (psi)

$$= [0.2f'_m \text{ or } \tfrac{1}{2}(0.2f'_m)] \text{ psi} \times R \quad \text{(Walls)}$$

R = h'/t reduction factor, decimal ≤ 1.00

$$= \left[1 - \left(\frac{h'}{42t}\right)^3\right]$$

P = actual load on column

$$P_a = \left(0.20 f'_m A_e + 0.65 A_s F_{sc}\right)\left[1 - \left(\frac{h'}{42t}\right)^3\right] \text{(Columns)}$$

F_b = maximum allowable flexural stress if members were carrying bending load only

$= 0.33 f'_m$ (psi) or $^1/_2(0.33f'_m)$ if no inspection is provided

f_b = actual computed bending stress not to exceed:

$$f_b \le \left(1 - \frac{f_a}{F_a}\right) F_b \text{ (psi)}$$

In the case of temporary loads due to wind or earthquake a one third increase may be included per UBC Section 2303(d). An example of this is, if the moment on the wall is caused by a wind then

$$\frac{\text{vertical} f_a}{\text{allowable } F_a} + \frac{\text{wind} f_b}{\text{allowable } F_b} \le 1.33 \quad \text{(Walls)}$$

$$\frac{P}{P_a} + \frac{f_b}{F_b} \le 1.33 \quad \text{(Columns)}$$

This is a simple and acceptable technique provided the resulting design is not less than the design determined using only dead and live loads.

5-8.B.1(a) Uncracked Section

The cross-section of the element is uncracked when the vertical stress is equal to or more than the flexural stress. This occurs when the eccentricity, e, of the load, P, is less than or equal to the kern distance:

$$e_k = \frac{l}{6} \text{ or } \frac{t}{6} \quad \text{(rectangular section)}$$

$$e_k = \frac{I}{A_y} = \frac{r^2}{y} \quad \text{(irregular section)}$$

Where:

I = moment of inertia of section

A = area section

y = distance from neutral axis to extreme edge

r = radius of gyration, or $\sqrt{\dfrac{I}{A}}$

The stress can also be determined by the equation

$$f_a = \frac{P}{bt}; \quad f_b = \frac{Mc}{I} = \frac{6M}{bt^2}$$

$$f = \frac{P}{A} \pm \frac{Mc}{I} = \frac{P}{bt} \pm \frac{6M}{bt^2}$$

When f_a is greater than or equal to f_b the section is always under compression.

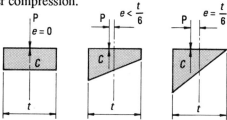

Figure 5-48 Wall under compression.

5-8.B.1(b) Cracked Section

If the virtual eccentricity is greater than the kern distance, there is tension on the face of the wall. Since it is assumed in reinforced masonry that the masonry does not resist tension, then the section is to be reinforced to resist the tension as if there was no vertical force to reduce it. This is a good approximation when the steel is located within the middle third of the wall. The design condition is as shown in Figure 5-49.

If credit is given to the tension bond between the mortar and the masonry, the comparative distance e_k may be increased from $t/6$ to $t/5$, or $t/4$ depending on the value given to the tension bond.

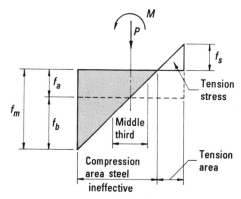

Figure 5-49 Wall under combined stresses with flexural stress exceeding axial stress.

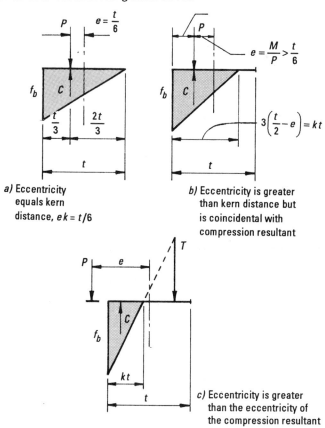

a) Eccentricity equals kern distance, $ek = t/6$

b) Eccentricity is greater than kern distance but is coincidental with compression resultant

c) Eccentricity is greater than the eccentricity of the compression resultant

Figure 5-50 Conditions of increasing eccentricity of load on wall.

When the eccentricity exceeds $t/6$ or $l/6$ and the tension capacity of the masonry is ignored, the section may be under compression only until it becomes necessary to provide reinforcing steel to resist the tension forces. This condition of compression stress only, may be assumed similar to an eccentrically loaded footing, which is capable of imposing only compression forces. (Figure 5-50b).

The limit of the condition where only compression forces exist is when the eccentric load is no longer coincidental with the resultant of the compression force in the stress block and the allowable compression stress on the masonry is not exceeded.

If the force in the reinforcing steel is to be included in the evaluation for the sum of the moments and sum of the forces, it may be necessary to reduce the assumed masonry compressive stress thus decreasing the eccentricity of the resultant compression in the masonry.

The resultant compression force will be balanced, $\Sigma F_v = 0$, by the eccentric vertical load and the tension force in the steel, see Figure 5-50c.

The maximum compressive stress on the masonry is determined based on satisfying the unity equation:

$$\frac{f_a}{F_a} + \frac{f_b}{F_b} = 1.00 \text{ or } 1.33$$

$$f_m = f_a + f_b$$

EXAMPLE 5-W   Combined Loading.

Determine whether steel is required for tension in an 8″ concrete masonry wall which is 13′ - 4″ high and is subjected to a wind pressure of 30 psf. $f'_m = 1500$ psi, $n = 25.8$, no special inspection, $F_s = 24{,}000$ psi, Vertical load, $P = 4000$ plf and Distance to steel, $d = 5.3$ in.

Solution 5-W

Moment perpendicular to wall due to wind, M:

$$M = \frac{wl^2}{8} = \frac{30 \times (13.33)^2}{8} = 667 \text{ ft lbs / ft}$$

Virtual eccentricity $e = \dfrac{M}{P} = \dfrac{667 \times 12}{4000} = 2$ in.

Kern distance $e_k = \dfrac{t}{6} = \dfrac{7.63}{6} = 1.27$ in. < 2 in.

Eccentricity exceeds kern distance

Length of compression area $= 3\left(\dfrac{t}{2} - e\right)$

$$= 3\left(\frac{7.63}{2} - 2\right) = 5.45 \text{ in.}$$

This indicates that $(7.63 - 5.45) = 2.18''$ of wall will have no stress on it and that steel located 5.3″ from the compression face would not be stressed in tension.

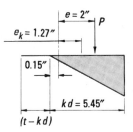

Actual compression stress due to eccentric vertical load

$$C = \tfrac{1}{2} f_b \times b \times kt$$

$$f_m = \frac{2C}{bkt} = \frac{2 \times 4000}{12 \times 5.45} = 122 \text{ psi}$$

Actual axial compression stress;

$$f_a = \frac{P}{b \times t} = \frac{4000}{12 \times 7.63} = 43.7 \text{ psi}$$

Allowable axial stress

$$F_a = \tfrac{1}{2} \times 0.2 f'_m \left[1 - \left(\frac{h'}{42t}\right)^3 \right]$$

$$= \tfrac{1}{2} \times 0.2(1500)\left[1 - \left(\frac{13.33 \times 12}{42 \times 7.63}\right)^3 \right]$$

$$= 131 \text{ psi} > 43.7 \text{ psi   O.K.}$$

Maximum allowable flexural compression stress

$$F_b = \tfrac{1}{2} \times 0.33 \times f'_m$$

$$= \tfrac{1}{2} \times 0.33 \times 1500$$

$$= 250 \text{ psi} > 122 \text{ psi   O.K.}$$

Check the interaction equation

Allowable bending stress

$$f_b = F_b\left(1.33 - \frac{f_a}{F_a}\right)$$

$$= 250\left(1.33 - \frac{43.7}{131}\right)$$

$$= 249 \text{ psi} > 122 \text{ psi   O.K.}$$

Maximum allowable compression stress

$$f_m = f_a + f_b = 43.7 + 249 = 292.7 \text{ psi}$$

$$122.0 \text{ psi} < 292.7 \text{ psi   O.K.}$$

The eccentric vertical load P is coincidental with the resultant compressive force C. No tension steel is required. Provide minimum steel as required by code.

EXAMPLE 5-X Steel Requirement.

Using Example 5-W, check the requirement for tension steel for the wind load only.

Solution 5-X

Design reinforcing for lateral wind load moment of 667 ft lbs/ft; $d = 5.3''$

$$K = \frac{M}{bd^2} = \frac{667 \times 12}{12 \times 5.3^2}$$

$$= 23.7 \text{ due to wind}$$

From Table E1a for $K = 23.7$

read $p = 0.0008$

$A_s = pbd = 0.0008 \times 12 \times 5.3$

$\quad = 0.051 \text{ sq. in./ft}$

Check against minimum A_s requirement

$A_s = 0.0013bt$

$\quad = 0.0013 \times 12 \times 7.63$

$\quad = 0.119 \text{ sq. in./ft} \leftarrow \text{Controls}$

Use #5 @ 32 o.c.

By the above equation, there is no tension on the wall and only minimum required reinforcing steel would be needed.

5-8.C Method 1. Vertical Load and Moment Considered Independently

The Method 1 analysis for interaction, particularly when the moment is perpendicular to the plane of the wall, is to consider each force independently. The stress for the vertical load is calculated and then the stress due to the moment based on a cracked section is calculated. The combination of compressive stresses should not exceed the unity equation.

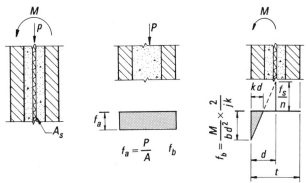

Figure 5-51 Unity equation assumed stress distribution; bending stress greater than axial compressive stress; $f_a < f_b$.

This handbook presents a direct method of designing a wall subjected to load and bending without the need to make assumptions for the amount of steel and then checking all stresses. The amount of reinforcing if needed can be directly determined for a wall subjected to bending perpendicular to the plane of the wall.

Calculate or assume:

M; P; f'_m; b (normally 12 in.); F_s

t (wall thickness); d (distance from compression face to center of steel); and h' (effective or actual height of wall

Solve for

1. Kern distance, $e_k = \dfrac{t}{6}$ or $e_k = \dfrac{I}{Ay} = \dfrac{S}{A}$

2. Virtual eccentricity, $e = \dfrac{M}{P}$

3a. If $e < e_k$ minimum reinforcement required

3b. If $e > e_k$ design for bending stress

4a. Actual axial stress, $f_a = \dfrac{P}{bt}$

   Note: Use actual cross-sectional area of masonry. For partially grouted walls use Table B-3 to find equivalent solid thicknesses (*EST*).

4b. Flexural stress assuming uncracked section

$$f_b = \frac{Mc}{I} = \frac{M}{S} = \frac{6M}{bt^2}$$

4c. If $f_a \geq f_b$, section under compression minimum reinforcement required, condition 3a.

   If $f_a < f_b$, section under tension, design reinforcing for flexural stress, condition 3b.

   However, if tensile stress does not exceed the allowable tensile stress for plain masonry, Table A-9, only minimum steel need be used.

5. $\dfrac{h'}{t}$ reduction factor, $R = \left[1 - \left(\dfrac{h'}{42t}\right)^3\right]$

   (See Table Q-1).

6. Maximum allowable axial stress
   $F_a = [0.2 f'_m \text{ or } \frac{1}{2}(0.2f'_m)]R$ (see Table Q-2).

7. Ratio of axial stresses f_a/F_a

8. Maximum allowable flexural compression stress
   $F_b = 0.33f'_m \text{ or } \frac{1}{2}(0.33f'_m)$; Max. 2000 psi

9. Maximum allowable flexural compression stress that will satisfy the unity equation

$$f_b = (1 - f_a/F_a)F_b$$

or $f_b = (1.33 - f_a/F_a)F_b$ if loads are temporary such as wind or earthquake.

Diagram F-4   *K* vs *p* for Various Masonry and Steel Stresses:
f'_m = 3000 psi, *n* = 12.9

For use of Diagram,
see Example 5-Y

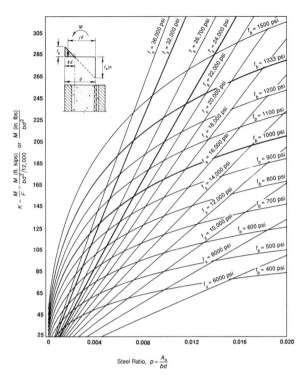

Steel Ratio, $p = \dfrac{A_s}{bd}$

Typical Diagram F

10. Compute the flexural coefficient,

$$K = \frac{M}{bd^2} = \text{ or } K = \frac{12,000M}{bd^2}$$

11. With K and f_b, determine the steel ratio, p, using Diagrams F-1 through F-8.

12. With steel ratio p determined from above and the given d, from Table C-4 select the reinforcing bars and spacing.

EXAMPLE 5-Y Load and Moment on Brick Wall.

A 9 inch solid grouted reinforced brick wall supports a vertical load of 3200 plf and a moment of 825 ft lbs/ft due to earth load.

Design the required steel if $d = 5$ inches and the effective height of wall = 10 feet, 6 inches.

Solution 5-Y

Assume $f'_m = 1500$ psi; no special inspection provided

$$n = 25.8; \; F_s = 24,000 \text{ psi}$$

1. Kern distance, $e_k = \dfrac{t}{6} = \dfrac{9}{6} = 1.5$ in.

2. Virtual eccentricity, $\; e = \dfrac{M}{P} = \dfrac{825 \times 12}{3200}$
$$= 3.1 \text{ in.}$$

3. $e > e_k$, therefore there is tension on section, assume cracked

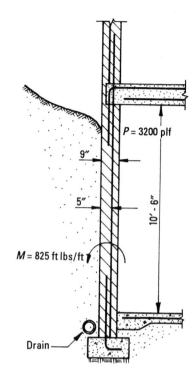

Figure 5-52 Cross section of brick wall with loads shown.

4. Actual axial compression stress

$$f_a = \frac{P}{bt} = \frac{3200}{12 \times 9} = 29.6 \text{ psi}$$

Check for $f_b = \dfrac{6M}{bt^2} = \dfrac{6 \times 825 \times 12}{12 \times 9^2} = 61 \text{ psi}$

Since the tensile stress of 61 psi exceeds the compression stress, 29.6 psi, assume section is cracked. It also exceeds allowable tension for plain masonry; $f_t = 40$ psi, Table A-9.

5. $\dfrac{h'}{t}$ reduction factor, $= \dfrac{10.5 \times 12}{9} = 14.0$

Enter Table Q-1 to find $R = 0.963$

6. Use Table Q-2 to find allowable axial stress, $F_a = 144$ psi

7. Ratio of axial stress $\dfrac{f_a}{F_a} = \dfrac{29.6}{144} = 0.205$

8. Maximum allowable flexural compression stress

$$f_b = \frac{1}{2}\left(\frac{1}{3}f'_m\right) = \frac{1}{2}\left(\frac{1}{3} \times 1500\right)$$
$$= 250 \text{ psi, Table A-3}$$

9. Maximum allowable flexural compression stress to satisfy the unity equation

$$f_b = \left(1 - \frac{f_a}{F_a}\right)F_b$$

$$= (1 - 0.205)250 = 198 \text{ psi}$$

10. The flexural coefficient, $K = \dfrac{M}{bd^2}$, for

$b = 12''$, $d = 5''$ and $M = 825$ ft lbs/ft

$$K = \frac{M}{bd^2} = \frac{825 \times 12}{12 \times 5^2} = 33$$

11. In diagram F-1, K vs p for $n = 25.8$

Enter $K = 33$ move right to intersect $f_b = 198$ psi.

Move down and read $p = 0.0046$

Diagram F-1 K vs p for Various Masonry and Steel Stresses:
$f'_m = 1500$ psi, $n = 25.8$

For use of Diagram,
see Example 5-Y

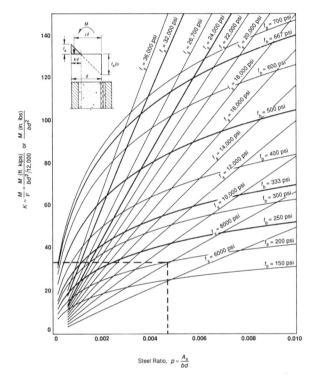

Steel Ratio, $p = \frac{A_s}{bd}$

Diagram F-1

12. Select reinforcing steel

$A_s = pbd = 0.0046 \times 12 \times 5 = 0.276$ sq. in./ft

See Table C-4a

Use #5 at 12″ o.c. ($A_s = 0.307$ sq. in./ft)

or #6 at 18″ o.c. ($A_s = 0.295$ sq. in./ft)

Alternate method to determine steel requirement:

After determining the maximum allowable flexural compressive stress that will satisfy the unity of equation as shown in step 9

$$f_b = \left(1 - \frac{f_a}{F_a}\right) F_b$$

or $f_b = \left(1.33 - \dfrac{f_a}{F_a}\right) F_b$ For wind or seismic loads

Equate to flexural formula

$$f_b = \frac{M}{bd^2} \times \frac{2}{jk}$$

$$\frac{2}{jk} = f_b \frac{bd^2}{M}$$

Solve for $\dfrac{2}{jk}$

From Table E-9 for $\dfrac{2}{jk}$; read np

Solve for $p = \dfrac{np}{n}$

Solve for $A_s = pbd$

From Example 5-Y , Step 9, $f_b = 198$ psi

Solve for $\dfrac{2}{jk}$

$$\frac{2}{jk} = \frac{198 \times 12 \times 5^2}{825 \times 12}$$

$$= 6.00$$

From Table E-9b for $\dfrac{2}{jk} = 6.00$

read $np = 0.118$

Use #5 at 12″ o.c. or #6 at 18″ o.c.

EXAMPLE 5-Z Reinforced Brick Column with an Eccentric Load.

Design a hollow clay brick column, 12 feet high, to support a live load of 4 kips and a dead load of 6 kips. The loads have an eccentricity of 6 inches from the center line of the column.

Assume: $f'_m = 2500$ psi, $f_y = 60,000$ psi, $n = 15.5$

No special inspection, use half stresses.

Solution 5-Z

Maximum allowable axial column stress on masonry is

$F_a = {}^1/2(0.2 f'_m) R$

For initial design assume $R = 0.95$

$F_a = {}^1/2(0.2 \times 2500) 0.95 = 238$ psi

Initial column area

$$A_e = \frac{(4 + 6) 1000}{238} = 42 \text{ sq. in.}$$

Use two hollow clay brick, 6″ × 4″ × 12″ actual $5^1/2'' \times 3^1/2'' \times 11^1/2''$

$A_e = 11.5 \times 11.5 = 132$ sq. in.

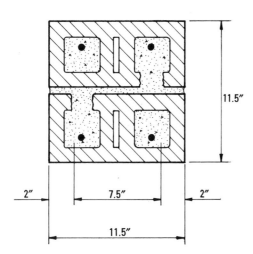

2″   **7.5″**   **2″**

11.5″

Figure 5-53 Cross section of hollow clay brick column.

Use minimum area of vertical steel $p = 0.05$

$$A_s = pbt$$
$$= 0.005 \times 11.5 \times 11.5$$
$$= 0.66 \text{ sq. in.}$$

Try 4 - #5 bars

$$A_s = 1.24 \text{ sq. in.}$$ (excess steel for moment consideration)

Check reduction coefficient, R

$$R = \left[1 - \left(\frac{h}{42t}\right)^3\right]$$
$$= \left[1 - \left(\frac{12 \times 12}{42 \times 11.5}\right)^3\right] = 0.97$$

Maximum allowable load with 4 - #5 bars

$$P_a = [^1/_2 \times 0.20 f'_m A_e + 0.65 A_s F_{sc}]R$$
$$= [^1/_2 \times 0.20(2500)(11.5)^2$$
$$+ 0.65(1.24)24,000]0.97$$
$$= 50,834 \text{ lbs}$$

Ratio of vertical loads,

$$\frac{P}{P_a} = \frac{10,000}{50,834} = 0.20$$

Based on unity equation the maximum allowable flexural compression masonry stress is:

$$f_b = \left(1 - \frac{P}{P_a}\right)F_b$$
$$F_b = ^1/_2(0.33)(2500)$$
$$= 417 \text{ psi for masonry}$$
$$\text{without special inspection}$$
$$f_b = (1 - 0.20)417$$
$$= 334 \text{ psi}$$

Determine the area of reinforcing steel required for the moment and limiting stress condition by the *npj* method.

Moment due to eccentric load

$$M = (4000 + 6000)6$$
$$= 60,000 \text{ in. lbs}$$

Solve for $\dfrac{2}{jk}$

$$\frac{2}{jk} = \frac{bd^2 f_b}{M}$$
$$= \frac{11.5 \times 9.5^2 \times 334}{60,000} = 5.778$$

From Table E-9b for $2/jk = 5.778$

$$np = 0.133$$

Solve for *npj*

$$npj = \frac{nM}{bd^2 f_s}$$
$$= \frac{15.5 \times 60,000}{11.5 \times 9.5^2 \times 24,000}$$
$$= 0.0373$$

From Table E-9a for $npj = 0.0373$

$$np = 0.041$$

Use maximum np value, masonry controls

$$p = \frac{np}{n} = \frac{0.133}{15.5} = 0.0086$$
$$A_s = pbd$$
$$= 0.0086 \times 11.5 \times 9.5 = 0.94 \text{ sq. in.}$$

Use 2 - #6 bars on each side; $A_s = 0.88$ sq. in.

Alternate Solution:

Flexural coefficient $K = \dfrac{M}{bd^2} = \dfrac{60,000}{11.5 \times 9.5^2} = 57.8$

From Diagram F-3

for $K = 57.8$ and $f_b = 334$ psi,

Read $p = 0.008$ (approximately same as above)

Ties in column

No special conditions since moment is not due to seismic forces.

From Tables 7-11 and 7-12

Use $^1/_4″$ ties at 12″ o.c.

Closer tie spacing would be required for Seismic Zone Nos. 3 and 4.

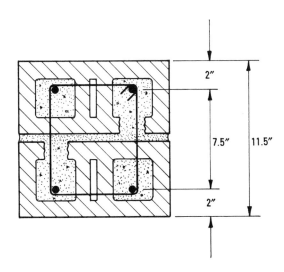

Figure 5-54 Cross section of column showing reinforcing and ties.

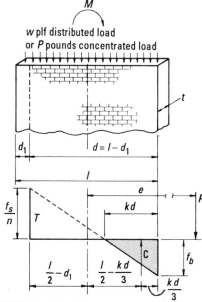

Figure 5-55 Load and moment on wall.

5-8.D Method 2. Evaluation of Forces Based on Static Equilibrium of $\sum F_v = 0$ and $\sum M = 0$

Given:

Length of wall $= l$ inches

Thickness of wall $= t$ inches

Distance to steel $= d$ inches

Distance to steel $= d_1$ inches

Axial load $= P$ pound or w plf

Compression force $= C$ pounds

Tension force $= T$ pounds

Moment $= M$ foot pounds

Steel stress $= f_s$ psi

Masonry stress $= f_m$ psi

Height of wall $= h$ feet

Compression force, $C = \frac{1}{2}tkdf_b$

Tension force, $T = C - P$

Taking the sum of the moments about the center line or axis of the vertical load:

$$C\left(\frac{l}{2} - \frac{kd}{3}\right) + T\left(\frac{l}{2} - d_1\right) - M = 0$$

but $T = C - P$

$$C\left(\frac{l}{2} - \frac{kd}{3}\right) + (C - P)\left(\frac{l}{2} - d_1\right) - M = 0$$

substituting for $C = \frac{1}{2}tkdf_m$

$$\left(\frac{1}{2}tkdf_m\right)\left(\frac{l}{2} - \frac{kd}{3}\right) + \left(\frac{1}{2}tkdf_m - P\right)\left(\frac{l}{2} - d_1\right) - M = 0$$

$$\frac{1}{4}tf_m lkd - \frac{1}{6}tf_m(kd)^2 + \frac{1}{2}tkd\left(\frac{l}{2} - d_1\right)f_m$$

$$-P\left(\frac{l}{2} - d_1\right) - M = 0$$

$$\frac{1}{4}tf_m lkd - \frac{1}{6}tf_m(kd)^2 + \frac{1}{4}tf_m lkd - \frac{1}{2}tf_m d_1 kd$$

$$-P\left(\frac{l}{2} - d_1\right) - M = 0$$

Change signs and combine terms

$$\frac{1}{6}tf_m(kd)^2 - \frac{1}{2}tf_m(l - d')kd + P\left(\frac{l}{2} - d_1\right) + M = 0$$

Solving this quadratic equation

$$\mathbf{a}x^2 + \mathbf{b}x + \mathbf{c} = 0$$

let $x = kd$

$$\mathbf{a} = \frac{1}{6}tf_m$$

$$\mathbf{b} = -\frac{1}{2}tf_m(l - d_1)$$

$$\mathbf{c} = P\left(\frac{l}{2} - d_1\right) + M$$

but $d = l - d$, so the equation for b simplifies to:

$$\mathbf{b} = -\frac{1}{2}tf_m d$$

Using the binominal formula to solve the quadratic equation,

$$kd = \frac{-\mathbf{b} \pm \sqrt{\mathbf{b}^2 - 4\mathbf{ac}}}{2\mathbf{a}}$$

Note:

The term $-\dfrac{\mathbf{b}}{2\mathbf{a}} = \dfrac{\frac{1}{2}tdf_m}{2 \times \frac{1}{6}tf_m}$

$= 1.5d$ which would result in a negative kd distance

$$kd = \frac{-\mathbf{b} - \sqrt{\mathbf{b}^2 - 4\mathbf{ac}}}{2\mathbf{a}}$$

$$kd = \frac{\tfrac{1}{2} t f_m d - \sqrt{\left(\tfrac{1}{2} t f_m d\right)^2 - 4\left(\tfrac{1}{6} t f_m\right)\left[P\left(\tfrac{l}{2} - d_1\right) + M\right]}}{2 \times \tfrac{1}{6} t f_m}$$

Determine the maximum allowable masonry stress, f'_m.

$$f_a = \frac{P}{lt}$$

$$F_a = 0.20 f'_m \left[1 - \left(\frac{h'}{42t}\right)^3\right]$$

$F_b = 0.33 f'_m$ but not to exceed 2000 psi

$$f_b = F_b\left(1.33 - \frac{f_a}{F_a}\right)$$

$$f_m = f_a + f_b$$

$$C = \tfrac{1}{2} t k d f_m$$

$$T = C - P$$

$$k = \frac{kd}{d}$$

$$f_s = \left(\frac{1-k}{k}\right) n f_m \quad \text{or}$$

f_s = allowable steel stress plus allowable increases.

$$A_s = \frac{T}{f_s}$$

If f_s exceeds allowable F_s, decrease f_m, and recompute values.

EXAMPLE 5-AA   Determine the Reinforcing for a Shear Wall (Method 2).

An 8 inch concrete masonry shear wall in a high rise building is subjected to a vertical load, P of 845 kips and a seismic moment M of 5840 ft kips. The wall is 9 ft 4 inches between floors, it is 30 ft long and d_1 is assumed 8 inches. $f'_m = 3000$ psi, $n = 12.9$.

Solution 5-AA

$$f_a = \frac{P}{lt} = \frac{845 \times 1000}{12 \times 30 \times 7.63}$$

$$= 308 \text{ psi}$$

$$F_a = 0.2(3000)\left[1 - \left(\frac{12 \times 9.33}{42 \times 7.63}\right)^3\right]$$

$$= 600(0.957)$$

$$= 574 \text{ psi}$$

$$F_b = 0.33 f'_m$$

$$F_b = 0.33(3000) = 1000 \text{ psi}$$

$$f_b = 1000\left[1.33 - \left(\frac{308}{574}\right)\right]$$

$$= 793 \text{ psi}$$

$$f_m = f_a + f_b$$

$$= 308 + 793 = 1101 \text{ psi maximum}$$

$$= 1.101 \text{ ksi}$$

Solve values kd, f_s, C, T and A_s

$$\mathbf{a} = \tfrac{1}{6} t f_m$$

$$= \tfrac{1}{6} \times 7.63 \times 1.1$$

$$= 1.40$$

$$\mathbf{b} = -\tfrac{1}{2} t f_m (t - d')$$

$$= -\tfrac{1}{2} \times 7.63 \times 1.1(360 - 8)$$

$$= -1477$$

$$\mathbf{c} = P\left(\frac{l}{2} - d_1\right) + M$$

$$= 845\left(\frac{360}{2} - 8\right) + 5840 \times 12$$

$$= 215,420$$

$$kd = \frac{-\mathbf{b} - \sqrt{\mathbf{b}^2 - 4\mathbf{ac}}}{2\mathbf{a}}$$

$$= \frac{+1477 - \sqrt{(-1477)^2 - 4(1.40)(215,420)}}{2 \times 1.40}$$

$$= 175 \text{ inches}$$

$$k = \frac{kd}{d} = \frac{175}{360 - 8}$$

$$= 0.497$$

$$C = \tfrac{1}{2} t k d f_m$$

$$= \tfrac{1}{2} \times 7.63 \times 175 \times 1.1$$

$$= 734 \text{ kips}$$

$$T = C - P$$

$$= 734 - 845 = -110 \text{ kips}$$

The negative sign indicates that no tension reinforcing steel is required and the eccentric axial load can be coincidental with the resultant compression force.

Virtual eccentricity $e = \dfrac{M}{P} = \dfrac{5840 \times 12}{845}$

$$= 83 \text{ in.}$$

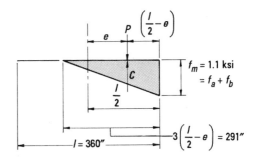

$$\frac{l}{2} - e = \frac{360}{2} - 83$$
$$= 97 \text{ in.}$$

$$3\left(\frac{l}{3} - e\right) = 3 \times 97$$
$$= 291 \text{ in.} > kd = 175$$

No tension steel required

Use minimum steel

$A_s = 0.0013\, bt \times \tfrac{1}{2}$ (each side)
$$= 0.0013 \times 360 \times 7.63 \times \tfrac{1}{2}$$
$$= 1.79 \text{ sq. in.}$$

Use 2 - #9 bars each side.

$$A_s = 2.00 \text{ sq. in.}$$

and #4 at 48″ o.c. in. between

Note: The stress in the masonry will actually be less than the maximum allowable stress of 1.1 ksi. Then the stress block will be 291 inches and the applied eccentric load, P, will be colinear with the resultant force C.

EXAMPLE 5-AB   Overturning Steel in a Wall (Method 2).

Determine the overturning steel for the wall shown.

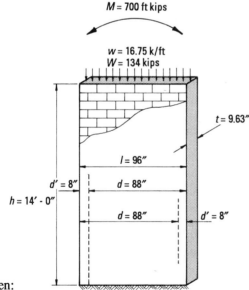

M = 700 ft kips

w = 16.75 k/ft
W = 134 kips

t = 9.63″

l = 96″

d' = 8″ d = 88″

h = 14′ - 0″

d = 88″ d' = 8″

Given:

Wall thickness nominal 10″ block

t = 9.63 inches

f_m' = 3000 psi

n = 12.9

F_s = 24,000 psi

Moment due to seismic forces

Solution 5-AB

Actual axial stress

$$f_a = \frac{P}{tl} = \frac{134}{9.63 \times 96}$$
$$= 0.145 \text{ ksi}$$

$$F_a = 0.2 \times 3\left[1 - \left(\frac{12 \times 14}{42 \times 9.63}\right)^3\right]$$
$$= 0.557 \text{ ksi}$$

$$F_b = 1.0 \text{ ksi}$$

$$f_b = F_b\left(1.33 - \frac{f_a}{F_a}\right)$$
$$= 1.0\left(1.33 - \frac{0.145}{0.557}\right)$$
$$= 1.07 \text{ ksi}$$

$$f_m = f_a + f_b$$
$$= 0.145 + 1.07$$
$$= 1.215 \text{ ksi}$$

Solve for kd, f_s, C, T and A_s

$\mathbf{a} = \tfrac{1}{6} t f_m$
$$= \tfrac{1}{6} \times 9.63 \times 1.215$$
$$= 1.95$$

$\mathbf{b} = -\tfrac{1}{2} t f_m (l - d_1)$
$$= -\tfrac{1}{2} \times 9.63 \times 1.215(96 - 8)$$
$$= -515$$

$\mathbf{c} = P\left(\frac{l}{2} - d_1\right) + M$
$$= 134\left(\frac{96}{2} - 8\right) + 700 \times 12$$
$$= 13760$$

$$kd = \frac{-\mathbf{b} - \sqrt{\mathbf{b}^2 - 4\mathbf{ac}}}{2\mathbf{a}}$$
$$= \frac{+515 - \sqrt{(515)^2 - 4(1.95)(13760)}}{2 \times 1.95}$$
$$= 30.2 \text{ inches}$$

$C = \tfrac{1}{2} t kd f_m$
$$= \tfrac{1}{2} \times 9.63 \times 30.2 \times 1.215$$
$$= 176.7 \text{ kips}$$

$T = C - P$
$$= 176.7 - 134$$
$$= 42.7 \text{ kips}$$

$$k = \frac{kd}{d} = \frac{30.2}{(96-8)} = 0.343$$

$$
\begin{aligned}
f_s &= \left(\frac{1-k}{k}\right) n f_m \\
&= \left(\frac{1-0.343}{0.343}\right) 12.9 \times 1.215 \\
&= 30 \text{ ksi}
\end{aligned}
$$

$$
\begin{aligned}
A &= \frac{T}{f_s} = \frac{42.7}{30.0} \\
&= 1.42 \text{ sq. in.}
\end{aligned}
$$

Use 2 - #8 bars each side ($A_s = 1.58$ sq. in.)

5-8.E Method 3. Section Assumed Homogeneous for Combined Loads, Vertical Load with Bending Moment Parallel to Wall

Walls and piers which resist forces parallel to the wall are subjected to overturning moments. The vertical load and the overturning moment cause combined stresses on the wall or pier. These overturning moments may be caused by wind, seismic or other lateral forces.

a. If the compressive stress, f_a, due to vertical load exceeds the flexural tension stress, f_b, due to overturning moment, the section is under compression and only minimum jamb steel is required.

b. If the tension stress due to the overturning moment exceeds the compression stress due to vertical load, determine the total net tension force and provide reinforcing steel to accommodate the tension force.

This method is also presented in the National Concrete Masonry Association Design Manual, *The Application of Reinforced Concrete Masonry Load Bearing Walls in Multi-Story Structures*, in the Concrete Masonry Association of California and Nevada publication, *Reinforced Load Bearing Concrete Block Walls for Multistory Construction* and in the *Recommended Practice for Engineered Brick Masonry* by the Brick Institute of America.

This method assumes that the section is homogeneous but the tension is taken by reinforcing steel.

1. $f_m = \dfrac{P}{A} \pm \dfrac{M}{S}$

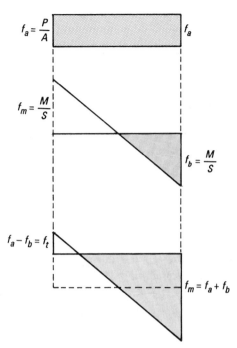

2. Check unity equation

$$\frac{f_a}{F_b} + \frac{f_b}{F_b} \le 1.00 \text{ or } 1.33$$

3. Determine the total net tension force

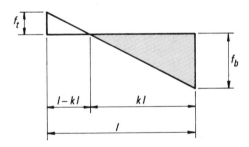

$$kl = \frac{f_m}{f_m + f_t} \times l$$

$$(l - kl) = \frac{f_t}{f_m + f_t} \times l$$

Tension Force $= \frac{1}{2} f_t \times b (l - kl)$

4. Area of steel

The area of steel may be determined by dividing the tension force by the allowable tension stress which may be increased by one third if the force is due to wind or earthquake.

$$A_s = \frac{T}{F_s} \text{ or } \frac{T}{1.33 F_s}$$

Using the allowable tensile stress for steel in the above equation is assuming that it will be strained sufficiently to produce a stress in the steel equal to the allowable stress.

An analysis in which the basic assumptions of:

a. Plane section remain plane after bending

b. Strain is proportional to the distance from the neutral axis.

may give results that indicate the strains may be of such a value that the actual steel stresses are less than allowable values.

It might be assumed that the steel is stressed to its allowable value because it may be assumed that:

c. Plane sections may not remain plane after bending

d. The section is cracked and the localized cracks will open up and cause a strain and thus a stress equivalent to the assumed stress level. This will then provide the required tension force.

5. Moment resistance of tension steel

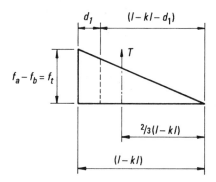

The moment of the tension force, T, about the neutral axis is:

$$M_{N.A.} = T\left[\tfrac{2}{3}(l-kl)\right]$$

If the reinforcing steel is moved from the centroid of the stress triangle, two thirds of the distance from the neutral axis, to the actual location, d', from the edge of the wall to the jamb steel, then the tension force can be reduced because the moment arm is increased.

The equivalent tension force, T_{eq}, required is:

$$T_{eq} = T\left[\tfrac{2}{3}(l-kl)\right] \times \frac{1}{\left(1-kl-d_1\right)}$$

The adjusted area of steel would be

$$\text{Equivalent } A_s = \frac{T_{eq}}{1.33 f_s}$$

6. The section is then investigated to insure that the sum of the vertical forces equals zero and that the internal resisting moment equals the external applied moment.

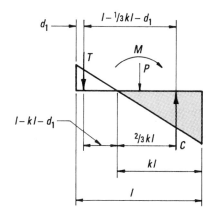

$$\Sigma F_v = C - T - P = 0$$
$$\Sigma M_c = M - T\left(l - \tfrac{1}{3}kl - d_1\right) = 0$$

EXAMPLE 5-AC Interaction Design (Method 3).

Using Method 3 determine the stress and reinforcing steel required for a nominal 8 inch solid load bearing reinforced concrete block wall. The wall is 12'-0" long and spans vertically 10'-0" high between horizontal supports. The wall carries a total load of 2500 plf and an overturning moment due to seismic forces of 500 ft kips.

Solution 5-AC

Assume $f'_m = 1500$ psi, $n = 25.8$, continuous special inspection is provided therefore full stress are permitted. Allowable steel stress, $F_s = 24,000$ psi.

Following the procedure outlined:

1. $f_a = \dfrac{P}{A} = \dfrac{2500 \times 12}{7.63 \times 12 \times 12}$

    $= 27.3$ psi

    $F_a = 0.2 f'_m \times \left[1 - \left(\dfrac{h'}{42t}\right)^3\right]$

    $\dfrac{h'}{t} = \dfrac{10 \times 12}{7.63} = 15.7$

From Table Q-2, $F_a = 284$ psi > 27.3 O.K.

$f_b = \dfrac{M}{S} = \dfrac{6M}{bd^2}$

$\quad = \dfrac{6 \times 500,000 \times 12}{7.63 \times (12 \times 12)^2}$

$\quad = 227.5$ psi

$F_b = 0.33 f'_m = 0.33 (1500)$

$\quad\quad = 500$ psi > 227.5 O.K.

$f_t = f_a - f_b = 27.3 - 227.5 = -200.3$ psi

$f_m = f_a + f_b = 27.3 + 227.5 = 254.8$ psi

2. Unity check

$$\frac{f_a}{F_a} + \frac{f_b}{F_b} = \frac{27.3}{284} + \frac{227.5}{500}$$

$$= 0.096 + 0.455$$

$$= 0.551 < 1.33$$

3. Tension force

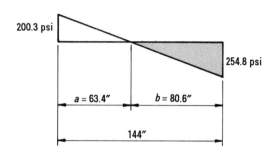

$$a = \frac{200.3}{200.3 + 254.8} \times 144 = 63.4''$$

$$b = \frac{254.8}{200.3 + 254.8} \times 144 = 80.6''$$

Tension force $= \frac{1}{2} f_t \times t \times a$

$$= \frac{1}{2} \times 200.3 \times 7.63 \times 63.4$$

$$= 48,446 \text{ lbs}$$

4. Area of steel

$$A_s = \frac{T}{1.33 F_s} = \frac{48,446}{1.33 \times 24,000} = 1.51 \text{ sq. in.}$$

Use 2 - #8 bars ($A_s = 1.58$ sq. in.)

5. Equivalent tension force

$$T_{eq} = T \left[\frac{2}{3}(l - kl) \right] \frac{1}{(l - kl - d_1)}$$

$$= 48,446 \left[\frac{2}{3}(144 - 80.6) \right] \frac{1}{(140 - 80.6 - 8)}$$

$$= 39,837 \text{ lbs}$$

Equiv. $A_s = \dfrac{39,837}{1.33 \times 24,000} = 1.24$ sq. in.

Use 2 - #7 bars ($A_s = 1.20$ sq. in.)

6. Compression force

$$C = \frac{1}{2}(254.8)7.63 \times 80.6 = 78,348 \text{ lbs}$$

7. Check sum of vertical forces

$$C - T = P$$

$$78,348 - 39,837 = 12 \times 2500$$

$$38,511 \neq 30,000$$

The sum of vertical forces are not in equilibrium, adjust the size of the compression stress block and magnitude of compression stress.

8. Adjust stress block by iteration (trial and error) assume $kl = 68$ in.

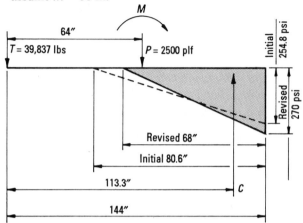

$$C = \frac{1}{2} \times 270 \times 7.63 \times 68 = 70,043 \text{ lbs}$$

$$\Sigma F_v = C - T = 70,043 - 39,837$$

$$= 30,206 \cong 30,000 \text{ lbs}$$

$$\Sigma M_T = 64 \times 30,000 - 113.3 \times 70,043 = M$$

$$1,920,000 - 7,938,206 = M$$

$$6,018,206 \text{ in. lbs} = M$$

$$501.5 \text{ ft kips} = M$$

External moment = 500 ft kips

9. Alternate method to check section for equilibrium of forces and moment

$M = 500$ ft kips

$P = 2.5$ k/ft $\times 12 = 30$ kips

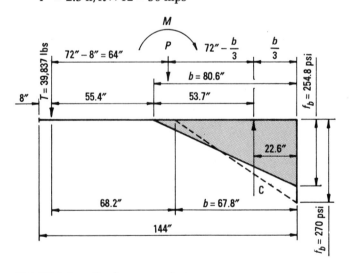

Establish length of compression area

$$\Sigma M = 0$$

$$M_c = 39.84 \left(64 + 72 - \frac{b}{3} \right) + 30 \left(72 - \frac{b}{3} \right)$$

$$500 \times 12 = 5420 - 13.3b + 2160 - 10b$$

$$6000 = 7580 - 23.3b$$

$$b = 67.8''$$

$$\Sigma F_v = 0$$

$$= 39,837 + 30,000 - C = 0$$

$$C = 69,837 \text{ lbs}$$

$$C = 69,837 = \tfrac{1}{2} f_m \times 7.63 \times 67.8$$

$$f_m = 270 \text{ psi}$$

The length of the compressive stress block and the magnitude of the compressive stress is adjusted to satisfy equilibrium of forces and moment.

5-9 WALLS WITH FLANGES AND RETURNS, INTERSECTING WALLS

5-9.A General

The design and analysis for combined stresses, axial and moment has been given in Section 5-8 for uniform rectangular sections. However, many walls intersect other walls and form I, U, C, Z and T sections. The sections provide greater moments of inertia and section moduli than a regular rectangular section.

Any reasonable assumption may be adopted for computing relative flexural stiffness of walls for the distribution of moment due to wind load. T-Beam action may be assumed where a shear wall intersects another wall or walls, using the effective flange for calculations width as one sixth of the total wall height above the level being analyzed and its overhanging width on either side of the shear wall up to six times the thickness of the intersected wall. Where shear walls intersect a wall or walls to form L or C sections, the effective overhanging flange width for calculations should not exceed one sixteenth of the total wall height above the level being analyzed nor six times the thickness of the intersected wall.

The vertical shear stress at the intersection may not exceed the allowable shear stress.

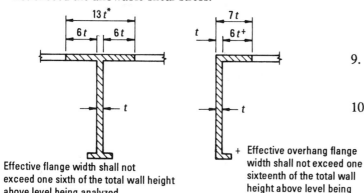

* Effective flange width shall not exceed one sixth of the total wall height above level being analyzed

\+ Effective overhang flange width shall not exceed one sixteenth of the total wall height above level being analyzed

Figure 5-56 Flanges on an intersection wall.

5-9.B Design Procedure

Proceed as follows:

Given, calculate or assume

M; P; f'_m; l (length of wall); t (wall thickness); I (moment of inertia of wall); S (Section modulus of wall to each side); d (distance from compression face to center of steel, each direction); flange width if applicable; h (effective or actual height of wall).

Solve for:

1. Effective width of flange at each end; $^1/_{16}$ to $^1/_6$ of the wall height, $6t$ maximum each side.

2. Moment of inertia, uncracked section

3. Section modulus to each side

4. Kern distance $e_k = \dfrac{I}{Ac} = \dfrac{S}{A}$

5. Virtual eccentricity $e = \dfrac{M}{P}$

6. If $e \le e_k$ minimum reinforcement required

   If $e > e_k$, consider tension bond capability or design the reinforcing for flexural stresses if the tension stress exceeds flexural bond.

7. Actual axial stress $f_a = \dfrac{P}{A} = \dfrac{P}{bt}$

Use actual cross-sectional area of masonry, web and flanges, and equivalent solid thickness for partially grouted walls.

8. h'/t Reduction factor

   $$R = \left[1 - \left(\dfrac{h'}{42t} \right)^3 \right], \text{ Table Q-1}$$

   The distance between points of support may be either horizontal, length of wall between the flanges, or the vertical, height between the floor and the roof.

9. Maximum allowable axial stress

   $$F_a = [0.2 \, f'_m \text{ or } \tfrac{1}{2}(0.2 \, f'_m)]R, \text{ Table Q-2}$$

10. Flexural stress assuming an uncracked section

    $$f_b = Mc/I \text{ for each side.}$$

11. Maximuim flexural compression stress

$$F_b = 0.33 \, f'_m \text{ or } \tfrac{1}{2}(0.33 \, f'_m); \, 2000 \text{ psi max.}$$

12. Unity equation check

$$\frac{f_a}{F_a} + \frac{f_b}{F_b} \le 1.00 \text{ or } \le 1.33$$

13. Combine stresses, f_a and f_b to establish the stress distribution on the wall.

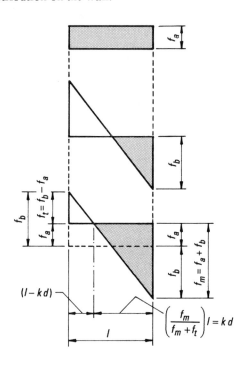

14. Tension force, T, obtained by the average tension stress times the tension area.

15. Compression force, C, obtained by taking moments about centroid of tension steel. The moment of load P times moment arm must equal the compression force times the moment arm.

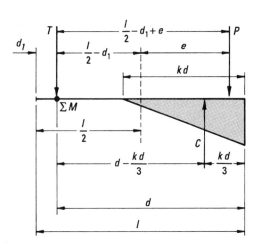

$$\Sigma M_T = \left(\frac{l}{2} - d_1 + e\right)P = \left(d - \frac{kd}{3}\right)C$$

16. Sum of the vertical forces must equal zero

$$\Sigma F_v = T + P - C = 0$$

If not in balance adjust compression force and moment arm accordingly.

17. The steel area using maximum steel stress values.

$$A_s = \frac{T}{F_s} \text{ or } \frac{T}{F_s \times 1.33}$$

If centroid of steel is not at previously assumed location adjust the value of T and moment arm.

18. Select reinforcing steel to satisfy the area requirement.

19. Select balance of steel for wall.

EXAMPLE 5-AD   Reinforcing for Moment in a Flanged Wall.

Design the flanged wall section shown which is in a high rise building subjected to a wind moment of 4000 ft. kips and an axial load of 400 kips. The wall is 8 inches nominal thickness concrete block with a clear height between lateral supports of 16' - 0".

Assume solid grouted reinforced hollow unit masonry, $f'_m = 2500$ psi, $F_s = 24,000$ psi; special continuous inspection will be provided, full stresses permitted.

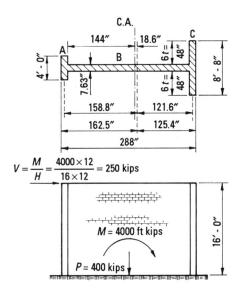

Figure 5-57 Shear wall with intersectiong walls forming I section.

Solution 5-AD

1. Flanges are as shown = 48" on one end of wall, 104" on the other end of wall, maximum overhang of $6t$ (48") on each side of wall.

2. Locate centroidal axis and determine moment of inertia.

$$\bar{x} = \frac{\text{Moment area}}{\text{area}} = \frac{525,850}{3235} = 162.5 \text{ in.}$$

TABLE 5-3 Location of Centroidal Axis and Determination of Moment Inertia

Section	Area	Arm	Moment Area	$I = \dfrac{bd^3}{12}$	$d =$ $(x - \text{Arm})$	Ad^2	$I + Ad^2$
A	366	3.81	1,395	1,772	158.8	9,229,583	9,231,360
B	2,075	144.0	298,800	12,795,286	18.6	717,867	13,513,153
C	794	284.2	225,655	3,850	121.6	11,740,530	11,744,380

$\Sigma = 3{,}235$ sq. in. $\Sigma = 525{,}850$ in.^{3} $\Sigma(I = Ad^2) = 34{,}488{,}893$
$\cong 34{,}489{,}000$

3. Section modulus, $S = I/c$

$$S = \frac{34{,}489{,}000}{162.6} = 212{,}109 \text{ in.}^3 \text{ to flange A}$$

$$S = \frac{34{,}489{,}000}{125.4} = 275{,}031 \text{ in.}^3 \text{ to flange C}$$

4. Kern distance

$$e_k = \frac{S}{A} = \frac{212{,}109}{3235} = 65.6'' \text{ to flange C}$$

$$e_k = \frac{S}{A} = \frac{275{,}031}{3235} = 85.0'' \text{ to flange } A$$

5. Virtual eccentricity

$$e = \frac{M}{P} = \frac{4000 \times 12}{400}$$
$$= 120'' > 65.6'' > 85.0''$$

6. Virtual eccentricity exceeds the kern distance for each direction from the neutral axis, therefore there will be tension on the section. Provide reinforcing steel to resist tension.

7. Actual axial stress

$$f_a = \frac{P}{A} = \frac{400 \times 1000}{3235} = 123.6 \text{ psi}$$

8. h'/t Reduction factor

$$\frac{h'}{t} = \frac{16 \times 12}{7.63} = 25.2$$
$$R = 0.784 \quad (\text{Table Q-1})$$

9. Maximum allowable axial stress

$$F_a = 0.2 f'_m R$$
$$= 0.2(2500)0.784 = 392 \text{ psi}$$

10. Flexural stress

$$f_b = \frac{M}{S} = \frac{4000 \times 12 \times 1000}{212{,}109}$$
$$= 226.3 \text{ psi on narrow end, tension or compression}$$

$$f_b = \frac{M}{S} = \frac{4000 \times 12 \times 1000}{275{,}031}$$
$$= 174.5 \text{ psi on wide end, tension or compression}$$

11. Maximum flexural compression stress

$$F_b = \tfrac{1}{3} f'_m = \tfrac{1}{3}(2500) = 833 \text{ psi}$$

12. Check unity equation

$$\frac{f_a}{F_a} + \frac{f_a}{F_b} \leq 1.33 \text{ (Wind forces)}$$
$$\frac{123.6}{392} + \frac{226.3}{833} = 0.315 + 0.272$$
$$= 0.587 < 1.33 \quad \text{O.K.}$$

13. Combine stress

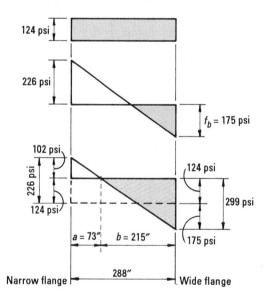

$$a = \frac{102}{102 + 299} \times 288 = 73''$$

$$b = \frac{299}{102 + 299} \times 288 = 215''$$

14. Tension force

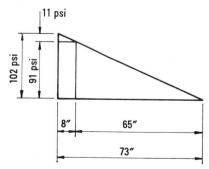

Maximum tension stress $= \dfrac{65}{73} \times 102 = 91$ psi

Tension force $= \left(\frac{1}{2} \times 91 \times 7.63 \times 65\right)$

$+(91 \times 7.63 \times 48) + \left(\frac{1}{2} \times 11 \times 7.63 \times 48\right)$

$= 22{,}566 + 33{,}328 + 2{,}015$

$= 57{,}910$ lbs

$A_s = \dfrac{57{,}910}{1.33 \times 24{,}000}$

$= 1.81$ in.^{2}

Use 2 - # 9 bars ($A_s = 2.0$ sq. in.)

15. Calculate compression force

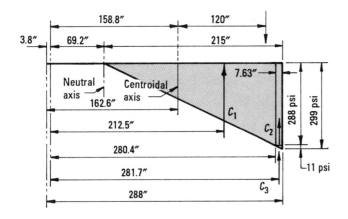

$\dfrac{215 - 7.6}{215} \times 299 = 288$ psi $= 0.288$ ksi

$C_1 = \frac{1}{2}(0.288)(7.63)(215 - 7.6)\ \ = 227.9$ kips

$C_2 = (0.288)(7.63)(104)\ \ = 228.5$

$C_3 = \frac{1}{2}(0.011)(7.63)(104)\ \ = \ \ 4.4$

$\overline{ 460.8 \text{ kips}}$

$\begin{aligned} \Sigma M_T &= 212.5(227.9) + 280.4(228.5) + 281.7(4.4) \\ &= (158.8 + 120)400 \\ &= 48{,}429 + 64{,}071 + 1239 = 111{,}570 \text{ ft k} \end{aligned}$

$113{,}739 \cong 111{,}520$ ft kips

$2219 \neq 0$

16. Sum of vertical forces

$\Sigma F_v = T + P - C = 0$

$= 57{,}910 + 400{,}000 \cong 460{,}800$ lbs

$- 2890 \neq 0$

17. The values above are within 1% range of error, and are acceptable.

The moment compression force and compression forces can be considered in equilibrium with the mo-

ment of the load and the tension force plus load, respectively.

18. Horizontal shear

$V = \dfrac{M}{h} = \dfrac{4000}{16} = 250$ psi

$f_v = \dfrac{V}{bd} = \dfrac{250 \times 1000}{7.63 \times 288}$

$= 113$ psi

$\dfrac{M}{Vd} = \dfrac{4000}{250 \times 24} = 0.67$

For $f'_m = 2500$ psi and $\dfrac{M}{Vd} = 0.67$

From Tables and Diagrams A-5 and A-6

Allowable shear on masonry =

$= 50 \times 1.33$

$= 66.7$ psi < 113 psi   N.G.

Allowable shear with reinforcing =

$= 84 \times 1.33$

$= 112$ psi $\cong 113$ psi

From Diagram K-2

For $t = 7.63''$, $v = 113$ psi, $F_s = 32{,}000$ psi

Use #8 at 24" o.c. spaced vertically

19. Consider moment in other direction. Flange A in compression

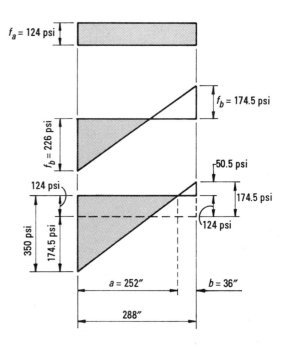

$a = \dfrac{350}{350 + 50.5} \times 288 = 252''$

$b = \dfrac{50.5}{350 + 50.5} \times 288 = 36''$

20. Tension in flange C.

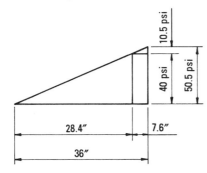

Maximum tension stress $= \dfrac{28.4}{36} \times 50.5$

$\qquad\qquad\qquad\quad = 40$ psi

Tension force =

$\qquad = \frac{1}{2}(40)(7.63)(28.4) + (40)(7.63)(104) +$

$\qquad\quad \frac{1}{2}(10.5)(7.63)(104)$

$\qquad = 4334 + 31,741 + 4166 = 40,241$ lbs

$A_s = \dfrac{40,241}{1.33 \times 24,000} = 1.26$ sq. in.

Use 2 - #8 ($A_s = 1.58$ sq. in.)

5-9.C Connections of Intersecting Walls

When cross walls are considered as flanges to walls that resist overturning moments, the connections between them must be properly designed. The intersection of the flange or cross wall element to the web section is the critical location for stress concentrations. This stress is a vertical shear stress for it is delivering compression forces to the masonry or tension forces to the flange steel.

These connections should be evaluated to determine flange masonry or the amount and location of reinforcing required to permit the connection to function as desired. This evaluation is based on calculating the shear stress at the intersection:

Where

$$v = \frac{VQ}{It} = \frac{VA_f y}{It}$$

v = Vertical shear stress

V = Total shear

A_f = Area of flange

y = Distance from centroidal axis of the section to the centroid of the flange

I = Moment of inertia

t = Thickness of web

The limiting allowable shear stress is based on either the masonry or the reinforcing steel resisting all shear and is governed by the M/Vd or h/d value. See Tables A-5 and A-6 for limiting values.

If the shear stress is equal to or less than the allowable value for masonry, no reinforcing is required. If it is equal to or less than the allowable value for reinforcing to resist the shear forces, provide shear steel. If it exceeds the allowable value for reinforcing steel, increase the thickness of the wall and recompute all stresses.

The shear steel shall be determined by the equation:

$$A_v = \frac{Vs}{F_s d}$$

Where

A_v = Area of shear steel

V = Total shear

s = Spacing of shear steel

F_s = Allowable tensile stress for shear steel. May be increased one third for wind and seismic forces.

d = Depth or length of shear wall

EXAMPLE 5-AE Intersecting Walls — Vertical Shear

Calculate the vertical shear at the intersection of the web and the flange from Example 5-AD.

$$V = \frac{M}{h} = \frac{4000}{16} = 250 \text{ kips}$$

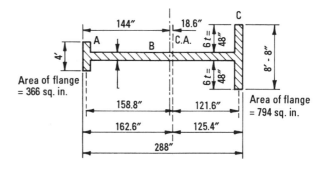

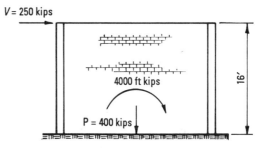

Figure 5-58 Flanged shear wall.

Vertical shear $v = \dfrac{V A_f y}{I t}$

$V = 250$ kips

$A_{fa} = 366$ sq. in. $y_a = 158.8$ in.

$A_{fc} = 794$ sq. in. $y_c = 121.6$ in.

$I = 34,489,000$ in.[4] $t = 7.63$ in.

$v_{fa} = \dfrac{V A_f y}{I t}$

$\quad = \dfrac{250 \times 1000 \times 366 \times 158.8}{34,489,000 \times 7.63}$

$\quad = 55.2$ psi

$v_{fc} = \dfrac{250 \times 1000 \times 794 \times 121.6}{34,489,000 \times 7.63}$

$\quad = 91.7$ psi

$\dfrac{M}{Vd} = \dfrac{4000}{250 \times 24} = 0.67$

for $f_m' = 2500$ psi and $\dfrac{M}{Vd} = 0.67$

From Tables and Diagrams A-5 and A-6

Allowable shear stress, masonry resisting shear,

35 psi $\times 1.33 = 44.7$ psi

Allowable shear stress, reinforcing resisting shear,

75 psi $\times 1.33 = 100$ psi

Provide shear reinforcing for vertical shear forces

Vertical shear $= v A_w$

$\quad = 91.7(7.63)(16 \times 12)$

$\quad = 134.3$ kips

$A_v = \dfrac{Vs}{F_s d} = \dfrac{134.3 \times 1000 \times 24}{32,000 \times 288} = 0.35$ sq. in.

Use #6 at 24 in. o.c. spaced vertically ($A_s = 0.44$ in^2)

The tension steel provided at the end will be adequate to resist and transfer the vertical shear between the web (cross wall) and the flanges (end walls).

Use 2 - #9 bars at wall A and 2 - #8 bars at wall C.

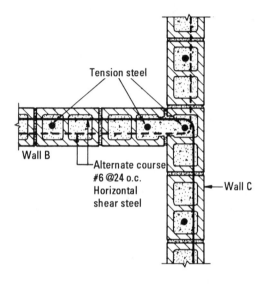

Figure 5-59 Detail of connection of intersecting walls.

5-10 QUESTIONS AND PROBLEMS

5-1   What are the basic assumptions in elastic design of a flexural member?

5-2   Is strain compatible with stress? What is its significance with respect to compression steel?

5-3   What is the modular ratio? What is its significance?

5-4   Explain the function of the flexural coefficient, K. How does it vary from an under-reinforced section to an over-reinforced section.

5-5   Given, a ten inch (nominal) concrete masonry cantilever retaining wall reinforced with vertical steel #6 bars 24 inches on center. What is the maximum d value that this wall could be designed for? Locate the neutral axis by means of transformed areas if this wall is solid grouted and f_m' is 2500 psi. Use full design stresses (special inspection). If the reinforcing steel has a maximum allowable stress

of 24,000 psi, what is the allowable moment for the section?

5-6   From basic principles, establish the following values for a rectangular section for $f_m' = 2250$ psi, $f_s = 18,000$ psi

a) balanced steel ratio, p

b) balanced flexural coefficient, K

c) j, k values for balanced condition

5-7   What is the limiting stress in compressive reinforcement? Explain in terms of n, f_m', f_s'. What are the limiting features?

5-8   Determine the moment capacity and maximum size reinforcing bar that can be placed in a 8″ CMU and still have the neutral axis in the face shell. Given face shell thickness $= 1^{1}/_{4}″$; $d = 5.3″$; bar spacing $= 24, 32$ and $48″$ o.c. and modular ratio, $n = 25.8$; 15.5 and 9.7, $F_s = 24,000$ psi.

5-9   A two wythe clay brick lintel beam is 10″ wide by 32″ deep. It spans over an opening 20 feet wide. What is the maximum uniform load that can be placed on this lintel beam if shear is the governing stress? The f_m' of the masonry is 2000 psi.

Determine the allowable super-imposed load for

a) masonry resists all shear, no special inspection

b) masonry resists all shear, with special inspection

c) steel resists all shear, no special inspection

d) steel resists all shear, with special inspection

Shear steel is #6 vertical bars at 14 inches on center, F_s = 24,000 psi.

5-10   What is the allowable shear stress parallel to a 10 inch thick brick shear wall if the wall is 20 feet long and 40 feet high and if the wall is 50 feet long and 15 feet high? Consider that the masonry is to resist all of the shear, f_m' = 2500 psi. Consider that the masonry is to resist none of the shear and that horizontal steel reinforcing (#6 bars 24 inches on center, F_s = 24,000 psi) resists all the shear. Consider both inspected and non-inspected masonry and give appropriate allowable stresses.

5-11   What is the shear resistance per linear foot of a 10 inch thick brick wall at the floor joint both parallel to and perpendicular to the wall if the axial stress is 135 psi?

5-12   What is the anchorage length required for a #7 bar in masonry (f_m' = 1500 psi) and in concrete (f_c' = 2000 psi)?

5-13   Design the tension reinforcing and specify the minimum allowable strength of masonry, f_m', for a wall subjected to axial load and seismic overturning moment. The wall is a nominal 10 inches thick, 10 feet long and 12 feet high. F_s = 24,000 psi. Axial load = 100 kips, overturing moment = 300 ft kips parallel to the wall.

5-14   An 8 inch concrete masonry wall, solid grouted is 12 feet high and is reinforced with #7 bars at 24 inches on center. Axial load is 3 kips per foot, f_m' = 1500 psi and special inspection is provided. What is the maximum moment that can be applied perpendicular to the wall if d is 3.75 inches and if d is 5.25 inches?

5-15   What is the reinforcing required for a wall subjected to vertical load of 100 kips and an overturing moment of 200 ft k. The masonry is 8 inches

solid grouted, f_m' = 2500 psi, F_s = 24,000 psi, h = 10 ft, and special inspection is provided.

a) Ignore Tee flange

b) Include Tee flange

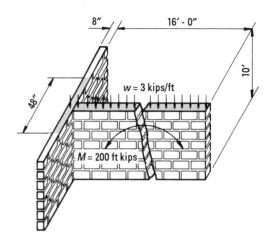

5-16   A 10 inch thick CMU beam spans 25 feet. The beam has a total depth of 48 inches and is continuous on both ends. It carries a live load of 1000 lbs per linear foot. Design the reinforcing both in the center and over the supports and the shear reinforcing, if required. f_m' = 1500 psi, special inspection is used in its construction. If the masonry strength is not sufficient, what f_m' should be used?

5-17   What is the moment capacity of a grouted concrete block beam 8 inches thick, total depth 32 inches, d = 26 inches and reinforced with two #8 bars? Use f_m' = 1500 psi and F_s = 24,000 psi.

5-18   A two wythe masonry wall 8½ inches thick is reinforced vertically with #6 bars @ 30 inches on centers in the center of the wall. It is subjected to a bending moment of 1000 ft pounds per foot. Assume that f_m' = 1800 psi and n = 21.5. What are the stresses in the masonry and steel? If the bending moment is 1.5 kip ft/ft what are the stresses? Are they within the allowable range?

5-19   A masonry beam 12 inches wide and 30 inches deep (d = 24 inches) spans 20 feet. It carries a live and dead load of 1000 plf. For f_m' = 2000 psi and F_s = 24,000 psi, design the tension reinforcement and compression reinforcement if needed, and the shear reinforcing. Also design reinforcing if the LL plus DL is 2000 plf.

5-20   Design a 13″ wide double reinforced brick beam for a total moment, M = 500 ft kips using f_m' = 2500 psi, Grade 60 reinforcing steel and a cover of 4 inches to center of steel. Assume d' = 4 inches and d = 60 inches. Determine the required steel.

5-21 A 12″ x 48″ concrete block beam has $d = 40$ inches, $d' = 4$ inches, and A 615-60 steel. What is the moment capacity if

   1) $A_s = 2 - \#9$, and $A'_s = 2 - \#6$

   2) $A_s = 2 - \#10$, and $A'_s = 2 - \#11$

   3) $A_s = 3 - \#11$, and $A'_s = 2 - \#11$

5-22 Calculate the allowable load on the following columns if $h = 13$ feet 4 inches and the columns have $^3/_8$ inch head joints.

Size (in.)	Reinforcing	f'_m (psi)	Inspection
8×32	4 - #6	1500	no
10×16	4 - #7	1500	yes
16×16	4 - #8	1500	no
24×32	8 - #9	2000	yes

5-23 A 20 ft high interior column supports and axial load of 200 kips. Determine the size of the column, vertical reinforcing steel, and the tie spacing, if the column is constructed without special inspection in Seismic Zone No. 4 and

   a) Reinforced brick, $f'_m = 2500$ psi

   b) Reinforced concrete masonry $f'_m = 1500$ psi

   Assume $F_s = 24,000$ psi.

5-24 A concrete masonry column 16″ x 16″ (nominal) is 14 ft high and is reinforced with four No. 9, grade 60 bars. What vertical load at an eccentricity of 12 inches can it support? Assume full allowable stress.

5-25 Design a 22 feet high reinforced brick wall to carry an axial load of 5 kips/ft and a moment perpendicular to the wall of 2 ft kips/ft. Use $f'_m = 2500$ psi, $F_s = 24,000$ psi and full design stresses.

5-26 Select the reinforcing required for a 10 inch brick wall which is subjected to an axial load of 2000 plf and a moment perpendicular to the wall of 2000 ft lbs/ft. Use $f'_m = 4000$ psi, $F_s = 24,000$ psi, $h = 18' - 0''$, steel in center of wall and full allowable stress.

5-27 For the concrete masonry beam shown below, $f'_m = 1500$ psi (use half stresses) and $F_s = 24,000$ psi. Neglecting the weight of beam, calculate the depth, d, and total depth of the beam for these items individually.

   a) depth without stirrups

   b) depth with stirrups

   c) depth for bond

   d) depth for stress in steel

   e) depth for maximum stress in steel or masonry

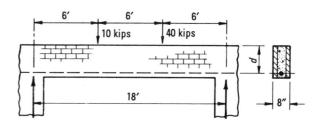

5-28 Design the flexural tension reinforcing, compression reinforcing, if needed, and shear reinforcing for the lintel beam shown below. Use $f'_m = 3000$ psi, 8″ CMU, normal weight, solid grouted and $F_s = 24,000$ psi.

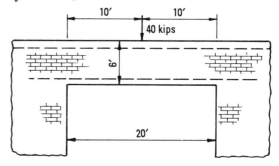

5-29 Design the shear reinforcing and calculate the embedment length for the cantilevered beam shown assuming $f'_m = 2000$ psi, $F_s = 24,000$ psi and (a) with special inspection (b) without special inspection.

5-30 Determine the shear in the 8 inch concrete masonry piers shown below. Determine the shear stress and shear reinforcing, if necessary. Assume $f'_m = 1500$ psi, $F_s = 24,000$ psi and use half stresses.

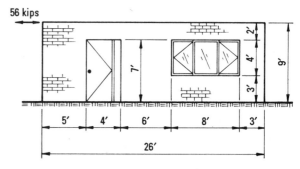

5-31 Design a reinforced masonry wall for a commercial building with walls 14′ - 0″ high from floor to roof ledger beam. Walls are 6 inches thick and the building is in Seismic Zone No. 2A. Wind = 15 psf, $f'_m = 1500$ psi and $F_s = 24,000$ psi

5-32 Determine the shear reinforcing and overturning steel for an 8″ CMU shear wall which is 10′ - 0″ long and 12 ft. high. Assume $f'_m = 2000$ psi, $F_s = 24,000$ psi and the lateral seismic force at the top of the wall is 90 kips.

SECTION 6

Design of Structural Members—Strength Design

6-1 GENERAL

The structural design of reinforced masonry is changing from the elastic working stress method to strength design procedures.

Charles Whitney pioneered the concept of strength design in his technical paper *Plastic Theory of Reinforced Concrete* published in the 1942 ASCE Transaction 107. His theory states that when a reinforced concrete section is subjected to high flexural moments, the concrete stress from the neutral axis to the extreme compression fibers would conform to the stress strain curve of the materials as if it were tested in compression.

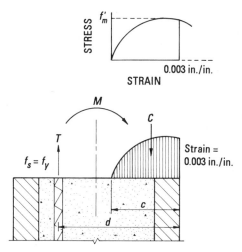

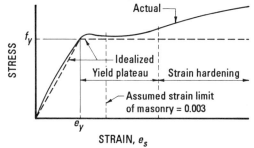

Figure 6-1 Stress due to flexural moment at balanced condition.

Whitney also states that when the tension reinforcing reaches its yield stress, it will continue to elongate without an increase in moment or force. This condition occurs at the yield plateau of the steel as shown on the stress-strain curve in Figure 6-2.

Figure 6-2 Idealized stress strain diagram for reinforcing steel.

The compressive stress block of the concrete, as shown in Figure 6-3, is simplified from the curved or parabolic shape to a rectangular configuration. This rectangular stress block, which is now often called Whitney's stress block, is approximated as having a length of ad and a height of $0.85 f'_m$.

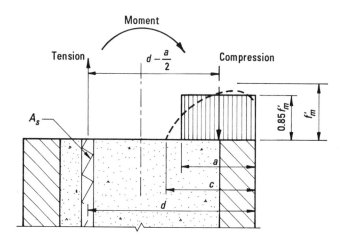

Figure 6-3 Assumed stress block at yield condition.

Masonry systems have compression stress-strain curves similar to those of concrete, in that the curves are curved or parabolic shaped and that they reach a strain of at least 0.003. Accordingly, the parameters of reinforced concrete strength design are being adopted with minor changes for masonry design.

6-2 DEVELOPMENT OF STRESS CONDITIONS

As a structural element is loaded in flexure, one side is stressed in tension while the other is stressed in compression. When the modulus of rupture is reached, the tension side of the element cracks and the reinforcing steel resists the tension force. As the moment is increased, the stress in the steel and masonry also increases. The shape of the stress block for the masonry parallels a stress-strain curve (Figure 6-4).

For safety, concrete and masonry sections are designed to be under-reinforced so the reinforcing steel is stressed to its yield strength well before the masonry

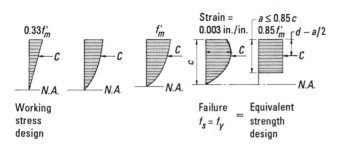

Figure 6-4 Variation in stress block as moment increases and the steel yields.

reaches its full strength capacity. This prevents the masonry from failing suddenly in compression.

When the steel is stressed to yield (which for Grade 60 steel is assumed to be 60,000 psi at an initial strain of 0.002 in./in.), it continues to stretch without a significant increase in stress as can be seen in Figures 6-2 and 6-5.

As the steel stretches, the depth of the masonry stress block decreases and the stress and strain increase until the masonry is strained to the assumed maximum strain of 0.003 in./in. at which point the masonry fails in a crushing compression failure.

The maximum masonry compression strain actually ranges from 0.003 to 0.005 in./in. The value 0.003 in./in. is conservatively used in UBC Section 2411(b)3 D.

Items 1, 2 and 3 below describe the conditions that occur on the stress and strain diagrams shown in Figure 6-5.

1. Allowable flexural tension stress for steel,

   $f_s = 0.4f_y = 24,000$ psi,

   Allowable flexural compression stress in masonry,

   $f_b = 0.33 f'_m$.

2. Steel is stressed to yield,

   $f_s = f_y$

   Masonry is stressed from 0.7 to 0.8 f'_m

3. Steel stretches,

   Strain increases in steel until the strain in masonry is 0.003 in./in.

6-3 STRENGTH DESIGN PROCEDURE

There are two conditions included in strength design. They are the load and the design parameters.

6-3.A Load Parameters

6-3.A.1 Load Factors

Service loads or actual loads are generally used for working stress design procedures. For strength design

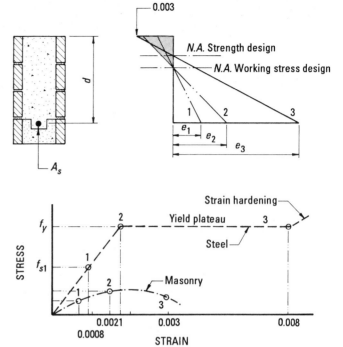

Figure 6-5 Development of stress and strain in a flexural member. (Leet, 1982)

procedures, however, the actual or specified code loads are increased by prescribed load factors. These load factors which are given in UBC Sections 2411 and 2412, consider live load, dead load, wind, earthquake, temperature, settlement and earth pressure. The appropriate or most severe loading condition is used to design the structural element.

6-3.A.1(a) Load Factors for Tall Slender Wall Design

UBC Sec.2411(b) for Slender Wall Design
3. **Strength design. A. Load factors.** Factored loads shall be based on:
$U = 1.4 D + 1.7 L$, or .... (11-2)
$U = 0.75 (1.4 D + 1.7 L + 1.87 E)$, or .... (11-3)
$U = 0.75 (1.4 D + 1.7 L + 1.7 W)$, or .... (11-4)
$U = 0.9 D + 1.43 E$, or .... (11-5)
$U = 0.9 D + 1.3 W$ .... (11-6)

Where:

D = dead loads, or related internal moments and forces.

E = load effects of earthquake, or related internal moments and forces.

L = live loads, or related internal moments and forces.

U = required strength to resist factored loads, or related internal moments and forces.

W = wind load, or related internal moments and forces.

6-3.A.1(b) Load factor for Shear Wall Design

UBC Sec.2412 for Shear Wall Design

(c) Shear Wall Design Procedure. 1. Required Strength. The required strength shall be determined as follows:

A. For earthquake loading, the load factors shall be:

$$U = 1.4(D + L + E) \quad\text{(12-1)}$$

$$U = 0.9D \pm 1.4E \quad\text{(12-2)}$$

B. Required strength U to resist dead load D and live load L shall be at lease equal to:

$$U = 1.4D + 1.7L \quad\text{(12-3)}$$

C. If resistance to structural effects of a specified wind load W are included in design, the following combinations of D, L and W shall be investigated to determine the greatest required strength U.

$$U = 0.75(1.4D + 1.7L + 1.7W) \quad\text{(12-4)}$$

where load combinations shall include both full value and zero value of L to determine the more severe condition, and

$$U = 0.9D + 1.3W \quad\text{(12-5)}$$

but for any combination of D, L and W, required strength U shall be not less than formula (12-3).

D. If resistance to earth pressure H is included in design, required strength U shall be at lease equal to

$$U = 1.4D + 1.7L + 1.7H \quad\text{(12-6)}$$

except that where D or L reduces the effect of H, $0.9D$ shall be substituted for $1.4D$ and zero value of L shall be used to determine the greatest required strength U. For any combination of D, L and H, required strength U shall be not less than Formula (12-3).

E. If resistance to loadings due to weight and pressure of fluids with well-defined densities and controllable maximum heights F is included in design, such loading shall have a load factor of 1.4 and be added to all loading combinations that include live load.

F. If resistance to impact effects is taken into account in design, such effects shall be included with live load, L.

G. Where structural effects T of differential settlement, creep, shrinkage or temperature change may be significant in design, required strength U shall be at least equal to

$$U = 0.75(1.4D + 1.4T + 1.7L) \quad\text{(12-7)}$$

but required strength U shall be not less than

$$U = 1.4(D + T) \quad\text{(12-8)}$$

6-3.A.2 Capacity Reduction Factor, ϕ

No material is precisely as specified and no construction is exactly in accordance with the plans. In each case, there are variations in the strength, size, and placement of the materials that will change, and possibly reduce the capacity of the section.

Accordingly, a capacity reduction factor, ϕ, is used to lower the capacity of the ideally constructed member to a realistic capacity that can be assured.

For masonry systems constructed with special inspection, the capacity reduction factor, ϕ, for flexural capacity is 0.80. Note that masonry systems may no longer be designed by strength procedures if special inspection of the construction will not be provided.

6-3.B Design Parameters

The parameters for strength design are:

a) The steel is at yield stress.

b) The masonry stress block is rectangular.

c) The masonry strain is limited to 0.003 in./in.

d) The steel ratio, p, is limited to 50% of the balanced reinforcing ratio, p_b, to assure that a ductile mechanism forms prior to brittle, crushing, behavior.

6-4 DERIVATION OF FLEXURAL STRENGTH DESIGN EQUATIONS

6-4.A Strength Design for Sections with Tension Steel Only

As stated above, the limits for flexural design using strength methods are that the stress in the steel is at yield strength and that the strain in the masonry is at 0.003. When these conditions occur at the same moment, the section is considered to be at balanced design.

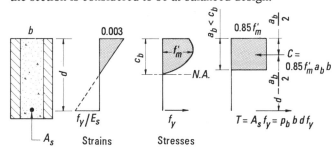

Figure 6-6 Strain and stress distribution on a flexural member balanced design.

The depth to the neutral axis, c_b, for a balanced design is:

$$c_b = \frac{0.003}{0.003 + \dfrac{f_y}{E_s}}d = \frac{87,000}{87,000 + f_y}d$$

For $f_y = 60,000$ psi

$$c_b = \frac{87,000}{87,000 + 60,000} d = 0.592d$$

The depth of the stress block for a balanced design, a_b is

$$a_b = \beta_1 c_b$$

Where:

$\beta_1 = 0.85$ for $f'_m \leq 4000$ psi

$\beta_1 = 0.825$ for $f'_m = 4500$ psi

$\beta_1 = 0.80$ for $f'_m = 5000$ psi

Therefore when $f'_m \leq 4000$ psi:

$$a_b = 0.85 c_b = 0.85 \times 0.592d$$

$$= 0.503d$$

This is the depth of the stress block for balanced conditions.

When design conditions are not at balanced conditions, the depth of the stress block will be less than a_b. The designation for the resulting depth of the stress block is a.

Equating the compression and tension forces

Compression force = $0.85 f'_m a b$

Tension force = $A_s f_y = p b d f_y$

$$C = T$$

$$0.85 f'_m a b = p b d f_y$$

Solve for a

$$a = \frac{p b d f_y}{0.85 f'_m b}$$

$$= \frac{p d f_y}{0.85 f'_m}$$

$$= p \left(\frac{f_y}{f'_m} \right) \left(\frac{d}{0.85} \right)$$

The steel quotient is defined as

$$q = p \left(f_y / f'_m \right)$$

Therefore

$$a = \left(\frac{p f_y}{f'_m} \right) \left(\frac{d}{0.85} \right) = \frac{q d}{0.85}$$

The moment capacity of the section is then found as:

$$M_n = C \left(d - \frac{a}{2} \right) = T \left(d - \frac{a}{2} \right)$$

$$M_n = 0.85 f'_m a b \left(d - \frac{a}{2} \right) \quad \text{(Masonry capacity)}$$

$$M_n = A_s f_y \left(d - \frac{a}{2} \right) \quad \text{(Steel capacity)}$$

Substituting $a = \dfrac{p f_y d}{0.85 f'_m}$ in the masonry capacity equation yields:

$$M_n = 0.85 f'_m b \left(\frac{p f_y d}{0.85 f'_m} \right) \times \left(d - \frac{p f_y d}{2 \times 0.85 f'_m} \right)$$

$$= p f_y b d^2 \left(\frac{1 - p f_y}{f'_m} \right)$$

Substituting $q = p f_y / f'_m$ and $p f_y = q f'_m$

$$M_n = b d^2 f'_m q (1 - 0.59 q)$$

The flexural coefficient K is then

$$M_n = b d^2 f'_m q (1 - 0.59 q)$$

$$M_n = K_n b d^2$$

$$K_n = f'_m q (1 - 0.59 q)$$

Introducing the capacity reduction factor, ϕ, the equations are

Nominal moment, $M_n \geq M_u / \phi$ and

$$M_u \leq \phi b d^2 f'_m q (1 - 0.59)$$

$$M_u \leq \phi K_n b d^2 = K_u b d^2$$

6-4.A.1 Balanced Steel Ratio

In order to insure that the reinforcing steel will be stressed to yield before the masonry achieves the strain limitation of 0.003 in./in., it is necessary to limit the amount of reinforcing steel in the section. For concrete design, this limitation is 75% of the reinforcing required for balanced design condition while the UBC currently limits reinforcing for masonry design to 50% p_b.

As stated previously, the definition of balanced design for strength design is that the steel is stressed to its yield strength just as the masonry achieves a strain of 0.003 in./in.

The balanced steel ratio is

$$p_b = \frac{0.85 \beta_1 f'_m}{f_y} \left(\frac{87,000}{87,000 + f_y} \right)$$

For $f_y = 60,000$ psi and $\beta_1 = 0.85$ the balanced steel ratio is

$$p_b = \frac{0.85 (0.85) f'_m}{60,000} \left(\frac{87,000}{87,000 + 60,000} \right)$$

$$= 0.00000713 f'_m$$

Therefore, the maximum steel ratio for $f_y = 60,000$ psi is

$$0.5 p_b = 0.00000356 f'_m$$

Table 6-1 shows the values of p_b, $0.5 p_b$, and $0.75 p_b$ for various f'_m values.

The balanced steel ratio, $p_b = A_{sb}/bd$, can also be determined by balancing the tension and compression forces.

$$\text{Compression force} = 0.85 f'_m (0.503d) b$$
$$= 0.428 f'_m bd$$

$$\text{Tension force} = A_{sb} f_y = p_b bd f_y$$

$$C = T$$

$$0.428 f'_m bd = p_b bd f_y$$

$$p_b = \frac{0.428 f'_m}{f_y}$$

TABLE 6-1　Maximum Allowable Steel Ratio p; $f_y = 60,000$ psi

f'_m	p_b	$0.50 p_b$[1]	$0.75 p_b$[2]
1500	0.0107	0.0053	0.0080
2000	0.0142	0.0071	0.0107
2500	0.0178	0.0089	0.0134
3000	0.0214	0.0107	0.0161
3500	0.0250	0.0125	0.0188
4000	0.0286	0.0143	0.0215
4500	0.0311	0.0156	0.0233
5000	0.0335	0.0168	0.0252

1. UBC Section 2411(b) limits p for slender wall design to $0.5 p_b$.
2. UBC Section 2610(d)3 limits p for concrete design to $0.75 p_b$.

EXAMPLE 6-A　Balance Steel Ratio, p_b.

Determine the steel ratio for a balanced design condition for strength design and compare to working stress design.

Given:

Strength of masonry, $f'_m = 1500$ psi;
Grade 60 steel, $f_y = 60,000$ psi

Solution 6-A

For strength design, balanced steel variable from Section 6-4.A.1

$p_b = 0.00000713(1500) = 0.0107$

For working stress design for $f'_m = 1500$ psi and $f_s = 24,000$ psi from Table E-1b; $p_b = 0.0036$.

Note: The balanced condition for strength design requires three times the amount of steel as for working stress design. Therefore, it is acceptable to "over reinforce" structural elements designed by WSD principles. This condition of an over reinforced section for WSD means that the masonry is fully stressed but the steel is not fully stressed.

EXAMPLE 6-B　Comparison of STR. DES. and WSD Balanced Steel Ratio.

Determine the balanced steel ratios by the strength design and working stress design methods when $f'_m = 3000$ psi and $f_y = 60,000$ psi.

Solution 6-B

For strength design, the balance steel ratio from Section 6-4.A.1 or Table 6-1 is

$p_b = 0.0000073(3,000) = 0.0214$

For working stress design when $f'_m = 3000$ psi and $f_s = 24,000$, $p_b = 0.0073$ (from Table E-4b).

From the above examples it is evident that it requires far more reinforcing to achieve balanced conditions for strength design than for working stress design.

EXAMPLE 6-C　Depth of Beam and Reinforcing Steel.

Determine the beam depth and reinforcing steel for a nominal 8″ concrete masonry beam to support a factored bending moment, M_u, of 95 ft kips.

Solution 6-C

Assume

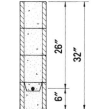

$f'_m = 1500$ psi

$f_y = 60,000$ psi

Maximum steel ratio, $p = 0.5 p_b$

From STR. DES. Table A-1 for steel ratio of $0.5 p_b$, $p_{max} = 0.0053$; $a_u = 3.50$ and $K_u = 220.8$

$$bd^2 = \frac{M_u}{K_u} = \frac{95 \times 1000 \times 12}{220.8}$$

$$= 5163$$

$$b = 7.63 \text{ inches}$$

$$d^2 = \frac{5163}{7.63} = 677$$

$$d = 26.0 \text{ inches}$$

Use total depth of 32 inches

Determine reinforcement

$$A_s = \frac{M_u}{a_u d} = \frac{95}{3.5 \times 26} = 1.04 \text{ sq. in.}$$

Alternate procedure

$$A_s = pbd$$
$$= 0.0053 \times 7.6 \times 26$$
$$= 1.05 \text{ sq. in.}$$

Use 1 - #9 bar ($A_s = 1.00$ sq. in.)

or 1 - #6 and 1 - #7 bar ($A_s = 1.04$ sq. in.)

EXAMPLE 6-D Area of Steel, Strength Design.

What is the area of reinforcement required for a beam subjected to a factored moment of 150 ft kips.

The beam is grouted brick 9.5 inches wide by 48 inches deep. The d distance is 42 inches, $f'_m = 2500$ psi, and $f_y = 60,000$ psi.

Solution 6-D

Determine the flexural coefficient

$$K_u = \frac{M_u}{bd^2} = \frac{150 \times 1000 \times 12}{9.5 \times 42^2}$$
$$= 107.4$$

From STR. DES. Table A-3 for $K_u = 107.4$; $p = 0.0023$ and $a_u = 3.87$

$$A_s = pbd = 0.0023 \times 9.5 \times 42$$
$$= 0.92 \text{ sq. in.}$$

Use 1 - # 9 bar ($A_s = 1.00$ sq. in.)

EXAMPLE 6-E Moment Capacity.

If the beam in Example 6-D was reinforced with 2 - #8 bars, what would be its factored moment capacity.

Solution 6-E

$$p = \frac{A_s}{bd} = \frac{2 \times 0.79}{9.5 \times 42}$$
$$= 0.0040$$

From STR. DES. Table A-3 for $p = 0.0040$

$$K_u = 179.3$$
$$M_u = K_u bd^2$$
$$= \frac{179.3 \times 9.5 \times 42^2}{12,000}$$
$$= 250 \text{ ft kips}$$

EXAMPLE 6-F Design Aid Strength Design Table B-1.

Using STR. DES. Table B-1 determine the required steel area for a nominal 8 inch concrete masonry solid grouted beam carrying a live load of 3000 plf and dead load including the weight of the beam of 2000 plf, $f'_m = 1350$ psi, $f_y = 60,000$ psi, $d = 58$ inches and overall depth = 64 inches. The beam spans 20 ft and special inspection is provided.

Solution 6-F

Factored loads:

$$U = 1.4D + 1.7L$$
$$= 1.4(2000) + 1.7(3000)$$
$$= 7900 \text{ plf}$$

Factored moment, M_u

$$M_u = \frac{wl^2}{8} = \frac{7900 \times 20^2}{8}$$
$$= 395,000 \text{ ft lbs}$$
$$= 395 \text{ ft kips}$$

Determine the steel requirement using STR. DES. Table B-1

$$q(1 - 0.59q) = \frac{M_u}{\phi bd^2 f'_m}$$
$$= \frac{395,000 \times 12}{0.8 \times 7.63 \times 58^2 \times 1350}$$
$$= 0.1710$$

From STR. DES. Table B-1 for $q(1 - 0.59q) = 0.1710$

$$q = 0.193$$

Steel ratio

$$p = \frac{qf'_m}{f_y}$$
$$p = \frac{0.193 \times 1350}{60,000}$$
$$= 0.0043$$

$$A_s = pbd = 0.0043 \times 7.63 \times 58$$
$$= 1.92 \text{ sq. in.}$$

Use 2 - # 9 bars ($A_s = 2.00$ sq. in.)

Check for $0.5p_b$

$$p_b = \frac{0.85 \beta f'_m}{f_y}\left(\frac{87,000}{87,000 + f_y}\right)$$
$$= \frac{0.85 \times 0.85 \times 1350}{60,000}\left(\frac{87,000}{87,000 + 60,000}\right)$$
$$= 0.0096$$

$0.5p_b = 0.0048 > 0.0043$ Satisfactory

6-4.B Strength Design for Sections with Tension and Compression Steel

The use of compression steel is very seldom required in masonry design. However, when there is steel in the compression stress block, it will contribute to the compression capacity of the section.

If more factored moment capacity is required than available by using the maximum allowable steel ratio of $0.5p_b$, additional tension and compression steel can be added to provide the increased moment capacity.

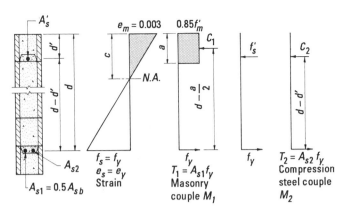

Figure 6-7 Strain, stress and moment diagram for flexural member with compression steel.

Factored moment capacity $M_u = \phi M_n = \phi(M_1 + M_2)$

Where: $\phi = 0.8$

$$M_1 = T_1\left(d - \frac{a}{2}\right)$$

$$M_2 = T_2(d - d')$$

Calculate maximum moment, M_1, for maximum steel ratio, $0.5p_b$ for a member with tension steel only.

$$A_{s1} = 0.5p_bbd \qquad T_1 = A_{s1}f_y$$

c/d from STR. DES. A Tables

$$c = d\left(\frac{c}{d}\right) \qquad a = 0.85c$$

Determine the value of M_2 as the difference between M_n and M_1. The moment arm is $(d - d')$.

The area of steel is based on the stress in the steel.

Tension steel $f_s = f_y$

Compression steel $f'_s \le f_y$

The stress in the compression steel can be determined by the geometry of the maximum masonry strain of 0.003, c distance to the neutral axis and the d' or $(c - d')$ value. The distance c is based on a flexural member with tension steel only.

Stress in compression steel

$$f'_s = \text{strain} \times E_s = e'_s \times E_s$$

$$= 0.003\left(\frac{c - d'}{c}\right)(29,000,000)$$

$$= 87,000\left(\frac{c - d'}{c}\right)$$

Where:

$$c = \frac{dpf_y}{0.85\,\beta_1 f'_m}; \quad \text{for } \beta_1 = 0.85:$$

$$c = \frac{dpf_y}{0.7225 f'_m}$$

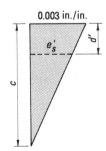

for $f_y = 60,000$ psi

$$c = \frac{83,045d\,p}{f'_m}$$

$$f'_s = 87,000\left[\frac{(83,045d\,p\,/\,f'_m) - d'}{(83,045d\,p\,/\,f'_m)}\right]$$

$$= 87,000\left(1 - \frac{d'f'_m}{83,045d\,p}\right)$$

$$= 87,000 - \frac{1,048d'f'_m}{d\,p}$$

The additional tension steel is based on the yield stress, f_y.

$$A_{s2} = \frac{T_2}{f_y}$$

The compression steel area is based on either f_y or f_s if it is below yield strain.

$$A'_s = \frac{M_2}{(d - d')f'_s} = \frac{C_2}{f'_s}$$

EXAMPLE 6-G Area of Tension and Compression Steel.

Given an 8" CMU beam with 32" total depth, $d = 26"$, $d' = 4"$ and subjected to a factored moment; M_u, of 150 ft kips. Determine the area of tension steel and compression steel if required. $f'_m = 2000$ psi, $f_y = 60,000$ psi, $\phi = 0.8$

Solution 6-G

$$M_u = \phi(M_1 + M_2)$$

$$150 = 0.8(M_1 + M_2)$$

$$M_1 + M_2 = 187.5 \text{ ft. kips}$$

The maximum allowable steel ratio is

$$0.5p_b = p_{max} = 0.0071$$

$$A_{s1} = pbd = 0.0071 \times 7.63 \times 26$$
$$= 1.4 \text{ sq. in.}$$

$$a = \frac{T_1}{0.85f'_m b} = \frac{1.4 \times 60,000}{0.85 \times 2000 \times 7.63}$$
$$= 6.48 \text{ inches}$$

$$M_1 = T_1\left(d - \frac{a}{2}\right)$$

$$= 1.4\left(\frac{60,000\left(26 - \dfrac{6.48}{2}\right)}{12,000}\right)$$

$$= 159.3 \text{ ft kips}$$

$$M_2 = M_n - M_1$$
$$= 187.5 - 159.3 = 28.2 \text{ ft kips}$$

Additional tension steel

$$A_{s2} = \frac{T_2}{f_y} = \frac{M_2}{(d-d')f_y}$$

$$= \frac{28.2 \times 12,000}{(26-4)60,000} = 0.26 \text{ sq. in.}$$

Tension steel $= A_1 + A_2$

$$= 1.4 + 0.26 = 1.66 \text{ sq. in.}$$

Use 2 - #8 ($A_s = 1.59$ sq. in.)

Compression steel

Check stress in compression steel

$$c = \frac{a}{0.85} = \frac{6.48}{0.85}$$

$$= 7.62 \text{ inches}$$

$$f_s' = 87,000\left(\frac{7.62-4}{7.2}\right)$$

$$= 41,330 \text{ psi}$$

The additional compression force C_2 is

$$M_2 = M_n - M_1 = 187.5 - 157.3 = 28.2 \text{ ft kips}$$

$$C_2 = \frac{M_2}{d-d'} = \frac{28.2 \times 12,000}{26-4}$$

$$= 15,382 \text{ lbs}$$

$$A_s' = \frac{C_2}{f_s'} = \frac{15,382}{41,330}$$

$$= 0.37 \text{ sq. in.}$$

Compression steel $= A_s'$

$$= 0.37 \text{ sq. in.}$$

Use 2 - #4 ($A_s = 0.40$ sq. in.)

6-4.C Strength Design for Combined Load and Moment.

Many walls are subjected to combined vertical loads and moments due to dead and live loads plus lateral forces either in plane or out of plane.

Accordingly, they are designed based on the parameters of strength design for factored loads, maximum allowable steel ratio and limitation of masonry strain.

6-4.C.1 Derivation for *P-M* Loading

The following derivation is based on simple statics by summing the moments and the vertical forces to equal zero.

Derivation:

$$C\left(\frac{l}{2}-\frac{a}{2}\right) + T\left(\frac{l}{2}-d_1\right) - M = 0$$

Sum of the moments about centroid of the load *P*.

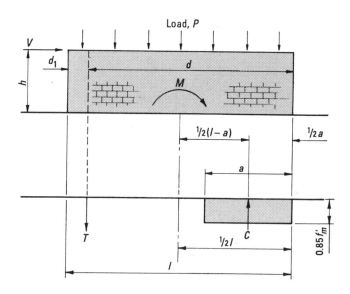

Figure 6-8 Shear wall with vertical and lateral load. Stress conditions shown.

Sum of the vertical forces

$$T = C - P$$

Substituting for *T*

$$C\left(\frac{l}{2}-\frac{a}{2}\right) + (C-P)\left(\frac{l}{2}-d_1\right) - M = 0$$

but

$$C = 0.85 f_m' a t$$

substituting for *C*

$$(0.85 f_m' a t)\left(\frac{l}{2}-\frac{a}{2}\right) + (0.85 f_m' a t - P)\left(\frac{l}{2}-d_1\right)$$

$$-M = 0$$

$$0.425 f_m' a t l - 0.425 f_m' a^2 t + 0.425 f_m' a t l$$

$$-0.85 f_m' a t d_1 - P\left(\frac{l}{2}-d_1\right) - M = 0$$

Change signs and combine terms

$$\underbrace{0.425 f_m' t a^2}_{\mathbf{a}} - \underbrace{0.85 f_m' t (l-d_1) a}_{\mathbf{b}} + \underbrace{P\left(\frac{l}{2}-d_1\right) + M}_{\mathbf{c}} = 0$$

Solving this quadratic equation for *a*

$$\mathbf{a} a^2 + \mathbf{b} a + \mathbf{c} = 0$$

Let $\mathbf{a} = 0.425 f_m' t$

$$\mathbf{b} = -0.85 f_m' t (l-d_1)$$

Note $(l-d_1) = d$

$$= 0.85 f_m' t d$$

$$\mathbf{c} = P\left(\frac{l}{2}-d_1\right) + M$$

Using the binomial formula to solve the quadratic equation

$$a = \frac{-b - \sqrt{b^2 - 4ac}}{2a}$$

$$a = \frac{0.85 f'_m t d - \sqrt{\left(-0.85 f'_m t d\right)^2 - 4\left(0.425 f'_m t\right)\left(P\left(\frac{l}{2} - d_1\right) + M\right)}}{2\left(0.425 f'_m t\right)}$$

Determining the size of the stress block a, calculate the compression force.

$$C = 0.85 f'_m a t$$

Determine the tension force

$$T = C - P$$

If the value is zero or negative, no tension steel is required. Use minimum steel as per code requirements.

Calculate the area of steel

$$A_s = \frac{T}{f_y}$$

EXAMPLE 6-H Shear wall design by strength methods; vertical load, overturning and shear.

A nominal 8 inch solid grouted concrete masonry shear wall carries a dead load of 4 kips/ft, live load of 1.5 kips/ft and a lateral force of 50 kips due to wind. $f'_m = 1500$ psi, $f_y = 60,000$ psi.

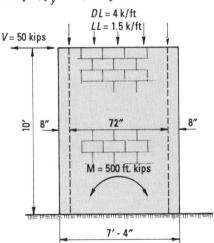

Determine the required tension and shear steel using factored loads and strength design procedures developed above.

Solution 6-H

Determine factored loads.

$$U = 0.9D + 1.3W \quad \text{(UBC Chapter 24, Equation 12-5)}$$

$$U = 0.9D = 0.9(4)7.33$$

$$= 26.4 \text{ kips}$$

Factored overturning moment

$$M = 1.3W = 1.3(50)10$$

$$= 650 \text{ ft kips}$$

Solve for length of stress block a

Determine the constants for the coefficients for the quadratic equations

$$\mathbf{a} = 0.425 f'_m t$$

$$= 0.425(1.15)7.63 = 4.9$$

$$\mathbf{b} = -0.85 f'_m t d$$

$$= -0.85(1.5)7.63(80) = -778$$

$$\mathbf{c} = P\left(\frac{l}{2} - d_1\right) + M$$

$$= 26.4\left(\frac{88}{2} - 8\right) + 650 \times 12$$

$$= 8750$$

Solve for length of stress block a

$$a = \frac{-b - \sqrt{b^2 - 4ac}}{2a}$$

$$a = \frac{-(-778) - \sqrt{(-778)^2 - 4(4.9)8750}}{2(4.9)}$$

$$= 12.2 \text{ inches}$$

Compression forces

$$C = 0.85 f'_m a t$$

$$= 0.85(1.5)12.2(7.63)$$

$$= 118.7 \text{ kips}$$

Tension force

$$T = C - P$$

$$= 118.7 - 26.4$$

$$= 92.3 \text{ kips}$$

Area of overturning tension steel

$$\phi = 0.65$$

$$A_s = \frac{T}{\phi f_y} = \frac{92.3}{0.65 \times 60}$$

$$= 2.4 \text{ sq. in.}$$

Use 2 - #10 bars each side ($A_s = 2.54$ sq. in.)

Shear design

$$V_n = V_m + V_s \quad \text{(UBC Chapter 24, Equation 12-13)}$$

Assume wall is in critical hinge area, all shear to be resisted by reinforcing steel

$$\phi = 0.60$$

Factored lateral load

$$U = 1.3 \times 50 = 65 \text{ kips}$$

$$V_n = \phi A_v F_y$$

$$A_v = \frac{65}{0.60 \times 10 \times 60} = 0.18 \text{ sq. in./ft}$$

Use #6 at 24 in. o.c.

$A_s = 0.22$ sq. in./ft

Out of plane reinforcing between OTM jamb steel

Use minimum steel $0.0013bt$.

$$A_s = 0.0013 \times 88 \times 7.63$$
$$= 0.87 \text{ sq. in.}$$

Use 1 - #5 bars between the #10 bars

$$A_s = 4(1.27) + 0.31 = 5.39 \text{ sq. in.}$$

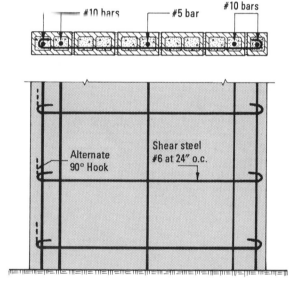

Figure 6-9 Layout of steel in shear wall.

6-5 TALL SLENDER WALLS

6-5.A General

In 1980 and 1981, the Structural Engineers Association of Southern California and the Southern California Chapter of the American Concrete Institute conducted a major research testing program to develop criteria for the design of tall, slender walls.

A total of 32 test panels were built with h'/t ranging from 30 to 57.

Panels were tested with a typical eccentric vertical roof load applied to a steel ledger at the top. Lateral pressure was applied through an air bag which loaded the wall for its full height and width.

Based on the test results, design techniques were developed and code requirements included in the UBC to reflect the performance of the walls in the test program.

This design criteria limits the deflection under service loads and requires ductile yield strength with factored loads. An acceptable design must satisfy both criteria.

Slender wall masonry panels ready to be tested.

6-5.B Slender Wall Design Requirements

The Uniform Building Code parameters for slender walls are as follows:

a) Vertical load stress is limited to a maximum of $0.04f'_m$.

b) Minimum nominal thickness of wall is 6 inches.

c) Maximum reinforcement ratio is $0.5p_b$.

d) Maximum lateral deflection due to service loads is $0.007h$.

e) The nominal resisting moment multiplied by the capacity reduction factor must equal or exceed the required factored moment.

f) Walls designed by strength design methods must have special inspection during construction.

g) Maximum usable strain in the masonry is 0.003 in./in.

h) Minimum $f'_m = 1500$ psi and maximum $f'_m = 4000$ psi.

6-5.B.1 Reinforcement

The balanced reinforcement ratio is calculated as shown in Section 6-4.A.1 using the equation:

$$p_b = \frac{0.85\beta f'_m}{f_y}\left(\frac{87,000}{87,000 + f_y}\right)$$

The maximum allowable steel ratio is $0.5p_b$.

6-5.B.2 Effective Steel Area

The vertical load on a wall acts as a reinforcing force and is therefore transformed into an equivalent steel area. The resulting effective steel area may be determined as:

$$A_{se} = \frac{P_u + A_s f_y}{f_y}$$

6-5.B.3 Nominal Moment Strength

The nominal moment strength, M_n, of the wall is determined based on the following formulas:

$$M_n = A_{se} f_y \left(d - \frac{a}{2} \right)$$

$$a = \frac{P_u + A_s f_y}{0.85 f'_m b}$$

6-5.C Design or Factored Strength of Wall Cross-Section

The design strength provided by a reinforced masonry wall cross section is computed as the nominal strength multiplied by a strength reduction force, ϕ, i.e.

$$M_u \le \phi M_n$$

and $\quad P_u \le \phi P_n$

Where

P_n = nominal axial strength for a cross section subjected to combined flexural and axial load

$P_u = P_{uw} + P_{uf}$

= factored wall load plus factored tributary floor or roof loads.

M_n = nominal moment strength for a cross section subjected to combined flexural and axial load.

M_u = factored moment on a section due to lateral loads and eccentric roof and wall loads.

ϕ = 0.8

6-5.C.1 Deflection Criteria

The mid-height deflection is limited so that a serviceable wall is designed. The maximum deflection permitted by the UBC is

$\Delta_s = 0.007h$ (UBC Chapter 24, Equation 11-10)

The maximum deflections allowed are thus directly proportional to the height of the wall.

This limitation was based on the capability of the wall to deflect elasticity to at least Δ_s and still rebound to its original vertical position. This recognizes that the wall will crack but will not impair its structural capacity. The SEAOSC/ACI committee recommended a deflection criteria of $0.01\,h$ but this was reduced when it was adopted by the UBC.

6-5.C.2 Deflection of Wall

Lateral and vertical service loads (unfactored) are used in computing the maximum horizontal deflection, which generally occurs at the mid-height of the wall.

Secondary moments induced by deflections at the mid-height of the wall are represented in the deflection calculation.

ϕ factors are not used in the deflection calculation since deflections results from unfactored loads and moments. The load-deflection relation for walls is assumed to follow a curve similar to the one in Figure 6-10.

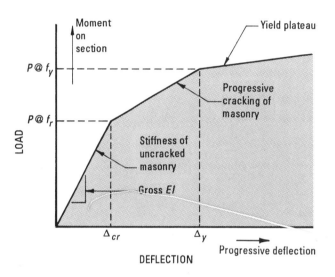

Figure 6-10 Load - deflection curve for a wall.

The slopes of the straight line parts of the load-deflection curve are as follows: (a) up to cracking load, the gross section moment of inertia, I_g, is used to compute deflection from the load; (b) additional deflection beyond the cracking load is computed using the cracked moment of inertia, I_{cr}.

The deflection of the wall at mid-height is determined by the following formula or an equivalent procedure.

Deflection at service load, Δ_s

$$\Delta_s = \frac{5 M_{cr} h^2}{48 E_m I_g} + \frac{5 \left(M_{ser} - M_{cr} \right) h^2}{48 E_m I_{cr}} \quad \begin{array}{l} \text{(UBC Ch. 24,} \\ \text{Eq. 11-12)} \end{array}$$

Where

M_{ser} = service moment on the masonry wall

M_{cr} = cracking moment strength of the masonry wall

$$M_{cr} = S f_r \quad \text{(UBC Chapter 24, Equation 11-13)}$$

S = section modulus

f_r = modulus of rupture as given in Table 6-2 and STR DES Table D-1.

TABLE 6-2   Modulus of Rupture of Masonry Walls Perpendicular to Bed Joint

	Fully Grouted	Partially Grouted
Hollow-unit Masonry	$4.0\sqrt{f'_m}$, 235 psi max.	$2.5\sqrt{f'_m}$, 125 psi max.
Two-Wythe Brick Masonry	$2.0\sqrt{f'_m}$, 125 psi max	Not allowed

Gross moment of inertia, I_g, Solid Grouted

$$I_g = \frac{bt^3}{12}$$

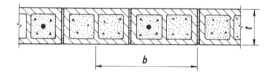

Cracked moment of inertia, I_{cr}

$$I_{cr} = nA_{se}(d-c)^2 + \frac{bc^3}{3}$$

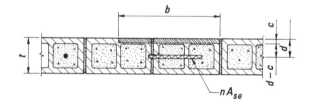

Distance to neutral axis, $c = \dfrac{a}{0.85}$

Service moment, M_s

$$M_s = \frac{wh^2}{8} + P_f\left(\Delta + \frac{e}{2}\right) + P_w\frac{\Delta}{2}$$

Where

w = unfactored lateral service load

P_f = unfactored load on the ledger from tributary floor or roof loads

e = eccentricity of the ledger load

P_w = unfactored weight of wall

Δ = deflection due to load and weight of wall

6-5.D   Determination of Moments at the Mid-Height of the Wall

The moment at the mid-height of the wall can be determined using statics. Consider the wall support and free body diagrams shown in Figure 6-11.

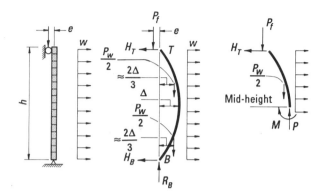

Figure 6-11   Wall support and free body diagrams.

The horizontal force at the roof line, H_t is found by summing moments about B.

Where

H_t = horizontal force at the roof line

w = lateral load acting on the wall

P_w = weight of the wall

P_f = load at the roof line

e = eccentricity of the roof load

By summing moments about the wall mid-height, the relation for mid-height moment, M, is obtained.

$$M = \frac{wh^2}{8} + \frac{P_w\Delta}{2} + P_f\Delta + \frac{P_f e}{2}$$

6-6   SLENDER WALL DESIGN EXAMPLE

6-6.A   General

The design example given below considers a partially grouted 8″ CMU wall is Seismic Zone No. 4. Because of partial grouting, the wall has lower lateral earthquake loads imposed on it as compared to a solid grouted wall.

The key to slender wall design is the assumption for the required steel reinforcing. The use of design aids as given in the Reference List will significantly reduce design time. Computer programs are also available which make slender wall design fast and simple.

EXAMPLE 6-I   Strength Design of wall, $h/t = 36.7$.

Using the slender wall design method given in UBC Sec. 2411, design the reinforcing steel and check the wall for compliance to service load deflection and factored strength requirements.

Given: 8″ nominal CMU wall, partially grouted in Seismic Zone No. 4. Special inspection is provided.

$\phi = 0.8$

$f'_m = 1500$ psi

$f_y = 60,000$ psi

The wall spans 23′ - 4″ between lateral supports. Roof load, $P_f = 500$ plf at an eccentricity of 7.3″

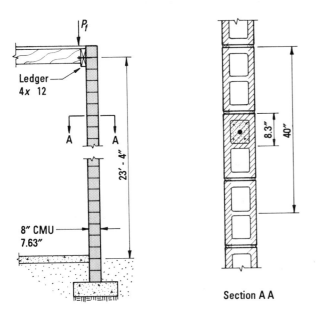

Figure 6-12 Slender wall cross-sections.

Section A A

Solution 6-I Using the $P\Delta$ Method

Assume steel is spaced 40″ o.c. and grouted only at steel, this is based on the estimating curves given in the references.

1. Loads

   a) Weight of wall: assume medium weight CMU grouted at 40″ o.c.; $Wt = 53$ psf from Table B-3

   $$P_w = \frac{53 \times 23.33}{2}$$
   $$= \frac{618 \text{ plf} \times 40}{12} = 2058 \text{ lbs}/40″$$

   Roof load $= \frac{500 \text{ plf} \times 40}{12} = 1667 \text{ lbs}/40″$

   Check axial load limitation

   $$A_g = (40 - 8.3)2 \times 1.25 +$$
   $$8.3 \times 7.63 = 142.6 \text{ sq. in.}$$

   $$\frac{P_w + P_f}{A_g} \le 0.04 f'_m \qquad \text{UBC Ch. 24, Eq. 11-1)}$$

   $$\frac{2058 + 1667}{142.6} < 0.04 \times 1500$$
   $$26.1 \text{ psi} < 60 \text{ psi O.K.}$$

b) Lateral load

   Seismic load $F_p = ZIC_pW_p$ (UBC Chapter 23 Equation 36-1)

   For Seismic Zone No. 4, $Z = 0.40$ from Table 3-11 (UBC Table 23-I)

   Assume building is a standard occupancy; $I = 1.0$.

   From Table 3-15 (UBC Table No. 23-P), $C_p = 0.75$ for walls

   $$F_p = 0.4 \times 1.00 \times 0.75 \times 53 \text{ psf}$$
   $$= 15.9 \text{ psf} = w_s$$

c) Factored loads; $U = 0.9D + 1.43E$

   Factored wall load

   $$P_{uw} = 0.9 \times 2058 = 1852 \text{ lbs}/40″$$

   Factored roof load

   $$P_{uf} = 0.9 \times 1667 = 1500 \text{ lbs}/40″$$

   Factored vertical loads

   $$P_u = P_{uf} + P_{uw}$$
   $$= 1500 + 1852$$
   $$= 3352 \text{ lbs}/40″$$

   Factored seismic load

   $$w_u = \frac{1.43 \times 15.9 \times 40}{12} = 75.8 \text{ lbs}/40″ \text{ width}$$

2. Assume vertical steel

   Maximum allowable steel ratio, $p = 0.5 \, p_b$

   From Table 6-1, $p_{max} = 0.0053$

   Assume #6 bars at 40″ o.c.

a) Gross steel ratio   (see Table C-9)

   $$p_g = \frac{A_s}{bt} = \frac{0.44}{40 \times 7.63}$$
   $$= 0.0014 < 0.0032 \quad (1988 \text{ UBC})$$

b) Structural steel ratio

   $$p = \frac{A_s}{bd} = \frac{0.44}{40 \times 3.81} = 0.0028 < 0.0053 \ (1991 \text{ UBC})$$

3. Determine E_m, n, f_r and I_g

a) Modulus of Elasticity, E_m

   $$E_m = 750 f'_m \quad \text{(see Table A-3)}$$
   $$= 750 \times 1500$$
   $$= 1,125,000 \text{ psi}$$

b) Modulus ratio, n (see Table A-3)

   $$n = \frac{E_s}{E_m}$$
   $$= \frac{29,000,000}{1,125,000} = 25.8$$

c) Modulus of rupture, f_r (see STR DES TableE-1)

$$f_r = 2.5\sqrt{f'_m} \quad \text{(UBC Sec. 2411(b)4)}$$

$$= 2.5\sqrt{1500} = 97 \text{ psi}$$

d) Gross moment of Inertia, I_g

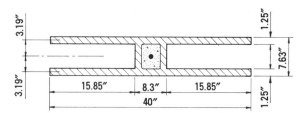

$$I_g = \frac{8.3 \times 7.63^3}{12} +$$

$$2\left(\frac{31.7 \times 1.25^3}{12} + 31.7 \times 1.25 \times 3.19^2\right)$$

$$= 307.2 + 2(5.2 + 403.2)$$

$$= 1124 \text{ in.}^4 / 40'' \text{ wide}$$

4. Moment at cracking, M_{cr}

$M_{cr} = Sf_r$ (UBC Chapter 24, Equation 11-13)

but $S = I/c$ where $c = t/2$

$$M_{cr} = \frac{2I_g f_r}{t} = \frac{2 \times 1124 \times 97}{7.63}$$

$$= 28,579 \text{ in. lbs}/40''$$

5. Cracked moment of Inertia

Calculate I_{cr} using the modular ratio, n, to transform the effective reinforcing steel into an equivalent area of masonry, and by using the expression:

$$I = \frac{bh^3}{12} + Ad^2$$

$$I_{cr} = \frac{(b - b_w)t_s^3}{12}$$

$$+ t_s(b - b_w)\left(c - \frac{t_s}{2}\right)^2 + \frac{b_w c^3}{3} + nA_{se}(d - c)^2$$

a)

$$A_{se} = \frac{P_u + A_s f_y}{f_y}$$

$$= \frac{3352 + 0.44 \times 60,000}{60,000}$$

$$= 0.50 \text{ sq. in.}/40''$$

b) Depth of rectangular stress block a

$$a = \frac{P_u + A_s f_y - 0.85 f'_m (b - b_w) t_s}{0.85 f'_m b_w}$$

$$= \frac{3352 + 0.44(60,000) - 0.85(1500)(40 - 8.3)1.25}{0.85(1500)8.3}$$

$$= -1.96$$

This results in a negative value. Therefore the stress block is completely in the shell.

$$a = \frac{P_u + A_s f_y}{0.85 f'_m b}$$

$$= \frac{3352 + 0.44(60,000)}{0.85(1500)40}$$

$$= 0.58 \text{ in.}$$

c) Distance to Neutral Axis, c

$$c = \frac{a}{0.85} = \frac{0.58}{0.85}$$

$$= 0.68''$$

Therefore — Cracked moment of inertia

$$I_{cr} = nA_{se}(d - c)^2 + \frac{bc^3}{3}$$

$$= 25.8 \times 0.50(3.81 - 0.68)^2 - \frac{40(0.68)^3}{3}$$

$$= 126.4 + 4.2$$

$$130.6 \text{ in.}^4 / 40'' \text{ width}$$

6. Calculate mid-height moment, M_s, and lateral deflection, Δ_s, due the service loads by iteration method

a) First iteration, assume $\Delta_s = 0''$

$$M_{s1} = \frac{wh^2}{8} + P_o\left(\frac{e}{2}\right) + (P_o + P_w)\Delta_s$$

$$= \frac{15.9 \times \left(\frac{40}{12}\right) \times 23.33^2 \times 12}{8} + 1667\left(\frac{7.3}{2}\right) +$$

$$(1667 + 2058)0$$

$$= 43,271 + 6085 + 0$$

$$= 49,356 \text{ in. lbs}/40'' > M_{cr}$$

$$\Delta_{s1} = \frac{5M_{cr}h^2}{48E_m I_g} + \frac{5(M_{ser} - M_{cr})h^2}{48E_m I_{cr}} \quad \begin{array}{l}\text{(UBC}\\\text{Ch. 24,}\\\text{Eq. 11-12)}\end{array}$$

$$= \frac{5 \times 28,579(23.33 \times 12)^2}{48 \times 1,125,000 \times 1124} +$$

$$\frac{5(49,356 - 28,579)(23.33 \times 12)^2}{48(1,125,000 \times 130.6)}$$

$$= 0.185 + 1.155$$

$$= 1.34''$$

b) Second iteration; $\Delta_s = 1.34''$

$$M_{s2} = 43,271 + 6085 + (1667 + 2058)1.34$$
$$= 54,347 \text{ in. lbs} / 40'' \text{ width}$$

$$\Delta_{s2} = 0.185 + \frac{5(23.33 \times 12)^2 (54,348 - 28,579)}{48 \times 1,125,000 \times 130.6}$$
$$= 0.185 + 5557 \times 10^{-5} (25,769)$$
$$= 0.185 + 1.432 = 1.617''$$

c) Third iteration $\Delta_s = 1.617''$

$$M_{s3} = 43,271 + 6085 + 3725(1.617)$$
$$= 55,379 \text{ in. lbs}$$

$$\Delta_{s3} = 0.185 + 5.557 \times 10^{-5}(55,379 - 28,579)$$
$$= 0.185 + 1.489$$
$$= 1.67''$$

d) Convergence based on deflection

$$\left(\frac{\Delta_{s3} - \Delta_{s2}}{\Delta_{s3}}\right)100 = \left(\frac{1.67 - 1.617}{1.67}\right)100$$
$$= 3.2\% \text{ Satisfactory}$$

7. Check lateral deflection allowance at service load

Allowable $\Delta_s = 0.007h$
$$= 0.007(23.33 \times 12)$$
$$= 1.96''$$

Actual $\Delta_{s3} = 1.67'' < 1.96''$ o.k.

The service load deflection of 1.67″ is less than the maximum allowable deflection of 1.96″. Therefore the deflection criteria is satisfied.

8. Strength calculation — based on a 40″ width

Calculate mid-height moment under factored loads

$$M_u = \frac{w_u h^2}{8} + P_{uf}\left(\frac{e}{2}\right) + P_u \Delta_u \quad \begin{array}{l}\text{(UBC Ch. 24,}\\\text{Eq. 11-7,}\\\text{Revised)}\end{array}$$

a) First iteration; Assume $\Delta_u = 0$

$$M_{u1} = \frac{75.8 \times 23.3^2 \times 12}{8} + \frac{1500 \times 7.3}{2} + (3352)0$$
$$= 61,727 + 5475 + 0$$
$$= 67,202 \text{ in lbs} / 40'' \text{ width}$$

$$\Delta_{u1} = \frac{5M_{cr}h^2}{48E_m I_g} + 5\left(\frac{\left(M_{u1} - M_{cr}\right)h^2}{48E_m I_{cr}}\right)$$

$$= \frac{5 \times 28,579(23.33 \times 12)^2}{48(1,125,000 \times 1124)} +$$

$$\frac{5(67,202 - 28,579)(23.33 \times 12)^2}{48 \times 1,125,000 \times 130.6}$$

$$= 0.185 + 5.56 \times 10^{-5}(67,202 - 28,579)$$
$$= 0.185 + 2.155$$
$$= 2.33''$$

b) Second iteration $\Delta_{u1} = 2.33''$

$$M_{u2} = 61,727 + 5475 + 3352(2.33)$$
$$= 75,012 \text{ in. lbs}/40'' \text{ width}$$

$$\Delta_{u2} = 0.185 + 5.56 \times 10^{-5}(75,012 - 28,579)$$
$$= 0.185 + 2.580$$
$$= 2.76''$$

c) Third iteration $\Delta_{u2} = 2.76''$

$$M_{u3} = 61,727 + 5,475 + 3,352(2.76)$$
$$= 76,470 \text{ in. lbs}/40'' \text{ width}$$

$$\Delta_{u3} = 0.185 + 5.56 \times 10^{-5}(76,470 - 28,579)$$
$$= 0.185 + 2.660$$
$$= 2.85''$$

d) Convergence based on deflection

$$\left(\frac{\Delta_{u3} - \Delta_{u2}}{\Delta_{u3}}\right) \times 100 = \left(\frac{2.85 - 2.76}{2.85}\right) \times 100$$
$$= 3.0\% \text{ Satisfactory}$$

9. Determine nominal strength of wall, M_n

$$M_n = 0.85 f'_m ab\left(d - \frac{a}{2}\right)$$

$$= 0.85 \times 1500 \times 0.58 \times 40\left(3.81 - \frac{0.58}{2}\right)$$

$$= 104,122 \text{ in. lbs} / 40''$$

$$\phi M_n = 0.8(104,122)$$
$$= 83,297 \text{ in. lbs} / 40''$$

$$\phi M_n > M_u$$

83,297 in. lbs > 76,470 in. lbs

Therefore the section is adequate for strength.

6-6.B Alternate Method of Moment Distribution

Moment and deflection calculations shown in Example 6-I are based on UBC Sec. 2411 which assumes simple support conditions top and bottom with the maximum moment and deflection occurring at mid-height.

Other support and fixity conditions may be used and the resulting moments and deflections may be calculated using established principles of mechanics.

For instance, assume a wall is fully fixed at the bottom and designed as a pinned cantilever.

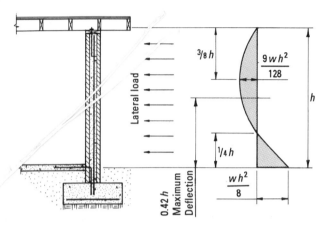

Figure 6-13 Slender wall fixed at bottom and pinned at top.

Under a uniform pressure, w, the moment at the base of the wall is $wh^2/8$. For this case the point of zero moment occurs at $0.25h$ and the maximum moment in wall is $9wh^2/128$ which occurs at $5/8h$. The maximum deflection occurs at $0.4215h$ from the top and is determined by the equation

$$\Delta_{max} = \frac{wh^4}{185EI}$$

This deflection is less than that of a simple span which is $5wh^4/48EI$ or about nineteen times as great.

Using this method the lower section of the wall can be reinforced for the maximum moment while significantly less reinforcing steel is required in the upper part of the wall.

6-7 STRENGTH DESIGN OF SHEAR WALLS

6-7.A General

Load bearing masonry walls support vertical and lateral loads. These loads create an interaction of load and moment on a wall. The strength design techniques for this condition are outlined in Section 2412 of the Uniform Building Code.

It provides appropriate load factors to be used and it prescribes the conditions for reinforcing, the hinge region and the required confinement for overturning steel.

Strength design procedures for shear walls allow masonry and reinforcing steel to resist shear forces even when the shear stress exceeds the capacity of the masonry. This is only permitted in the portion of the shear wall above the potential hinge region. In the hinge region, which is defined by the base of the shear wall and a plane at a height equal to the length of the wall, the shear must be entirely resisted by the reinforcing steel.

C. All continuous reinforcement shall be anchored or spliced in accordance with Section 2409(e)1, 2, 3A (with $f_s = 0.5 f_y$), 3B, 3D, 3F and 3G (with $F_s = f_y$).

D. The minimum amount of vertical reinforcement shall not be less than one half the horizontal reinforcement.

E. Maximum spacing of horizontal reinforcement within the region defined in Section 2412(c)6 C(i) shall not exceed three times nominal wall thickness or 24 inches, whichever is less.

5. Axial strength. The nominal axial strength of the shear wall supporting axial loads only shall be calculated by Formula (12-11).

$$P_o = 0.85 f_m'(A_e - A_s) + f_y A_s \qquad (12\text{-}11)$$

Axial design strength provided by the shear wall cross section shall satisfy Formula (12-12).

$$P_u \leq \phi(0.80)P_o \qquad (12\text{-}12)$$

6. Shear strength. Shear strength shall be as follows:

A. The nominal shear strength shall be determined using either Section 2412(c) 6 B or C. Table No. 24-N gives the maximum nominal shear strength values.

B. The nominal shear strength of the shear wall shall be determined from Formula (12-13), except as provided in Section 2412(c)6 C.

$$V_n = V_m + V_s \qquad (12\text{-}13)$$

Where:

$$V_m = C_d A_{mv} (f_m')^{1/2} \qquad (12\text{-}14)$$

and

$$V_s = A_{mv} \rho_n f_y \qquad (12\text{-}15)$$

C. For a shear wall whose nominal shear strength exceeds the shear corresponding to development of its nominal flexural strength two shear regions exist.

(i) For all cross sections within the region defined by the base of the shear wall and a plane at a distance L_w above the base of the shear wall, the nominal shear strength shall be determined from Formula (12-16)

$$V_n = A_{mv} \rho_n f_y \qquad (12\text{-}16)$$

The required shear strength for this region shall be calculated at a distance $L_w/2$ above the base of the shear wall but not to exceed one-half story height.

(ii) For the other region, the nominal shear strength of the shear wall shall be determined from Formula (12-13).

7. Boundary member. Boundary members shall be as follows:

A. The need for boundary members at boundaries of the shear wall shall be determined using either Section 2412(c)7 B or C.

B. Boundary members shall be provided when the failure mode is flexure and when the maximum extreme fiber stress, corresponding to factored forces, exceeds $0.2 f_m'$. The boundary member may be discontinued where the calculated compressive stress is less than $0.15 f_m'$. Stresses may be calculated for the factored forces using a linearly elastic model and gross-section properties.

C. Boundary members shall be provided to confine all vertical reinforcement whose corresponding masonry compressive stress, corresponding to factored forces, exceeds $0.4 f_m'$ when the failure mode is flexure.

D. The minimum length of the boundary member shall be three times the thickness of the wall.

E. Boundary members shall be confined with a minimum of No. 3 bars at a maximum of 8-inch spacing or equivalent within the grouted core and within the region defined by the base of the shear wall and a plane at a distance L_w above the base of the shear wall.

6-8   DESIGN EXAMPLE — SHEAR WALL

This section provides a detailed design example based on strength design requirements of UBC Section 2412.

It is suggested that a shear wall computer program be used to estimate the location of the neutral axis and determination of stresses, loads and moments. The Concrete Masonry Association of California and Nevada, 6060 Sunrise Drive, Suite 1875, Citrus Heights, CA 95610, has a program available.

EXAMPLE 6-J   Strength Design of a Shear Wall.

Determine the reinforcing steel for the overturning moment, load and shear force on the solid grouted 8 inch concrete masonry wall shown. Verify that the wall meets the requirements of UBC Section 2412.

V_s = 100 kips   (Earthquake Load)

P_s = 200 kips   (Dead Load)

M_s = 1100 ft kips   (Earthquake Load)

Wall properties

8″ CMU = 7.625 inches actual

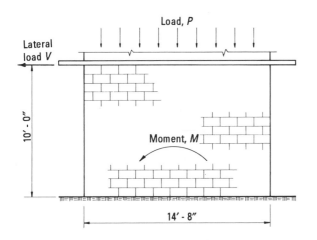

Figure 6-14 Masonry shear wall subjected to combined loading and moment.

Given $f'_m = 3000$ psi; $f_y = 60,000$ psi

From Table A-3,
$$E_m = 2,250,000 \text{ psi};   n = 12.9$$

From STR DES Table E-1:

Modulus of rupture = $4\sqrt{f'_m} = 219$ psi

Maximum usable masonry strain, $E_m = 0.003$ in./in.

Load factors

$$U = 1.4D + 1.7L$$

$$U = 1.4D + 1.7L + 1.4E$$

$$U = 0.9D + 1.4E$$

Strength reduction factors, ϕ

$\phi = 0.65$ Axial load only

$\phi = 0.65$ Axial load and moment

$\phi = 0.80$ Moment only

$\phi = 0.60$ Shear

Estimate vertical steel requirement for overturning moment.

$$A_s = \frac{M}{d f'_y} = \frac{1100 \times 1.4 \times 12}{14 \times 12 \times 60} = 1.8 \text{ sq. in.}$$

Try 8 - # 6 bars ($A_s = 3.52$ sq. in.)

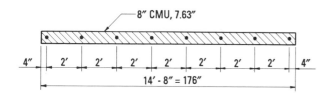

Figure 6-15 Distribution of steel in wall.

Analyze the shear wall by:

1. Plotting the interaction diagram for the wall.

2. Determining the cracking moment, $M_n \geq M_{cr}$.

3. Checking loading conditions for vertical load and moment.

4. Checking the requirements for boundary members and confinement.

5. Determining the shear reinforcing.

6. Comparing the design to wall designed by the working stress method.

Solution 6-J

1. Plot interaction diagram

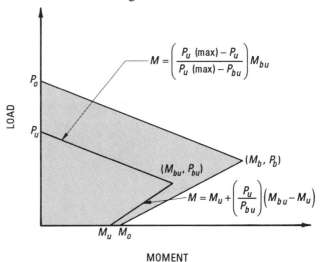

Figure 6-16 Simplified generic interaction diagram.

Where

P_o = Nominal axial strength

P_u = ϕ times the nominal axial strength or the factored axial load on the wall

M_o = Nominal moment strength

M_u = ϕ times the nominal moment strength or the factored moment on the wall

P_b = Balanced axial strength

P_{bu} = ϕ times the balanced axial strength

M_b = Balanced moment strength

M_{bu} = ϕ times the balanced moment strength

a)   Nominal axial load P_o

$$P_o = 0.85 f'_m (A_e - A_s) + f_y A_s \quad \text{(UBC Ch. 24, Eq. 12-11)}$$

$$= 0.85(3)\,[(14.67 \times 12 \times 7.625) - (8 \times 0.44)] + 60(8 \times 0.44)$$

$$= 3413 + 211 = 3624 \text{ kips}$$

b) Factored axial load, P_u

P_u = 1.4 × 200 = 280 k

Check $P_u \leq \phi(0.8)P_o$

280 = 0.65(0.80)3624

280 ≤ 1885 kips   O.K.

c) Nominal moment strength, M_o

Solve for location of the neutral axis (NA) so that sum of vertical forces equals zero.

Assume location for NA; c = 10 inch.

Use maximum allowable CMU strain = 0.003.

Solution by iteration.

Take the sum of the moments about the extreme compression fiber at the end of wall.

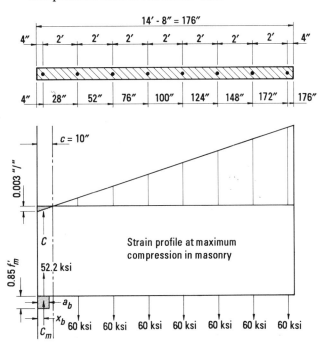

Figure 6-17 Steel location, strain condition and force equilibrium diagrams.

a = Depth of compression stress block

= 0.85c = 0.85(10) = 8.5 inch

x = 0.5a = 4.25 inch

Tension force, $T = A_s f_y$
= 0.44 (7) (60)
= 185 kips

Compression force = $A_s f_s + 0.85 f'_m b a$

= 0.44(52.2) +
0.85(3)(7.625)(8.5)

= 23 + 165 = 188 kips

T – C = 185 – 188 = – 3 kips

Close enough use c = 10 in.

Nominal bending moment, M_o

Sum of moments about left edge of wall

$M_o = T$ (moment arm) – C (moment arm)

= $A_s f_s$(moment arm) – [0.85$f'_m b a_b$(moment arm)
+ $A_s f_y$(moment arm)]

= 0.44[60(28 + 52 + 76 + 100 +
124 +148 + 172)] –
[0.85(3)7.625(10)(5) + 0.44(52.2)(4)]

= 18,480 – 972 – 92

= 17,416 in. kips = 1451 ft kips

d) Design bending moment, M_u

$M_u = \phi M_o = 0.80(1451)$

= 1161 ft kips

e) Nominal balanced design axial load, P_b

Compression capacity, $C_m = 0.85 f'_m b a_b$

Where balanced stress block, $a_b = 0.85c$

$$c_b = \left(\frac{e_{mu}}{e_{mu} + \dfrac{f_y}{E_s}} \right) d$$

$$= \left(\frac{0.003}{0.003 + \dfrac{60,000}{29,000,000}} \right) d = 0.592d$$

$c_b = 0.592(172) = 102''$ Neutral axis for balanced design

$a_b = 0.85c_b = 0.85(0.592)d$

= 0.503d

$a_b = 0.503(172)$

= 86.5 inches, depth of compression stress block

Tension force = $T = A_s f_s$

= 0.44(18.9 + 39.4 + 60)

= 0.44(118.3) = 52 kips

Compression force $C = A_s f_s + 0.85 f'_m b a_b$

= 0.44(1.7 + 22.2 +
42.6 + 60 + 60) +
0.85(3)(7.625)(86.5)

= 82 + 1682 = 1764 kips

Sum of vertical forces

$P_b = C - T$
= 1764 – 52
= 1712 kips

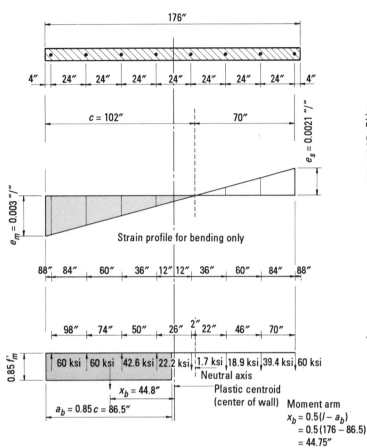

Figure 6-18 Balanced design load condition.

f) Design balanced axial load, P_{bu}

$$P_{bu} = \phi P_b$$
$$= 0.65(1712) = 1113 \text{ kips}$$

g) Nominal balanced design moment strength, M_b

Take moments about plastic centroid which is the center of the wall as it is symmetrical for masonry and steel

$$M_b = A_s f_s (\text{moment arm}) + 0.85 f'_m a_b b x_b$$

$$= 0.44[60(84) + 39.4(60) + 18.9(36) -$$
$$1.7(12) + 22.2(12) + 42.6(36) + 60(60) +$$
$$60(84)] + 0.85(3)(86.5)(7.625)(44.8)$$

$$= 0.44(18,504) + 75,348 = 8,142 + 75,348$$

$$= 83,490 \text{ in. kips} = 6958 \text{ ft kips}$$

h) Design balanced moment strength, M_{bu}

$$M_{bu} = \phi M_b$$
$$= 0.65(6958) = 4522 \text{ ft kips}$$

i) Plot the interaction diagram

2. Cracking moment, M_{cr}

Using gross section properties and linear elastic theory:

$$f_r = \frac{M_{cr}}{S} - \frac{P}{A}$$

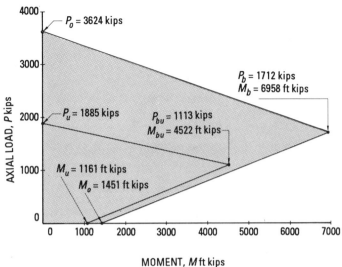

Figure 6-19 Interaction diagram for wall, Example 6-J for the assumed reinforcing steel.

Where

A = area of cross-section, bl
$= 7.625 \times 176 = 1342$ sq. in.

S = section modulus = $\dfrac{bl^2}{6}$

$= \dfrac{7.625 \times 176^2}{6} = 39,365$ in.^{3}

$f_r = 4\sqrt{f'_m} = 4\sqrt{3000} = 219$ psi

P = dead load = 200 kips

$$M_{cr} = S\left(\frac{P}{A} + f_r\right)$$

$$= 39,365\left(\frac{200,000}{1342} + 219\right)\frac{1}{1000}$$

$$= 14,488 \text{ in. kips} = 1207 \text{ ft kips}$$

3. Analyze two loading conditions for combined loading, vertical load and moment

a) Load condition 1

$$U = 1.4DL + 1.7LL + 1.4E$$
$$= 1.4(200) + 1.7(0) + 1.4(1100)$$
$$= 280 \text{ kips} + 0 + 1540 \text{ ft kips}$$

$$P_u = 280 \text{ kips} < P_{bu} = 1113 \text{ kips}$$

Determine the nominal moment strength, M_n, for $P_u = 280$ k

$$\frac{P_{bu}}{M_{bu} - M_u} = \frac{P_u}{M_x}$$

$$M_x = \left(\frac{P_u}{P_{bu}}\right)(M_{bu} - M_u)$$

$$M_n = M_u + M_x$$

$$= 1161 + \left(\frac{280}{1113}\right)(4522 - 1161)$$

$$= 2006 \text{ ft kips}$$

Check ratio of nominal moment to cracking moment

$$\frac{M_n}{M_{cr}} = \frac{2006}{1207} = 1.7$$

The nominal moment is 70% greater than the cracking moment.

b) Load condition 2

$$U = 0.9D + 1.4E$$

$$= 0.9(200) + 1.4(1100)$$

$$= 180 \text{ kips (vertical load)} +$$
$$1540 \text{ ft kips (seismic moment)}$$

Use factored load and compare to value on interaction diagram.

From interaction diagram

$$P_u = 180 \text{ kips} < P_{bu} = 1113 \text{ kips}$$

Nominal moment strength, M_n for $P_u = 180$ k

$$M_n = M_u + \left(\frac{P_u}{P_{bu}}\right)(M_{bu} - M_u)$$

$$= 1161 + \left(\frac{180}{1113}\right)(4522 - 1161)$$

$$= 1705 \text{ ft kips}$$

Check ratio of nominal moment to cracking moment

$$\frac{M_n}{M_{cr}} = \frac{1705}{1207} = 1.4$$

The nominal moment is 40% greater than the cracking moment

4. Check requirements for boundary and confinement conditions

a) Check for stress in masonry, $f_m > 0.2 f_m'$

Assume linear elastic conditions and an uncracked section

$$f_m = \text{maximum extreme compression fiber stress}$$

$$f_m = \frac{P_u}{A} + \frac{M_u}{S}$$

$$= \frac{280,000}{1342} + \frac{(1100 \times 1.4) \times 12,000}{39,365}$$

$$= 209 + 469 = 678 \text{ psi} > 0.2 f_m' < 0.4 f_m'$$

$$0.2 f_m' = 600 \text{ psi}$$
$$0.4 f_m' = 1200 \text{ psi}$$

As f_m is greater than $0.2 f_m'$ (UBC Sec. 2412(c)7) use boundary members to confine outside elements

b) Loading condition 1

$$U = 1.4D + 1.7L + 1.4E$$

$$P_u = 1.4(200) = 280 \text{ kips}$$

$$M_u = 1.4(1100) = 1540 \text{ ft kips}$$

Assume $c = 68$ inches

$$C - T = P = 280 \text{ kips}$$

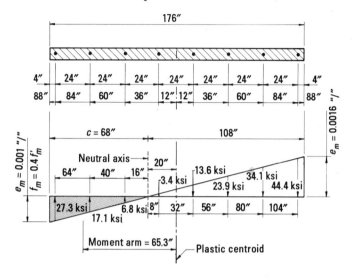

Figure 6-20 Stresses due to loading condition 1.

$$T = A_s f_s = 0.44(3.4 + 13.6 + 23.9 + 34.1 + 44.4)$$

$$= 0.44(119.4) = 52.5 \text{ kips}$$

$$C = A_s f_s + 0.4 f_m' b \times (\tfrac{1}{2} \times c)$$

$$= 0.44(6.8 + 117.1 + 27.3) +$$
$$0.4(3)(7.625)(\tfrac{1}{2} \times 68)$$

$$= 22.5 + 311.1 = 334 \text{ k}$$

$$C - T = 334 - 53 = 281 \text{ kips}$$

$$P_u = 280 \text{ kips; O.K. for location of neutral axis}$$

To determine the moment corresponding to $0.4 f_m'$, take moments about the plastic centroid.

If the moment, M_u, is more than the moment corresponding to $0.4 f_m'$, confinement of the vertical steel is required.

$$\text{Moment} = A_s f_s \text{ (moment arm)} +$$
$$0.4 f_m' (\tfrac{1}{2} \times c)b \text{ (moment arm)}$$

$$= 0.44[(13.6)(12) + 23.9(36) + 34.1(60)$$
$$+ 44.4(84) - 3.4(12) + 6.8(36)$$
$$+ 17.1(60) + 27.3(84)]$$
$$+ 0.4(3)(\tfrac{1}{2} \times 68)(7.625)(65.3)$$

$$= 4542 + 20,315$$

$$= 24,857 \text{ in. kips} = 2071 \text{ ft kips}$$

$$M_u = 1540 \text{ ft kips} < 2071 \text{ ft kips}$$

Total confinement of vertical steel NOT required.

c) Loading condition 2

$$U = 0.9D + 1.4E$$

$$P_u = 0.9\,(200) = 180 \text{ kips}$$

$$M_u = 1.4\,(1100) = 1540 \text{ ft kips}$$

$$C - T = P_u = 180 \text{ kips}$$

Assume neutral axis = 54 inches

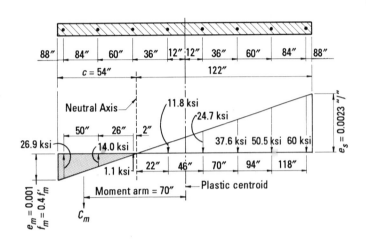

Figure 6-21 Stress due to loading condition 2.

$$T = A_s f_s = 0.44\,(11.8 + 24.7 + 37.6 + 50.5 + 60)$$
$$= 0.44\,(184.6) = 81.2 \text{ k}$$

$$C = A_s f_s + 0.4 f'_m b \times ({}^{1}/_2 \times c)$$

$$= 0.44\,(1.1 + 14 + 26.9) +$$
$$0.4\,(3)\,(7.625)\,(0.5)\,54$$

$$= 18.5 + 247.1 = 265.5$$

$$C - T = 265.5 - 81.2 = 184.3 \text{ k}$$

$$P_u = 180 \text{ kips; O.K. for location of neutral axis}$$

d) Determine moment for loading condition 2

Sum of moments about plastic centroid

$$\text{Moment} = A_s f_s \,(\text{moment arm}) +$$
$$(0.4 f'_m)({}^{1}/_2 \times c)\,b \,(\text{moment arm})$$

$$= 0.44\,[(60)\,(84) + 50.5\,(60) + 37.6\,(36$$
$$+ 24.7\,(12) - 11.8\,(12) + 1.(36) +$$
$$14\,(60) + 26.9\,(84)] +$$
$${}^{1}/_2\,(0.4)\,(3)\,(54)\,(7.625)\,(70)$$

$$= 5596 + 17,294 = 22,897 \text{ in. kips}$$

$$= 1907 \text{ ft kips}$$

$$M_u = 1540 \text{ ft kips} < 1907 \text{ ft kips}$$

Confinement of all vertical steel is NOT required.

Use confinement devices in boundary elements for 32″ on each side at 8″ vertical spacing.

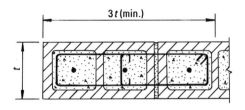

a) *#3 Confinement ties spaced at 8″ vertically.*

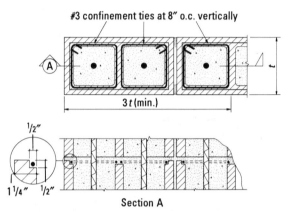

b) *#3 Confinement ties at 8″ o.c. vertically. (Detail of confinement ties used on the 28 story Excalibur Hotel, Las Vegas, Nevada.)*

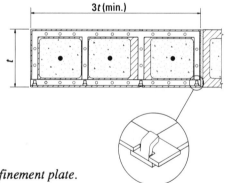

c) *Confinement plate.*

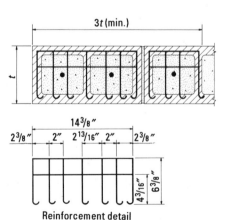

d) *Open wire mesh bedjoint reinforcement.*

Figure 6-22 Confinement devices for masonry boundary members.

5. Shear Design

a) Shear requirement

$$V_u = 1.4V_s$$
$$= 1.4(100) = 140 \text{ kips}$$

b) Shear strength of masonry only

$$V_n = V_m + V_s \quad \text{(UBC Ch. 24, Eq. 12-13)}$$

$$V_s = 0$$

$$V_m = C_d A_{mv} \sqrt{f'_m} \quad \text{(UBC Ch. 24, Eq. 12-14)}$$

C_d is proportional to M/Vd

$$d = 14.67' - 4'' = 14.3 \text{ ft}$$

$$M_u = 1.4(1100) = 1540 \text{ ft kips}$$

$$\frac{M_u}{V_{ud}} = \frac{1540}{140 \times 14.3} = 0.77$$

$$A_{mv} = l \times b = 176 \times 7.63 = 1342 \text{ sq. in.}$$

From STR. DES. Table and Diagram E-1

for $M/Vd = 0.77$; $C_d = 1.58$

or $C_d = 1.6(1 - M/Vd) + 1.2$
$$= 1.6(1 - 0.77) + 1.2 = 1.57$$

$$V_m = 1.57 \times 1342\sqrt{3000}$$
$$= 115.4 \text{ kips}$$

$$\phi V_m = 0.60 \times 115.4 = 69.2 \text{ kips}$$

$$V_u = 140 \text{ kips} > 69.2 \text{ kips}$$

Shear reinforcement is required

c) Design shear reinforcement to resist all the shear load since it is in the potential plastic hinge area

$$\phi V_n = \phi V_s; \quad V_m = 0$$
$$\phi V_s \geq 140 \text{ kips}$$
$$V_s = \phi A_{mv} p_n f_y \quad \text{(UBC Ch. 24, Eq. 12-15)}$$

$$p_n = \frac{V_s}{\phi A_{mv} f_y} = \frac{140}{0.60 \times 1342 \times 60}$$
$$= 0.0029$$

$$A_{sv} = p_n bh$$
$$= 0.0029 \times 7.625 \times 12$$
$$= 0.265 \text{ sq. in./ft}$$

Use #6 bars at 16" o.c.

$$A_{sv} = 0.33 \text{ sq. in./ft} > 0.265 \text{ sq. in./ft}$$

6. Alternate Design Using Working Stress Method

Special inspection provided - Use full stresses

Moment, $M = 1100$ ft kips

Load, $P = 200$ kips

Lateral shear, $V = 100$ kips

$f'_m = 3000$ psi, $F_s = 24,000$ psi, $n = 12.9$

Length of wall = 176" Assume $d = 164''$

a) Use unity equation

$$\frac{f_a}{F_a} + \frac{f_b}{F_b} \leq 1.33 \quad \text{(lateral seismic load)}$$

$$f_a = \frac{P}{A} = \frac{200,000}{7.63 \times 176}$$
$$= 149 \text{ psi}$$

$$F_a = 0.2f'_m \left[1 - \left(\frac{h'}{42t}\right)^3\right]$$

From Tables Q-1 and Q-2 find $F_a = 568$ psi.

$$f_b = \left(1.33 - \frac{f_a}{F_a}\right)F_b$$
$$= \left(1.33 - \frac{149}{568}\right)600$$
$$= 640 \text{ psi}$$

$$\frac{2}{jk} = \frac{bd^2 f_b}{M}$$
$$= 7.63 \times 164^2 \times \frac{640}{1100} \div 12,000$$
$$= 9.950$$

From Table E-9 for $2/jk = 9.950$

$$np = 0.030$$

$$npj = \frac{nM}{bd^2 f_s}$$
$$= \frac{12.9 \times 1100 \times 12,000}{7.63 \times 164^2 \times 32,000}$$
$$= 0.0259$$

From Table E-9 for $npj = 0.0259$

$$np = 0.028$$

Use $np = 0.030$ — masonry controls

$$p = \frac{np}{n} = \frac{0.030}{12.9}$$
$$= 0.0023$$

$$A_s = pbd = 0.0023 \times 7.63 \times 164$$
$$= 2.91 \text{ sq. in.}$$

Use 3 - #9 bars on each side

Use #5 bars at 48" o.c. in between

b) Shear stress, $f_v = \frac{V}{bjd} = \frac{100,000}{7.63 \times 0.95 \times 164}$
$$= 84 \text{ psi}$$

$$\frac{M}{Vd} = \frac{h}{d} = 10 \times \frac{12}{164} = 0.73$$

$$f_v = \frac{1}{2}\left(4 - \frac{M}{Vd}\right)\sqrt{f'_m}$$

From Table A-6:

$$f_v = 89.5 \text{ psi} \times \frac{4}{3}$$

$$= 119 \text{ psi} \quad \text{Maximum allowable for reinforced shear}$$

From Table K-2b:

Try #6 bars; $A_s = 0.44$ sq. in.

Use #6 bars at 16" o.c.

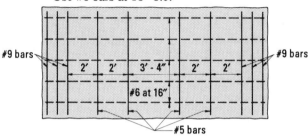

Figure 6-23 Shear wall reinforcing steel layout, design based on W.S.D.

6-9   WALL FRAMES

6-9.A   General

Masonry walls are generally considered solid elements with few openings.

Figure 6-24 Shear walls with few small openings.

As the openings in the walls are increased in size, a system of vertical load carrying elements (columns) and horizontal spandrel elements (beams) is created. As the proportions of the piers and connecting elements are changed, the system approaches the concept of a building wall frame.

Research has been conducted by Dr. Nigel Priestly at the University of Canterberry in Christ Church, New Zealand and at the University of California, San Diego proving the capability of masonry wall frames. As a result of this research, requirements have been formulated and are under consideration for inclusion into the building code.

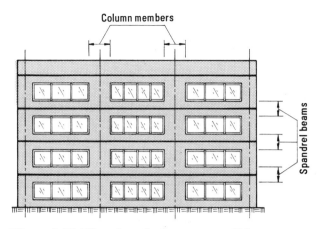

Figure 6-25 Elevation of a four story wall frame building.

Masonry wall frames have demonstrated their ability to transmit shear and moment. They function in a ductile manner when properly proportioned and detailed. The system must be under reinforced based on strength design requirements and the concept of a strong column and weak beam is used. This is to insure a ductile mechanism forming in the beam and maintaining a strong column to support vertical load.

The masonry frame must be solid grouted using open-end concrete or clay units. Special inspection is required.

Given below is a restatement of the proposed requirements by the Structural Engineers Association of California Seismology Committee for reinforced ductile masonry wall frames with some drawings to explain these requirements.

6-9.B   Proportion Requirements

Shown in Figure 6-26 are the proportional requirements for the spandrel beam.

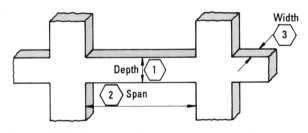

Figure 6-26 Horizontal spandrel beam framing member.

$\langle 1 \rangle$ Depth of spandrel, horizontal beam between columns not less that 32 inches or four masonry units which ever is greater. The nominal depth to width ratio shall be 6 or less.

$\langle 2 \rangle$ The clear span for the beam shall be three times its depth or more.

$\langle 3 \rangle$ The nominal width shall be 8 inches or $1/26$ of the clear span whichever is greater.

The pier or vertical column proportional requirements are shown in Figure 6-27.

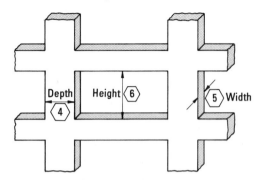

Figure 6-27 Vertical column/pier framing member.

⟨4⟩ The nominal depth of the column should not be more than 96 inches nor less than 32 inches.

⟨5⟩ The nominal width of the column shall not be less than the nominal width of the beam and not less than eight inches or $^1/_{14}$ of the clear height between beam faces whichever is greater.

⟨6⟩ The clear height to depth ratio should not exceed five.

6-9.C Analysis of Masonry Wall Frames

The design and analysis of masonry wall frames shall use strength design requirements and load factors to determine the cross-section size and reinforcing steel requirements. It shall take into consideration the relative stiffness of columns and beams including the stiffening influence of the joints and the contribution of floor slab reinforcing if any.

6-9.D Design Strength Reduction Factor, φ

All members shall have a strength greater than the required strength. The design strength for flexure, shear and axial load shall be the nominal strength multiplied by the strength reduction factor, φ.

Flexural reduction factor with or without axial load shall be

$$0.65 \le \phi = 0.85 - 2\left(\frac{P}{A_n f'_m}\right) \le 0.85$$

Shear reduction factor shall be

$$\phi = 0.80$$

6-9.E Reinforcement Details

6-9.E.1 General

a. The nominal moment strength at any section along a member shall not be less than one fourth of the higher moment strength provided at the two ends of the members.

b. Lap splices shall be as defined in UBC Section 2412(c)4C. The center of the lap splice shall be at the center of the member clear length.

c. Welded splices and mechanical connections conforming to UBC Sections 2612(o)3.A through D may be used for splicing the reinforcement at any section provided not more than alternate longitudinal bars are spliced at a section, and the distance between splices on alternate bars is at least 24 inches along the longitudinal axis.

d. Reinforcement shall have a specified yield strength of 60 ksi. The actual yield strength based on mill tests shall not exceed 78 ksi.

6-9.F Spandrel Beams

These requirements apply to beams proportioned primarily to resist flexure. Factored axial compression force on the beam shall not exceed $0.10\, A_n f'_m$.

6-9.F.1 Longitudinal Reinforcement

a. At any section of a beam, each masonry unit through the beam depth shall contain longitudinal reinforcement.

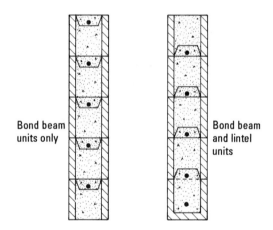

Figure 6-28 Uniform distribution of steel throughout the depth of the spandrel beam.

b. The variation in the longitudinal reinforcement area between units at any section shall not be greater than 50 percent, except multiple #4 bars shall not be greater than 100 percent, of the minimum area of longitudinal reinforcement contained by any one unit except where splices occur.

c. Minimum reinforcement ratio calculated over the gross cross section shall be 0.002.

d. Maximum reinforcement ratio calculated over the gross cross section shall be $0.15 f'_m / f_y$.

6-9.F.2 Transverse Reinforcement — Beams

a. Transverse reinforcement shall be hooked around top and bottom longitudinal bars with a standard 180-degree hook and shall be single pieces.

b. Within an end region extending one beam depth from pier faces and at any region at which beam plastic hinges may form during seismic or wind loading, maximum spacing of transverse reinforcement shall not exceed one fourth the nominal depth of the beam.

c. The maximum spacing of transverse reinforcement shall not exceed one half the nominal depth of the beam.

d. Minimum reinforcement ratio shall be 0.0015.

e. The first transverse bar shall be not more than 4 inches from the face of the pier.

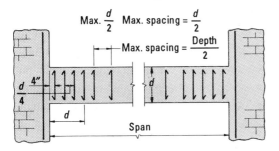

Figure 6-29 Maximum spacing of transverse shear reinforcing in spandrel beam.

6-9.G Piers Subjected to Axial Force and Flexure

These requirements apply to piers proportioned to resist flexure in conjunction with axial load.

6-9.G.1 Longitudinal Reinforcement

a. A minimum of 4 longitudinal bars shall be provided at all sections of every pier.

b. Flexural reinforcement shall be essentially uniformly distributed across the member depth.

c. Minimum reinforcement ratio shall be 0.002.

d. Maximum reinforcement ratio shall be $0.15 f'_m / f_y$.

e. Maximum bar diameter shall be $1/8$ nominal width of the pier.

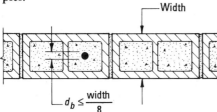

Figure 6-30 Maximum diameter of vertical reinforcing steel.

f. Maximum reinforcement in cells shall be 4 percent of the cell area without splices and 8 percent of the cell area with splices.

6-9.G.2 Transverse Reinforcement — Piers

a. Transverse reinforcement shall be hooked around the extreme longitudinal bars with standard 180-degree hook.

b. Within an end region extending one pier depth from the end of the beam, and at any region at which plastic hinges may form during seismic or wind loading, the maximum spacing of transverse reinforcement shall not exceed one fourth of the nominal depth of the pier.

c. The maximum spacing of transverse reinforcement shall not exceed one-half the nominal depth of pier.

d. Minimum transverse reinforcement ratio shall be 0.0015.

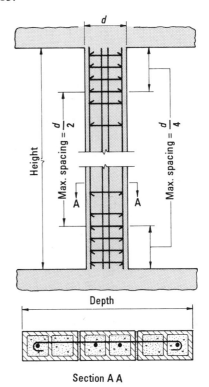

Figure 6-31 Spacing of transverse steel in pier.

6-9.G.3 Lateral Reinforcement

Reinforcement shall be provided to confine all vertical reinforcement with compressive strains greater than 0.0015, corresponding to factored forces and with R_w equal to 1.5.

Confinement reinforcing shall be a minimum of No. 3 bars at a maximum of 8-inch spacing within the grouted core or equivalent confinement which can develop an ultimate compressive strain of at least 0.006.

6-9.H Pier Design Forces

6-9.H.1 Nominal Moment Strength

Pier nominal moment strength shall not be less than 1.6 times the pier moment corresponding to the development of beam plastic hinges except at the foundation level.

6-9.H.2 Axial Load

Pier axial load based on the development of beam plastic hinges and including factored dead and live loads shall not exceed $0.15\, A_n f'_m$.

6-9.H.3 Plastic Hinge

The base plastic hinge of the pier must form immediately adjacent to the level of lateral support provided at the base or foundation.

6-9.I Shear Design

6-9.I.1 General

Beam and pier nominal shear strength shall be not less than 1.4 times the shears corresponding to the development of beam plastic hinges.

It shall be assumed in the calculation of member shear force that moments of opposite sign act at the joint faces and that the member is loaded with the tributary gravity load along its span.

6-9.I.2 Vertical Member Shear Strength

The nominal shear strength of the section shall be determined as follows:

$$V_n = V_m + V_s$$

Where:

$$V_m = C_d A_{mv} \sqrt{f'_m}$$

(For values of C_d see STR DES Table D-1)

and

$$V_s = A_{mv}\, p_n\, f_y$$

V_m shall be zero within an end region extending one pier depth from beam faces, and at any region where pier plastic hinges may form during seismic loading, and at piers subjected to net tension factored loads.

The nominal pier shear strength, V_n, shall not exceed the value given in UBC Table No. 24-K.

6-9.I.3 Beam Shear Strength

The nominal shear strength shall be determined as follows:

$$V_n = V_m + V_s$$

Where:

$$V_m = 1.2 A_{mv} \sqrt{f'_m}$$

V_m shall be zero with an end region extending one beam depth from pier faces and at any region at which beam plastic hinges may form during seismic loading.

$$V_s = A_{mv}\, p_n\, f_y$$

The nominal beam shear strength V_n, shall not exceed $4 A_{mv} \sqrt{f'_m}$.

6-9.J Joints

6-9.J.1 General Requirements

a. Where reinforcing bars extend through a joint, the joint dimensions shall be proportioned such that

$$h_p = \frac{4800 d_{bb}}{\sqrt{f'_g}}$$

and

$$h_b = \frac{1800 d_{bp}}{\sqrt{f'_g}}$$

The grout strength shall not exceed 5000 psi.

b. Joint shear forces shall be calculated on the assumption that the stress in all flexural tension reinforcement of the beams at the pier faces is $1.4 f_y$.

c. Strength of joint shall be governed by the appropriate strength reduction factors.

d. Beam longitudinal reinforcement terminating in a pier shall be extended to the far face of the pier and anchored by a standard 90-degree or 180-degree hook bent back into the pier.

e. Pier longitudinal reinforcement terminating in a beam shall be extended to the far face of the beam and anchored by a standard 90-degree or 180-degree hook bent back to the beam.

6-9.J.2 Transverse Reinforcement

a. Special horizontal joint shear reinforcement crossing a potential corner to corner diagonal joint shear crack, and anchored by standard hooks around the extreme pier reinforcing bars shall be provided such that

$$A_{jh} = \frac{0.5 V_{jh}}{f_{yh}}$$

b. Vertical shear forces may be considered to be carried by a combination of masonry shear resisting mechanisms and truss mechanisms involving intermediate pier reinforcing bars.

6-9.J.3 Shear Strength

The nominal horizontal shear stress of the joint shall not exceed $7\sqrt{f'_m}$ or 350 psi, whichever is less.

6-9.K Seismic Coefficient

It is suggested that for ductile masonry moment resistant wall frames the seismic coefficient $R_w = 9$ and the height be limited to 160 ft.

6-10 THE CORE METHOD OF DESIGN

6-10.A Core Method

Grouting between masonry wythes provides a vertical element, called a core, which is concrete. This concrete core can be considered the structural member which resists both vertical and lateral loads due to wind, earthquake, or more commonly, earth pressure for a retaining wall. The brick or block serves as a form for the concrete grout and also provides the color, texture and architectural features of the wall.

There are concrete masonry face shell units designed to act as forms and provide the look of masonry. Figure 6-32 shows how the shells are tied together with rectangular 9 gauge wire. The walls can be made to whatever width desired. These components are lightweight or medium weight concrete bricks conforming to ASTM C-55 with a minimum strength of 2500 psi and may be specified for higher strengths such as 3750 psi. The components are given a 4-hour rating when used in 8″ walls. Because the face shells are separate until tied in the wall, different units may be used on each side of the wall.

The system can have the units laid in mortar allowing the full width to be used in calculating masonry stresses. Both the masonry and the concrete core can be designed based on strength design methods. When the design is based on using only the concrete core, the requirement of UBC Chapter 26, applies.

The prime advantage of this method of construction and design is that high strength concrete can be utilized.

After the units are laid, the core is filled with concrete grout. The wall thickness for concrete design purposes is measured from inside face to inside face. Ties are commonly made for walls 6″ to 24″ thick in 1″ increments.

Component or expandable units are ideal for subterranean walls, retaining walls, and shear walls. They are also very useful when there is a congestion of reinforcing steel such as at the end of shear walls.

Component units are used where there is steel congestion

To add texture to exposed portions of walls, split face or patterned units can be used or standard units may be sandblasted.

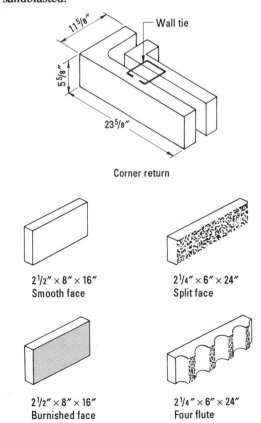

Figure 6-32 Typical component units.

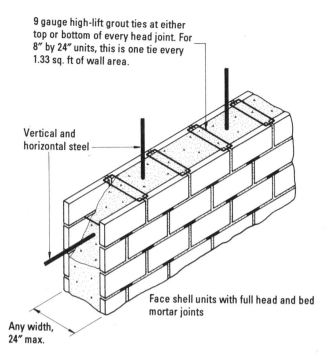

9 gauge high-lift grout ties at either top or bottom of every head joint. For 8″ by 24″ units, this is one tie every 1.33 sq. ft of wall area.

Vertical and horizontal steel

Face shell units with full head and bed mortar joints

Any width, 24″ max.

Figure 6-33 Wires that tie the masonry components together.

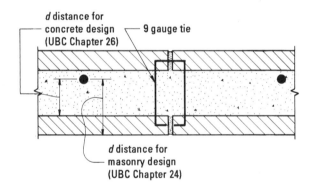

d distance for concrete design (UBC Chapter 26)

9 gauge tie

d distance for masonry design (UBC Chapter 24)

Figure 6-34 Component wall showing tie and d distance for either concrete or masonry design calculations.

For solid grouted masonry that is constructed without special inspection, the assumed strength is f'_m = 1500 psi and the maximum allowable working stress is $F_b = 0.5(0.33)f'_m = 250$ psi.

The grout in the core space between wythes must have a minimum strength $f'_c = 2000$ psi. This core may be considered as a concrete member and designed by the strength design methods of UBC Chapter 26.

The use of strength design whether for a masonry section or concrete section varies only in the coefficients. Load factors are the same for each material but the flexural strength reduction factor is 0.80 for masonry and 0.90 for concrete. Additionally, the limitation on the maximum allowable steel ratio is $0.75p_b$ for concrete and $0.50p_b$ for masonry.

6-10.B Comparison of the Design of a Wall Section with Component Units Using Masonry Design and Concrete Core Design

EXAMPLE 6-K Component Design.

Compare the cross-section requirements, *d* distance, and area of steel for a 12 ft high cantilever retaining wall using form or component units which are held in position by wire ties. Use a) working stress design method for masonry; b) strength design method for masonry; c) strength design method for the concrete core.

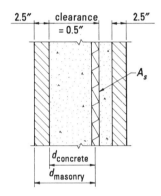

Assume f'_m = 1500 psi, $f_g = f'_c$ = 3000 psi and Grade 60 reinforcing.

Given:

Backfill is on a slope of 3 to 1, equivalent fluid pressure, EFP = 38 pcf.

$$\text{Moment} = \frac{1}{6}wh^3$$

$$= \frac{1}{6} \times 38 \times 12^3$$

$$= 10,944 \text{ ft lbs/ft}$$

6-10.B.1 Masonry — Working Stress Design

Assume inspected; solid grouted

f'_m = 1500 psi, f_s = 24,000 psi; n = 25.8

From Table E-1b

Balanced K = 77.2, p = 0.0036

$$bd^2 = \frac{M}{K} = \frac{10,944 \times 12}{77.2}$$

$$bd^2 = 1701$$

$$b = 12''$$

$$d^2 = 142$$

$$d = 11.9 \approx 12''$$

Total thickness = 12 + 0.5(clearance) +

0.5(to center of bar) +

2.5(Shell thickness) = 15.5″

Space units for $d = 12''$

$$K = \frac{M}{bd^2} = \frac{10,944 \times 12}{12 \times 12^2}$$

$$= 76.0$$

From Table E-1b   for $K = 76.0$

$p = 0.0035$

$A_s = pbd$
$= 0.0035 \times 12 \times 12$
$= 0.50$ sq. in./ft

Use #9 at 24" o.c. ($A_s = 0.50$ sq. in./ft)

Horizontal steel $= 0.0007bt$
$= 0.0007 \times 12 \times 15$
$= 0.126$ sq. in./ft

Use #5 @ 24" o.c. ($A_s = 0.15$ sq. in./ft)

6-10.B.2 Masonry - Strength Design

$f'_m = 1500$ psi, $f_y = 60,000$ psi

Load factor $= 1.7$

Strength reduction Factor $\phi = 0.8$

Maximum steel ratio $p = 0.5p_b$

Factored moment, $M_u = 1.7 \times 10,944$

$$= 18,604 \text{ ft lbs/ft}$$

From Table 6-1

$p_{max} = 0.5p_b = 0.0053$

From STR DES Table A-1

for $p = 0.0053$, $K_u = 220.8$

$$bd^2 = \frac{M_u}{K_u} = \frac{18,604 \times 12}{220.8} = 1011$$

Where $b = 12$ inches

$d^2 = 84.3$

$d = 9.2$ inches, use $d = 9.5''$

Total thickness $= 9.5 + 1 + 2.5 = 13$ inches

From Strength Design Table A-1

for $K_u = 225$ read $p = 0.0053$

$A_s = pbd = 0.0053 \times 12 \times 9.5$
$= 0.60$ sq. in./ft

Use #8 at 16" o.c. ($A_s = 0.59$ sq. in./ft)

Horizontal steel $= 0.0007bt$
$= 0.0007 \times 12 \times 13$
$= 0.109$ sq. in./ft

Use #5 @ 32" o.c. ($A_s = 0.116$ sq. in./ft)

6-10.B.3 Concrete Strength Design

$f'_c = 3000$ psi, $f_y = 60,000$ psi

Load factor $= 1.7$

ϕ factor $= 0.9$

Max $p = 0.75p_b$

From Table 6-1, $p_{max} = 0.0161$

$$K_u = \phi p f_y \left[1 - 0.59 \left(\frac{p f_y}{f'_c} \right) \right]$$

$$= 0.9(0.0161)60,000 \left[1 - 0.59 \left(\frac{0.0161 \times 60,000}{3000} \right) \right]$$

$$= 704$$

$$bd^2 = \frac{M_u}{K_u} = \frac{18,604 \times 12}{704} = 317$$

Where:

$b = 12$ inches

$d^2 = 26.4$

$d = 5.1$ inches, use $5.0''$

Total thickness $= (2 \times 2.5) + 5.0 + 0.5 + 0.5$
$= 11$ inches

$A_s = pbd = 0.0161 \times 12 \times 5.0$
$= 0.97$ sq. in./ft

Use # 9 at 12" o.c.

Horizontal steel $= 0.002bt$
$= 0.002 \times 12 \times 6$
$= 0.14$ sq. in./ft

Use #5 @ 24" o.c. ($A_s = 0.15$ sq. in./ft)

6-10.B.4 Comparison to Masonry Strength Design

Section 6-10.B.2 *Masonry-Strength Design* uses a total thickness of 13" and vertical reinforcing of #8 at 16" o.c. ($A_s = 0.59$ sq. in./ft).

For comparative concrete core design, use total thickness of 13" and $d = 7''$.

Determine steel ratio.

Using STR DES Table A-9, substitute f'_c for f'_m.

$$\frac{M_u}{\phi f'_c bd^2} = q(1 - 0.59q)$$

$$= \frac{18,604 \times 12}{0.9 \times 3000 \times 12 \times 7^2} = 0.141$$

From STR DES Table A-9

For 0.141 read $q = 0.155$

$$p = \frac{qf'_c}{f_y} = \frac{0.155 \times 3000}{60,000} = 0.0078$$

$$A_s = pbd = 0.0078 \times 12 \times 7$$
$$= 0.65 \text{ sq. in./ft}$$

Use #9 at 18″ o.c. ($A_s = 0.67$ sq. in./ft)

Horizontal steel $= 0.002bt$

$$= 0.002 \times 12 \times 8$$
$$= 0.192 \text{ sq. in./ft}$$

Use #5 at 20″ o.c. ($A_s = 0.19$ sq. in./ft)

TABLE 6-3 Summary of Comparison of Designs for Moment = 10.9 ft kips/ft

	Masonry WSD	Masonry STR. DES.	Concrete STR. DES.	Concrete STR. DES.
	0.5″ / 2.5″ / 12″ / 15.5″	1″ / 2.5″ / 9.5″ / 13″	2.5″ / 5″ / 3.5″ / 11″	2.5″ / 7″ / 1″ / 2.5″ / 13″
f'_m or f_c; psi	1500	1500	3000	3000
Depth d″ Total Thickness t″	12.0 15.5	9.5 13.0	5.0 11.0	7.0 13.0
Vertical Reinforcing sq. in./ft	#9 @ 24″ 0.50	#8 @ 16″ 0.60	#9 @ 12″ 0.97	#9 @ 18″ 0.65
Horizontal Reinforcing sq. in./ft	#5 @ 24″ 0.13	#5 @ 32″ 0.111	#5 @ 24″ 0.14	#5 @ 20″ 0.19

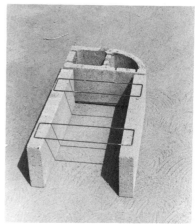

Shrine Auditorium garage built with concrete component units, 6 levels, 645 car capacity — Los Angeles, CA.

6-11 LIMIT STATE

6-11.A General

Design of masonry is based on several states that limit its use or stress conditions. The qualification of these limit states may be based on the loading, the stress or the strain conditions imposed on either the reinforcing steel or masonry or on the deflection of the members.

The concept of limit state conditions were recognized by the 1963 ACI Code in a minor way and were later stated in the 1971 ACI Code as moment redistribution. The 1971 Code included the concept of changing moment pattern, stress conditions, curvature and deflection conditions.

For a reinforced masonry structural member subjected to an ever increasing bending moment, there are three distinctive limit states that may be considered as the stress on the section changes.

The following subsections outline these basic limit states as shown in Figure 6-35.

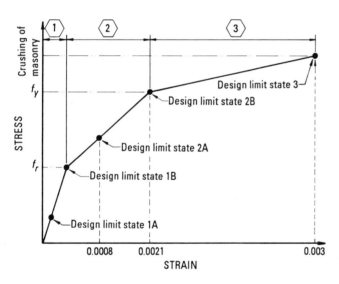

1️⃣ Behavior state 1

2️⃣ Behavior state 2

3️⃣ Behavior state 3

Figure 6-35 Limit and behavior states of a flexural member.

6-11.B Behavior State 1

Within this behavior state, the masonry system is not cracked. The mortar joint, the bond between mortar and unit, and the masonry unit itself resist the tensile forces caused by moment on the section. The stresses range from zero to less than the modulus of rupture.

The limit of behavior state 1 is reached when the moment on the section stresses the masonry to the modulus of rupture.

6-11.B.1 Design Limit State 1A

At the design limit state 1A, the tensile stress of the masonry is limited, based on Table A-9 (UBC Section 2406(c)4 and ACI/ASCE Table 6.3.1.1.) which form the basis for the design of unreinforced masonry systems.

6-11.B.2 Design Limit State 1B

At design limit state 1B, the modulus of rupture is reached and the section cracks. The modulus of rupture value is approximated in STR. DES. Table D-1 based on UBC Section 2411(b)4.

The cracking moment strength of the wall is determined by the equation:

$$M_{cr} = S f_r \quad \text{(UBC Chapter 24, Equation 11-13)}$$

6-11.C Behavior State 2

When the moment on the section exceeds the modulus of rupture, the masonry will crack and behavior state 2 is reached. The reinforcing steel in the system resists the tensile forces and the masonry resists the compression forces. This is the basis for reinforced masonry.

6-11.C.1 Design Limit State 2A

At design state limit 2A the stresses in the steel and the masonry are limited to maximum allowable values as given in UBC Sec 2406 and ACI/ASCE Sec. 7.3. These values are well within the elastic range of the materials and are used for working stress design procedures (See Section 5).

Masonry compressive stress, flexural.

$$F_b = 0.33 f'_m, \quad 2,000 \text{ psi maximum}$$

Steel tension stress, deformed bars

$$F_s = 0.4 f_y, \quad 24,000 \text{ psi maximum}$$

6-11.C.2 Design Limit State 2B

As the moment on the section increases, the stresses in the reinforcing steel and masonry increase.

To assure a ductile failure of a member, the reinforcing steel ratio is limited so that it will yield well before the masonry crushes. Limit state 2 occurs at the point where the steel first reaches its yield strength.

$$f_y = 60,000 \text{ psi specified min.}$$

$$< 78,000 \text{ psi actual max.}$$

$$e_y = 0.0021 \text{ in./in. for } f_y = 60 \text{ ksi}$$

$$e_{sh} = 0.008 \text{ in./in. for } f_y = 60 \text{ ksi}$$

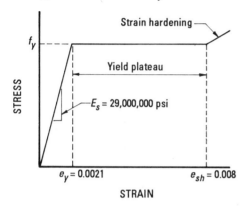

Figure 6-36 Stress-strain relationship for Grade 60 reinforcing steel.

6-11.D Behavior State 3

After limit state 2 is achieved, the reinforcing steel stretches without significantly increasing the moment on the section. The strain in the masonry increases throughout behavior state 3 until the limiting strain in the masonry is exceeded at which point the masonry will fail in compression. The limit state for the maximum masonry compression strain ranges from 0.0025 to 0.005. Building codes, however, limit the maximum masonry compression strain to 0.003 in./in. (UBC Section 2411(b).3.D).

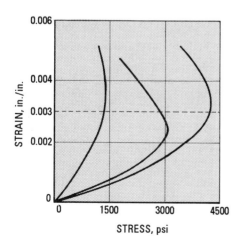

Figure 6-37 Stress-strain relationship for masonry.

6-11.D.1 Limit State 3

At limit state 3, the steel is at yield stress and the masonry reaches its crushing strain which is defined as 0.003 in./in.

This condition is the basis for strength design procedures of a member as given in UBC Section 2411 for tall slender walls and in UBC Section 2412 for shear walls.

6-11.D.2 Proposed Masonry Limit States Design Standards

The Masonry Society has developed a proposed Limit State Design Standard which is based on limit state 3 including serviceability limits, and strength limits.

The fundamental concept for the proposed standard is that of using statistically determined expected values of the physical properties, strength, moduli of elasticity, etc. for the materials, assemblies and systems. Similar to the requirements for strength design, load factors and strength reduction factors are used to provide satisfactory factors of safety.

Research is continuing on using this inelastic method of structural design as a standard.

6-12 QUESTIONS AND PROBLEMS

6-1 You wish to use 8 inch concrete masonry units for a 24 foot high bearing wall. Explain how you would do this in order to comply with the code.

6-2 An 8 inch thick non-load bearing concrete masonry wall is 20 feet high. Design the vertical and horizontal reinforcing steel if the wind load is 20 psf, $f_y = 60,000$ psi, $f'_m = 2000$ psi. Use strength design procedures.

6-3 A 6″ nominal (5¹/₂″ actual) hollow clay masonry beam has an overall depth of 36 inches. The beam is continuous at the supports and has a clear span of 24 feet. $f'_m = 2500$ psi, $f_y = 60,000$ psi, $LL = 1000$ lbs/ft, $DL = 740$ lbs/ft plus the weight of the beam. Use strength design methods to determine tension steel and if necessary compression steel. Check if shear reinforcing is required.

6-4 What is the live load capacity for a 8″ CMU solid grouted beam spanning 16 ft if it is 32 inches deep with $d = 26″$, $A_s = 2.00$ sq. in.; $d' = 3″$, $A'_s = 0.62$ sq. in., $f'_m = 1500$ psi; $f_y = 60,000$ psi; $LL = 1200$ plf; $DL = 800$ lbs/ft. Use strength design methods.

6-5   Design a 10 ft high reinforced 6″ clay block wall for a vertical load of 4 kips/ft and a moment perpendicular to the plane of the wall of 2 ft kip/ft. Assume that the wall is fixed at the bottom and pinned at the top. Use $f'_m = 3000$ psi and $f_y = 60,000$ psi. Specify reinforcing size and spacing.

6-6   Using the slender wall method of design, check the adequacy of an 8 inch concrete masonry wall having vertical reinforcing steel of #5 @ 24″. Assume that the wall is grouted at @ 24″ o.c. and is located in Seismic Zone 4, wind pressure 12 psf.

The wall is 20′ between pinned lateral supports at the floor and roof diaphragm.

The axial dead load on the wall is 450 plf, $f'_m = 1500$ psi, and $f_y = 60$ ksi.

6-7   A solid grouted reinforced brick wall is 26 ft high between the lateral supports of the floor and roof diaphragm. It is located in Seismic Zone No. 3 where the wind pressure is 20 psf. It supports a roof live load of 370 plf with an eccentricity of 7″ to the center of the wall.

Given:

$$t = 10'', d = 5'' \qquad f_y = 60,000 \text{ psi}$$
$$f'_m = 2500 \text{ psi} \qquad \phi = 0.8$$

Determine the reinforcing steel size and spacing, and check for adequacy using the slender wall method of design. Wall is assumed pinned at the top and bottom.

6-8   Given a nominal 8″ hollow clay masonry shear wall, solid grouted. Wall is 12 ft high; 8 ft 6 in. long, $f'_m = 2500$ psi; $f_y = 60,000$ psi; Units are $3\frac{1}{2}'' \times 7\frac{1}{2}'' \times 11\frac{1}{2}''$.

Lateral seismic shear	V	= 60 kips
Vertical dead load	P_{DL}	= 100 kips
Vertical live load	P_{LL}	= 90 kips
Seismic Moment	M	= 720 ft kips

It is reinforced with 9 - #8 bars.

Using UBC Section 2412 and Section 6-7 plot the interaction diagram and determine if wall and reinforcing is adequate for the loads and moments imposed? Try for nominal moment strength, M_n, neutral axis at 18.5″; for $D.l$, load condition 1, $N.A. = 57''$; For load condition 2, $N.A. = 37''$.

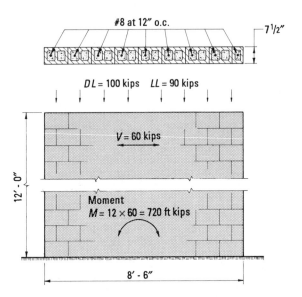

6-9   Compare the moment capacity of a component wall system by strength design and working stress design methods. $f'_m = 2000$ psi; $f_g = 3000$ psi; $f_s = 24,000$ psi; $f_y = 60,000$ psi; $\phi = 0.8$; special inspection provided.

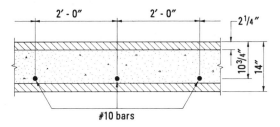

6-10   Using the cross-section and material properties of problem 6-9 compare the moment capacity for $d = 10.75$ in. using masonry working stress design to $d = 8.75$ in. using concrete strength design, $\phi = 0.9$.

SECTION 7

Details of Reinforcing Steel

7-1 MINIMUM REINFORCING STEEL

The minimum area of steel to be used in masonry depends on the seismic zone in which the structure is to be constructed.

7-1.A Seismic Zone Nos. 0 and 1

There are no requirements for reinforcing steel in masonry for Seismic Zone Nos. 0 and 1.

7-1.B Seismic Zone No. 2

In Seismic Zone No. 2 the minimum wall reinforcement is required as follows in UBC Sec. 2407(h)3B.

Reinforcing in a concrete masonry wall located in Seismic Zone No. 4.

UBC Sec. 2407(h)3

B. Wall Reinforcement. Vertical reinforcement of at least 0.20 square inch in cross-sectional area shall be provided continuously from support to support at each corner, at each side of each opening, at the ends of walls and at a maximum spacing of 4 feet apart, horizontally throughout the wall.

Horizontal reinforcement not less than 0.2 square inch in cross-sectional area shall be provided

(1) at the bottom and top of wall openings and shall not extend less than 24 inches nor less than 40 bar diameters past the opening,

(2) continuously at structurally connected roof and floor levels and at the top of walls,

(3) at the bottom of the wall or in the top of the foundations when dowelled to the wall,

(4) at maximum spacing of 10 feet unless uniformly distributed joint reinforcement is provided.

Reinforcement at the top and bottom of openings when continuous in the wall may be used in determining the maximum spacing specified in Item (1) above.

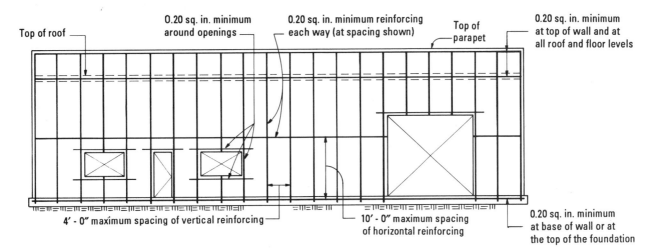

Figure 7-1 Minimum reinforcing for Seismic Zone No. 2 (coordinate this figure with Figure 7-4).

7-1.C Seismic Zone Nos. 3 and 4

In Seismic Zone Nos. 3 and 4, walls must be reinforced in accordance with the requirements for Seismic Zone No. 2 and with the additional requirements given in UBC Sec. 2407(h)4B.

UBC Sec. 2407(h)4

B. Wall Reinforcement. All walls shall be reinforced with both vertical and horizontal reinforcement. The sum of the areas of horizontal and vertical reinforcement shall be at least 0.002 times the gross cross-sectional area of the wall, and the minimum area of reinforcement in either direction shall not be less than 0.0007 times the gross cross-sectional area of the wall. The minimum steel requirements for Seismic Zone No. 2 in Section 2407(h)3B may be included in the sum.

The spacing of reinforcement shall not exceed 4 feet. The diameter of reinforcement shall not be less than 3/8 inch except that joint reinforcement may be considered as part or all of the requirement for minimum reinforcement. Reinforcement shall be continuous around wall corners and through intersections. Only reinforcement which is continuous in the wall or element shall be considered in computing the minimum area of reinforcement. Reinforcement with splices conforming to Section 2409(e)6 shall be considered as continuous reinforcement.

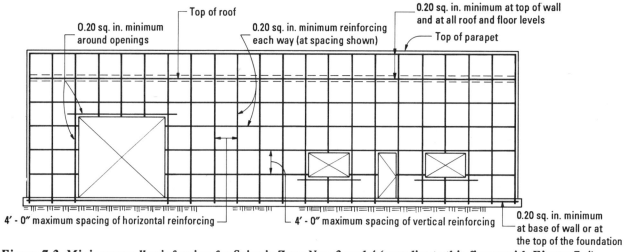

Figure 7-2 Minimum wall reinforcing for Seismic Zone Nos. 3 and 4 (coordinate this figure with Figure 7-4).

7-1.D Calculation of Minimum Steel Area

The Uniform Building Code states in Section 2407(h)4B that in order to use stresses permitted for reinforced masonry design, a wall must be reinforced both vertically and horizontally with a required minimum amount of reinforcing. The minimum area of reinforcement for Seismic Zone Nos. 3 and 4, in one direction either vertically or horizontally, may not be less than 0.0007 times the gross cross-sectional area of the wall.

The gross cross-sectional area is the width of the wall (usually 1-foot) times a given length.

EXAMPLE 7-A Minimum Areas of Steel.

Based on the 1991 UBC, what is the minimum size and spacing of reinforcing steel in each direction for:

(a) 9″ solid grouted brick wall

(b) 8″ concrete block wall

Assume these walls are located in either Seismic Zone No. 3 or 4.

Solution 7-A

UBC Section 2407(h)4.B requires at least $A_s = 0.0007bt$ in both directions with a minimum total area of steel of $0.002bt$ for all reinforced masonry structures located in Seismic Zone Nos. 3 and 4. Generally, $0.0007bt$ is placed in the wall opposite of the direction the wall spans. The balance of the reinforcement ($0.002bt - 0.0007bt = 0.0013bt$) is placed in the direction the wall is principally spanning.

(a) 9″ Solid Grouted Brick Wall

From Table C-6, *Maximum Spacing of Minimum Steel*, the principal reinforcement should be at least:

$$A_s = 0.00013bt = 0.076 \text{ sq.in./ft}$$

Use #5 @ 26″ in the principal direction of stress

Use Table C-5 to find the reinforcement in the opposite direction.

$$A_s = 0.0007bt = 0.076 \text{ sq.in./ft}$$

Choose #5 @ 48" o.c. in other direction

(b) 8" Concrete block wall

From Table C-6,

$$A_s = 0.0013bt = 0.119 \text{ sq.in./ft}$$

Select either #5 @ 32" o.c. or #7 @ 48" o.c. in the direction of principal stress

Similarly, from Table C-5

$$A_s = 0.0007bt = 0.065 \text{ sq.in./ft}$$

Choose either #4 @ 32" o.c. or #5 @ 48" o.c. in opposite direction

EXAMPLE 7-B   Minimum Steel Requirements Utilizing Joint Reinforcement.

Select the minimum vertical and horizontal reinforcement for an 8" block wall which spans 12 feet between the foundation and the roof bond beam. Note that the wall is located in Seismic Zone No. 4. Joint reinforcement with #9 gauge longitudinal wires will be placed every 16" vertically in the wall bed joints.

Solution 7-B

Use $A_s = 0.0013bt$ vertically and $A_s = 0.0007bt$ horizontally to satisfy the requirements of UBC Section 2407(h)4.B.

Therefore:

Minimum vertical $A_s = 0.119$ sq.in./ft (Table C-6)

Minimum horizontal $A_s = 0.064$ sq.in./ft (Table C-5)

Also from Table C-6, choose the vertical reinforcing of #5 @ 32" o.c.

To find the additional horizontal area of steel required to meet the $A_s = 0.064$ sq.in./ft, the contribution of the joint reinforcement must first be determined.

Total required horizontal steel, $A_s = 0.064 \times 12$
$$= 0.769 \text{ sq. in.}$$

Place the joint reinforcing in every other mortar joint or at 16" o.c.

Therefore:

$$\text{The number of joints reinforced} = \frac{12 \text{ ft} \times 12 \text{ in./ft}}{16 \text{ in.}} - 1$$
$$= 8 \text{ joints}$$

From Table C-4a, the area of 2 - #9 longitudinal joint reinforcing wires is 0.035 sq.in. Therefore, the area of steel provided by the joint reinforcing is:

$$A_s = 0.035 \times 8 \text{ joints reinforced} = 0.28 \text{ in.}^2$$

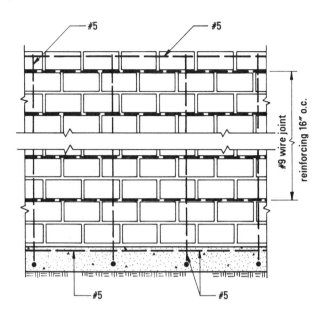

Figure 7-3 Wall with joint reinforcing.

Area of steel needed in the bond beam and the top of the footing is:

$$\frac{0.769 - 0.28}{2} = 0.24 \text{ in.}^2$$

Use #5 bar in the bond beam and top of the footing

It is general practice for the principal steel which resists the design stresses, to be the larger amount of steel, ($A_s = 0.0013bt$), and perpendicular to it would be the minimum amount of steel (A_s min. = 0.0007bt). Thus, if a wall spans vertically, between floors, or between the floor and the roof, the principal steel would be vertical and would be 0.0013bt or, as required by engineering calculations. The minimum horizontal steel could then be 0.0007bt as required.

Many times, however, the same amount of steel is used both vertically and horizontally. In that case, the area of steel would be 0.001bt placed in both directions.

These minimum steel area requirements are somewhat arbitrary and are an outgrowth of the minimum requirements initially used for reinforced concrete. Concrete requires a fairly large amount of minimum steel because it is cast in a plastic state and is subject to significant shrinkage during hydration. Masonry units, on the other hand, are for the most part, dimensionally stable when the wall is constructed. Only plastic mortar and grout are added to the masonry structure. Because there is far less material to shrink in a masonry wall than in a concrete wall, the minimum steel requirements have been set at half that of concrete.

7-2 REINFORCING STEEL AROUND OPENINGS

In reinforced masonry walls there must be not less than a #4 bar or two #3 bars on all sides of, and adjacent to, every opening which exceeds 24 inches in either direction. These bars must extend not less than 40 diameters, but in no case less than 24 inches, beyond the corner of the opening. These bars are required in addition to the minimum reinforcement, unless they are continuous throughout the length of the wall.

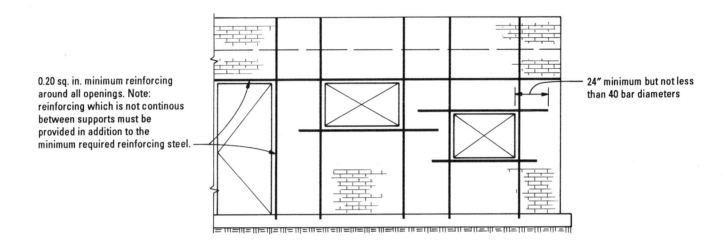

0.20 sq. in. minimum reinforcing around all openings. Note: reinforcing which is not continous between supports must be provided in addition to the minimum required reinforcing steel.

24″ minimum but not less than 40 bar diameters

Figure 7-4 Typical reinforcing steel around openings (Coordinate this figure with Figures 7-1 and 7-2 for minimum wall reinforcing requirements).

7-3 PLACEMENT OF STEEL

7-3.A Positioning of Steel

The placement of reinforcing bars should conform to the recommended practice of placing reinforcing bars in concrete. Principal steel should be properly located and secured in position so that it will resist the forces for which it was designed. This is particularly important in cantilever retaining walls, beams, columns, etc.

To insure correct location of principal steel, vertical bars should be held in place at top and bottom and at intervals not exceeding 200 diameters of the reinforcement. For bars no more than 6″ out of position, they may be bent at a slope of 1 to 6 to properly align them (Figure 7-5).

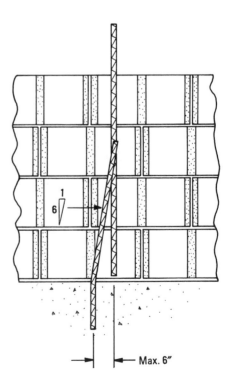

Figure 7-5 Slope for bending reinforcing steel into position.

7-3.B Tolerances for Placement of Steel

For reinforced masonry to perform as designed, it is important that the reinforcing steel be in the proper location.

The proper placement of reinforcing steel is governed by UBC Section 2404(e). These allowable tolerances are shown in Figure 7-6 and in Table 7-1.

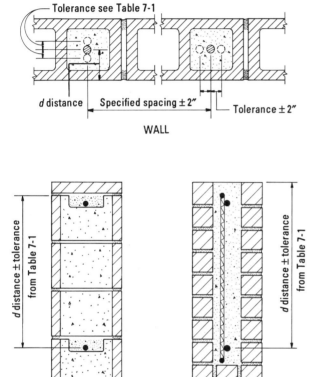

Figure 7-6 Illustration of tolerances for steel placement.

TABLE 7-1   Tolerances for Placing Reinforcement

Distance, *d*, from face of CMU to the center of reinforcing	Allowable tolerance
$d \le 8''$	$\pm \, ^1/_2''$
$8'' < d \le 24''$	$\pm \, 1''$
$d \le 24''$	$\pm \, 1^1/_4''$

7-3.C Clearances

7-3.C.1 Clearance Between Reinforcing Steel and Masonry Units

To be effective, reinforcing steel must be surrounded by grout. Reinforcing steel bars must have a minimum of $^1/_4$-inch of grout between the steel and the masonry when fine grout is used. When coarse grout (pea gravel grout) is used, the clearance between the steel and the masonry units must be at least $^1/_2$-inch. This assures proper bond so that stresses may be transferred between the steel to the masonry.

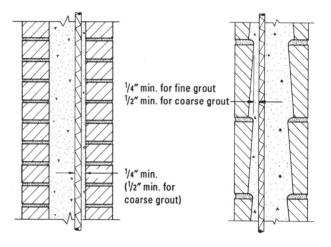

Figure 7-7 Reinforcing steel clearances.

7-3.C.2 Clear Spacing Between Reinforcing Bars

UBC Sec. 2409(e)2

2. Spacing of longitudinal reinforcement. The clear distance between parallel bars, except in columns, shall be not less than the nominal diameter of the bars or 1 inch, except that bars in a splice may be in contact. This clear distance requirement applies to the clear distance between a contact splice and adjacent splices or bars. The minimum clear distance between parallel bars in columns shall be two and one-half times the bar diameter.

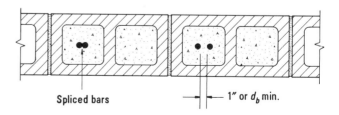

Figure 7-8 Minimum spacing of vertical reinforcing in cell.

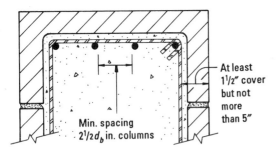

Figure 7-9 Minimum clearance between bars in a column.

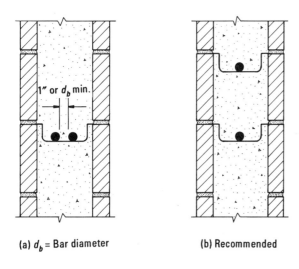

(a) d_b = Bar diameter   (b) Recommended

Figure 7-10 Spacing of horizontal reinforcing in a concrete masonry wall.

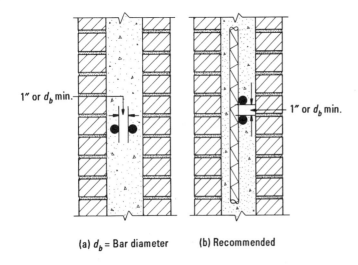

(a) d_b = Bar diameter   (b) Recommended

Figure 7-11 Spacing of horizontal reinforcing in a brick wall.

7-3.D   Cover Over Reinforcing

7-3.D.1   Steel Bars

> **UBC Sec. 2409(e)2**
> **(Third Paragraph)**
>
> All reinforcing bars, except joint reinforcing, shall be completely embedded in mortar or grout and have a minimum cover, including the masonry unit, of at least $3/4$ inch, $1\frac{1}{2}$ inches of cover when exposed to weather and 2 inches of cover when exposed to soil. [Refer to Figure 7-11 for minimum cover requirements.]

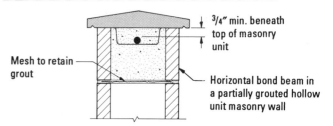

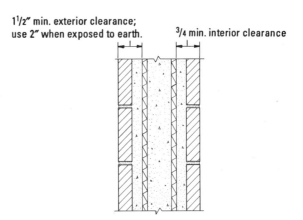

Figure 7-12 Minimum cover over reinforcing steel.

7-3.D.2   Cover for Joint Reinforcing and Ties

> **UBC Sec.2407(f)**
>
> **(f) Protection of Ties, and Joint Reinforcement.** A minimum of $5/8$-inch mortar cover shall be provided between ties or joint reinforcement and any exposed face. The thickness of grout or mortar between masonry units and joint reinforcement shall not be less than $1/4$ inch, except that $1/4$ inch or smaller diameter reinforcement or bolts may be placed in bed joints which are at least twice the thickness of the reinforcement.

Joint reinforcing steel can be used in mortar joints that are at least twice as thick as the joint reinforcement. There must be a minimum of $5/8$ inch of mortar coverage from the outside of the masonry wall to the face of the joint reinforcement (see Figure 7-13).

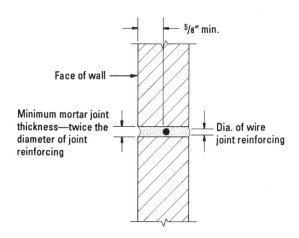

Figure 7-13 Cover over joint reinforcing.

7-3.D.3   Cover for Column Reinforcing

Lateral ties and longitudinal bars in columns must be placed at least $1\frac{1}{2}''$ but not more than $5''$ from the surface of the column per UBC Sec. 2409(b)5B (see Figure 7-9 for minimum column cover requirements).

7-4   EFFECTIVE DEPTH, *d*, IN A WALL

Determination of the *d* distance to the reinforcing steel perpendicular to the plane of the wall is given in Tables 7-2, 7-3, 7-4 and 7-5:

7-4.A   Hollow Masonry Unit Walls

TABLE 7-2   Steel in Center of Cell, Block

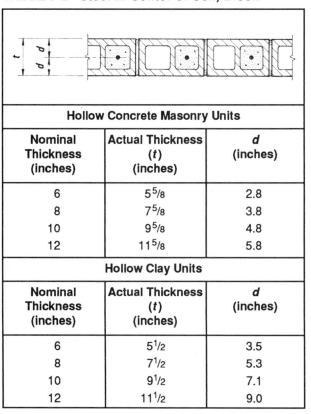

Hollow Concrete Masonry Units		
Nominal Thickness (inches)	Actual Thickness (*t*) (inches)	*d* (inches)
6	$5^5/8$	2.8
8	$7^5/8$	3.8
10	$9^5/8$	4.8
12	$11^5/8$	5.8
Hollow Clay Units		
Nominal Thickness (inches)	Actual Thickness (*t*) (inches)	*d* (inches)
6	$5^1/2$	3.5
8	$7^1/2$	5.3
10	$9^1/2$	7.1
12	$11^1/2$	9.0

TABLE 7-3   Steel Place for Maximum *d*, Block

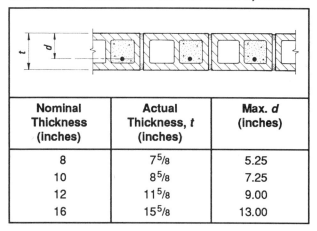

Nominal Thickness (inches)	Actual Thickness, *t* (inches)	Max. *d* (inches)
8	$7^5/8$	5.25
10	$8^5/8$	7.25
12	$11^5/8$	9.00
16	$15^5/8$	13.00

7-4.B   Brick Walls

TABLE 7-4   Steel in Center of Grout Space, Brick

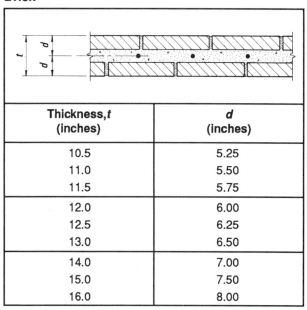

Thickness, *t* (inches)	*d* (inches)
10.5	5.25
11.0	5.50
11.5	5.75
12.0	6.00
12.5	6.25
13.0	6.50
14.0	7.00
15.0	7.50
16.0	8.00

TABLE 7-5   Steel Placed for Maximum *d*, Brick

Actual Thickness, *t* (inches)	*d* (inches)	Actual Thickness, *t* (inches)	*d* (inches)
9.0	5.00	12.5	8.00
9.5	5.25	13.0	8.50
10.0	5.50	14.5	9.50
10.5	6.00	15.0	10.50
11.0	6.50	16.0	11.50
11.5	7.00	18.0	13.50
12.0	7.50	20.0	15.50

7-4.C Effect of *d* Distance in a Wall (Location of Steel)

If a wall is subjected to lateral forces from either face, it is generally more economical to place the steel in the center of the wall rather than $1/2$ the amount of steel against each outside face.

EXAMPLE 7-C Moment Capacity vs. *d* Distance.

Assume: $f'_m = 1500$ psi; $n = 25.8$; no special inspection

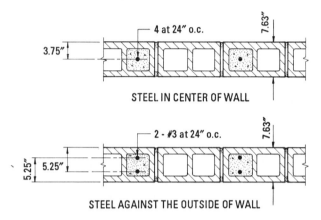

STEEL IN CENTER OF WALL

STEEL AGAINST THE OUTSIDE OF WALL

Steel in Center of Wall

A_s = #4 at 24″ o.c. = 0.10 sq. in./ft

From Table C-8b for $d = 3.75″$; $p = 0.0022$

From Table E-1a for $f'_m = 1500$ psi and no special inspection;

$p = 0.0022$ $K = 32.5$

$$\text{Moment} = Kbd^2$$
$$= 32.5 \times 12 \times 3.75^2$$
$$= 5484 \text{ in. lbs/ft}$$

Steel Placed for Maximum *d* Distance

A_s = #3 at 24″ o.c. each face; = 0.11 sq. in./ft

From Table C-8c for $d = 5.25″$
$p = 0.0009$ #3 at 24″

From Table E-1a for $f'_m = 1500$ psi; half stresses
$p = 0.0009$, $K = 20.5$

$$\text{Moment} = Kbd^2$$
$$= 20.5 \times 12 \times 5.25^2$$
$$= 6780 \text{ in. lbs/ft}$$

Although the moment capacity is greater with the steel against the outside face, it requires twice the number of bars and increased steel placement costs plus the congestion with added steel, negates the benefit of increased moment. To compensate increase the steel to #5 at 24″ o.c. in center of wall.

A_s = #5 at 24″ o.c. = 1.55 sq. in./ft

From Table C-8b for $d = 3.75$; $p = 0.0034$

From Table E-1a for $f'_m = 1500$ psi, half stresses

for $p = 0.0034$, $K = 37.5$

$$\text{Moment} = Kbd^2 = 37.5 \times 12 \times 3.75^2$$
$$= 6328 \text{ in. lbs/ft}$$

It is easier, faster and more economical to place one bar of steel in the center rather than a bar of steel on each side.

7-5 ANCHORAGE OF REINFORCING STEEL

7-5.A Development Length, Bond

To develop a reinforcing bar, adequate development length, l_d, is required.

The development length is based on the allowable bond stress, the bar diameter, and the stress to be developed in the steel bar.

TABLE 7-6 Development Length, l_d (inches)[1]
Grade 60, $F_s = 24,000$ psi

Bar Size		l_d (inches) for Deformed Bars		l_d for Smooth Bars
No	Dia, d_b (inches)	Bars in Tension[2]	Bars in Compression[3]	Tension or Compression[4]
3	0.375	18	14	36
4	0.500	24	18	48
5	0.625	30	23	60
6	0.750	36	27	72
7	0.875	42	32	84
8	1.000	48	36	96
9	1.128	54	41	108
10	1.270	61	46	122
11	1.410	68	51	132

1. Based on UBC Sec. 2409(e)3.
2. $l_d = 0.002 d_b f_s = 48 d_b$.
3. $l_d = 0.0015 d_b f_s = 36 d_b$.
4. $l_d = 2(0.002 d_b f_s) = 96 d_b$.

7-5.B Hooks

A standard 90 degree and 180 degree hook has a tension capacity to develop 7500 psi stress in a bar per UBC Section 2409(e)5E.

A hook has the benefit of developing full stress within a very short distance.

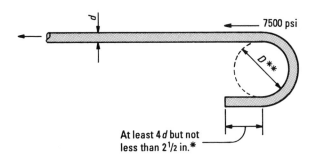

Figure 7-14a Standard 180° hook.

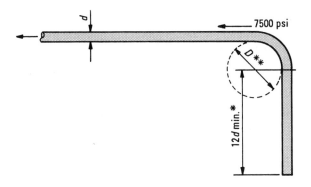

Figure 7-14b Standard 90° hook.

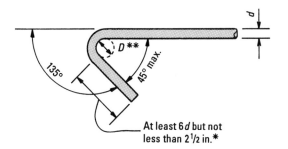

Figure 7-14c Standard 135° hook.

* See also the additional requirements in UBC Sec. 2407(h)4G for Seismic Zone Nos. 3 and 4.

** See UBC Sec. 2409(e)5B and C for dimension D.

TABLE 7-7 Tension Capacity, T (lbs) and Dimensions, D (inches) of Standard Hooks Grade 40 and 60

Bar Size			$T = A_s F_t$ [1] (lbs)	D_i (inside dia) (inches)	D_o (outside dia) (inches)	Remarks [2]
No.	Dia., d_b (inches)	Cross Sect. A_s (sq. in.)				
3	0.375	0.11	825	2¼	3	
4	0.500	0.20	1,500	3	4	
5	0.625	0.31	2,325	3¾	5	$D_i = 6d_b$
6	0.750	0.44	3,300	4½	6	
7	0.875	0.60	4,500	5¼	7	
8	1.000	0.79	5,925	6	8	
9	1.128	1.00	7,500	9	11¼	
10	1.270	1.27	9,525	10	12¾	$D_i = 8d_b$
11	1.410	1.56	11,700	11¼	14	

1. Hooks max. tensile stress, F_t = 7,500 psi per UBC Section 2409(e)5F.

2. Based on UBC Table No. 24-F.

7-6 DEVELOPMENT LENGTH IN CONCRETE

For bars anchored in concrete, the development length is based on UBC Sec. 2612(b)(c) and (d).

> **UBC Sec. 2612(b), (c) & (d)**
>
> (b) **Development of Reinforcement— General.** [12.1] Calculated tension or compression in reinforcement at each section of reinforced concrete members shall be developed on each side of that section by embedment length, hook or mechanical device, or a combination thereof. Hooks may be used in developing bars in tension only.
>
> The values of $\sqrt{f_c'}$ used in Section 2612 shall not exceed 100 psi.
>
> (c) **Development of Deformed Bars and Deformed Wire in Tension.** 1. Development length, l_d, in inches, for deformed bars and deformed wire in tension shall be computed as the product of the basic development length l_{db} of Section 2612(c)2 and the applicable modification factors of Section 2612(c)3 through 5, but l_d shall not be less than 12 inches.
>
> 2. Basic development length l_{db} shall be:
>
> No. 11 bar and smaller
>
> and deformed wire .... $0.04 A_1 f_y / \sqrt{f_c'}$ *
>
> No. 14 bar .... $0.085 f_y / \sqrt{f_c'}$ **
>
> No 18 bar .... $0.125 f_y / \sqrt{f_c'}$ **

*The constant carries the unit of one/inch.

**The constant carries the unit of inch.

3. To account for bar spacing, amount of cover and enclosing transverse reinforcement, the basic development length shall be multiplied by a factor from Section 2612(c)3A, B or C which may be modified by Section 2612(c)3D or 2612(c)3E, but shall not be less than provided by Section 2612(c)3F.

A. For bars satisfying any one of the following conditions 1.0

(i) Bars in beams or columns with (1) minimum cover as specified in Section 2607(h), (2) with transverse reinforcement satisfying tie requirements of Section 2607(k)3 or minimum stirrup requirements of Section 2611(f)4 and 5C along the development length, and (3) with clear spacing of not less than $3d_b$.

(ii) Bars in beams or columns with (1) minimum cover as specified in Section 2607(h), and (2) enclosed within transverse reinforcement A_{tr}, along the development length satisfying Formula (12-1).

$$A_{tr} \geq \frac{d_b s N}{40} \qquad (12\text{-}1)$$

where d_b is the diameter of the bar being developed.

(iii) Bars in the inner layer of slab or wall reinforcement and with clear spacing of not less than $3d_b$.

(iv) Any bars with cover of not less than $2d_b$ and with clear spacing of not less than $3d_b$.

B. For bars with cover of d_b or less or with clear spacing of $2d_b$ or less .... 2.0.

C. For bars not included in Section 2612(c)3A or 3B .... 1.4.

D. For No. 11 bars and smaller with clear spacing not less than $5d_b$ and with cover from face of member to edge bar, measured in the plane of the bars, not less than $2.5d_b$, the factors in Section 2612(c)3A through C may be multiplied by 0.8.

E. For reinforcement enclosed within spiral reinforcement not less than $1/4$-inch diameter and not more than 4 inches pitch, within No. 4 or larger circular ties spaced at not more than 4 inches on center, or within No. 4 or larger ties or stirrups spaced not more than 4 inches on center and arranged such that alternate bars shall have support provided by the corner of a tie or hoop with an included angle of not more than 135 degrees, the factors in Section 2612(c)3A through C may be multiplied by 0.75.

F. The basic development length multiplied by the applicable factor of Section 2612(c)A through 3C with modifiers of Section 2612(c)3D and/or E shall not be taken less than ....

.... $0.03 d_b f_y / \sqrt{f'_c}$

4. Basic development length l_{db} as modified by Section 2612(c)3 shall also be multiplied by applicable factor or factors for:

A. **Top reinforcement.** Horizontal reinforcement so placed that more than 12 inches of fresh concrete is cast in the member below the development length or splice 1.3

B. **Lightweight aggregate concrete** 1.3

Or when f_{ct} is specified .... $6.7 \sqrt{f'_c} / f_{ct}$

but not less than 1.0.

C. **Epoxy-coated reinforcement.** Bars with cover less than $3d_b$ or clear spacing between bars less than $6d_b$ .... 1.5.

All other conditions .... 1.2.

The product of the factor for top reinforcement of Section 2612(c)4A and the factor for epoxy-coated reinforcement of this section need not be taken greater than 1.7.

5. **Excess reinforcement.** Development length may be reduced where reinforcement in a flexural member is in excess of that required by analysis except where anchorage or development for f_y is specifically required or the reinforcement is designed under provisions of Section 2621(c)1E ....
.... [$(A_s$ required) $/ (A_s$ provided)]

(d) **Development of Deformed Bars in Compression.** [12.3] Development length l_d, in inches, for deformed bars in compression shall be computed as the product of the basic development length and applicable modification factors as defined in this section, but l_d shall be not less than 8 inches.

Basic development length shall be

.... $0.02 d_b f_y / \sqrt{f'_c}$

but not less than .... $0.0003 d_b f_y$

Basic development length may be multiplied by applicable factors for:

Reinforcement in excess of that required

by analysis .... $(A_s$ required) $/ (A_s$ provided)

Reinforcement enclosed within spiral reinforcement not less than $1/4$-inch diameter and not more than 4-inch pitch or within No. 4 ties in conformance with Section 2607(k)3 and spaced not more than 4 inches on center 0.75

7-7 LAP SPLICES FOR REINFORCING STEEL

It is not reasonable to build a reinforced masonry wall using a single continuous length of reinforcing steel. Instead, the steel is placed using bars that have been cut to manageable lengths. For these shorter lengths of steel to function as continuous reinforcement, they must be connected in some fashion.

The usual method is to lap the bars at specified length. The Uniform Building Code requires that reinforcing steel bars in tension have a lapped length of 40 bar diameters for Grade 40 steel and 48 bar diameters for Grade 60 steel. When the steel is in compression, the required lap length is 30 bar diameters for Grade 40 steel and 36 bar diameters for Grade 60 steel.

TABLE 7-8 Length of Lap (inches)[1] Grade 60 Steel, F_s = 24,000 psi

Bar Size		Laps for Compression Bars			Laps for Tension Bars		
No.	Dia., d_b (inches)	36 Dia.[2] Min.	36 x 1.3[3] = 47 Dia.	36 x 1.5[4] = 54 Dia.	48 Dia.[2] Min.	48 x 1.3[3] = 62 Dia.	48 x 1.5[4] = 72 Dia.
3	0.375	14	18	20	18	23	27
4	0.500	18	24	27	24	31	36
5	0.625	23	29	34	30	39	45
6	0.750	27	35	41	36	47	54
7	0.875	32	41	47	42	54	63
8	1.000	36	47	54	48	62	72
9	1.128	41	53	61	54	70	81
10	1.270	46	60	69	61	79	91
11	1.410	51	66	76	68	87	102

1. Based on UBC Sec. 2409(e)6 and 2409(e)3A and G.
2. Minimum development length, l_d, and lap splice.
3. Use when bars are separated by 3 inches or less.
4. Use where $f_s > 0.80 F_s$ per UBC Sec. 2409(e)3G.

Splices may be made only at certain locations and in such manner that the structural strength of the member will not be reduced.

The Uniform Building Code Sec.2409(e)6 requires that when adjacent bars (2 or more) are spaced 3 inches or closer together, the required lap length must be increased 30%. (See Table 7-8 and Figure 7-15.)

If the lapped splices are staggered, the lap lengths need not be increased.

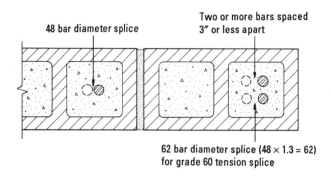

Figure 7-15 Lap splice of steel in cell.

7-8 SPLICES FOR JOINT REINFORCING

Properly designed splices are a key element in a well executed design. For joint reinforcing in principal framing elements, end anchorage terminating at the edges of a structure often must be made in joints where space is limited and the fit is complicated by crossing bars. Design and layout of anchorages and splices involve a unique combination of art and science in any complete reinforced masonry structural design. Practical considerations of cost, construction time and feasibility under normal construction conditions are of equal importance in meeting theoretical or building code requirements.

Splices for joint reinforcing are shown in Table 7-9 and Figure 7-16 based on UBC Section 2404(h). (See Table C-2 for additional properties of steel reinforcing wire.)

TABLE 7-9 Length of Lap of Wire (inches)[1,2]

Wire Size		54 Dia Lap[3]	75 Dia Lap[4]	Alternate Bed Joint Lap	
No.	Dia			Vertical Spacing of Joint Reinforcing	
				8"	16"
9	0.1483	8	11	24	40
8	0.1620	9	12	25	41
3/16"	0.1875	10	14	26	42

1. Based on UBC Sec. 2404(h).
2. Coordinate this Table with Figure 7-16.
3. Grouted cell.
4. Mortared bed joint.

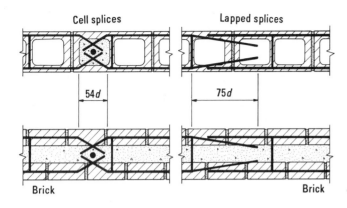

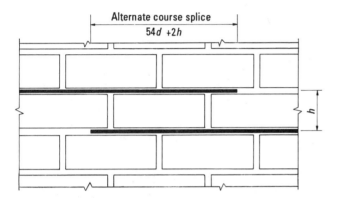

Figure 7-16 Typical splice arrangement for Ladder Type joint reinforcements. When calculations show that shorter splices are adequate in areas of minimum stress (e.g. inflection points), shorter splices may be used.

7-9 ANCHOR BOLTS

7-9.A Anchor Bolts in Masonry

Anchor bolts are used to tie masonry to structural supports and to transfer loads from masonry attachments such as ledgers, sill plates, etc. Some examples where anchor bolts may be used are, the connections between masonry walls and roofs, floors, ledger beams, and signs.

Conventional embedded anchor bolts are commonly specified as bent bar anchor bolts, plate anchor bolts and headed anchor bolts. They are available in standard sizes (diameters and lengths) or can be fabricated to meet specific project requirements.

Anchor bolts are commonly embedded at:

1. The surface of walls—for connecting relief angles and wood or steel ledger beams to the walls.

2. The top of walls—for attaching sill plates and base plates to the walls.

3. The top of columns—for anchoring steel bearing plates onto the columns.

Anchor bolts are generally divided into two categories:

1. Embedded anchor bolts which are placed in the grout during construction, and

2. Drilled in anchors which are placed after construction of the masonry.

Anchor bolts are subjected to shear and tension forces resulting from gravity loads, earthquakes, wind forces, differential movements, dynamic vibrations, etc. The magnitude of these loads vary significantly.

The values for shear and tension given in the code are generalized and in some cases very conservative. Tables A-7a, A-7b, and A-8 give allowable shear and tension capacities of typical size anchor bolts based on UBC Section 2406(h) and UBC Table Nos. 24-D-1, 24-D-2 and 24-E.

Note that anchor bolts subjected to combined shear and tension forces must be designed per UBC Section 2406h(4), Equation 6-33:

$$\frac{b_t}{B_t} + \frac{b_v}{B_v} \leq 1.0$$

7-9.B Effective Embedment Length

The minimum embedment depth l_b. per UBC Section 2406(h)6 must be 4 bolt diameters but not less than 2 inches. Table 7-10 lists minimum embedment depths for common size anchor bolts.

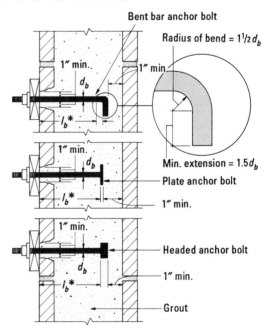

✱ Minimum embedment length $l_b = 4d_b$ but l_b may not be less than 2″. See Table 7-10 for minimum embedment lengths for common diameter anchor bolts.

Figure 7-17 Effective embedment.

TABLE 7-10 Minimum Embedment Depth[1]
(inches)

Diameter (inches)	Minimum Embedment (inches)
3/8	2
1/2	2
5/8	2 1/2
3/4	3
7/8	3 1/2
1	4
1 1/8	4 1/2
1 1/4	5

1. Based on UBC Section 2406(h)6 with a minimum embedment of 4 bolt diameters but not less than 2".

7-9.C Minimum Edge Distance and Spacing Requirements

The minimum edge distance, l_{be}, measured from the edge of the masonry parallel with the anchor bolt to the surface of the anchor bolt must be 1 1/2 inches per UBC Sec. 2406(h)5. Additionally, UBC Section 2406(h)7 requires the minimum center-to-center spacing between anchor bolts to be four bolt diameters.

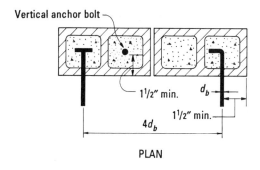

PLAN

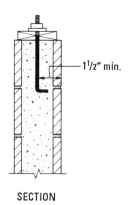

SECTION

Figure 7-18 Minimum edge distance.

7-10 BEAMS

7-10.A General

The Uniform Building Code does not specify a minimum amount of steel or steel ratio, p, for flexural beams. Engineering practice generally recommends that for masonry beams, the minimum reinforcement ratio, p, be not less than $80/F_y$. Therefore, for grade 60 steel, the minimum steel ratio should be $p = 80/60,000 = 0.0013$.

7-10.B Continuity of Reinforcing Steel in Flexural Members.

Continuity requirements for reinforcing steel in flexural members is given in the UBC Sec. 2409(e)3B, C, D, E, F and G.

UBC Sec. 2409(e)3B, C, D, E, F and G.

B. Except at supports or at the free end of cantilevers, every reinforcing bar shall be extended beyond the point at which it is no longer needed to resist tensile stress for a distance equal to 12 bar diameters or the depth of the beam, whichever is greater. No flexural bar shall be terminated in a tensile zone unless one of the following conditions is satisfied:

(i) The shear is not over one half that permitted, including allowance for shear reinforcement, if any.

(ii) Additional shear reinforcement in excess of that required is provided each way from the cutoff a distance equal to the depth of the beam. The shear reinforcement spacing shall not exceed $d/8r_b$, where r_b is the ratio of the area of bars cut off to the total area of bars at the section.

(iii) The continuing bars provide double the area required for flexure at that point or double the perimeter required for reinforcing bond.

C. At least one third of the total reinforcement provided for negative moment at the support shall be extended beyond the extreme position of the point of inflection a distance sufficient to develop one half the allowable stress in the bar, not less than one sixteenth of the clear span, nor the depth d of the member, whichever is greater.

D. Tensile reinforcement for negative moment in any span of a continuous restrained or cantilever beam, or in any member of a rigid frame, shall be adequately anchored by reinforcing bond, hooks or mechanical anchors in or through the supporting member.

E. At least one third of the required positive moment reinforcement in simple beams or at the freely supported end of continuous beams shall extend along the same face of the beam into the support at least 6 inches. At least one fourth of the required positive moment reinforcement at the continuous end of continuous beams shall extend along the same face of the beam into the support at least 6 inches.

F. Compression reinforcement in flexural members shall be anchored by ties or stirrups not less than 1/4 inch in diameter, spaced not farther apart than 16 bar diameters or 48 tie diameters. Such ties or stirrups shall be used throughout the distance where compression steel is required.

G. In regions of moment where the design tensile stresses in the steel are greater than 80 percent of the allowable steel tensile stress (F_s), the lap length of splices shall be increased not less than 50 percent of the minimum required

length. Other equivalent means of stress transfer to accomplish the same 50 percent increase may be used.

Figure 7-19 shows the UBC minimum requirements for continuity in flexural members. Note that it is often good engineering practice to continue much of the beam reinforcing through the length of the beam. Continuous bars which are adequately anchored and lapped provide a certain amount of redundancy and added safety into the structure. Continuous reinforcement eliminates much of the concern over whether the bars are properly placed in the field.

Similarly, ending bars in tension zones may allow cracks to form at the ends of the bars. Although the Uniform Building Code requires additional precautions for shear near the ends of such terminated bars, it is recommended that these bars be extended and anchored into the compression zone of the beam.

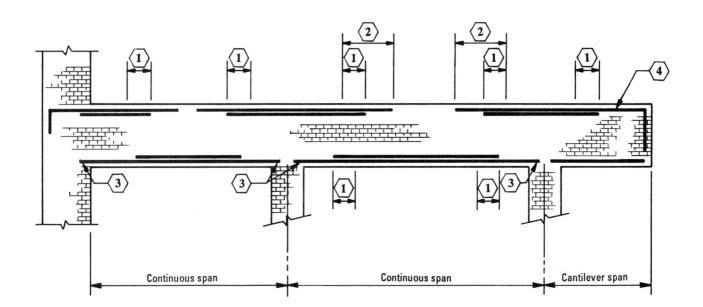

① Extend steel at least 12 bar diameters beyond the point where it is no longer required per UBC Sec. 2409(e)3B.

② Extend at least one third of negative moment reinforcing beyond the inflection point for the distance required in UBC Sec. 2409(e)3C.

③ Extend at least one fourth of the positive reinforcement from continuous beams into the support a distance of 6 inches per UBC Sec. 2409(e)3E.

④ No flexural bars shall be terminated in a tension zone unless the requirements of UBC Sec. 2409(e)3B are met.

Figure 7-19 Steel detailing for continuity.

7-11 TIES FOR BEAM STEEL IN COMPRESSION

Compression reinforcement in flexural members must be tied to secure it in position and to prevent it from buckling. UBC Section 2409(e)3F states the requirements for ties:

UBC Sec. 2409(e)3F

F. Compression reinforcement in flexural members shall be anchored by ties or stirrups not less than ¼ inch in diameter, spaced not farther apart than 16 bar diameters or 48 tie diameters. Such ties or stirrups shall be used throughout the distance where compression steel is required.

Compression steel in top of beam

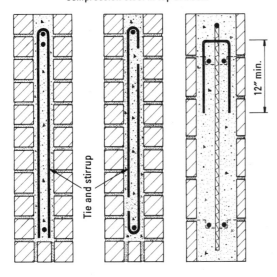

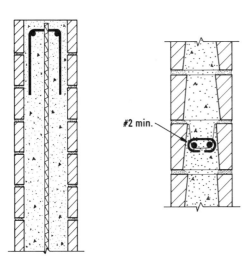

Figure 7-20 Ties for compression steel in beams.

7-12 SHEAR REINFORCEMENT REQUIREMENTS IN BEAMS

7-12.A General

Section 2409(d)1 of the Uniform Building Code requires that shear reinforcement be provided when the computed shear stress exceeds the allowable shear stress and that the shear reinforcement be designed to resist the entire shear force.

For beams, the maximum shearing forces are generally at the end of the beams, with less shear force near the middle. It follows that shear reinforcing bars will be required to be spaced more closely near the beam end. As a minimum, the Uniform Building Code requires in Sec. 2409(d)2 that web reinforcement be spaced so that every potential 45-degree crack extending from a point at $d/2$ of the beam to the longitudinal tension steel be crossed by at least one shear reinforcing bar.

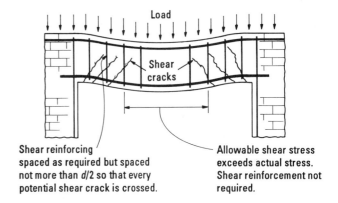

Figure 7-21 Beam showing potential shear cracks and shear reinforcing bars.

7-12.B Types of Shear Reinforcement

Web reinforcement may consist of:

1. Stirrups or web reinforcement bars perpendicular to the longitudinal steel.

2. Stirrups or web reinforcement bars rigidly attached to the longitudinal steel and making an angle of 30 degrees or more thereto.

3. Longitudinal bars bent so that the axis of the inclined portion of the bar makes an angle of 15 degrees or more with the axis of the longitudinal portion of the bar.

4. Special arrangements of bars with adequate provisions to prevent slip of bars or splitting of masonry by the reinforcement.

7-12.C Anchorage of Shear Reinforcement

Bars used as shear reinforcement must be anchored at each end by one of the methods in UBC Sec. 2409(e).

UBC Sec. 2409(e)4A, B and C.

4. Anchorage of shear reinforcement. A. Single separate bars used as shear reinforcement shall be anchored at each end by one of the following methods:

(i) Hooking tightly around the longitudinal reinforcement through 180 degrees.

(ii) Embedment above or below the mid-depth of the beam on the compression side a distance sufficient to develop the stress in the bar for plain or deformed bars.

(iii) By a standard hook [see Section 2409(e)5] considered as developing 7,500 psi, plus embedment sufficient to develop the remainder of the stress to which the bar is subjected. The effective embedded length shall not be assumed to exceed the distance between the mid-depth of the beam and the tangent of the hook.

B. The ends of bars forming a single U or multiple U stirrup shall be anchored by one of the methods of Section 2409(e)4A or shall be bent through an angle of at least 90 degrees tightly around a longitudinal reinforcing bar not less in diameter than the stirrup bar, and shall project beyond the bend at least 12 diameters of the stirrup.

C. The loops or closed ends of simple U or multiple U stirrups shall be anchored by bending around the longitudinal reinforcement through an angle of at least 90 degrees and project beyond the end of the bend at least 12 diameters of the stirrup bar.

7-12.D Shear Reinforcing Details

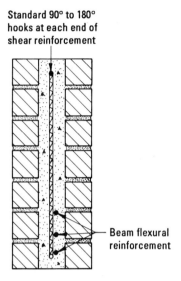

Standard 90° to 180° hooks at each end of shear reinforcement

Beam flexural reinforcement

Figure 7-22 Cross section of beam showing vertical shear reinforcing steel.

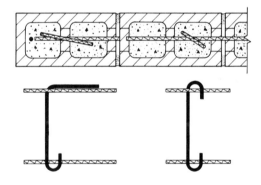

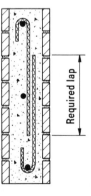

Required lap

Figure 7-23 Anchorage details for shear reinforcement.

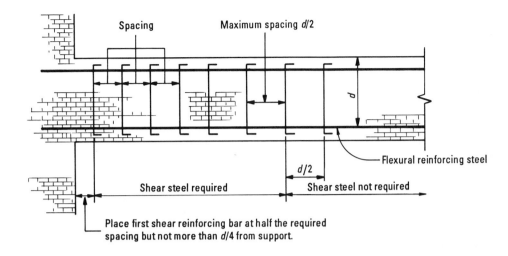

Spacing

Maximum spacing $d/2$

d

Flexural reinforcing steel

$d/2$

Shear steel required

Shear steel not required

Place first shear reinforcing bar at half the required spacing but not more than $d/4$ from support.

Figure 7-24 Vertical web or shear reinforcing steel arrangement for beams.

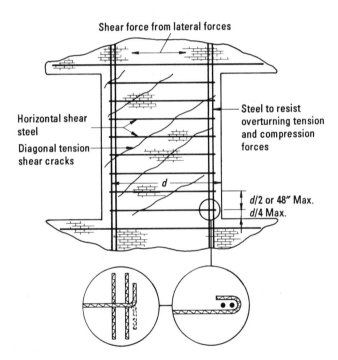

Shear force from lateral forces

Horizontal shear steel

Diagonal tension shear cracks

Steel to resist overturning tension and compression forces

d

$d/2$ or 48″ Max.
$d/4$ Max.

Figure 7-25 Shear wall reinforced with horizontal steel to resist lateral shear forces induced by wind or seismic forces.

7-13 COMPRESSION JAMB STEEL AT THE END OF PIERS AND SHEAR WALLS

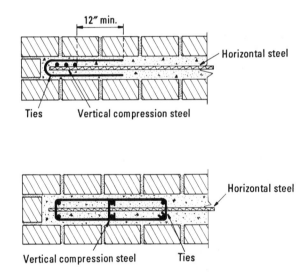

12″ min.

Horizontal steel

Ties   Vertical compression steel

Horizontal steel

Vertical compression steel   Ties

Figure 7-26 Door jamb reinforcing at the ends of brick walls or piers.

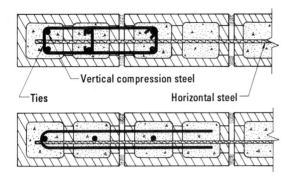

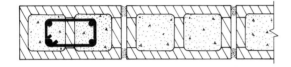

Figure 7-27 Door jamb reinforcing at the ends of concrete masonry walls.

7-14 COLUMNS

7-14.A General

In the design of columns, the vertical reinforcing steel significantly contributes to the load-carrying capacity of the member because the ties prevent the reinforcing from buckling.

Columns may be categorized by their location; they may be isolated (free standing), projecting from a wall, or flush in a wall. The least dimension of columns should not be less than 12 inches, unless they are designed for $1/2$ of the allowable stresses, then they may be as small as 8 inches square.

> ### UBC Sec. 2407(h)4 E(ii)
>
> (ii) **Columns.** The least nominal dimension of a reinforced masonry column shall be 12 inches except that if the allowable stresses are reduced to one half the values given in Section 2406, the minimum nominal dimension shall be 8 inches.

> ### UBC Sec. 2409(b)5A
>
> 5. **Reinforcement for columns. A. Vertical reinforcement.** The area of vertical reinforcement shall be not less than $0.005A_e$ and not more than $0.04A_e$. At least four No. 3 bars shall be provided.

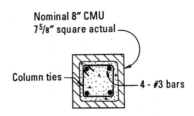

Figure 7-28 Minimum column size and reinforcing.

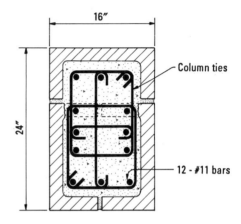

Figure 7-29 Maximum amount of steel in a 16″ × 24″ column.

7-14.B Projecting Wall Columns or Pilasters

Heavily loaded girders which frame into a wall may require substantial base plates and columns. In order to provide a convenient girder seat and adequate column capacity, columns called pilasters are often built projecting out from the face of the wall.

Projecting pilasters also serve to stiffen the wall if they are adequately supported at the top and bottom. The wall between pilasters can then be designed to span horizontally allowing very high walls to be built using only nominal masonry thicknesses.

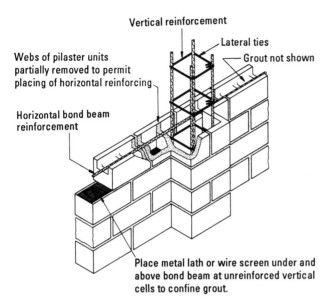

Figure 7-30 *Construction of reinforced concrete masonry pilaster with continuous bond beams.*

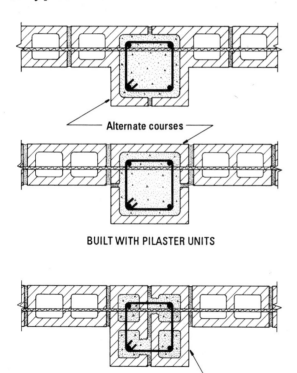

BUILT WITH PILASTER UNITS

BUILT WITH TWO CORE STANDARD MASONRY UNITS

Figure 7-31 Projecting concrete masonry wall column details.

7-14.C Flush Wall Columns

If engineering design permits, it will economically benefit the owner and the contractor to build columns that are contained in the wall and are flush with the wall. Wall-contained columns permit faster construction, since there are no projections from the wall and no special units are required. The reinforcing steel must be tied in accordance with the code requirements.

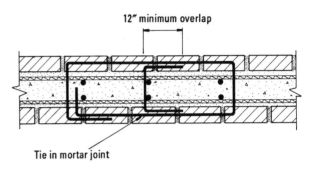

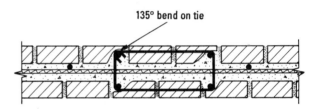

Figure 7-32 Flush wall brick columns with ties in mortar joint.

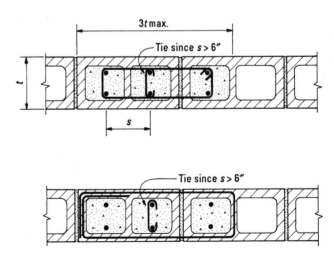

Figure 7-33 Flush wall concrete masonry wall columns.

7-14.D Column Tie Requirements

UBC Sec. 2409(b)5B covers the requirements for column ties based on the longitudinal bar and tie diameters.

UBC Sec. 2409(b)5B.

B. **Lateral ties.** All longitudinal bars for columns shall be enclosed by lateral ties. Lateral support shall be provided to the longitudinal bars by the corner of a complete tie having an included angle of not more than 135 degrees or by a hook at the end of a tie. The corner bars shall have such support provided by a complete tie enclosing the longitudinal bars. Alternate longitudinal bars shall have such lateral support provided by ties and no bar shall be farther than 6 inches from such laterally supported bar.

Lateral ties and longitudinal bars shall be placed not less than 1½ inches and not more than 5 inches from the surface of the column. Lateral ties may be against the longitudinal bars or placed in the horizontal bed joints if the requirements of Section 2407(f) are met. Spacing of ties shall be not more than 16 longitudinal bar diameters, 48 tie diameters or the least dimension of the column but not more than 18 inches.

Ties shall be at least ¼ inch in diameter for No. 7 or smaller longitudinal bars and No. 3 for larger longitudinal bars. Ties less than ³⁄₈ inch in diameter may be used for longitudinal bars larger than No. 7, provided the total cross-sectional area of such smaller ties crossing a longitudinal plane is equal to that of the larger ties at their required spacing.

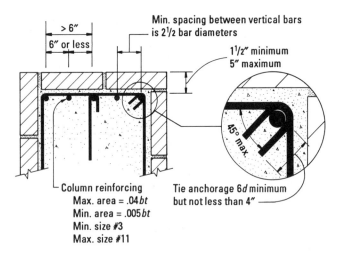

Min. spacing between vertical bars is 2½ bar diameters

1½" minimum
5" maximum

45° max.

Column reinforcing
Max. area = .04*bt*
Min. area = .005*bt*
Min. size #3
Max. size #11

Tie anchorage 6*d* minimum but not less than 4"

Figure 7-34 Reinforcing tie details.

TABLE 7-11 Maximum Tie Spacing Based on Longitudinal Bar Size[1]

Compression Steel Bar No.	Maximum Tie Spacing (inches)
#3	6
#4	8
#5	10
#6	12
#7	14
#8	16
#9	18
#10	18
#11	18

1. Based on UBC Section 2409(b)5B. Maximum tie spacing may not exceed 16 longitudinal bar diameters, 48 tie diameters, 18" nor the least column dimension. Coordinate this Table with Table 7-12.

TABLE 7-12 Maximum Tie Spacing Based on Tie Size[1]

Tie Steel Size	Maximum Tie Spacing (inches)
¼"	12
#3	18
#4	18
#5	18

1. Based on UBC Section 2409(b)5B. Maximum tie spacing may not exceed 16 longitudinal bar diameters, 48 tie diameters, 18" nor the least column dimension. Coordinate this Table with Table 7-11.

Note: #2 (¼") ties at 8" spacing is equivalent to #3 (³⁄₈") tie at 16" spacing.

7-14.E Lateral Tie Spacing

7-14.E.1 Lateral Tie Spacing, Seismic Zone Nos. 0, 1 and 2.

There are no special tie spacing requirements for Seismic Zone Nos. 0, 1 or 2. Therefore, the normal spacing of 16 bar diameters, 48 tie diameters, 18 inches or least column dimension which ever is less applies. See Figure 7-35.

Specific lateral tie spacing requirements for buildings located in Seismic Zone Nos. 3 and 4 are shown in Figure 7-36.

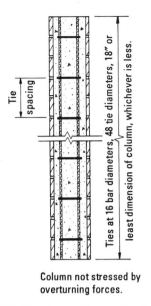

Column not stressed by
overturning forces.

*Figure 7-35 Maximum tie spacing in columns in
Seismic Zone Nos. 0, 1 & 2.*

7-14.E.2 Lateral Tie Spacing in Seismic Zone Nos. 3 and 4.

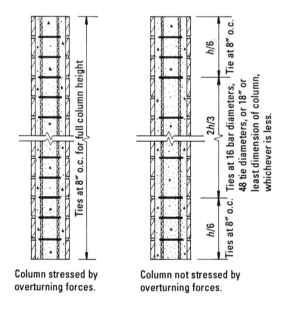

Column stressed by
overturning forces.

Column not stressed by
overturning forces.

*Figure 7-36 Maximum tie spacing in columns in
Seismic Zone Nos. 3 and 4.*

7-14.F   Ties Around Anchor Bolts on Columns

UBC Sec. 2409(b)5C.

C. **Anchor bolt ties.** Additional ties shall be provided around anchor bolts which are set in the top of the column. Such ties shall engage at least four bolts or, alternately, at least four vertical column bars or a combination of bolts and bars totaling four in number. Such ties shall be located within the top 5 inches of the column and shall provide a total of 0.4 square inch or more in cross-sectional area. The uppermost tie shall be within 2 inches of the top of the column.

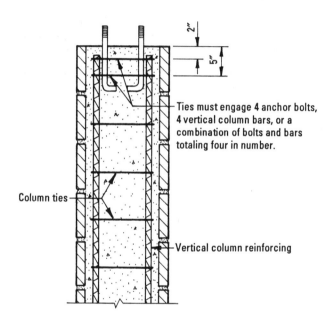

Figure 7-37 Ties at anchor bolts in the top of columns.

7-15 QUESTIONS AND PROBLEMS

7-1   What are the minimum reinforcing steel requirements for Seismic Zones Nos. 0, 1, 2A and 2B.

7-2   What are two reasons to provide steel around openings in a wall?

7-3   Under what conditions or uses is joint reinforcing more desirable than reinforcing bars?

7-4   Detail the reinforcing required for a two-story building located in Seismic Zone No. 2B. Show the reinforcing at the corners, floor and roof, around openings, etc.

7-5   What is the minimum amount of reinforcing required for walls in Seismic Zone No. 3? If the vertical steel is in the center of a 9″ brick wall and the steel ratio, p, = 0.004, how much steel must be used horizontally. Specify an appropriate size and spacing of reinforcing bars. If p = 0.002, what is the size, spacing and steel ratio of the horizontal steel?

7-6   A 10 inch solid grouted masonry wall has #6 bars 18 inches o.c. vertically. How much steel must be placed horizontally to comply with the minimum code requirements for Seismic Zone No. 4?

7-7   Determine the minimum steel required for a 10 inch brick wall, 18 ft high located in Seismic Zone No. 3. The parapet extends 30 inches above the roof line. Use joint reinforcement between the footing and bond beam. Assume two #4 bars are used in the bond beam and at the top of the footing. Also assume the wall spans vertically. Use minimum steel requirements without structural calculations.

SECTION 8

Building Details

8-1 GENERAL

One of the most important considerations of any structure subjected to seismic forces is the connections. If connections hold together and make the structure perform as a total system there is an excellent chance for the structure to survive even great earthquakes.

Details of reinforcing bar size and spacing is dependent on engineering requirements. All connections must be satisfactory to transmit the forces due to lateral and vertical loads. The elements must be sufficiently tied together to cause them to act as a unit.

This section shows some of the more typical wall connections and building details.

8-2 WALL TO WALL CONNECTIONS

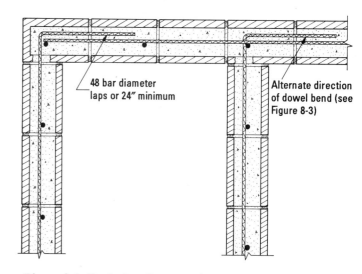

Figure 8-1 Typical wall connections.

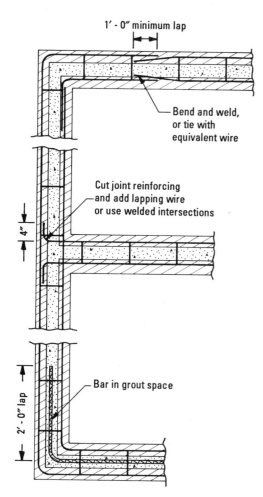

Figure 8-2 Typical wall connections—plan of joint reinforcing showing intersection and alternate lapping.

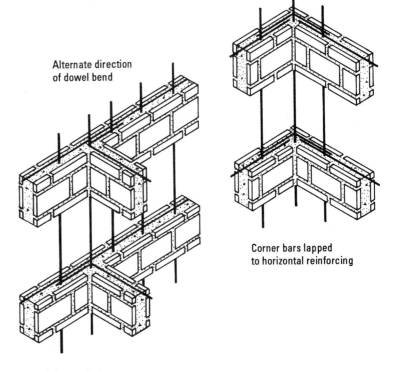

Figure 8-3 Exploded isometric view of reinforcing steel for intersecting walls.

8-3 LINTEL AND BOND BEAM CONNECTION TO WALL

Masonry beam over an opening.

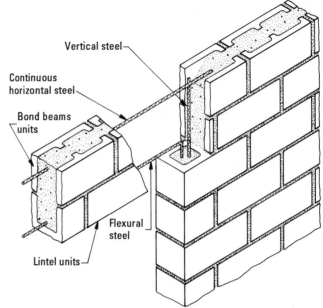

Figure 8-4 Lintel and bond beam detail.

8-4 WALL TO WOOD DIAPHRAGM CONNECTIONS

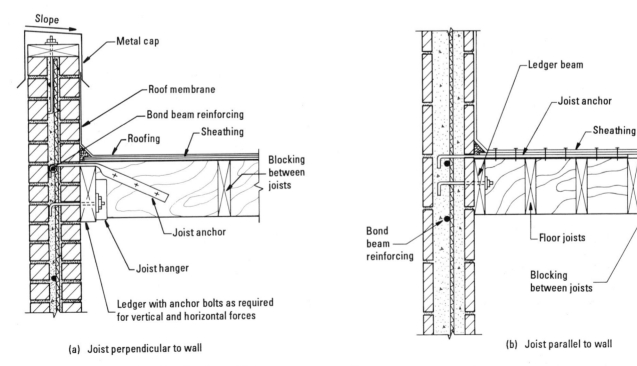

(a) Joist perpendicular to wall

(b) Joist parallel to wall

Figure 8-5 Connection details of wood joists to masonry walls.

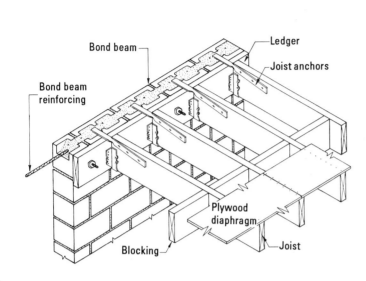

Figure 8-6 Isometric view of connection of wood diaphragm to masonry wall.

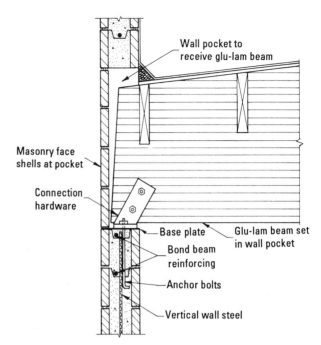
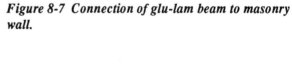

Figure 8-7 Connection of glu-lam beam to masonry wall.

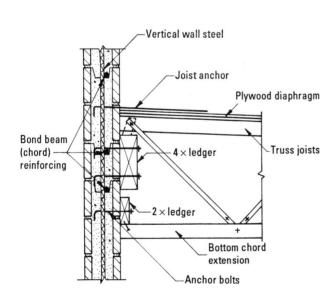
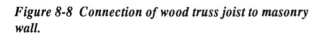

Figure 8-8 Connection of wood truss joist to masonry wall.

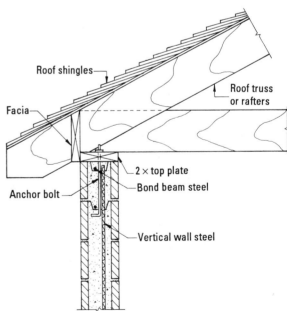

Figure 8-9 Connection of roof rafters or truss to masonry wall.

8-5 WALL TO CONCRETE DIAPHRAGM CONNECTIONS

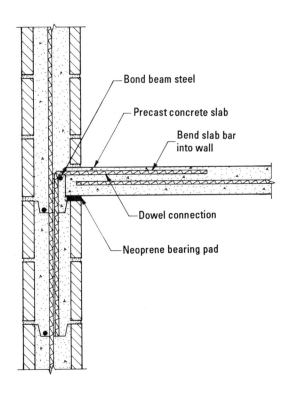

Figure 8-10 Precast slab bearing on masonry wall.

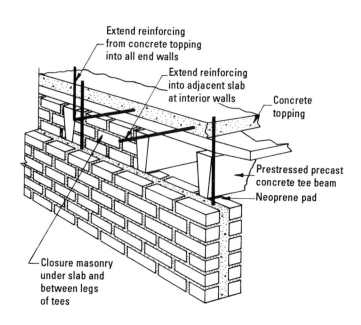

Figure 8-11 Precast tee beam bearing on masonry wall.

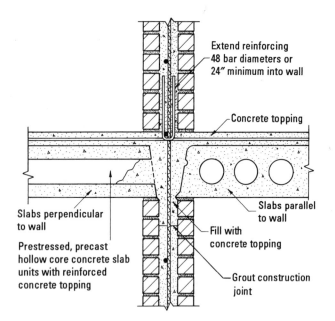

Figure 8-12 Precast, prestressed hollow core slabs with concrete topping on masonry wall.

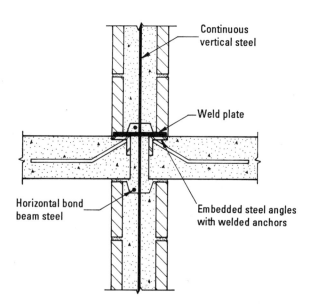

Figure 8-13 Precast concrete slabs connected to masonry wall with welded connections.

8-6 WALL TO STEEL DIAPHRAGM CONNECTIONS

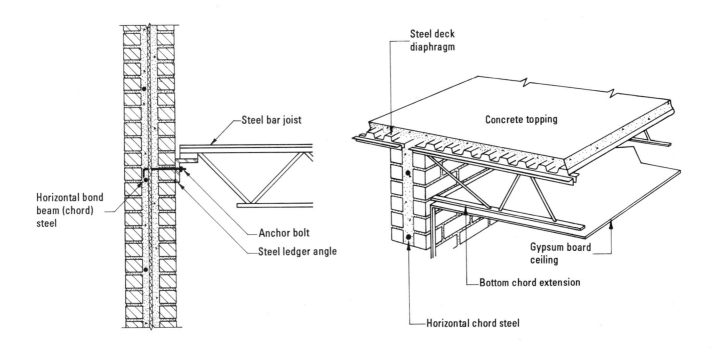

Figure 8-14 Steel bar joist floor or roof system connected to masonry wall with a ledger angle.

Figure 8-15 Isometric view of connection of steel bar joist floor system to masonry wall.

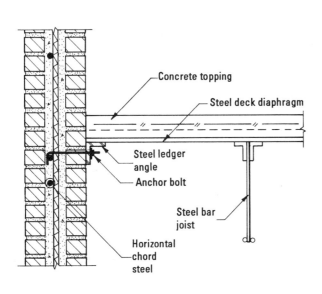

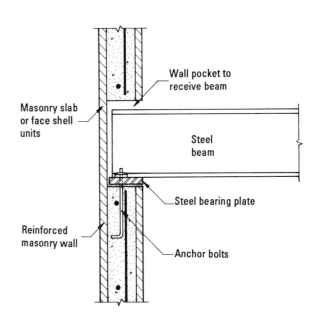

Figure 8-16 Steel bar joist and roof deck connection with bar joists parallel to wall.

Figure 8-17 Steel beam bearing on masonry wall.

8-7   WALL FOUNDATION DETAILS

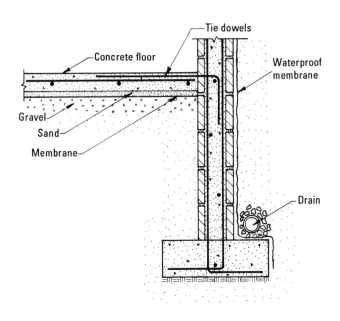

Figure 8-18 Exterior bearing wall with earth backfill.

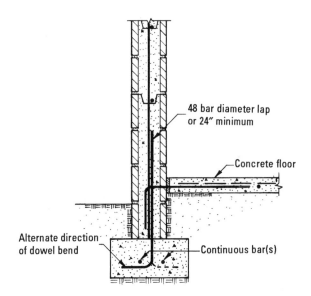

Figure 8-19 Typical footing detail for exterior masonry wall.

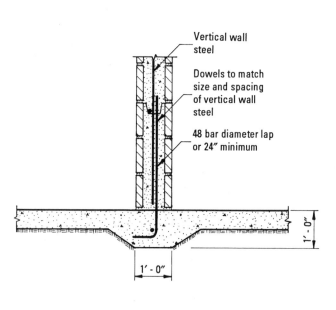

Figure 8-20 Typical interior, non-load bearing wall and footing detail.

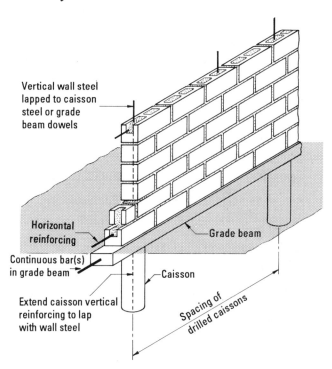

Figure 8-21 Grade beam and caisson system for supporting masonry wall.

Special Topics

9-1 MOVEMENT JOINTS

9-1.A General

All structures move when they are subjected to changes in moisture, temperature and loads. The movements can cause damage or cracks especially when no provisions are made to allow the structure to move.

The type, location and spacing of movement joints depend on the masonry materials, climatic conditions, size and type of structure, local factors and experience.

Movement joints should be located where they will least impair the strength of the finished structure, where they will not adversely affect the architectural design, and where they can facilitate the construction of the walls. They should never be located by chance or convenience without regard for their effect on the strength or the appearance of the completed structure.

Movement joints in a wall, whether they are control joints or expansion joints, should match any joints built into the roof system, the floor system, the spandrel beams or other elements intended to accommodate the overall movement of the building.

9-1.B Movement Joints for Clay Masonry Structures

9-1.B.1 General

Clay masonry units are normally smallest just after firing. As they gain moisture they may expand 0.02 percent for each percent of moisture increase. Thus, a 100 foot long brick wall may increase in length approximately $1/4''$ for each percent of moisture increase. If the wall is restrained from movement, high compressive stresses may develop, often high enough to crush the brick or push an adjoining wall out of plumb.

Clay masonry units also expand approximately 0.036% per 100°F temperature increase.

9-1.B.2 Vertical Expansion Joints

Expansion joints are used to accommodate an increase in length and height of a masonry wall due to thermal expansion or swelling of the clay masonry from a moisture increase. The need for expansion joints depends upon:

1. The climatic area in which the structure is located
2. Dimensions and configuration of the building
3. Temperature change and provisions for temperature control
4. Type of structural frame, connection to the foundation, and symmetry of stiffness against lateral displacement
5. Materials of construction

The Brick Institute of America's Technical Note 18A estimates unrestrained expansion of clay masonry as:

$$\Delta L = [0.0005 + 0.000004(\Delta T)]L$$

Where ΔL = unrestrained expansion, inches

ΔT = difference in change of temperature of the wall from minimum to maximum, degrees Farenheit

L = Length of wall in inches

Note however that the BIA recognizes the total amount of expansion as somewhat less due to indeterminate factors such as restraint, shrinkage and plastic flow of mortar and variations in workmanship.

9-1.B.3 Location and Spacing of Expansion Joints

Spacing of expansion joints should be between 150 to 200 feet for a straight length of wall and should be located with consideration to the shape, plan and elevation of the structure. The expansion joints should be filled with a caulking material that will expand and compress to prevent moisture penetration through the joint.

Vertical control joints are used to divide long walls or to separate changes in height or thickness at the junctions in T, U or L shaped buildings (Figure 9-1). They may be installed at the abutment of walls and columns, and at one or both sides of wall openings.

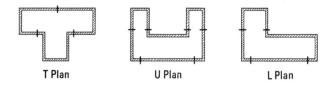

T Plan   U Plan   L Plan

Figure 9-1 Location of expansion joints in irregular shaped buildings.

Because of the moisture expansion of clay masonry, brick veneer walls may have narrow movement joints spaced 25 to 30 feet apart. Figure 9-2 shows typical types of vertical joints.

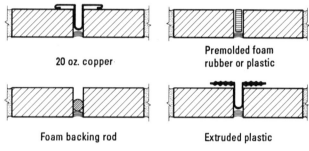

20 oz. copper   Premolded foam rubber or plastic

Foam backing rod   Extruded plastic

Figure 9-2 Details of vertical expansion joints.

9-1.B.4   Horizontal Expansion Joints

Horizontal expansion joints, or soft joints (Figure 9-3), are especially important where brick veneer is placed on high-rise buildings. These joints allow vertical short-

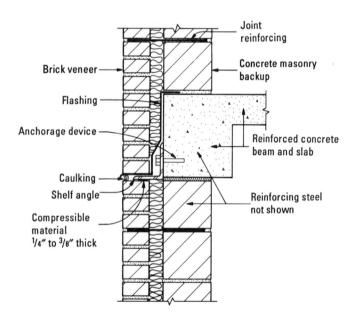

Joint reinforcing
Concrete masonry backup
Brick veneer
Flashing
Anchorage device
Reinforced concrete beam and slab
Caulking
Shelf angle
Reinforcing steel not shown
Compressible material 1/4″ to 3/8″ thick

Figure 9-3 Horizontal expansion joint to allow movement of the wall.

ening of the building frame, vertical deflection of the supporting members and expansion of the brick veneer. Their absence can create severe problems in both reinforced concrete frame buildings and in masonry buildings with exterior clay-brick wythes. The combined effect of drying shrinkage, creep and plastic flow in a structural frame reduces the building's floor-to-floor height. Any expansion of the clay brick veneer facing adds to the problem. Without the horizontal soft joints between the bottom of a shelf angle and the top course of the masonry panel below it, cracking or crushing is likely to occur.

To properly account for these vertical movements, a compressible material should be provided at the top of the panel just below the supporting shelf angle (Figures 9-3 and 9-4).

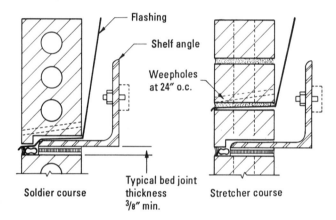

Flashing
Shelf angle
Weepholes at 24″ o.c.
Soldier course
Typical bed joint thickness 3/8″ min.
Stretcher course

Figure 9-4 Cut units to reduce height of exposed movement joint at support angle.

9-1.C   Movement Joints in Concrete Masonry Structures

9-1.C.1   General

Concrete masonry units are subjected to significant shrinkage due to moisture loss and/or temperature decrease. Of particular concern is the drying shrinkage of the concrete masonry during the early curing and drying of masonry walls which introduces tension stresses into the masonry units. During the construction of concrete masonry walls, wet fluid grout adds moisture to the masonry wall, causing it to expand. As the grout hydrates and hardens and as the masonry units dry out, the wall tries to shrink. Since the face shells lose moisture rapidly in a dry climate, they shrink quickly, thus putting them into tension while the interior of the wall is subjected to compression (Figure 9-5). Cracking of concrete masonry units will occur if these tensile stresses exceed the tensile strength of the materials.

Several major factors affect the volume change of concrete masonry generated by moisture fluctuations including the type of aggregate and the curing method. Generally concrete masonry units made with normal weight sand and gravel aggregate exhibit less volume change than those made with light weight aggregate or cinders. Similarly, units cured by steam pressure or autoclaving show a significant decrease in volume change characteristics compared to masonry cured by air. Coefficients for the moisture related volume change of concrete masonry units vary from about 0.01% to 0.1%.

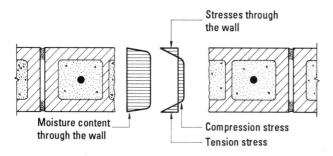

Figure 9-5 Moisture content and shrinkage stresses in a concrete masonry wall.

9-1.C.2   Crack Control for Concrete Masonry

Major methods employed to control cracking in concrete masonry structures are:

1. Specifying materials which limit the drying-shrinkage potential, (i.e., Type I moisture controlled units).

2. Reinforcing the masonry to resist tension stresses and thus increase crack resistance.

3. Providing control joints to accommodate the expected movement.

Although the use of Type I moisture controlled units are more satisfactory in their relative shrinkage characteristics than Type II non-moisture controlled units, their possible increase in cost or their unavailability can make specifying Type I units undesirable.

9-1.C.3   Control Joints in Concrete Masonry Walls

Shrinkage control joints panelize a wall, allowing shrinkage to take place within a small, relatively unrestrained panel. The panel can shorten in length allowing potential cracks to occur at the control joint.

These control joints are basically weakened head joints which extend vertically straight up and down the wall through the use of full and half masonry units. The mortar at the control joints is either left out entirely or deeply raked back. Joint reinforcing is generally termi-

nated at the control joint, although chord reinforcing steel at floors and roofs must continue through the control joints. Likewise any of the horizontal reinforcing required for structural considerations such as lintel reinforcing should continue through the joints.

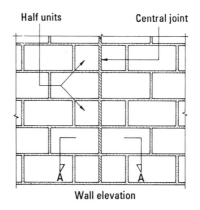

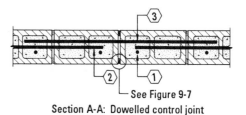

Section A-A: Dowelled control joint

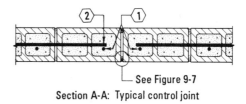

Section A-A: Typical control joint

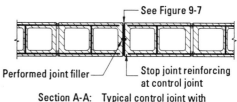

Section A-A: Typical control joint with joint reinforcing

1   Additional vertical bars on each side of all control joints.

2   Terminate all non-structural reinforcing 2″ from control joints.

3   Provide 4′ - 0″ long smooth dowels across the joint. Prevent bond between bar and grout with grease or a plastic sleeve. Cap all dowels to allow 1″ of movement.

Figure 9-6 Typical control joint detail in concrete masonry walls.

9-1.C.4 Spacing of Vertical Control Joints

Vertical control joints are usually spaced at close intervals so that when shortening takes place, the resulting crack will occur at the joint. It is important to provide sufficient control joints so that the movement occurs at the joint rather than midway between the control joints.

The maximum horizontal spacing of control joints in unreinforced concrete masonry walls should be approximately 32 feet, but a length (spacing) to height of wall ratio is a better measure. See Table 9-1 for the maximum recommended control joint spacing of horizontally reinforced walls.

TABLE 9-1 Spacing of Control Joints for Type I Concrete Masonry Units with Horizontal Reinforcing[1,2]

Recommended Spacing of Control Joints	Vertical Spacing of Joint or Horizontal Reinforcement (Inches)			
	48	24	16	8
Expressed as Ratio of Panel Length to Height, *L/H*	2	2½	3	4
With Panel Length (*L*) Not to Exceed:	40 ft	45 ft	50 ft	60 ft

1. The spacing of control joints should be reduced for Type II non-moisture controlled units. It may be reduced by one-half or an amount based on experience or practice in the area where the project is located.
2. Based on NCMA TEK Note 3, Table 1.

9-1.C.5 Vertical Expansion Joints in Concrete Masonry Walls

Generally, concrete masonry buildings less than 200 feet long do not require expansion joints if adequate control joints have been provided. However, if a concrete masonry structure is of unusual size or length or if it is subjected to severe conditions, expansion joints are advisable.

Additionally, the need for thermal expansion joints in long buildings should be determined based on local practice.

9-1.D Caulking Details

Movement joints should be constructed as continuous vertical head joints, by using full and half masonry units, and by raking back the mortar at least one inch deep. The raked vertical head joint should then be caulked to keep it weather-proof. A backer rod should be provided in the joint to limit the depth of the caulking and to limit the adhesion of the caulk to the ends of the block.

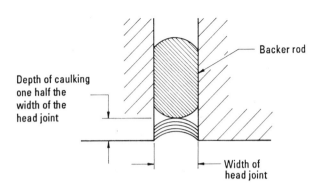

Figure 9-7 Caulking detail.

9-2 WATERPROOFING MASONRY STRUCTURES

9-2.A General

Masonry materials are relatively porous and may absorb water under certain conditions. Since water infiltration can deteriorate the masonry as well as damage a building's appearance and interior finishes, every effort should be taken to repel water infiltration.

Normally masonry structures are protected from water damage by one of two methods. The first method consists of constructing the walls with an exterior masonry veneer separated from the structural back-up by an air gap. Any water which penetrates the veneer runs down the back side of the veneer since it can not cross the air gap. Flashing and weep holes at the base of the cavity direct the water back out the wall, thus keeping the interior of the building dry. This cavity wall system is quite effective and has been used extensively in the past. The Brick Institute of America's *Technical Notes*, as well as other publications provide excellent design and detailing procedures for this type of wall.

The second method to limit water damage is to repel water infiltration through proper design techniques, material selection, construction methods, surface treatments and maintenance. Single wythe masonry walls and most reinforced masonry walls must be waterproofed in this fashion. The remainder of this section will provide general guidelines to effectively waterproof these types of walls.

9-2.B Design Considerations

Thoughtful design and careful detailing of a masonry building can significantly reduce potential leaks. Special attention should be given to leak-prone areas as described hereafter.

9-2.B.1   Mortar Joints

Certain types of mortar joints, such as concave and V type joints, are much more weather resistant than others, as indicated in Figure 1-10. Well tooled joints compact the mortar, filling voids and cracks which could lead to water migration.

9-2.B.2   Parapets and Fire Walls

Exposed on both faces, parapets and firewalls are subjected to high wind forces, extreme rain and snow, and severe temperature fluctuations. Providing a well constructed wall cap and a positive membrane water-proofing on the roof side of these walls can effectively eliminate water penetration.

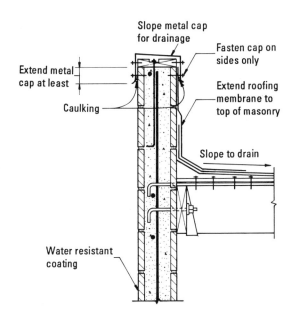

Figure 9-8 Parapet wall detail.

9-2.B.3   Movement Joints

Generally too few movement joints are provided in masonry structures to properly accommodate moisture and temperature fluctuations. Shrinkage and temperature hairline cracks which can develop without these joints allow water passage through the masonry. Additionally, leakage can occur at the movement joints themselves, through cracked, unbonded or misapplied caulks and sealants.

For more information regarding movement joints, see Section 9-1.

9-2.B.4   Horizontal Surfaces - Projections, Ledges and Sills

Horizontal surfaces contribute to the possibility of water penetration. Ledges and sills are particularly vulnerable as water may penetrate the top of the mortar joints, causing cracking and spalling. If possible, slope all projections, ledges and sills or provide a sloped flashing above them (Figure 9-9).

Poor ledge detail

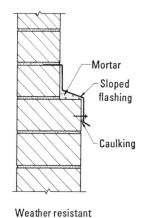

Weather resistant ledge detail

Figure 9-9 Ledge detail.

9-2.B.5   Copings and Wall Caps

Ample slope should also be provided on the top of all copings and wall caps so that water may be shed quickly. Masonry and precast copings should extend past the face of the wall to reduce water penetration through the joint between the coping and the wall. Additionally, all overhangs should have drip edges to prevent water migration along the bottom of the wall cap and then down the wall (Figure 9-10).

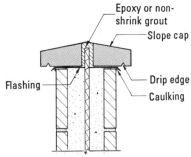

Figure 9-10 Typical drip edge on precast concrete wall cap.

Mortar caps should be avoided since they crack easily and are quite porous. If a mortar cap must be used, add a latex admixture to the mortar to reduce cracking and to increase the tensile strength and bond of the mortar.

Metal wall caps can prevent water penetration effectively, provided joints between cap pieces are lapped or sealed and provided the cap extends down the face of the masonry sufficiently. Because walls are often topped with 2 × 6 or 8 wood nailers and because the metal caps slope, the extension of the skirt should be 4 to 6 inches in order to extend sufficiently over the masonry (Figure 9-8).

9-2.B.6   Wall Penetrations

Possibly the most overlooked areas subjected to leaking are wall openings. Door and window frames must be installed and caulked properly to eliminate leaking. Likewise, penetrations for plumbing and electrical conduits will leak profusely if not properly flashed and caulked.

9-2.C   Material Selection

Because of the numerous types and diversity of masonry materials and products it is often difficult to select the best materials for a particular application. Each material has its own traits and characteristics making it useful in only certain types of construction. This section covers only the basic concerns in selecting materials for water resistant structures. For additional information, see Section 1 of this book which discusses common masonry materials along with their properties and uses. Also refer to manufacturer literature to select masonry materials which will provide the best resistance to water infiltration.

Common waterproofing products are also discussed in Section 9-2.E and can be used for general information on these products. Under all circumstances, the product or material manufacturer should be consulted to obtain specific product information.

Specified materials should be of high quality meeting all the appropriate standards of the industry. Choose products from reputable manufacturers who have a history of successful use of the desired product. Where appropriate, require guarantees of at least 5 years, especially for applied waterproofing products and sealants. Always completely follow the manufacturer's installation instructions.

Concrete masonry materials should be in climatic balance at the time of installation to limit the possibility of drying shrinkage cracks. "Green" block which has not cured thoroughly or which is wet and has not achieved climatic balance, shrinks substantially and can develop numerous cracks despite proper control joints or reinforcing steel. In areas of heavy precipitation, Type I (ASTM C 90-90) or Grade N Type I (ASTM C 90-85) concrete masonry units are warranted if available. Only Grades SW (severe weathering) or MW (moderate weathering) brick should be used in exterior applications.

9-2.D   Construction Procedures and Application Methods

Workmanship is very important and critical to provide good water resistance. Quality work with proper materials helps insure weathertight walls. Because of this, choose qualified, well established contractors for all aspects of construction.

Masonry industry standards and procedures should be followed throughout the construction process to help eliminate the potential for water penetration. Special care should be taken to provide adequate bond between the masonry units and the mortar since leaks often occur at the bed joints. Masonry materials should be properly stored, generally off the ground and away from detrimental materials. If exposed to rain or snow, they should be covered because excessively wet units may not adequately bond to mortar and grout. Additionally, drying shrinkage cracks and efflorescence can develop if masonry materials become saturated.

Mortar and grout must be mixed thoroughly. As previously mentioned, tooled mortar joints compact the mortar, reducing cracks and improving bond. Additionally full head and bed joints are often prudent.

Grout should contain sufficient water for a slump of 8 to 10 inches to flow readily into small voids and cavities. Thoroughly consolidating the grout eliminates voids and provides better bond to the masonry units and the reinforcing steel.

Prior to applying waterproofing products, the masonry surface must be clean and properly prepared. Oil, dust, efflorescence and other detrimental substances must be removed from the surface of the masonry so the applied coatings will adhere properly. Since few waterproofing products effectively span over cracks, all cracks should be repaired.

Some coatings require the surface to be dry prior to the application of the coating while others require damp substrates. Therefore, the product manufacturer should confirm that the surface is properly prepared prior to the use of their products. Always follow the manufacturer's recommendations fully and ensure the products are applied at the appropriate coverage rates.

9-2.E Waterproofing Products

There are numerous waterproofing products available, each with their own special characteristics and attributes. The following information briefly describes the major types of waterproofing products which are readily available. Note that no product works equally well on all substrates and the manufacturer should always be consulted to determine the best product for the job.

9-2.E.1 Bituminous Waterproofing Products

Used primarily below grade to resist moisture penetration through basement and retaining walls, bituminous waterproofing products have a long history of success. To perform effectively, a system to remove seepage and/or groundwater must be provided. (Figure 9-11). These products can be combined with felts or fabrics to form a built-up membrane.

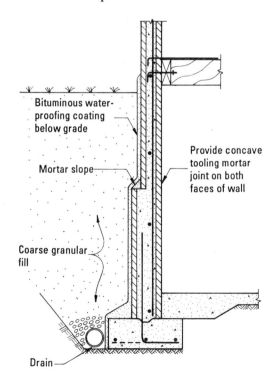

Bituminous water-proofing coating below grade

Mortar slope

Coarse granular fill

Provide concave tooling mortar joint on both faces of wall

Drain

Figure 9-11 Bituminous waterproofing system.

9-2.E.2 Clear Water Repellents

Clear water repellent products may be used on brick and block walls to shield the masonry from rainwater. Their main advantage over other waterproofing products is that they are clear so that the color and texture of the masonry can be seen. Some repellents can also include colored stains to enhance the masonry.

Most of these coatings repel water by producing high capillary pore angles so the masonry will no longer readily absorb water. They span over only the smallest cracks and every effort should be made to fill cracks and beeholes.

Breathable water repellents are recommended so internal moisture can escape. If the moisture becomes trapped in the wall it can freeze causing severe cracking and spalling. Unfortunately, salts cannot as readily escape through some waterproofing materials. As these salts build up within the wall, they may cause cracking and spalling of the brick.

Water repellents should not yellow with age nor should they abnormally darken the masonry surface. Repellents which do not give sheens are generally considered more acceptable. Select a repellent effective in resisting wind driven rain. Note that no known water repellents withstand water under pressure and therefore they should not be used below grade.

9-2.E.2(a) Types of Clear Water Repellents

Four generic types of clear water repellents are prevalent: acrylics, silicones, silanes, and siloxanes.

Acrylics and Silicones

The first two, acrylics and silicones, are deposited on the surface of the masonry, forming a thin film as the solvent evaporates. Generally they are applied with a low-pressure, airless sprayer on an air dry surface. Some acrylics may slightly darken the color of the masonry.

Silanes and Siloxanes

Silanes and siloxanes are characterized as penetrating repellents which, by undergoing a chemical reaction, form a water repellent barrier in the pores of the masonry. Some of these products, especially many silanes, react more completely in the presence of moisture and alkalies. Since concrete is by nature an alkaline material, these products often form an effective barrier on moist concrete block.

9-2.E.3 Paints

Paints can also provide a relatively low cost method to achieve water resistance. They have a long history of success when applied properly and can be very durable.

They can add a variety of color to a masonry structure although their opaque nature can also be a disadvantage since it hides the beauty and texture of the masonry.

Like water repellents, paints should normally be breathable, so internal moisture will not be trapped within the wall. Since moisture vapor enters through the interior surface of walls in cold climates and tries to exit through the exterior face of the wall, exterior paints should generally be more permeable than the internal paints. If an impermeable paint is applied on the outside face of such a wall, the trapped water may cause blistering and peeling of the paint or even worse, cracking and spalling of the masonry. Because of this, impermeable paints are generally recommended only for surfaces which are constantly subjected to moisture, such as swimming pools.

9-2.E.3(a) Types of Paints

The two most common types of paint are cement-based and latex-based paints. Oil-based paints are also sometimes used but are generally reserved for interior applications.

Cement-based Paints

Cement-based paints are very durable and form a hard, flat, breathable coating. They are not normally harmed by alkalies, allowing them to be placed on new concrete masonry as soon as the mortar dries. Unfortunately, these paints often chalk and fade with time and will crack and chip if applied too thickly.

Latex Paints

Latex paints are also breathable and quite durable under normal conditions. They have excellent color retention, and are easy to use. Although latex paints are permeable to water, some trap salts within the wall as the water vapor escapes. Since salt build-up within the wall generates extreme pressures, which can cause spalling and cracking of the masonry, it is important to use materials relatively free of salts when using these paints.

9-2.E.4 Elastomeric Coatings

These coatings are extremely water resistant although they can have a high first time cost. They have excellent flexibility, allowing them to bridge over hairline cracks when properly applied. They can be blended into a variety of colors but, unfortunately, like paint, they can not yet be made to be clear and transparent.

9-2.E.5 Integral Water Repellents

Used primarily in concrete masonry construction, integral water repellents provide an effective alternative to clear water repellents. These products are added directly into the concrete mix used to make the block units and must also be added into the mortar. They coat the pores of the concrete masonry units and mortar, making them more water resistant. Because they are added directly into the concrete and mortar, they should not wear off like other repellents. Concrete masonry walls which have integral water repellents also have the advantage over normal concrete masonry of being easier to clean or paint.

The largest drawback of integral water repellents is their inability to span over cracks or gaps in the masonry. If the mortar does not bond well to the units, water will pass through the cracks just as in any other concrete masonry wall. Therefore, whenever using these products, special care should be taken to assure mortar joints are well tooled and ample movement joints provided.

Integral water repellents cannot withstand water under pressure and should not be relied on in most below grade situations. Note also that integral water repellents must be added to the concrete mix and mortar in precise dosages, as given by the manufacturer, adding excessive amounts of these products may increase the water repellency of the wall, but it can also decrease the bond between the units and the mortar. Similarly, excessive amounts of integral water repellents have been reported to retard the mortar set.

9-2.E.6 Membrane Waterproofing

Continuous waterproofing membranes can effectively resist water penetration under most circumstances. Designed and installed correctly, they can even withstand water under pressure, and therefore are often applied against basement walls. By using asphalt for water resistance, and plastic polymers for added ultra violet radiation durability, these membranes can also effectively resist moisture penetration through the roof side of parapets (Figure 9-8).

9-2.F Maintenance of Waterproofing Systems

Throughout the life of a structure, maintenance must be performed to keep the waterproofing system working as intended. Periodic inspections of the structure should be performed to define areas requiring attention. Any work required should be performed promptly as waiting often allows significant damage to occur.

Roof drains, gutters, and weep holes must be kept cleaned and free from clogs. Cracks in the masonry should be filled as they form. Paint and other applied waterproofing products require periodic applications in order to remain effective. Likewise, caulking and sealants should be removed and replaced as they crack or separate from the substrates. In severely deteriorated structures, broken or cracked masonry units should be replaced and deteriorated mortar joints should be re-pointed.

9-3 FIRE RESISTANCE

9-3.A General

It is imperative to provide walls that are not only structurally adequate but are also fire resistant. Masonry walls excel in resisting the passage of heat or flames and can also be used to effectively contain most fires.

Fire resistance is determined by a series of fire tests conducted in accordance with the ASTM E 119, *Standard Methods for Fire Tests of Building Construction and Materials*. These fire tests require that a wall specimen be subjected to fire having the time/temperature curve shown in Figure 9-12.

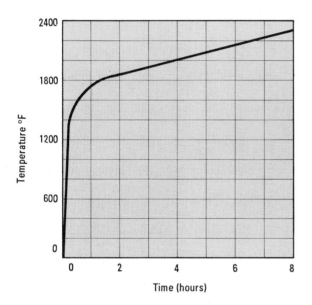

Figure 9-12 The ASTM E 119 standard time-temperature curve.

9-3.A.1 Temperature Rise Test

The termination of the fire test, or end point is reached, 1) when the passage of flame or gases are hot enough to ignite cotton waste on the opposite side of the wall, or 2) when the average temperature rises more than 250°F, based on temperatures recorded at no less than nine points on the unexposed side of the wall. Additionally, the temperature at any single recording point on the unexposed side of the wall may not rise to more than 325°F. The fire test rating is then given initially as an hourly rating.

9-3.A.2 Hose Stream Test

A similar wall specimen to be used for the hose stream test is then subjected to a fire exposure of one half the time determined by the time/temperature curve but not to exceed one hour.

Immediately after the second fire test, the fire-exposed side of the wall specimen is subjected to the hose stream tests. The impact, erosion and cooling effects of the hose stream is first directed at the center of the wall and then at all parts of the exposed face. The water pressure and the duration of application of the hose stream are based on the rating classification time period. For example, if a wall achieves a one-hour rating, the water pressure must be 30 pounds per square inch applied to the wall for one minute per 100 square inch of exposed area.

Hose Stream Test

9-3.A.3 End of Test

If a wall reaches the end point either by increased average temperature or single point increase, for one hour and 59 minutes, it is rated as a one hour wall. If it reaches the end point at 2 hours and 1 minute, the wall is rated as a 2 hour wall. Note: both walls must pass the hose stream test for these ratings.

Because masonry walls resist fire penetration extremely well, the masonry wall specimen that was subjected to the time temperature fire test is also sometimes subjected to the hose steam test. This is far more severe than the ASTM E 119 requirements, but it eliminates the need to fire test two walls.

TABLE 9-2   ASTM E 119 Acceptance Criteria for Walls

Acceptance Criteria Applicable to:					
Structural Fire Resistance			Barrier Fire Resistance		
	Sustain Load	Hose Stream	Hose Stream	Unexposed Surface Temp.	Cotton Waste
Bearing	X	X	X	X	X
Non-Bearing	NA	X	X	X	X

9-3.A.4   Fire Ratings as per UBC Chapter 43

Table 9-3 shows fire ratings from UBC Table No. 43-B, *Rated Fire-Resistive Periods for Various Walls and Partitions*. This table provides a classification of masonry walls based on the required wall thickness for a specified fire rating time.

The specified wall thickness for masonry shown in Table 9-3 is the equivalent solid thickness. For solid clay units, the equivalent solid thickness is the actual thickness of the unit or wall. However, for hollow clay or concrete units, the wall is considered either ungrouted or solid grouted (partial grouted walls are considered as ungrouted). Footnote 2 provides an equation to calculate the equivalent solid thickness of ungrouted walls. Tables B-3a and B-3b, *Average Weight of Completed Walls and Equivalent Solid Thickness*, can also be used to find the equivalent solid thicknesses of ungrouted hollow unit walls.

As an alternative to UBC Table No. 43-B, fire-resistive construction may be approved by the building official, based on evidence submitted showing that the construction meets the required fire-resistive classification.

TABLE 9-3   Rated Fire-Resistive Periods for Various Walls and Partitions[a,1,17]

Material	Item Number	Construction	Minimum Finished Thickness Face-to-Face[2] (In Inches)			
			4 Hr	3 Hr	2 Hr	1 Hr
1. Brick of Clay or Shale	1-1.1	Solid units (at least 75 percent solid).	8		6[3]	4
	1-1.2	Solid units plastered each side with 5/8" gypsum or portland cement plaster. Portland cement plaster mixed 1:2½ by weight, cement to sand.			4¾[4]	
	1-2.1	Hollow brick units[5] at least 71 percent solid.		8		
	1-2.2	Hollow brick units[5] at least 71 percent solid, plastered each side with 5/8" gypsum plaster.	8¾			
	1-3.1	Hollow (rowlock[6]).	12		8	
	1-3.2	Hollow (rowlock[6]) plastered each side with 5/8" gypsum or portland cement plaster. Portland cement plaster mixed 1:2½ by weight, cement to sand.	9			
	1-4.1	Hollow cavity wall consisting of two 4" nominal clay brick units with air space between.	10			
	1-4.2	Cavity wall consisting of two 3-inch nominal thick solid clay units with air space.		8		
	1-5.1	Hollow brick, not filled.	5.0	4.3	3.4	2.3
	1-5.2	Hollow brick unit wall, grouted or filled with perlite, vermiculite, expanded shale aggregate.	6.6	5.5	4.4	3.0
	1-6.1	4" nominal thick units at least 75 percent solid backed with a hat-shaped metal furring channel 3/4 inch thick formed from 0.021-inch sheet metal attached to the brick wall on 24-inch centers with approved fasteners; and 1/2 inch Type X gypsum wallboard[7] attached to the metal furring strips with 1-inch-long Type S screws spaced 8 inches on center.			5[4]	

(continued on following page)

TABLE 9-3   Rated Fire-Resistive Periods for Various Walls and Partitions[a,1,17] (Continued)

Material	Item Number	Construction	Minimum Finished Thickness Face-to-Face[2] (In Inches)			
			4 Hr	3 Hr	2 Hr	1 Hr
2. Hollow Clay Tile, Non-load-bearing (End or Side Construction)	2-1.1	One cell in wall thickness, units at least 50 percent solid, plastered each side with $5/8''$ gypsum plaster.				$4\frac{1}{4}$
	2-1.2	Two cells in wall thickness, units at least 45 percent solid.				6
	2-1.3	Two cells in wall thickness, units at least 45 percent solid. Plastered each side with $5/8''$ gypsum plaster.			7	
	2-1.4	Two cells in wall thickness, units at least 60 percent solid. Plastered each side with $5/8''$ gypsum plaster.			5	
3. Hollow Clay Tile, Load-bearing (End or Side Construction)	3-1.1	Two cells in wall thickness, units at least 40 percent solid.				8
	3-1.2	Two cells in wall thickness, units at least 40 percent solid. Plastered one side with $5/8''$ gypsum plaster.			$8\frac{1}{2}$	
	3-1.3	Two cells in wall thickness, units at least 49 percent solid.			8	
	3-1.4	Three cells in wall thickness, units at least 40 percent solid.			12	
	3-1.5	Two units and three cells in wall thickness, units at least 40 percent solid.		12		
	3-1.6	Two units and four cells in wall thickness, units at least 45 percent solid.	12			
	3-2.1	Two units and three cells in wall thickness, units at least 40 percent solid. Plastered one side with $5/8''$ gypsum plaster.	$12\frac{1}{2}$			
	3-2.2	Three cells in wall thickness, units at least 43 percent solid. Plastered one side with $5/8''$ gypsum plaster.		$8\frac{1}{2}$		
	3-2.3	Two cells in wall thickness, units at least 40 percent solid. Plastered each side with $5/8''$ gypsum plaster.		9		
	3-2.4	Three cells in wall thickness, units at least 43 percent solid. Plastered each side with $5/8''$ gypsum plaster.	9			
	3-2.5	Three cells in wall thickness, units at least 40 percent solid. Plastered each side with $5/8''$ gypsum plaster.	13			
	3-3.1	Hollow cavity wall consisting of two $4''$ nominal clay tile units (at least 40 percent solid) with air space between. Plastered one side (exterior) with $3/4''$ portland cement plaster and other side with $5/8''$ gypsum plaster. Portland cement plaster mixed 1:3 by volume, cement to sand.	10			
4. Combination of Clay Brick and Load bearing Hollow Clay Tile	4-1.1	$4''$ brick and $8''$ tile.	12			
	4-1.2	$4''$ brick and $4''$ tile.		8		
	4-1.3	$4''$ brick and $4''$ tile plastered on the tile side with $5/8''$ gypsum plaster.	$8\frac{1}{2}$			
5. Concrete Masonry Units[8]	5-1.1[13,14]	Expanded slag or pumice.	4.7	4.0	3.2	2.1
	5-1.2[13,14]	Expanded clay, shale or slate.	5.1	4.4	3.6	2.6
	5-1.3[13]	Limestone, cinders or air-cooled slag.	5.9	5.0	4.0	2.7
	5-1.4[13,14]	Calcereous or siliceous gravel.	6.2	5.3	4.2	2.8

(See next page for applicable footnotes.)

Table 9-3 Footnotes

a. Generic fire-resistance ratings (those not designated by a company code letter) as listed in the Fire Resistance Design Manual, Twelfth Edition, dated August 1988, as published by the Gypsum Association, may be accepted as if herein listed.

1. Staples with equivalent holding power and penetration may be used as alternate fasteners to nails for attachment to wood framing.

2. Thickness shown for brick and clay tile are nominal thicknesses unless plastered, in which case thicknesses are net. Thickness shown for concrete masonry and hollow clay or shale brick is equivalent thickness defined as the average thickness of solid material in the wall and is represented by the formula:

$$T_E = \frac{V_n}{L \times H}$$

WHERE:

T_E = Equivalent thickness, in inches

V_n = Net volume (gross volume less volume of voids), in cubic inches.

L = Length of block or brick using specified dimensions as defined in UBC Chapter 24, in inches.

H = Height of block or brick using specified dimensions as defined in UBC Chapter 24, in inches.

Thickness includes plaster, lath and gypsum wallboard, where mentioned, and fill or grout when all cells are solid grouted or when all cells are filled with silicone-treated perlite loose-fill insulation; vermiculite loose-fill insulation; or expanded clay, shale or slate lightweight aggregate.

3. Single-wythe brick.

4. Shall be used for nonbearing purposes only.

5. Hollow brick units 4-inch by 8-inch by 12-inch nominal with two interior cells having a $1\frac{1}{2}$-inch web thickness between cells and $1\frac{3}{4}$-inch-thick face shells.

6. Rowlock design employs clay brick with all or part of bricks laid on edge with the bond broken vertically.

7. For all of the construction with gypsum wall board described in Table No. 43-B, gypsum base for veneer plaster of the same size, thickness and core type may be substituted for gypsum wallboard, provided attachment is identical to that specified for the wall board and the joints on the face layer are reinforced and the entire surface is covered with a minumum of $\frac{1}{16}$-inch gypsum veneer plaster.

8. See also Footnote No. 2. The equivalent thickness may include the thickness of portland cement plaster or 1.5 times the thickness of gypsum plaster applied in accordance with the requirements of UBC Chapter 47 of the code.

13. The fire-resistive time period for concrete masonry units meeting the equivalent thicknesses required for a two-hour fire-resistive rating in Item 5, and having a thickness of not less that $7\frac{5}{8}$ inches is four-hour when cores which are not grouted are filled with silicone-treated perlite loose-fill insulation; vermiculite loose-fill insulation; or expanded clay, shale or slate lightweight aggregate; sand or slag having a maximum particle size of $\frac{3}{8}$ inch.

14. For determining equivalent thickness of concrete masonry units made from unblended aggregates, see Footnote No. 2. The equivalent thickness of units composed of blends of two or more aggregate categories shall be determined by interpolating between the equivalent thickness values specified in UBC Table No. 43-B in proportion to the percentage by volume of each aggregate.

The equivalent thickness required to provide a desired fire-resistive time period for concrete masonry composed of units manufactured with fine aggregates passing a No. 4 sieve listed in Item 5-1.4 of UBC Table No. 43-B blended with aggregates listed in Items 5-1.1 or 5-1.2 shall be determined by interpolating between the equivalent thickness values specified in UBC Table No. 43-B in proportion to the percentage by volume of each aggregate as follows:

1. $ET_{required} = ET_{1.4} \times V_{1.4} + ET_{1.1} \times V_{1.1}$
2. $ET_{required} = ET_{1.4} \times V_{1.4} + ET_{1.2} \times V_{1.2}$

The required equivalent thickness of concrete masonry units manufactured with aggregates listed in Items 5-1.1, 5-1.2 and 5-1.4 of Table No. 43-B shall be determined as follows:

3. $ET_{required} = ET_{1.4} \times V_{1.4} + ET_{1.2} \times (V_{1.1} + V_{1.2})$

WHERE:

$ET_{1.1}, ET_{1.2}, ET_{1.4}$ = specified equivalent thickness for Items 5-1.1, 5-1.2 and 5-1.4 of UBC Table No. 43-B

$V_{1.1}, V_{1.2}, V_{1.4}$ = volume of aggregates expressed as a percentage of the total aggregate volume for Item 5-1.1, 5-1.2 or 5-1.4 of UBC Table No. 43-B

17. Portion of UBC Table No. 43-B.

EXAMPLE 9-A Fire Resistive Period Calculation.

Calculate the fire-resitive period of a wall constructed of standard, 8″ lightweight concrete masonry units based on the ratings in UBC Table 43-B (see Table 9-3).

Approximate the net volume based on the typical dimensions of an 8″ standard block.

$$V_n = A_n \times \text{height} = \begin{bmatrix} 2(15.625 \times 1.25) + \\ 2(5.125 \times 1.25) + \\ (5.125 + \times 1) \end{bmatrix} \times 7\frac{5}{8}''$$

$$= 57'' \times 7\frac{5}{8}''$$

$$T_E = \frac{V_n}{L \times H} = \frac{57 \times 7\frac{5}{8}}{15.625 \times 7\frac{5}{8}} = 3.65''$$

Since lightweight units are often made with expanded slag or pumice, the rated fire-resistive period from Table 9-3 is 2 hours.

Alternately T_E could be found assuming the block is approximately 50% solid. Thus:

$$T_E = \frac{0.50(7.625 \times 15.625) \times 7.625}{15.625 \times 7.625}$$

$$= 3.8''$$

From Table 9-3, the wall is still rated as 2 hours.

9-4 INTERNATIONAL SYSTEM OF UNITS (SI, SYSTEM)

9-4.A General

The American system of measurement is gradually changing from the English system of pounds, kips, inches and feet to the International System of Measurement (SI system) as adopted in 1960 by the General Conference of Weights and Measures. Based on earlier metric systems, the SI system standardized several units of measurements which are scaled in multiples of 1000. To avoid confusion with other measurement system, numbers are arranged in groups of three with respect to the decimal point by spaces. Thus a long block of numbers such as 12345.6789 is grouped as 12 345.678 9, not as 12,345.6789.

9-4.B Measurement Conversion Factors

The U Tables have been provided in the back of this book to give a comprehensive list of common conversion factors between the English and SI systems. For a complete list of conversion factors along with a discussion on SI formatting guidelines, see ASTM E 380.

For convenience, common units of measurement are discussed briefly hereafter with appropriate conversion factors.

Length or Distance Measurements

Under the SI system, the basic distance measurement unit is the metre which is approximately equal to 3'-3" or 39^3/$_8$". A millimetre, or ^{1}/$_{1000}$ of a metre, is equivalent to 0.0394 inch (about ^{1}/$_{32}$"). Thus one inch equals 25.4 mm.

In the SI system, only metres and millimetres are used for length measurements. However, in areas where the metric system has been used for a long period, the designation of centimetres (ten millimetres) is commonly used. In fact, construction accuracy is often given to the one half centimetre. Similarly mortar joint thicknesses are often given as one centimetre not ten millimetres.

A standard U.S. concrete block with nominal dimensions of 4 × 8 × 16 inches is shown in Figure 9-13 with the actual dimensions and SI conversions. Metric blocks, however, are manufactured to actual dimensions of 90 × 190 × 390 mm making them slightly smaller than U.S. standard concrete masonry units. The inclusion of 10 mm mortar joints standardizes SI nominal dimensions as 100 × 200 × 400 mm.

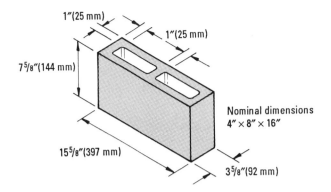

Figure 9-13 Standard U.S. hollow concrete unit with SI conversions shown.

Under the SI system an area is measured in square metres (m^2) where one square metre is approximately equal to 10.75 square feet (3.28^2). Thus a rough approximation is that there are 10 square feet in a square metre.

Mass

Under the SI system, mass is measured in kilograms (kg). Since one kilogram is equal to 2.2 pounds, 1000 pounds or 1 kip is equal to 453.5 kilograms, and 1 ton is equal to 907.2 kilograms.

Force

Force is mass times the acceleration of gravity, g, and although it can be stated as kilograms force, kgf, it should actually be given as newtons, N, or kilogram metre/second^2, kg·m/s^2.

Since one pound mass equals 0.454 kg, one pound force equals 0.454 kg times the acceleration of gravity (9.807 m/s^2). Thus one pound force equals 4.45 N and 1000 N equals about 225 pounds force.

Pressure

In the English system, pressure is commonly measured in pounds per square inch or pounds per square foot. In the SI system, it is measured in newtons per square metre. N/m^2 or pascals, Pa, where one pascal equals 1000 N/m^2. Pressure may also be measured in kilograms per metre second^2, kg/m·s^2. The conversion factors from psi and psf to pascals are respectively 6,894.8 psi/Pa and 47.88 psf/Pa as determined as follows:

$$1 \text{ psi} = (1 \text{ kg}/2.2 \text{ lbs}) \times 9.807 \text{ m/s}^2 \times (1 \text{ in.}/0.0254 \text{ m})^2$$

$$6900 \text{ kg/m·s}^2 = 6900 \text{ Pa}$$

$$1 \text{ psf} = (1 \text{ kg}/2.2 \text{ lbs}) \times 9.807 \text{ m/s}^2 \times (1 \text{ ft}/0.3048 \text{ m})^2$$

$$= 47.88 \text{ kg/m·s}^2 = 47.88 \text{ Pa}$$

EXAMPLE 9-B   Change $f'_m = 1500$ psi into pascals or mega pascals.

From Table U-1,
1 psi = 6 894.9 Pa
1 ksi = 6.8948 MPa
Thus,
$$f'_m = 1500 \times 6\ 895$$
$$= 10\ 342\ 500 \text{ Pa or}$$
$$= 1.5 \times 6.895$$
$$= 10.34 \text{ MPa}$$
Alternately, use Table U-4 to find:
$f'_m = 10.34$ MPa

EXAMPLE 9-C   Convert the modulus of elasticity of steel into SI units.

E_s = 29,000,000 psi
Since 1 psi = 6895 Pa,
E_s = 29 000 000 × 6 895
   = 199 955 000 000 Pa
   = 199 955 MPa
   = 199.955 GPa

EXAMPLE 9-D   Change the UBC equation for the modulus of elasticity of masonry into SI units.

E_m = 750 f'_m (psi)
E_m = 750 × 6.895 f'_m
   = 5171.25 f'_m Pa (for f'_m in psi)
   = 5.17 f'_m MPa (for f'_m in psi)

EXAMPLE 9-E   Check the modular ratio for $f'_m = 1500$ psi.

f'_m = 1500 psi = 10.34 MPa (from Example 9-B)
E_m = 750 × 10.34 = 7 755 MPa
(Alternately E_m = 5.17(1500) = 7 755 MPa)

$$n = \frac{E_s}{E_m} = \frac{199\ 955 \text{ MPa}}{7\ 755 \text{ MPa}} = 25.8$$

EXAMPLE 9-F   For $w = 200$ lbs/ft and $l = 25$ ft, find the simple beam moment in SI units.

Since
1 lb/ft = 14.585 N/m;
w = 200 lbs/ft × 14.6 N/m
   = 2917 N/m
l = 25 ft. × 0.305 m/ft = 7.625 m

$$M = \frac{wl^2}{8} = \frac{2917 \times 7.625^2}{8} = 21\ 200 \text{ N} \cdot \text{m}$$
$$= 21.2 \text{ kN} \cdot \text{m}$$

Alternate solution:

$$M = \frac{200 \times 25^2}{8}$$
$$M = 15,625 \text{ ft lbs}$$
Use Table U-10 to estimate:
$M = 21\ 200$ N·m

EXAMPLE 9-G   Determine the section modulus of a section 8″ wide by 18″ deep.

$$S = \frac{bd^2}{6}$$
b = 8″ = 8 × 25.4 = 203.2 mm
d = 18″ = 18 × 25.4 = 457.2 mm
$$S = \frac{bd^2}{6} = \frac{203.2 \times 457.2^2}{6}$$
$$= 7\ 079\ 212 \text{ mm}^3$$
$$= 7.08 \times 10^3 \text{ m}^3$$

9-5 QUESTIONS AND PROBLEMS

9-1 Name four factors which affect the size and spacing of movement joint.

9-2 Why are movement joints in clay masonry structures generally called expansion joints while most joints in concrete masonry structures are called control or shrinkage joint?

9-3 What is the estimated expansion of a 150 ft long brick wall which undergoes a temperature change of 60° F?

9-4 Name four areas where leakage can easily occur if not properly designed, detailed or constructed.

9-5 What type of waterproofing system would be prudent for (a) a brick wall, (b) a concrete block wall and (c) a concrete block basement wall.

9-6 Based on the UBC, calculate the fire-resistive period rating of a 6″ hollow concrete masonry wall if

    a) it is ungrouted

    b) it is grouted at 48″ o.c.

    c) it is grouted at 24″ o.c.

    d) it is solid grouted.

Assume pumice aggregate was used to make the block.

9-7 Using the 1991 UBC, find the fire-resistive period rating of a 6″ hollow clay msonry wall if it is:

    a) it is ungrouted

    b) it is grouted at 48″ o.c.

    c) it is grouted at 24″ o.c.

    d) it is solid grouted.

9-8 Convert the following English units to SI units

    a) 13.5 inches

    b) 12 feet $7^{1}/_{4}$ inches

    c) 367 square feet

    d) 163 feet $11^{3}/_{8}$ inches

    e) 237 pounds

    f) 43.23 kips

    g) 1742 foot pounds

    h) 42.7 foot kips per foot

    i) 150 pounds per cubic foot

    j) 3740 pounds per cubic yard

    k) 1200 pounds per square inch

    l) 2000 pounds per square foot

    m) 26,667 pounds per square inch.

9-9 Calculate the maximum negative and positive moments in SI units for the beam shown.

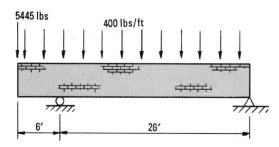

9-10 Calculate the unit compressive stress in SI units for an $8^{1}/_{2}$ inch wall, 16 ft high with a load of 10,000 pounds per foot. If $f'_m = 2000$ psi what is the allowable compressive stress, F_a.

SECTION 10

Formulas For Reinforced Masonry Design

10-1 GENERAL

This section is presented in two main subsections to show the formulas for reinforced masonry designed by (1) the Working Stress Design (WSD) Method, and (2) the Strength Design Method.

The Working Stress Design subsection is organized in two tables:

a) Table 10-1, *Allowable stress equations*

b) Table 10-2, *Design formulas*

The strength design subsection is also organized in two tables:

a) Table 10-3, *Formulas for strength design of reinforced masonry — 1991 Uniform Building Code equations*

b) Table 10-4, *Derived strength design formulas*

Where applicable, UBC and ACI/ASCE code references are provided along with reference to any applicable tables in this book.

Refer to page vii for definitions of symbols and notations.

10-2 WORKING STRESS DESIGN FORMULAS

TABLE 10-1 Allowable Stress Equations — WSD

Type of Stress	Allowable Stress Equation	Code References		Applicable Tables
		UBC[1]	ACI/ASCE	
Tensile Stress in Steel Reinforcement	Deformed bars, $F_s = 0.5 f_y$, 24,000 psi maximum	Eq. 6-16	Sec. 7.2.1.1 limits F_s to 24 ksi	Table A-4
	Wire reinforcement, $F_s = 0.5 f_y$, 30,000 psi maximum	Eq. 6-17	Sec. 7.2.1.1(c) limits F_s to 30,000 psi	Table A-4
	Ties, anchors and smooth bars, $F_s = 0.4 f_y$, 20,000 psi maximum	Eq. 6-18	—	—
Compressive Stress in Steel Reinforcing	Deformed bars in columns, $F_{sc} = 0.4 f_y$, 24,000 psi maximum	Eq. 6-19	Sec. 7.2.1.2(b)	Table A-4
	Deformed bars in flexural members, $F_s = 0.5 f_y$, 24,000 psi maximum	Eq. 6-20	—	Table A-4
	Confined, deformed bars in shear walls, $F_{sc} = 0.4 f_y$, 24,000 psi maximum	Eq. 6-21	—	Table A-4
Modulus of Elasticity	Reinforcing steel, $E_s = 29,000,000$ psi	Eq. 6-25	Sec. 5.5.1.1	Table A-4
	Masonry (clay or concrete), $E_m = 750 f'_m$, 3,000,000 psi maximum	Eq. 6-23 & 6-24	—	Table A-3
	Grout[2], $E_g = 57,000 \sqrt{f_g}$	—	Sec. 5.5.1.4	—

1. UBC equations listed are from UBC Chapter 24.

2. This equation will be modified in the 1992 edition of the ACI/ASCE Code to: $E_g = 500 f_g$.

(Continued on following 2 pages)

TABLE 10-1 Allowable Stress Equations — WSD (continued)

Type of Stress	Allowable Stress Equation	Code References		Applicable Tables
		UBC[1]	ACI/ASCE	
Shear Modulus	$G = 0.4E_m$ (UBC)	Eq. 6-26	—	Table A-3
	$E_v = 0.4E_m$ (ACI / ASCE)	—	Sec. 5.5.1.2(b) and 5.5.1.3(b)	A/A Tables A-1, C-1
Axial Compressive Stress	Walls designed per UBC: $$F_a = 0.20f'_m\left[1 - \left(\frac{h'}{42t}\right)^3\right]$$	Eq. 6-3	—	Q Tables
	Tied columns designed per UBC: $$P_a = \left(0.20f'_m A_e + 0.65A_s F_{sc}\right)\left[1 - \left(\frac{h'}{42t}\right)^3\right]$$	Eq. 6-4 & 6-5	—	Q and S Tables Table A-3
	Walls or columns designed per ACI / ASCE: When $h/r < 99$, $$F_a = \frac{1}{4}f'_m\left[1 - \left(\frac{h}{140r}\right)^2\right]$$	—	Eq. 7-1	—
	When $h/r \geq 99$, $$F_a = \frac{1}{4}f'_m\left(\frac{70r}{h}\right)^2$$	—	Eq. 7-2	—
Flexural Compressive Stress	$F_b = 0.33f'_m$, (UBC limits F_b to a maximum of 2000 psi)	Eq. 6-6	Sec. 7.3.1.2	Table A-3
Combined Compressive Stresses (Unity Equation)	$\dfrac{f_a}{F_a} + \dfrac{f_b}{F_b} \leq 1$	Eq. 6-22	—	—
Shear Stress	Flexural members without shear reinforcing, $F_v = 1.0\sqrt{f'_m}$ $F_{v(max)} = 50$ psi	Eq. 6-7	Eq. 7-4	Table A-3
	Flexural members with reinforcing steel carrying shear forces, $F_v = 3.0\sqrt{f'_m}$ $F_{v(max)} = 150$ psi	Eq. 6-8	Eq. 7-7	Table A-3

1. UBC equations listed are from UBC Chapter 24.

(Continued on following page)

TABLE 10-1   Allowable Stress Equations — WSD   (continued)

| Type of Stress | Allowable Stress Equation | Code References | | Applicable Tables |
		UBC[1]	ACI/ASCE	
Shear Stress — Shear Walls	Shear walls with masonry designed to carry all the shear force,			Tables & Diagram A-5
	When $\dfrac{M}{Vd} < 1$, $F_v = \frac{1}{3}\left[4 - \left(\dfrac{M}{Vd}\right)\right]\sqrt{f'_m}$	Eq. 6-10	Eq. 7-5	
	$F_{v(\max)} = 80 - 45\left(\dfrac{M}{Vd}\right)$ psi			
	When $\dfrac{M}{Vd} \geq 1$, $F_v = 1.0\sqrt{f'_m}$	Eq. 6-11	Eq. 7-6	
	$F_{v(\max)} = 35$ psi			
	Shear walls with reinforcing steel designed to carry all the shear force,			Tables & Diagram A-6
	When $\dfrac{M}{Vd} < 1$, $F_v = \frac{1}{2}\left[4 - \left(\dfrac{M}{Vd}\right)\right]\sqrt{f'_m}$	Eq. 6-12	Eq. 7-8	
	$F_{v(\max)} = 120 - 45\left(\dfrac{M}{Vd}\right)$ psi			
	When $\dfrac{M}{Vd} \geq 1$, $F_v = 1.5\sqrt{f'_m}$	Eq. 6-13	Eq. 7-9	
	$F_{v(\max)} = 75$ psi			
Bearing Stress	On full area, $F_{br} = 0.26 f'_m$	Eq. 6-14	—	Table A-3
	On one - third area or less, $F_{br} = 0.38 f'_m$	Eq. 6-15	—	Table A-3
	For ACI / ASCE Code, $F_{br} = 0.25 f'_m$	—	Sec. 5.12	
Tension on Embedded Anchor Bolts	The lesser of,			Tables A-7a, A-7b & A-7c
	$B_t = 0.5 A_p \sqrt{f'_m}$	Eq. 6-27	Eq. 5-1	
	$B_t = 0.2 A_b f_y$	Eq. 6-28	Eq. 5-2	
	(Note ACI / ASCE uses B_a to denote the allowable bolt tensile capacity.)			
Shear on Embedded Anchor Bolts	The lesser of,			Tables A-8a, A-8b & A-8c
	$B_v = 350\sqrt[4]{f'_m A_b}$	Eq. 6-31	Eq. 5-5	
	$B_v = 0.12 A_b f_y$	Eq. 6-32	Eq. 5-6	
Combined Shear and Tension on Anchor Bolts	$\dfrac{b_t}{B_t} + \dfrac{b_v}{B_v} \leq 1.0$	Eq. 6-33	Eq. 5-7	—
	(Note ACI / ASCE uses B_a to denote the allowable bolt tensile capacity.)			

1. UBC equations listed are from UBC Chapter 24.

TABLE 10-2 Design Formulas — WSD

Item	Design Formula	Standard Units	Useful Tables and Diagrams
Modular Ratio, n	$n = \dfrac{E_s}{E_m}$	—	A-2, A/A Tables A-1 & C-1
Tension Steel Reinforcing Ratio, p	$p = \dfrac{A_s}{bd} = \dfrac{K}{f_s j}$	—	Table C-9
Area of Tension Steel, A_s	$A_s = pbd = \dfrac{M}{f_s jd} = \dfrac{T}{f_s}$	sq. in.	Table C-3, C-4 & C-9
Compression Steel Reinforcing Ratio, p'	$p' = \dfrac{K - K_b}{(2n-1)\left(\dfrac{k - d'/d}{k}\right)\left(1 - \dfrac{d'}{d}\right)f_b} = \dfrac{A'_s}{bd}$	—	Tables N-1 to N-8 Diagrams P-1 to P-8
Area of Compression Steel, A'_s	$A'_s = \dfrac{M - KF}{cd}$	sq. in.	—
Perimeter of Circular Reinforcing, Bar Σ_o	$\Sigma_o = \pi d$	in.	Table C-3 & C-4
Moment Capacity of the Masonry, M_m	$M_m = \tfrac{1}{2} F_b k j b d^2 = K b d^2$	in. lb ft k	—
Moment Capacity of the Tension Steel	$M_s = F_s A_s jd = K b d^2$	in. lb ft k	—
Flexural Coefficient, K	$K = \tfrac{1}{2} f_b k j = \dfrac{M}{bd^2} = f_s p j$	—	E Tables & A/A Tables B & D
Coefficient, k	For members with tension steel only, $k = \sqrt{(np)^2 + 2np} - np$ $k = \dfrac{1}{1 + \dfrac{f_s}{nf_b}}$ Members with tension and compression reinforcement, $k = \sqrt{[np + (2n-1)]^2 + 2(2n-1)p'\dfrac{d'}{d}}$ $\quad - [np + (2n-1)p']$	—	E Tables & A/A Tables B & D
Coefficient, j	Members with tension steel only, $j = 1 - \dfrac{k}{3}$ Members with tension and compression steel, $j = 1 - \dfrac{z}{3}$	—	E Tables & A/A Tables B & D
Coefficient, z	$z = \dfrac{\tfrac{1}{6} + \dfrac{(2n-1)A'_s}{kbd} \times \dfrac{d'}{kd} \times \left(1 - \dfrac{d'}{kd}\right)}{\tfrac{1}{2} + \dfrac{(2n-1)A'_s}{kbd} \times \left(1 - \dfrac{d'}{kd}\right)}$	—	—

(Continued on following page)

TABLE 10-2 Design Formulas — WSD (Continued)

Item	Design Formula	Standard Units	Useful Tables and Diagrams
Dimensional Coefficient, F	$F = \dfrac{bd^2}{12,000}$	—	—
Resultant Compression Force, C	$C = \frac{1}{2} f_b k d b$	lbs, kips	
Resultant Tension Force, T	$T = A_s f_s$	lbs, kips	
Tension Steel Stress, f_s	$f_s = \dfrac{M}{A_s j d}$	psi	E Tables
Compression Steel Stress, f_{sc}, f_s'	$f_{sc} = 2n f_b \left(\dfrac{kd - d'}{kd} \right)$	psi	—
Masonry Stress, f_b	$f_b = \dfrac{2M}{bd^2 jk} = \dfrac{2}{jk} K$	psi	E Tables
Shear Stress, f_v or v, for beams and shear walls	$f_v = \dfrac{V}{bjd} \approx \dfrac{V}{bd}$ or $\dfrac{V}{bl}$	psi	K Diagrams & Tables
Spacing of Shear Steel, s	$s = \dfrac{A_v F_s d}{V}$	in.	K Diagrams & Tables
Shear Strength provided by the Reinforcing Steel, F_v	$F_v = \dfrac{A_v F_s}{bjs}$ or conservatively, $F_v = \dfrac{A_v F_s}{bs}$	psi	K Tables
Area of Shear Steel, A_v	$A_v = \dfrac{Vs}{F_s d}$	sq. in.	K Diagrams & Tables
Bond Stress, u	$u = \dfrac{V}{\Sigma_o j d}$	psi	Tables A-3, C1a, C-3 & C-4
Effective Height to thickness reduction factor, R	$R = \left[1 - \left(\dfrac{h'}{42t} \right)^3 \right]$	—	Tables Q-1 & Q-2
Interaction of Load and Moment	$\dfrac{f_a}{F_a} + \dfrac{f_b}{F_b} \leq 1.00$ or 1.33 $f_b = \left(1 - \dfrac{f_a}{F_a} \right) F_b$ $F_b = 0.33 f_m'$ $f_a = \dfrac{P}{Ae} = \dfrac{P}{bd}$ $f_m = f_a + f_b$ $kd = \dfrac{-\mathbf{b} \pm \sqrt{\mathbf{b}^2 - 4\mathbf{ac}}}{2\mathbf{a}}$ $\mathbf{a} = \frac{1}{6} t f_m$ $\mathbf{b} = -\frac{1}{2} t f_m d$ $\mathbf{c} = P\left(\dfrac{l}{2} - d_1 \right) + M$	—	—

10-3 STRENGTH DESIGN FORMULAS

TABLE 10-3 Formulas for Strength Design of Reinforced Masonry — 1991 UBC Equations

Item	Formula	UBC Chapter 24 Reference
Limiting vertical stress equation for slender wall design	$\dfrac{P_w + P_f}{A_g} \le 0.04 f'_m$	Eq. 11-1
Maximum Reinforcement Ratio	$p_{\max} = 0.5 p_b$	Sec. 2411(b)
Total Factored Load Equations	$U = 1.4D + 1.7L(+I)^1$	Eq. 11-2 & Eq. 12-3
	$U = 0.75\left(1.4D + 1.7L(+I)^1 + 1.87E\right)$, or	Eq. 11-3
	$U = 0.75\left(1.4D + 1.7L(+I)^1 + 1.7W\right)$, or Also let $L = 0$ and use most severe case	Eq. 11-4 & Eq. 12-4
	$U = 0.9D + 1.43E$	Eq. 11-5
	$U = 0.9D + 1.3W$	Eq. 11-6 & Eq. 12-5
	$U = 1.4\left(D + L(+I)^1 + E\right)$	Eq. 12-1
	$U = 0.9D \pm 1.4E$	Eq. 12-2
	$U = 1.4D + 1.7L(+I)^1 + 1.7H$ but use $0.9D$ and $L = 0$ if D or L reduces the effect of H	Eq. 12-6
	$U = 1.4F + $ all other loading combinations	Sec. 2412(c)1.E
	$U = 0.75\left(1.4D + 1.4T + 1.7L(+I)^1\right)$	Eq. 12-7
	$U = 1.4(D + T)$	Eq. 12-8
Factored Moment at midheight of slender walls M_u	$M_u = \dfrac{w_u h^2}{8} + P_{uf}\dfrac{e}{2} + P_u\,\Delta_u$	Eq. 11-7
Axial Load at midheight of slender walls, P_u	$P_u = P_{uw} + P_{uf}$	Eq. 11-8
Limiting moment strength equation	$M_u \le \phi M_n$	Eq. 11-9
Nominal Moment Strength, M_n	$M_n = A_{se}\, f_y\left(d - \dfrac{a}{2}\right)$	Sec. 2411(b)3C
Effective Area of Reinforcing, A_{se}	$A_{se} = \dfrac{A_s f_y + P_u}{f_y}$	Sec. 2411(b)3C

1. Add effects of Impact (I) to L per UBC Sec. 2412(c)1.F.   **(Continued on following page)**

**TABLE 10-3 Formulas for Strength Design of Reinforced Masonry
— 1991 UBC Equations (Continued)**

Item	Formula	UBC Chapter 24 Reference
Depth of Stress block, a	$a = \dfrac{P_u + A_s f_y}{0.85 f'_m b}$	Sec. 2411(b)3C
Midheight Deflection Limitation for Slender Walls	$\Delta_s = 0.007h$ maximum	Eq. 11-10
Calculated Midheight Deflection, Δ_s	$M_{ser} \leq M_{cr}$, $\Delta_s = \dfrac{5 M_s h^2}{48 E_m I_g}$ $M_{cr} < M_{ser} < M_n$, $\Delta_s = \dfrac{5 M_{cr} h^2}{48 E_m I_g} + \dfrac{5\left(M_{ser} - M_{cr}\right) h^2}{48 E_m I_{cr}}$	Eq. 11-11 Eq. 11-12
Cracking Moment Strength, M_{cr}	$M_{cr} = S f_r$	Eq.11-13
Modulus of Rupture, f_r (See STR. DES. Table D-1)	Fully grouted hollow unit masonry, $f_r = 4.0\sqrt{f'_m}$, 235 psi max Partially grouted hollow unit masonry, $f_r = 2.5\sqrt{f'_m}$, 125 psi max Fully grouted two - wythe brick masonry, $f_r = 2.0\sqrt{f'_m}$, 125 psi max	Sec. 2411(b)4 — —
Nominal Balanced Axial Strength, P_b	$P_b = 0.85 f'_m b a_b$	Eq. 12-9
Balanced depth of stress block, a_b	$a_b = 0.85 \left[\dfrac{e_{mu}}{\left(e_{mu} + \dfrac{f_y}{E_s}\right)} \right] d$	Eq. 12-10
Nominal Axial Strength of a Shear Wall supporting only Axial Loads, P_o	$P_o = 0.85 f'_m \left(A_e - A_s\right) + f_y A_s$	Eq. 12-11
Nominal Shear Strength of a Shear Wall, V_n	$V_n = V_m + V_s$ or (See STR. DES. $V_n = A_{mv} p_n f_y$ Diagram and Table E-2)	Eq. 12-13 Eq. 12-16
Nominal Shear Strength provided by the Masonry, V_n	$V_m = C_d A_{mv} \left(f'_m\right)^{1/2}$ (See STR. DES. Diagram and Table E-1)	Eq. 12-14
Nominal Shear Strength provided by the Shear Reinforcement, V_s	$V_s = A_{mv} p_n f_y$	Eq. 12-15

TABLE 10-4 Derived Strength Design Formulas

Item	Design Formula	Standard Units
Balanced Design Reinforcing Steel Ratio, p_b	$p_b = \dfrac{0.85\,\beta_1 f'_m}{f_y}\left(\dfrac{87,000}{87,000+f_y}\right)$ $= 7.13 \times 10^{-6} f'_m$ (for $f_y = 60,000$ psi and $\beta_1 = 0.85$)	—
Steel Quotient, q	$q = \dfrac{p f_y}{f'_m}$	—
Reinforcing Steel Ratio, p	$p = \dfrac{A_s}{bd} = q\dfrac{f'_m}{f_y}$	—
Area of Tension Steel, A_s	$A_s = pbd = \dfrac{M_u}{a_u d}$	sq. in.
Flexural Coefficient, K_u	$K_u = f'_m q(1 - 0.59q)$ (See STR. DES. Table A-1 to A-8)	—
Coefficient, c_b	$c_b = \dfrac{0.0003d}{0.005 + \dfrac{f_y}{E_s}} = \dfrac{a_b}{0.85} = \dfrac{87,000d}{87,000+f_y}$	—
Coefficient, a	$a = \dfrac{pbd f_y}{0.85 b f'_m} = 0.85c = \dfrac{qd}{0.85}$	—
Coefficient, a_b	$a_b = 0.85 c_b$	—
Coefficient, a_u	$a_u = \dfrac{K_u}{12,000\,p} = \dfrac{M_u}{A_s d} = \dfrac{\phi f_y(1 - 0.59q)}{12,000}$	—
Coefficient, c	$c = \dfrac{a}{0.85}$	—
$q(1 - 0.59q)$	$q(1 - 0.59q) = \dfrac{M_u}{\phi f'_m bd^2}$ (See STR. DES. Table A-9)	—
Interaction Coefficient, a	$a = \dfrac{-\mathbf{b} \pm \sqrt{\mathbf{b}^2 - 4\mathbf{ac}}}{2\mathbf{a}}$ $\mathbf{a} = 0.425 f'_m t$ $\mathbf{b} = -0.85 f'_m t(l - d_1),$ note $l - d_1 = d$ $\mathbf{b} = -0.85 f'_m t d$ $\mathbf{c} = P\left(\dfrac{l}{2} - d_1\right) + M$	—
Tension Force, T	$T = A_s f_y$ $= C - P$	lbs, kips
Compression Force, C	$C = 0.85 f'_m a t$	lbs, kips

SECTION 11

Design of One-Story Industrial Building

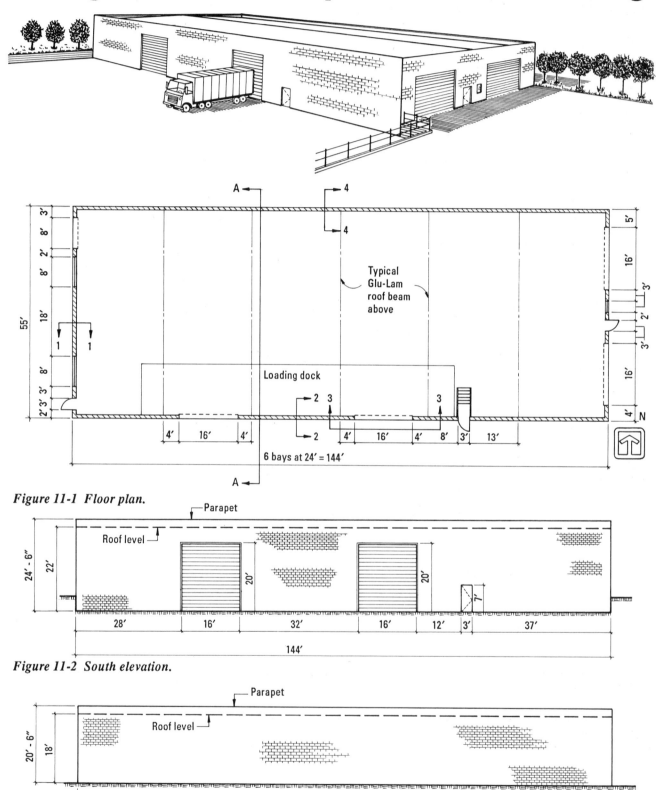

Figure 11-1 Floor plan.

Figure 11-2 South elevation.

Figure 11-3 North elevation.

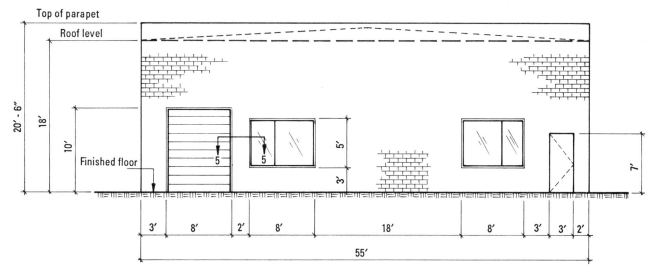

Figure 11-4 West elevation.

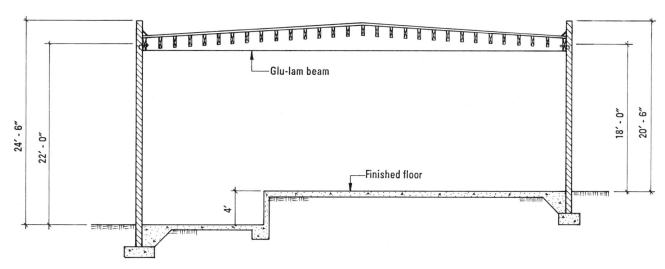

Figure 11-5 East elevation.

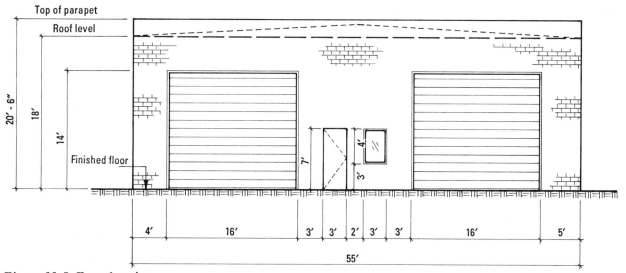

Figure 11-6 Typical section AA.

11-1 DESIGN CRITERIA: WORKING STRESS DESIGN

11-1.A Materials and Allowable Stresses

Design to be based on the 1991 UBC.

Walls are to be two wythe brick masonry walls, solid grouted. Total thickness = 9″ with 3″ grout space.

Brick:　　　Minimum Strength = 4000 psi

Mortar:　　Type S with 1 part portland cement, $\frac{1}{2}$ part lime, and $4\frac{1}{2}$ parts sand.

Grout:　　　Coarse pea gravel grout with a minimum strength of 2700 psi.

Masonry:　Strength of masonry, f'_m, = 2000 psi. Special inspection will not be provided (use half stresses for design per UBC Sec. 2406(c)).

Reinforcing
Steel:　　　Grade 60

$$f_y = 60{,}000 \text{ psi}$$

$$F_s = 24{,}000 \text{ psi}$$

$$E_s = 29{,}000{,}000 \text{ psi}$$

Concrete:　f'_c = 2000 psi at 28 days

From Table A-3:

$$\text{Modular ratio, } n = \frac{E_s}{E_m} = \frac{29{,}000{,}000}{750 f'_m}$$
$$= 19.3$$

Maximum allowable compressive stresses:

$$F_b = \frac{1}{2}(0.33 f'_m) = 333 \text{ psi}$$

Maximum allowable axial compressive strength:

$$F_a = \frac{1}{2}(0.2 f'_m)R = 200 \text{ psi} \times R$$

All allowable stresses may be increased by $\frac{1}{3}$ when considering wind or seismic forces (UBC Sec. 2303(d))

For members subjected to combined axial and flexural stresses due to vertical, seismic or wind forces.

$$\frac{f_a}{F_a} + \frac{f_b}{F_b} \le 1.0 \text{ or } 1.33 \quad \text{(UBC Chapter 24, Eq. 6-22)}$$

The maximum bending stress required to satisfy the unity equation when wind or seismic forces are considered is:

$$f_b = \left(1.33 - \frac{f_a}{F_a}\right)F_b$$

11-1.B Loads

11-1.B.1 Lateral Loads (Wind and Seismic)

11-1.B.1(a) Seismic Loads (per UBC Chapter 23, Part III)

Location: Seismic Zone No. 2B

Seismic load on total building:

$$V = \frac{ZIC}{R_w} W \quad \text{(UBC Chapter 23, Eq. 34-1)}$$

Where:　$Z = 0.20$ from Table 3-11 (UBC Table No. 23-I)

　　　　$I = 1.0$ from Table 3-5 (UBC Table No. 23-L)

$$C = \frac{1.25 S}{T^{2/3}} \le 2.75$$

　　　Use $S = 1.5$ for Type S_3 soil per Table 3-13 (UBC Table No. 23-J, footnote 1).

$$T = C_t \left(h_n\right)^{3/4}$$

　　　$C_t = 0.020$ for a masonry bearing wall building (UBC Sec. 2334(b)2.A)

$$T = 0.020(18)^{3/4} = 0.17 \text{ sec.}$$

$$C = \frac{1.25(1.5)}{(0.17)^{2/3}} = 6.11 > 2.75$$

　　　Use $C = 2.75$

　$R_w = 6$ from Table 3-14 (UBC Table No. 23-O)

　　Seismic base shear, $V = \dfrac{0.20 \times 1.0 \times 2.75}{6} W$
$$= 0.092 W$$

Seismic load on portions of the building:

$$F_p = ZIC_p W_p \quad \text{(UBC Chapter 23, Eq. 36-1)}$$

For the walls, $C_p = 0.75$

$$F_p = 0.20(1.0)(0.75)W_p = 0.15 W_p$$

since the walls weigh 90 psf, $F_p = 0.15(90) = 13.5$ psf

For the parapet, $C_p = 2.0$ and;

$$F_p = 0.20(1.0)(2.0)(90) = 36 \text{ psf}$$

11-1.B.1(b) Wind Loads (per UBC Chapter 23, Part II)

Basic wind speed = 70 mph
Exposure B
H ≤ 20 to 25 ft

Wind pressure on the primary structure by the projected area method:

$$P = C_e \times C_q \times q_s \times I \quad \text{(UBC Chapter 23, Eq. 16-1)}$$

C_e = 0.72 from Table 3-6 (UBC Table No. 23-G)

C_q = 1.3 from Table 3-7 (UBC Table No. 23-H)

q_s = 12.6 from Table 3-3 (UBC Table No. 23-F)

I = 1.0 from Table 3-5 (UBC Table No. 23-L)

$P = 0.72 \times 1.3 \times 12.6 \times 1 = 12$ psf

Note: Seismic pressure of 13.5 psf governs the design of the building.

11-1.B.2 Vertical Loads

No snow load considered

Roof load — Slope roof $^1/_4$ inch/ft for drainage

Roofing (Built up)	=	4.0 psf
Plywood, $^1/_2''$ thick	=	1.5
Framing	=	2.5
Sprinklers	=	1.5
Miscellaneous	=	1.5
Roof framing	=	11.0 psf
Roof beams	=	3.0
Total roof-dead load	=	14.0 psf
Roof live load (reduceable)	=	20.0

Total dead and live loads for
load bearing wall design   = 34.0 psf

Roof live load for pilaster
design (Live load reduced
as allowed in UBC Sec. 2306.
See bearing angle design
(Sec. 11-5) for calculation)   = 12.0 psf

Total dead and live loads
for pilaster design (14 + 12)   = 26.0 psf

Weight of wall (From Table B-4 based
on 9″ × 10 psf/1″ thickness)   = 90.0 psf

11-2 DESIGN OF WEST MASONRY BEARING WALL — SECTION 1-1

11-2.A   Vertical Loads on Wall

Tributary width of roof = $^1/_2$ × 24 ft roof span
= 12 ft per foot length of wall

Roof live load = 20 psf × 12 ft = 240 plf

Roof dead load = 14 psf × 12 ft = 168 plf

Dead load on wall at mid-height
(between footing and ledger beam)

DL parapet = 90 psf × 2.5   = 225 plf

DL wall = 90 psf × $^1/_2$(18 ft) =   810 plf

Total wall DL at mid-height = 1035 plf

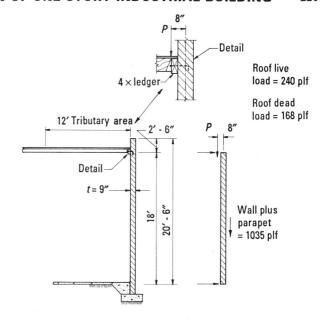

Figure 11-7 West wall Section 1-1.

11-2.B Lateral Forces on Wall

From section 11-1.B, Lateral loads:

Wind force, P = 12.0 psf
Seismic force, F_p = 13.5 psf > 12 psf

Seismic force governs. Use this value for the design of the wall between the ledger and the footing.

Seismic force of parapet, F_p = 36 psf.

Lateral seismic moment on wall assuming pin connection at top and bottom with no seismic load on parapet.

$$M = {}^1/_8 wh^2 = {}^1/_8 \times 13.5 \times 18^2$$
$$= 547 \text{ ft lbs/ft}$$

Moment due to eccentric roof dead load

$$M = 168 \times \frac{8}{12} = 112 \text{ ft lbs / ft}$$

Moment at mid-height

$$M = {}^1/_2 \times 112 = 56 \text{ ft lbs/ft}$$

Moment due to eccentric roof live load

$$M = 240 \times \frac{8}{12} = 160 \text{ ft lbs / ft}$$

Moment at mid-height

$$M = {}^1/_2 \times 160 = 80 \text{ ft lbs/ft}$$

Total moment, lateral plus roof DL at mid-height

$$M = 547 + 56 = 603 \text{ ft lbs/ft}$$

Maximum moment at mid-height

$$M = 547 + 56 + 80 = 683 \text{ ft lbs/ft}$$

Total moment at top of wall due to live and dead loads

$$M = 112 + 160 = 272 \text{ ft lbs/ft}$$

11-2.C Vertical Load on Wall at Mid-Height

Vertical dead load on wall at mid-height
= roof + parapet + ½ wall
= 168 + 90(2.5 + ½ × 18) = 1203 plf

Vertical dead and live load on wall at mid-height
= 1203 + 240 = 1443 plf

> NOTE: When lateral wind or seismic forces are considered, the roof live load is generally ignored except for snow load.

11-2.D Design Wall for Condition at Mid-Height — Section 1-1

Axial dead load = 1203 plf

$$t = 9'' \qquad h' = 18 \text{ ft}$$

from Tables Q-1 and Q-2.

$$\frac{h'}{t} \text{ ratio} = 24 \text{ and}$$

Reduction factor, $R = 0.789$

Maximum allowable axial stress, for $f'_m = 2000$ psi with no inspection:

$$F_a = 163 \text{ psi} \quad \text{(From Table Q-2)}$$

Actual axial stress, $f_a = \dfrac{P}{A}$

$$= \frac{1203}{9 \times 12} = 11.1 \text{ psi}$$

Bending moment at mid-height = 603 ft lbs. Design wall to span vertically between footing and diaphragm.

Steel is located in the center of the wall:

$$d = \frac{t}{2} = 4.5''$$

Check flexural stress as if section was uncracked

$$f_b = \frac{6M}{bt^2} = \frac{6 \times 603 \times 12}{12 \times 9^2}$$
$$= 45 \text{ psi} > 11.1 \text{ psi}$$

The bending stress is greater than the dead load axial stress — therefore design for tension as a cracked section.

Maximum moment at mid-height,

$$M = 683 \text{ ft lbs/ft}$$

Flexural coefficient $K = \dfrac{M}{bd^2} = \dfrac{683 \times 12}{12 \times 4.5^2} = 33.7$ psi

Maximum allowable bending stress for combined stress conditions.

$$f_b = \left[1.33 - \frac{f_a}{F_a} \right] F_b$$

$$f_b = \left[1.33 - \frac{1443/(12 \times 9)}{163} \right] 333$$

$$= 416 \text{ psi maximum bending stress}$$

$$< 333 \times 1.33 = 443 \text{ psi}$$

From Diagram F-2, K vs p, enter with :

$K = 33.7$; $f_b = 416$ psi; and $f_s = 32,000$ psi
Read $p = 0.0012$, steel stress governs

From Table C-8c Steel Ratio

for $p = 0.0012$ and $d = 4.5$ inches
use #5 at 48″ o.c. vertical bars, $p = 0.0014$

From Table C-4c, Area of Steel

#5 at 48″ o.c. $A_s = 0.077$ sq. in./ft

UBC Sec. 2407(h)3.B requires #4 horizontal reinforcing bars to be spaced no more than 10′ on centers for buildings in Seismic Zone No. 2. Therefore place #4 bar at ledger, bottom of wall and at mid-height (s ≈ 9 ft). Alternately use joint reinforcement as part of the horizontal steel.

11-2.E Design of Parapet Wall — Section 1-1

Seismic lateral force on parapet wall = 36 psf

Cantilever moment = ½wh^2
$$= ½ × 36 × 2.5^2 = 113 \text{ ft lbs/ft}$$

$$K = \frac{M}{bd^2} = \frac{113 \times 12}{12 \times 4.5^2} = 5.6$$

Minimum steel will be satisfactory for this small moment. Use #4 bars to match the vertical steel or continue the #5 vertical bars into the parapet.

Also provide 1 - #4 horizontal bar at the top of the parapet.

11-3 DESIGN OF SOUTH MASONRY WALL — SECTION 2-2

11-3.A Loads

The wall is 22 ft high between footing and ledger beam.

$$\frac{h}{t} = \frac{22 \times 12}{9} = 29.3$$

For a simple beam span, the

However, if the wall is assumed fixed at the bottom, the inflection point can be approximated at $0.2h$ from the bottom. The effective height of the wall, h', is between the ledger beam at the top and the inflection point.

Effective height, $h' = 0.8h = 0.8 \times 22$
$$= 17.6 \text{ ft}$$

$$\frac{h'}{t} = \frac{17.6 \times 12}{9} = 23.5$$

Since the roof spans east-west, the south wall is essentially a non-load bearing wall except at the roof beams. However, to be conservative, 4 ft of roof load is included in the wall design.

Live load from roof = 20 psf × 4 = 80 plf
Dead load from roof = 11 psf × 4 = 44 plf

Calculate the dead load at mid-height between ledger and inflection point

$$DL = \text{(roof)} + \text{(parapet)} + (\tfrac{1}{2} \text{ wall})$$
$$DL = 44 + (2.5 \times 90) + (\tfrac{1}{2} \times 17.6 \times 90)$$
$$= 1061 \text{ plf}$$

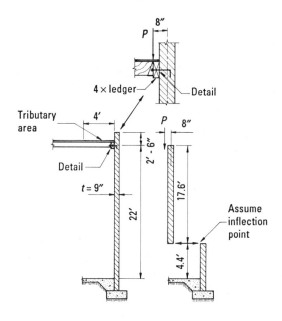

Figure 11-8 South wall Section 2-2

Lateral seismic moment $= \dfrac{wh^2}{8} = \dfrac{13.5 \times 17.6^2}{8}$
$$= 522.7 \text{ ft lbs / ft}$$

Moment due to eccentric roof load at mid height

$$= 44 \times \frac{8}{12} \times \frac{1}{2} = 14.7 \text{ ft lbs / ft}$$

Total moment $= 522.7 + 14.7 = 537.4 \text{ ft lbs / ft}$

The design procedure for the upper part of the wall is similar to Section 1-1.

11-3.B Design of Lower Part of Wall

Lateral load from upper wall

Reaction $= 13.5 \times \tfrac{1}{2} \times 17.6$
$$= 118.8 \text{ plf} \approx 119 \text{ plf}$$

Vertical dead load from upper wall, parapet and roof

$$P = 90(2.5 + 17.5) + 44$$
$$= 1844 \text{ plf}$$

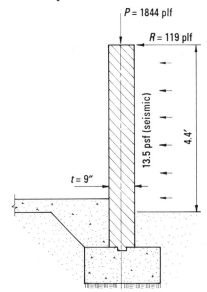

Figure 11-9 Load on lower section of wall.

Assume lower part of wall acts as a cantilever and therefore the effective height is twice the actual height:

$$h' = 2h = 2(4.4) = 8.8 \text{ ft}$$

From Table Q-1, $R = 0.98$

Use Table Q-2 to find; $F_a = 196$ psi

Total dead load at bottom of wall

$$= 90(2.5 + 22) + 44 = 2249 \text{ plf}$$

Axial dead load stress $f_a = \dfrac{P}{bt} = \dfrac{2249}{12 \times 9}$
$$= 20.8 \text{ psi} < 196 \text{ psi O.K.}$$

Total *DL* and *LL* axial stress:

$$f_a = 20.8 + \frac{80}{12 \times 9} = 21.5 \text{ psi} < 196 \text{ O.K.}$$

Moment at bottom of wall =

$$= (119 \times 4.4) + \frac{13.5 \times 4.4^2}{2} = 654 \text{ ft lbs}$$

Flexural coefficient

$$K = \frac{M}{bd^2} = \frac{654 \times 12}{12 \times 4.5^2} = 32.3$$

Check flexural stress as if section were uncracked

$$f_b = \frac{6M}{bt^2} = \frac{6 \times 654 \times 12}{12 \times 9^2}$$
$$= 48 \text{ psi} > f_a$$

Bending stress is greater than the axial dead load stress and therefore the wall will probably crack.

Determine the allowable flexural compressive stress to satisfy the unity equation

$$\frac{f_a}{F_a} + \frac{f_b}{F_b} \leq 1.33$$

$$f_b = \left(1.33 - \frac{f_a}{F_a}\right)F_b$$

$$F_b = \tfrac{1}{2}(0.33 f'_m) = 333 \text{ psi}$$

$$f_b = \left(1.33 - \frac{21.5}{196}\right)333$$

$$= 406.3 \text{ psi}$$

From Diagram F-2, K vs p, for $K = 32.3$, $f_b = 406$ psi and $f_s = 32{,}000$ psi, determine p:

Steel ratio $p = 0.001$ (f_s governs)

From Table C-8, Steel Ratio, for $d = 4.5''$ and $p = .001$, use same vertical reinforcement as for the upper wall (#5 at 48'' o.c vertical bars and #5 @ 9 ft horizontal bars).

Use the same steel in all walls plus reinforcing around the openings as outlined in UBC Sec. 2407 (h)3.B.

11-3.C   Design of Footing — Section 2-2

Assume footing is 2' - 6'' wide by 12'' thick

LL on roof	= 4' × 20 psf	=	80	plf
Wt. of roof	= 4' × 11 psf	=	44	plf
Wt. of wall	= 90 psf (2.5 + 22')	=	2205	plf
Wt. of ftg. + conc. slab (estimated)		=	450	plf
	LL + DL	=	2779	plf
	DL only	=	2699	plf
		Use	2700	plf

Σ Moments about centerline of footing
   (see Figure 11-10):

$$M = 119(6.4) + 13.5 \times 4.4 \times \left(\frac{4.4}{2} + 2\right)$$

$$= 1011 \text{ ft lbs / ft}$$

$$e = \frac{M}{P} = \frac{1011}{2700} = 0.37 < \frac{2.5}{6} = 0.42$$

Resultant is in middle third,

Check soil bearing $= f_{sb} = \dfrac{P}{A} \pm \dfrac{Pe}{S} = \dfrac{P}{A}\left(1 \pm \dfrac{6e}{d}\right)$

$$f_{sb} = \frac{2700}{(2.5)(1)}\left(1 \pm \frac{6(0.37)}{2.5}\right)$$

$$= 2039 \text{ psf (maximum)}$$

$$< 2000 \text{ psf} \times 1.33 = 2667 \text{ psf   O.K.}$$

$$= 121 \text{ psf minimum}$$

Footing steel $d = 12'' - 3.0'' - 0.5'' = 8.5''$

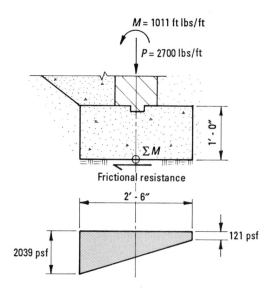

Figure 11-10 Soil bearing pressure under service loads.

Factored Loads

Check the footing for the factored loads and moments obtained from UBC Chapter 26, Eq. 9-2.

$$U = 0.75(1.4D + 1.7L + 1.7W)$$

However since seismic forces controlled the design, substitute 1.1E for W per UBC Sec. 2609(c)3. Therefore

$$U = 0.75(1.4D + 1.7L + 1.7(1.1E))$$

$$= 1.1D + 1.4L + 1.4E$$

For this design the load, P is almost entirely composed of dead load and the moment is attributed primarily to seismic forces. Thus: estimate the load factor for P as 1.1 and for M as 1.4.

$$P_u = 1.1P = 1.1(2780) = 3058 \text{ lbs/ft}$$

$$M_u = 1.4M = 1.4(1011) = 1415 \text{ ft lbs/ft}$$

Net soil pressure under load factors (f_{sbn})

$$e = \frac{M}{P} = \frac{1415}{3058} = 0.46 \approx \frac{2.5}{6} = 0.42$$

Approximately at the edge of kern area

$$f_{sbn} \approx \frac{3058}{2.5 \times 1}\left(1 \pm \frac{6 \times 0.46}{2.5}\right)$$

$$= 2574 \text{ psf at toe}$$

$$= 0 \text{ psf at heel}$$

Distance to the point about which moments are to be taken, ΣM_a. In accordance with UBC Sec. 2615(e) 2.B, the moment on a footing under a masonry wall shall be taken half way between the middle and the edge of the wall. This recognizes the possibility of edge softness before full fixity is achieved.

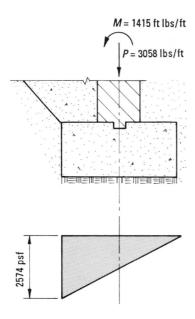

Figure 11-11 Footing design under factored loads.

Distance to $\Sigma M = \frac{1}{2}(2.5) - \dfrac{0.75}{4} = 1.06$ ft

Soil pressure at point of moment

$$f_{sb} = 2574 \times \frac{2.5 - 1.06}{2.5}$$

$$= 1483 \text{ psf}$$

Moment on footing (M_u) about the quarter point axis.

$$M_u = (1483 \times 1.06 \times \frac{1}{2} \times 1.06)$$
$$+ (\frac{1}{2}(2574 - 1480)\frac{2}{3} \times 1.06^2)$$
$$= 833 + 410 = 1243 \text{ ft lbs/ft}$$

Compute the moment of inertia (I_g) of the section. The bottom 1 or 2 inches of concrete placed against the ground is usually assumed to be of poor quality and neglected. Neglecting the bottom 2 inches.

$$I_g = \frac{12(10)^3}{12} = 1,000 \text{ in.}^4$$

$$f_t = \frac{M_u(h/2)}{\phi I_g} = \frac{1243 \times 12(10/2)}{0.65 \times 1000}$$
$$= 115 \text{ psi}$$

Allowable $f_t = 5\phi\sqrt{f'}$ (UBC Sec. 2615(l)3) for extreme fiber stress in tension.

$$f_t = 5 \times 0.65\sqrt{2000}$$
$$= 145 \text{ psi} > 115 \text{ psi} \quad \text{O.K.}$$

Provide alternate bend to the wall main reinforcing bars to furnish minimum allowable bond length and to insure wall fixity at floor and foundation.

Put 2 - #4 bars longitudinally; one at the top of foundation and one at the bottom.

Moment on footing

$$= (119 \times 6.4) + (13.5 \times 4.4)\left(\frac{4.4}{2} + 2\right)$$

$$= 1011 \text{ ft lbs/ft}$$

Dead load on footing = 2700 plf

Friction on bottom of footing = $\mu \times DL$

From Table 3-20 (UBC Table No. 29-B) or Table 5-1, choose $\mu = 0.35$

Frictional resistance = $0.35 \times 2700 = 945$ plf

Required depth from steel floor tie to bottom of footing

$$= \frac{1011 \times 12}{945} = 12.8'' < 24'' \text{ assumed depth} \quad \text{O.K.}$$

Area of floor tie, $A_s = \dfrac{M}{f_s d}$

$$A_s = \frac{1011 \times 12}{32,000 \times 24} = 0.016 \text{ sq. in./ft}$$

Use #4 at 48" o.c. ($A_s = 0.050$ sq. in./ft)

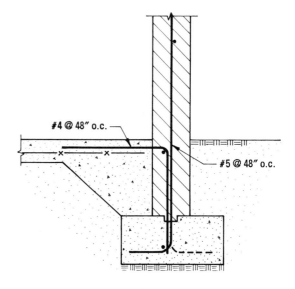

Figure 11-12 Tie of footing to slab.

Alternate footing design

This method develops fixity and lateral resistance by flag pole action.

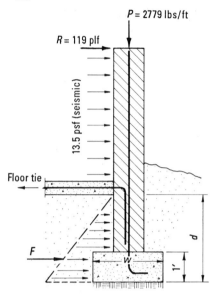

Figure 11-13 Wall footing.

Allowable bearing soil pressure 2000 psf

$$w = \frac{P}{S.B.} = \frac{2779}{2000} \times 12 = 16.7'' < 2'-6'' \quad \text{O.K.}$$

Allowable lateral soil bearing value from Table 3-20 (UBC Table No. 29-B), assuming soil to be gravel, equals 200 psf/ft depth. A one third increase is permitted for wind or earthquake loading.

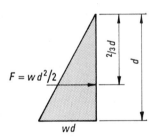

$$M = \frac{wd^2}{2} \times \frac{2d}{3} = \frac{wd^3}{3}$$

$$d = \sqrt[3]{\frac{3M}{w \times 1.33}}$$

$$M = 119 \times 4.4 + 13.5 \times \frac{4.4^2}{2} = 654 \text{ ft lbs}$$

$$w = 200 \text{ psf / ft depth} \times 1.67 \text{ ft deep} = 334 \text{ psf}$$

$$d = \sqrt[3]{\frac{3 \times 654}{334 \times 1.33}} = 1.6 < 1.67 \text{ assumed depth   O.K.}$$

11-4 DESIGN OF LINTEL BEAM SOUTH WALL — SECTION 3-3

11-4.A Flexural Design

Dead load from roof	= 11 psf × 4′	=	44 plf
Live load from roof	= 20 psf × 4′	=	80 plf
Dead load of wall	= 90 psf × 4.5′	=	405 plf

$$\overline{ 529 \text{ plf}}$$

$$\text{Moment} = \frac{wl^2}{12} = \frac{529 \times 16^2}{12}$$
$$= 11,285 \text{ ft lbs}$$

$f_s = 32,000$ psi;

$f_m = 333$ psi; $n = 19.3$

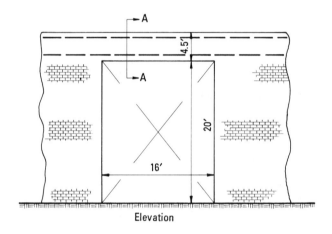

Elevation

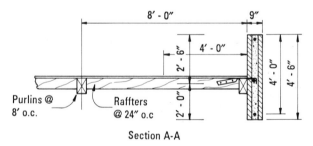

Section A-A

Figure 11-14 Lintel beam.

Determine the minimum depth of beam required for an allowable shear stress of 20 psi and no reinforcing (see exception of UBC Sec. 2406(c)6.A).

$$V = \frac{1}{2} \times 529 \times 16 = 4232 \text{ lbs}$$

$$d = \frac{V}{F_v j b} \quad (\text{estimate } j = 0.88)$$

$$= \frac{4232}{(20)(0.88)(9)} = 26.7 \text{ psi}$$

Use $d = (4.5' \times 12) - 6 = 48''$

Determine the flexural steel.

Moment $M = 11,285$ ft lbs

Flexural coefficient

$$K = \frac{M}{bd^2} = \frac{11,285 \times 12}{9 \times 48^2} = 6.5$$

From Table E-2a for $K = 6.5$, read

$p = 0.0003$

$A_s = 0.0003(9)(48) = 0.13$ sq. in.

The #4 bar at the top of the parapet is adequate. However, for safety, use at least:

$A_s = 0.0007bt = 0.0007 \times 9 \times 54 = 0.34$ sq. in.

Therefore provide an additional #4 bar at the top and bottom of beams over openings.

Also provide at least 1 - #4 in all jambs around openings.

11-4.B Lateral Seismic Load on Beam

From Table E-2a for $1.33K = 14.2$, read $p = 0.0005$.

$$\text{Moment } M = \frac{wl^2}{12} = \frac{13.5 \times 16^2 \times 4.5}{12}$$
$$= 1296 \text{ ft lbs (seismic)}$$

$$\text{Flexural coefficient } K = \frac{M}{bd^2} = \frac{1296 \times 12}{54 \times 4.5^2} = 14.2$$

$A_s = 0.0005 \times 54 \times 4.5 = 0.12$ sq. in.

The horizontal #4 bars are adequate.

11-4.C Deep Lintel Beams

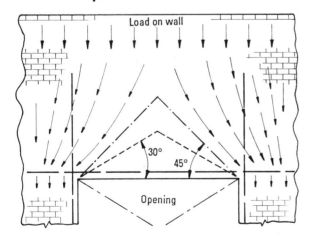

Figure 11-15 Distribution of load over an opening.

    The wall over a lintel will tend to create an arch over the opening leaving the area under the arch to be carried by the lintel beam. A conservative arch angle would be 45°; although an angle of 30° would probably be more realistic to the true arch action of modern reinforced masonry.

11-5 DESIGN OF FLUSH WALL PILASTER NORTH WALL — SECTION 4-4. DESIGNED AS A WALL NOT A COLUMN

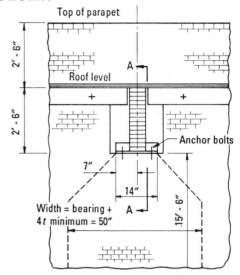

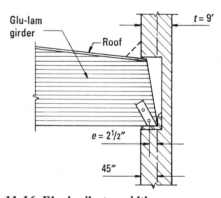

Figure 11-16 Flush pilaster width.

11-5.A Loads

$$\text{Tributary area} = 24\left(\frac{55}{2} - \frac{9}{12}\right) = 642 \text{ sq. ft}$$

Live load = 20 psf. Reduce as permitted in UBC Sec. 2306.

From UBC Table 23-C for a flat roof, $r = 0.08$.

Therefore:

$R = 0.08(642 - 150) = 39.4\% < 40\%$ O.K.

Check UBC Chapter 23, Eq. 6-2

$$R = 23.1\left(1 + \frac{D}{L}\right) = 23.1\left(1 + \frac{14}{20}\right) = 39.3\%$$

Design $LL = 20(1 - 0.39) = 12$ psf

Live load from roof $= 12 \times 642 = 7704$ lbs

Dead load from roof $= 14 \times 642 = 8988$ lbs

Total load $(DL + LL) =$ $16,692$ lbs

11-5.B   Bearing Angle Design

Bearing plate under Glu-Lam roof girder

Allowable masonry bearing stress on the full area per UBC Sec. 2406(c)8 (see Table A-3).

$$F_{bv} = \tfrac{1}{2}(0.26f'_m) = \tfrac{1}{2}(0.26 \times 2000)$$
$$= 260 \text{ psi}$$

$$\text{Required area of bearing plate} = \frac{16,692}{260}$$
$$= 64.2 \text{ sq. in.}$$

Glu-Lam beam assumed $6\tfrac{3}{4}'' \times 30''$

Width of bearing plate, assume 5''

$$\text{Required length of plate} = \frac{64.2}{5}$$
$$= 12.8''$$
$$\text{Use } 14''$$

Use steel angle for bearing plate.

To determine the bending moment on the bearing angle assume that the maximum moment or the point of fixity is at the center of the Glu-Lam beam, half way between the ends of the bearing angle. This is to recognize the compression or softness of the edges of the Glu-Lam beam.

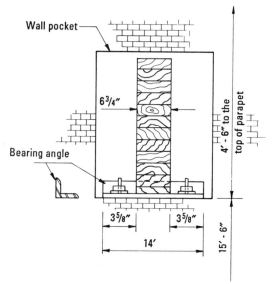

Figure 11-17 Glu-Lam beam bearing angle.

Uniform pressure under the bearing angle

$$= \frac{16,692}{14 \times 5} = 238 \text{ psi}$$

$$\text{Moment} = 238 \times 5 \times \frac{7^2}{2}$$
$$= 29,211 \text{ in. lbs}$$

Assume $F_s = 0.66\, F_y = 0.66(36,000) = 24,000$ psi

$$\text{Section modulus} = \frac{M}{F_s} = \frac{29,200}{24,000}$$
$$= 1.22 \text{ in.}^3$$

Use $5'' \times 3\tfrac{1}{2}'' \times \tfrac{1}{2}'' \times 1' - 2''$ angle $(S = 1.56 \text{ in.}^3)$

Anchor the angle to the wall with a minimum of two $\tfrac{1}{2}''$ dia. anchor bolts, 12'' long with 2 - #3 ties around bolts.

11-5.C   North Wall Pilaster Design

The north wall under the glue laminated beam is designed as a load bearing wall not as a column. If it were designed as a pilaster column the minimum steel ratio would be $\tfrac{1}{2}\%$ and column ties would be required.

The minimum wall steel required in Seismic Zone Nos. 3 and 4 is $0.0007bt$ and #4 @ 48'' in Seismic Zone No. 2. Because this wall steel is not tied, it is assumed that it does not carry any compression load. The steel is considered to resist only tension force due to bending. For the pilaster, $t = 9''$ and $h = 15' - 6''$.

Design at the top of the pilaster for the roof load, the parapet weight and the moment due to an eccentric load from the Glu-Lam beam.

Maximum width of flush wall pilaster (UBC Sec. 2407(c)1.A)

$$b = (4 \times 9) + 14 = 50''$$

Weight on pilaster at beam level

$$P = \text{live load} + \text{dead load} + \text{wt. of pilaster}$$
$$= 7704 + 8988 + \left(90 \times 4.5 \times \frac{50}{12}\right)$$
$$= 18,380 \text{ lbs}$$

$$\text{Actual} f_a = \frac{P}{bt} = \frac{18,380}{50 \times 9} = 40.8 \text{ psi}$$

Find the allowable axial stress, F_a.

From Table Q-1 for $\dfrac{h'}{t} = \dfrac{15.5 \times 12}{9} = 20.67$;

$$R = 0.880$$

Use Table Q-2 for $f'_m = 2000$ psi and no inspection to find:

$$F_a = 176 \text{ psi}$$

$$\frac{f_a}{F_a} + \frac{f_b}{F_b} = 1; \qquad F_b = 333 \text{ psi}$$

$$f_b = \left(1 - \frac{f_a}{F_a}\right)F_b = \left(1 - \frac{40.8}{176}\right)333$$
$$= 256 \text{ psi}$$

$$\text{Moment} = Pe = (7704 + 8988)(2.5'')$$
$$= 41,730 \text{ in. lbs}$$

Flexural Coefficient

$$K = \frac{M}{bd^2} = \frac{41,730}{50 \times 4.5^2} = 41.2$$

From Diagram F-2, K vs p, for $K = 41.2$ and $f_b = 256$ psi,

read $p = 0.055$ and $f_s = 8000$ psi

Area of steel

$$A_s = pbd = 0.0055 \times 50 \times 4.5$$
$$= 1.24 \text{ sq. in.}$$

Use #5 at 16″ o.c., therefore there are 4 - #5 bars in the 50″ width.

$$4 \times 0.31 = 1.24 \text{ sq. in.}$$

Check pilaster at mid-height and include lateral seismic force

Moment due to seismic force $= \dfrac{wl^2}{8} = \dfrac{13.5 \times 15.5^2}{8} \times \dfrac{50}{12}$
$$= 1689 \text{ ft lbs} / 50″$$

Moment at mid-height due to eccentric dead load

$$M_{DL} = 8988 \times \frac{2.5 \times \frac{1}{2}}{12} = 936 \text{ ft lbs}$$

Total Moment at mid-height, $M_T = 1689 + 936$
$$= 2625 \text{ ft lbs}$$

Vertical dead load = Glu-Lam + ½ wall pilaster + wall and parapet above.

$$P = 8988 + \left(\frac{15.5}{2} + 5.5\right)(90)\left(\frac{50}{12}\right)$$
$$= 13,957 \text{ lbs}$$

$$f_a = \frac{P}{bt} = \frac{13,957}{50 \times 9} 31 \text{ psi}$$

$$f_b = \left(1.33 - \frac{31}{176}\right)333$$
$$= 384 \text{ psi}$$

Flexural coefficient, $K = \dfrac{M_T}{bd^2}$

$$K = \frac{2625 \times 12}{50 \times 4.5^2} = 31.1$$

From Diagram F-2 for $K = 31.1$, $f_b = 385$ psi and $f_s = 32,000$ psi

read $p = 0.001$

$$A_s = 0.001 \times 50 \times 4.5 = 0.23 \text{ sq. in.}$$

#5 bars at 16″ o.c. are adequate.

11-6 DESIGN OF SECTION 5-5 FOR VERTICAL AND LATERAL LOADS

Tributary loading width $= \dfrac{8}{2} + 2 + \dfrac{8}{2}$
$$= 10 \text{ ft}$$

Vertical load at top of window = roof + parapet + wall

$$= (10 \times 12 \times 34) + (10 \times 2.5 \times 90) +$$
$$(8 \times 4 \times 90) + (10 \times 6 \times 90)$$
$$= 4080 + 2250 + 2880 + 5400$$
$$= 14,610 \text{ lbs/ft}$$

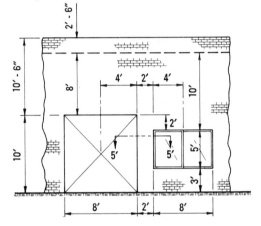

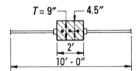

Figure 11-18 Detail elevation of west wall.

Axial stress, $f_a = \dfrac{P}{bt} = \dfrac{14,610}{24 \times 9} = 67.6$ psi

Allowable axial stress, $F_a = \frac{1}{2} \times 0.2 f'_m \left[1 - \left(\dfrac{h'}{42}\right)^3\right]$

From Tables Q-1 and Q-2 for $h'' = 18$ ft, $t = 9$ in., $f'_m = 2000$ psi and no special inspection, $F_a = 163$ psi.

Check combined stresses

Maximum allowable flexural stress, $F_b = \frac{1}{2} \times 0.33 f'_m$

$$F_b = \frac{1}{2} \times 0.33 \times 2000 = 333 \text{ psi}$$

Allowable flexural stress, f_b

$$f_b = \left(1.33 - \frac{f_a}{F_a}\right)F_b$$
$$= \left(1.33 - \frac{67.6}{163}\right)333$$
$$= 304 \text{ psi}$$

Lateral load on the wall from the earthquake = 13.5 psf.

Moment on wall at the top of the window (assuming the wall is pinned at the top and bottom and conservatively neglecting the seismic force on force on the parapet:

$$M = \frac{wh^2}{8} = \frac{(13.5 \times 10)(18)^2}{8}$$
$$= 5468 \text{ ft lbs}$$

Flexural coefficient, $K = \dfrac{M}{bd^2}$

$$K = \frac{5468 \times 12}{24 \times 4.5^2} = 135$$

This is in excess of allowable values (see Table E-2 and Diagram F-2).

Increase wall thickness at pier for full height to 15 in.

Recompute stresses and K.

From Tables Q-1 and Q-2, for $t = 15''$, $h' = 18$ ft, $f'_m = 2000$ psi and no special inspection;

$F_a = 192$ psi

Actual axial stress, f_a

$$f_a = \frac{P}{A} = \frac{14,610}{24 \times 15} = 40.6 \text{ psi}$$

Allowable flexural stress, f_b

$$f_b = \left(1.33 - \frac{f_a}{F_a}\right)F_b$$
$$= \left(1.33 - \frac{40.6}{192}\right)333$$
$$= 373 \text{ psi}$$

Flexural coefficient $K = \dfrac{M}{bd^2}$; assume $d = 10''$

$$K = \frac{5468 \times 12}{24 \times 10^2} = 27.3$$

From Diagram F-2

for $K = 27.3$, $f_b = 373$ psi, and $f_s = 32,000$ psi,

read $p = 0.0009$ (steel governs)

$A_s = pbd = 0.0009 \times 24 \times 10$
$\quad\quad = 0.22$ sq. in.

Use 2 - #5 bars each side ($A_s = 0.62$ sq. in.).

Note: Because the length of the pier is less than three times the thickness, it should be reinforced and tied as a column.

Use $^1/_4$ in. dia. ties at 12" o.c.

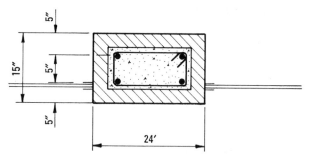

Figure 11-19   Cross-section of pier Section 5-5.

11-7 WIND AND SEISMIC FORCES ON TOTAL BUILDING

After each of the individual structural elements such as walls, spandrels, piers, columns, etc. are designed, the total building must be analyzed and designed for lateral seismic or wind forces.

By inspection, it is apparent that the east-west walls will have the highest shear stresses since they are relatively short but receive large loads from a N-S earthquake force. Therefore, check the shear capacity of these walls.

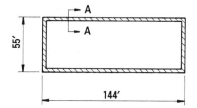

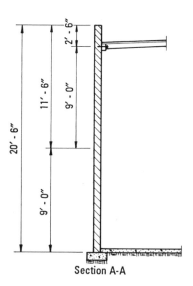

Figure 11-20 Plan of building and section thru wall.

11-7.A Loads

Load in North-South Direction

Consider the area of $\frac{1}{2}$ wall height from footing to ledger beam plus parapet $= \frac{1}{2}(18) + 2.5 = 11.5$ ft

Wind load $= 12$ psf $\times 11.5$ ft $= 138$ plf

Weight of structure acting on roof diaphragm

$= 2(\frac{1}{2}$ wall ht. + parapet) + roof dead load
$= 2(2.5 + \frac{1}{2} \times 18)(90$ psf$) + (14$ psf $\times 55$ ft$)$
$= 2070 + 770 = 2840$ plf

Seismic forces from longitudinal walls to roof diaphragm

$v = 0.092W$ (from Section 11-1.B.1(a))
$= 0.092(2840) = 262$ plf on the building

Seismic forces govern over wind forces

262 plf > 138 plf, and
262 plf > 200 plf (min. code force, UBC Sec. 2310)

Use 262 plf.

The roof diaphragm acts as horizontal beam with the end shear walls serving as the reactions. The ledger and bond beams act as the flanges that resist the tension and compression forces in this beam.

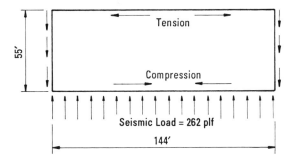

Figure 11-21 Seismic load to roof diaphragm.

$$\text{Moment} = \frac{wl^2}{8} = \frac{262 \times 144^2}{8}$$
$$= 679,104 \text{ ft lbs}$$

$$\text{Flange stress} = \frac{M}{d} = \frac{679,104}{54.25} = 12,518 \text{ lbs}$$

Chord steel or tension steel in bond beam

$$A_s = \frac{T}{f_s} = \frac{12,518}{32,000} = 0.39 \text{ sq. in.}$$

Use 1 - #6 or 2 - #4 bars in bond beam at ledger

Seismic force to end transverse shear walls

$$\text{Seismic force} = 262 \times \frac{144}{2} = 18,864 \text{ lbs}$$

Seismic shear per foot on transverse shear walls at roof

$$\text{ledger} = \frac{18,864}{55} = 343 \text{ plf} > 200 \text{ lb/ft}   \text{(UBC minimum)}$$

Additional seismic force due to weight of end walls. For simplicity assume no openings in the walls.

Weight of walls $= 20.5 \times 90$ psf $= 1845$ plf
Seismic force $= 0.092 \times 1845 = 170$ plf

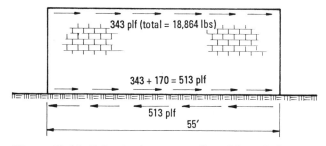

Figure 11-22 Seismic shear on wall and foundation.

Shear per linear foot at connection between wall and footing

$$343 + 170 = 513 \text{ plf}$$
$$\text{Unit shear} = \frac{513}{9 \times 12} = 4.75 \text{ psi}$$

Shear friction and steel dowels will resist this stress. (See Section 5-4.D, Table 5-1 and Table A-8.)

11-7.B Ledger Bolt and Ledger Beam Design

Ledger bolts and joist anchors tie the roof and/or floor diaphragms to the masonry walls. They transmit the vertical and lateral shear loads from the roof and/or floor system to the load bearing, shear resisting wall elements. They also resist loads perpendicular to the wall. UBC Sec. 2310 requires that the diaphragm connection be designed for a minimum force of 200 pounds per linear foot both parallel and perpendicular to the wall.

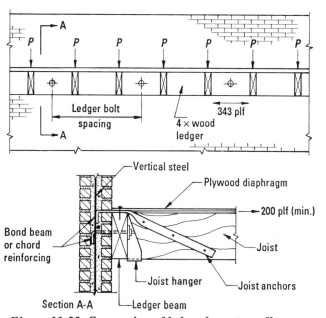

Figure 11-23 Connection of ledger beam to wall.

The diaphragm must be connected in a positive manner by means of joist anchors to the wall. The consideration of cross grain tension on the ledger to resist forces perpendicular to the wall is not permitted.

Seismic force: On wall = 13.5 psf
On parapet = 36 psf

Conservatively, the maximum wall reaction at the roof is

$$R = 13.5 \text{ psf} \left(\frac{22}{2}\right) + 36 \text{ psf} (2.5) = 238.5 \text{ plf}$$

This exceeds the minimum force of 200 plf required by the code.

Joist anchors must be spaced 48″ o.c. to resist 239 plf × 4 ft = 956 lbs. These anchors must be properly fastened by means of bolts or lag screws into the joists.

Shear forces parallel to the shear wall are transmitted from the wood diaphragm to the ledger beam by means of proper nailing. The ledger beam then transmits the shear force to the wall through the ledger bolts.

Horizontal shear force on ledger parallel to wall = 343 plf.

From UBC Table No. 25-J-1 *Allowable Shear In Pounds Per Foot For Horizontal Plywood Diaphragms With Framing Of Douglas Fir-larch Or Southern Pine,* determine: (a) plywood grade, (b) blocked or unblocked diaphragm, (c) nail size, (d) spacing of nails at various locations on the plywood panel, (e) minimum width of framing member, (f) plywood thickness.

From UBC Table No. 25-J-1 use:

   a. Structural I plywood grade

   b. Blocked diaphragm

   c. 10d nails

   d. Space nails 6″ o.c. at diaphragm boundaries and at continuous panel edges parallel to load

      Space nails 6″ o.c. at other plywood panel edges

      Space nails at 12″ o.c. for field nailing

   e. Minimum nominal width of framing member is 3″

   f. Minimum nominal plywood thickness is $^{15}/_{32}$ in.

Allowable shear = 360 plf > 343 plf   O.K.

Ledger bolts
   Shear parallel to wall = 343 plf
   Vertical dead load = 11 psf × 12 ft = 132 plf

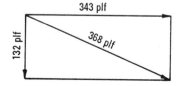

From Table A-8 Allowable Shear on Bolts

   Use $^5/_8$″ diameter bolts;

   Allowable shear = 1330 lbs × 1.33
                    = 1770 lbs / bolt

   Spacing of bolts $= \dfrac{1770}{368} \times 12 = 58″$

Space bolts 48″ o.c. or as shown below and locate them 4″ below top of member.

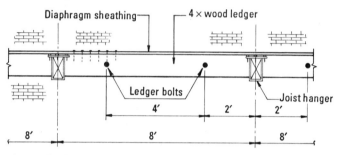

Figure 11-24 Bolt pattern in ledger.

The wood bearing stress must also be checked to assure the members are adequate.

11-8 DISTRIBUTION OF SHEAR FORCE IN END WALLS

The lateral forces on a building are distributed by the flexible wood diaphragm to the shear resisting walls. These walls carry the forces to the foundation and if the wall has openings due to doors and windows, the piers are subjected to shear forces in proportion to their respective rigidities. Rigidity of a pier is inversely proportional to its flexibility and deflection. This deflection is made up of both moment and shear deflection for cantilever and fixed piers.

Total shear and moment deflection for a cantilever pier is:

$$\Delta_C = \frac{P}{E_m t}\left[4\left(\frac{h}{d}\right)^3 + 3\left(\frac{h}{d}\right)\right]$$

Total shear and moment deflection for a fixed pier is:

$$\Delta_C = \frac{P}{E_m t}\left[\left(\frac{h}{d}\right)^3 + 3\left(\frac{h}{d}\right)\right]$$

The modulus of elasticity of masonry is based on the value 1,000,000 psi. Table T-1 gives values per inch of thickness for Δ_F; Δ_C; $\dfrac{1}{\Delta_F}$ and $\dfrac{1}{\Delta_C}$ for various $\dfrac{h}{d}$ ratios to facilitate determining the deflection and rigidity of piers in a wall.

If the unit shear stress exceeds the allowable shear for masonry with no shear reinforcement, the pier must be reinforced and the horizontal shear reinforcing must be capable of resisting all the shear stress.

11-8.A Shear Distribution in West Wall

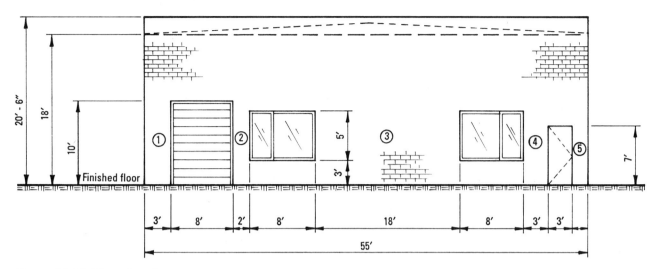

Figure 11-25 West elevation.

Seismic lateral force
at top of wall from ledger = 18,864 lbs

Calculate the seismic force from the weight of the wall.

Dead weight of wall, w

$w = 90[(20.5 \times 55) - (10 \times 8) - (2 \times 5 \times 8) - (7 \times 3)]$
$= 90[946.5] = 85,185$ lbs

$V_{wall} = 0.092(85,185) = 7837$ lbs

Total base shear at the foundation $= 18,864 + 7837$
$= 26,701$ lbs

Seismic lateral force at 8 ft above the foundation

$= 18,864 + 0.092(90 \times 10.5 \times 55) = 23,646$ lbs

The shear stresses in the piers (see Table 11-1) do not exceed the allowable values for masonry from Table A-5 or Diagram A-5 and only require nominal minimum

reinforcing steel. It is good practice to place 2 - #4 bars or 1 - #5 bar on each side, above and below all openings.

If this building were located in Seismic Zone No. 3 or 4, the shear forces due to seismic action would need to be increased by 1.5 (UBC Sec. 2407(h)4.F). The overturning moments due to seismic forces need not be increased 1.5 times.

Only the shear walls need be designed for 1.5 times the seismic forces. The diaphragm, diaphragm connections and chord steel need only be designed for the seismic force or wind, whichever governs. However, it would be good practice to design the connections for 1.5 times the seismic force also.

For this building the wind force is 138 plf and the seismic force is 262 plf.

TABLE 11-1 Relative Rigidities of Piers — West Wall

Pier No.	Height h (ft)	Length l (ft)	h/l Ratio (all piers fixed)	Relative Rigidity Table T-1 Fixed Piers[2]	Percentage Lateral Force to Each Pier	Force V to each Pier[3] (pounds)	Unit Shear f_v, in each pier = $\frac{V}{tl}$ (psi)
1	10	3	3.33	0.213	1.5	355	1.1
2	5	2	2.50	0.72	5.0	1,182	5.5
3	5	18	0.28	11.602	81.2	19,200	9.9
4	4	3	1.33	1.577	11.0	2,601	8.0
5	7	2	3.50	0.187	1.3	308	1.4
				$\Sigma = \overline{14.299}$	100%	$\Sigma = \overline{23,646}$ pounds[1]	

1. It would be conservative and quite usual to use the base shear, $V = 26,700$ lbs as the force on the wall and distribute this amount to all piers. The approach here is a little more detailed.

2. In Section 11-6 pier number 2 was increased in thickness to 15″. The stiffness is increased in the above table accordingly; $0.432 \times 15/9 = 0.72$.

3. UBC Sec. 2407(h)4.F requires for Seismic Zone Nos. 3 and 4 that when calculating shear or diagonal tension stresses, the shear walls which resist seismic forces must be designed to resist 1.5 times the seismic shear force V.

TABLE 11-2 Lateral Force to Piers — West Wall; 1.5 × Lateral Seismic Force = 110,000 pounds. Assume Piers Fixed Top and Bottom

Pier No.	M/Vd or $h/2l$	% Lateral force to each pier	Force $1.5V$ to each Pier (pounds)	Unit Shear[2], f_v in each Pier = $\dfrac{1.5V}{bjd}$ (psi)	Allowable Shear[1] × 1.33 No Reinforcing (psi)	Allowable Shear × 1.33 with Reinforcing (psi)
1	1.67	1.5	1,650	6.6	23.9	45.2
2	1.25	5.0	5,500	22.2	23.9	45.2
3	0.14	81.2	89,320	51.4	37.9	57.2
4	0.67	11.0	12,100	48.7	32.4	49.6
5	1.75	1.3	1,430	9.6	23.9	45.2

$$\Sigma = \overline{110,000} \text{ lbs}$$

1. From Diagrams A-5 and A-6, *Maximum Allowable Shear Stress.*

2. It has been assumed that the piers will crack. Therefore the UBC equation $f_v = V/bjd$ was used for design. The distance, d, was assumed to equal the length of the pier minus 6" ($d = l - 6''$) to allow for a brick end closure and tolerance for the placement of the reinforcing bars. The term b equals the thickness of the wall (9") and j was assumed to be 0.92.

11-8.B West Wall Pier Design for Seismic Forces

To demonstrate the techniques of pier design when the shear stress, f_v, exceeds the allowable shear, assume the structure is located in Seismic Zone No. 3 and the total seismic shear is:

$$V = 73,333 \text{ lbs}$$
$$1.5V = 110,000 \text{ lbs}$$

See Table 11-3 for distribution of forces.

Pier 2 has ties from the previous design (Section 11-5).

Piers 1 and 5 require only nominal horizontal reinforcing steel. Use #5 at 32" o.c. or #4 at 20" o.c.

Provide horizontal shear reinforcing for piers 3 and 4.

11-8.C Design of Shear Reinforcing in Piers 3 and 4

Pier 3

Calculate the spacing of shear steel for $f_v = 51.4$ psi, $V = 89,320$ lbs, $F_s = 32,000$ lbs and $d = 210''$

For #5 shear reinforcing bars:

$$s = \frac{A_v F_s d}{V} = \frac{0.31(32,000)(210)}{89,320}$$
$$= 23.3''$$

For #4 bars:

$$s = \frac{0.20(32,000)(210)}{89,320}$$
$$= 15''$$

Provide #5 at 22" o.c. or #4 at 12" o.c. horizontal shear reinforcing.

Check overturning moment. Assume pier is fixed at the top and the bottom

$$OTM = V \times h/2$$

The lateral force for overturning $= \dfrac{89,320}{1.5}$
$$= 59,550 \text{ lbs}$$

Overturning moment, $OTM = 59.55 \times \dfrac{5}{2} = 149$ ft kips

Combined stresses: Unity equation

$$\frac{f_a}{F_b} + \frac{f_b}{F_b} \leq 1 \text{ or } 1.33$$

Allowable axial stress, $F_a = \dfrac{1}{2} \times 0.2 f'_m \left[1 - \left(\dfrac{h'}{42t} \right)^3 \right]$

From Table Q-1 and Q-2 for:

$f'_m = 2000$ psi (No inspection)

$h' = 18$ ft and $t = 9$ inches:

$R = 0.813$

$F_a = 163$ psi

Actual axial stress, $f_a = \dfrac{P}{tl}$

$P = $ wall weight + roof load
$$= 90[(4 + 18 + 4)12 + (5 \times 18)]$$
$$+ 11[12(4 + 18 + 4)$$
$$= 37,350 + 3432 = 40,782 \text{ lbs}$$

$$f_a = \frac{P}{tl} = \frac{40,782}{9 \times 216} = 21.0 \text{ psi}$$

Allowable flexural stress, $F_b = \frac{1}{2}(0.33 f'_m)$
$$= 333 \text{ psi}$$

Interaction design using Method 3

Moment on pier 3, $M = 149$ ft k

Actual flexural stress, $f_b = \dfrac{M}{S} = \dfrac{6M}{tl^2}$

$$f_b = \frac{6 \times 149,000 \times 12}{9 \times 216^2}$$
$$= 25.5 \text{ psi}$$

$$\frac{f_a}{F_a} + \frac{f_b}{F_b} = \frac{21.0}{163} + \frac{25.5}{333}$$
$$= 0.13 + 0.08$$
$$= 0.21 << 1.33$$

Stress on section

$$f = \frac{P}{A} \pm \frac{M}{S}$$
$$= 21 \pm 25.5$$
$$= 4.5 \text{ psi tension}$$
$$= 46.5 \text{ psi compression}$$

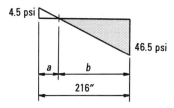

$$a = \frac{4.5}{4.5 + 46.5} \times 216 = 19 \text{ in.}$$
$$b = 216 - 19 = 197 \text{ in.}$$

Tension force $= \frac{1}{2} \times 4.5 \times 9 \times 19$
$$= 385 \text{ lbs}$$

By inspection, use 1 - #5 bar each side and, similar to the standard design for the wall, use #5 bars @ 48″ in between.

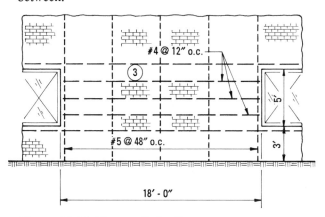

Figure 11-26 West wall pier 3.

Design of pier 4 will be similar.

11-8.D Shear Distribution in East Wall

Seismic shear force in East Wall

Use base shear,
Dead weight of wall $= 90[(20.5 \times 55) - (2 \times 16 \times 14)$
$$- (3 \times 7) - (3 \times 4)]$$
$$= 90[646.5]$$
$$= 58,185 \text{ lbs}$$

$$V = 0.092(58,185)$$
$$= 5353 \text{ lbs}$$

Base shear $= 18,864 + 5353$
$$= 24,217 \text{ lbs}$$

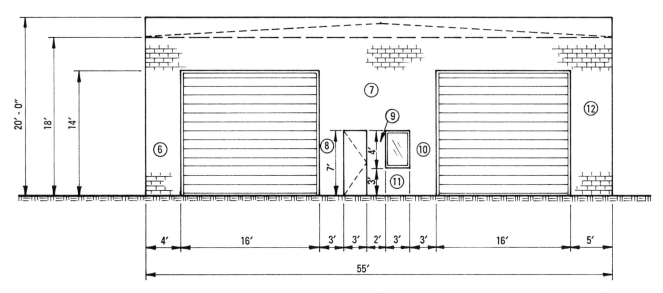

Figure 11-27 East wall — detailed method.

TABLE 11-3 Rigidity of Piers Using Table T-1

Pier No.	Height h (ft)	Length l, (ft)	h/l Ratio	Δ_y Deflection	R_y Rigidity
6	14	4	3.50	5.338	0.187
7	7	14	0.50	0.163	6.154
8	7	3	2.33	1.964	0.509
9	4	2	2.00	1.400	0.714
10	4	3	1.33	0.634	1.577
11	3	8	0.38	0.119	8.369
12	14	5	2.80	3.035	0.329

Determine the ridigity of the wall panel made up of piers 7, 8, 9, 10 and 11.

Panel 9, 10 and 11

$$\Delta\, 9,\,10,\,11 = \Delta_{11} + \cfrac{1}{\cfrac{1}{\Delta_9} + \cfrac{1}{\Delta_{10}}}$$

$$= \Delta_{11} + \frac{1}{R_9 + R_{10}}$$

$$= 0.119 + \frac{1}{0.714 + 1.577}$$

$$= 0.119 + 0.436$$

$$\Delta_{9,\,10,\,11} = 0.555 \text{ Deflection Coefficient}$$

Rigidity coefficient $R_{9,\,10,\,11} = \dfrac{1}{\Delta_{9,\,10,\,11}} = \dfrac{1}{0.555}$

$$= 1.80$$

Middle panel 7, 8, 9, 10 and 11

$$\Delta_{7,\,8,\,9,\,10,\,11} = \Delta_7 + \cfrac{1}{\cfrac{1}{\Delta_8} + \cfrac{1}{\Delta_{11} + \cfrac{1}{\cfrac{1}{\Delta_9} + \cfrac{1}{\Delta_{10}}}}}$$

$$= \Delta_7 + \cfrac{1}{R_8 + \cfrac{1}{\Delta_{11} + \cfrac{1}{R_9 + R_{10}}}}$$

$$= \Delta_7 + \cfrac{1}{R_8 + \cfrac{1}{\Delta_{11} + \Delta_{9,\,10}}}$$

$$= \Delta_7 + \cfrac{1}{R_8 + R_{9,\,10,\,11}}$$

$$\Delta_7 + \Delta_{8,\,9,\,10,\,11} = 0.16 + \frac{1}{0.509 + 1.80}$$

$$\Delta_{7,\,8,\,9,\,10,\,11} = 0.16 + 0.433$$

$$= 0.595 \text{ Deflection coefficient}$$

Rigidity coefficient $= \dfrac{1}{\Delta_{7,\,8,\,9,\,10,\,11}} = \dfrac{1}{0.596} = 1.68$

Seismic base shear force in East wall;

$$V = 24,217 \text{ lbs}$$

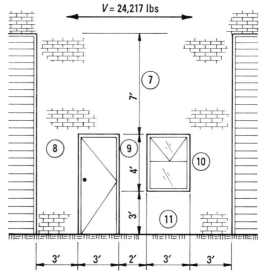

Figure 11-28 Middle panel, east wall piers 7, 8, 9, 10 and 11.

TABLE 11-4 Lateral Force Distribution

Pier	Δ_F Deflection	R_F Rigidity	Percent of Lateral Force to Each Pier	Lateral Force to Each Pier (pounds)
6	5.337	0.187	8.5	2058
Middle Panel	0.596	1.680	76.5	18,526
12	3.035	0.329	15.0	3,633
	$\Sigma = 2.196$		100.0%	24,217

Distribution of shear force to elements in middle panel

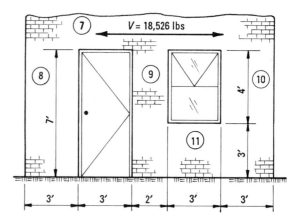

Figure 11-29 Middle panel, east wall piers 8, 9, 10 and 11.

TABLE 11-5 Lateral Force Distribution

Pier	R_F Rigidity	Percent of Lateral Force to Each Pier	Lateral Force to Each Pier (pounds)
8	0.509	22.0	4,076
9, 10,11	1.80	78.0	14,450
$\Sigma = \overline{2.309}$		$\overline{100.0\%}$	$\overline{18,526}$

TABLE 11-6 Lateral Force Distribution

Pier	R_F Rigidity	Percent of Lateral Force to Each Pier	Lateral Force to Each Pier (pounds)
9	0.714	31.2	4,503
10	1.577	68.8	9,947
$\Sigma = \overline{2.291}$		$\overline{100.0\%}$	$\overline{14,450}$

Distribution of force to each pier in panel 9, 10 and 11

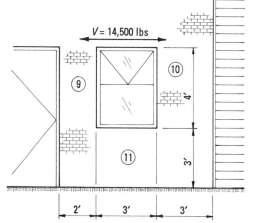

Figure 11-30 Middle panel, east wall piers 9, 10 & 11.

The shear stress in all piers is less than the allowable shear values for masonry resisting shear stresses and therefore, only minimum or nominal shear reinforcing steel is required.

Check all piers and sections for minimum steel. Try to maintain uniformity of bar size and spacing throughout the wall, wherever possible.

TABLE 11-7 Shear in Piers, East Wall. Total Lateral Force = 24,217 lbs

Pier	$h/2l$	l (inches)	t (inches)	V (pounds)	Unit Shear[1], $f_v = V/lt$ (psi)	1.33 x Allowable Shear[2] No Reinforcing (psi)	1.33 x Allowable Shear[2] w/ Reinforcing (psi)
6	1.75	48	9	2,058	4.8	23.9	45.2
7	0.25	168	9	18,526	12.3	37.2	55.1
8	1.17	36	9	4,076	12.6	23.9	45.2
9	1.00	24	9	4,503	20.8	23.9	45.2
10	0.67	36	9	9,947	30.7	32.6	49.7
11	0.19	96	9	14,450	16.7	37.2	55.9
12	1.40	60	9	3,633	6.7	23.9	45.2

1. The full cross-sectional area is used here. For piers where f_v exceeds the capacity of the masonry with no reinforcing, recalculate f_v as V/bjd.

2. From Tables A-5 and A-6, *Maximum Allowable Shear Stress.*

11-8.E Alternate Approximate Method Considering Only Piers 6, 8, 9, 10 & 12

TABLE 11-8 Rigidity of Piers and Lateral Force Distribution. Base Shear = 24,217 lbs

Pier No.	Height h (ft)	Length l (ft)	h/l Ratio	R_F Rigidity	% Lateral Force to Each Pier	Lateral Force to Each Pier, (pounds)
6	14	4	3.50	0.187	5.64	1,366
8	7	3	2.33	0.509	15.34	3,717
9	4	2	2.00	0.714	21.53	5,214
10	4	3	1.33	1.577	47.54	11,517
12	14	5	2.80	0.329	9.95	2,403
				$\Sigma = \overline{3.316}$	$\overline{100.00}$	$\overline{24,217}$

TABLE 11-9   Comparison Between Detailed Method and Approximate Method

Pier No.	V (pounds)		1.33 × Allowable Shear No Reinforcing	% of Difference against Detailed Method
	Detailed Method	Approx-imate Method		
6	2,058	1,366	22.6	−33.6%
8	4,076	3,717	22.6	−8.8%
9	4,503	5,214	22.6	+15.8%
10	9,947	11,517	25.3	+15.8%
12	3,633	2,403	22.6	−33.9%

For piers 6 and 12, the percentages are not really representing a large difference, that is because of the small amount of shear force on the piers, and the values do not exceed the allowable shear stress without reinforcing.

Piers 9 and 10 are more critical using the approximate method.

In conclusion, both methods are acceptable, and it is the decision of the designer which method should be used considering the general assumptions made for loads and strengths of materials.

11-9   QUESTIONS AND PROBLEMS

11-1   Determine the required reinforcing steel for the shear and overturning forces on pier 4 of the West wall of the industrial building in this Section.

The lateral force at the top of the pier is 15 kips, vertical dead load is 20 kips. f'_m = 2000 psi and f_y = 60,000 psi. No special inspection will be provided. Use interaction method 3.

11-2   Design and detail the flexural and shear reinforcing steel for the solid grouted continuous masonry beam shown below.

The 8″ concrete masonry units have a strength of 2800 psi and type S mortar will be used. Determine f'_m from Table 2-3. Assume the wind load is 20 psf and the structure is located in Seismic Zone 4. Vertical LL = 1400 plf, DL = 600 plf. No special inspection is provided.

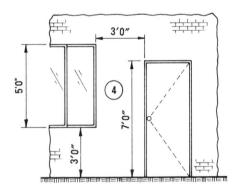

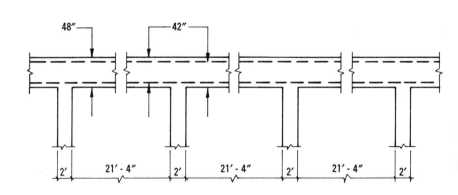

11-3 Determine the shear stress in each 9 inch thick grouted brick pier. If $f'_m = 2000$ psi and if there is no special inspection, will shear reinforcing steel be necessary in any of the piers?

11-4 Design the piers for the end wall shown below if $f'_m = 1500$ psi and Grade 60 reinforcing steel is used. Assume special inspection will be provided. Distribute the lateral force to each of the piers in relation to their rigidity. The lateral force is 30 kips at the roof level and 40 kips at the second floor level. Note the lateral force for the piers on the first floor must resist the total lateral force.

11-5 Using the dimensions and details for the South wall of the industrial building in this Section, determine the vertical and horizontal reinforcing steel using the slender wall design method.

Assume the structure is in Seismic Zone 4, the live load is 250 plf and the dead load is 300 plf at an eccentricity of 8 inches. The wall is pin connected at the top and bottom.

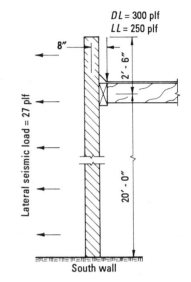

Problem 11-5

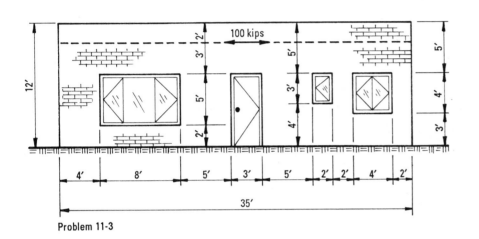

Problem 11-3

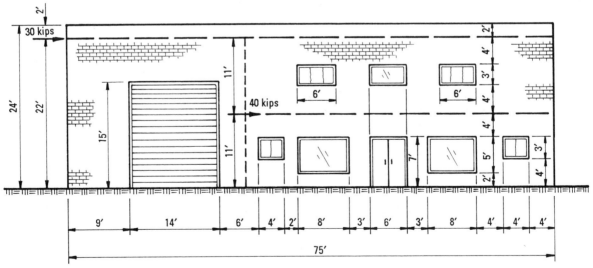

Problem 11-4

SECTION 12

Design of Seven-Story Masonry Load Bearing Wall Apartment Building

12-1 DESIGN CRITERIA, ELEVATION AND PLAN

Figure 12-1 Seven story apartment building.

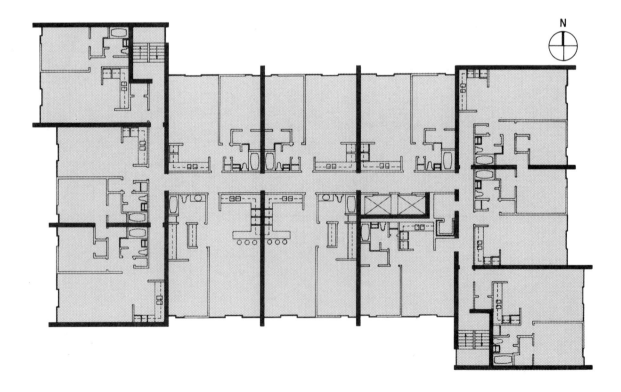

Figure 12-2 Typical architectural floor plan.

12-1.A General

Calculate the centers of mass and rigidity for the second and fifth floors and determine the design eccentricities. Find the shear forces and overturning moments in each wall for the controlling wind or seismic forces in the transverse direction. Also, determine the required reinforcing in the first and fourth story walls **f** and **j**.

12-1.B Floor and Roof Systems

Lightweight concrete precast slabs are used for the floor and roof systems (110 pcf).

Floor Loads:

Slab, 8″	= 73 psf
Partition	= 20 psf
Misc.	= 5
Dead load	= 98 psf

Roof Loads:

Slabs, 6″	= 55 psf
Fill	= 12
Roofing	= 4
Dead load	= 71 psf

Live Loads:

40 psf Apartments   20 psf Roof

100 psf Corridor   Snow load not considered

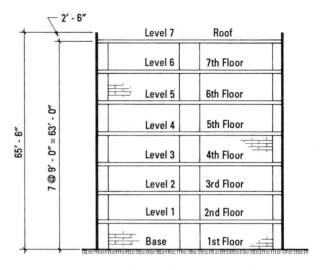

Floor to ceiling height = 8′ - 4″

Figure 12-3 Transverse cross-section.

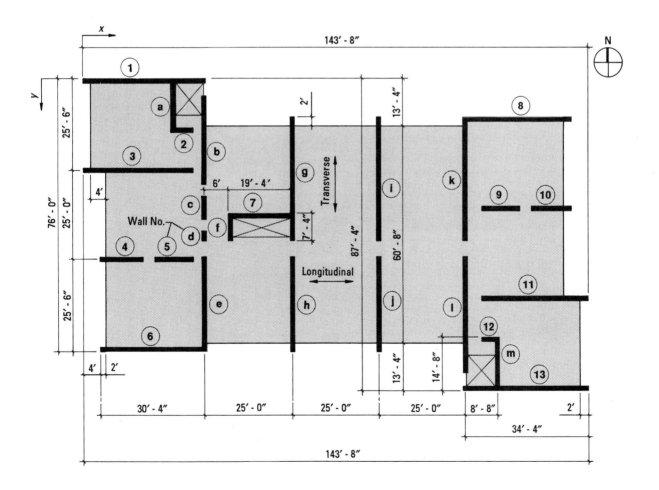

Figure 12-4 Typical structural floor plan.

12-1.C   Structural Wall System

8″ medium weight concrete block masonry units — 8″ × 4″ × 16″ units.

Wall weight = 78 psf (See Table B-3, Weight of Walls)

Masonry walls are to be solid grouted for maximum STC (Sound Transmission Coefficient) values and fire ratings.

Masonry walls will be continually inspected.

Values of f'_m, specified strength of masonry, are to be determined based on the magnitude of vertical and lateral stresses.

12-1.D   Lateral Force Design

Based on 1991 UBC seismic requirements Seismic Base Shear is:

$$V = \frac{ZIC}{R_w} \times W$$

Where:

Z = 0.40 (Seismic Zone No. 4)

I = 1.0

C = 2.75 (max.)

R_w = 6 (shear wall building)

Shear walls must be designed for shear to resist 1.5 times the base shear force, V (UBC Sec. 2407(h)4F(i)).

12-2   LOCATION OF CENTER OF MASS

The center of mass is the mathematical point at which it is assumed all the weight or mass of the floor or structure is concentrated. It is located by taking moments about a base line of each element weight and then dividing the sum of the moments by the total weight. The location of the center of mass will change if the floor or wall layout changes.

Center of mass of walls (values from Table 12-1)

$$\bar{x} = \frac{\Sigma L \times x}{\Sigma L} = \frac{39,191}{551.7} = 71.04 \text{ ft}$$

$$\bar{y} = \frac{\Sigma L \times y}{\Sigma L} = \frac{23,963}{551.7} = 43.43 \text{ ft}$$

Total *DL* of wall = 551.7 × 8.33 × 0.078 ksf
    = 358.5 kips/floor

Center of mass of floor and roof

Floor dead load = 98 psf
Roof dead load = 71 psf

TABLE 12-1   Center of Mass of Walls

NOTE: All walls are the same height, thickness and weight per square foot; therefore, only the length of the walls need be used in this table to determine the center of mass.

Wall No.	Direction of Wall	Length, L (feet)	Center of Gravity of Wall x distance (feet)	Center of Gravity of Wall y distance (feet)	L × x	L × y
1	x	34.67	17.33	0.33	600.8	11.4
2	x	4.67	28.33	14.33	132.3	66.9
3	x	30.00	15.00	25.50	450.0	765.0
4	x	12.00	10.00	50.50	120.0	606.0
5	x	11.33	24.67	50.50	279.5	572.2
6	x	30.67	19.33	75.67	592.9	2,320.8
7	x	19.33	50.00	38.33	966.5	740.9
8	x	30.67	124.33	11.67	3,813.2	357.9
9	x	11.33	118.67	36.83	1,344.5	417.3
10	x	12.00	133.33	36.83	1,600.0	442.0
11	x	30.00	128.00	61.83	3,840.0	1,854.9
12	x	4.67	116.00	73.00	541.7	340.9
13	x	34.67	126.33	87.00	4,379.9	3,016.3
a	y	14.67	25.67	7.33	376.6	107.5
b	y	24.67	34.33	17.33	846.9	427.5
c	y	6.67	34.33	36.33	229.0	242.3
d	y	2.67	34.33	44.00	91.7	117.5
e	y	26.67	34.33	62.67	915.6	1,671.4
f	y	7.33	40.67	41.67	298.1	305.4
g	y	34.00	59.33	28.33	2,017.2	963.2
h	y	26.67	59.33	62.67	1,582.3	1,671.4
i	y	34.00	84.33	28.33	2,867.2	963.2
j	y	26.67	84.33	62.67	2,249.1	1,671.4
k	y	34.00	109.33	28.33	3,717.2	963.2
l	y	33.00	109.33	65.83	3,607.9	2,172.4
m	y	14.67	118.00	80.00	1,731.1	1,173.6

Σ = 551.7′   Σ = 39,191

Σ = 23,963

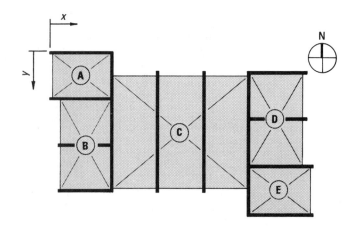

Figure 12-5 Typical floor plan.

TABLE12-2 Dead Weight of Floor

Area	Size		Area (sq. ft)
A	32.33 x	25.5	824.4
B	28.33 x	50.5	1,430.7
C	75.0 x	60.67	4,550.3
D	28.33 x	50.5	1,430.7
E	32.33 x	25.5	824.4

$$\Sigma = 9,060.5 \text{ sq. ft}$$

Dead weight of floor = 9060.5 sq. ft × 0.098 ksf
= 887.9 kips per floor

Due to symmetry, the center of floor and roof mass is:

$$\bar{x} = \frac{143.67}{2} = 71.84 \text{ ft}$$

$$\bar{y} = \frac{87.33}{2} = 43.67 \text{ ft}$$

Combined center of mass of walls and floor or roof

$$\bar{x} = \frac{\overset{\text{(walls)}}{(358.5 \times 71.04)} + \overset{\text{(floor)}}{(887.9 \times 71.84)}}{358.5 + 887.9}$$

= 71.61 ft

$$\bar{y} = \frac{\overset{\text{(walls)}}{(358.5 \times 43.43)} + \overset{\text{(floor)}}{(887.9 \times 43.67)}}{358.5 + 887.9}$$

= 43.61 ft

20-story loadbearing masonry Park Lane Towers, Denver, Colorado.

12-3 LOCATION OF CENTER OF RIGIDITY AND DETERMINATION OF DESIGN ECCENTRICITIES

The center of rigidity is the mathematical point that concentrates all of the rigidity of a system at that level. It is the axis about which the center of mass and the lateral forces rotate.

For a shear wall system in which the floors provide little vertical resistance to deflection or rotation of the walls, the walls may be assumed cantilevering from the base. The h/d value for the walls will change at each floor level and thus the stiffness or rigidity of the walls will change at each floor level.

Example of increasing flexibility or decreasing stiffness as a wall increases in height. Use wall **k**, length = 34 ft.

TABLE 12-3 Effect of Wall Height on Rigidity

Level	Base	1	2	3	4	5	6	7
Height above base (ft)	0	9	18	27	36	45	54	63
$h/_l = h/_d$	0	0.265	0.529	0.794	1.059	1.323	1.588	1.853
Rigidity R_c		11.507	4.576	2.303	1.259	0.760	0.480	0.324

```
7
6
5
4
3
2
1
Base
```

Accidental Torsion

Because the location and amount of some loads are unknown or vary and since the materials, construction and other influencing factors are not ideal, UBC Section 2334(e) requires the center of mass at each floor level to be displaced five percent from the calculated center of mass.

TABLE 12-4 Location of Center of Rigidity; Second Floor Walls, Level One

Wall Number	Direction of Wall	Length, l (feet)	$\dfrac{h}{l} = \dfrac{h}{d}$ (h = 8.33 ft)	Rigidity, R_x (Table T-1)	Rigidity, R_y (Table T-1)	x Distance (feet)	y Distance (feet)	$R_y \times x$	$R_x \times y$
1	x	34.67	0.24	12.898	—	—	0.33	—	4.26
2	x	4.67	1.78	0.358	—	—	14.33	—	5.13
3	x	30.00	0.28	10.778	—	—	25.50	—	274.84
4	x	12.00	0.69	2.955	—	—	50.50	—	149.23
5	x	11.33	0.74	2.604	—	—	50.50	—	131.50
6	x	30.67	0.27	11.252	—	—	75.67	—	851.44
7	x	19.33	0.43	6.219	—	—	38.33	—	238.37
8	x	30.67	0.27	11.252	—	—	11.67	—	131.31
9	x	11.33	0.74	2.604	—	—	36.83	—	95.91
10	x	12.00	0.69	2.955	—	—	36.83	—	108.83
11	x	30.00	0.28	10.778	—	—	61.83	—	666.40
12	x	4.67	1.78	0.358	—	—	73.00	—	26.13
13	x	34.67	0.24	12.898	—	—	87.00	—	1,122.13
a	y	14.67	0.57	—	4.080	25.67	—	104.73	—
b	y	24.67	0.34	—	8.495	34.33	—	291.63	—
c	y	6.67	1.25	—	0.865	34.33	—	29.70	—
d	y	2.67	3.12	—	0.076	34.33	—	2.61	—
e	y	26.67	0.31	—	9.531	34.33	—	327.20	—
f	y	7.33	1.14	—	1.070	40.67	—	43.52	—
g	y	34.00	0.25	—	12.308	59.33	—	730.23	—
h	y	26.67	0.31	—	9.531	59.33	—	565.47	—
i	y	34.00	0.25	—	12.308	84.33	—	1,037.93	—
j	y	26.67	0.31	—	9.531	84.33	—	803.75	—
k	y	34.00	0.25	—	12.308	109.33	—	1,345.63	—
l	y	33.00	0.25	—	12.308	109.33	—	1,345.63	—
m	y	14.67	0.57	—	4.080	118.00	—	481.44	—

$$\Sigma R_x = \overline{87.91}$$

$$\Sigma R_y = \overline{96.49}$$

$$\Sigma R_x \times y = \overline{3,805.5}$$

$$\Sigma R_x \times x = \overline{7,109.5}$$

Center of rigidity, CR, level one

$$\overline{x}_{CR} = \frac{\Sigma R_y \times x}{\Sigma R_y} = \frac{7109.5}{96.49} = 73.7 \text{ ft}$$

$$\overline{y}_{CR} = \frac{\Sigma R_x \times y}{\Sigma R_x} = \frac{3805.5}{87.91} = 43.3 \text{ ft}$$

Eccentricity at first level, center of mass to center of rigidity.

$$\overline{e}_x = \overline{x}_{CM} - \overline{x}_{CR} = 71.6 - 73.7 = 2.1 \text{ ft}$$

$$\overline{e}_y = \overline{y}_{CM} - \overline{y}_{CR} = 43.6 - 43.3 = 0.3 \text{ ft}$$

Minimum eccentricity from accidental torsion

$$\text{Min } e_x = 0.05 \times 143.67 = 7.2 \text{ ft}$$

$$\text{Min } e_y = 0.05 \times 87.33 = 4.37 \text{ ft}$$

$$\text{Total } e_x = 7.2 + 2.1 = 9.3 \text{ ft}$$

$$\text{Total } e_y = 4.37 + 0.3 = 4.67 \text{ ft}$$

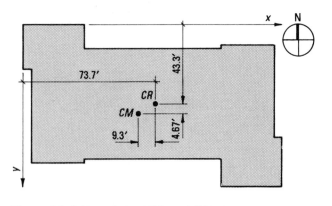

Figure 12-6 Location of CR and CM, level one.

TABLE 12-5 Location of Center of Rigidity; Fifth Floor Walls, Level Four

Wall Number	Direction of Wall	Length, l (feet)	$\frac{h}{l}=\frac{h}{d}$ (h = 44.33 ft)	Rigidity, R_x (Table T-1)	Rigidity, R_y (Table T-1)	x Distance (feet)	y Distance (feet)	$R_y \times x$	$R_x \times y$
1	x	34.67	1.28	0.818	—	—	0.33	—	0.27
2	x	4.67	9.49	0.003	—	—	14.33	—	0.04
3	x	30.00	1.48	0.574	—	—	25.50	—	14.64
4	x	12.00	3.69	0.047	—	—	50.50	—	2.37
5	x	11.33	3.91	0.040	—	—	50.50	—	2.02
6	x	30.67	1.45	0.604	—	—	75.67	—	45.70
7	x	19.33	2.29	0.182	—	—	38.33	—	6.98
8	x	30.67	1.45	0.604	—	—	11.67	—	7.05
9	x	11.33	3.91	0.040	—	—	36.83	—	1.47
10	x	12.00	3.69	0.047	—	—	36.83	—	1.73
11	x	30.00	1.48	0.574	—	—	61.83	—	35.49
12	x	4.67	9.49	0.003	—	—	73.00	—	0.22
13	x	34.67	1.28	0.818	—	—	87.00	—	71.17
a	y	14.67	3.02	—	0.084	25.67	—	2.16	—
b	y	24.67	1.80	—	0.348	34.33	—	11.95	—
c	y	6.67	6.65	—	0.008	34.33	—	0.27	—
d	y	2.67	16.60	—	0.001	34.33	—	0.03	—
e	y	26.67	1.66	—	0.430	34.33	—	14.76	—
f	y	7.33	6.05	—	0.011	40.67	—	0.45	—
g	y	34.00	1.30	—	0.788	59.33	—	46.75	—
h	y	26.67	1.66	—	0.430	59.33	—	25.51	—
i	y	34.00	1.30	—	0.788	84.33	—	66.45	—
j	y	26.67	1.66	—	0.430	84.33	—	36.26	—
k	y	34.00	1.30	—	0.788	109.33	—	86.15	—
l	y	33.00	1.34	—	0.733	109.33	—	80.14	—
m	y	14.67	3.02	—	0.084	118.00	—	9.91	—

$$\Sigma R_x = \overline{4.35} \qquad \Sigma R_x \times y = \overline{189.2}$$

$$\Sigma R_y = \overline{4.92} \qquad \Sigma R_y \times x = \overline{380.7}$$

Center of rigidity, CR, level four

$$\bar{x}_{CR} = \frac{\Sigma R_y \times x}{\Sigma R_y} = \frac{380.7}{4.92} = 77.4 \text{ ft}$$

$$\bar{y}_{CR} = \frac{\Sigma R_x \times y}{\Sigma R_x} = \frac{189.2}{4.35} = 43.6 \text{ ft}$$

Eccentricity at fourth level, center of mass to center of rigidity

$$\bar{e}_x = \bar{x}_{CM} - \bar{x}_{CR} = 71.6 - 77.4 = 5.8 \text{ ft}$$

$$\bar{e}_y = \bar{y}_{CM} - \bar{y}_{CR} = 43.6 - 43.6 = 0.00 \text{ ft}$$

Min. design eccentricity including accidental eccentricity (see previous page):

$$e_x = 7.2 + 5.8 = 13.0 \text{ ft}$$

$$e_y = 4.37 + 0 = 4.37 \text{ ft}$$

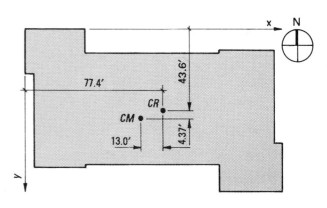

Figure 12-7 Location of CR and CM, level four.

12-4   LATERAL FORCE DESIGN

12-4.A   General

It is assumed in this example that for purposes of lateral force distribution, the floor slabs between walls are considered flexible in the vertical direction.

Consideration of coupled walls connected by spandrel beams can be investigated to determine the stresses, moments and shears induced in these members due to lateral forces.

12-4.B   Base Shear

12-4.B.1   Seismic Base Shear

The total base shear according to UBC Sec. 2334(b) is

$$V = \frac{ZIC}{R_w} W$$

Lateral force coefficient, C

$$C = \frac{1.25S}{T^{2/3}}$$

Assume type S_3 soil and therefore $S = 1.5$ from Table 3-13 (UBC Table No. 23-J: see footnote 1).

Period, $T = C_t \left(h_n \right)^{3/4}$

Where $C_t = 0.020$ and $h_n = 63$ ft

$$T = 0.020(63)^{3/4} = 0.45$$

$$C = \frac{1.25(1.5)}{0.45^{2/3}} = 3.19$$

But C need not exceed 2.75 without a soils report (UBC Sec. 2334(b)1).

Thus the base shear is:

$$V = \frac{0.4 \times 1 \times 2.75}{6} W$$
$$= 0.183W$$

Determine effective dead load; W_{DL}

At roof level

Roof slab $= \left(\frac{71}{98} \right) 887.9$ $= 643.3$ kips

Tributary wall load $= \left(\dfrac{2.5 + \dfrac{8.33}{2}}{8.33} \right) 358.5 = 287.1$ kips

At floor levels

Floors $= 6 \times 887.9$ $= 5327.4$ kips

Wall load $= 6 \times 358.5$ $= 2151.0$ kips

Total W_{DL} $= 8408.8$ kips

Base shear, V (Transverse)

$$V = 0.183 \times 8408.8 = 1538.8 \text{ kips}$$

The same method may be used to determine the base shear in longitudinal direction.

12-4.B.2   Wind Base Shear

Determine the wind pressure on the building using UBC Sec. 2316 and Section 3-5 of this text. Use the projected area method to find the design wind pressure, P.

$$P = C_e C_q q_s I$$

Table 12-6 summarizes the calculation of design wind pressures based on the following:

Basic wind speed = 70 mph

Exposure C for the determination of C_e from Table 3-6 (UBC Table No. 23-G)

Pressure coefficients, C_q, from Table 3-7 (UBC Table No. 23-H):

$C_q = 1.3$ for Height from 0 to 40 feet

$C_q = 1.4$ for Height from 40 to 65 feet

$q_s = 12.6$ psf from Table 3-3 (UBC Table No. 23-F) for $V = 70$ mph.

$I = 1.0$ from Table 3-5 (UBC Table No. 23-L).

TABLE 12-6   Wind Pressure, P

Height (feet)	C_e	C_q	q_s	I	P (psf)
0 - 15	1.06	1.3	12.6	1.0	17.4
15 - 20	1.13	1.3	12.6	1.0	18.5
20 - 25	1.19	1.3	12.6	1.0	19.5
25 - 30	1.23	1.3	12.6	1.0	20.1
30 - 40	1.31	1.3	12.6	1.0	21.5
40 - 60	1.43	1.4	12.6	1.0	25.2
60 - 80	1.53	1.4	12.6	1.0	27.0

Check base shear due to wind in the transverse direction:
Length = 143.67 ft
Wind base shear

$$V = \begin{bmatrix} 15(17.4) + 5(18.5) + 5(19.5) + 5(20.1) \\ + 10(21.5) + 20(25.2) + 5.5(27.0) \end{bmatrix} \left(\frac{143.67}{1000} \right)$$
$$= 203.9 \text{ kips}$$

Therefore seismic governs for base shear,

$$V = 1539 \text{ kips}$$

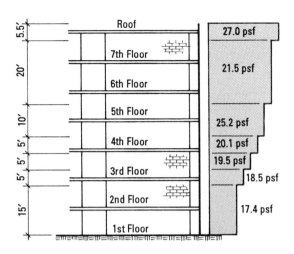

Figure 12-8 Wind forces on the building.

12-5 LATERAL FORCE DISTRIBUTION

Dead load at roof level:

Roof load	= 643.3 kips
Wall load	= 287.1 kips
Load at top level, roof w_n	= 930.4 kips

Dead load to typical floor:

Floor load	= 887.9 kips
Wall load	= 358.5 kips
Load at floor level, w_i	= 1246.4 kips

Dead load at base:

Consider only weight of half the walls since remaining portions of walls and slabs are massed at each floor level.

$$w = 358.5/2 = 179.3 \text{ kips}$$

Determine whiplash force at top, F_t

Transverse direction:
$F_t = 0$ since $T < 0.7$ sec. (UBC 2334 (d))

12-6 OVERTURNING MOMENT

The Overturning Moment at the base, M_B, is governed by the controlling lateral forces due to wind or seismic forces.

$$V = \frac{ZIC}{R_w} W$$

$$M_{Base} = \sum_{x=1}^{n} F_x h_x$$

Where:

$$F_x = \frac{(V - F_t) w_x h_x}{\sum_{i=1}^{n} w_i h_i} \qquad \text{(See Table 12-7)}$$

$$M_x = F_t(h_n - h_x) + \sum_{i=x+1}^{n-1} F_i(h_i - h_x)$$

However since $F_t = 0$:

$$M_x = \sum_{i=x+1}^{n-1} F_i(h_i - h_x)$$

$$M_{Base} = \sum_{x=1}^{n} F_x h_x$$

TABLE 12-7   Lateral Seismic Force Distribution to Each Floor Level in Transverse Direction

Level[1]	w_i (kips)	h_i (feet)	$w_i h_i$ (ft kips)	F_x (kips)	ΣF_x (kips)
7	930.4	63	58,615	307	307
6	1,246.4	54	67,306	352	659
5	1,246.4	45	56,088	293	952
4	1,246.4	36	44,870	235	1,187
3	1,246.4	27	33,653	176	1,363
2	1,246.4	18	22,435	117	1,480
1	1,246.4	9	11,218	59	1,539
Base[2]	179.3	0	0	0	1,539

1. Roof is level 7.
2. First floor is base.

$$\sum_{i=1}^{n} w_i h_i = \overline{294,185} \text{ kips}$$

Base Shear, $V = \Sigma F_x = 1539$ kips

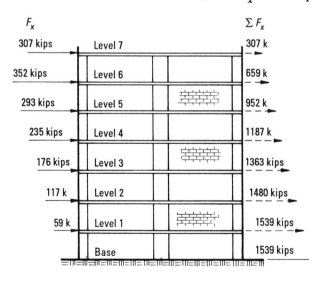

Figure 12-9 Lateral seismic forces on the building.

TABLE 12-8 Overturning Moment on Building at Each Floor Level in Transverse Direction

Level	F_x (kips)	h_x (feet)	$f_x h_x$	M_x[1] (ft kips)
7	307	63	19,341	0
6	352	54	19,008	2,763
5	293	45	13,185	8,694
4	235	36	8,460	17,262
3	176	27	4,752	27,945
2	117	18	2,106	40,212
1	59	9	531	53,532
Base	0	0	0	67,383

1. Calculation of M_x shown below.

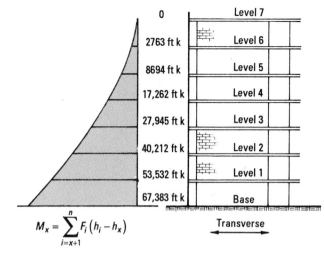

$$M_x = \sum_{i=x+1}^{n} F_i \left(h_i - h_x \right)$$

Figure 12-10 Overturning moment diagram, ft kips.

M_6 = 307 x 9 = 2763 ft kips

M_5 = 307 x 18 + 352 x 9 = 8694 ft kips

M_4 = 307 x 27 + 352 x 18 + 293 x 9 = 17,262 ft kips

M_3 = 307 x 36 + 352 x 27 + 293 x 18 + 235 x 9 = 27,945 ft kips

M_2 = 307 x 45 + 352 x 36 + 293 x 27 + 235 x 18 + 176 x 9 = 40,212 ft kips

M_1 = 307 x 54 + 352 x 45 + 293 x 36 + 235 x 27 + 176 x 18 + 117 x 9 = 53,532 ft kips

M_{Base} = 307 x 63 + 352 x 54 + 293 x 45 + 235 x 36 + 176 x 27 + 117 x 18 + 59 x 9 = 67,383 ft kips

Check overturning due to wind (see Figure 12-8).

$$M_{Base} = (27.0 \times 5.5 \times 62.75) + (25.2 \times 20 \times 50) + (21.5 \times 10 \times 35) + (20.1 \times 5 \times 27.5) + (21.5 \times 10 \times 35) + (20.1 \times 5 \times 27.5) + (19.5 \times 5 \times 22.5) + (18.5 \times 5 \times 17.5) + (17.4 \times 15 \times 7.5)$$

$$= 9318 + 25,200 + 7525 + 2764 + 2194 + 1619 + 1958$$

$$= 50,578 \text{ ft lbs}$$

Seismic controls,

$$M_{Base} = 67,383 \text{ ft kips}$$

12-7 LOAD AND MOMENT DISTRIBUTION TO FIRST STORY, BASE LEVEL, SHEAR WALLS (TRANSVERSE DIRECTION)

Distribute shears and overturning moments to resisting walls according to relative rigidities (see Table 12-9).

1st story shear, V = 1539 kips

Overturning moment, $M_x = M_{Base} = 67,383$ ft kips

Eccentricity, $e_x = 9.3$ ft

Torsional moment, $M_T = 1539 \times 9.3$

$$= 14,313 \text{ ft kips}$$

Force to wall due to story shear

$$F_v = \frac{R}{\Sigma R} \times V$$

Force to wall due to story torsional moment

$$F_t = \frac{Rd}{\Sigma Rd^2} \times M_T$$

Design shear force = $1.5(F_v + F_t)$

Overturning moment to wall due to story overturning moment

$$OTM = \frac{R}{\Sigma R} \times M_x + F_t \times h_{wall}$$

$$OTM = OTM_x + OTM_T$$

TABLE 12-9 Load and Moment Distribution to Shear Walls on First Floor, Base Level, (Transverse Direction)

Shear V = 1539 kips

Torsional Moment M_T = 14,313 ft kips x_{CR} = 73.7 ft $d_x = (x - x_{CR})$

Overturning Moment, M_{Base} = 67,383 ft kips y_{CR} = 43.3 ft $d_y = (y - y_{CR})$

Col. No. 1	2	3	4	5	6	7	8	9	10	11	12	13
Wall No.	Rigidity R_c	x (ft)	d_x (ft)	Rd_x	Rd_x^2	Lateral Force, F_v $\frac{R}{\Sigma R}V$	Torsional Force, F_t $\frac{Rd}{\Sigma Rd^2}M_T$	Shear $F = F_v + F_t$ (kips)	Shear, 1.5F (kips)	$OTM_T = F_t \times 8.33$ (ft kips)	$OTM_x = \frac{R}{\Sigma R}M_{Base}$ (ft kips)	$OTM = OTM_T + OTM_x$ (ft kips)
Transverse Walls, y direction												
a	4.080	25.67	−48.03	−196.0	9,412.1	65.1	−16.9	65.1	97.7	−141	2,849	2,849
b	8.495	34.33	−39.37	−334.4	13,167.2	135.5	−28.8	135.5	203.3	−240	5,932	5,932
c	0.865	34.33	−39.37	−34.1	1,340.7	13.8	−2.9	13.8	20.7	−24	604	604
d	0.076	34.33	−39.37	−3.0	117.8	1.2	−0.3	1.2	1.8	−2	53	53
e	9.531	34.33	−39.37	−375.2	14,773.0	152.0	−32.3	152.0	228.0	−269	6,656	6,656
f	1.070	40.67	−33.03	−35.3	1,167.3	17.1	−3.0	17.1	25.7	−25	747	747
g	12.308	59.33	−14.37	−176.9	2,541.6	196.3	−15.2	196.3	294.5	−127	8,595	8,595
h	9.531	59.33	−14.37	−137.0	1,968.1	152.0	−11.8	152.0	228.0	−98	6,656	6,656
i	12.308	84.33	10.63	130.8	1,390.8	196.3	11.3	207.6	311.4	94	8,595	8,689
j	9.531	84.33	10.63	101.3	1,077.0	152.0	8.7	160.7	241.1	72	6,656	6,728
k	12.308	109.33	35.63	438.5	15,625.0	196.3	37.8	234.1	351.2	315	8,595	8,910
l	12.308	109.33	35.63	438.5	15,625.0	196.3	37.8	234.1	351.2	315	8,595	8,910
m	4.080	118.00	44.30	180.7	8,007.0	65.1	15.6	80.7	121.1	130	2,849	2,979

ΣR = 96.491

Col. No. 1	2	y (feet)	d_y (feet)	Rd_y	Rd_y^2	7	8	9	10	11	12	13
Longitudinal Walls, x direction												
1	12.898	0.33	−42.97	−554.2	23,815.1	—	−47.8	47.8	71.7	—	—	—
2	0.358	14.33	−28.97	−10.4	300.5	—	−0.9	0.9	1.4	—	—	—
3	10.778	25.50	−17.80	−191.8	3,414.9	—	−16.5	16.5	24.8	—	—	—
4	2.955	50.50	7.20	21.3	153.2	—	1.8	1.8	2.7	—	—	—
5	2.604	50.50	7.20	18.7	135.0	—	1.6	1.6	2.4	—	—	—
6	11.252	75.67	32.37	364.2	11,790.0	—	31.4	31.4	47.1	—	—	—
7	6.219	38.33	−4.97	−30.9	153.6	—	−2.7	2.7	4.1	—	—	—
8	11.252	11.67	−31.63	−355.9	11,257.1	—	−30.7	30.7	46.1	—	—	—
9	2.604	36.83	−6.47	−16.8	109.0	—	−1.5	1.5	2.3	—	—	—
10	2.955	36.83	−6.47	−19.1	123.7	—	−1.6	1.6	2.4	—	—	—
11	10.778	61.83	18.53	199.7	3,700.7	—	17.2	17.2	25.8	—	—	—
12	0.358	73.00	29.70	10.6	315.8	—	0.9	0.9	1.4	—	—	—
13	12.898	87.00	43.70	563.6	24,631.2	—	48.6	48.6	72.9	—	—	—

ΣR = 87.909 ΣRd^2 = 166,112.4

Column 1 — Wall Identification.

Column 2 — Rigidity from Table 12-4 or 12-5 and from Table T-1.

Column 3 — x distance from the Y axis (or y distance from X axis).

Column 4 — $d_x = x$ distance minus the distance to the center of rigidity (or $d_y = y - y_{CR}$).

Column 5 — Column 2 times Column 4.

Column 6 — Column 5 times Column 4.

Column 7 — Rigidity of Column 2 divided by sum of the rigidities of Column 2 times the base shear, V.

Column 8 — Column 5 divided by Column the sum of Column 6 times the torsional moment, M_T.

Column 9 — Total shear force is the direct shear force F_v plus the torsional force F_t which is column 7 plus Column 8.

Column 10 — 1.5 × Column 9.

Column 11 — Torsional overturning moment equals the torsional force, F_t, from Column 8 multiplied by the height of the wall (8.33 ft).

Column 12 — Overturning moment on the wall at the level x is the rigidity of Column 2 divided by the sum of the rigidities in Column 2 multiplied by the overturning moment at that level.

Column 13 — Total overturning moment on the wall at the level considered is the sum of Column 11 plus Column 12. Negative moments are not deducted and the maximum moment is used.

12-8 LOAD AND MOMENT DISTRIBUTION TO FIFTH STORY, LEVEL 4, SHEAR WALLS (TRANSVERSE DIRECTION)

Distribute shears and overturning moments to resisting walls according to relative rigidities.

The walls act as cantilever beams from the foundation and have a height at the fifth floor level of 44.33 ft.

Fifth story shear, $V = 952$ kips (Table 12-7)
Overturning moment at level 4,
$$M_x = M_4 = 17,262 \text{ ft kips (Table 12-8)}$$

Eccentricity, $e_x = 13.0$

Torsional moment, $M_T = Ve_x = 952 \times 13.0$
$$= 12,376 \text{ ft kips}$$

TABLE 12-10 Load and Moment Distribution to Shear Walls on Fifth Floor, Level 4 (Transverse Direction)

Shear $V = 952$ kips			$x_{CR} = 77.4$ ft		$d_x = (x - x_{CR})$							
Torsional Moment $M_T = 12,376$ ft kips			$y_{CR} = 43.6$ ft		$d_y = (y - y_{CR})$							
Overturning Moment, $M_4 = 17,262$ ft kips												

Col. No. 1[1]	2	3	4	5	6	7	8	9	10	11	12	13
	Wall No.	Rigidity R_c	x (ft)	d_x (ft)	Rd_x	Rd_x^2	Lateral Force, F_v $\dfrac{R}{\Sigma R}V$	Torsional Force, F_t $\dfrac{Rd}{\Sigma Rd^2}M_T$	Shear $F = F_v + F_t$ (kips)	Shear, $1.5F$ (kips)	$OTM_T = F_t \times 8.33$ (ft kips)	$OTM_x = \dfrac{R}{\Sigma R}M_4$ (ft kips)	$OTM = OTM_T + OTM_x$ (ft kips)
Transverse Walls, y direction	a	0.084	25.67	−51.73	−4.3	224.4	16.2	−6.2	16.2	24.3	−52	294	294
	b	0.348	34.33	−43.07	−15.0	645.5	67.3	−21.7	67.3	101.0	−181	1,220	1,220
	c	0.008	34.33	−43.07	−0.3	14.8	1.5	−0.4	1.5	2.3	−3	28	28
	d	0.001	34.33	−43.07	0.4	1.9	0.2	0.0	0.2	0.3	0	4	4
	e	0.430	34.33	−43.07	−18.5	797.7	83.2	−26.7	83.2	124.8	−222	1,507	1,507
	f	0.011	40.67	−36.73	−0.4	14.8	2.1	−0.6	2.1	3.2	−5	39	39
	g	0.788	59.33	−18.07	−14.2	257.3	152.4	−20.5	152.4	228.6	−171	2,763	2,763
	h	0.430	59.33	−18.07	−7.8	140.3	83.2	−11.3	83.2	124.8	−94	1,507	1,507
	i	0.788	84.33	6.93	5.5	37.8	152.4	8.0	160.4	240.6	67	2,763	2,830
	j	0.430	84.33	6.93	3.0	20.7	83.2	4.3	87.5	131.3	36	1,507	1,543
	k	0.788	109.33	31.93	25.2	803.4	152.4	36.4	188.8	283.2	303	2,763	3,066
	l	0.733	109.33	31.93	23.4	747.3	141.7	33.8	175.5	263.3	282	2,570	2,852
	m	0.084	118.00	40.60	3.4	138.5	16.2	4.9	21.1	31.7	41	294	335
	$\Sigma R = 4.923$												
			y (feet)	d_y (feet)	Rd_y	Rd_y^2							
Longitudinal Walls, x direction	1	0.818	0.33	−43.27	−35.4	1,531.5	—	−51.2	51.2	76.8	—	—	—
	2	0.003	14.33	−29.27	−0.1	2.6	—	−0.1	0.1	0.2	—	—	—
	3	0.574	25.50	−18.10	−10.4	188.0	—	−15.0	15.0	22.5	—	—	—
	4	0.047	50.50	6.90	0.3	2.2	—	0.4	0.4	0.6	—	—	—
	5	0.040	50.50	6.90	0.3	1.9	—	0.4	0.4	0.6	—	—	—
	6	0.604	75.67	32.07	19.4	621.6	—	28.0	28.0	42.0	—	—	—
	7	0.182	38.33	−5.27	−1.0	5.1	—	−1.4	1.4	2.1	—	—	—
	8	0.604	11.67	−31.93	−19.3	615.8	—	−27.9	27.9	41.9	—	—	—
	9	0.040	36.83	−6.77	−0.3	1.8	—	−0.4	0.4	0.6	—	—	—
	10	0.047	36.83	−6.77	−0.3	2.2	—	−0.4	0.4	0.6	—	—	—
	11	0.574	61.83	18.23	10.5	190.9	—	15.2	15.2	22.8	—	—	—
	12	0.003	73.00	29.40	0.1	2.6	—	0.1	0.1	0.2	—	—	—
	13	0.818	87.00	43.40	35.5	1,540.8	—	51.3	51.3	77.0	—	—	—
	$\Sigma R = 4.354$				$\Sigma Rd^2 = 8,551.4$								

1. See Table 12-9 Footnotes for explanation of columns.

12-9 UNIT SHEAR AND FLEXURAL STRESSES

l = length of wall, ft

V = shear force, kips

t = wall thickness = 7.63 in.

UBC Sec. 2407(h)4.F requires that when calculating shear stresses, shear walls which resist seismic forces must be designed to resist 1.5 times the calculated force.

$$f_v = \text{unit shear stress} = \frac{1000(1.5F)}{12lt}$$

$$= \frac{10.92(1.5F)}{l} \text{ psi}$$

I = moment of inertia of wall = $\dfrac{tl^3}{12}$

$$= \frac{7.63 \times 1728}{12} l^3 \text{ in.}^4$$

h = clear height of walls = 8' - 4"

OTM = overturning moment, ft kips

f_b = flexural stress due to overturning based on uncracked section

$$= \frac{M_c}{I} = \frac{1200(OTM)(l \times 12)}{2 \times 1098 \times l^3} \qquad c = \frac{l}{2}$$

$$f_b = \frac{65.57 \times OTM}{l^2} \text{ psi (8" thick walls)}$$

TABLE 12-11 Unit Shear and Flexural Stresses on 1st Story, Base Level, 5th Story, Level 4, Transverse Walls in _y_ Direction

	Wall Number	Length, L (Feet)	$\frac{h}{l} = \frac{h}{d}$ (h = 8.33)	Shear, 1.5F (kips)	Shear Stress, f_v (psi)	Max. Allow. Shear Stress (psi) × 1.33 w/o Reinforcing per Diagram A-5	Max. Allow. Shear Stress (psi) × 1.33 w/ Reinforcing per Diagram A-6	OTM = OTM_T + OTM_x (ft kips)	f_b (psi)
Fifth Story Walls, Level 4 f'_m = 1500 psi	a	14.67	0.57	24.3	18.1	59	88	294	89.6
	b	24.67	0.34	101.0	44.7	63	94	1220	131.4
	c	6.67	1.25	2.3	3.8	47	77	28	41.3
	d	2.67	3.12	0.3	1.2	47	77	4	36.8
	e	26.67	0.31	124.8	51.1	64	96	1507	138.9
	f	7.33	1.14	3.2	4.8	47	77	39	47.6
	g	34.00	0.25	228.6	73.4	64	97	2763	156.7
	h	26.67	0.31	124.8	51.1	64	96	1507	138.9
	i	34.00	0.25	240.6	77.3	64	97	2830	160.5
	j	26.67	0.31	131.3	53.8	64	96	1543	142.2
	k	34.00	0.25	283.2	91.0	64	97	3066	173.9
	l	33.00	0.25	263.3	87.1	64	97	2852	171.7
	m	14.67	0.57	31.7	23.6	59	88	335	102.1
First Story Walls, Base Level f'_m = 3000 psi	a	14.67	0.57	97.7	72.7	72	125	2849	868.0
	b	24.67	0.34	203.3	90.0	86	133	5932	639.1
	c	6.67	1.25	20.7	33.9	47	100	604	890.2
	d	2.67	3.12	1.8	7.4	47	100	53	487.5
	e	26.67	0.31	228.0	93.4	88	133	6656	613.6
	f	7.33	1.14	25.7	38.3	47	100	747	911.6
	g	34.00	0.25	294.5	94.6	91	137	8595	487.5
	h	26.67	0.31	228.0	93.4	88	133	6656	613.6
	i	34	0.25	311.4	100.0	91	137	8689	492.9
	j	26.67	0.31	241.1	98.7	88	133	6728	620.2
	k	34	0.25	351.2	112.8	91	137	8910	505.4
	l	33	0.25	351.2	116.2	91	137	8910	536.5
	m	14.67	0.57	121.0	90.1	72	125	2979	907.6

12-10   DESIGN OF WALL f ON FIRST STORY, BASE LEVEL

12-10.A   General

Wall **f** carries relatively low vertical loads and therefore overturning controls the design. Neglect any elevator loads imposed on wall **f**.

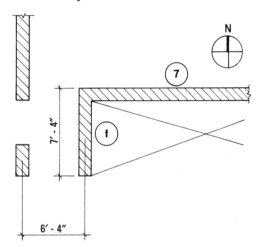

Figure 12-11 Wall f.

From the evaluation of the shear stresses and overturning stresses on the transverse walls on the first floor in Table 12-11, it is evident that high strength masonry will be required. The high shear stresses; (i.e. for wall **k**, f_v = 112.8 psi and for wall **l**, f_v = 116.2 psi) and the high flexural compression stresses based on an uncracked section; (i.e. for wall **a**, f_b = 868 psi and for wall **f**, f_b = 912 psi) require that the f'_m of the masonry be 3000 psi.

Specify strength of masonry f'_m = 3000 psi for the lower floors.

See Table A-3, Allowable Stresses

n = 12.9

Allowable flexural stress, F_b = 0.33 f'_m = 1000 psi

Allowable shear stress, from Diagrams A-5 and A-6 (See Table 12-11)

Allowable steel stress, F_s = 24,000 psi

Allowable axial wall stress, F_a, from Tables Q-1 and Q-2 for h = 8′ - 4″ and actual thickness, t = 7.63″ (nominal thickness = 8″):

$$\frac{h'}{t} = 13.1 \qquad\qquad R = 0.970$$

$$F_a = 0.2 f'_m \left[1 - \left(\frac{h'}{42t} \right)^3 \right] = 582 \text{ psi}$$

Note: Allowable unit stresses may be increased one third for combined lateral and vertical loads due to wind or seismic forces.

12-10.B   Loads

Load on wall **f**

Clear span of slab supported by wall **f** = 5′ - 8″.

Tributary width to wall = 2.83 ft

Dead load

Roof load	2.83×71 psf	=	201 plf
Floor load	$6 \times 2.83 \times 98$	=	1664 plf
Wall load	$7 \times 9 \times 78$ psf	=	4914 plf
	Total dead load	=	6779 plf

Live load reduction;

Corridor LL = 100 psf

Area supported by wall = $6 \times 2.83 \times 7.33$ = 124.5 sq ft
                                                                    < 150 sq ft

No reduction permitted, UBC Sec. 2306.

Live load

Roof load	2.83×20 psf	=	57 plf
Floor load	$6 \times 2.83 \times 100$ psf	=	1698 plf
	Total live load	=	1755 plf

Actual axial stress

$$\text{Dead load } f_a = \frac{P_{DL}}{bt} = \frac{6779}{12 \times 7.63} = 74 \text{ psi}$$

$$\text{Live load } f_a = \frac{P_{LL}}{bt} = \frac{1755}{12 \times 7.63} = 19 \text{ psi}$$

Floor live load (w/o roof LL)

$$f_a = \frac{1698}{12 \times 7.63} = 19 \text{ psi}$$

Total load w/o roof LL = 74 + 19
                                        = 93 psi

12-10.C   Design

(a) Lateral shear stress on wall

From Table 12-11 for wall **f**, f_v = 38.3 psi

$$\frac{h}{l} = \frac{8.33}{7.33} = 1.14 > 1$$

From Diagram A-5 the maximum allowable shear stress in the masonry without reinforcing is:

F_{vm} = 35 psi x 1.33 = 47 psi > 38.3 psi   O.K.

No shear reinforcing is required.

Provide minimum steel.

(b) Vertical load axial stress

$f_a = 93$ psi $< F_a = 584$ psi O.K.

(c) Vertical dead load + live load + seismic load perpendicular to wall

$f_a = 93$ psi

$F_a = 584$ psi

Lateral seismic force,

$F_p = ZIC_pW_p$; where $Z = 0.4$

$I = 1$; $C_p = 0.75$; $W_p = 78$ psf

$F_p = 0.4 \times 1 \times 0.75 \times 78 = 23.4$ psf

Lateral seismic moment

$M = \dfrac{1}{12}wl^2$

$= \dfrac{1}{12} \times 23.4 \times 8.33^2$

$= 135.3$ ft lbs / ft

Note: Wind loads may be more critical normal to exterior walls, than seismic loads at high floor levels.

Check flexural stress assuming section is uncracked

$f_b = \dfrac{6M}{bt} = \dfrac{6 \times 135.3 \times 12}{12 \times 7.63^2} = 13.9$ psi

Axial stress is greater than flexural stress

93 psi > 13.9 psi

Minimum required steel is adequate; jamb steel may be included.

Minimum steel ($0.0013bt$) required from Table C-6 for $t = 7.63''$ is #5 at 24" o.c. vertical bars. Jamb steel can serve as vertical reinforcing.

(d) Vertical dead load + floor live load (w/o roof LL) + seismic load parallel to wall

$t = 88''$, Assume $d = 80''$

Axial stress$\dfrac{P}{A} = f_a = 93$ psi

Allowable axial stress

$F_a = 584$ psi (Table Q-2)

Overturning moment on wall **f** at the first floor (from Table 12-11):

$M_{OT} = 747$ ft kips

(e) Using interaction design method 2, assume wall with no return.

Calculate f_m, **a**, **b**, **c** and kd

$f_b = \left[1.33 - \dfrac{f_a}{F_a}\right]F_b$

$= \left[1.33 - \dfrac{93}{584}\right]900$

$= 1{,}170$ psi

$f_m = f_a + f_b$

$= 93 + 1{,}170$

$= 1263$ psi

$\mathbf{a} = \dfrac{1}{6}\,t\,f_m$

$= \dfrac{1}{6} \times 7.63 \times 1263$

$= 1607$

$\mathbf{b} = -\dfrac{1}{2}\,t\,d\,f_m$

$= -\dfrac{1}{2} \times 7.63 \times 80 \times 1263$

$= -385{,}470$

$\mathbf{c} = P\left(\dfrac{l}{2} - d_1\right) + M$

$P = 93 \text{ psi} \times 7.63 \times 88 = 62{,}444$ lbs

$M = 747$ ft kips

$\mathbf{c} = 62{,}444\left(\dfrac{88}{2} - 8\right) + 747 \times 12{,}000$

$= 11{,}212{,}000$

$kd = \dfrac{-\mathbf{b} - \sqrt{\mathbf{b}^2 - 4\mathbf{ac}}}{2\mathbf{a}}$

$= \dfrac{385{,}470 - \sqrt{(385{,}470)^2 - 4 \times 1607 \times 11{,}212{,}000}}{2 \times 1607}$

$= 33.8$ in.

Calculate f_s, C, T and A_s

$f_s = \left(\dfrac{d - kd}{kd}\right)n f_m$

$= \left(\dfrac{88 - 33.8}{33.8}\right)12.9 \times 1263$

$= 26{,}126$ psi $< 32{,}000$ psi O.K.

$C = \dfrac{1}{2}\,t\,k\,d\,f_m$

$= \dfrac{1}{2} \times 7.63 \times 33.8 \times 1263$

$C = 162{,}860$ lbs

$P = -62{,}444$ lbs

$T = 100{,}416$ lbs

$A_s = \dfrac{T}{f_s} = \dfrac{100{,}416}{26{,}126} = 3.8$ sq. in.

Use 3 - #10 bars ($A_s = 3.81$ sq. in.)

12-10.D Alternate Design

(a) Design of wall **f** considering the effect of the flange return wall, using interaction Method 3.

Vertical dead load + floor live load (without roof live load) + seismic load parallel to wall.

Effective width of overhanging flange = $6t$ = 48".

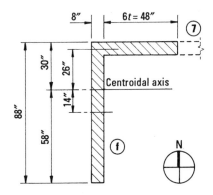

Figure 12-12 Location of centroidal axis.

TABLE 12-12 Location of Centroidal Axis

Wall No.	Area, A (sq. in.)	Arm, x (inches)	Moment Area = A_x
f	7.63 x 88 = 671	44	29,524
7	7.63 x 48 = 366	84	30,744

Σ areas = 1037 sq. in. ΣM = 60,268 in.³

$$x = \frac{Ax}{A} = \frac{60,268}{1037} = 58''$$

TABLE 12-13 Moment of Inertia of Section

Wall No.	$\frac{bh^3}{12}$	Area, A (sq. in.)	d (in.)	Ad^2	$I_{CA} = I + Ad^2$
f	$\frac{7.63 \times 88^3}{12} = 433{,}300$	671	14	131,516	564,816
7	$\frac{48 \times 7.63^3}{12} = 1{,}777$	366	26	247,416	249,193

ΣI_{CA} = 814,009 in.⁴

(b) Section modulus

To flange $S = \dfrac{I}{c} = \dfrac{814,009}{30}$

$\qquad\qquad = 27{,}134$ in.³

To narrow end $S = \dfrac{I}{c} = \dfrac{814,009}{58}$

$\qquad\qquad = 14{,}035$ in.³

From previous calculations

$\qquad f_a$ = 93 psi

$\qquad F_a$ = 584 psi

Overturning moment = 747 kips   (from Table 12-11)

(c) Flexural stresses and steel requirement

1) Moment subjecting flange to compression

$$f_b = \frac{M}{S} = \frac{747 \times 12,000}{27,134}$$
$$= 330 \text{ psi Compression}$$

$$f_t = \frac{M}{S} = \frac{747 \times 12,000}{14,035}$$
$$= 637 \text{ psi Tension}$$

Unity check

$$\frac{f_a}{F_a} + \frac{f_b}{F_b} = \frac{93}{584} + \frac{330}{1000}$$
$$= 0.16 + 0.33 = 0.49 < 1.33$$

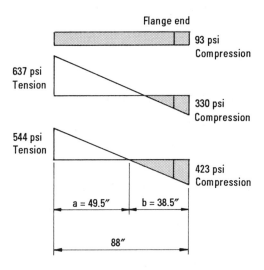

$$a = \frac{544}{544 + 423} \times 88$$
$$= 49.5 \text{ in.}$$

$b = 88 - 49.5 = 38.5$ in.

Tension force = $\frac{1}{2} \times 544 \times 7.63 \times 49.5$
$\qquad\qquad = 102,730$ lbs

$$A_s = \frac{T}{f_s} = \frac{102,730}{1.33 \times 24,000}$$
$$= 3.2 \text{ sq. in.}$$

Use 3 - #10 bars (A_s = 3.81 sq. in.)

(Same as previous design)

2) Moment subjecting narrow end to compression

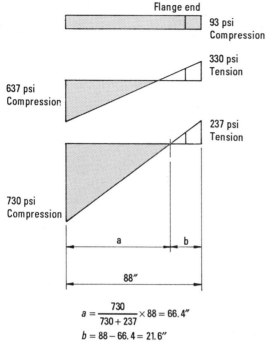

$$a = \frac{730}{730 + 237} \times 88 = 66.4''$$

$$b = 88 - 66.4 = 21.6''$$

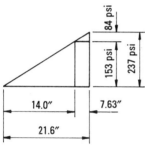

Tension force = $\frac{1}{2} \times 237 \times 7.63 \times 21.6$
$+ 153 \times 7.63 \times 48$
$+ \frac{1}{2} \times 84 \times 7.63 \times 48$
$= 90,947$ lbs

$$A_s = \frac{T}{F_s} = \frac{90,947}{37,000}$$
$$= 2.84 \text{ sq. in.}$$

Use 3 - #9 bars ($A_s = 3.0$ sq. in.) or 3 - #10 bars to match other end of wall.

(d) Check vertical shear at the intersection of wall **f** to flange

$$v = \frac{VAy}{It}$$

Where V = Shear force
$= 25.7$ kips (Table 12-11)

A = Area of flange
$= 7.63 \times 56 = 427$ sq. in.

y = Distance from neutral axis to the centroid of the flange area = 26″

I = Moment of inertia of section
$= 814,009$ in.⁴

t = thickness of web = 7.63 in.

$$v = \frac{25.7 \times 1000 \times 427 \times 26}{814,000 \times 7.63} = 45.9 \text{ psi}$$

The allowable shear stress for $f'_m = 3000$ psi, $M/Vd > 1$, special inspection and no shear reinforcing is found from Table A-5b as:

$$F_{vm} = 1.33(35)$$
$$= 46.6 \text{ psi} > 45.9 \quad \text{O.K.}$$

No vertical shear is reinforcing required

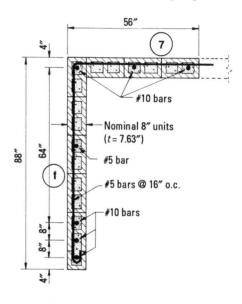

*Figure 12-13 Detail of reinforcing steel in wall **f** at first story, base level. (Coordinate this plan with Figure 12-15.)*

12-11 DESIGN OF WALL f ON FIFTH STORY, LEVEL 4

12-11.A General

From an evaluation of the shear stresses and over-turning flexural compression stresses on the transverse walls on the fifth story (see Table 12-11), it is evident that the high strength masonry needed for the lower stories is not required for this level and above.

An f'_m of 1500 psi will accommodate stresses imposed at this level and above.

Specify $f'_m = 1500$ psi with special inspection; $n = 25.8$

Allowable stresses, Table A-3

Flexural compression stress, $F_b = \frac{1}{3} f'_m$
$= 500$ psi

Axial compression stress, $F_a = 0.2 f'_m R$

$$\frac{h}{t} = \frac{8.33 \times 12}{7.63} = 13.1$$

From Table Q-1, $R = 0.970$

From Table Q-2, $F_a = 291$ psi

12-11.B Loads

Dead load

Roof load	2.83×71 psf =	201 plf
Floor loads	$2 \times 2.83 \times 98$ psf =	555 plf
Wall loads	$3 \times 9 \times 78$ psf =	2106 plf
	Total dead load =	2862 plf

Live load

Roof load	2.83×20 psf =	57 plf
Floor loads	$2 \times 2.83 \times 100$ psf =	566 plf
	Total live load =	623 plf

Axial stress at fifth story

Dead load stress

$$f_a = \frac{P_{DL}}{bt} = \frac{2862}{12 \times 7.63} = 31 \text{ psi}$$

Live load stress

$$f_a = \frac{P_{LL}}{bt} = \frac{623}{12 \times 7.63} = 7 \text{ psi}$$

Floor live load stress (w/o roof live load)

$$f_a = \frac{P_{LL}}{bt} = \frac{566}{12 \times 7.63} = 6 \text{ psi}$$

12-11.C Design

(a) From Table 12-11, unit shear stress on wall **f**:

$$f_v = 4.8 \text{ psi}$$
$$\frac{h}{l} = \frac{8.33}{7.33} = 1.14$$

Maximum allowable shear w/o shear reinforcement (Diagram A-5):

$$F_{vm} = 35 \times 1.33$$
$$= 47 \text{ psi} > 4.8 \text{ psi   O.K.}$$

No shear reinforcement required — provide minimum steel horizontally

Use #5 at 48″ o.c.
or #4 at 32″ o.c.

(b) Axial compression stress

$$f_a = 31 + 7$$
$$= 38 \text{ psi} < F_a = 291 \text{ psi   O.K.}$$

(c) Combined forces

Vertical dead load + live load + seismic load perpendicular to wall

$$f_a = 38 \text{ psi}$$
$$F_a = 291 \text{ psi}$$

Lateral seismic force perpendicular to wall

$$F_p = ZIC_p W_p$$
$$= 0.4 \times 1 \times 0.75 \times 78$$
$$= 23.4 \text{ psf}$$

Moment on wall

$$= \frac{1}{10} wl^2 \text{ (assume some fixity)}$$
$$= \frac{1}{10} \times 23.4 \times 8.33^2$$
$$= 162.4 \text{ ft lbs/ft}$$

Check flexural stress assuming an uncracked section

$$f_b = \frac{6M}{bt^2} = \frac{6 \times 162.4 \times 12}{12 \times 7.63^2} = 16.7 \text{ psi}$$

Axial stress is greater than flexural stress

$$f_a > f_b$$
$$38 \text{ psi} > 16.7 \text{ psi}$$

Minimum steel is adequate.

(d) Combined forces; Use Method 1

Vertical dead load + live load + seismic load parallel to wall

From Table 12-11, the overturning moment on Wall **f** = 39 ft kips

Length of wall = 88 inches

Axial stress = 38 psi

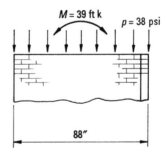

Ignore return on wall

Maximum allowable flexural compressive stress

$$f_b = \left(1.33 - \frac{f_a}{F_a}\right) F_a$$
$$F_b = 500 \text{ psi}$$
$$f_b = \left(1.33 - \frac{38}{291}\right) 500$$
$$= 600 \text{ psi}$$

Calculate the flexural coefficient, K, assuming $d = 80''$

$$K = \frac{M}{bd^2} = \frac{39 \times 12,000}{7.63 \times 80^2}$$
$$= 9.6$$

From Diagram F-1 for $K = 9.6$, $f_b = 600$ psi and $f_s = 32,000$ psi, it is apparent that the minimum required reinforcing will control. Provide $0.0007bt$ horizontal steel and $0.0013bt$ vertical steel.

$$A_s = pbd = 0.0013 \times 7.63 \times 88$$
$$= 0.87 \text{ sq. in.}$$

Use 1 - #5, at each end of the wall and one in the center of the wall as shown in Figure 12-14.

Total vertical steel

$$A_s = 3 \times 0.31 = 0.93 \text{ sq. in.}$$

More than required minimum steel

Use #5 at 48" o.c. horizontal steel.

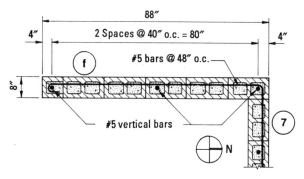

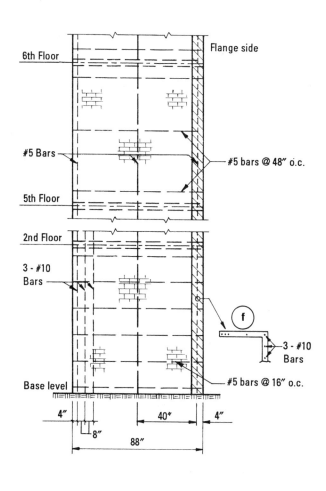

Figure 12-15 Looking west at wall f.

Figure 12-14 Detail of reinforcing steel in wall f at fifth story, level 4. (Coordinate this plan with Figure 12-15.)

12-12 DESIGN OF WALL j ON FIRST STORY, BASE LEVEL

12-12.A General

Wall **j** is a typical cross bearing wall carrying relatively high axial loads. Length of wall = 26' - 8" = 320" and the actual wall thickness = 7.63".

12-12.B Loads

Load on wall

Tributary floor width = 25'

Dead load

Roof load	25' × 71 psf	= 1,775 plf
Floor load	6 × 24.33 × 98 psf	= 14,306 plf
Wall load	7 × 9 × 78 psf	= 4,914 plf
	Total dead load	= 20,995 plf

Live load

Live load reduction for roof

Area supported by wall = 25 × 26.67
= 667 sq. ft

From UBC Table 23-C: Live load = 12 psf

Live load reduction for floors

Area supported by wall = 6 × 25 × 26.67
= 4000 sq. ft

$$R = r(A - 150) = 0.08(4000 - 150)$$
$$= 308\%$$
Maximum allowable
reduction = 60%

$$R = 23.1\left(1+\frac{DL}{LL}\right)$$

$DL = 98$ psf

$LL = 40$ psf

$$R = 23.1\left(1+\frac{98}{40}\right)$$

$\quad = 79.7\%$, maximum
allowable reduction = 60%

Live load = $40(1 - 0.6) = 16$ psf

Roof load	$25' \times 12$ psf =	300 plf
Floor load	$6 \times 25' \times 16$ psf =	2400 plf
	Total live load =	2700 plf

Actual axial stress

$$\text{Dead load } f_a = \frac{P_{DL}}{bt} = \frac{20,995}{12 \times 7.63} = 229 \text{ psi}$$

$$\text{Live load } f_a = \frac{P_{LL}}{bt} = \frac{2,700}{12 \times 7.63} = 29 \text{ psi}$$

Floor live load (w/o roof LL)

$$f_a = \frac{2,400}{12 \times 7.63} = 26 \text{ psi}$$

Total $DL + LL$ (w/o roof LL)

$$f_a = 229 + 26 = 255 \text{ psi}$$

12-12.C Design

(a) Lateral shear on wall

From Table 12-11 for wall **j**, Shear = 241 k.

$$f_v = \frac{1000 \times 241}{320 \times 7.63} = 98.7 \text{ psi}$$

Determine allowable shear

$$\frac{h}{l} = \frac{8.33}{26.7} = 0.31$$

From Diagram A-5, the maximum allowable shear without shear reinforcing

$F_{vm} = 66$ psi $\times 1.33$
$\quad = 88$ psi < 98.7 psi   N.G.

Shear reinforcing is required.

From Diagram A-6:

$F_{vs} = 100 \times 1.33$
$\quad = 133$ psi > 98.7 psi   O.K.

Use Diagram K-2b to select the required shear reinforcing steel ($F_s = 1.33 \times 24,000 = 32,000$ psi).

Select #6 @ 16" ($F_{vs} = 116$ psi)

(b) Vertical load axial stress (dead load + live load + seismic load perpendicular to wall)

$$f_a = 229 + 26$$
$$\quad = 255 \text{ psi}$$

The allowable axial stress using Tables Q-1 and Q-2 is:

$$F_a = 582 \text{ psi}$$

Lateral seismic force perpendicular to wall,

$F_p = ZIC_pW_p$
$\quad = 0.40 \times 1.0 \times 0.75 \times 78$
$\quad = 23.4$ psf

Lateral seismic moment

$M = \frac{1}{10}wl^2 = \frac{1}{10} \times 23.4 \times 8.33^2$
$\quad\quad = 162$ ft lbs/ft

Check flexural stress assuming uncracked section

$$f_b = \frac{6M}{bt^2} = \frac{6 \times 162 \times 12}{12 \times 7.63^2} = 16.7 \text{ psi}$$

Actual axial stress is much greater than the flexural stress:

$\quad$ 255 psi > 16.7 psi, no tension stress

Minimum required steel is adequate

(c) Vertical dead load + floor live load + seismic load parallel to wall

From Table 12-11, Overturning moment on wall **j**

$\quad M_{OT} = 6728$ ft kips

Combined axial load and moment

$\quad l = 26'$ - $8'' = 320''$ $\quad\quad h = 8'$ - $4''$

$\quad$ Estimate $d_1 = 16''$ and $d = 320 - 16 = 304''$

$\quad F_a = 582$ psi (Table Q-1 and Q-2)

Check flexural stress assuming uncracked section

$$f_b = \frac{6M}{bl^2} = \frac{6 \times 6728 \times 12}{7.63 \times 320^2}$$
$$\quad = 620 \text{ psi} > 255 \text{ psi}$$

$$f_b > f_a$$

Design as a cracked section
Use Method 2, Interaction design

Maximum allowable flexural stress

$$f_b = \left(1.33 - \frac{f_a}{F_a}\right)F_b$$

$$\quad = \left(1.33 - \frac{255}{582}\right)1000$$

$$\quad = 892 \text{ psi}$$

Maximum allowable compressive stress

$$f_m = f_a + f_b$$
$$= 255 + 892$$
$$= 1147 \text{ psi} = 1.147 \text{ ksi}$$

$$\mathbf{a} = \tfrac{1}{6} t f_m$$
$$= \tfrac{1}{6} \times 7.63 \times 1.147$$
$$= 1.46$$

$$\mathbf{b} = -\tfrac{1}{2} t f_m (l - d_1)$$
$$= -\tfrac{1}{2} \times 7.63 \times 1.147 (320 - 16)$$
$$= -1330$$

$$\mathbf{c} = P\left(\frac{l}{2} - d_1\right) + M$$
$$P = 0.255 \times 7.63 \times 320$$
$$= 623,000 \text{ lbs}$$

$$M = 6728 \text{ ft k}$$

$$\mathbf{c} = 623\left(\frac{320}{2} - 16\right) + 6728 \times 12$$
$$= 170,448$$

$$kd = \frac{-\mathbf{b} - \sqrt{\mathbf{b}^2 - 4\mathbf{ac}}}{2\mathbf{a}}$$
$$= \frac{1330 - \sqrt{(1330)^2 - 4 \times 1.46 \times 170,448}}{2 \times 1.46}$$
$$= 154 \text{ in.}$$

Stress in steel

$$f_s = \left(\frac{d - kd}{kd}\right) n f_m$$
$$= \left(\frac{304 - 154}{154}\right) 12.9 \times 1,147$$
$$= 14,412 \text{ psi}$$

$$\text{Compression} = \tfrac{1}{2} t k d f_m$$
$$= \tfrac{1}{2} \times 7.63 \times 154 \times 1,147$$
$$= 673,874 \text{ lbs}$$
$$C = 674 \text{ kips}$$
$$-P = \underline{623 \text{ kips}}$$
$$\text{Tension} = 51 \text{ kips}$$

Area of steel

$$A_s = \frac{T}{f_s} = \frac{51,000}{14,412} = 3.54 \text{ sq. in.}$$

Use 4 - #9 bars at each end and minimum steel between boundary members.

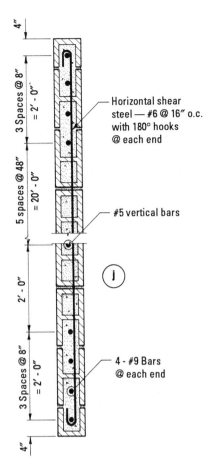

Figure 12-16 Detail of reinforcing steel in wall j at first floor, base level. (Coordinate this plan with Figure 12-18.)

12-13 DESIGN OF WALL j ON FIFTH STORY, LEVEL 4

12-13.A General

Use $f'_m = 1500$ psi with special continuous inspection for fifth floor and above. See Section 12-11.A General.

12-13.B Loads

Dead load

Roof load	25×71 psf =	1775 plf
Floor loads	$2 \times 24.33 \times 98$ psf =	4769 plf
Wall loads	$3 \times 9 \times 78$ psf =	2106 plf
	Total dead load =	8650 plf

Live load (see Section 12-12.B for reduced live loads)

Roof load	25×12 psf =	300 plf
Floor load	$2 \times 24.33 \times 16$ psf =	779 plf
	Total live load =	1079 plf

Axial stress at fifth story

Dead load stress

$$f_a = \frac{P_{DL}}{bt} = \frac{8650}{12 \times 7.63} = 94 \text{ psi}$$

Live load stress

$$f_a = \frac{P_{LL}}{bt} = \frac{779}{12 \times 7.63} = 12 \text{ psi}$$

Floor live load stress (w/o roof live load)

$$f_a = \frac{P_{LL}}{bt} = \frac{779}{12 \times 7.63} = 8.5 \text{ psi}$$

Total load w/o roof live load = 94 + 8.5
$$= 102.5 \approx 103 \text{ psi}$$

12-13.C Design

(a) From Table 12-11, Unit shear stress, f_v, = 53.8 psi imposed.

$$\frac{h}{l} = \frac{8.33}{26.67} = 0.31$$

Use Diagram A-5, for $h/l = 0.31$, and $f'_m = 1500$ psi to find the maximum allowable shear stress w/o shear reinforcing:

$$F_{vm} = 48 \times 1.33 = 64 \text{ psi} > 53.8 \text{ psi} \text{ O.K.}$$

Provide minimum horizontal steel

Use #5 at 48" o.c.

(b) Axial compression stress

$$f_a = 103 \text{ psi}; F_a = 291 \text{ psi (Tables Q-1 and Q-2)}$$

(c) Vertical dead load + live load + seismic load perpendicular to wall

Lateral seismic force

$$F_p = ZIC_pW_p$$
$$= 0.40 \times 1.0 \times 0.75 \times 78 = 23.4 \text{ psf}$$

Moment on wall,

$$M = {}^1\!/_{10}wl^2 \text{ (assume some fixity)}$$
$$= {}^1\!/_{10} \times 23.4 \times 8.33^2 = 162.4 \text{ ft lbs/ft}$$

Check flexural stress assuming an uncracked section

$$f_b = \frac{6M}{bt^2} = \frac{6 \times 162.4 \times 12}{12 \times 7.63^2} = 16.7 \text{ psi} < 103 \text{ psi}$$

Flexural stress is less than axial stress

Minimum steel is adequate for lateral load perpendicular to wall.

(d) Vertical dead load + live load + seismic load parllel to wall.

From Table 12-11, the overturning moment on Wall **j** = 1543 ft kips

Length of wall = 320 inches

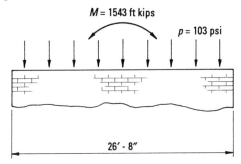

Check flexural compression stress assuming an uncracked section.

Overturning moment = 1543 ft kips

$$f_b = \frac{6M}{bl^2} = \frac{6 \times 1543 \times 12,000}{7.63 \times 320^2}$$
$$= 142 \text{ psi} > 103 \text{ psi}$$

Flexural stress is greater than axial compression stress

Design Tension Reinforcement

Interaction design, Method 2

Maximum allowable compressive stress, f_m

$$f_m = \left(1.33 - \frac{f_a}{F_a}\right)F_b + f_a$$
$$= \left(1.33 - \frac{103}{291}\right)500 + 103$$
$$= 488 + 103 = 591 \text{ psi}$$

Calculate a, b, c and kd

$$\mathbf{a} = {}^1\!/_6\, t f_m$$
$$= {}^1\!/_6 \times 7.63 \times 0.591 = 0.752$$

$$\mathbf{b} = -{}^1\!/_2\, t f_m (l - d_1)$$
$$= -{}^1\!/_2 \times 7.63 \times 0.591 (320 - 8) = -703$$

$$\mathbf{c} = -P\left(\frac{l}{2} - d_1\right) + M$$
$$= 251\left(\frac{320}{2} - 8\right) + 1542 \times 12 = 56,730$$

$$kd = \frac{-\mathbf{b} - \sqrt{\mathbf{b}^2 - 4\mathbf{ac}}}{2\mathbf{a}}$$

$$= \frac{703 - \sqrt{703^2 - 4 \times 0.752 \times 56,730}}{2 \times 0.752}$$

$$= 89''$$

Equivalent eccentricity of vertical load

$$e = \frac{M}{P} = \frac{1543 \times 12}{251} = 73.7 \text{ in.}$$

Calculate the compressive force, C:

$$C = \tfrac{1}{2}tkdf_m = \tfrac{1}{2}(7.63)(89)(591)$$
$$= 200,665 \text{ lbs}$$

The resulting tension force is:

$$T = C - P = 200,665 - 251,000$$
$$= -50,335 \text{ lbs}$$

The negative sign indicates that no tension reinforcement is required and the resultant compressive force can be made coincidental with the eccentric axial load by reducing the masonry compressive stress.

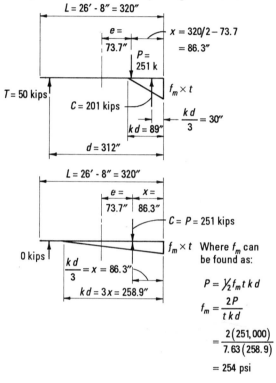

Use minimum steel to satisfy code requirements.

Minimum steel is provided horizontally: #5 at 48" o.c.

$$p = \frac{0.31}{7.63 \times 48} = 0.0008$$

Vertical steel ratio required

$$p_g = 0.0020 - 0.0008$$
$$= 0.0012$$

Vertical steel,

$$A_s = 0.0012 \times 7.63 \times 320$$
$$= 2.93 \text{ sq. in.}$$

Provide #7 bars at each end and six #5 bars in between.

$$A_s = 2 \times 0.60 + 6 \times 0.31$$
$$= 3.06 \text{ sq. in.}$$
$$p_g = \frac{3.06}{7.63 \times 320}$$
$$= 0.00125 > 0.0012 \quad \text{O.K.}$$

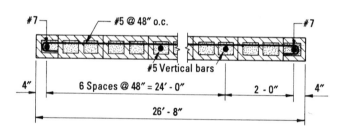

Figure 12-17   Detail of reinforcing steel in Wall j at Fifth Story, Level 4. (Coordinate this plan with Figure 12-18.)

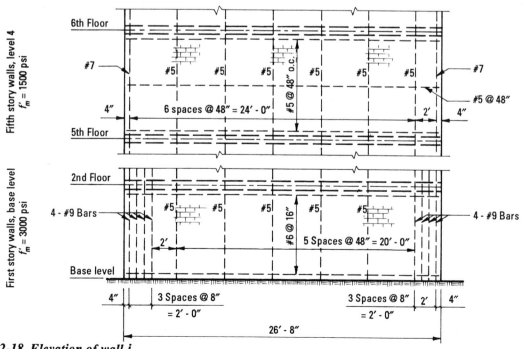

Figure 12-18 Elevation of wall j.

12-14 ADDITIONAL CONSIDERATIONS IN THE DESIGN OF MULTI-STORY SHEAR WALL STRUCTURES

The concepts outlined in this book provide information on the design and detailing of a multi-story shear wall structure. In addition, there are alternate concepts and considerations that should be examined.

Repetition of floor layout from floor to floor will speed construction. The contractor and mason will become familiar with the layout and production will increase as the building is constructed.

If possible, walls should be continuous to the foundation. This will eliminate discontinuities or the need to develop a shear platform to transfer the shear to other walls.

It is advantageous to have as uniform and symmetrical floor plans as possible. This will reduce torsional forces in the building and simplify design, detailing and construction.

Buildings with irregular plans such as T, Z, U and L layouts, should be designed with floor diaphragms sufficiently stiff to transmit the torsional forces to the various shear wall elements or there should be a separation of each wing so that each section will act independently. The separation should be adequate to prevent impact between building sections and should be at least the computed drift due to wind or earthquake forces plus a clearance allowance. It is recommended that the seismic clearance between buildings be not less than 1″ per story for Seismic Zones 3 and 4, 1/2″ per story for Seismic Zone 2 and 1/4″ per story for Seismic Zone 1 and that the clearance for wind forces be not less than 1/4″ per story. (See UBC Sec. 2334(h) for addition code requirements).

Buildings can react as shear structures in which the major drift, lateral deflection, due to lateral loads is caused by shear deformation of the walls rather than moment deflection of the frame and shortening of the column or boundary members due to overturning forces. Shear structures in which this phenomenon dominates generally have a height to width ratio of 1.5 or less. The shear walls in these buildings may be designed on a floor to floor basis. The height to width ratio of the walls may be less than 1.0 and accordingly the shear capacity and the allowable shear stress are increased as shown in Diagrams A-5 and A-6.

When the height to width ratio of a building is 1.20 or more it can react as a moment or flexural structure in which the major drift due to lateral forces is caused by moment deformation, joint rotation (in the case of a frame building) and column or boundary member shortening due to overturning moment and flange stresses. The shear walls in these relatively flexible structures should be designed for their full height as walls loaded laterally and cantilevering from the foundation as in the design example in this section. The height to width ratio would be large, greater than 1, and the shear capacity and allowable shear stresses reduced as shown in Diagrams A-5 and A-6.

Shear wall buildings are generally stiff buildings and do not distort as much as frame buildings. This means that there will be little or no damage to non-structural, architectural elements, and no damage to the structural walls in buildings subjected to small earthquakes, and only slight damage in medium earthquakes.

Shear walls which resist the lateral forces, can vary greatly in stiffness depending on the placement of openings in the wall. Figure 12-20 compares the rigidity of a wall with no openings to walls of the same size but with various opening patterns.

By staggering the openings, the wall acts as a solid wall with scattered openings. Walls that have the openings stacked on top of each other act as independent elements and the total rigidity is significantly reduced. This comparison assumes that the floor system connecting the walls has no resistance to transfer shear or moment.

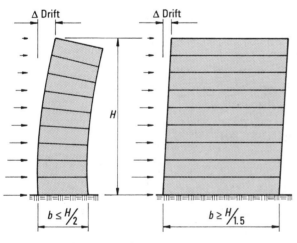

Cantilever or moment system   Shear system

Figure 12-19 Drift of a flexible moment frame structure and a rigid shear wall structure.

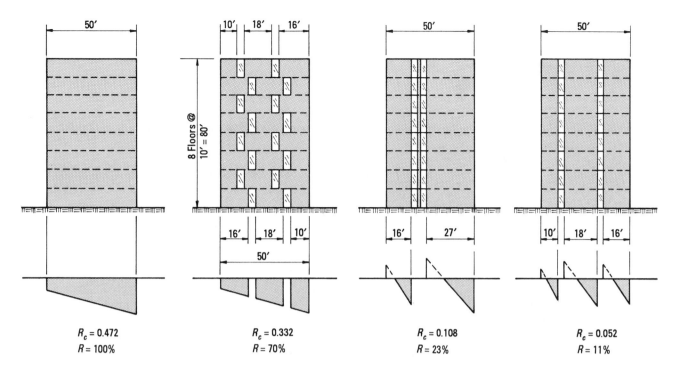

$R_c = 0.472$
$R = 100\%$

$R_c = 0.332$
$R = 70\%$

$R_c = 0.108$
$R = 23\%$

$R_c = 0.052$
$R = 11\%$

Figure 12-20 Relative rigidities of walls with various opening patterns.

The strength of the connecting member between the wall elements can also significantly influence the total strength of the wall. Figure 12-21 demonstrates the effect of the stiffness of a connecting member on the rigidity of the system and produces a coupled wall mechanism. The connecting members, floor slabs or spandrel or lintel beams in coupled shear walls should be investigated and designed for the stresses, moments and shear forces induced in them.

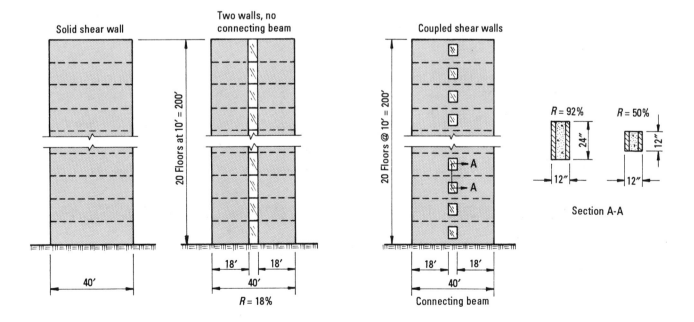

Figure 12-21 Influence of coupling shear walls.

The moment capacity of shear walls can be greatly increased when acting as vertical cantilever elements by using returns at the ends which function as flanges and increased area to resist compression forces. The flanges will increase the moment of inertia of the wall section, decrease the flexural stresses and facilitate the placement and efficiency of reinforcing steel. These flanges can be readily incorporated at door returns as shown in Figure 12-22. They also significantly reduce shear and moment stresses and thus reduce the amount of steel required for overturning forces.

If the total shear, V, is the same for the 26 ft distance of 2 motel rooms, shown in Figure 12-21, the unit shear, f_v, for Figure 12-22(b) would be only 76% of the unit shear, f_v, for Figure 12-22(a). The compressive bending stress, f_b, in the flanges of Figure 12-22(b) is only 42% of the compressive bending stress, f_b, for Figure 12-22(a) and in turn only 59% of the steel is required for the plan with the flanges.

This same concept can be utilized in cross walls as well as longitudinal walls (Figure 12-23). The flanges can be single or double returns and can include corners and intersections of walls.

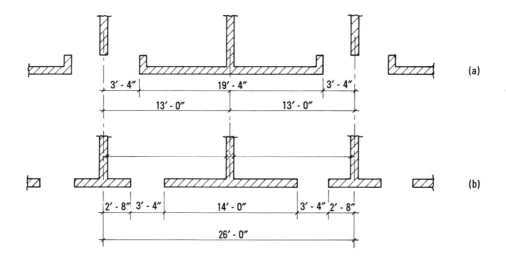

Figure 12-22 Wall returns increase rigidity of shear walls.

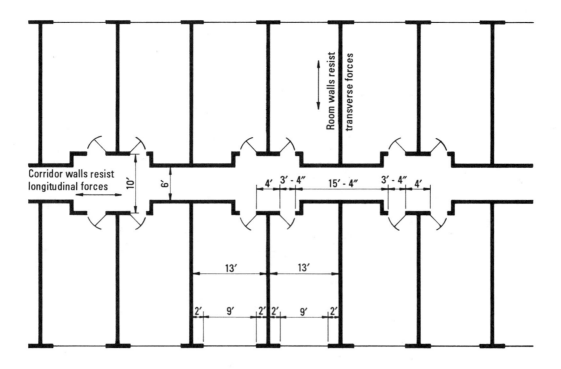

Figure 12-23 Floor layout to increase rigidity and shear capacity of a building.

12-15 QUESTIONS AND PROBLEMS

12-1 What is the overturning moment and shear on the 7 story building at each floor level given the force distribution shown?

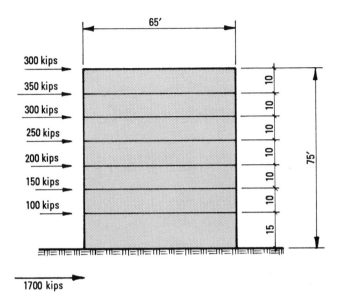

12-2 An 8-story building, 88 feet high and 70 feet wide is subjected to a base shear of 600 kips. What is the shear distribution on the building? All floors are 11 feet apart and floors and roof have same weight.

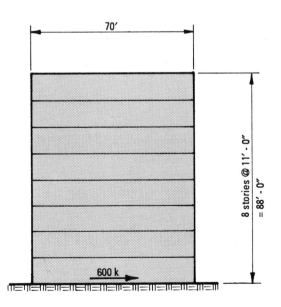

12-3 Given the first floor plan and the second through fifth floor plans in an office building, locate the center of mass and the center of rigidity for the first and fifth floor levels. What is the eccentricity to be used in the design for each direction.

Assume the dead load of the floor is 85 lbs per square foot. The walls are 12 feet high floor to floor and are made of 8 inch thick normal weight solid grouted masonry walls. The window walls weigh 15 lbs per square foot. Make whatever additional assumptions are necessary. What are the forces to walls 2, 6, B and I on the first floor and to walls 6, 12, B and I on the fifth floor?

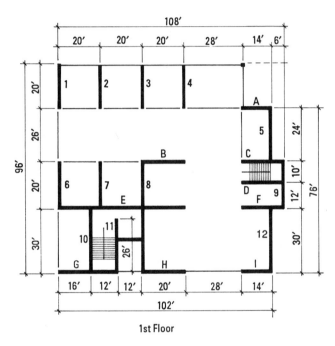

1st Floor

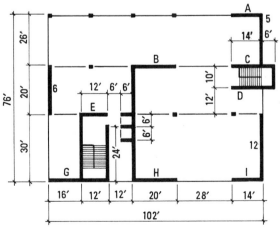

2nd to 5th Floor

12-4 Design shear walls 2 and 1 for the 2nd floor and 6th floor of the 8-story apartment building shown below. Assume it is located in Seismic Zone No. 3. The floor and roof dead load is 600 kips each and the live plus partition load is 45 lbs per square foot. Compute the center of rigidity and mass and the direct shear and torsional force to each wall.

Use two strengths of masonry; high strength for the first five floors and lower strength for the upper three floors. Make any other engineering assumptions necessary, assume all walls are the same thickness for calculation of center of rigidity and center of mass.

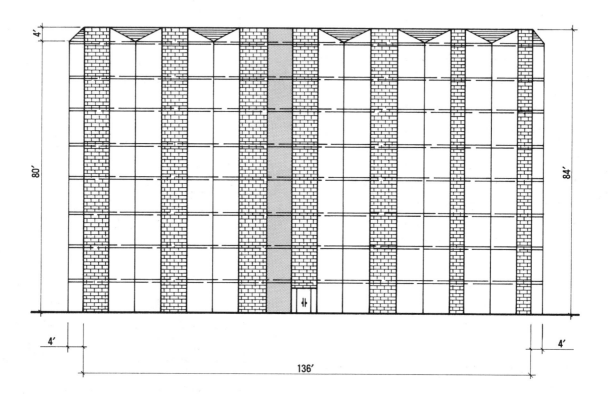

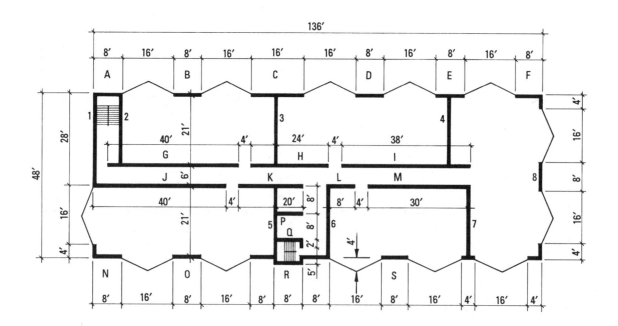

SECTION 13

Retaining Walls

13-1 GENERAL

One of the many applications of masonry is for retaining walls. Masonry walls inherently provide visual beauty along with the required structural strength to resist the imposed vertical and lateral loads.

13-2 TYPES OF RETAINING WALLS

There are four basic types of retaining walls: gravity walls, counterfort or buttressed walls, cantilever walls, and supported walls. The selection of the particular type of wall will depend on the site, size of wall, loads, soil conditions, use and economics of construction.

13-2.A Gravity Walls

These walls are of mass masonry designed so that there are no tension stresses developed in the wall under most loading conditions (Figure 13-1). In some instances, low tension stresses are permitted by providing reinforcing in the wall. These partially reinforced walls are considered as semi-gravity walls. Neither masonry gravity walls nor semi-gravity walls are used to any great extent in new structures.

Reinforced masonry wall retaining eight feet of backfill

the vertical supports are exposed in front of the wall, they are called buttresses and are compression members (Figure 13-3). In either case, the main wall is considered as a continuous member supported at each cross wall.

Counterfort and buttressed retaining walls are used to retain high fills up to 25 ft.

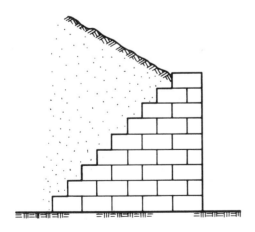

Figure 13-1 Gravity retaining wall.

13-2.B Counterfort or Buttressed Walls

These walls span horizontally between vertical support members. If the vertical supports are behind the wall and buried in the earth backfill, they are called counterforts and are tension members (Figure 13-2). If

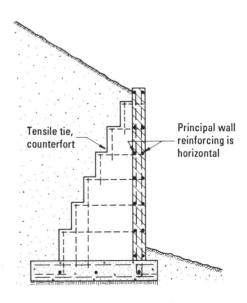

Figure 13-2 Counterfort retaining wall.

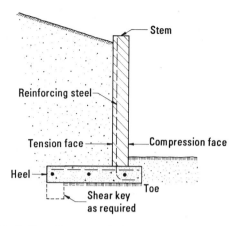

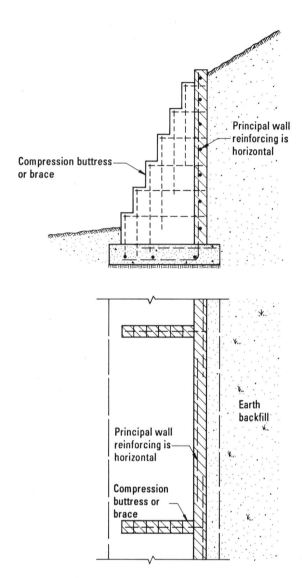

Figure 13-3 Buttress retaining wall.

13-2.C Cantilever Retaining Walls

These walls are so named because the vertical stem wall is designed to cantilever from the base (Figure 13-4). The tension stresses which develop on the wall stem are resisted by reinforcing steel placed on the soil side (tension face) of the wall. The base resists sliding, overturning and rotating due to the lateral loading and must also be large enough to ensure that the bearing capacity of the soil is not exceeded.

Cantilever retaining walls can be designed without a footing heel and are especially useful in limited space areas such as near property lines and existing utility lines (Figure 13-5). These walls require special attention to assure they can resist the lateral earth forces and overturning moment through their own weight and strength. Since there is no footing heel on most property walls, there is also no resisting soil mass and thus the wall foundation may be especially large and heavy.

Figure 13-4 Cantilever retaining wall.

To reduce the need for such a large foundation, adjacent slabs are often tied into the foundation (Figure 13-5(b)). These slabs help resist sliding forces and contribute somewhat to reducing the soil bearing pressure and the overturning forces.

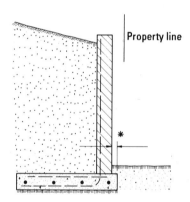

(a) Retaining wall adjacent to property line.

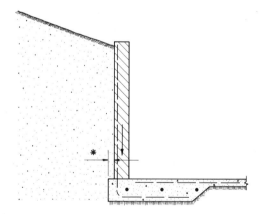

(b) Retaining wall adjacent to property line with foundation tied into floor slab to increase the sliding resistance.

Figure 13-5 "Property" line type retaining walls.

* If possible, provide at least a 2″ footing extension to allow room for construction tolerances.

13-2.D Supported Walls

Walls of basements and subterranean garages are often laterally supported at their tops by floor systems (Figure 13-6).

Depending on the type of support provided by the floor and foundation systems, a supported wall could be considered having either a fixed top and bottom, a fixed base with simply supported top or a simply supported top and bottom (Figure 13-7). Each wall type must be designed and reinforced accordingly. Continuity of the connections at the top and the bottom must be developed by proper reinforcing in order to provide the required degree of fixity.

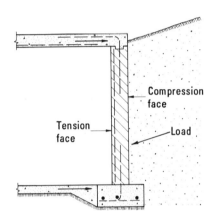

Figure 13-6 Supported retaining wall.

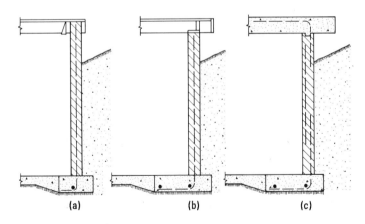

(a) Simply supported top and bottom

(b) Simply supported top: fixed at bottom

(c) Fixed at top and bottom

Figure 13-7 Supported retaining walls with various end conditions.

Basement or subterranean garage walls are often subjected to both vertical and lateral loads since these perimeter walls support the building above as well as resist the earth pressure (Figure 13-8). The combined loading on the wall, vertical load plus lateral load, must be considered in the design of the wall.

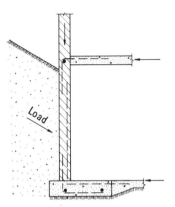

Figure 13-8 Vertical and lateral loads on a supported retaining wall.

13-3 DESIGN OF RETAINING WALLS

Masonry retaining walls may be designed by two methods. They can be designed as masonry walls using allowable masonry stresses as previously discussed in this book, or they can be designed assuming that the masonry is a form for the concrete grout and allowable concrete stresses are used in the stress analysis (see Section 6-10).

Both design methods depend on the loads imposed, the height of the wall, the limiting thickness of wall, the construction procedure and the economics of the finished wall.

It is usually recommended that the resultant of the forces on a retaining wall fall within the middle third of the footing. This results in the most efficient use of the footing. However, this general recommendation is not mandatory and the resultant force can fall outside of the middle third, provided the resulting soil pressure does not exceed the allowable soil bearing pressure and the overturning moment does not exceed the resisting moment within the allowable factor of safety.

13-3.A Effect of Corners on Lateral Supporting Capacity of Retaining Walls

Retaining walls that change direction create a condition of increased strength at the corner. Each direction of wall will tend to support the opposite wall for a certain distance (Figure 13-9). This condition is common in basements and underground garages.

If the length of the wall is not more than about 3 times the height, the wall can be designed by a procedure similar to the one presented by the Portland Cement Association in its publication *Rectangular Concrete Tanks* (IS003). Using this procedure, walls may have four conditions of support:

1) fixed both sides, hinged top and bottom,

2) fixed both sides and bottom, free at top,

3) fixed both sides, hinged at bottom and free at top, and

4) hinged at all four edges.

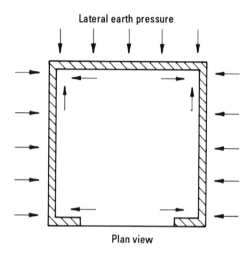

Figure 13-9 Intersecting walls brace retaining walls.

The loading on the wall is assumed to be a lateral triangular earth load plus any surcharge load and perhaps a vertical load due to the structure above. If the length of the wall is greater than about 3 times its height, the influence of the corner is significantly reduced and it is usually considered effective over a distance equal to the height of the wall.

13-3.B Preliminary Proportioning of Retaining Walls

Retaining walls are designed by initially selecting tentative dimensions which are analyzed for stability and structural requirements and then revised as required. Since this is trial and error process, various solutions to the problem may be obtained, all of which may be satisfactory.

Dimensions for a retaining wall must be adequate for the structural stability of the wall and must satisfy local building code requirements. Tentative dimensions of cantilever retaining walls are shown in Figure 13-10 and are based in part on experience of satisfactorily constructed walls. They may be used for the initial design proportions, although they sometimes result in overly conservative designs.

The stem should generally be made of typical masonry units and the base of the stem should be thick enough to satisfy the shear requirements without the use of shear reinforcing steel.

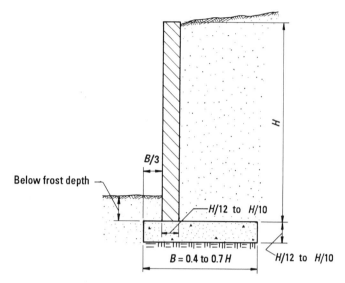

Figure 13-10 Tentative design dimensions for a cantilever retaining wall.

The base-slab dimensions should be such that the resultant of the vertical loads falls within the middle one-third (Figure 13-11). If the resultant falls outside the middle one-third, the toe pressures may be excessively large and only a part of the footing will be effective.

When the resultant falls within the middle third, no uplift will occur and the following equations may be used for design:

$$\text{Eccentricity, } e = \frac{l}{2} - \left(\frac{M_R - OTM}{\Sigma W} \right), \text{ and}$$

$$\text{Soil Pressure} = \frac{W}{A} \pm \frac{We}{S}$$

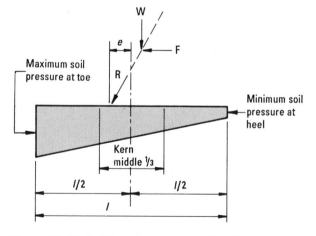

Figure 13-11 Soil bearing pressure distribution, resultant within middle third.

When the resultant of the forces falls outside the middle third, the above equations do not apply since only a portion of the footing is effective (Figure 13-12). For this condition, moments may be taken about the toe of the footing to find:

$$\text{Eccentricity, } e = \frac{l}{2} - \left(\frac{M_R - OTM}{\Sigma W}\right)$$

$$\text{Length of compression area} = l' = 3\left(\frac{l}{2} - e\right)$$

$$\text{Maximum soil pressure} = 2 \times \frac{\Sigma W}{l'}$$

The stabilizing resisting moment should be at least $1\frac{1}{2}$ times greater than the overturning moment and the allowable soil bearing pressure must not be exceeded. If the maximum soil pressure does exceed the allowable soil pressure, the toe or heel may be extended to decrease the eccentricity, e, and to increase the length of compression area.

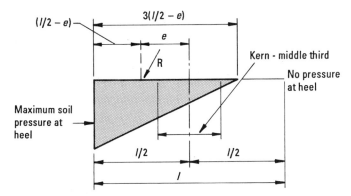

Figure 13-12 *Soil bearing pressure distribution, resultant outside of middle third.*

13-4   CANTILEVER RETAINING WALL DESIGN EXAMPLE

13-4.A   Design Criteria

The design for the cantilever retaining wall is based on a per foot length of wall.

Retaining wall — 10 ft high

Backfill slope — 1:3

Backfill soil — course-grained soil of low permeability due to admixture of particles of silt size, $w_s = 110$ pcf (Type 2 Soil)

Masonry: Design of wall using two different masonry requirements.

Steel $f_s = 24,000$ psi, $E_s = 29,000,000$ psi

Footing Concrete, $f'_c = 2500$ psi, no special inspection required.

Lateral Pressure — Equivalent Fluid Pressure.

From Tables 3-18 and 3-19 for Type 2 soil with a 1:3 backfill slope:

$$k_h = 40 \text{ pcf and } k_v = 12 \text{ pcf}$$

In order to use these values, weepholes or a drainage system must be provided to prevent hydrostatic head from building up behind the wall. These holes or systems must be located near the bottom of the wall or at the ground surface.

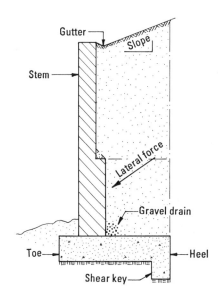

Figure 13-13 *Cross-section of cantilever retaining wall.*

Determine the required wall thickness and reinforcing in the wall for balanced design conditions if it is constructed with:

Part (a): Two wythe, grouted clay brick masonry wall without special inspection; $f'_m = 2500$ psi.

Part (b): Concrete block masonry with special inspection; $f'_m = 1500$ psi.

Also check the footing and other conditions of design for the clay brick masonry stem as determined in part (a).

13-4.B   Stem Design

13-4.B.1   Part (a) — Brick Wall Stem

Since no inspection will be provided, use half stresses. For balanced design:

$$f_b = \frac{1}{2}(0.33f'_m) = \frac{1}{2}(0.33 \times 2500) = 417 \text{ psi}$$

$$f_s = 24,000 \text{ psi}$$

Use a minimum width at the top of 10 inches to provide a 4 inch grout space.

$t = 10$ inches, use $d = 5.5''$

Enter Table E-3a for $f'_m = 2500$ psi; half stresses and $n = 15.5$. Balanced conditions occur when $f_b = 417$ psi and $f_s = 24,000$ psi:

$$K_b = 41.1; \; p_b = 0.0018$$

Balanced moment

$$M_b = \frac{K_b \times bd^2}{12}$$
$$= \frac{41.1 \times 12 \times 5.5^2}{12}$$
$$= 1,243 \text{ ft lbs / ft}$$

(Note, Table H-3 could also have been used to find $M_b = 1.24$ ft K/ft).

Find the height of backfill where the actual moment will exceed M_b. For a cantilevered wall with a triangular (equivalent fluid) soil surcharge:

$$\text{Moment} = k_h \frac{h^3}{6} = 1,243 \text{ ft lbs / ft}$$

$$h = \left(\frac{6M}{k_h} \right)^{1/3} = \left(\frac{6 \times 1243}{40} \right)^{1/3}$$

$$h = 5.71 \text{ ft} \approx 5'\text{-}8''$$

This is equivalent to 17 courses of brick with a 4" vertical module.

$$A_s = p_b bd = 0.0018 \times 12 \times 5.5$$
$$= 0.119 \text{ sq. in./ft}$$

From Table C-4b

Use #5 at 30" o.c.

$A_s = 0.123$ sq. in./ft

Minimum horizontal steel = $0.0007 \, bt$

From Table C-5

Select #4 bars spaced no more than 29 inches o.c.

Use a more practical spacing of 24 inches

Determine the required wall thickness and reinforcing at the base of the wall where the moment is maximum. The moment at base is:

$$M = \frac{k_h h^3}{6} = \frac{40 \times 10^3}{6}$$
$$= 6,670 \text{ ft kips / ft}$$

$$bd^2 = \frac{M}{K_b} = \frac{6670 \times 12}{41.1} = 1,947$$

Since $b = 12''$,

$$d = \sqrt{\frac{1947}{12}} = 12.7''$$

Try $t = 17.0''$.

$$A_s = p_b bd = 0.0018 \times 12 \times 12.7$$
$$= 0.274 \text{ sq. in./ft}$$

From Table C-4b:

Choose #8 at 30" o.c. $A_s = 0.314$ sq. in./ft

(This spacing matches the spacing of the reinforcing in the upper section of the wall.)

Horizontal steel = $0.0007 \, bt$
$$= 0.0007 \times 12 \times 17$$
$$= 0.143 \text{ sq. in./ft}$$

Use #5 bars at 24" o.c. $A_s = 0.154$ sq. in./ft

If the lateral force was large enough, it could split the wall vertically at the point of change of section. Determine the maximum shear force at that point.

Shear on wall

Shear 5'-8" from top

$$V = \frac{k_h h^2}{2} = \frac{40 \times 5.67^2}{2}$$
$$= 643 \text{ lbs / ft}$$

Assume the shear is resisted by the grout and masonry where $b = 12''$, $d = 5.50''$ and $t = 10''$. Also assume $j = 0.9$.

$$v = \frac{V}{bjd} = \frac{643}{12 \times 0.9 \times 5.5}$$
$$= 10.8 \text{ psi}$$

From Table A-3 *Allowable Stresses*:

Allowable $v_m = 25$ psi > 10.8 psi   O.K.

Shear at base of stem at footing

$$V = \frac{k_h h^2}{2} = \frac{40 \times 10^2}{2}$$
$$= 2000 \text{ lbs / ft}$$

Assume shear is resisted by grout and masonry where $b = 12''$, $t = 18''$ and $d = 13.5''$. Assume $j = 0.9$.

$$v = \frac{V}{bjd} = \frac{2000}{12 \times 0.9 \times 13.5}$$
$$= 13.7 \text{ psi}$$

Allowable $v_m = 25$ psi > 13.7 psi   O.K.

If the actual shear exceeds the allowable shear strength of the masonry, provide a shear key similar to those shown in Figure 13-14 and use concrete shear capacity values.

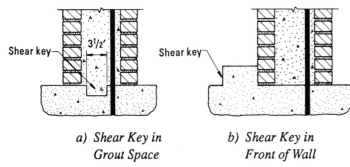

a) *Shear Key in*
Grout Space

b) *Shear Key in*
Front of Wall

Figure 13-14 Connection of wall with footing.

The shear on a key shown in Figure 13-14a is:

$$v = \frac{V}{bd} = \frac{2000}{12 \times 3.5}$$

$$= 47.6 \text{ psi}$$

The allowable $v_c = 1.1\sqrt{f_g} = 1.1\sqrt{2000} = 49$ psi

Determine the area of steel ties required. Tie each vertical bar to prevent vertical splitting. Therefore for each 30″ of wall,

$$V = \frac{30}{12}(643) = 1608 \text{ lbs}$$

$$A_s = \frac{V}{f_s} = \frac{1608}{24,000}$$

$$= 0.067 \text{ sq. in.}$$

Provide 2 - $^1/_4$″ ties at 6″ o.c. vertically at every vertical bar

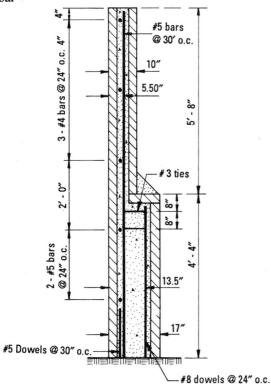

#5 bars @ 30′ o.c.

10″

5.50″

4″

4″

3 - #4 bars @ 24″ o.c.

5′ - 8″

#3 ties

8″

8″

2′ - 0″

4′ - 4″

2 - #5 bars @ 24″ o.c.

13.5″

17″

#5 Dowels @ 30″ o.c.

#8 dowels @ 24″ o.c.

Figure 13-15 Detail of wall for part (a), brick masonry stem, balanced design without special inspection.

13-4.B.2   Part (b) — Concrete Masonry Stem

Determine the reinforcing for a concrete masonry wall with f'_m = 1500 psi and special continuous inspection (use full allowable stresses).

Use balanced design conditions:

f_b = 500 psi and f_s = 24,000 psi.

From Table E-1b,

n = 25.8, K_b = 77.2 and p_b = 0.0036

Try 8 in. CMU for top section.

t = 7.63 in. and d = 5.25 in.

Balanced moment

$$M_b = K_b \times bd^2$$

$$= 77.2 \times 12 \times \frac{5.25^2}{12}$$

$$= 2128 \text{ ft lbs / ft}$$

$$\text{Moment} = k_h \frac{h^3}{6} = 2128 \text{ ft lbs / ft}$$

$$h = \left(\frac{6M}{k_h}\right)^{1/3} = \left(\frac{6 \times 2128}{40}\right)^{1/3}$$

$$h = 6.8 \text{ ft} \approx 82 \text{ inches}$$

This is equivalent to 10 courses of 8″ high CMU (80″)

$$A_s = p_b bd = 0.0036 \times 12 \times 5.25$$

$$= 0.227 \text{ sq. in./ft}$$

From Table C-4a

Use #5 at 16″ o.c. (A_s = 0.230 sq. in./ft)

Horizontal steel

From Table C-5, Minimum A_s = 0.064 sq. in./ft

Select #4 bars @ 37″ o.c. is adequate but since this is not modular, use #4 bars @ 32″ o.c.

Moment at base

$$M = 6670 \text{ ft lbs / ft [From Part (a)]}$$

$$bd^2 = \frac{M}{K_b} = \frac{6670 \times 12}{77.2} = 1037$$

Since b = 12″,

$$d = \sqrt{\frac{1037}{12}} = 9.3$$

Use 12″ CMU, d = 9″

$$K = \frac{M}{bd^2} = \frac{6670 \times 12}{12 \times 9^2}$$

$$= 82.3$$

From Table E-1b

n = 25.8 for K = 8.4, read p = 0.0045

From Table C-8g

　for d = 9.0″ and p = 0.0045

　　Use #7 at 16″ o.c.

Horizontal Steel

From Table C-5, Minimum A_s = 0.098 sq. in./ft

　　Use #4 at 24′ o.c.

13-4.C Footing Design

13-4.C.1 Soil Bearing

　The clay brick stem, as designed in Sec. 13-4.B.1, will be used to complete the design of the retaining wall.

　For the most efficient footing design, it is advisable to have the resultant of forces fall within the middle third of the footing.

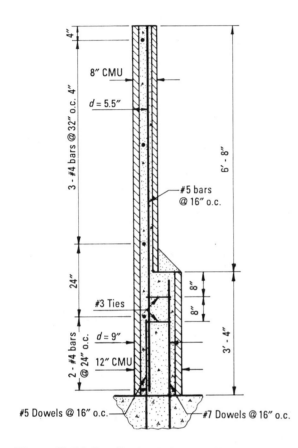

Figure 13-16 *Detail of wall for part (b), concrete block masonry with special inspection.*

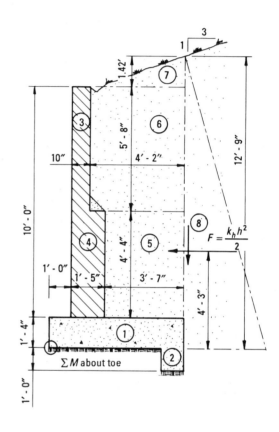

Figure 13-17 *Forces on wall.*

TABLE 13-1 Resisting Moment (Σ Moments at Toe of Footing)

Item								Weight lbs/ft		Moment Arm, ft		Resisting Moment, ft lbs/ft
(1) Footing		1.33	×	6	×	150	=	1,200	×	3.00	=	3,600
(2) Key		1	×	1.0	×	150	=	150	×	5.50	=	825
(3) Wall		0.83	×	5.67	×	120	=	565	×	1.42	=	802
(4) Wall		1.42	×	4.33	×	120	=	738	×	1.71	=	1,262
(5) Earth		3.58	×	4.33	×	110	=	1,705	×	4.13	=	7,042
(6) Earth		4.17	×	5.67	×	110	=	2,601	×	3.91	=	10,169
(7) Earth	½ ×	4.17	×	1.42	×	110	=	326	×	4.61	=	1,501
						Total of weight = 7,285 lb/ft						
(8) Vertical component of lateral earth pressure = ½ $k_v h^2$ = ½ × 12 × 12.75² = 975 ×									6.00		=	5,852
					Total vertical force = 8,260 lb/ft							
							Total Resisting Moment ΣM			=		31,053 ft lb/ft

Lateral earth pressure

$$F = \frac{k_h h^2}{2} = \frac{40 \times 12.75^2}{2}$$
$$= 3251 \text{ lbs/ft}$$

Overturning moment, assuming the shear key is one foot deep:

$$OTM = \frac{k_h h^3}{6} + \frac{p_p \text{ Shear Key}}{3}$$
$$= \frac{40 \times 12.75^3}{6} + \frac{200 \times 1^3}{3}$$
$$= 13,885 \text{ ft lbs/ft}$$

$$\text{Overturning safety factor} = \frac{M_R}{OTM}$$
$$= \frac{31,053}{13,885} = 2.24 \quad \text{O.K.}$$

$$\text{Eccentricity} = \frac{l}{2} - \left[\frac{M_R - OTM}{\Sigma W}\right]$$
$$= \frac{6}{2} - \left[\frac{31,053 - 13,885}{8260}\right]$$
$$= 0.92 \text{ ft}$$

$$\text{Third point} = \frac{l}{6} = \frac{6}{6} = 1.00 \text{ ft}$$

The eccentricity, $e = 0.92$ ft, is within the middle third. Thus there will be no uplift on the footing.

$$\text{Soil pressure} = \frac{W}{l} \pm \frac{6We}{l^2}$$
$$= \frac{7,285}{6} \pm \frac{6 \times 7285 \times 0.92}{6^2}$$
$$= 1214 \pm 1117$$
$$= 2331 \text{ psf maximum}$$
$$= 97 \text{ psf minimum}$$

Allowable soil bearing = 3000 psf O.K.

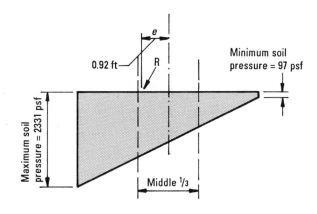

Figure 13-18 Soil pressure under footing.

13-4.C.2 Design of Footing Heel

The moment on the footing under the masonry wall is to be taken at a point halfway between the middle and edge of the wall (UBC 2615 (e) 2.B.)

$$\text{Soil pressure at } A = 97 + (2331 - 97)\frac{3.94}{6}$$
$$= 1564 \text{ psf}$$

$$\text{Soil Pressure at } B = 97 + (2331 - 97)\frac{4.64}{6}$$
$$= 1825 \text{ psf}$$

13-4.C.2(a) Moment Design of Heel

Steel design for Design Moment at AA (From Table 13-2). Ignore soil reaction and therefore add terms 7 and 8 back into M_A to find M_u.

$$M_u = 14,026 + (1280 + 6436)$$
$$M_u = 21,742 \text{ ft lbs/ft}$$

Moment resisted by minimum area of steel required by UBC Sec. 2610(f)

$$A_{s(min)} = \frac{200bd}{f_y} = \frac{200 \times 12 \times 12}{60,000}$$
$$= 0.48 \text{ sq. in./ft}$$

Since wall steel is spaced at 30" on centers, keep slab steel modular. Try #7 @ 15" ($A_s = 0.48$ sq. in./ft)

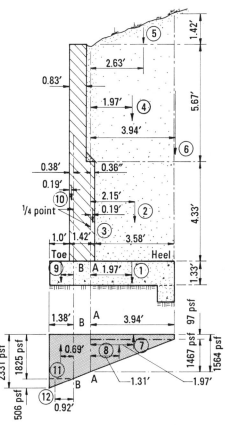

Figure 13-19 Design of heel and toe.

TABLE 13-2 Moment at AA, Quarter Point Under Wall (Strength Design Method)

Item								Load Factor		Wt. (lbs/ft)		Arm (ft)		Moment (ft lbs/ft)
(1) Footing			1.33	x	3.94	x	150	x 1.4	=	1,100	x	1.97	=	2,168
(2) Earth			4.33	x	3.58	x	110	x 1.4	=	2,387	x	2.15	=	5,133
(4) Earth			5.67	x	3.94	x	110	x 1.4	=	3,440	x	1.97	=	6,777
(5) Earth	½	x	1.42	x	3.94	x	110	x 1.4	=	431	x	2.63	=	1,133
(6) Vertical soil component							975	x 1.7	=	1,658	x	3.94	=	6,531
(7)*Soil reaction					3.94	x	− 97	x 1.7	=	− 650	x	1.97	=	− 1,280
(8)*Soil reaction	½	x			3.94	x	−1,467	x 1.7	=	− 4,913	x	1.31	=	− 6,436
								$\Sigma V_A =$		3,453		$\Sigma M_A =$		14,026

* A common conservative approach is to ignore the soil reaction which increases the moment and the shear on the heel of the footing. Therefore these two items (7 and 8) will be added to the values above to determine M_u and V_u as shown below.

$$a = \frac{A_s f_y}{0.85 f'_c b} = \frac{0.48 \times 60,000}{0.85 \times 2500 \times 12}$$
$$= 1.13''$$

$$M_n = A_s f_y \left(d - \frac{a}{2} \right)$$

$$M_n = 0.48 \times 60,000 \left(12 - \frac{1.13}{2} \right)$$

$$M_n = 329,336 \text{ in. lbs / ft}$$
$$= 27,445 \text{ ft lbs / ft}$$

$$M_n \geq \frac{M_u}{0.9}$$

$$27,445 \geq \frac{21,742}{0.9}$$

$$27,384 > 24,157 \text{ ft lbs / ft O.K.}$$

Minimum steel requirements are sufficient.

Use #7 @ 15 o.c.

13-4.C.2(b) Shear Design of Heel

For the heel, the critical section for shear is at the face of the wall (UBC Sec. 2615(f)2).

$$97 + (2331 - 97) \frac{3.58}{6} = 1430 \text{ psf}$$

Calculate the ultimate shear by conservatively adding back in the soil reaction (see Table 13-3).

$$V_u = 3915 + 4647 = 8562 \text{ lbs / ft}$$
$$V_c = 2\sqrt{f'_c} \, bd = 2\sqrt{2500} \times 12 \times 12$$
$$= 14,400 \text{ lbs / ft}$$

$$V_u \leq \phi V_c$$

$$8562 \leq 0.85 \times 14,400$$
$$8562 < 12,240 \text{ lbs / ft}$$

Therefore concrete can resist shear stress adequately.

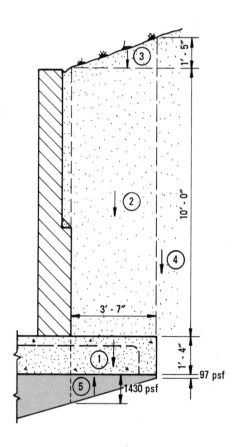

Figure 13-20 Loads on heel of footing.

TABLE 13-3 Shear at Face of Wall (Strength Design)

Item												Shear (lbs/ft)
(1) Footing			1.33	x	3.58	x	150	x	1.4	=		1,000
(2) Earth			3.58	x	10.00	x	110	x	1.4	=		5,513
(3) Earth	$^1/_2$	x	3.58	x	1.42	x	110	x	1.4	=		391
(4) Vert. Comp.							975	x	1.7	=		1,658
(5)* Soil Reaction			3.58 x 97 + $^1/_2$ x 3.58(1,430-97)					x	1.7	=		− 4,647
									Shear	=		3,915 lbs/ft

* When calculating the ultimate shear, V_u, it is a common conservative approach to neglect the upward soil reaction. Therefore, the soil reaction will be added to the shear above to determine V_u.

13-4.C.3 Design of Toe

13-4.C.3(a) Moment Design of Toe

Steel design for moment at BB (Figure 13-19)

$M_u = 3234$ ft lbs/ft (from Table 13-4)

By inspection compared to heel design, use minimum steel.

Use #7 at 15″ o.c.; $A_s = 0.48$ sq. in./ft

Temperature Steel parallel to wall

Minimum Steel, $A_s = 0.0018bt$ [UBC Sec. 2607(m)2.A(ii)]

$A_s = 0.0018 \times 12 \times 16$

$\quad = 0.35$ sq. in./ft $\times 6$ ft $= 2.07$ sq. in.

Use 5 - #6 bars at 16″ with 4″ cover at end $A_s = 2.20$ sq. in.

13-4.C.3(b) Shear Design of Toe

The critical section for shear is taken at d distance from the face of the wall (Figure 13-21).

No shear requirement for toe design.

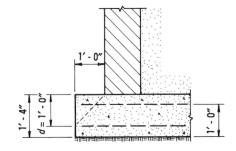

Figure 13-21 Location of critical shear section for toe.

13-4.C.4 Design of Shear Key

13-4.C.4(a) Lateral Force and Resisting Force

Sliding force at the base of the footing =

$\quad = ^1/_2 \, k_h \times h^2$

$\quad = ^1/_2(40)(1.42 + 10 + 1.33)^2 = 3252$ lbs/ft

Frictional resistance = μN

Frictional resistance coefficient $\mu = 0.35$ (see Table 3-20 or Table 5-i).

Only dead load can be used to determine the frictional sliding resistance. (Values from Table 13-1)

$N \quad = 8260 - 975 = 7285$ lbs/ft

$F.R. = 0.35 \times 7285$

$\qquad = 2550$ lbs/ft < 3252 lbs/ft N.G.

Therefore use a shear key and passive soil pressure to resist the remaining shear force.

TABLE 13-4 Moment at BB, Quarter Point Under Wall (Strength Design Method)

Item								Load Factor		Wt. (lbs/ft)		Arm (ft)		Moment (ft lbs/ft)
(9) Footing	1.33	x	1.38	x	150	x	1.4	=		385	x	0.69	=	266
(11) − Soil reaction			− 1.38	x	1,825	x	1.7	=		− 4,281	x	− 0.69	=	− 2,954
(12) − Soil reaction	− $^1/_2$	x	1.38	x	506	x	1.7	=		− 594	x	0.92	=	− 546
												M_n =		− 3,234

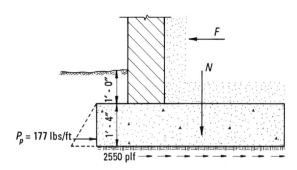

Figure 13-22 Lateral resistance on footing.

Neglect the 1 ft 0 in. cover above the footing toe

Passive lateral soil resistance = 200 psf/ft of depth and may be increased 200 psf for each additional foot of depth to a maximum of 15 times the designated value.

The average soil pressure under the footing is

$\frac{1}{2}(2331 + 97) = 1214$ psf   (UBC Table No. 29-B, Footnote 4)

This would be equivalent to a depth of soil

$$\text{Depth} = \frac{1215}{110} = 11.0 \text{ ft}$$

Lateral passive soil resistance at end of toe

$$P_p = 200 \times 11.0 = 2200 \text{ psf}$$
$$P_p = \frac{1}{2} \times wd^2$$
$$= \frac{1}{2}(200)\, 1.33^2$$
$$= 177 \text{ lb/ft}$$

It would be conservative to ignore passive resistance of soil at the end of the toe because of possible soil erosion.

Required additional passive soil resistance =

= Shear Frictional Resistance

= 3252 – 2550 = 702 lbs/ft

but this does not include a factor of safety. Therefore, for a factor of safety against sliding of 1.5, the required total passive soil resistance is:

1.5(3252) = 4878

Frictional Resistance	= 2550
Provide shear key for	2328 lbs/ft

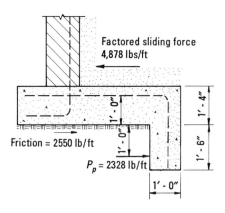

Figure 13-23 Shear key detail

13-4.C.4(b) Shear Key Design

Lateral resistance of shear key; h = depth of shear key

$$\frac{1}{2} \times 2200 \times h^2 = 2328 \text{ lbs/ft}$$

$$h = \sqrt{2.12 \text{ ft}} = 1.5 \text{ ft} = 18 \text{ in.}$$

$$\frac{V}{bd} = \frac{2328}{12 \times 9} = 21.6 \text{ psi}$$

Note: Shear stress on key is approximately:

Very small — therefore shear capacity is adequate.

Moment on key

$$M = \frac{2}{3} \times 12 \times 2328$$
$$= 18,624 \text{ in. lbs}$$
$$f = \frac{6M}{bt^2} = \frac{6 \times 18,624}{12 \times 12^2}$$
$$f_c = f_t = 65 \text{ psi, compression or tension}$$

UBC Sec. 2609(f)2.c

$$f_r = 7.5\sqrt{f_c'}$$
$$= 7.5\sqrt{2500}$$
$$= 375 \text{ psi} > 65 \text{ psi}$$

No reinforcing required — Provide minimum steel for footing.

13-5 SUPPORTED RETAINING WALL DESIGN

13-5.A Design Example (Basement Wall)

A 12-inch thick residential basement wall is to be constructed of 1500 psi masonry to support the loads shown in Figure 13-24.

Determine the size and spacing of grade 60 reinforcement required if special inspection is not provided. Additionally, the contractor has requested backfilling to Elevation 713′ prior to placing the first floor diaphragm in order to facilitate nearby construction.

Assume the wall stability has already been checked against overturning, sliding, etc.

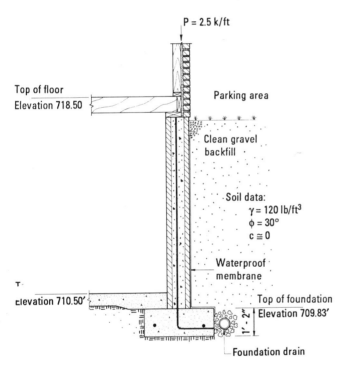

Figure 13-24 Residential basement wall.

13-5.B Full Soil Load on Completed Structure

Conservatively neglect any moment restraint at the foundation and assume the wall is simply supported at the first floor and at the basement slab with a span length of 8′-0″.

There is some question as to whether the wall should be designed for active soil pressures or for the larger at-rest lateral earth pressures since the first floor diaphragm prevents free translation of the top of the wall.

For active pressures to effectively develop, the top of the wall must translate approximately $0.002H$ in cohesionless soils (Bowles, 1977). Therefore, the top of an 8′-0″ wall should translate approximately $0.002 \times (8 \times 12 \text{ in/ft}) = 0.19″$, or about $3/16″$.

For this example, it is assumed that this small amount of movement can occur through a combination of foundation rotation, diaphragm deflection, and wall deflection. In actual design however, these assumptions should be examined carefully. Always consult a geotechnical engineer to determine design soil pressures, bearing capacities and expected settlement data.

For active (Rankine) earth pressures:

$K_a = \tan^2(45 - \phi/2)$ (Das, 1984).

Therefore, for $\phi = 30$ degrees:

$K_a = \tan^2(45 - 30/2) = 0.333.$

Based on this value of K_a, an equivalent fluid pressure can be approximated as:

$K_a \times \gamma_s = 0.333 \times 120 = 40 \text{ pcf}$

To account for the parking lot, it is customary to approximate the surcharge as 2 feet of soil. Therefore, the parking lot surcharge, q, can be taken as:

$0.333 \times (2′ \times 120 \text{ pcf}) = 80 \text{ psf}$

Figure 13-25 shows the wall with these design loads.

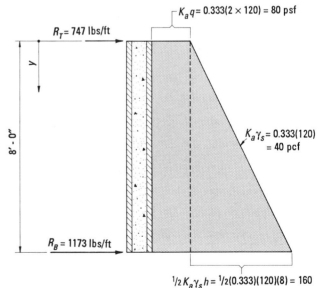

Figure 13-25 Applied forces on basement wall.

Calculate the reactions at the top and the base by using moment and force equilibrium equations.

Taking the sum of the moments about the base of the wall, find the reaction at the top of the wall.

$$8 \times R_t = \tfrac{1}{2}(80)(8^2) + \tfrac{1}{6}(40)(8^3)$$

$$8R_t = 5973$$

$$R_t = 747 \text{ lbs/ft}$$

To find the reaction at the base of the wall, sum lateral forces.

$$R_b + 747 = 80(8) + \tfrac{1}{2}(40)(8^2)$$

$$R_b = 1173 \text{ lbs/ft}$$

Find the maximum wall moment by first finding where the shear is zero. Take a free body diagram of the top of the wall to a distance y. Sum the lateral forces and solve for y using the quadratic equation:

$$y = \frac{-b \pm \sqrt{b^2 - 4ac}}{2a}$$

Thus:

$$747 = \tfrac{1}{2}(40)\left(y^2\right) + 80y$$

$$0 = 20y^2 + 80y - 747$$

$$y = \frac{-80 \pm \sqrt{80^2 - 4(20)(-747)}}{2(20)}$$

$$y = 4.43' \text{ or } -8.43'$$

(4.43′ is the only practical solution).

Since the maximum moment occurs where the shear is zero:

$$M_{\text{max.}} = 747(4.43) - 80\left(\frac{4.43^2}{2}\right) - \tfrac{1}{6}(40)\left(4.43^3\right)$$

$$= 1945 \text{ ft lb per ft length of wall}$$

Determine if the extreme face of the masonry is in tension by finding if the bending stress exceeds the axial compressive stress. If the wall is uncracked and assumed to be solidly grouted:

$$f_b = \frac{M}{S} = \frac{6M}{bt^2} = \frac{6(1945 \times 12 \text{ in./ft})}{12 \times 11.63^2}$$

$$= 86 \text{ psi}$$

Likewise, if the wall is assumed to be solidly grouted, the axial compressive stress at the point of maximum moment will be no more than:

$$f_{a(\text{max})} = \frac{P}{A} = \frac{2500 + 133(4.43)}{12 \times 11.63}$$

$$= 22.1 \text{ psi}$$

where 133 psf = weight of a solid grouted 12″ normal weight concrete masonry wall from Table B-3a and 2500 lb/ft is the vertical load per foot of wall as shown in Figure 13-24.

Since $f_b > f_{a(max.)}$, tension will exist on the face of the wall and the masonry will most likely crack. Reinforce the wall for tension stresses.

To calculate the required reinforcing, assume a more practical spacing of grouted cores and reinforcing steel of 32 in. From Table B-3a, the estimated wall weight is then 89 psf and the equivalent solid thickness (EST) is 7.0 in.

$$f_{a(\text{max})} = \frac{P_{\text{max}}}{A} = \frac{P_{\text{max}}}{EST \times L}, \text{ where } L = 12 \text{ in.}$$

$$= \frac{2500 + 89(8 \text{ ft})}{7.0 \times 12} = 38.2 \text{ psi (small)}$$

and at the point of maximum moment:

$$f_a = \frac{2500 + 89(4.43)}{7.0 \times 12} = 34.5 \text{ psi}$$

The allowable compressive stress can be found from Tables Q-1 and Q-2. From Table Q-1, h'/t is about 8.3 and R is about 0.992, for $t = 11.63$ in. and $h' = 8$ ft. Therefore from Table Q-2, F_a is 149 psi for non-inspected with $f'_m = 1500$ psi. Since F_a exceeds $f_{a(max)}$, the base of the wall is adequate for axial compression.

Check combined compressive stresses at the point of maximum moment by limiting the maximum value of f_b with the unity equation.

Note the allowable maximum bending stress in compression must be reduced by one half since inspection will not be provided. Therefore

$$F_b = \tfrac{1}{2} \times \tfrac{1}{3} \times f'_m = 250 \text{ psi.}$$

$$\frac{f_a}{F_a} + \frac{f_b}{F_b} \leq 1, \text{ or}$$

$$f_{b(\text{max})} = \left(1 - \frac{f_a}{F_a}\right)F_b = \left(1 - \frac{35}{149}\right)250$$

$$f_{b(\text{max})} = 191 \text{ psi}$$

Calculate the flexural coefficient, K. Estimate the effective depth, d, as $11.63 - 1.5$ in. face shell thickness $- 1$ in. clear = 9.13 in. Use 9 in.

$$K = \frac{M}{bd^2} = \frac{1945 \times 12 \text{ in./ft}}{12 \times 9^2} = 24$$

Using Diagram F-1 with $K = 24$ and $f_b = 191$ psi, the required reinforcing ratio, p, can be found as approximately 0.0020. Therefore the required area of reinforcing steel is:

$$A_s = pbd = 0.0020 \times 12 \times 9 = 0.22 \text{ sq. in./ft}$$

From Table C-4b, use #7 bars @ 32″ ($A_s = 0.24$ sq. in./ft)

13-5.C Cantilevered Wall with Backfill

Check the freestanding wall without the top diaphragm for backfill to El. 713 ft. Conservatively include a surcharge in the calculations since machinery will probably be driven near the wall.

Design the wall as a cantilever as shown in Figure 15-26 and find the additional reinforcing required in the outside face of the wall. Neglect vertical loads since Sec. 13-5.B of this problem showed them to be so small.

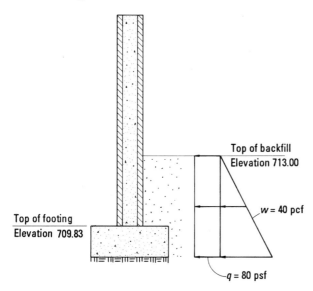

Figure 13-26 Cantilevered basement wall with partial backfill loading.

For a uniform surcharge load of 80 psf and an equivalent fluid pressure of 40 pcf:

$$M = \tfrac{1}{2}(80)\left(4^2\right) + \tfrac{1}{6}(40)\left(4^3\right)$$

$$M = 640 + 426 = 1067 \text{ ft lbs / ft}$$

$$K = \frac{M}{bd^2} = \frac{1067 \times 12 \text{ in./ft}}{12 \times 9^2} = 13.2$$

Use Diagram F-1 or Table E-1a to find $p = 0.0006$. Therefore, $A_s = 0.0006 \times 12 \times 9 = 0.065$ sq. ft/ft. From Table C-4b choose #4 @ 32″ on the outside face of the wall.

Check minimum reinforcing steel requirements of UBC Sec. 2407(h)4.B. Use:

$$\text{Vertical Steel} \geq 0.0013bt = 0.0013(12)(11.63)$$
$$= 0.18 \text{ sq. in./ft}$$

$$\text{Horizontal Steel} \geq 0.0007bt = 0.0007(12)(11.63)$$
$$= 0.098 \text{ sq. in./ft}$$

The vertical steel of #7 @ 32″ on the inside face and #4 @ 32″ on the inside face and #4 @ 32″ on the outside face provide:

$$A_s = (0.60 + 0.20)\left(\frac{12}{32}\right) = 0.30 \text{ sq. in./ft} > 0.18 \quad \text{O.K.}$$

For minimum horizontal reinforcing, refer to Table C-5 and select #6 horizontal bars @ 48″ o.c. See Figure 13-27 for the final reinforcing layout.

Note this design assumes the wall is stable against overturning and sliding as given in the problem statement. Likewise the design assumes the concrete foundation, the first floor diaphragm and all connections have also be designed properly.

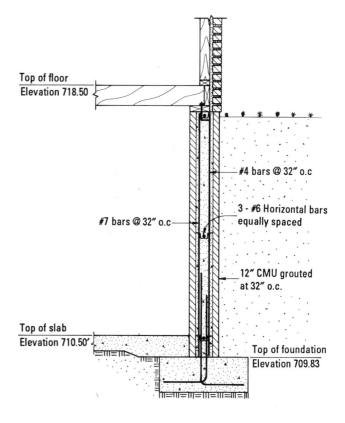

Figure 13-27 Basement wall reinforcing layout.

13-6 QUESTIONS AND PROBLEMS

13-1 Design a brick property line retaining wall and footing without a toe. Assume h = 6 ft, f'_m = 2500 psi (use half stresses), F_s = 24,000 psi, lateral earth pressure, E.F.P. = 30 pcf, f'_c = 2500 psi for footing, allowable soil bearing = 3000 psf, and weight of earth = 125 psf.

13-2 Design an 8 ft high supported concrete block wall for a subterranean garage to support an equivalent fluid pressure of 45 pounds per square foot. The wall is shown below and will be constructed with f'_m = 2000 psi, no special inspection, F_s = 24,000 psi, and an allowable soil bearing = 2500 psf.

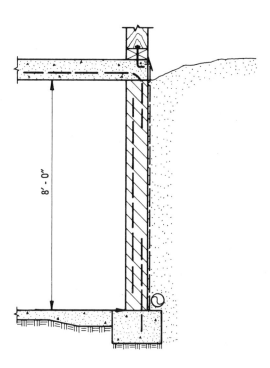

13-3 Design a 5 ft high retaining wall using (a) concrete block masonry solid grouted with no special inspection, f'_m = 1500 psi and (b) solid grouted brick masonry, f'_m = 2000 psi (special inspection provided). For both parts (a) and (b) assume f'_c = 2500 psi, F_s = 24,000 psi and allowable soil bearing = 2000 psf. Backfill is level, E.F.P. = 40 psf with no surcharge. Footing extends under backfill.

13-4 Design a 16 feet high buttress retaining wall with the buttresses located 12 feet on centers. The backfill against the wall is on a slope of 2 to 1. Design the wall using grouted brick, f'_m = 2500 psi, special inspection provided, F_s = 24,000 psi, and f'_c = 2500 psi. Weight of soil is 110 pounds per cubic foot and the allowable soil bearing is 4000 psf.

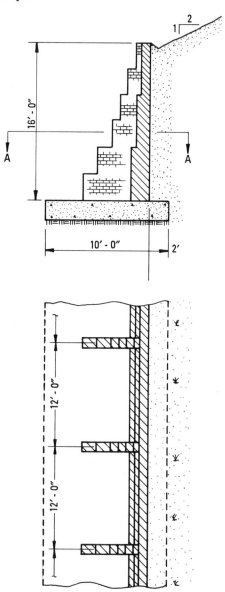

Section A-A

Explanation of Tables and Diagrams

14-1 WORKING STRESS DESIGN TABLES AND DIAGRAMS

The tables listed in this section are primarily based on the Uniform Building Code requirements. However, many of these tables can be used for both designs conforming to the UBC and the ACI/ASCE requirements.

Unless noted otherwise, all design tables are based on grade 60 steel reinforcement with $F_s = 24,000$ psi. Where applicable, stress and load values may be increased by one-third when considering wind or seismic forces per UBC Section 2303(d).

TABLE A-1a Specified Compressive Strength of Clay Masonry Assemblages, f'_m

This table provides a value for the specified compressive strength of a clay masonry assemblage, f'_m, based on the strength of the clay units and the type of mortar. It is based on UBC Table No. 24-C with appropriate footnotes.

TABLE A-1b Clay Masonry f'_m, E_m, n and G Values Based on the Clay Masonry Unit Strength and the Mortar Type

Values for the modulus of elasticity, E_m, the modular ratio, n, and the shear modulus, G, are given based on the specified strength of the clay masonry, f'_m, which is found from the clay masonry unit strength and the mortar type.

TABLE A-2a Specified Compressive Strength of Concrete Masonry Assemblages, f'_m

Values for the specified strength, f'_m, of concrete masonry systems are listed based on the net area strength of the units and the type of mortar.

TABLE A-2b Concrete Masonry f'_m, E_m, n and G Values Based on the Concrete Masonry Unit Strength and the Mortar Type

Based on the specified strength of the concrete masonry system, f'_m, which is found from the concrete masonry unit strength and the mortar type, the values for the modulus of elasticity, E_m, the modular ratio, n, and the shear modulus, G, are given.

TABLE A-3 Maximum Allowable Working Stresses for Reinforced Solid and Hollow Unit Masonry

This table lists the allowable working stresses of masonry systems for f'_m values between 1500 to 5000 psi. It also includes half stresses for f'_m values from 1500 to 3000 psi. The footnotes provide clarifications, explanations, and UBC references.

TABLE A-4 Allowable Steel Working Stresses

This table gives the allowable steel stresses in tension and compression of various grades of steel.

DIAGRAMS AND TABLES A-5 and A-6 Allowable Shear Wall Stresses

These diagrams plot the maximum allowable shear stress values for a shear wall assuming, in Diagram A-5, that the masonry alone resists the shear force and, for Diagram A-6, that the reinforcing steel alone resists the shear force.

Values for half stresses (no special inspection) are also included.

The tables list the allowable shear values based on strength of masonry, f'_m, and M/Vd.

As the tables and diagrams show, the smaller the M/Vd (or h/l) value, the higher the allowable shear stress, up to the allowable maximum stress. Therefore, it is reasonable and advantageous to design shear walls for their story-to-story height since higher allowable shear stresses may be used.

TABLES A-7a, b and c Allowable Tension, B_t, for Embedded Anchor Bolts in Clay and Concrete Masonry

The allowable pull out or tension values of anchor bolts is based on the length of embedment or the edge distance, the strength of masonry, f'_m, and the diameter of the anchor bolt. The lesser value from Table A-7a (based on the masonry capacity) or Table A-7b (based on the bolt capacity) must be used for design. These values may be increased one-third for wind or earthquake forces as allowed in UBC Sec. 2303(d) (except as noted in UBC Sec. 2337(b)9E for some diaphragm connections on buildings in Seismic Zone Nos. 3 and 4).

Table A-7c shows the percentage capacity or percentage reduction based on bolt spacing as per UBC Sec. 2406(h)2 and ACI/ASCE Sec. 5.14.2.1.

TABLES A-8a, b and c   Allowable Shear, B_v, for Embedded Anchors Bolts in Clay and Concrete Masonry

Table A-8a gives the shear capacity of embedded anchors and is useful in determining the number and size of bolts required to secure ledger beams to walls. It may also be used to determine the additional shear capacity provided by dowels extending from the foundation into a masonry wall.

The allowable shear values are based on the strength of masonry, f'_m, and the anchor bolt diameter. For wind or seismic forces, table values may be increased one-third per UBC Sec. 2303(d) (except as noted in UBC Sec. 2337(b)9E for some diaphragm connections on buildings in Seismic Zone Nos. 3 and 4).

Table A-8b provides the percentage shear capacity of anchor bolts when the bolt spacing is less than $12d_b$ per the reductions found in UBC Sec. 2406(h)3. Similarly, Table A-8c shows the percentage shear capacity of anchor bolts when the edge distance is less than $8d_b$.

TABLES A-9 and A-10   Allowable Stresses for Unreinforced Masonry

These tables are used for rational design of unreinforced masonry.

In the table for allowable tensile stresses (Table A-10), PCL represents mortars consisting of portland cement-lime and sand mixed in accordance with UBC Standard No. 24-20 (ASTM C 270). MC represents masonry cement conforming to UBC Standard No. 24-16 (ASTM C 91). Mortar cement is a new classification of mortar in the UBC and is governed by UBC Standard No. 24-19. Mortar cement is essentially masonry cement in composition. However, UBC Standard No. 24-19 requires flexural bond strength testing for Types M, S, and N mortar cements used in high seismic zone regions. Presently, PCL and mortar cements may be used in all seismic zones, while masonry cement is limited in use to only buildings in Seismic Zones Nos. 0 and 1.

Rational designs permit flexural tensile stresses in unreinforced masonry. The designer can choose the type of mortar and unit needed to satisfy design conditions. As the table shows, PCL mortars and mortar cements have higher allowable stresses than masonry cement mortars. If the calculated flexural tensile strength exceeds the allowable stresses in Table A-10, the UBC requires that the masonry element be reinforced to resist flexural tension.

TABLE B-1   Weights of Building Materials

Self explanatory.

TABLES B-2 and B-3a and b   Average Weights of Concrete Masonry Units and Weight of Completed Masonry Walls

The weights of lightweight, medium weight and normal weight masonry units are given based on the unit thickness and the percent of cells grouted solid.

TABLE B-4   Average Weights of Reinforced Grouted Brick Walls

Self explanatory.

TABLES B-5a and B-5b   Approximate Grout Quantities Required in Block and Brick Walls

Table B-5a gives the amount of grout required in 100 square feet of wall or 100 concrete units. It also shows how many blocks will be filled with one cubic yard of grout.

Table B-5b gives the amount of grout required per 100 square feet of wall and the square feet of brick wall per cubic yard of grout. It assumes the brick walls are solid grouted and are based on the width of the grout space.

TABLE B-6   Approximate Measurements of Masonry Materials

Standard measures for mortar materials such as sack, ton and cubic yard are shown providing an easy understanding of the industry's volume or weight measures. Because the table is in common terms, used by the industry, it can be used to determine the amount of materials required for a mortar batch.

TABLES C-1a and b   Properties of Standard Steel Reinforcing Bars

The nominal weights, dimensions, and determination requirements are given for the various steel bar sizes.

TABLE C-2   Properties of Steel Reinforcing Wire

This table lists the dimensions, weights and allowable tensile capacities, typical steel wire sizes based on UBC Standard No. 24-15.

TABLE C-3   Areas and Perimeters of Various Combinations of Bars

For a required area of steel or specified perimeter, combinations of bars can be selected from this table, or, for combinations of bars, areas and perimeters can easily be obtained.

TABLES C-4a and b   Areas and Perimeters of Reinforcing Steel per Foot

For various bar sizes and spacing, the area and perimeter can be obtained on a per foot length of section basis.

TABLE C-5   Maximum Spacing in Inches of Minimum Reinforcing Steel, $A_s = 0.0007bt$

The maximum spacing for a particular bar may be obtained based on minimum code requirements. The table limits the maximum spacing to 48 inches per UBC Sec. 2407(h)4.B.

TABLE C-6   Maximum Spacing in Inches of Reinforcing Steel, $A_s = 0.0013bt$

Maximum spacing may be obtained for selected bar sizes and is based on minimum code requirements. Maximum spacing is limited to six times the wall thickness or a maximum of 48 inches.

TABLE C-7   Maximum Spacing in Inches of Reinforcing Steel, $A_s = 0.001bt$

Minimum reinforcing may be evenly divided horizontally and vertically, $0.001bt$ each way. Maximum spacing is limited to six times the wall thickness or a maximum of 48 inches.

TABLES C-8a to m   Steel Ratio, $p = A_s/bd$

For given p and d values, the bar size and spacing can be immediately obtained, or alternately, for a given bar size, spacing and specified distance d, the steel ratio, p, can be readily found.

TABLE C-9   Ratio of Steel Area to Gross Cross-Sectional Area

Steel ratios are shown based on various steel sizes, spacing and wall thicknesses. The minimum ratio given is just below the allowable minimum $0.0007bt$ for reinforced masonry walls.

TABLE C-10   Maximum Area of Steel per CMU Cell

UBC Sec. 2409(e) limits the area of steel per cell to 6% of the cell area. This table lists the maximum areas of steel for 4″, 6″, 8″, 10″ and 12″ units based on this limit. Values based on a recommended 5% of the cell area are also given.

TABLE C-11   Maximum Number of Reinforcing Bars per Cell

Based on the maximum area of steel per cell as given in Table C-10, the number of bars per cell is listed. Values are based on 5% and 6% of cell area.

TABLES E-1 to E-8   Flexural Design Coefficients

These tables give the flexural coefficients K, p, np, k, j, $2/jk$, based on the straight line working stress theory for various values of f'_m and f_s. The required steel ratio, p, can also be determined from these tables based on the value $K = M/bd^2$.

The tables are limited to the maximum allowable masonry stress and/or the maximum allowable steel stress. At the condition of maximum allowable masonry stress together with maximum allowable steel stress, balanced design is achieved, and the maximum efficiency of the materials is obtained.

Tables are given for specified masonry strengths, f'_m, from 1500 psi to 5000 psi for full allowable stresses. Tables based on f'_m of 1500 psi to 4000 psi for half allowable stresses are also given.

TABLES E-9a and b   Flexural Coefficients Based on np Values

This table lists values for np, $2/jk$, j, k and npj. It is used for analysis as follows:

A steel size and steel spacing is assumed for a member and the steel ratio, p, is computed as: $p = A_s/bd$. The modular ratio, n, is established based on the specified materials. With the np value, values of $2/jk$, j and k, may be obtained from the table. The masonry stress, $f_b = \dfrac{M}{bd^2} \times \dfrac{2}{jk}$ and the steel stress, $f_s = \dfrac{M}{A_s jd}$ or $\dfrac{M}{bd^2} \times \dfrac{1}{pj}$ may then be computed and compared to the allowable stress values.

For design, $2/jk$ and npj are calculated from the applied moment, b, d and the allowable stresses as follows:

$$\frac{2}{jk} = \frac{f_b bd^2}{M}; \quad npj = \frac{nM}{F_s bd^2}$$

With these values, the table may be entered to determine the value of np. The steel ratio, p, is then calculated by $p = np/n$, and the area of steel by the equation $A_s = pbd$.

DIAGRAMS F-1 to F-8 K vs p for Various Masonry and Steel Stresses

These diagrams are a series of curves for various values of masonry and steel stresses based upon the coefficient, K, and the steel ratio, p. The modular ratio, n, determines which diagram to use.

By calculating the value $K = M/bd^2$, the diagram can be entered horizontally to the allowable steel stress or masonry stress. Whichever stress value is intersected farthest to the right controls. Proceeding vertically downward from the controlling stress value yields the steel ratio, p.

If the steel ratio, p, is known, and K is calculated as M/bd^2, the steel and masonry stress levels can be immediately determined at the intersection of the K and p lines.

A wide range of stresses for masonry and steel are given in these charts. The allowable stress curves shown in the diagrams as extra heavy lines.

DIAGRAM F-9   *K* vs *np* for masonry stresses

With this diagram, the masonry stress may be determined using only K and np values.

Alternately, for given f_b and np values, the flexural coefficient, K, may be determined from which the moment capacity can be calculated.

TABLES H-1 to H-6   Moment Capacity of Walls and Beams for Balanced Design Conditions

These tables give the moment capacity for 12 inch wide walls or beams based upon balanced design conditions with steel and masonry stresses at maximum allowable values.

For given values of d and f_m', the moment capacity in foot kips is given along with the required area of steel. Moment capacities may be increased one third for wind or earthquake conditions.

TABLES J-1 to J-3   Moment Capacity of Walls with Minimum Areas of Steel

These tables give the moment capacity for sections reinforced with $A_s = 0.0007bt$, $A_s = 0.0013bt$ and $A_s = 0.001bt$ sq. in. of steel. Moment capacities are shown for eight values of allowable masonry stresses. For a given wall thickness, the area of reinforcing steel based on the table's reinforcement ratio, p, is also given.

DIAGRAMS AND TABLES K-1 to K-4   Spacing of Shear Reinforcing Steel

Plots of the shear capacity of reinforcing bars versus the spacing of the shear steel are given in these diagrams for 6″, 8″, 10″ and 12″ sections.

For analysis enter the appropriate wall thickness diagram with the size and the spacing of shear reinforcing steel to determine the allowable unit shear stress, v.

For design, locate the calculated unit shear stress, v, and select the size and spacing of reinforcing that are adequate. These tables list the shear capacity values from the K diagrams based on the size and the spacing of the reinforcing steel.

TABLES N-1 to N-8   Coefficients *p* and *p'* for Tension and Compression Steel in Flexural Members

These design tables are based upon the allowable stresses in the masonry and the tensile steel under balanced design conditions. For a given moment, b, d and d', the flexural coefficient $K = M/bd^2$ may be determined. With K and the d'/d ratio, the steel ratio p and p' can be selected from the table. This immediately gives the steel ratios required for the moment imposed upon the section when it is in excess of the balanced capacity of the section.

Eight tables are provided for various values of f_b. The compression steel ratio, p', is based on $(2n - 1)$ to take into account the creep and plastic flow of the masonry that, with time, will increase the load on the steel.

DIAGRAMS P-1 to P-8   Steel Ratios *p* and *p'* versus *K*

Plots of the N tables are shown in these diagrams in which p and p' can be determined in the appropriate stress diagram from K and d'/d values.

TABLE Q-1   Stress Reduction Coefficients, *R*, Based on the *h'/t* Ratio

This table gives the stress reduction factor, R, equal to $\left[1 - (h'/42t)^3\right]$, based on the wall's effective height to the thickness ratio, h'/t.

TABLE Q-2   Allowable Wall Axial Compressive Stresses

The maximum allowable axial compressive stress, F_a, can be found using this table with the h'/t ratio and the specified masonry strength, f_m'.

TABLES S-1 to S-3   Tied Masonry Column Capacity

These tables provide the masonry compressive strength capacity for columns from $8″ \times 8″$ to $32″ \times 60″$. The column capacity is computed according to the actual dimension of the column which is reduced by the thickness of the head joints. For columns less than 12 inches in either dimension, half loads are given according to UBC Sec. 2407(h)4E(ii). Values for eight masonry strengths for special inspection are listed. For non-inspected masonry, the table P_m values must be reduced by one half. Also given are the capacities of steel reinforcement for the minimum and maximum reinforcement ratios of 1/2 and 4%.

TABLE S-4   Capacity of Reinforcing Steel in Tied Masonry Columns, Load on Bars

For a given load on the reinforcing steel in a tied masonry column, the bar size and number of bars can be selected. Alternately, for a given size and number of bars, the load capacity is given.

Table S-5   Maximum Spacing of Column Ties

This table shows the maximum spacing of ties permitted by UBC Sec. 2409(B)5B based on the tie diameter, the longitudinal bar diameter and a maximum spacing of 18″. The spacing determined from this table may not exceed the least column dimension.

TABLES T-1a to g   Coefficients for Deflection and Rigidity Walls and Piers for the Distribution of Horizontal Forces

These tables can be used to determine the shear in piers and walls subjected to a lateral load, parallel to the length of the wall. These coefficients are based on a modulus of elasticity, E_m, of 1,000,000 psi, a modulus of rigidity, G (or E_v) of $0.4 E_m = 400,000$ psi, a 1 inch wall thickness and a lateral force of 100,000 pounds.

To obtain deflection or rigidity coefficients for the masonry's actual modulus of elasticity, multiply the values in the tables by the ratio $E_m / 1,000,000$.

If the wall is composed of elements with various thicknesses, divide the deflection coefficients by the actual thicknesses and multiply the coefficients of rigidity by the thickness of each element of the wall.

TABLES U-1 to U-14   Conversion Tables From English Measurements to SI (or Metric) Measurements

These tables provide both the coefficients to convert from the English system to the SI (Systems International), system, and the actual converted values for common measurements used in structural design.

14-2   STRENGTH DESIGN TABLES AND DIAGRAMS

STR. DES. TABLES A-1 to A-8   Coefficients for Flexural Strength Design

These tables give the flexural coefficients for strength design theory, which assumes a rectangular stress block. Values for K_u, q, a_u, p, c/d, a/d and j are given for masonry strengths from 1500 to 5000 psi.

STR. DES. TABLE B-1   Moment Coefficients on Rectangular Sections

Based on the value $M_u / f'_m bd^2$, the value q can be found on the left side of the table to hundredths, and on the top of the table for added thousandths.

With q, the steel ratio, p, can be calculated from the formula $p = q f'_m / f_y$.

STR. DES. TABLES C-1 to C-8   Moment Capacity of Rectangular Sections

These tables provide the moment capacity of a section based on the tension steel ratio, p, depth to the steel, d, and strength of masonry, f'_m. They are predicated on steel yield stress of $F_y = 60,000$ psi and a section width of 12 inches.

The moment capacities given are in foot kips.

STR. DES. TABLE D-1   Modulus of Rupture, f_r, of Masonry

Modulus of rupture values are listed for various types and strengths of masonry based on the equations found in UBC Sec. 2411(b)4.

STR. DES. TABLE D-2   Strength Reduction Factor, ϕ, for Axial Loads or Combined Axial and Flexural Loads

This table gives ϕ factors to be used in the design of masonry shear walls subjected to either axial loads or axial and flexural loads based on UBC Sec. 2412(c)2A.

STR. DES. DIAGRAM AND TABLE E-1   Maximum Nominal Shear Stress provided by the Masonry, v_m.

STR. DES. Table E-1 provides C_d values for shear walls with various M/Vd ratios per UBC Table No. 24-O. Maximum nominal shear stress values provided by the masonry, v_m, are shown in both the table and the diagram based on C_d, M/Vd, and f'_m values.

STR. DES. DIAGRAM AND TABLE E-2   Maximum Nominal Shear Stress Values, v_n.

The coefficient $V_n / A_e \sqrt{f'_m}$ is shown in Table E-2 for shear walls with various M/Vd values per UBC Table No. 24-N. Maximum nominal shear stress values, v_n, are shown in the table and diagram based on the coefficient and the M/Vd and f'_m values.

14-3   ACI/ASCE TABLES

A/A TABLE A-1   Clay Masonry, f'_m, E_m, n and E_v Values Based on the Clay Masonry Unit Strength and the Mortar Type

Based on the ACI/ASCE Standard Table 5.5.1.2 and ACI/ASCE Specification Table 1.6.2.1, the specified strength of clay masonry, f'_m, the modulus of elasticity, E_m, the modular ratio, n, and the shear modulus, E_v, are given.

A/A TABLE A-2 Correction Factors for Clay Masonry Prism Strength

Clay masonry strengths are based on a standard prism with a height to thickness ratio of 5. The strengths of prisms constructed with other h/t ratios must be normalized to relate to the standard size with the values given in this table.

A/A TABLES B-1 to B-6 Flexural Design Coefficients for Clay Masonry

These tables provide values for design or analysis and are similar to the WSD E tables which are based on UBC values.

The type of mortar, N, S or M, changes the modulus of elasticity and the modular ratio, n. Note some of these tables have the same f'_m values but differ for the type of mortar and thus the modular ratio is different.

A/A TABLE C-1 Concrete Masonry f'_m, E_m, n and E_v Values Based on the Concrete Masonry Unit Strength and the Mortar Type

Based on the ACI/ASCE Standard Table 5.5.1.3 and ACI/ASCE Specification Table 1.6.2.2, the specified strength of concrete masonry, f'_m, the modulus of elasticity, E_m, the modular ratio, n, and the shear modulus, E_v, are given.

A/A TABLE C-2 Correction Factors for Concrete Masonry Prism Strength

Concrete masonry correction factors for the h/t ratio of the specimens are the same as the correction factors as given in the UBC.

A/A TABLES D-1 to D-4 Flexural Design Coefficients for Concrete Masonry

Concrete masonry working stress design flexural coefficients are given based on the type of mortar, the specified strength of masonry, f'_m, and the modular ratio n.

These tables provide values for design or analysis and are similar to the WSD E tables which are based on UBC values.

A/A TABLES E-1, 2 and 3 Radius of Gyration, r

These tables provide the radius of gyration, r, of hollow concrete masonry walls and hollow clay masonry walls based on the thickness of the units and the spacing of the grouted cells.

A/A TABLE E-4 Maximum Allowable Axial Stress on Walls and Columns, F_a.

This table shows the maximum allowable axial stress, F_a, based upon the h/r ratio and the f'_m value. Values are determined from ACI/ASCE Equations 7-1 and 7-2.

WORKING
STRESS
DESIGN
TABLES
and
DIAGRAMS

**Based on the
Uniform Building
Code Requirements**

Use judgment when using tables
to the 4th decimal when the initial
data is based on an estimate.

Don't be so precise that you forget
to be accurate.

JAMES E. AMRHEIN
Civil & Structural Engineer

TABLE A-1a   Specified Compressive Strength of Clay Masonry Assemblages, f'_m (psi)[2,4]

Compressive Strength of Clay Masonry Units[1] (psi)	Specified Compressive Strength of Masonry, f'_m	
	Type M or S Mortar[3] (psi)	Type N Mortar[3] (psi)
14,000 or more	5,300	4,400
12,000	4,700	3,800
10,000	4,000	3,300
8,000	3,350	2,700
6,000	2,700	2,200
4,000	2,000	1,600

1. Compressive strength of solid masonry units is based on the gross area. Compressive strength of hollow clay masonry units is based on minimum net area. Values may be interpolated. When hollow clay masonry units are grouted, the grout shall conform to the proportions of UBC Table 24-B (See Table 1-19).

2. Assumed assemblage. The specified compressive strength of masonry, f'_m is based on gross area strength when using solid units or solid grouted masonry and net area strength when using ungrouted hollow units.

3. Mortar for unit masonry, proportion specification, as specified in UBC Table No. 24-A. These values apply to portland cement-lime mortars without added air-entraining materials.

4. Based on UBC Table No. 24-C.

TABLE A-1b   Clay Masonry f'_m, E_m, n and G Values Based on the Clay Masonry Unit Strength and the Mortar Type

Type N Mortar				
Compressive Strength of Clay Masonry[1] (psi)	Specified Compressive Strength of Clay Masonry Assemblage[2], f'_m (psi)	Modulus of Elasticity[3] $E_m = 750 f'_m$ (psi) E_m (max) = 3,000,000 psi	Modular Ratio $n = E_s/E_m$ Where E_s = 29,000,000 psi[4]	Modulus of Rigidity[5] $G = 0.4 E_m = 300 f'_m$ (psi) G (max) = 1,200,000 (psi)
14,000 or more	4,400	3,000,000	9.7	1,200,000
12,000	3,800	2,850,000	10.2	1,140,000
10,000	3,300	2,475,000	11.7	990,000
8,000	2,700	2,025,000	14.3	810,000
6,000	2,200	1,650,000	17.6	660,000
4,000	1,600	1,200,000	24.2	480,000
Type M or S Mortar				
Compressive Strength of Clay Masonry[1] (psi)	Specified Compressive Strength of Clay Masonry Assemblage[2], f'_m (psi)	Modulus of Elasticity[3] $E_m = 750 f'_m$ (psi) E_m (max) = 3,000,000 psi	Modular Ratio $n = E_s/E_m$ Where E_s = 29,000,000 psi[4]	Modulus of Rigidity[5] $G = 0.4 E_m = 300 f'_m$ (psi) G (max) = 1,200,000 (psi)
14,000 or more	5,300	3,000,000	9.7	1,200,000
12,000	4,700	3,000,000	9.7	1,200,000
10,000	4,000	3,000,000	9.7	1,200,000
8,000	3,350	2,512,500	11.5	1,005,000
6,000	2,700	2,025,000	14.3	810,000
4,000	2,000	1,500,000	19.3	600,000

1. Compressive strength of solid masonry units is based on the gross area. Compressive strength of hollow clay masonry units is based on minimum net area. Values may be interpolated.

2. Based on UBC Table No. 24-C.

3. Based on UBC Chapter 24, Eq. 6-23.

4. Based on UBC Chapter 24, Eq. 6-25.

5. Based on UBC Chapter 24, Eq. 6-26.

TABLE A-2a Specified Compressive Strength of Concrete Masonry Assemblages, f'_m (psi)[2,4]

Compressive Strength of Concrete Masonry Units[1] (psi)	Specified Compressive Strength of Masonry, f'_m	
	Type M or S Mortar[3] (psi)	Type N Mortar[3] (psi)
4,800 or more	3,000	2,800
3,750	2,500	2,350
2,800	2,000	1,850
1,900	1,500	1,350
1,250	1,000	950

1. Compressive strength of solid masonry units is based on the gross area. Compressive strength of hollow concrete masonry units is based on minimum net area. Values may be interpolated.

2. Assumed assemblage. The specified compressive strength of masonry, f'_m is based on gross area strength when using solid units or solid grouted masonry and net area strength when using ungrouted hollow units.

3. Mortar for unit masonry, proportion specification, as specified in UBC Table No. 24-A. These values apply to portland cement-lime mortars without added air-entraining materials.

4. Based on UBC Table No. 24-C.

TABLE A-2b Concrete Masonry f'_m, E_m, n and G Values Based on the Concrete Masonry Unit Strength and the Mortar Type

Type N Mortar				
Compressive Strength of Concrete Masonry[1] (psi)	Specified Compressive Strength of Concrete Masonry Assemblage[2], f'_m(psi)	Modulus of Elasticity[3] $E_m = 750 f'_m$ (psi) E_m(max) = 3,000,000 psi	Modular Ratio $n = E_s/E_m$ Where $E_s =$ 29,000,000 psi[4]	Modulus of Rigidity[5] $G = 0.4 E_m = 300 f'_m$ (psi) G(max) = 1,200,000 (psi)
4,800 or more	2,800	2,100,000	13.8	840,000
3,750	2,350	1,762,500	16.5	705,000
2,800	1,850	1,387,500	20.9	555,000
1,900	1,350	1,012,500	28.6	405,000
1,250	950	712,500	40.7	285,000
Type M or S Mortar				
Compressive Strength of Concrete Masonry[1] (psi)	Specified Compressive Strength of Concrete Masonry Assemblage[2], f'_m(psi)	Modulus of Elasticity[3] $E_m = 750 f'_m$ (psi) E_m(max) = 3,000,000 psi	Modular Ratio $n = E_s/E_m$ Where $E_s =$ 29,000,000 psi[4]	Modulus of Rigidity[5] $G = 0.4 E_m = 300 f'_m$ (psi) G(max) = 1,200,000 (psi)
4,800 or more	3,000	2,250,000	12.9	900,000
3,750	2,500	1,875,000	15.5	750,000
2,800	2,000	1,500,000	19.3	600,000
1,900	1,500	1,125,000	25.8	450,000
1,250	1,000	750,000	38.7	300,000

1. Compressive strength of solid masonry units is based on the gross area. Compressive strength of hollow clay masonry units is based on minimum net area. Values may be interpolated.

2. Based on UBC Table No. 24-C.

3. Based on UBC Chapter 24, Eq. 6-24.

4. Based on UBC Chapter 24, Eq. 6-25.

5. Based on UBC Chapter 24, Eq. 6-26.

TABLE A-3 Maximum Allowable Working Stresses (psi), for Reinforced Solid and Hollow Unit Masonry[1]

Type of Stress	Allowable Stress or Stress Coefficient		Specified Strength of Masonry, f'_m (psi)											
	f'_m (psi)		1500		2000		2500		3000		3500	4000	4500	5000
Special Inspection Required[8]	NO	YES	NO[8]	YES	NO[8]	YES	NO[8]	YES	NO[8]	YES	YES	YES	YES	YES
Compression-Axial Columns[7,10]	$0.2f'_m$	$0.2f'_m$	150	300	200	400	250	500	300	600	700	800	900	1000
Walls[6,10]	$0.2f'_m$	$0.2f'_m$	150	300	200	400	250	500	300	600	700	800	900	1000
Compression-Flexural[10]	$0.33f'_m$; 1000 psi max.	$0.33f'_m$; 2000 psi max.	250	500	333	667	417	833	500	1000	1167	1333	1500	1667
Shear No Shear Reinforcing														
Flexural[2,10]	$\frac{1}{2}(1.0\sqrt{f'_m})$; 25 psi max.	$1.0(\sqrt{f'_m})$; 50 psi max.	20	39	23	45	25	50	25	50	50	50	50	50
Shear Walls[3,10]														
$M/Vd > 1$[4,10]	$\frac{1}{2}(\sqrt{f'_m})$; 18 psi max.	$1.0(\sqrt{f'_m})$; 35 psi max.	18	35	18	35	18	35	18	35	35	35	35	35
$M/Vd = 0$[4,10]	$\frac{1}{2}(\frac{1}{3}(4 - M/Vd)(\sqrt{f'_m}))$; $\frac{1}{2}(80 - 45\,M/Vd)$ max.	$\frac{1}{3}(4 - M/Vd)(\sqrt{f'_m})$; $(80 - 45\,M/Vd)$ max.	26	52	30	60	33	67	37	73	79	80	80	80
Reinforcing taking all shear														
Flexural[2,10]	$\frac{1}{2}(3.0\sqrt{f'_m})$; 75 psi max.	$3.0(\sqrt{f'_m})$; 150 psi max.	58	116	67	134	75	150	75	150	150	150	150	150
Shear Walls[3,10]														
$M/Vd > 1$[4,10]	$\frac{1}{2}(1.5\sqrt{f'_m})$; 38 psi max.	$1.5(\sqrt{f'_m})$; 75 psi max.	29	58	34	67	38	75	38	75	75	75	75	75
$M/Vd = 0$[4,10]	$\frac{1}{2}(\frac{1}{2}(4 - M/Vd)\sqrt{f'_m})$; $\frac{1}{2}(120 - 45M/Vd)$ max.	$\frac{1}{2}(4 - M/Vd)\sqrt{f'_m}$; $(120 - 45M/Vd)$ max.	39	77	45	89	50	100	55	110	118	120	120	120
Mod. of Elasticity $E_m \times 10^6$ (psi)	$750f'_m$; 3×10^6 psi max.	$750f'_m$; 3×10^6 psi max.	1.125	1.125	1.500	1.500	1.875	1.875	2.250	2.250	2.625	3.000	3.000	3.000
Modular Ratio $= n$ $n = E_s/E_m$	$29{,}000{,}000/750f'_m$; 9.7 min.	$29{,}000{,}000/750f'_m$; 9.7 min.	25.8	25.8	19.3	19.3	15.5	15.5	12.9	12.9	11.0	9.7	9.7	9.7
Mod. of Ridigity $G = 0.4\,E_m \times 10^5$ (psi)	$0.4E_m$; 12.0×10^5 psi max.	$0.4E_m$; 12.0×10^5 psi max.	4.50	4.50	6.00	6.00	7.50	7.50	9.0	9.0	10.5	12.0	12.0	12.0
Bearing on full area[5,10]	$\frac{1}{2}(0.26f'_m)$	$0.26f'_m$	195	390	260	520	325	650	390	780	910	1040	1170	1300
Bearing on 1/3 or less area[5,10]	$\frac{1}{2}(0.38f'_m)$	$0.38f'_m$	285	570	380	760	475	950	570	1140	1330	1520	1710	1900
Bond[9,10]	See Footnotes 9 and 10.													

TABLE A-3 FOOTNOTES

1. Stresses for hollow unit masonry are based on the net section.

2. Web reinforcement must be provided to carry the entire shear in excess of 20 pounds per square inch whenever there is required negative reinforcement and for a distance of one-sixteenth the clear span beyond the point of inflection. (Sec. 2406(c)6.A. Exception.)

3. Equations based on UBC Sec. 2406(c)7.

   UBC Section 2407(h)4.F(i) requires shear walls located in seismic zones 3 and 4 to be designed for increased loads when they resist seismic forces. These shear walls must be designed to resist 1.5 times the forces required by UBC Section 2333 and 2334(b) when calculating shear or diagonal tension. Therefore:

   $V = 1.5 \times (ZIC/R_w)W$ (UBC Ch. 23, Eq. 34-1)

   Note that this 1.5 factor should not be included in V when calculating M/Vd.

4. M is the maximum bending moment occurring simultaneously with the shear load V at the section under consideration.

   For allowable shear stresses when $0 < M/Vd < 1$, see Tables A-5a, A-5b, A-6a, and A-6b.

5. Allowable bearing stresses are determined from UBC Chapter 24, Equations 6-14 and 6-15. Note that the allowable stress increase for bearing on one-third the area or less is permitted only when the least distance between the edges of the loaded and unloaded areas is a minimum of one-fourth of the parallel side dimension of the loaded area. The allowable bearing stress on a reasonable concentric area greater than one-third, but less that the full area, may be interpolated between the values given. (UBC Sec. 2406(c)8.)

6. The axial stress in reinforced masonry bearing walls may not exceed the value determined by the following formula:

   $F_a = 0.20 f'_m [1 - (h'/42t)^3]$ (UBC Ch. 24, Eq. 6-3)

   Table A-3 gives only the terms $0.20 f'_m$ for non-inspected masonry or $\frac{1}{2}(0.20 f'_m)$ for inspected masonry as shown in the *Allowable Stress or Stress Coefficient* column of the Table. Multiply the values from Table A-3 by the reduction factor, $[1 - (h'/42t)^3]$, as shown in the Q Tables.

7. The axial load on columns may not exceed:

   $P = (0.20 f'_m A_e + 0.65 A_s F_{sc})[1 - (h'/42t)^3]$ (UBC Ch. 24, Eq. 6-4)

   Table A-3 gives only the terms $0.20 f'_m$ for inspected masonry or $\frac{1}{2}(0.20 f'_m)$ for non-inspected masonry as shown in the *Allowable Stress or Stress Coefficient* column of the Table. Coordinate the Table A-3 values with the Q and S Tables.

8. When special inspection will not be provided during construction, the allowable design masonry stresses must be reduced by one half per UBC Section 2406 (c).

9. Allowable Bond Strengths (psi) per UBC Section 2406(c)5 are as follows:

Bar Type	No Special Inspection	Special Inspection
Plain	30	60
Deformed 1988 UBC	70	140
1991 UBC	100	200

10. UBC Section 2303(d) permits an increase in the allowable stress values when considering wind or seismic forces.

TABLE A-4 Allowable Steel Working Stresses[1]

Item:	Pounds Per Square Inch
Tensile Strength, F_s: For deformed bars with a yield strength of 40,000 psi (Grade 40)	20,000[2]
For deformed bars with a yield strength of 60,000 psi or more and in sizes No. 11 and smaller	24,000[2]
Joint reinforcement, 50 percent of the minimum yield point specified in UBC Standards for the particular kind and grade of steel used, but in no case to exceed	30,000[2]
Compressive Stress in Column Verticals, F'_s: 40 percent of the minimum yield strength, but not to exceed	24,000[2]
Compressive Stress in Flexural Members: For compression reinforcement in flexural members, the allowable stress shall not be taken as greater than the allowable tensile stress shown above.	
The modulus of elasticity of steel reinforcement, E_s	29,000,000

1. Based on UBC Section 2406(d).

2. UBC Section 2303(d) permits an increase of one-third in the allowable stress values when considering wind or seismic forces.

DIAGRAM A-5   Allowable Shear Wall Stresses (psi) with the Masonry Designed to Carry the Entire Shear Load[1,2]

For use of Diagram and Tables, see Example 5-M

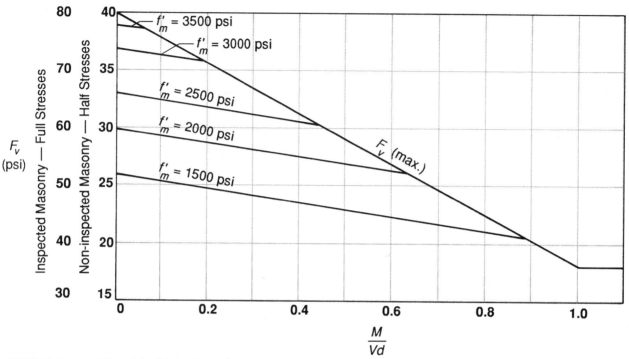

TABLE A-5a   Allowable Shear Wall Stresses (psi) for Non-inspected Masonry Designed to Carry the Entire Shear Load[1,2]

f'_m (psi)	M/Vd										
	0.0	0.1	0.2	0.3	0.4	0.5	0.6	0.7	0.8	0.9	1.0+
1500	26	25	25	24	23	23	22	21	21	20	18
2000	30	29	28	28	27	26	25	24	22	20	18
2500	33	33	32	31	30	29	27	24	22	20	18
3000	37	36	35	33	31	29	27	24	22	20	18
3500	39	38	36	33	31	29	27	24	22	20	18
4000+	40	38	36	33	31	29	27	24	22	20	18

TABLE A-5b   Allowable Shear Wall Stresses (psi) for Inspected Masonry Designed to Carry the Entire Shear Load[1]

f'_m (psi)	M/Vd										
	0.0	0.1	0.2	0.3	0.4	0.5	0.6	0.7	0.8	0.9	1.0+
1500	52	50	49	48	46	45	44	43	41	40	35
2000	60	58	57	55	54	52	51	49	44	40	35
2500	67	65	63	62	60	58	53	49	44	40	35
3000	73	71	69	67	62	58	53	49	44	40	35
3500	79	76	71	67	62	58	53	49	44	40	35
4000+	80	76	71	67	62	58	53	49	44	40	35

1. Based on UBC Chapter 24, Equation 6-10 and 6-11 or ACI/ASCE Equations 7-5 and 7-6:

   For $M/Vd < 1$, $F_v = \frac{1}{2}\left(4 - M/Vd\right)\sqrt{f'_m}$ with $F_{v(max)} = 120 - 45\,M/Vd$,   and

   For $M/Vd \geq 1$, $F_v = 1.5\sqrt{f'_m}$ with $F_{v(max)} = 75$ psi

2. Allowable stress values are reduced by one-half for non-inspected masonry per UBC Section 2406(c)1.

DIAGRAM A-6 Allowable Shear Wall Stresses (psi) with the Reinforcing Steel Designed to Carry the Entire Shear Load[1,2]

For use of Diagram and Tables, see Example 5-M

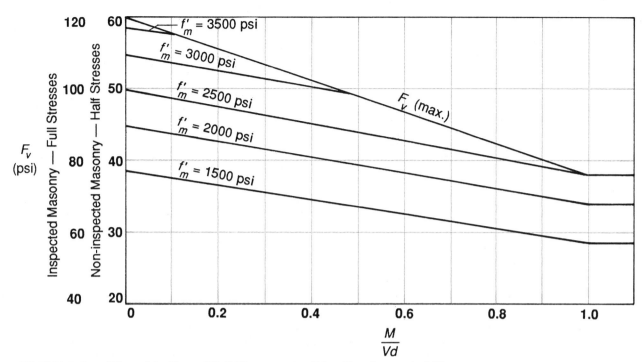

TABLE A-6a Allowable Shear Wall Stresses (psi) for Non-inspected Masonry with the Reinforcing Steel Designed to Carry the Entire Shear Load[1,2]

f'_m (psi)	M/Vd										
	0.0	0.1	0.2	0.3	0.4	0.5	0.6	0.7	0.8	0.9	1.0+
1500	39	38	37	36	35	34	33	32	31	30	29
2000	45	44	42	41	40	39	38	37	36	35	34
2500	50	49	48	46	45	44	43	41	40	39	38
3000	55	53	52	51	49	48	47	44	42	40	38
3500	59	58	56	53	51	49	47	44	42	40	38
4000+	60	58	56	53	51	49	47	44	42	40	38

TABLE A-6b Shear Wall Allowable Shear Stresses (psi) for Non-inspected Masonry with the Reinforcing Steel Designed to Carry the Entire Shear Load[1]

f'_m (psi)	M/Vd										
	0.0	0.1	0.2	0.3	0.4	0.5	0.6	0.7	0.8	0.9	1.0+
1500	77	76	74	72	70	68	66	64	62	60	58
2000	89	87	85	83	81	78	76	74	72	69	67
2500	100	98	95	93	90	88	85	83	80	78	75
3000	110	107	104	101	99	96	93	89	84	80	75
3500	118	115	111	107	102	98	93	89	84	80	75
4000+	120	116	111	107	102	98	93	89	84	80	75

1. Based on UBC Chapter 24, Equation 6-10 and 6-11 or ACI/ASCE Equations 7-8 and 7-9:

   For $M/Vd < 1$, $F_v = \frac{1}{2}\left(4 - M/Vd\right)\sqrt{f'_m}$ with $F_{v(max)} = 120 - 45\,M/Vd$, and

   For $M/Vd \geq 1$, $F_v = 1.5\sqrt{f'_m}$ with $F_{v(max)} = 75$ psi

2. Allowable stress values are reduced by one-half for non-inspected masonry per UBC Section 2406(c)1.

TABLE A-7a Allowable Tension, B_t, for Embedded Anchor Bolts in Clay and Concrete Masonry based on the Masonry Strength (pounds)[1,2,3,4,5]

For use of Tables see Example 5-V

f'_m (psi)	Embedment Length, l_b, or Edge Distance l_{be} (inches)						
	2	3	4	5	6	8	10
1500	240	550	970	1520	2190	3890	6080
1800	270	600	1070	1670	2400	4260	6660
2000	280	630	1120	1760	2520	4500	7020
2500	310	710	1260	1960	2830	5030	7850
3000	340	770	1380	2150	3100	5510	8600
4000	400	890	1590	2480	3580	6360	9930
5000	440	1000	1780	2780	4000	7110	11100
6000	480	1090	1950	3040	4380	7790	12200

1. The allowable tension values in Table are based on compressive strength of masonry assemblages. Where yield strength of anchor bolt steel governs, the allowable tension in pounds is given in Table A-7b.
2. Values are for bolts of at least A 307 quality. Bolts shall be those specified in UBC Section 2406(h)1A.
3. Values shown are for work with or without special inspection.
4. Values based on UBC Table No. 24-D-1or ACI/ASCE Equation 5-1 (except $B_a = B_t$).
5. Values may be increased by one-third when considering wind or seismic forces per UBC Section 2303(d).

TABLE A-7b Allowable Tension, B_t, for Embedded Anchor Bolts for Clay and Concrete Masonry based on the Anchor Bolt Strength (pounds)[1,2,3,4]

Bent Bar Anchor Bolt Diameter (inches)							
1/4	3/8	1/2	5/8	3/4	7/8	1	1 1/8
350	790	1410	2210	3180	4330	5650	7160

1. Values are for bolts of at least A 307 quality. Bolts shall be those specified in UBC Section 2406(h)1A.
2. Values shown are for work with or without special inspection.
3. Values based on UBC Table No. 24-D-2 or ACI/ASCE Equation 5-2 (except $B_a = B_t$).
4. Values may be increased by one-third when considering wind or seismic forces per UBC Section 2303(d).

TABLE A-7c Percent Tension Capacity of Anchor Bolts based on Bolt Spacing[1,2]

Per UBC Sec. 2406(h)2 or ACI/ASCE Sec. 5.14.2.1, the tension capacity of anchors bolts must be reduced if the areas of their tension (pullout) cones, A_p, overlap. The tensile capacity of such bolts must be determined by reducing A_p of the bolts by one half the overlapping area. The values in this table show the appropriate percent capacity or percent capacity reduction based on the spacing of the anchor bolts (see figure below).

Area of Segment, adb = Area of Sector, adbc – Area of Triangle, abc

$$= \pi l_b^2 \cos^{-1}\left(\frac{s}{2l_b}\right) - \left(\frac{s}{2}\right)\sqrt{l_b^2 - \left(\frac{s}{2}\right)^2}$$

Tension Cone Area, $A_p = \pi l_b^2$

Reduction % = Area of Segment, adb $\times 100 / A_p$

Note to find the percent reduction, set $l_b = 1.0$

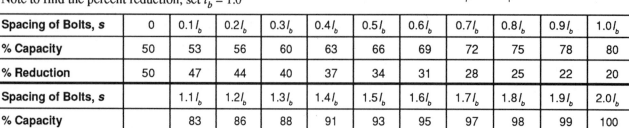

Spacing of Bolts, s	0	$0.1l_b$	$0.2l_b$	$0.3l_b$	$0.4l_b$	$0.5l_b$	$0.6l_b$	$0.7l_b$	$0.8l_b$	$0.9l_b$	$1.0l_b$
% Capacity	50	53	56	60	63	66	69	72	75	78	80
% Reduction	50	47	44	40	37	34	31	28	25	22	20
Spacing of Bolts, s		$1.1l_b$	$1.2l_b$	$1.3l_b$	$1.4l_b$	$1.5l_b$	$1.6l_b$	$1.7l_b$	$1.8l_b$	$1.9l_b$	$2.0l_b$
% Capacity		83	86	88	91	93	95	97	98	99	100
% Reduction		17	14	12	9	7	5	3	2	1	0

1. l_b = Embedment depth of anchor bolts, inches.
2. UBC Sec. 2406(h)7 limits the maximum spacing, S, between anchor bolts to four bolt diameters ($4d_b$).

TABLE A-8a Allowable Shear, B_v, for Embedded Anchor Bolts in Clay and Concrete Masonry (pounds)[1,2,3,4,5]

For use of Table see Example 5-V

For use of Table see Example 5-V

f'_m (psi)	Bent Bar Anchor Bolt Diameter (inches)						
	3/8	1/2	5/8	3/4	7/8	1	1 1/8
1500	480	850	1330	1780	1920	2050	2170
1800	480	850	1330	1860	2010	2150	2280
2000	480	850	1330	1900 [4]	2060	2200	2340
2500	480	850	1330	1900	2180	2330	2470
3000	480	850	1330	1900	2280	2440	2590
4000	480	850	1330	1900	2450	2620	2780
5000	480	850	1330	1900	2590	2770	2940
6000	480	850	1330	1900	2600	2900	3080

1. Values are for bolts of at least A 307 quality. Bolts shall be those specified in UBC Section 2406(h)1A.
2. Values shown are for work with or without special inspection.
3. Values may be increased by one-third when considering wind or seismic forces per UBC Section 2303(d).
4. Values based on UBC Table No. 24-E and UBC Eqs. 6-31 and 6-32 (ACI/ASCE Eqs. 5-5 and 5-6 similar). Shaded values are controlled by the capacity of the bolt as given by UBC Eq. 6-32 (ACI/ASCE Eq. 5-6).
5. Refer to Tables A-8b and A-8c for the percent capacity of anchor bolts based on edge distance and bolt spacing.

TABLE A-8b Percentage Shear Capacity of Anchor Bolts based on Edge Distance, l_{be}[1,2]

Bolt Diameter (inches)		3/8		1/2		5/8		3/4		7/8		1		1 1/8	
		l_{be}	% Capacity	l_{be}	% Capacity	l_{be}	% Capacity	l_{be}	% Capacity	l_{be}	% Capacity	l_{be}	% Capacity	l_{be}	% Capacity
Edge Distance in Bolt Diameters	$12 d_b$	4.5	100.0	6.0	100.0	7.5	100.0	9.0	100.0	10.5	100.0	12.0	100.0	13.5	100.0
	$10 d_b$	3.8	75.0	5.0	77.8	6.3	79.2	7.5	80.0	8.8	80.6	10.0	81.0	11.3	81.3
	$8 d_b$	3.0	50.0	4.0	55.6	5.0	58.3	6.0	60.0	7.0	61.1	8.0	61.9	9.0	62.5
	$6 d_b$	2.3	25.0	3.0	33.3	3.8	37.5	4.5	40.0	5.3	41.7	6.0	42.9	6.8	43.8
	$4 d_b$	1.5	0.0	2.0	11.1	2.5	16.7	3.0	20.0	3.5	22.2	4.0	23.8	4.5	25.0
	$2 d_b$	—	0.0	—	0.0	—	0.0	1.5	0.0	1.8	2.8	2.0	4.8	2.3	6.3
Minimum l_{be}	1.5"	1.5	0.0	1.5	0.0	1.5	0.0	1.5	0.0	1.5	0.0	1.5	0.0	1.5	0.0

1. UBC Section 2406(h)3 requires the capacity of anchor bolts determined by UBC Equation 6-31 be reduced when the edge distance is less than $12 d_b$.
2. l_{be} = Edge distance in inches.

TABLE A-8c Percentage Shear Capacity of Anchor Bolts based on the Bolt Spacing, s[1,2]

Bolt Diameter (inches)		3/8		1/2		5/8		3/4		7/8		1		1 1/8	
		s	% Capacity	s	% Capacity	s	% Capacity	s	% Capacity	s	% Capacity	s	% Capacity	s	% Capacity
Bolt Spacing in Bolt Diameters	$8 d_b$	3.0	100.0	4.0	100.0	5.0	100.0	6.0	100.0	7.0	100.0	8.0	100.0	9.0	100.0
	$7 d_b$	2.6	93.8	3.5	93.8	4.4	93.8	5.3	93.8	6.1	93.8	7.0	93.8	7.9	93.8
	$6 d_b$	2.3	87.5	3.0	87.5	3.8	87.5	4.5	87.5	5.3	87.5	6.0	87.5	6.8	87.5
	$5 d_b$	1.9	81.3	2.5	81.3	3.1	81.3	3.8	81.3	4.4	81.3	5.0	81.3	5.6	81.3
	$4 d_b$	1.5	75.0	2.0	75.0	2.5	75.0	3.0	75.0	3.5	75.0	4.0	75.0	4.5	75.0

1. UBC Section 2406(h)3 requires the capacity of anchor bolts determined by UBC Equation 6-31 be reduced when the bolt spacing is less than $8 d_b$.

WSD

TABLE A-9 Allowable Compressive Stresses for Empirical Design of Masonry Based on UBC, Table No. 24-H

Construction: Compressive Strength of Unit; Gross Area	Allowable Compressive Stresses[1] Gross Cross-sectional Area	
	Type M or S Mortar	Type N Mortar
Solid masonry of brick and other solid units of clay or shale; sand-lime or concrete brick:		
8,000 plus, psi	350	300
4,500 psi	225	200
2,500 psi	160	140
1,500 psi	115	100
Grouted masonry of clay or shale; sand-lime or concrete:		
4,500 plus, psi	275	200
2,500 psi	215	140
1,500 psi	175	100
Solid masonry of solid concrete masonry units;		
3,000 plus, psi	225	200
2,000 psi	160	140
1,200 psi	115	100
Masonry of hollow load-bearing units;		
2,000 plus, psi	140	120
1,500 psi	115	100
1,000 psi	75	70
700 psi	60	55
Hollow walls (cavity or masonry bonded)[2] solid units;		
2,500 plus, psi	160	140
1,500 psi	115	100
Hollow units	75	70
Stone ashlar masonry;		
Granite	720	640
Limestone or marble	450	400
Sandstone or cast stone	360	320
Rubble stone masonry		
Coarse, rough or random	120	100
Unburned clay masonry	30	—

1. Linear interpolation may be used for determining allowable stresses for masonry units having compressive strengths which are intermediate between those given in the table.

2. Where floor and roof loads are carried upon one wythe, the gross cross-sectional area is that of the wythe under load. If both wythes are loaded, the gross cross-sectional area is that of the wall minus the area of the cavity between the wythes.

TABLE A-10 Allowable Tensile Stresses for Unreinforced Masonry Walls in Flexure[1,2]

Tension Normal to the Bed Joints (psi)[1,2]				
Mortar Type \ Unit Type	Type M or S PCL or Mortar Cement	Type N PCL or Mortar Cement	Type M or S Masonry Cement	Type N Masonry Cement
Solid Units	40	30	24	15
Hollow Units	25	19	15	9

Tension Normal to the Head Joints (psi)[1,2]				
Mortar Type \ Unit Type	Type M or S PCL or Mortar Cement	Type N PCL or Mortar Cement	Type M or S Masonry Cement	Type N Masonry Cement
Solid Units	80	60	48	30
Hollow Units	50	38	30	19

1. Based on UBC Section 2406(c)4 with the following reductions:

    (a) For Type N Mortar made with portland cement and lime (PCL) or mortar cement, reduce UBC Table values by 25 percent.

    (b) For Type M os S masonry cement mortars, reduce the UBC Table values by 40 percent.

    (c) For Type N masonry cement mortars, reduce the UBC Table values by 63 percent.

2. Values for tension normal to the head joints are for running bond only. For stack bond, no tension is permitted across the head joints.

TABLE B-1 Weights of Building Materials

FLOORS: — Weight (psf) Pounds per Square Foot

Material	Weight (psf)
Concrete finish, per inch of thickness	12
Light weight concrete fill, per inch of thickness	9
7/8" Hardwood floor on sleepers clipped to concrete without fill	5
1½" Terrazzo floor finish directly on slab	19
1½" Terrazzo floor finish on 1" mortar bed	30
1" Terrazzo finish on 2" concrete bed	38
3/4" Ceramic or quarry tile on ½" mortar bed	16
3/4" Ceramic or quarry tile on 1" mortar bed	22
1/4" Linoleum or asphalt tile directly on concrete	1
1/4" Linoleum or asphalt tile on 1" mortar bed	12
3/4" Mastic floor	9
Hardwood flooring, 7/8" thick	4
Sub-flooring (soft wood), 3/4" thick	2½
Gypsum Slab, per inch of thickness	6
Asphalt mastic finish, 1½" thick	18
½" Douglas Fir plywood	1½
1" Douglas Fir plywood	3

ROOFS:

Material	Weight (psf)
Five-ply felt and gravel (or slag)	6½
Three-ply felt and gravel (or slag)	5½
Five-ply composition roof, no gravel	4
Three-ply felt composition roof, no gravel	3
Asphalt strip shingles	3
Concrete tile	16
Slate, 1/4" thick (laid)	10
Slate, 1/2" thick (laid)	20
Sheathing, 3/4" thick, yellow pine	3½
Sheathing, 3/4" thick, spruce or hemlock	2½
Skylight with galvanized iron frame, 1/4" wire glass	7
Gypsum, per inch of thickness	4
Poured gypsum on steel rails, per inch of thickness	5
Light weight fill or insulation, porous glass, vermiculite, etc, per inch of thickness	1 to 2
Spanish tile (laid)	9 to 12
Shingle-type clay tile	12 to 14
Metal deck (20 gauge)	2
Metal deck (18 gauge)	3
Corrugated metal (20 gauge)	2

CEILINGS:

Material	Weight (psf)
3/4" Plaster directly on concrete, blocks or tile	5
3/4" Plaster on metal lath furring	8
3/4" Gypsum plater on metal lath and channel suspended ceiling construction	10
Plaster on rock lath and channel ceiling construction	6
Acoustical fiber tile directly on concrete blocks or tile	1
Acoustical fiber tile on rock lath and channel ceiling construction	5
Acoustical fiber tile on suspended wood furring strips	3

WALLS:[1]

Material	Weight (psf)
Windows, Glass, frame and sash	8
Porcelain enamel on sheet steel	3
Structural glass, per inch of thickness	15
Stone 4" thick	55
Glass block 4" thick	18

PARTITIONS:[2]

Material	Weight (psf)
3" clay tile	17
4" clay tile	18
6" clay tile	25
8" clay tile	31
10" clay tile	35
3" gypsum block	10
4" gypsum block	13
5" gypsum block	16
6" gypsum block	17
2" solid plaster	20
2 × 4 studs, or metal studs, lath and 3/4" plaster	18
Steel partitions	4
Gypsum plaster per 1/8" thick	1

1. See Tables B-2, B-3, and B-4 for masonry walls.
2. UBC Section 2304(d) specifies uniformly distributed deadload for moveable partitions of 20 pounds per square foot.

TABLE B-2   Average Weight[1] of Concrete Masonry Units, Pounds Per Unit (16″ Long Units)

Thickness of Units		Lightweight Units: 103 pcf					Medium Weight Units: 115 pcf					Normal Weight Units: 135 pcf				
		4″	6″	8″	10″	12″	4″	6″	8″	10″	12″	4″	6″	8″	10″	12″
Individual Block	4″ high units	8	11	13	15	20	9	13	15	17	22	10	16	18	20	26
	8″ high units	16	23	27	32	42	18	28	32	36	47	21	33	37	42	55

1. ASTM C 90 classifies masonry units as follows: Lightweight: Less than 105 pcf. Medium Weight: 105 pcf to 125 pcf. Normal Weight: 125 pcf or more.

TABLE B-3a   Average Weight of Completed Walls,[1] Pounds per Square Foot, and Equivalent Solid Thickness, Inches (Weight of Grout = 140 pcf)

Wall Thickness		Hollow Concrete Block												Hollow Clay Block 120 pcf			Equivalent Solid Thickness[2] (Inches)			
		Lightweight 103 pcf				Medium Weight 115 pcf				Normal Weight 135 pcf										
		6″	8″	10″	12″	6″	8″	10″	12″	6″	8″	10″	12″	4″	6″	8″	6″	8″	10″	12″
Solid Grouted Wall		52	75	93	118	58	78	98	124	63	84	104	133	38	56	77	5.6	7.6	9.6	11.6
Vertical Cores Grouted at	16″ o.c.	41	60	69	88	47	63	80	94	52	66	86	103	33	45	59	4.5	5.8	7.2	8.5
	24″ o.c.	37	55	61	79	43	58	72	85	48	61	78	94	31	42	54	4.1	5.2	6.3	7.5
	32″ o.c.	36	52	57	74	42	55	68	80	47	58	74	89	30	40	51	4.0	4.9	5.9	7.0
	40″ o.c.	35	50	55	71	41	53	66	77	46	56	72	86	29	39	49	3.8	4.7	5.7	6.7
	48″ o.c.	34	49	53	69	40	45	64	75	45	55	70	83	28	38	48	3.7	4.6	5.5	6.5
No Grout in Wall		26	33	36	47	32	36	41	53	37	42	47	62	25	30	35	3.4	4.0	4.7	5.5

1. The above table gives the average weight of completed walls of various thicknesses in pounds per square foot of wall face area. An average amount has been added into these values to include the weight of bond beams and reinforcing steel.

2. Equivalent solid thickness means the calculated thickness of the wall if there were no hollow cores, and is obtained by dividing the volume of the solid material in the wall by the face area of the wall. This Equivalent Solid Thickness (EST) is for the determination of area for structural design only, e.g., $f_a = P/(EST)b$. A fire rating thickness is based either on equivalent solid thickness of ungrouted units or solid grouted walls (partial grouted walls are considered as ungrouted for fire ratings).

TABLE B-3b   Average Weight of Completed Walls,[1] Pounds Per Square Foot and Equivalent Solid Thickness (Weight of Grout = 105 pcf)

Wall Thickness		Hollow Concrete Block												Hollow Clay Block 120 pcf			Equivalent Solid Thickness[2] (Inches)			
		Lightweight 103 pcf				Medium Weight 115 pcf				Normal Weight 135 pcf										
		6″	8″	10″	12″	6″	8″	10″	12″	6″	8″	10″	12″	4″	6″	8″	6″	8″	10″	12″
Solid Grouted Wall		45	65	79	100	51	68	84	106	56	74	90	115	35	49	66	5.6	7.6	9.6	11.6
Vertical Cores Grouted at	16″ o.c.	37	51	61	78	43	54	66	84	48	60	72	93	31	39	49	4.5	5.8	7.2	8.5
	24″ o.c.	35	47	55	71	41	50	60	77	46	56	66	86	30	39	49	4.1	5.2	6.3	7.5
	32″ o.c.	33	45	52	67	39	48	57	73	44	54	63	82	29	37	47	4.0	4.9	5.9	7.0
	40″ o.c.	32	43	50	65	38	46	55	71	43	52	61	80	28	36	45	3.8	4.7	5.7	6.7
	48″ o.c.	31	42	49	63	37	45	54	69	42	51	60	78	27	35	44	3.7	4.6	5.5	6.5
No Grout in Wall		26	33	36	47	32	36	41	53	37	42	47	62	25	30	35	3.4	4.0	4.7	5.5

1. The above table gives the average weight of completed walls of various thicknesses in pounds per square foot of wall face area. An average amount has been added into these values to include the weight of bond beams and reinforcing steel.

2. Equivalent solid thickness means the calculated thickness of the wall if there were no hollow cores, and is obtained by dividing the volume of the solid material in the wall by the face area of the wall. This Equivalent Solid Thickness (EST) is for the determination of area for structural design only, e.g., $f_a = P/(EST)b$. A fire rating thickness is based either on equivalent solid thickness of ungrouted units or solid grouted walls (partial grouted walls are considered as ungrouted for fire ratings).

TABLE B-4   Average Weight of Reinforced Grouted Brick Walls[1]

Wall Thickness	Weight (psf)	Wall Thickness	Weight (psf)	Wall Thickness	Weight (psf)	Wall Thickness	Weight (psf)
8″	80	9″	90	10″	100	12″	120
8½″	85	9½″	95	11″	110	13″	130

1. Based on an average weight of completed wall of 10 psf per 1″ thickness.

TABLE B-5a   Approximate Grout Quantities in Concrete Masonry Walls

Thickness of Standard Two Cell Concrete Masonry Units[1]	Spacing of Grouted Cells and Vertical Reinforcing Bars	Cubic Yards of Grout[2] per 100 Block Units[3] Square Feet of Wall	Cubic Yards of Grout[2] per 100 Block Units[3]	Number of Block Units[3] Filled per Cubic Yard of Grout
6″	All Cells Grouted Solid	0.93	0.83	120
	16″	0.55	0.49	205
	24″	0.42	0.37	270
	32″	0.35	0.31	320
	40″	0.31	0.28	360
	48″	0.28	0.25	396
8″	All Cells Grouted Solid	1.12	1.00	100
	16″	0.65	0.58	171
	24″	0.50	0.44	225
	32″	0.43	0.38	267
	40″	0.37	0.33	300
	48″	0.34	0.30	330
10″	All Cells Grouted Solid	1.38	1.23	80
	16″	0.82	0.73	137
	24″	0.63	0.56	180
	32″	0.53	0.47	214
	40″	0.47	0.42	240
	48″	0.43	0.38	264
12″	All Cells Grouted Solid	1.73	1.54	65
	16″	1.01	0.90	111
	24″	0.76	0.68	146
	32″	0.64	0.57	174
	40″	0.57	0.51	195
	48″	0.53	0.47	215

1. For open ended block increase the approxiamate quanties of grout required by about 10 percent. For slumped block reduce the above grout quantities by about 5 percent. Note, table assumes horizontal bond beams at 4 ft o.c.
2. Table includes a 3 percent allowance for grout loss and various job conditions.
3. Based on standard 8″ high by 16″ long concrete masonry units.

TABLE B-5b   Approximate Grout Quantities Needed in 2 Wythe Brick Wall Construction

Width of Grout Space (Inches)	Cubic Yards of Grout[1] per 100 Square Feet of Wall	Square Feet of Wall Filled per Cubic Yard of Grout[1]
2.0	0.64	157
2.5	0.79	126
3.0	0.96	105
3.5	1.11	90
4.0	1.27	79
4.5	1.43	70
5.0[2]	1.59	63
5.5[2]	1.75	57
6.0[2]	1.91	52
6.5[2]	2.07	48
7.0[2]	2.23	45
8.0[2]	2.54	39

1. Table includes a 3 percent allowance for grout loss and various job conditions.
2. When the width of the grout space is 5″ or more, it is advisable to use floaters during low lift grouting.

TABLE B-6 Approximate Measurements of Masonry Materials

Item	Weight	Volume
Portland Cement		
1 Bag of Portland Cement	94 lbs	1.0 cu. ft
1, 12-quart Bucket of Portland Cement	38 lbs	0.4 cu. ft
Lime		
1 Cubic foot of Lime Putty[1]	80 lbs	1 cu. ft
1, 12 quart Bucket of Lime Putty	30 lbs	0.37 cu. ft
6.5 Full No. 2 Shovels of Lime Putty	80 lbs	1 cu. ft
1 Cubic foot of Hydrated Lime	40 lbs	1 cu. ft
100 pounds of Hydrated Lime make the following volume of Lime Putty	100 lbs	2.18 cu. ft
1 Cubic foot of Quick-Lime	60 lbs	1 cu. ft
100 pounds of Quick-lime makes the following volume of Lime Putty	100 lbs	3.69 cu. ft
Sand		
1 Cubic Yard of Sand	2700 lbs	1 cu. yd
1 Ton of Sand	2000 lbs	3/4 cu. yd
1 Cubic Foot of Sand	100 lbs	1 cu. ft
1, 12 Quart Bucket of Sand	40 lbs	0.4 cu. ft

1. Made from approximately 45.8 lbs of hydrated lime or 27.3 lbs of quicklime.

TABLE C-1a Properties of Standard Steel Reinforcing Bars

Bar designation number[2]	Nominal weight (lbs/ft)	Nominal dimensions[1]			Deformation requirements (inches)		
		Diameter (inches)	Cross-Sectional area (sq. in.)	Perimeter (inches)	Maximum average spacing	Minimum average height	Maximum gap (chord of 12½% of nominal perimeter)
#3	0.376	0.375	0.11	1.178	0.262	0.015	0.143
#4	0.668	0.500	0.20	1.571	0.350	0.020	0.191
#5	1.043	0.625	0.31	1.963	0.437	0.028	0.239
#6	1.502	0.750	0.44	2.356	0.525	0.038	0.286
#7	2.044	0.875	0.60	2.749	0.612	0.044	0.334
#8	2.670	1.000	0.79	3.142	0.700	0.050	0.383
#9	3.400	1.128	1.00	3.544	0.790	0.056	0.431
#10	4.303	1.270	1.27	3.990	0.889	0.064	0.487
#11	5.313	1.410	1.56	4.430	0.987	0.071	0.540
#14	7.650	1.693	2.25	5.320	1.185	0.085	0.648
#18	13.600	2.257	4.00	7.090	1.580	0.102	0.864

1. The nominal dimension of a deformed bar is equivalent to that of a plain round bar having the same weight per foot as the deformed bar.

2. Bar numbers are based on the number of eighths of an inch included in the nominal diameter of the bars.

Bar diameters are nominal with the actual diameter outside of deformations being somewhat greater. The outside diameter may be important when punching holes in structural steel members to accommodate bars or when allowing for the out-to-out width of a group of beam bars crossing and in contact with column bars. Approximately $1/16$ in. for #3, #4, #5 bars, $1/8$ in. for #6, #7, #8, #9 bars and $3/16$ in. for #10 and #11 bars should be added to the nominal bar diameter for the height of the deformations. See Table C-1b below.

TABLE C-1b Overall Diameter of Bars

Bar Size	Approx. dia. to outside of deformations[1] (inches)	Bar Size	Approx. dia. to outside of deformations[1] (inches)
#3	$7/16$	#8	$1\tfrac{1}{8}$
#4	$9/16$	#9	$1\tfrac{1}{4}$
#5	$11/16$	#10	$1\tfrac{7}{16}$
#6	$7/8$	#11	$1\tfrac{5}{8}$
#7	1	#14	$1\tfrac{7}{8}$
		#18	$2\tfrac{1}{2}$

1. Diameters tabulated are the approximate dimension to the outside of the deformations.

TABLE C-2 Properties of Steel Reinforcing Wire[1]

Steel wire gauge	Diameter (inches)	Perimeter (inches)	Area (square inches)	Ultimate strength (lbs)	Weight (plf)	Allowable tension load @ 30,000 psi (lbs)	Allowable tension load @ 40,000 psi[2] (lbs)
14	.080	.251	.005	375	.017	150	200
13	.092	.287	.007	525	.022	198	264
12	.106	.330	.009	675	.030	261	348
11	.120	.377	.011	825	.039	342	456
10	.135	.424	.014	1050	.049	429	572
9 [3]	.148	.465	.017	1275	.059	519	692
8 [3]	.162	.509	.021	1575	.070	618	824
7	.177	.556	.025	1875	.084	738	984
3/16" [3]	.188	.587	.028	2100	.094	828	1104
6	.192	.603	.029	2175	.098	867	1156
5	.207	.650	.034	2550	.114	1008	1344
4	.225	.707	.040	3000	.135	1197	1596
3	.244	.763	.047	3525	.158	1398	1864
1/4" [3]	.250	.785	.049	3675	.167	1473	1964
2	.262	.823	.054	4050	.184	1623	2163
1	.283	.889	.063	4725	.214	1887	2515
1-0	.306	.961	.074	5550	.251	2214	2952
5/16" [3]	.313	.980	.077	5775	.261	2301	3068
2-0	.331	1.040	.086	6450	.292	2580	3440
3-0	.362	1.137	.103	7725	.351	3096	4128

1. Based on the United States Steel Wire Gauge and UBC Standard No. 24-15 with F_{su} = 75,000 psi min., F_y = 60,000 psi min. and F_s allowable = 30,000 psi.

2. Allowable tension loads increased $1/3$ for wind and seismic loads.

3. Used for joint reinforcing.

WSD

TABLE C-3   Areas and Perimeters of Various Combinations of Bars

Areas, A_s (or A'_s) (top) sq. in.

Perimeters, Σ_o (bottom) in.

Columns headed $\boxed{0}\ \boxed{5}$ contain data for bars of one size in groups of one to ten.

Columns headed $\boxed{1}\ \boxed{2}\ \boxed{3}\ \boxed{4}\ \boxed{5}$ contain data for bars of two sizes with one to five bars of each size.

For bars of one size:
Σ_o = sum of perimeters

For bars of two sizes:
$$\Sigma_o = \frac{4A_s}{D}$$
where D = largest bar dia.

Each cell is given as **area (top) / perimeter (bottom)**.

Block #4 (pair #4)

#4	n	0	5	#3 · 1	#3 · 2	#3 · 3	#3 · 4	#3 · 5
	1	0.20 / 1.6	1.20 / 9.4	0.31 / 2.5	0.42 / 3.4	0.53 / 4.2	0.64 / 5.1	0.75 / 6.0
	2	0.40 / 3.1	1.40 / 11.0	0.51 / 4.1	0.62 / 5.0	0.73 / 5.8	0.84 / 6.7	0.95 / 7.6
#4	3	0.60 / 4.7	1.60 / 12.6	0.71 / 5.7	0.82 / 6.6	0.93 / 7.4	1.04 / 8.3	1.15 / 9.2
	4	0.80 / 6.3	1.80 / 14.1	0.91 / 7.3	1.02 / 8.2	1.13 / 9.0	1.24 / 9.9	1.35 / 10.8
	5	1.00 / 7.9	2.00 / 15.7	1.11 / 8.9	1.22 / 9.8	1.33 / 10.6	1.44 / 11.5	1.55 / 12.4

Block #5 (pair #5)

#5	n	0	5	#4 · 1	#4 · 2	#4 · 3	#4 · 4	#4 · 5	#3 · 1	#3 · 2	#3 · 3	#3 · 4	#3 · 5
	1	0.31 / 2.0	1.86 / 11.8	0.51 / 3.3	0.71 / 4.5	0.91 / 5.8	1.11 / 7.1	1.31 / 8.4	0.42 / 2.7	0.53 / 3.4	0.64 / 4.1	0.75 / 4.8	0.86 / 5.5
	2	0.62 / 3.9	2.17 / 13.7	0.82 / 5.2	1.02 / 6.5	1.22 / 7.8	1.42 / 9.1	1.62 / 10.4	0.73 / 4.7	0.84 / 5.4	0.95 / 6.1	1.06 / 6.8	1.17 / 7.5
#5	3	0.93 / 5.9	2.48 / 15.7	1.13 / 7.2	1.33 / 8.5	1.53 / 9.8	1.73 / 11.1	1.93 / 12.4	1.04 / 6.7	1.15 / 7.4	1.26 / 8.1	1.37 / 8.8	1.48 / 9.5
	4	1.24 / 7.9	2.79 / 17.7	1.44 / 9.2	1.64 / 10.5	1.84 / 11.8	2.04 / 13.1	2.24 / 14.3	1.35 / 8.6	1.46 / 9.3	1.57 / 10.0	1.68 / 10.8	1.79 / 11.5
	5	1.55 / 9.8	3.10 / 19.6	1.75 / 11.2	1.95 / 12.5	2.15 / 13.8	2.35 / 15.0	2.55 / 16.3	1.66 / 10.6	1.77 / 11.3	1.88 / 12.0	1.99 / 12.7	2.10 / 13.4

Block #6 (pair #6)

#6	n	0	5	#5 · 1	#5 · 2	#5 · 3	#5 · 4	#5 · 5	#4 · 1	#4 · 2	#4 · 3	#4 · 4	#4 · 5
	1	0.44 / 2.4	2.64 / 14.1	0.75 / 4.0	1.06 / 5.7	1.37 / 7.3	1.68 / 9.0	1.99 / 10.6	0.64 / 3.4	0.84 / 4.5	1.04 / 5.5	1.24 / 6.6	1.44 / 7.7
	2	0.88 / 4.7	3.08 / 16.5	1.19 / 6.3	1.50 / 8.0	1.81 / 9.7	2.12 / 11.3	2.43 / 13.0	1.08 / 5.8	1.28 / 6.8	1.48 / 7.9	1.68 / 9.0	1.88 / 10.0
#6	3	1.32 / 7.1	3.52 / 18.8	1.63 / 8.7	1.94 / 10.3	2.25 / 12.0	2.56 / 13.7	2.87 / 15.3	1.52 / 8.1	1.72 / 9.2	1.92 / 10.2	2.12 / 11.3	2.32 / 12.4
	4	1.76 / 9.4	3.96 / 21.2	2.07 / 11.0	2.38 / 12.7	2.69 / 14.3	3.00 / 16.0	3.31 / 17.7	1.96 / 10.5	2.16 / 11.5	2.36 / 12.6	2.56 / 13.7	2.76 / 14.7
	5	2.20 / 11.8	4.40 / 23.6	2.51 / 13.4	2.82 / 15.0	3.13 / 16.7	3.44 / 18.3	3.75 / 20.0	2.40 / 12.8	2.60 / 13.9	2.80 / 14.9	3.00 / 16.0	3.20 / 17.1

Block #7 (pair #7)

#7	n	0	5	#6 · 1	#6 · 2	#6 · 3	#6 · 4	#6 · 5	#5 · 1	#5 · 2	#5 · 3	#5 · 4	#5 · 5	#4 · 1	#4 · 2	#4 · 3	#4 · 4	#4 · 5
	1	0.60 / 2.7	3.60 / 16.5	1.04 / 4.8	1.48 / 6.8	1.92 / 8.8	2.36 / 10.8	2.80 / 12.8	0.91 / 4.2	1.22 / 5.6	1.53 / 7.0	1.84 / 8.4	2.15 / 9.8	0.80 / 3.7	1.00 / 4.6	1.20 / 5.5	1.40 / 6.4	1.60 / 7.3
	2	1.20 / 5.5	4.20 / 19.2	1.64 / 7.5	2.08 / 9.5	2.52 / 11.5	2.96 / 13.5	3.40 / 15.5	1.51 / 6.9	1.82 / 8.3	2.13 / 9.7	2.44 / 11.2	2.75 / 12.6	1.40 / 6.4	1.60 / 7.3	1.80 / 8.2	2.00 / 9.1	2.20 / 10.1
#7	3	1.80 / 8.2	4.80 / 22.0	2.24 / 10.2	2.68 / 12.3	3.12 / 14.3	3.56 / 16.3	4.00 / 18.3	2.11 / 9.6	2.42 / 11.1	2.73 / 12.5	3.04 / 13.9	3.35 / 15.3	2.00 / 9.1	2.20 / 10.1	2.40 / 11.0	2.60 / 11.9	2.80 / 12.8
	4	2.40 / 11.0	5.40 / 24.7	2.84 / 13.0	3.28 / 15.0	3.72 / 17.0	4.16 / 19.0	4.60 / 21.0	2.71 / 12.4	3.02 / 13.8	3.33 / 15.2	3.64 / 16.6	3.95 / 18.1	2.60 / 11.9	2.80 / 12.8	3.00 / 13.7	3.20 / 14.6	3.40 / 15.5
	5	3.00 / 13.7	6.00 / 27.5	3.44 / 15.7	3.88 / 17.7	4.32 / 19.7	4.76 / 21.8	5.20 / 23.8	3.31 / 15.1	3.62 / 16.5	3.93 / 18.0	4.24 / 19.4	4.55 / 20.8	3.20 / 14.6	3.40 / 15.5	3.60 / 16.5	3.80 / 17.4	4.00 / 18.3

Block #8 (pair #8)

#8	n	0	5	#7 · 1	#7 · 2	#7 · 3	#7 · 4	#7 · 5	#6 · 1	#6 · 2	#6 · 3	#6 · 4	#6 · 5	#5 · 1	#5 · 2	#5 · 3	#5 · 4	#5 · 5
	1	0.79 / 3.1	4.74 / 18.9	1.39 / 5.6	1.99 / 8.0	2.59 / 10.4	3.19 / 12.8	3.79 / 15.2	1.23 / 4.9	1.67 / 6.7	2.11 / 8.4	2.55 / 10.2	2.99 / 12.0	1.10 / 4.4	1.41 / 5.6	1.72 / 6.9	2.03 / 8.1	2.34 / 9.4
	2	1.58 / 6.3	5.53 / 22.0	2.18 / 8.7	2.78 / 11.1	3.38 / 13.5	3.98 / 15.9	4.58 / 18.3	2.02 / 8.1	2.46 / 9.8	2.90 / 11.6	3.34 / 13.4	3.78 / 15.1	1.89 / 7.6	2.20 / 8.8	2.51 / 10.0	2.82 / 11.3	3.13 / 12.5
#8	3	2.37 / 9.4	6.32 / 25.1	2.97 / 11.9	3.57 / 14.3	4.17 / 16.7	4.77 / 19.1	5.37 / 21.5	2.81 / 11.2	3.25 / 13.0	3.69 / 14.8	4.13 / 16.5	4.57 / 18.3	2.68 / 10.7	2.99 / 12.0	3.30 / 13.2	3.61 / 14.4	3.92 / 15.7
	4	3.16 / 12.6	7.11 / 28.3	3.76 / 15.0	4.36 / 17.4	4.96 / 19.8	5.56 / 22.2	6.16 / 24.6	3.60 / 14.4	4.04 / 16.2	4.48 / 17.9	4.92 / 19.7	5.36 / 21.4	3.47 / 13.9	3.78 / 15.1	4.09 / 16.4	4.40 / 17.6	4.71 / 18.8
	5	3.95 / 15.7	7.90 / 31.4	4.55 / 18.2	5.15 / 20.6	5.75 / 23.0	6.35 / 25.4	6.95 / 27.8	4.39 / 17.6	4.83 / 19.3	5.27 / 21.1	5.71 / 22.8	6.15 / 24.6	4.26 / 17.0	4.57 / 18.3	4.88 / 19.5	5.19 / 20.8	5.50 / 22.0

Block #9 (pair #9)

#9	n	0	5	#8 · 1	#8 · 2	#8 · 3	#8 · 4	#8 · 5	#7 · 1	#7 · 2	#7 · 3	#7 · 4	#7 · 5	#6 · 1	#6 · 2	#6 · 3	#6 · 4	#6 · 5
	1	1.00 / 3.5	6.00 / 21.3	1.79 / 6.3	2.58 / 9.1	3.37 / 12.0	4.16 / 14.8	4.95 / 17.6	1.60 / 5.7	2.20 / 7.8	2.80 / 9.9	3.40 / 12.1	4.00 / 14.2	1.44 / 5.1	1.88 / 6.7	2.32 / 8.2	2.76 / 9.8	3.20 / 11.3
	2	2.00 / 7.1	7.00 / 24.8	2.79 / 9.9	3.58 / 12.7	4.37 / 15.5	5.16 / 18.3	5.95 / 21.1	2.60 / 9.2	3.20 / 11.3	3.80 / 13.5	4.40 / 15.6	5.00 / 17.7	2.44 / 8.7	2.88 / 10.2	3.32 / 11.8	3.76 / 13.3	4.20 / 14.9
#9	3	3.00 / 10.6	8.00 / 28.4	3.79 / 13.4	4.58 / 16.2	5.37 / 19.0	6.16 / 21.8	6.95 / 24.6	3.60 / 12.8	4.20 / 14.9	4.80 / 17.0	5.40 / 19.1	6.00 / 21.3	3.44 / 12.2	3.88 / 13.8	4.32 / 15.3	4.76 / 16.9	5.20 / 18.4
	4	4.00 / 14.2	9.00 / 31.9	4.79 / 17.0	5.58 / 19.8	6.37 / 22.6	7.16 / 25.4	7.95 / 28.2	4.60 / 16.3	5.20 / 18.4	5.80 / 20.6	6.40 / 22.7	7.00 / 24.8	4.44 / 15.7	4.88 / 17.3	5.32 / 18.9	5.76 / 20.4	6.20 / 22.0
	5	5.00 / 17.7	10.00 / 35.5	5.79 / 20.5	6.58 / 23.3	7.37 / 26.1	8.16 / 28.9	8.95 / 31.7	5.60 / 19.9	6.20 / 22.0	6.80 / 24.1	7.40 / 26.2	8.00 / 28.4	5.44 / 19.3	5.88 / 20.9	6.32 / 22.4	6.76 / 24.0	7.20 / 25.5

Block #10 (pair #10)

#10	n	0	5	#9 · 1	#9 · 2	#9 · 3	#9 · 4	#9 · 5	#8 · 1	#8 · 2	#8 · 3	#8 · 4	#8 · 5	#7 · 1	#7 · 2	#7 · 3	#7 · 4	#7 · 5
	1	1.27 / 4.0	7.62 / 24.0	2.27 / 7.1	3.27 / 10.3	4.27 / 13.4	5.27 / 16.6	6.27 / 19.7	2.06 / 6.5	2.85 / 9.0	3.64 / 11.5	4.43 / 14.0	5.22 / 16.4	1.87 / 5.9	2.47 / 7.8	3.07 / 9.7	3.67 / 11.6	4.27 / 13.4
	2	2.54 / 8.0	8.89 / 27.9	3.54 / 11.2	4.54 / 14.3	5.54 / 17.4	6.54 / 20.6	7.54 / 23.7	3.33 / 10.5	4.12 / 13.0	4.91 / 15.5	5.70 / 18.0	6.49 / 20.4	3.14 / 9.9	3.74 / 11.8	4.34 / 13.7	4.94 / 15.6	5.54 / 17.4
#10	3	3.81 / 12.0	10.16 / 31.9	4.81 / 15.2	5.81 / 18.3	6.81 / 21.4	7.81 / 24.6	8.81 / 27.7	4.00 / 14.5	5.39 / 17.0	6.18 / 19.5	6.97 / 22.0	7.76 / 24.4	4.41 / 13.9	5.01 / 15.8	5.61 / 17.7	6.21 / 19.6	6.81 / 21.4
	4	5.08 / 16.0	11.43 / 35.9	6.08 / 19.2	7.08 / 22.3	8.08 / 25.4	9.08 / 28.6	10.08 / 31.7	5.87 / 18.5	6.66 / 21.0	7.45 / 23.5	8.24 / 26.0	9.03 / 28.4	5.68 / 17.9	6.28 / 19.8	6.88 / 21.7	7.48 / 23.6	8.08 / 25.4
	5	6.35 / 20.0	12.70 / 39.9	7.35 / 23.2	8.35 / 26.3	9.35 / 29.4	10.35 / 32.6	11.35 / 35.7	7.14 / 22.5	7.93 / 25.0	8.72 / 27.5	9.51 / 30.0	10.30 / 32.4	6.95 / 21.9	7.55 / 23.8	8.15 / 25.7	8.75 / 27.6	9.35 / 29.4

Block #11 (pair #11)

#11	n	0	5	#10 · 1	#10 · 2	#10 · 3	#10 · 4	#10 · 5	#9 · 1	#9 · 2	#9 · 3	#9 · 4	#9 · 5	#8 · 1	#8 · 2	#8 · 3	#8 · 4	#8 · 5
	1	1.56 / 4.4	9.36 / 26.6	2.83 / 8.0	4.10 / 11.6	5.37 / 15.2	6.64 / 18.8	7.91 / 22.4	2.56 / 7.3	3.56 / 10.1	4.56 / 12.9	5.56 / 15.8	6.56 / 18.6	2.35 / 6.7	3.14 / 8.9	3.93 / 11.1	4.72 / 13.4	5.51 / 15.6
	2	3.12 / 8.9	10.92 / 31.0	4.39 / 12.5	5.66 / 16.1	6.93 / 19.7	8.20 / 23.3	9.47 / 26.9	4.12 / 11.7	5.12 / 14.5	6.12 / 17.4	7.12 / 20.2	8.12 / 23.0	3.91 / 11.1	4.70 / 13.3	5.49 / 15.6	6.28 / 17.8	7.07 / 20.1
#11	3	4.68 / 13.3	12.48 / 35.4	5.95 / 16.9	7.22 / 20.5	8.49 / 24.1	9.76 / 27.7	11.03 / 31.3	5.68 / 16.1	6.68 / 19.0	7.68 / 21.8	8.68 / 24.6	9.68 / 27.5	5.47 / 15.5	6.26 / 17.8	7.05 / 20.0	7.84 / 22.9	8.63 / 21.5
	4	6.24 / 17.7	14.04 / 39.9	7.51 / 21.3	8.78 / 24.9	10.05 / 28.5	11.32 / 32.1	12.59 / 35.7	7.24 / 20.5	8.24 / 23.4	9.24 / 26.3	10.24 / 29.1	11.24 / 31.9	7.03 / 19.9	7.82 / 22.2	8.61 / 24.4	9.40 / 26.7	10.19 / 28.9
	5	7.80 / 22.2	15.60 / 44.3	9.07 / 25.7	10.34 / 29.3	11.61 / 32.9	12.88 / 36.6	14.15 / 40.1	8.80 / 25.0	9.80 / 27.8	10.8 / 30.6	11.80 / 33.5	12.80 / 36.3	8.59 / 24.4	9.38 / 26.6	10.17 / 28.9	10.96 / 31.1	11.75 / 33.3

TABLE C-4a Areas and Perimeters of Reinforcing Steel Per Foot

Bar				Bar Spacing								
Size	Diameter (inches)	Area (sq. in.)		8	10	12	14	16	18	20	22	24
		Perimeter (inches)		0' - 8"	0' - 10"	1' - 0"	1' - 2"	1' - 4"	1' - 6"	1' - 8"	1' - 10"	2' - 0"
2 - #9 wires	0.148	0.035	A_s	0.052	0.041	0.035	0.030	0.026	0.023	0.021	0.019	0.017
		0.930	Σ_o	1.395	1.116	0.930	0.797	0.698	0.620	0.558	0.507	0.465
2 - #8 wires	0.162	0.041	A_s	0.062	0.049	0.041	0.035	0.031	0.027	0.025	0.022	0.021
		1.018	Σ_o	1.527	1.222	1.018	0.873	0.764	0.679	0.611	0.555	0.509
2 - 3/16" wires	0.1875	0.055	A_s	0.083	0.066	0.055	0.047	0.041	0.037	0.033	0.030	0.028
		1.174	Σ_o	1.761	1.409	1.174	1.006	0.881	0.783	0.704	0.640	0.587
2 - 1/4" wires	0.25	0.098	A_s	0.147	0.118	0.098	0.084	0.074	0.065	0.059	0.053	0.049
		1.570	Σ_o	2.355	1.884	1.570	1.346	1.178	1.047	0.942	0.856	0.785
2 - 5/16" wires	0.3125	0.152	A_s	0.228	0.182	0.152	0.130	0.114	0.101	0.091	0.083	0.076
		1.960	Σ_o	2.940	2.352	1.960	1.680	1.470	1.307	1.176	1.069	0.980
#2	1/4	0.049	A_s	0.074	0.059	0.049	0.042	0.037	0.033	0.029	0.027	0.025
		0.786	Σ_o	1.179	0.943	0.786	0.674	0.590	0.524	0.472	0.429	0.393
#3	3/8	0.110	A_s	0.165	0.132	0.110	0.094	0.083	0.073	0.066	0.060	0.055
		1.178	Σ_o	1.767	1.414	1.178	1.010	0.884	0.785	0.707	0.643	0.589
#4	1/2	0.196	A_s	0.294	0.235	0.196	0.168	0.147	0.131	0.118	0.107	0.098
		1.571	Σ_o	2.357	1.885	1.571	1.347	1.178	1.047	0.943	0.857	0.786
#5	5/8	0.307	A_s	0.461	0.368	0.307	0.263	0.230	0.205	0.184	0.167	0.154
		1.963	Σ_o	2.945	2.356	1.963	1.683	1.472	1.309	1.178	1.071	0.982
#6	3/4	0.442	A_s	0.663	0.530	0.442	0.379	0.332	0.295	0.265	0.241	0.221
		2.356	Σ_o	3.534	2.827	2.356	2.019	1.767	1.571	1.414	1.285	1.178
#7	7/8	0.601	A_s	0.902	0.721	0.601	0.515	0.451	0.401	0.361	0.328	0.301
		2.749	Σ_o	4.124	3.299	2.749	2.356	2.062	1.833	1.649	1.499	1.375
#8	1.0	0.786	A_s	1.179	0.943	0.786	0.674	0.590	0.524	0.472	0.429	0.393
		3.142	Σ_o	4.713	3.770	3.142	2.693	2.357	2.095	1.885	1.714	1.571
#9	1-1/8	1.000	A_s	1.500	1.200	1.000	0.857	0.750	0.667	0.600	0.545	0.500
		3.544	Σ_o	5.316	4.253	3.544	3.038	2.658	2.363	2.126	1.933	1.772
#10	1-1/4	1.270	A_s	1.905	1.524	1.270	1.089	0.953	0.847	0.762	0.693	0.635
		3.990	Σ_o	5.985	4.788	3.990	3.420	2.993	2.660	2.394	2.176	1.995
#11	1-3/8	1.560	A_s	2.340	1.872	1.560	1.337	1.170	1.040	0.936	0.851	0.780
		4.430	Σ_o	6.645	5.316	4.430	3.797	3.323	2.953	2.658	2.416	2.215

TABLE C-4b   Areas and Perimeters of Reinforcing Steel Per Foot

Bar				Bar Spacing								
Size	Diameter (inches)	Area (sq. in.)		26	28	30	32	34	36	40	44	48
		Perimeter (inches)		2' - 2"	2' - 4"	2' - 6"	2' - 8"	2' - 10"	3' - 0"	3' - 4"	3' - 8"	4' - 0"
2 - #9 wires	0.148	0.035	A_s	0.016	0.015	0.014	0.013	0.012	0.012	0.010	0.009	0.009
		0.930	Σ_o	0.429	0.399	0.372	0.349	0.328	0.310	0.279	0.254	0.233
2 - #8 wires	0.162	0.041	A_s	0.019	0.018	0.016	0.015	0.015	0.014	0.012	0.011	0.010
		1.018	Σ_o	0.470	0.436	0.407	0.382	0.359	0.339	0.305	0.278	0.255
2 - 3/16" wires	0.1875	0.055	A_s	0.025	0.024	0.022	0.021	0.019	0.018	0.017	0.015	0.014
		1.174	Σ_o	0.542	0.503	0.470	0.440	0.414	0.391	0.352	0.320	0.294
2 - 1/4" wires	0.25	0.098	A_s	0.045	0.042	0.039	0.037	0.035	0.033	0.029	0.027	0.025
		1.570	Σ_o	0.725	0.673	0.628	0.589	0.554	0.523	0.471	0.428	0.393
2 - 5/16" wires	0.3125	0.152	A_s	0.070	0.065	0.061	0.057	0.054	0.051	0.046	0.041	0.038
		1.960	Σ_o	0.905	0.840	0.784	0.735	0.692	0.653	0.588	0.535	0.490
#2	1/4	0.049	A_s	0.023	0.021	0.020	0.018	0.017	0.016	0.015	0.013	0.012
		0.786	Σ_o	0.363	0.337	0.314	0.295	0.277	0.262	0.236	0.214	0.197
#3	3/8	0.110	A_s	0.051	0.047	0.044	0.041	0.039	0.037	0.033	0.030	0.028
		1.178	Σ_o	0.544	0.505	0.471	0.442	0.416	0.393	0.353	0.321	0.295
#4	1/2	0.196	A_s	0.090	0.084	0.078	0.074	0.069	0.065	0.059	0.053	0.049
		1.571	Σ_o	0.725	0.673	0.628	0.589	0.554	0.524	0.471	0.428	0.393
#5	5/8	0.307	A_s	0.142	0.132	0.123	0.115	0.108	0.102	0.092	0.084	0.077
		1.963	Σ_o	0.906	0.841	0.785	0.736	0.693	0.654	0.589	0.535	0.491
#6	3/4	0.442	A_s	0.204	0.189	0.177	0.166	0.156	0.147	0.133	0.121	0.111
		2.356	Σ_o	1.087	1.010	0.942	0.884	0.832	0.785	0.707	0.643	0.589
#7	7/8	0.601	A_s	0.277	0.258	0.240	0.225	0.212	0.200	0.180	0.164	0.150
		2.749	Σ_o	1.269	1.178	1.100	1.031	0.970	0.916	0.825	0.750	0.687
#8	1.0	0.786	A_s	0.363	0.337	0.314	0.295	0.277	0.262	0.236	0.214	0.197
		3.142	Σ_o	1.450	1.347	1.257	1.178	1.109	1.047	0.943	0.857	0.786
#9	1-1/8	1.000	A_s	0.462	0.429	0.400	0.375	0.353	0.333	0.300	0.273	0.250
		3.544	Σ_o	1.636	1.519	1.418	1.329	1.251	1.181	1.063	0.967	0.886
#10	1-1/4	1.270	A_s	0.586	0.544	0.508	0.476	0.448	0.423	0.381	0.346	0.318
		3.990	Σ_o	1.842	1.710	1.596	1.496	1.408	1.330	1.197	1.088	0.998
#11	1-3/8	1.560	A_s	0.720	0.669	0.624	0.585	0.551	0.520	0.468	0.425	0.390
		4.430	Σ_o	2.045	1.899	1.772	1.661	1.564	1.477	1.329	1.208	1.108

TABLE C-5   Maximum Spacing in Inches of Minimum Reinforcing Steel, $A_s = 0.0007\,bt$[1,2]

Actual Wall Thickness, t (inches)	Min. Area[1,2] (sq. in./ft)	Reinforcing Bar Size						
		#3	#4	#5	#6	#7	#8	#9
3.50	0.029	45	48					
3.63	0.030	43	48					
3.75	0.032	42	48					
4.00	0.034	39	48					
4.50	0.038	35	48					
5.00	0.042	31	48					
5.25	0.044	30	48			See		
5.50	0.046	29	48			Footnote		
5.63	0.047	28	48			3		
5.75	0.048	27	48					
6.00	0.050	26	48					
6.25	0.053	25	46	48				
6.50	0.055	24	44	48				
6.75	0.057	23	42	48				
7.00	0.059	22	41	48				
7.25	0.061	21	39	48				
7.50	0.063	20	38	48				
7.63	0.064	20	37	48				
7.75	0.065	20	36	48				
8.00	0.067	20	36	48				
8.50	0.071	18	34	48				
8.75	0.074	18	33	48				
9.00	0.076	17	32	48				
9.50	0.080	17	30	47	48			
9.63	0.081	16	30	46	48			
10.00	0.084	16	29	44	48			
10.50	0.088	15	27	42	48			
11.00	0.092	14	26	40	48			
11.50	0.097	14	25	39	48			
11.63	0.098	13	24	38	48			
12.00	0.101	13	24	37	48			
12.50	0.105	13	23	35	48			
13.00	0.109	12	22	34	48			
13.50	0.113	12	21	33	47	48		
14.00	0.118	11	20	32	45	48		
14.50	0.122	11	20	31	43	48		
15.00	0.126	10	19	30	42	48		
15.50	0.130	10	18	29	41	48		
15.63	0.131	10	18	28	40	48		
16.00	0.134	10	18	28	39	48		

1. Reinforcing steel spacing shown will provide required area of steel based on $p_{min.} = 0.0007$. To be conservative, all spacing values shown were truncated to the nearest lower inch.

2. Minimum area of principal reinforcement may be $0.0013\,bt$ but may not be less than $0.0007\,bt$ per UBC Section 2407(h)4.B.

3. Values shown to the right of the heavy zigzag line are limited to a maximum spacing of 4' (48") o.c. per UBC Section 2407(h)4.B.

TABLE C-6 Maximum Spacing in Inches of Reinforcing Steel, $A_s = 0.0013bt^{1,2}$

Actual Wall Thickness, t (inches)	Min. Area[1,2] (sq. in./ft)	Reinforcing Bar Size						
		#3	#4	#5	#6	#7	#8	#9
3.50	0.055	21						
3.63	0.057	22						
3.75	0.059	23	23					
4.00	0.062	21	24					
4.50	0.070	19	27					
5.00	0.078	17	30	30				
5.25	0.082	16	29	31			See	
5.50	0.086	15	28	33			Footnote	
5.63	0.088	15	27	34			3	
5.75	0.090	14	27	35				
6.00	0.094	14	26	36				
6.25	0.098	14	25	37				
6.50	0.101	13	24	37	39			
6.75	0.105	13	23	35	40			
7.00	0.109	12	22	34	42			
7.25	0.113	11	21	33	44			
7.50	0.117	11	21	32	45	48		
7.63	0.119	11	20	31	44	48		
7.75	0.121	11	20	31	44	48		
8.00	0.125	11	19	30	42	48		
8.50	0.133	10	18	28	40	48		
8.75	0.137	9	18	27	39	48		
9.00	0.140	9	17	26	38	48		
9.50	0.148	9	16	25	36	48		
9.63	0.150	8	15	24	35	48	48	
10.00	0.156	8	15	24	34	46	48	
10.50	0.164	8	15	23	32	44	48	
11.00	0.172	8	14	22	31	42	48	
11.50	0.179	7	13	21	29	40	48	
11.63	0.181	7	13	20	29	39	48	
12.00	0.187	7	13	20	28	38	48	
12.50	0.195	7	12	19	27	37	48	
13.00	0.203	7	12	18	26	36	47	48
13.50	0.211	6	11	18	25	34	45	48
14.00	0.218	6	11	17	24	33	43	48
14.50	0.226	6	11	16	23	32	42	48
15.00	0.234	6	10	16	23	31	41	48
15.50	0.242	5	10	15	22	30	39	48
15.63	0.244	5	10	15	22	30	38	48
16.00	0.250	5	10	15	21	29	38	48

1. Reinforcing steel spacing shown will provide required area of steel based on $p_{min.} = 0.0013$. To be conservative, all spacing values shown were truncated to the nearest lower inch.

2. Minimum area of principal reinforcement may be $0.0013bt$ but may not be less than $0.007bt$ per UBC Section 2407(h)4.B.

3. Values shown to the right of the heavy zigzag line are limited to a maximum spacing of 4' (48") o.c. per UBC Section 2407(h)4.B. Values to the right of the zigzag line which are less than 48" are limited to 6 times the thickness per UBC Section 2409(c)4.

TABLE C-7   Maximum Spacing in Inches of Reinforcing Steel, $A_s = 0.001\,bt^{1,2}$

Actual Wall Thickness, t (inches)	Min. Area[1,2] (sq. in./ft)	Reinforcing Bar Size						
		#3	#4	#5	#6	#7	#8	#9
3.50	0.042	21						
3.63	0.044	22						
3.75	0.045	23						
4.00	0.048	24						
4.50	0.054	24	27					
5.00	0.060	22	30					
5.25	0.063	21	31				See	
5.50	0.066	20	33				Footnote	
5.63	0.068	19	34				3	
5.75	0.069	19	35					
6.00	0.072	18	33	36				
6.25	0.075	18	32	37				
6.50	0.078	17	31	39				
6.75	0.081	16	30	40				
7.00	0.084	16	29	42				
7.25	0.087	15	28	43	44			
7.50	0.090	15	27	41	45			
7.63	0.092	14	26	40	46			
7.75	0.093	14	26	40	47			
8.00	0.096	14	25	39	48			
8.50	0.102	13	24	36	48			
8.75	0.105	13	23	35	48			
9.00	0.108	12	22	34	48	48		
9.50	0.114	12	21	33	46	48		
9.63	0.116	11	21	32	46	48		
10.00	0.120	11	20	31	44	48		
10.50	0.126	10	19	30	42	48		
11.00	0.132	10	18	28	40	48		
11.50	0.138	10	17	27	38	48		
11.63	0.140	9	17	27	38	48		
12.00	0.144	9	17	26	37	48		
12.50	0.150	9	16	25	35	48	48	
13.00	0.156	8	15	24	34	46	48	
13.50	0.162	8	15	23	33	44	48	
14.00	0.168	8	14	22	31	43	48	
14.50	0.174	8	14	21	30	41	48	
15.00	0.180	7	13	21	29	40	48	
15.50	0.186	7	13	20	28	39	48	
15.63	0.188	7	13	20	28	38	48	
16.00	0.192	7	13	19	28	38	47	

1. Reinforcing steel spacing shown will provide required area of steel based on $p_{min.} = 0.001$. To be conservative, all spacing values shown were truncated to the nearest lower inch.

2. Minimum area of principal reinforcement may be $0.0013\,bt$ but may not be less than $0.007\,bt$ per UBC Section 2407(h)4.B.

3. Values shown to the right of the heavy zigzag line are limited to a maximum spacing of 4' (48″) o.c. per UBC Section 2407(h)4.B. Values to the right of the zigzag line which are less than 48″ are limited to 6 times the thickness per UBC Section 2409(c)4.

TABLE C-8a   Steel Ratio $p = \dfrac{A_s}{bd}$ A_s in square inches; *b* and *d* in inches

d (inches)	Bar Size	Steel Area (sq. in.)	8	12	16	20	24	28	32	36	40	44	48	*d* (inches)
	#3	0.11	0.0055	0.0037	0.0028	0.0022	0.0018	0.0016	0.0014	0.0012	0.0011	0.0010	0.0009	
	#4	0.20	0.0100	0.0067	0.0050	0.0040	0.0033	0.0029	0.0025	0.0022	0.0020	0.0018	0.0017	
	#5	0.31	0.0155	0.0103	0.0078	0.0062	0.0052	0.0044	0.0039	0.0034	0.0031	0.0028	0.0026	
2.50	#6	0.44	0.0220	0.0147	0.0110	0.0088	0.0073	0.0063	0.0055	0.0049	0.0044	0.0040	0.0037	2.50
	#7	0.60	0.0300	0.0200	0.0150	0.0120	0.0100	0.0086	0.0075	0.0067	0.0060	0.0055	0.0050	
	#8	0.79	0.0395	0.0263	0.0198	0.0158	0.0132	0.0113	0.0099	0.0088	0.0079	0.0072	0.0066	
	#9	1.00	0.0500	0.0333	0.0250	0.0200	0.0167	0.0143	0.0125	0.0111	0.0100	0.0091	0.0083	
	#10	1.27	0.0635	0.0423	0.0318	0.0254	0.0212	0.0181	0.0159	0.0141	0.0127	0.0115	0.0106	
	#11	1.56	0.0780	0.0520	0.0390	0.0312	0.0260	0.0223	0.0195	0.0173	0.0156	0.0142	0.0130	

d (inches)	Bar Size	Steel Area (sq. in.)	8	12	16	20	24	28	32	36	40	44	48	*d* (inches)
	#3	0.11	0.0050	0.0033	0.0025	0.0020	0.0017	0.0014	0.0013	0.0011	0.0010	0.0009	0.0008	
	#4	0.20	0.0091	0.0061	0.0045	0.0036	0.0030	0.0026	0.0023	0.0020	0.0018	0.0017	0.0015	
	#5	0.31	0.0141	0.0094	0.0070	0.0056	0.0047	0.0040	0.0035	0.0031	0.0028	0.0026	0.0023	
2.75	#6	0.44	0.0200	0.0133	0.0100	0.0080	0.0067	0.0057	0.0050	0.0044	0.0040	0.0036	0.0033	2.75
	#7	0.60	0.0273	0.0182	0.0136	0.0109	0.0091	0.0078	0.0068	0.0061	0.0055	0.0050	0.0045	
	#8	0.79	0.0359	0.0239	0.0180	0.0144	0.0120	0.0103	0.0090	0.0080	0.0072	0.0065	0.0060	
	#9	1.00	0.0455	0.0303	0.0227	0.0182	0.0152	0.0130	0.0114	0.0101	0.0091	0.0083	0.0076	
	#10	1.27	0.0577	0.0385	0.0289	0.0231	0.0192	0.0165	0.0144	0.0128	0.0115	0.0105	0.0096	
	#11	1.56	0.0709	0.0473	0.0355	0.0284	0.0236	0.0203	0.0177	0.0158	0.0142	0.0129	0.0118	

d (inches)	Bar Size	Steel Area (sq. in.)	8	12	16	20	24	28	32	36	40	44	48	*d* (inches)
	#3	0.11	0.0046	0.0031	0.0023	0.0018	0.0015	0.0013	0.0011	0.0010	0.0009	0.0008	0.0008	
	#4	0.20	0.0083	0.0056	0.0042	0.0033	0.0028	0.0024	0.0021	0.0019	0.0017	0.0015	0.0014	
	#5	0.31	0.0129	0.0086	0.0065	0.0052	0.0043	0.0037	0.0032	0.0029	0.0026	0.0023	0.0022	
3.00	#6	0.44	0.0183	0.0122	0.0092	0.0073	0.0061	0.0052	0.0046	0.0041	0.0037	0.0033	0.0031	3.00
	#7	0.60	0.0250	0.0167	0.0125	0.0100	0.0083	0.0071	0.0063	0.0056	0.0050	0.0045	0.0042	
	#8	0.79	0.0329	0.0219	0.0165	0.0132	0.0110	0.0094	0.0082	0.0073	0.0066	0.0060	0.0055	
	#9	1.00	0.0417	0.0278	0.0208	0.0167	0.0139	0.0119	0.0104	0.0093	0.0083	0.0076	0.0069	
	#10	1.27	0.0529	0.0353	0.0265	0.0212	0.0176	0.0151	0.0132	0.0118	0.0106	0.0096	0.0088	
	#11	1.56	0.0650	0.0433	0.0325	0.0260	0.0217	0.0186	0.0163	0.0144	0.0130	0.0118	0.0108	

d (inches)	Bar Size	Steel Area (sq. in.)	8	12	16	20	24	28	32	36	40	44	48	*d* (inches)
	#3	0.11	0.0042	0.0028	0.0021	0.0017	0.0014	0.0012	0.0011	0.0009	0.0008	0.0008	0.0007	
	#4	0.20	0.0077	0.0051	0.0038	0.0031	0.0026	0.0022	0.0019	0.0017	0.0015	0.0014	0.0013	
	#5	0.31	0.0119	0.0079	0.0060	0.0048	0.0040	0.0034	0.0030	0.0026	0.0024	0.0022	0.0020	
3.25	#6	0.44	0.0169	0.0113	0.0085	0.0068	0.0056	0.0048	0.0042	0.0038	0.0034	0.0031	0.0028	3.25
	#7	0.60	0.0231	0.0154	0.0115	0.0092	0.0077	0.0066	0.0058	0.0051	0.0046	0.0042	0.0038	
	#8	0.79	0.0304	0.0203	0.0152	0.0122	0.0101	0.0087	0.0076	0.0068	0.0061	0.0055	0.0051	
	#9	1.00	0.0385	0.0256	0.0192	0.0154	0.0128	0.0110	0.0096	0.0085	0.0077	0.0070	0.0064	
	#10	1.27	0.0488	0.0326	0.0244	0.0195	0.0163	0.0140	0.0122	0.0109	0.0098	0.0089	0.0081	
	#11	1.56	0.0600	0.0400	0.0300	0.0240	0.0200	0.0171	0.0150	0.0133	0.0120	0.0109	0.0100	

TABLE C-8b   Steel Ratio $p = \dfrac{A_s}{bd}$ A_s in square inches; b and d in inches

d (inches)	Bar Size	Steel Area (sq. in.)	Steel Spacing (inches)											d (inches)
			8	12	16	20	24	28	32	36	40	44	48	
3.50	#3	0.11	0.0039	0.0026	0.0020	0.0016	0.0013	0.0011	0.0010	0.0009	0.0008	0.0007	0.0007	3.50
	#4	0.20	0.0071	0.0048	0.0036	0.0029	0.0024	0.0020	0.0018	0.0016	0.0014	0.0013	0.0012	
	#5	0.31	0.0111	0.0074	0.0055	0.0044	0.0037	0.0032	0.0028	0.0025	0.0022	0.0020	0.0018	
	#6	0.44	0.0157	0.0105	0.0079	0.0063	0.0052	0.0045	0.0039	0.0035	0.0031	0.0029	0.0026	
	#7	0.60	0.0214	0.0143	0.0107	0.0086	0.0071	0.0061	0.0054	0.0048	0.0043	0.0039	0.0036	
	#8	0.79	0.0282	0.0188	0.0141	0.0113	0.0094	0.0081	0.0071	0.0063	0.0056	0.0051	0.0047	
	#9	1.00	0.0357	0.0238	0.0179	0.0143	0.0119	0.0102	0.0089	0.0079	0.0071	0.0065	0.0060	
	#10	1.27	0.0454	0.0302	0.0227	0.0181	0.0151	0.0130	0.0113	0.0101	0.0091	0.0082	0.0076	
	#11	1.56	0.0557	0.0371	0.0279	0.0223	0.0186	0.0159	0.0139	0.0124	0.0111	0.0101	0.0093	

d (inches)	Bar Size	Steel Area (sq. in.)	Steel Spacing (inches)											d (inches)
			8	12	16	20	24	28	32	36	40	44	48	
3.75	#3	0.11	0.0037	0.0024	0.0018	0.0015	0.0012	0.0010	0.0009	0.0008	0.0007	0.0007	0.0006	3.75
	#4	0.20	0.0067	0.0044	0.0033	0.0027	0.0022	0.0019	0.0017	0.0015	0.0013	0.0012	0.0011	
	#5	0.31	0.0103	0.0069	0.0052	0.0041	0.0034	0.0030	0.0026	0.0023	0.0021	0.0019	0.0017	
	#6	0.44	0.0147	0.0098	0.0073	0.0059	0.0049	0.0042	0.0037	0.0033	0.0029	0.0027	0.0024	
	#7	0.60	0.0200	0.0133	0.0100	0.0080	0.0067	0.0057	0.0050	0.0044	0.0040	0.0036	0.0033	
	#8	0.79	0.0263	0.0176	0.0132	0.0105	0.0088	0.0075	0.0066	0.0059	0.0053	0.0048	0.0044	
	#9	1.00	0.0333	0.0222	0.0167	0.0133	0.0111	0.0095	0.0083	0.0074	0.0067	0.0061	0.0056	
	#10	1.27	0.0423	0.0282	0.0212	0.0169	0.0141	0.0121	0.0106	0.0094	0.0085	0.0077	0.0071	
	#11	1.56	0.0520	0.0347	0.0260	0.0208	0.0173	0.0149	0.0130	0.0116	0.0104	0.0095	0.0087	

d (inches)	Bar Size	Steel Area (sq. in.)	Steel Spacing (inches)											d (inches)
			8	12	16	20	24	28	32	36	40	44	48	
4.00	#3	0.11	0.0034	0.0023	0.0017	0.0014	0.0011	0.0010	0.0009	0.0008	0.0007	0.0006	0.0006	4.00
	#4	0.20	0.0063	0.0042	0.0031	0.0025	0.0021	0.0018	0.0016	0.0014	0.0013	0.0011	0.0010	
	#5	0.31	0.0097	0.0065	0.0048	0.0039	0.0032	0.0028	0.0024	0.0022	0.0019	0.0018	0.0016	
	#6	0.44	0.0138	0.0092	0.0069	0.0055	0.0046	0.0039	0.0034	0.0031	0.0028	0.0025	0.0023	
	#7	0.60	0.0188	0.0125	0.0094	0.0075	0.0063	0.0054	0.0047	0.0042	0.0038	0.0034	0.0031	
	#8	0.79	0.0247	0.0165	0.0123	0.0099	0.0082	0.0071	0.0062	0.0055	0.0049	0.0045	0.0041	
	#9	1.00	0.0313	0.0208	0.0156	0.0125	0.0104	0.0089	0.0078	0.0069	0.0063	0.0057	0.0052	
	#10	1.27	0.0397	0.0265	0.0198	0.0159	0.0132	0.0113	0.0099	0.0088	0.0079	0.0072	0.0066	
	#11	1.56	0.0488	0.0325	0.0244	0.0195	0.0163	0.0139	0.0122	0.0108	0.0098	0.0089	0.0081	

d (inches)	Bar Size	Steel Area (sq. in.)	Steel Spacing (inches)											d (inches)
			8	12	16	20	24	28	32	36	40	44	48	
4.25	#3	0.11	0.0032	0.0022	0.0016	0.0013	0.0011	0.0009	0.0008	0.0007	0.0006	0.0006	0.0005	4.25
	#4	0.20	0.0059	0.0039	0.0029	0.0024	0.0020	0.0017	0.0015	0.0013	0.0012	0.0011	0.0010	
	#5	0.31	0.0091	0.0061	0.0046	0.0036	0.0030	0.0026	0.0023	0.0020	0.0018	0.0017	0.0015	
	#6	0.44	0.0129	0.0086	0.0065	0.0052	0.0043	0.0037	0.0032	0.0029	0.0026	0.0024	0.0022	
	#7	0.60	0.0176	0.0118	0.0088	0.0071	0.0059	0.0050	0.0044	0.0039	0.0035	0.0032	0.0029	
	#8	0.79	0.0232	0.0155	0.0116	0.0093	0.0077	0.0066	0.0058	0.0052	0.0046	0.0042	0.0039	
	#9	1.00	0.0294	0.0196	0.0147	0.0118	0.0098	0.0084	0.0074	0.0065	0.0059	0.0053	0.0049	
	#10	1.27	0.0374	0.0249	0.0187	0.0149	0.0125	0.0107	0.0093	0.0083	0.0075	0.0068	0.0062	
	#11	1.56	0.0459	0.0306	0.0229	0.0184	0.0153	0.0131	0.0115	0.0102	0.0092	0.0083	0.0076	

TABLE C-8c   Steel Ratio $p = \dfrac{A_s}{bd}$ *A_s* in square inches; *b* and *d* in inches

WSD

d (inches)	Bar Size	Steel Area (sq. in.)	Steel Spacing (inches)											*d* (inches)
			8	12	16	20	24	28	32	36	40	44	48	
4.50	#3	0.11	0.0031	0.0020	0.0015	0.0012	0.0010	0.0009	0.0008	0.0007	0.0006	0.0006	0.0005	4.50
	#4	0.20	0.0056	0.0037	0.0028	0.0022	0.0019	0.0016	0.0014	0.0012	0.0011	0.0010	0.0009	
	#5	0.31	0.0086	0.0057	0.0043	0.0034	0.0029	0.0025	0.0022	0.0019	0.0017	0.0016	0.0014	
	#6	0.44	0.0122	0.0081	0.0061	0.0049	0.0041	0.0035	0.0031	0.0027	0.0024	0.0022	0.0020	
	#7	0.60	0.0167	0.0111	0.0083	0.0067	0.0056	0.0048	0.0042	0.0037	0.0033	0.0030	0.0028	
	#8	0.79	0.0219	0.0146	0.0110	0.0088	0.0073	0.0063	0.0055	0.0049	0.0044	0.0040	0.0037	
	#9	1.00	0.0278	0.0185	0.0139	0.0111	0.0093	0.0079	0.0069	0.0062	0.0056	0.0051	0.0046	
	#10	1.27	0.0353	0.0235	0.0176	0.0141	0.0118	0.0101	0.0088	0.0078	0.0071	0.0064	0.0059	
	#11	1.56	0.0433	0.0289	0.0217	0.0173	0.0144	0.0124	0.0108	0.0096	0.0087	0.0079	0.0072	

d (inches)	Bar Size	Steel Area (sq. in.)	Steel Spacing (inches)											*d* (inches)
			8	12	16	20	24	28	32	36	40	44	48	
4.75	#3	0.11	0.0029	0.0019	0.0014	0.0012	0.0010	0.0008	0.0007	0.0006	0.0006	0.0005	0.0005	4.75
	#4	0.20	0.0053	0.0035	0.0026	0.0021	0.0018	0.0015	0.0013	0.0012	0.0011	0.0010	0.0009	
	#5	0.31	0.0082	0.0054	0.0041	0.0033	0.0027	0.0023	0.0020	0.0018	0.0016	0.0015	0.0014	
	#6	0.44	0.0116	0.0077	0.0058	0.0046	0.0039	0.0033	0.0029	0.0026	0.0023	0.0021	0.0019	
	#7	0.60	0.0158	0.0105	0.0079	0.0063	0.0053	0.0045	0.0039	0.0035	0.0032	0.0029	0.0026	
	#8	0.79	0.0208	0.0139	0.0104	0.0083	0.0069	0.0059	0.0052	0.0046	0.0042	0.0038	0.0035	
	#9	1.00	0.0263	0.0175	0.0132	0.0105	0.0088	0.0075	0.0066	0.0058	0.0053	0.0048	0.0044	
	#10	1.27	0.0334	0.0223	0.0167	0.0134	0.0111	0.0095	0.0084	0.0074	0.0067	0.0061	0.0056	
	#11	1.56	0.0411	0.0274	0.0205	0.0164	0.0137	0.0117	0.0103	0.0091	0.0082	0.0075	0.0068	

d (inches)	Bar Size	Steel Area (sq. in.)	Steel Spacing (inches)											*d* (inches)
			8	12	16	20	24	28	32	36	40	44	48	
5.00	#3	0.11	0.0028	0.0018	0.0014	0.0011	0.0009	0.0008	0.0007	0.0006	0.0006	0.0005	0.0005	5.00
	#4	0.20	0.0050	0.0033	0.0025	0.0020	0.0017	0.0014	0.0013	0.0011	0.0010	0.0009	0.0008	
	#5	0.31	0.0078	0.0052	0.0039	0.0031	0.0026	0.0022	0.0019	0.0017	0.0016	0.0014	0.0013	
	#6	0.44	0.0110	0.0073	0.0055	0.0044	0.0037	0.0031	0.0028	0.0024	0.0022	0.0020	0.0018	
	#7	0.60	0.0150	0.0100	0.0075	0.0060	0.0050	0.0043	0.0038	0.0033	0.0030	0.0027	0.0025	
	#8	0.79	0.0198	0.0132	0.0099	0.0079	0.0066	0.0056	0.0049	0.0044	0.0040	0.0036	0.0033	
	#9	1.00	0.0250	0.0167	0.0125	0.0100	0.0083	0.0071	0.0063	0.0056	0.0050	0.0045	0.0042	
	#10	1.27	0.0318	0.0212	0.0159	0.0127	0.0106	0.0091	0.0079	0.0071	0.0064	0.0058	0.0053	
	#11	1.56	0.0390	0.0260	0.0195	0.0156	0.0130	0.0111	0.0098	0.0087	0.0078	0.0071	0.0065	

d (inches)	Bar Size	Steel Area (sq. in.)	Steel Spacing (inches)											*d* (inches)
			8	12	16	20	24	28	32	36	40	44	48	
5.25	#3	0.11	0.0026	0.0017	0.0013	0.0010	0.0009	0.0007	0.0007	0.0006	0.0005	0.0005	0.0004	5.25
	#4	0.20	0.0048	0.0032	0.0024	0.0019	0.0016	0.0014	0.0012	0.0011	0.0010	0.0009	0.0008	
	#5	0.31	0.0074	0.0049	0.0037	0.0030	0.0025	0.0021	0.0018	0.0016	0.0015	0.0013	0.0012	
	#6	0.44	0.0105	0.0070	0.0052	0.0042	0.0035	0.0030	0.0026	0.0023	0.0021	0.0019	0.0017	
	#7	0.60	0.0143	0.0095	0.0071	0.0057	0.0048	0.0041	0.0036	0.0032	0.0029	0.0026	0.0024	
	#8	0.79	0.0188	0.0125	0.0094	0.0075	0.0063	0.0054	0.0047	0.0042	0.0038	0.0034	0.0031	
	#9	1.00	0.0238	0.0159	0.0119	0.0095	0.0079	0.0068	0.0060	0.0053	0.0048	0.0043	0.0040	
	#10	1.27	0.0302	0.0202	0.0151	0.0121	0.0101	0.0086	0.0076	0.0067	0.0060	0.0055	0.0050	
	#11	1.56	0.0371	0.0248	0.0186	0.0149	0.0124	0.0106	0.0093	0.0083	0.0074	0.0068	0.0062	

TABLE C-8d   Steel Ratio $p = \dfrac{A_s}{bd}$ A_s in square inches; b and d in inches

d (inches)	Bar Size	Steel Area (sq. in.)	Steel Spacing (inches) 8	12	16	20	24	28	32	36	40	44	48	d (inches)
5.50	#3	0.11	0.0025	0.0017	0.0013	0.0010	0.0008	0.0007	0.0006	0.0006	0.0005	0.0005	0.0004	5.50
	#4	0.20	0.0045	0.0030	0.0023	0.0018	0.0015	0.0013	0.0011	0.0010	0.0009	0.0008	0.0008	
	#5	0.31	0.0070	0.0047	0.0035	0.0028	0.0023	0.0020	0.0018	0.0016	0.0014	0.0013	0.0012	
	#6	0.44	0.0100	0.0067	0.0050	0.0040	0.0033	0.0029	0.0025	0.0022	0.0020	0.0018	0.0017	
	#7	0.60	0.0136	0.0091	0.0068	0.0055	0.0045	0.0039	0.0034	0.0030	0.0027	0.0025	0.0023	
	#8	0.79	0.0180	0.0120	0.0090	0.0072	0.0060	0.0051	0.0045	0.0040	0.0036	0.0033	0.0030	
	#9	1.00	0.0227	0.0152	0.0114	0.0091	0.0076	0.0065	0.0057	0.0051	0.0045	0.0041	0.0038	
	#10	1.27	0.0289	0.0192	0.0144	0.0115	0.0096	0.0082	0.0072	0.0064	0.0058	0.0052	0.0048	
	#11	1.56	0.0355	0.0236	0.0177	0.0142	0.0118	0.0101	0.0089	0.0079	0.0071	0.0064	0.0059	

d (inches)	Bar Size	Steel Area (sq. in.)	Steel Spacing (inches) 8	12	16	20	24	28	32	36	40	44	48	d (inches)
5.75	#3	0.11	0.0024	0.0016	0.0012	0.0010	0.0008	0.0007	0.0006	0.0005	0.0005	0.0004	0.0004	5.75
	#4	0.20	0.0043	0.0029	0.0022	0.0017	0.0014	0.0012	0.0011	0.0010	0.0009	0.0008	0.0007	
	#5	0.31	0.0067	0.0045	0.0034	0.0027	0.0022	0.0019	0.0017	0.0015	0.0013	0.0012	0.0011	
	#6	0.44	0.0096	0.0064	0.0048	0.0038	0.0032	0.0027	0.0024	0.0021	0.0019	0.0017	0.0016	
	#7	0.60	0.0130	0.0087	0.0065	0.0052	0.0043	0.0037	0.0033	0.0029	0.0026	0.0024	0.0022	
	#8	0.79	0.0172	0.0114	0.0086	0.0069	0.0057	0.0049	0.0043	0.0038	0.0034	0.0031	0.0029	
	#9	1.00	0.0217	0.0145	0.0109	0.0087	0.0072	0.0062	0.0054	0.0048	0.0043	0.0040	0.0036	
	#10	1.27	0.0276	0.0184	0.0138	0.0110	0.0092	0.0079	0.0069	0.0061	0.0055	0.0050	0.0046	
	#11	1.56	0.0339	0.0226	0.0170	0.0136	0.0113	0.0097	0.0085	0.0075	0.0068	0.0062	0.0057	

d (inches)	Bar Size	Steel Area (sq. in.)	Steel Spacing (inches) 8	12	16	20	24	28	32	36	40	44	48	d (inches)
6.00	#3	0.11	0.0023	0.0015	0.0011	0.0009	0.0008	0.0007	0.0006	0.0005	0.0005	0.0004	0.0004	6.00
	#4	0.20	0.0042	0.0028	0.0021	0.0017	0.0014	0.0012	0.0010	0.0009	0.0008	0.0008	0.0007	
	#5	0.31	0.0065	0.0043	0.0032	0.0026	0.0022	0.0018	0.0016	0.0014	0.0013	0.0012	0.0011	
	#6	0.44	0.0092	0.0061	0.0046	0.0037	0.0031	0.0026	0.0023	0.0020	0.0018	0.0017	0.0015	
	#7	0.60	0.0125	0.0083	0.0063	0.0050	0.0042	0.0036	0.0031	0.0028	0.0025	0.0023	0.0021	
	#8	0.79	0.0165	0.0110	0.0082	0.0066	0.0055	0.0047	0.0041	0.0037	0.0033	0.0030	0.0027	
	#9	1.00	0.0208	0.0139	0.0104	0.0083	0.0069	0.0060	0.0052	0.0046	0.0042	0.0038	0.0035	
	#10	1.27	0.0265	0.0176	0.0132	0.0106	0.0088	0.0076	0.0066	0.0059	0.0053	0.0048	0.0044	
	#11	1.56	0.0325	0.0217	0.0163	0.0130	0.0108	0.0093	0.0081	0.0072	0.0065	0.0059	0.0054	

d (inches)	Bar Size	Steel Area (sq. in.)	Steel Spacing (inches) 8	12	16	20	24	28	32	36	40	44	48	d (inches)
6.25	#3	0.11	0.0022	0.0015	0.0011	0.0009	0.0007	0.0006	0.0006	0.0005	0.0004	0.0004	0.0004	6.25
	#4	0.20	0.0040	0.0027	0.0020	0.0016	0.0013	0.0011	0.0010	0.0009	0.0008	0.0007	0.0007	
	#5	0.31	0.0062	0.0041	0.0031	0.0025	0.0021	0.0018	0.0016	0.0014	0.0012	0.0011	0.0010	
	#6	0.44	0.0088	0.0059	0.0044	0.0035	0.0029	0.0025	0.0022	0.0020	0.0018	0.0016	0.0015	
	#7	0.60	0.0120	0.0080	0.0060	0.0048	0.0040	0.0034	0.0030	0.0027	0.0024	0.0022	0.0020	
	#8	0.79	0.0158	0.0105	0.0079	0.0063	0.0053	0.0045	0.0040	0.0035	0.0032	0.0029	0.0026	
	#9	1.00	0.0200	0.0133	0.0100	0.0080	0.0067	0.0057	0.0050	0.0044	0.0040	0.0036	0.0033	
	#10	1.27	0.0254	0.0169	0.0127	0.0102	0.0085	0.0073	0.0064	0.0056	0.0051	0.0046	0.0042	
	#11	1.56	0.0312	0.0208	0.0156	0.0125	0.0104	0.0089	0.0078	0.0069	0.0062	0.0057	0.0052	

TABLE C-8e Steel Ratio $p = \dfrac{A_s}{bd}$ _A_$_s$ in square inches; _b_ and _d_ in inches

WSD

d (inches)	Bar Size	Steel Area (sq. in.)	Steel Spacing (inches) 8	12	16	20	24	28	32	36	40	44	48	d (inches)
6.50	#3	0.11	0.0021	0.0014	0.0011	0.0008	0.0007	0.0006	0.0005	0.0005	0.0004	0.0004	0.0004	6.50
	#4	0.20	0.0038	0.0026	0.0019	0.0015	0.0013	0.0011	0.0010	0.0009	0.0008	0.0007	0.0006	
	#5	0.31	0.0060	0.0040	0.0030	0.0024	0.0020	0.0017	0.0015	0.0013	0.0012	0.0011	0.0010	
	#6	0.44	0.0085	0.0056	0.0042	0.0034	0.0028	0.0024	0.0021	0.0019	0.0017	0.0015	0.0014	
	#7	0.60	0.0115	0.0077	0.0058	0.0046	0.0038	0.0033	0.0029	0.0026	0.0023	0.0021	0.0019	
	#8	0.79	0.0152	0.0101	0.0076	0.0061	0.0051	0.0043	0.0038	0.0034	0.0030	0.0028	0.0025	
	#9	1.00	0.0192	0.0128	0.0096	0.0077	0.0064	0.0055	0.0048	0.0043	0.0038	0.0035	0.0032	
	#10	1.27	0.0244	0.0163	0.0122	0.0098	0.0081	0.0070	0.0061	0.0054	0.0049	0.0044	0.0041	
	#11	1.56	0.0300	0.0200	0.0150	0.0120	0.0100	0.0086	0.0075	0.0067	0.0060	0.0055	0.0050	

d (inches)	Bar Size	Steel Area (sq. in.)	Steel Spacing (inches) 8	12	16	20	24	28	32	36	40	44	48	d (inches)
6.75	#3	0.11	0.0020	0.0014	0.0010	0.0008	0.0007	0.0006	0.0005	0.0005	0.0004	0.0004	0.0003	6.75
	#4	0.20	0.0037	0.0025	0.0019	0.0015	0.0012	0.0011	0.0009	0.0008	0.0007	0.0007	0.0006	
	#5	0.31	0.0057	0.0038	0.0029	0.0023	0.0019	0.0016	0.0014	0.0013	0.0011	0.0010	0.0010	
	#6	0.44	0.0081	0.0054	0.0041	0.0033	0.0027	0.0023	0.0020	0.0018	0.0016	0.0015	0.0014	
	#7	0.60	0.0111	0.0074	0.0056	0.0044	0.0037	0.0032	0.0028	0.0025	0.0022	0.0020	0.0019	
	#8	0.79	0.0146	0.0098	0.0073	0.0059	0.0049	0.0042	0.0037	0.0033	0.0029	0.0027	0.0024	
	#9	1.00	0.0185	0.0123	0.0093	0.0074	0.0062	0.0053	0.0046	0.0041	0.0037	0.0034	0.0031	
	#10	1.27	0.0235	0.0157	0.0118	0.0094	0.0078	0.0067	0.0059	0.0052	0.0047	0.0043	0.0039	
	#11	1.56	0.0289	0.0193	0.0144	0.0116	0.0096	0.0083	0.0072	0.0064	0.0058	0.0053	0.0048	

d (inches)	Bar Size	Steel Area (sq. in.)	Steel Spacing (inches) 8	12	16	20	24	28	32	36	40	44	48	d (inches)
7.00	#3	0.11	0.0020	0.0013	0.0010	0.0008	0.0007	0.0006	0.0005	0.0004	0.0004	0.0004	0.0003	7.00
	#4	0.20	0.0036	0.0024	0.0018	0.0014	0.0012	0.0010	0.0009	0.0008	0.0007	0.0006	0.0006	
	#5	0.31	0.0055	0.0037	0.0028	0.0022	0.0018	0.0016	0.0014	0.0012	0.0011	0.0010	0.0009	
	#6	0.44	0.0079	0.0052	0.0039	0.0031	0.0026	0.0022	0.0020	0.0017	0.0016	0.0014	0.0013	
	#7	0.60	0.0107	0.0071	0.0054	0.0043	0.0036	0.0031	0.0027	0.0024	0.0021	0.0019	0.0018	
	#8	0.79	0.0141	0.0094	0.0071	0.0056	0.0047	0.0040	0.0035	0.0031	0.0028	0.0026	0.0024	
	#9	1.00	0.0179	0.0119	0.0089	0.0071	0.0060	0.0051	0.0045	0.0040	0.0036	0.0032	0.0030	
	#10	1.27	0.0227	0.0151	0.0113	0.0091	0.0076	0.0065	0.0057	0.0050	0.0045	0.0041	0.0038	
	#11	1.56	0.0279	0.0186	0.0139	0.0111	0.0093	0.0080	0.0070	0.0062	0.0056	0.0051	0.0046	

d (inches)	Bar Size	Steel Area (sq. in.)	Steel Spacing (inches) 8	12	16	20	24	28	32	36	40	44	48	d (inches)
7.25	#3	0.11	0.0019	0.0013	0.0009	0.0008	0.0006	0.0005	0.0005	0.0004	0.0004	0.0003	0.0003	7.25
	#4	0.20	0.0034	0.0023	0.0017	0.0014	0.0011	0.0010	0.0009	0.0008	0.0007	0.0006	0.0006	
	#5	0.31	0.0053	0.0036	0.0027	0.0021	0.0018	0.0015	0.0013	0.0012	0.0011	0.0010	0.0009	
	#6	0.44	0.0076	0.0051	0.0038	0.0030	0.0025	0.0022	0.0019	0.0017	0.0015	0.0014	0.0013	
	#7	0.60	0.0103	0.0069	0.0052	0.0041	0.0034	0.0030	0.0026	0.0023	0.0021	0.0019	0.0017	
	#8	0.79	0.0136	0.0091	0.0068	0.0054	0.0045	0.0039	0.0034	0.0030	0.0027	0.0025	0.0023	
	#9	1.00	0.0172	0.0115	0.0086	0.0069	0.0057	0.0049	0.0043	0.0038	0.0034	0.0031	0.0029	
	#10	1.27	0.0219	0.0146	0.0109	0.0088	0.0073	0.0063	0.0055	0.0049	0.0044	0.0040	0.0036	
	#11	1.56	0.0269	0.0179	0.0134	0.0108	0.0090	0.0077	0.0067	0.0060	0.0054	0.0049	0.0045	

TABLE C-8f   Steel Ratio $p = \dfrac{A_s}{bd}$ A_s in square inches; b and d in inches

d (inches)	Bar Size	Steel Area (sq. in.)	Steel Spacing (inches)											d (inches)
			8	12	16	20	24	28	32	36	40	44	48	
7.50	#3	0.11	0.0018	0.0012	0.0009	0.0007	0.0006	0.0005	0.0005	0.0004	0.0004	0.0003	0.0003	7.50
	#4	0.20	0.0033	0.0022	0.0017	0.0013	0.0011	0.0010	0.0008	0.0007	0.0007	0.0006	0.0006	
	#5	0.31	0.0052	0.0034	0.0026	0.0021	0.0017	0.0015	0.0013	0.0011	0.0010	0.0009	0.0009	
	#6	0.44	0.0073	0.0049	0.0037	0.0029	0.0024	0.0021	0.0018	0.0016	0.0015	0.0013	0.0012	
	#7	0.60	0.0100	0.0067	0.0050	0.0040	0.0033	0.0029	0.0025	0.0022	0.0020	0.0018	0.0017	
	#8	0.79	0.0132	0.0088	0.0066	0.0053	0.0044	0.0038	0.0033	0.0029	0.0026	0.0024	0.0022	
	#9	1.00	0.0167	0.0111	0.0083	0.0067	0.0056	0.0048	0.0042	0.0037	0.0033	0.0030	0.0028	
	#10	1.27	0.0212	0.0141	0.0106	0.0085	0.0071	0.0060	0.0053	0.0047	0.0042	0.0038	0.0035	
	#11	1.56	0.0260	0.0173	0.0130	0.0104	0.0087	0.0074	0.0065	0.0058	0.0052	0.0047	0.0043	

d (inches)	Bar Size	Steel Area (sq. in.)	Steel Spacing (inches)											d (inches)
			8	12	16	20	24	28	32	36	40	44	48	
7.75	#3	0.11	0.0018	0.0012	0.0009	0.0007	0.0006	0.0005	0.0004	0.0004	0.0004	0.0003	0.0003	7.75
	#4	0.20	0.0032	0.0022	0.0016	0.0013	0.0011	0.0009	0.0008	0.0007	0.0006	0.0006	0.0005	
	#5	0.31	0.0050	0.0033	0.0025	0.0020	0.0017	0.0014	0.0013	0.0011	0.0010	0.0009	0.0008	
	#6	0.44	0.0071	0.0047	0.0035	0.0028	0.0024	0.0020	0.0018	0.0016	0.0014	0.0013	0.0012	
	#7	0.60	0.0097	0.0065	0.0048	0.0039	0.0032	0.0028	0.0024	0.0022	0.0019	0.0018	0.0016	
	#8	0.79	0.0127	0.0085	0.0064	0.0051	0.0042	0.0036	0.0032	0.0028	0.0025	0.0023	0.0021	
	#9	1.00	0.0161	0.0108	0.0081	0.0065	0.0054	0.0046	0.0040	0.0036	0.0032	0.0029	0.0027	
	#10	1.27	0.0205	0.0137	0.0102	0.0082	0.0068	0.0059	0.0051	0.0046	0.0041	0.0037	0.0034	
	#11	1.56	0.0252	0.0168	0.0126	0.0101	0.0084	0.0072	0.0063	0.0056	0.0050	0.0046	0.0042	

d (inches)	Bar Size	Steel Area (sq. in.)	Steel Spacing (inches)											d (inches)
			8	12	16	20	24	28	32	36	40	44	48	
8.00	#3	0.11	0.0017	0.0011	0.0009	0.0007	0.0006	0.0005	0.0004	0.0004	0.0003	0.0003	0.0003	8.00
	#4	0.20	0.0031	0.0021	0.0016	0.0013	0.0010	0.0009	0.0008	0.0007	0.0006	0.0006	0.0005	
	#5	0.31	0.0048	0.0032	0.0024	0.0019	0.0016	0.0014	0.0012	0.0011	0.0010	0.0009	0.0008	
	#6	0.44	0.0069	0.0046	0.0034	0.0028	0.0023	0.0020	0.0017	0.0015	0.0014	0.0013	0.0011	
	#7	0.60	0.0094	0.0063	0.0047	0.0038	0.0031	0.0027	0.0023	0.0021	0.0019	0.0017	0.0016	
	#8	0.79	0.0123	0.0082	0.0062	0.0049	0.0041	0.0035	0.0031	0.0027	0.0025	0.0022	0.0021	
	#9	1.00	0.0156	0.0104	0.0078	0.0063	0.0052	0.0045	0.0039	0.0035	0.0031	0.0028	0.0026	
	#10	1.27	0.0198	0.0132	0.0099	0.0079	0.0066	0.0057	0.0050	0.0044	0.0040	0.0036	0.0033	
	#11	1.56	0.0244	0.0163	0.0122	0.0098	0.0081	0.0070	0.0061	0.0054	0.0049	0.0044	0.0041	

d (inches)	Bar Size	Steel Area (sq. in.)	Steel Spacing (inches)											d (inches)
			8	12	16	20	24	28	32	36	40	44	48	
8.25	#3	0.11	0.0017	0.0011	0.0008	0.0007	0.0006	0.0005	0.0004	0.0004	0.0003	0.0003	0.0003	8.25
	#4	0.20	0.0030	0.0020	0.0015	0.0012	0.0010	0.0009	0.0008	0.0007	0.0006	0.0006	0.0005	
	#5	0.31	0.0047	0.0031	0.0023	0.0019	0.0016	0.0013	0.0012	0.0010	0.0009	0.0009	0.0008	
	#6	0.44	0.0067	0.0044	0.0033	0.0027	0.0022	0.0019	0.0017	0.0015	0.0013	0.0012	0.0011	
	#7	0.60	0.0091	0.0061	0.0045	0.0036	0.0030	0.0026	0.0023	0.0020	0.0018	0.0017	0.0015	
	#8	0.79	0.0120	0.0080	0.0060	0.0048	0.0040	0.0034	0.0030	0.0027	0.0024	0.0022	0.0020	
	#9	1.00	0.0152	0.0101	0.0076	0.0061	0.0051	0.0043	0.0038	0.0034	0.0030	0.0028	0.0025	
	#10	1.27	0.0192	0.0128	0.0096	0.0077	0.0064	0.0055	0.0048	0.0043	0.0038	0.0035	0.0032	
	#11	1.56	0.0236	0.0158	0.0118	0.0095	0.0079	0.0068	0.0059	0.0053	0.0047	0.0043	0.0039	

TABLE C-8g   Steel Ratio $p = \dfrac{A_s}{bd}$ A_s in square inches; b and d in inches

d (inches)	Bar Size	Steel Area (sq. in.)	Steel Spacing (inches)											d (inches)
			8	12	16	20	24	28	32	36	40	44	48	
8.50	#3	0.11	0.0016	0.0011	0.0008	0.0006	0.0005	0.0005	0.0004	0.0004	0.0003	0.0003	0.0003	8.50
	#4	0.20	0.0029	0.0020	0.0015	0.0012	0.0010	0.0008	0.0007	0.0007	0.0006	0.0005	0.0005	
	#5	0.31	0.0046	0.0030	0.0023	0.0018	0.0015	0.0013	0.0011	0.0010	0.0009	0.0008	0.0008	
	#6	0.44	0.0065	0.0043	0.0032	0.0026	0.0022	0.0018	0.0016	0.0014	0.0013	0.0012	0.0011	
	#7	0.60	0.0088	0.0059	0.0044	0.0035	0.0029	0.0025	0.0022	0.0020	0.0018	0.0016	0.0015	
	#8	0.79	0.0116	0.0077	0.0058	0.0046	0.0039	0.0033	0.0029	0.0026	0.0023	0.0021	0.0019	
	#9	1.00	0.0147	0.0098	0.0074	0.0059	0.0049	0.0042	0.0037	0.0033	0.0029	0.0027	0.0025	
	#10	1.27	0.0187	0.0125	0.0093	0.0075	0.0062	0.0053	0.0047	0.0042	0.0037	0.0034	0.0031	
	#11	1.56	0.0229	0.0153	0.0115	0.0092	0.0076	0.0066	0.0057	0.0051	0.0046	0.0042	0.0038	

d (inches)	Bar Size	Steel Area (sq. in.)	Steel Spacing (inches)											d (inches)
			8	12	16	20	24	28	32	36	40	44	48	
8.75	#3	0.11	0.0016	0.0010	0.0008	0.0006	0.0005	0.0004	0.0004	0.0003	0.0003	0.0003	0.0003	8.75
	#4	0.20	0.0029	0.0019	0.0014	0.0011	0.0010	0.0008	0.0007	0.0006	0.0006	0.0005	0.0005	
	#5	0.31	0.0044	0.0030	0.0022	0.0018	0.0015	0.0013	0.0011	0.0010	0.0009	0.0008	0.0007	
	#6	0.44	0.0063	0.0042	0.0031	0.0025	0.0021	0.0018	0.0016	0.0014	0.0013	0.0011	0.0010	
	#7	0.60	0.0086	0.0057	0.0043	0.0034	0.0029	0.0024	0.0021	0.0019	0.0017	0.0016	0.0014	
	#8	0.79	0.0113	0.0075	0.0056	0.0045	0.0038	0.0032	0.0028	0.0025	0.0023	0.0021	0.0019	
	#9	1.00	0.0143	0.0095	0.0071	0.0057	0.0048	0.0041	0.0036	0.0032	0.0029	0.0026	0.0024	
	#10	1.27	0.0181	0.0121	0.0091	0.0073	0.0060	0.0052	0.0045	0.0040	0.0036	0.0033	0.0030	
	#11	1.56	0.0223	0.0149	0.0111	0.0089	0.0074	0.0064	0.0056	0.0050	0.0045	0.0041	0.0037	

d (inches)	Bar Size	Steel Area (sq. in.)	Steel Spacing (inches)											d (inches)
			8	12	16	20	24	28	32	36	40	44	48	
9.00	#3	0.11	0.0015	0.0010	0.0008	0.0006	0.0005	0.0004	0.0004	0.0003	0.0003	0.0003	0.0003	9.00
	#4	0.20	0.0028	0.0019	0.0014	0.0011	0.0009	0.0008	0.0007	0.0006	0.0006	0.0005	0.0005	
	#5	0.31	0.0043	0.0029	0.0022	0.0017	0.0014	0.0012	0.0011	0.0010	0.0009	0.0008	0.0007	
	#6	0.44	0.0061	0.0041	0.0031	0.0024	0.0020	0.0017	0.0015	0.0014	0.0012	0.0011	0.0010	
	#7	0.60	0.0083	0.0056	0.0042	0.0033	0.0028	0.0024	0.0021	0.0019	0.0017	0.0015	0.0014	
	#8	0.79	0.0110	0.0073	0.0055	0.0044	0.0037	0.0031	0.0027	0.0024	0.0022	0.0020	0.0018	
	#9	1.00	0.0139	0.0093	0.0069	0.0056	0.0046	0.0040	0.0035	0.0031	0.0028	0.0025	0.0023	
	#10	1.27	0.0176	0.0118	0.0088	0.0071	0.0059	0.0050	0.0044	0.0039	0.0035	0.0032	0.0029	
	#11	1.56	0.0217	0.0144	0.0108	0.0087	0.0072	0.0062	0.0054	0.0048	0.0043	0.0039	0.0036	

d (inches)	Bar Size	Steel Area (sq. in.)	Steel Spacing (inches)											d (inches)
			8	12	16	20	24	28	32	36	40	44	48	
9.25	#3	0.11	0.0015	0.0010	0.0007	0.0006	0.0005	0.0004	0.0004	0.0003	0.0003	0.0003	0.0002	9.25
	#4	0.20	0.0027	0.0018	0.0014	0.0011	0.0009	0.0008	0.0007	0.0006	0.0005	0.0005	0.0005	
	#5	0.31	0.0042	0.0028	0.0021	0.0017	0.0014	0.0012	0.0010	0.0009	0.0008	0.0008	0.0007	
	#6	0.44	0.0059	0.0040	0.0030	0.0024	0.0020	0.0017	0.0015	0.0013	0.0012	0.0011	0.0010	
	#7	0.60	0.0081	0.0054	0.0041	0.0032	0.0027	0.0023	0.0020	0.0018	0.0016	0.0015	0.0014	
	#8	0.79	0.0107	0.0071	0.0053	0.0043	0.0036	0.0031	0.0027	0.0024	0.0021	0.0019	0.0018	
	#9	1.00	0.0135	0.0090	0.0068	0.0054	0.0045	0.0039	0.0034	0.0030	0.0027	0.0025	0.0023	
	#10	1.27	0.0172	0.0114	0.0086	0.0069	0.0057	0.0049	0.0043	0.0038	0.0034	0.0031	0.0029	
	#11	1.56	0.0211	0.0141	0.0105	0.0084	0.0070	0.0060	0.0053	0.0047	0.0042	0.0038	0.0035	

TABLE C-8h   Steel Ratio $p = \dfrac{A_s}{bd}$ A_s in square inches; b and d in inches

d (inches)	Bar Size	Steel Area (sq. in.)	Steel Spacing (inches)											d (inches)
			8	12	16	20	24	28	32	36	40	44	48	
9.50	#3	0.11	0.0014	0.0010	0.0007	0.0006	0.0005	0.0004	0.0004	0.0003	0.0003	0.0003	0.0002	9.50
	#4	0.20	0.0026	0.0018	0.0013	0.0011	0.0009	0.0008	0.0007	0.0006	0.0005	0.0005	0.0004	
	#5	0.31	0.0041	0.0027	0.0020	0.0016	0.0014	0.0012	0.0010	0.0009	0.0008	0.0007	0.0007	
	#6	0.44	0.0058	0.0039	0.0029	0.0023	0.0019	0.0017	0.0014	0.0013	0.0012	0.0011	0.0010	
	#7	0.60	0.0079	0.0053	0.0039	0.0032	0.0026	0.0023	0.0020	0.0018	0.0016	0.0014	0.0013	
	#8	0.79	0.0104	0.0069	0.0052	0.0042	0.0035	0.0030	0.0026	0.0023	0.0021	0.0019	0.0017	
	#9	1.00	0.0132	0.0088	0.0066	0.0053	0.0044	0.0038	0.0033	0.0029	0.0026	0.0024	0.0022	
	#10	1.27	0.0167	0.0111	0.0084	0.0067	0.0056	0.0048	0.0042	0.0037	0.0033	0.0030	0.0028	
	#11	1.56	0.0205	0.0137	0.0103	0.0082	0.0068	0.0059	0.0051	0.0046	0.0041	0.0037	0.0034	

d (inches)	Bar Size	Steel Area (sq. in.)	Steel Spacing (inches)											d (inches)
			8	12	16	20	24	28	32	36	40	44	48	
9.75	#3	0.11	0.0014	0.0009	0.0007	0.0006	0.0005	0.0004	0.0004	0.0003	0.0003	0.0003	0.0002	9.75
	#4	0.20	0.0026	0.0017	0.0013	0.0010	0.0009	0.0007	0.0006	0.0006	0.0005	0.0005	0.0004	
	#5	0.31	0.0040	0.0026	0.0020	0.0016	0.0013	0.0011	0.0010	0.0009	0.0008	0.0007	0.0007	
	#6	0.44	0.0056	0.0038	0.0028	0.0023	0.0019	0.0016	0.0014	0.0013	0.0011	0.0010	0.0009	
	#7	0.60	0.0077	0.0051	0.0038	0.0031	0.0026	0.0022	0.0019	0.0017	0.0015	0.0014	0.0013	
	#8	0.79	0.0101	0.0068	0.0051	0.0041	0.0034	0.0029	0.0025	0.0023	0.0020	0.0018	0.0017	
	#9	1.00	0.0128	0.0085	0.0064	0.0051	0.0043	0.0037	0.0032	0.0028	0.0026	0.0023	0.0021	
	#10	1.27	0.0163	0.0109	0.0081	0.0065	0.0054	0.0047	0.0041	0.0036	0.0033	0.0030	0.0027	
	#11	1.56	0.0200	0.0133	0.0100	0.0080	0.0067	0.0057	0.0050	0.0044	0.0040	0.0036	0.0033	

d (inches)	Bar Size	Steel Area (sq. in.)	Steel Spacing (inches)											d (inches)
			8	12	16	20	24	28	32	36	40	44	48	
10.00	#3	0.11	0.0014	0.0009	0.0007	0.0006	0.0005	0.0004	0.0003	0.0003	0.0003	0.0003	0.0002	10.00
	#4	0.20	0.0025	0.0017	0.0013	0.0010	0.0008	0.0007	0.0006	0.0006	0.0005	0.0005	0.0004	
	#5	0.31	0.0039	0.0026	0.0019	0.0016	0.0013	0.0011	0.0010	0.0009	0.0008	0.0007	0.0006	
	#6	0.44	0.0055	0.0037	0.0028	0.0022	0.0018	0.0016	0.0014	0.0012	0.0011	0.0010	0.0009	
	#7	0.60	0.0075	0.0050	0.0038	0.0030	0.0025	0.0021	0.0019	0.0017	0.0015	0.0014	0.0013	
	#8	0.79	0.0099	0.0066	0.0049	0.0040	0.0033	0.0028	0.0025	0.0022	0.0020	0.0018	0.0016	
	#9	1.00	0.0125	0.0083	0.0063	0.0050	0.0042	0.0036	0.0031	0.0028	0.0025	0.0023	0.0021	
	#10	1.27	0.0159	0.0106	0.0079	0.0064	0.0053	0.0045	0.0040	0.0035	0.0032	0.0029	0.0026	
	#11	1.56	0.0195	0.0130	0.0098	0.0078	0.0065	0.0056	0.0049	0.0043	0.0039	0.0035	0.0033	

d (inches)	Bar Size	Steel Area (sq. in.)	Steel Spacing (inches)											d (inches)
			8	12	16	20	24	28	32	36	40	44	48	
10.25	#3	0.11	0.0013	0.0009	0.0007	0.0005	0.0004	0.0004	0.0003	0.0003	0.0003	0.0002	0.0002	10.25
	#4	0.20	0.0024	0.0016	0.0012	0.0010	0.0008	0.0007	0.0006	0.0005	0.0005	0.0004	0.0004	
	#5	0.31	0.0038	0.0025	0.0019	0.0015	0.0013	0.0011	0.0009	0.0008	0.0008	0.0007	0.0006	
	#6	0.44	0.0054	0.0036	0.0027	0.0021	0.0018	0.0015	0.0013	0.0012	0.0011	0.0010	0.0009	
	#7	0.60	0.0073	0.0049	0.0037	0.0029	0.0024	0.0021	0.0018	0.0016	0.0015	0.0013	0.0012	
	#8	0.79	0.0096	0.0064	0.0048	0.0039	0.0032	0.0028	0.0024	0.0021	0.0019	0.0018	0.0016	
	#9	1.00	0.0122	0.0081	0.0061	0.0049	0.0041	0.0035	0.0030	0.0027	0.0024	0.0022	0.0020	
	#10	1.27	0.0155	0.0103	0.0077	0.0062	0.0052	0.0044	0.0039	0.0034	0.0031	0.0028	0.0026	
	#11	1.56	0.0190	0.0127	0.0095	0.0076	0.0063	0.0054	0.0048	0.0042	0.0038	0.0035	0.0032	

TABLE C-8i   Steel Ratio $p = \dfrac{A_s}{bd}$ *A_s* in square inches; *b* and *d* in inches

d (inches)	Bar Size	Steel Area (sq. in.)	Steel Spacing (inches)											*d* (inches)
			8	12	16	20	24	28	32	36	40	44	48	
	#3	0.11	0.0013	0.0009	0.0007	0.0005	0.0004	0.0004	0.0003	0.0003	0.0003	0.0002	0.0002	
	#4	0.20	0.0024	0.0016	0.0012	0.0010	0.0008	0.0007	0.0006	0.0005	0.0005	0.0004	0.0004	
	#5	0.31	0.0037	0.0025	0.0018	0.0015	0.0012	0.0011	0.0009	0.0008	0.0007	0.0007	0.0006	
	#6	0.44	0.0052	0.0035	0.0026	0.0021	0.0017	0.0015	0.0013	0.0012	0.0010	0.0010	0.0009	
10.50	#7	0.60	0.0071	0.0048	0.0036	0.0029	0.0024	0.0020	0.0018	0.0016	0.0014	0.0013	0.0012	10.50
	#8	0.79	0.0094	0.0063	0.0047	0.0038	0.0031	0.0027	0.0024	0.0021	0.0019	0.0017	0.0016	
	#9	1.00	0.0119	0.0079	0.0060	0.0048	0.0040	0.0034	0.0030	0.0026	0.0024	0.0022	0.0020	
	#10	1.27	0.0151	0.0101	0.0076	0.0060	0.0050	0.0043	0.0038	0.0034	0.0030	0.0027	0.0025	
	#11	1.56	0.0186	0.0124	0.0093	0.0074	0.0062	0.0053	0.0046	0.0041	0.0037	0.0034	0.0031	

d (inches)	Bar Size	Steel Area (sq. in.)	Steel Spacing (inches)											*d* (inches)
			8	12	16	20	24	28	32	36	40	44	48	
	#3	0.11	0.0013	0.0009	0.0006	0.0005	0.0004	0.0004	0.0003	0.0003	0.0003	0.0002	0.0002	
	#4	0.20	0.0023	0.0016	0.0012	0.0009	0.0008	0.0007	0.0006	0.0005	0.0005	0.0004	0.0004	
	#5	0.31	0.0036	0.0024	0.0018	0.0014	0.0012	0.0010	0.0009	0.0008	0.0007	0.0007	0.0006	
	#6	0.44	0.0051	0.0034	0.0026	0.0020	0.0017	0.0015	0.0013	0.0011	0.0010	0.0009	0.0009	
10.75	#7	0.60	0.0070	0.0047	0.0035	0.0028	0.0023	0.0020	0.0017	0.0016	0.0014	0.0013	0.0012	10.75
	#8	0.79	0.0092	0.0061	0.0046	0.0037	0.0031	0.0026	0.0023	0.0020	0.0018	0.0017	0.0015	
	#9	1.00	0.0116	0.0078	0.0058	0.0047	0.0039	0.0033	0.0029	0.0026	0.0023	0.0021	0.0019	
	#10	1.27	0.0148	0.0098	0.0074	0.0059	0.0049	0.0042	0.0037	0.0033	0.0030	0.0027	0.0025	
	#11	1.56	0.0181	0.0121	0.0091	0.0073	0.0060	0.0052	0.0045	0.0040	0.0036	0.0033	0.0030	

d (inches)	Bar Size	Steel Area (sq. in.)	Steel Spacing (inches)											*d* (inches)
			8	12	16	20	24	28	32	36	40	44	48	
	#3	0.11	0.0013	0.0008	0.0006	0.0005	0.0004	0.0004	0.0003	0.0003	0.0003	0.0002	0.0002	
	#4	0.20	0.0023	0.0015	0.0011	0.0009	0.0008	0.0006	0.0006	0.0005	0.0005	0.0004	0.0004	
	#5	0.31	0.0035	0.0023	0.0018	0.0014	0.0012	0.0010	0.0009	0.0008	0.0007	0.0006	0.0006	
	#6	0.44	0.0050	0.0033	0.0025	0.0020	0.0017	0.0014	0.0013	0.0011	0.0010	0.0009	0.0008	
11.00	#7	0.60	0.0068	0.0045	0.0034	0.0027	0.0023	0.0019	0.0017	0.0015	0.0014	0.0012	0.0011	11.00
	#8	0.79	0.0090	0.0060	0.0045	0.0036	0.0030	0.0026	0.0022	0.0020	0.0018	0.0016	0.0015	
	#9	1.00	0.0114	0.0076	0.0057	0.0045	0.0038	0.0032	0.0028	0.0025	0.0023	0.0021	0.0019	
	#10	1.27	0.0144	0.0096	0.0072	0.0058	0.0048	0.0041	0.0036	0.0032	0.0029	0.0026	0.0024	
	#11	1.56	0.0177	0.0118	0.0089	0.0071	0.0059	0.0051	0.0044	0.0039	0.0035	0.0032	0.0030	

d (inches)	Bar Size	Steel Area (sq. in.)	Steel Spacing (inches)											*d* (inches)
			8	12	16	20	24	28	32	36	40	44	48	
	#3	0.11	0.0012	0.0008	0.0006	0.0005	0.0004	0.0003	0.0003	0.0003	0.0002	0.0002	0.0002	
	#4	0.20	0.0022	0.0015	0.0011	0.0009	0.0007	0.0006	0.0006	0.0005	0.0004	0.0004	0.0004	
	#5	0.31	0.0034	0.0023	0.0017	0.0014	0.0011	0.0010	0.0009	0.0008	0.0007	0.0006	0.0006	
	#6	0.44	0.0049	0.0033	0.0024	0.0020	0.0016	0.0014	0.0012	0.0011	0.0010	0.0009	0.0008	
11.25	#7	0.60	0.0067	0.0044	0.0033	0.0027	0.0022	0.0019	0.0017	0.0015	0.0013	0.0012	0.0011	11.25
	#8	0.79	0.0088	0.0059	0.0044	0.0035	0.0029	0.0025	0.0022	0.0020	0.0018	0.0016	0.0015	
	#9	1.00	0.0111	0.0074	0.0056	0.0044	0.0037	0.0032	0.0028	0.0025	0.0022	0.0020	0.0019	
	#10	1.27	0.0141	0.0094	0.0071	0.0056	0.0047	0.0040	0.0035	0.0031	0.0028	0.0026	0.0024	
	#11	1.56	0.0173	0.0116	0.0087	0.0069	0.0058	0.0050	0.0043	0.0039	0.0035	0.0032	0.0029	

TABLE C-8j   Steel Ratio $p = \dfrac{A_s}{bd}$ A_s in square inches; b and d in inches

d (inches)	Bar Size	Steel Area (sq. in.)	\multicolumn{11}{c	}{Steel Spacing (inches)}	d (inches)
			8	12	16	20	24	28	32	36	40	44	48	
11.50	#3	0.11	0.0012	0.0008	0.0006	0.0005	0.0004	0.0003	0.0003	0.0003	0.0002	0.0002	0.0002	11.50
	#4	0.20	0.0022	0.0014	0.0011	0.0009	0.0007	0.0006	0.0005	0.0005	0.0004	0.0004	0.0004	
	#5	0.31	0.0034	0.0022	0.0017	0.0013	0.0011	0.0010	0.0008	0.0007	0.0007	0.0006	0.0006	
	#6	0.44	0.0048	0.0032	0.0024	0.0019	0.0016	0.0014	0.0012	0.0011	0.0010	0.0009	0.0008	
	#7	0.60	0.0065	0.0043	0.0033	0.0026	0.0022	0.0019	0.0016	0.0014	0.0013	0.0012	0.0011	
	#8	0.79	0.0086	0.0057	0.0043	0.0034	0.0029	0.0025	0.0021	0.0019	0.0017	0.0016	0.0014	
	#9	1.00	0.0109	0.0072	0.0054	0.0043	0.0036	0.0031	0.0027	0.0024	0.0022	0.0020	0.0018	
	#10	1.27	0.0138	0.0092	0.0069	0.0055	0.0046	0.0039	0.0035	0.0031	0.0028	0.0025	0.0023	
	#11	1.56	0.0170	0.0113	0.0085	0.0068	0.0057	0.0048	0.0042	0.0038	0.0034	0.0031	0.0028	

d (inches)	Bar Size	Steel Area (sq. in.)	\multicolumn{11}{c	}{Steel Spacing (inches)}	d (inches)
			8	12	16	20	24	28	32	36	40	44	48	
11.75	#3	0.11	0.0012	0.0008	0.0006	0.0005	0.0004	0.0003	0.0003	0.0003	0.0002	0.0002	0.0002	11.75
	#4	0.20	0.0021	0.0014	0.0011	0.0009	0.0007	0.0006	0.0005	0.0005	0.0004	0.0004	0.0004	
	#5	0.31	0.0033	0.0022	0.0016	0.0013	0.0011	0.0009	0.0008	0.0007	0.0007	0.0006	0.0005	
	#6	0.44	0.0047	0.0031	0.0023	0.0019	0.0016	0.0013	0.0012	0.0010	0.0009	0.0009	0.0008	
	#7	0.60	0.0064	0.0043	0.0032	0.0026	0.0021	0.0018	0.0016	0.0014	0.0013	0.0012	0.0011	
	#8	0.79	0.0084	0.0056	0.0042	0.0034	0.0028	0.0024	0.0021	0.0019	0.0017	0.0015	0.0014	
	#9	1.00	0.0106	0.0071	0.0053	0.0043	0.0035	0.0030	0.0027	0.0024	0.0021	0.0019	0.0018	
	#10	1.27	0.0135	0.0090	0.0068	0.0054	0.0045	0.0039	0.0034	0.0030	0.0027	0.0025	0.0023	
	#11	1.56	0.0166	0.0111	0.0083	0.0066	0.0055	0.0047	0.0041	0.0037	0.0033	0.0030	0.0028	

d (inches)	Bar Size	Steel Area (sq. in.)	\multicolumn{11}{c	}{Steel Spacing (inches)}	d (inches)
			8	12	16	20	24	28	32	36	40	44	48	
12.00	#3	0.11	0.0011	0.0008	0.0006	0.0005	0.0004	0.0003	0.0003	0.0003	0.0002	0.0002	0.0002	12.00
	#4	0.20	0.0021	0.0014	0.0010	0.0008	0.0007	0.0006	0.0005	0.0005	0.0004	0.0004	0.0003	
	#5	0.31	0.0032	0.0022	0.0016	0.0013	0.0011	0.0009	0.0008	0.0007	0.0006	0.0006	0.0005	
	#6	0.44	0.0046	0.0031	0.0023	0.0018	0.0015	0.0013	0.0011	0.0010	0.0009	0.0008	0.0008	
	#7	0.60	0.0063	0.0042	0.0031	0.0025	0.0021	0.0018	0.0016	0.0014	0.0013	0.0011	0.0010	
	#8	0.79	0.0082	0.0055	0.0041	0.0033	0.0027	0.0024	0.0021	0.0018	0.0016	0.0015	0.0014	
	#9	1.00	0.0104	0.0069	0.0052	0.0042	0.0035	0.0030	0.0026	0.0023	0.0021	0.0019	0.0017	
	#10	1.27	0.0132	0.0088	0.0066	0.0053	0.0044	0.0038	0.0033	0.0029	0.0026	0.0024	0.0022	
	#11	1.56	0.0163	0.0108	0.0081	0.0065	0.0054	0.0046	0.0041	0.0036	0.0033	0.0030	0.0027	

d (inches)	Bar Size	Steel Area (sq. in.)	\multicolumn{11}{c	}{Steel Spacing (inches)}	d (inches)
			8	12	16	20	24	28	32	36	40	44	48	
12.25	#3	0.11	0.0011	0.0007	0.0006	0.0004	0.0004	0.0003	0.0003	0.0002	0.0002	0.0002	0.0002	12.25
	#4	0.20	0.0020	0.0014	0.0010	0.0008	0.0007	0.0006	0.0005	0.0005	0.0004	0.0004	0.0003	
	#5	0.31	0.0032	0.0021	0.0016	0.0013	0.0011	0.0009	0.0008	0.0007	0.0006	0.0006	0.0005	
	#6	0.44	0.0045	0.0030	0.0022	0.0018	0.0015	0.0013	0.0011	0.0010	0.0009	0.0008	0.0007	
	#7	0.60	0.0061	0.0041	0.0031	0.0024	0.0020	0.0017	0.0015	0.0014	0.0012	0.0011	0.0010	
	#8	0.79	0.0081	0.0054	0.0040	0.0032	0.0027	0.0023	0.0020	0.0018	0.0016	0.0015	0.0013	
	#9	1.00	0.0102	0.0068	0.0051	0.0041	0.0034	0.0029	0.0026	0.0023	0.0020	0.0019	0.0017	
	#10	1.27	0.0130	0.0086	0.0065	0.0052	0.0043	0.0037	0.0032	0.0029	0.0026	0.0024	0.0022	
	#11	1.56	0.0159	0.0106	0.0080	0.0064	0.0053	0.0045	0.0040	0.0035	0.0032	0.0029	0.0027	

TABLE C-8k   Steel Ratio $p = \dfrac{A_s}{bd}$ *A_s in square inches; b and d in inches*

WSD

d (inches)	Bar Size	Steel Area (sq. in.)	8	12	16	20	24	28	32	36	40	44	48	d (inches)
			Steel Spacing (inches)											
12.50	#3	0.11	0.0011	0.0007	0.0006	0.0004	0.0004	0.0003	0.0003	0.0002	0.0002	0.0002	0.0002	12.50
	#4	0.20	0.0020	0.0013	0.0010	0.0008	0.0007	0.0006	0.0005	0.0004	0.0004	0.0004	0.0003	
	#5	0.31	0.0031	0.0021	0.0016	0.0012	0.0010	0.0009	0.0008	0.0007	0.0006	0.0006	0.0005	
	#6	0.44	0.0044	0.0029	0.0022	0.0018	0.0015	0.0013	0.0011	0.0010	0.0009	0.0008	0.0007	
	#7	0.60	0.0060	0.0040	0.0030	0.0024	0.0020	0.0017	0.0015	0.0013	0.0012	0.0011	0.0010	
	#8	0.79	0.0079	0.0053	0.0040	0.0032	0.0026	0.0023	0.0020	0.0018	0.0016	0.0014	0.0013	
	#9	1.00	0.0100	0.0067	0.0050	0.0040	0.0033	0.0029	0.0025	0.0022	0.0020	0.0018	0.0017	
	#10	1.27	0.0127	0.0085	0.0064	0.0051	0.0042	0.0036	0.0032	0.0028	0.0025	0.0023	0.0021	
	#11	1.56	0.0156	0.0104	0.0078	0.0062	0.0052	0.0045	0.0039	0.0035	0.0031	0.0028	0.0026	

d (inches)	Bar Size	Steel Area (sq. in.)	8	12	16	20	24	28	32	36	40	44	48	d (inches)
			Steel Spacing (inches)											
12.75	#3	0.11	0.0011	0.0007	0.0005	0.0004	0.0004	0.0003	0.0003	0.0002	0.0002	0.0002	0.0002	12.75
	#4	0.20	0.0020	0.0013	0.0010	0.0008	0.0007	0.0006	0.0005	0.0004	0.0004	0.0004	0.0003	
	#5	0.31	0.0030	0.0020	0.0015	0.0012	0.0010	0.0009	0.0008	0.0007	0.0006	0.0006	0.0005	
	#6	0.44	0.0043	0.0029	0.0022	0.0017	0.0014	0.0012	0.0011	0.0010	0.0009	0.0008	0.0007	
	#7	0.60	0.0059	0.0039	0.0029	0.0024	0.0020	0.0017	0.0015	0.0013	0.0012	0.0011	0.0010	
	#8	0.79	0.0077	0.0052	0.0039	0.0031	0.0026	0.0022	0.0019	0.0017	0.0015	0.0014	0.0013	
	#9	1.00	0.0098	0.0065	0.0049	0.0039	0.0033	0.0028	0.0025	0.0022	0.0020	0.0018	0.0016	
	#10	1.27	0.0125	0.0083	0.0062	0.0050	0.0042	0.0036	0.0031	0.0028	0.0025	0.0023	0.0021	
	#11	1.56	0.0153	0.0102	0.0076	0.0061	0.0051	0.0044	0.0038	0.0034	0.0031	0.0028	0.0025	

d (inches)	Bar Size	Steel Area (sq. in.)	8	12	16	20	24	28	32	36	40	44	48	d (inches)
			Steel Spacing (inches)											
13.00	#3	0.11	0.0011	0.0007	0.0005	0.0004	0.0004	0.0003	0.0003	0.0002	0.0002	0.0002	0.0002	13.00
	#4	0.20	0.0019	0.0013	0.0010	0.0008	0.0006	0.0005	0.0005	0.0004	0.0004	0.0003	0.0003	
	#5	0.31	0.0030	0.0020	0.0015	0.0012	0.0010	0.0009	0.0007	0.0007	0.0006	0.0005	0.0005	
	#6	0.44	0.0042	0.0028	0.0021	0.0017	0.0014	0.0012	0.0011	0.0009	0.0008	0.0008	0.0007	
	#7	0.60	0.0058	0.0038	0.0029	0.0023	0.0019	0.0016	0.0014	0.0013	0.0012	0.0010	0.0010	
	#8	0.79	0.0076	0.0051	0.0038	0.0030	0.0025	0.0022	0.0019	0.0017	0.0015	0.0014	0.0013	
	#9	1.00	0.0096	0.0064	0.0048	0.0038	0.0032	0.0027	0.0024	0.0021	0.0019	0.0017	0.0016	
	#10	1.27	0.0122	0.0081	0.0061	0.0049	0.0041	0.0035	0.0031	0.0027	0.0024	0.0022	0.0020	
	#11	1.56	0.0150	0.0100	0.0075	0.0060	0.0050	0.0043	0.0038	0.0033	0.0030	0.0027	0.0025	

d (inches)	Bar Size	Steel Area (sq. in.)	8	12	16	20	24	28	32	36	40	44	48	d (inches)
			Steel Spacing (inches)											
13.25	#3	0.11	0.0010	0.0007	0.0005	0.0004	0.0003	0.0003	0.0003	0.0002	0.0002	0.0002	0.0002	13.25
	#4	0.20	0.0019	0.0013	0.0009	0.0008	0.0006	0.0005	0.0005	0.0004	0.0004	0.0003	0.0003	
	#5	0.31	0.0029	0.0019	0.0015	0.0012	0.0010	0.0008	0.0007	0.0006	0.0006	0.0005	0.0005	
	#6	0.44	0.0042	0.0028	0.0021	0.0017	0.0014	0.0012	0.0010	0.0009	0.0008	0.0008	0.0007	
	#7	0.60	0.0057	0.0038	0.0028	0.0023	0.0019	0.0016	0.0014	0.0013	0.0011	0.0010	0.0009	
	#8	0.79	0.0075	0.0050	0.0037	0.0030	0.0025	0.0021	0.0019	0.0017	0.0015	0.0014	0.0012	
	#9	1.00	0.0094	0.0063	0.0047	0.0038	0.0031	0.0027	0.0024	0.0021	0.0019	0.0017	0.0016	
	#10	1.27	0.0120	0.0080	0.0060	0.0048	0.0040	0.0034	0.0030	0.0027	0.0024	0.0022	0.0020	
	#11	1.56	0.0147	0.0098	0.0074	0.0059	0.0049	0.0042	0.0037	0.0033	0.0029	0.0027	0.0025	

TABLE C-8I   Steel Ratio $p = \dfrac{A_s}{bd}$ A_s in square inches; b and d in inches

d (inches)	Bar Size	Steel Area (sq. in.)	Steel Spacing (inches)											d (inches)
			8	12	16	20	24	28	32	36	40	44	48	
13.50	#3	0.11	0.0010	0.0007	0.0005	0.0004	0.0003	0.0003	0.0003	0.0002	0.0002	0.0002	0.0002	13.50
	#4	0.20	0.0019	0.0012	0.0009	0.0007	0.0006	0.0005	0.0005	0.0004	0.0004	0.0003	0.0003	
	#5	0.31	0.0029	0.0019	0.0014	0.0011	0.0010	0.0008	0.0007	0.0006	0.0006	0.0005	0.0005	
	#6	0.44	0.0041	0.0027	0.0020	0.0016	0.0014	0.0012	0.0010	0.0009	0.0008	0.0007	0.0007	
	#7	0.60	0.0056	0.0037	0.0028	0.0022	0.0019	0.0016	0.0014	0.0012	0.0011	0.0010	0.0009	
	#8	0.79	0.0073	0.0049	0.0037	0.0029	0.0024	0.0021	0.0018	0.0016	0.0015	0.0013	0.0012	
	#9	1.00	0.0093	0.0062	0.0046	0.0037	0.0031	0.0026	0.0023	0.0021	0.0019	0.0017	0.0015	
	#10	1.27	0.0118	0.0078	0.0059	0.0047	0.0039	0.0034	0.0029	0.0026	0.0024	0.0021	0.0020	
	#11	1.56	0.0144	0.0096	0.0072	0.0058	0.0048	0.0041	0.0036	0.0032	0.0029	0.0026	0.0024	

d (inches)	Bar Size	Steel Area (sq. in.)	Steel Spacing (inches)											d (inches)
			8	12	16	20	24	28	32	36	40	44	48	
13.75	#3	0.11	0.0010	0.0007	0.0005	0.0004	0.0003	0.0003	0.0003	0.0002	0.0002	0.0002	0.0002	13.75
	#4	0.20	0.0018	0.0012	0.0009	0.0007	0.0006	0.0005	0.0005	0.0004	0.0004	0.0003	0.0003	
	#5	0.31	0.0028	0.0019	0.0014	0.0011	0.0009	0.0008	0.0007	0.0006	0.0006	0.0005	0.0005	
	#6	0.44	0.0040	0.0027	0.0020	0.0016	0.0013	0.0011	0.0010	0.0009	0.0008	0.0007	0.0007	
	#7	0.60	0.0055	0.0036	0.0027	0.0022	0.0018	0.0016	0.0014	0.0012	0.0011	0.0010	0.0009	
	#8	0.79	0.0072	0.0048	0.0036	0.0029	0.0024	0.0021	0.0018	0.0016	0.0014	0.0013	0.0012	
	#9	1.00	0.0091	0.0061	0.0045	0.0036	0.0030	0.0026	0.0023	0.0020	0.0018	0.0017	0.0015	
	#10	1.27	0.0115	0.0077	0.0058	0.0046	0.0038	0.0033	0.0029	0.0026	0.0023	0.0021	0.0019	
	#11	1.56	0.0142	0.0095	0.0071	0.0057	0.0047	0.0041	0.0035	0.0032	0.0028	0.0026	0.0024	

d (inches)	Bar Size	Steel Area (sq. in.)	Steel Spacing (inches)											d (inches)
			8	12	16	20	24	28	32	36	40	44	48	
14.00	#3	0.11	0.0010	0.0007	0.0005	0.0004	0.0003	0.0003	0.0002	0.0002	0.0002	0.0002	0.0002	14.00
	#4	0.20	0.0018	0.0012	0.0009	0.0007	0.0006	0.0005	0.0004	0.0004	0.0004	0.0003	0.0003	
	#5	0.31	0.0028	0.0018	0.0014	0.0011	0.0009	0.0008	0.0007	0.0006	0.0006	0.0005	0.0005	
	#6	0.44	0.0039	0.0026	0.0020	0.0016	0.0013	0.0011	0.0010	0.0009	0.0008	0.0007	0.0007	
	#7	0.60	0.0054	0.0036	0.0027	0.0021	0.0018	0.0015	0.0013	0.0012	0.0011	0.0010	0.0009	
	#8	0.79	0.0071	0.0047	0.0035	0.0028	0.0024	0.0020	0.0018	0.0016	0.0014	0.0013	0.0012	
	#9	1.00	0.0089	0.0060	0.0045	0.0036	0.0030	0.0026	0.0022	0.0020	0.0018	0.0016	0.0015	
	#10	1.27	0.0113	0.0076	0.0057	0.0045	0.0038	0.0032	0.0028	0.0025	0.0023	0.0021	0.0019	
	#11	1.56	0.0139	0.0093	0.0070	0.0056	0.0046	0.0040	0.0035	0.0031	0.0028	0.0025	0.0023	

d (inches)	Bar Size	Steel Area (sq. in.)	Steel Spacing (inches)											d (inches)
			8	12	16	20	24	28	32	36	40	44	48	
14.25	#3	0.11	0.0010	0.0006	0.0005	0.0004	0.0003	0.0003	0.0002	0.0002	0.0002	0.0002	0.0002	14.25
	#4	0.20	0.0018	0.0012	0.0009	0.0007	0.0006	0.0005	0.0004	0.0004	0.0004	0.0003	0.0003	
	#5	0.31	0.0027	0.0018	0.0014	0.0011	0.0009	0.0008	0.0007	0.0006	0.0005	0.0005	0.0005	
	#6	0.44	0.0039	0.0026	0.0019	0.0015	0.0013	0.0011	0.0010	0.0009	0.0008	0.0007	0.0006	
	#7	0.60	0.0053	0.0035	0.0026	0.0021	0.0018	0.0015	0.0013	0.0012	0.0011	0.0010	0.0009	
	#8	0.79	0.0069	0.0046	0.0035	0.0028	0.0023	0.0020	0.0017	0.0015	0.0014	0.0013	0.0012	
	#9	1.00	0.0088	0.0058	0.0044	0.0035	0.0029	0.0025	0.0022	0.0019	0.0018	0.0016	0.0015	
	#10	1.27	0.0111	0.0074	0.0056	0.0045	0.0037	0.0032	0.0028	0.0025	0.0022	0.0020	0.0019	
	#11	1.56	0.0137	0.0091	0.0068	0.0055	0.0046	0.0039	0.0034	0.0030	0.0027	0.0025	0.0023	

TABLE C-8m   Steel Ratio $p = \dfrac{A_s}{bd}$ *A_s* in square inches; *b* and *d* in inches

d (inches)	Bar Size	Steel Area (sq. in.)	Steel Spacing (inches)											*d* (inches)
			8	12	16	20	24	28	32	36	40	44	48	
14.50	#3	0.11	0.0009	0.0006	0.0005	0.0004	0.0003	0.0003	0.0002	0.0002	0.0002	0.0002	0.0002	14.50
	#4	0.20	0.0017	0.0011	0.0009	0.0007	0.0006	0.0005	0.0004	0.0004	0.0003	0.0003	0.0003	
	#5	0.31	0.0027	0.0018	0.0013	0.0011	0.0009	0.0008	0.0007	0.0006	0.0005	0.0005	0.0004	
	#6	0.44	0.0038	0.0025	0.0019	0.0015	0.0013	0.0011	0.0009	0.0008	0.0008	0.0007	0.0006	
	#7	0.60	0.0052	0.0034	0.0026	0.0021	0.0017	0.0015	0.0013	0.0011	0.0010	0.0009	0.0009	
	#8	0.79	0.0068	0.0045	0.0034	0.0027	0.0023	0.0019	0.0017	0.0015	0.0014	0.0012	0.0011	
	#9	1.00	0.0086	0.0057	0.0043	0.0034	0.0029	0.0025	0.0022	0.0019	0.0017	0.0016	0.0014	
	#10	1.27	0.0109	0.0073	0.0055	0.0044	0.0036	0.0031	0.0027	0.0024	0.0022	0.0020	0.0018	
	#11	1.56	0.0134	0.0090	0.0067	0.0054	0.0045	0.0038	0.0034	0.0030	0.0027	0.0024	0.0022	

d (inches)	Bar Size	Steel Area (sq. in.)	Steel Spacing (inches)											*d* (inches)
			8	12	16	20	24	28	32	36	40	44	48	
14.75	#3	0.11	0.0009	0.0006	0.0005	0.0004	0.0003	0.0003	0.0002	0.0002	0.0002	0.0002	0.0002	14.75
	#4	0.20	0.0017	0.0011	0.0008	0.0007	0.0006	0.0005	0.0004	0.0004	0.0003	0.0003	0.0003	
	#5	0.31	0.0026	0.0018	0.0013	0.0011	0.0009	0.0008	0.0007	0.0006	0.0005	0.0005	0.0004	
	#6	0.44	0.0037	0.0025	0.0019	0.0015	0.0012	0.0011	0.0009	0.0008	0.0007	0.0007	0.0006	
	#7	0.60	0.0051	0.0034	0.0025	0.0020	0.0017	0.0015	0.0013	0.0011	0.0010	0.0009	0.0008	
	#8	0.79	0.0067	0.0045	0.0033	0.0027	0.0022	0.0019	0.0017	0.0015	0.0013	0.0012	0.0011	
	#9	1.00	0.0085	0.0056	0.0042	0.0034	0.0028	0.0024	0.0021	0.0019	0.0017	0.0015	0.0014	
	#10	1.27	0.0108	0.0072	0.0054	0.0043	0.0036	0.0031	0.0027	0.0024	0.0022	0.0020	0.0018	
	#11	1.56	0.0132	0.0088	0.0066	0.0053	0.0044	0.0038	0.0033	0.0029	0.0026	0.0024	0.0022	

d (inches)	Bar Size	Steel Area (sq. in.)	Steel Spacing (inches)											*d* (inches)
			8	12	16	20	24	28	32	36	40	44	48	
15.00	#3	0.11	0.0009	0.0006	0.0005	0.0004	0.0003	0.0003	0.0002	0.0002	0.0002	0.0002	0.0002	15.00
	#4	0.20	0.0017	0.0011	0.0008	0.0007	0.0006	0.0005	0.0004	0.0004	0.0003	0.0003	0.0003	
	#5	0.31	0.0026	0.0017	0.0013	0.0010	0.0009	0.0007	0.0006	0.0006	0.0005	0.0005	0.0004	
	#6	0.44	0.0037	0.0024	0.0018	0.0015	0.0012	0.0010	0.0009	0.0008	0.0007	0.0007	0.0006	
	#7	0.60	0.0050	0.0033	0.0025	0.0020	0.0017	0.0014	0.0013	0.0011	0.0010	0.0009	0.0008	
	#8	0.79	0.0066	0.0044	0.0033	0.0026	0.0022	0.0019	0.0016	0.0015	0.0013	0.0012	0.0011	
	#9	1.00	0.0083	0.0056	0.0042	0.0033	0.0028	0.0024	0.0021	0.0019	0.0017	0.0015	0.0014	
	#10	1.27	0.0106	0.0071	0.0053	0.0042	0.0035	0.0030	0.0026	0.0024	0.0021	0.0019	0.0018	
	#11	1.56	0.0130	0.0087	0.0065	0.0052	0.0043	0.0037	0.0033	0.0029	0.0026	0.0024	0.0022	

d (inches)	Bar Size	Steel Area (sq. in.)	Steel Spacing (inches)											*d* (inches)
			8	12	16	20	24	28	32	36	40	44	48	
15.25	#3	0.11	0.0009	0.0006	0.0005	0.0004	0.0003	0.0003	0.0002	0.0002	0.0002	0.0002	0.0002	15.25
	#4	0.20	0.0016	0.0011	0.0008	0.0007	0.0005	0.0005	0.0004	0.0004	0.0003	0.0003	0.0003	
	#5	0.31	0.0025	0.0017	0.0013	0.0010	0.0008	0.0007	0.0006	0.0006	0.0005	0.0005	0.0004	
	#6	0.44	0.0036	0.0024	0.0018	0.0014	0.0012	0.0010	0.0009	0.0008	0.0007	0.0007	0.0006	
	#7	0.60	0.0049	0.0033	0.0025	0.0020	0.0016	0.0014	0.0012	0.0011	0.0010	0.0009	0.0008	
	#8	0.79	0.0065	0.0043	0.0032	0.0026	0.0022	0.0019	0.0016	0.0014	0.0013	0.0012	0.0011	
	#9	1.00	0.0082	0.0055	0.0041	0.0033	0.0027	0.0023	0.0020	0.0018	0.0016	0.0015	0.0014	
	#10	1.27	0.0104	0.0069	0.0052	0.0042	0.0035	0.0030	0.0026	0.0023	0.0021	0.0019	0.0017	
	#11	1.56	0.0128	0.0085	0.0064	0.0051	0.0043	0.0037	0.0032	0.0028	0.0026	0.0023	0.0021	

TABLE C-9 Ratio of Steel Area to Gross Cross-Sectional Area[1]

Wall Thick-ness (inches)	Bar Size	Area (sq. in.)	Spacing of Steel Reinforcing (inches)										
			8	12	16	20	24	28	32	36	40	44	48
5.625 6" Nom.	#3	0.11	0.0024	0.0016	0.0012	0.0010	0.0008	0.0007	0.0006	0.0005	0.0005	0.0004	0.0004
	#4	0.20	0.0044	0.0030	0.0022	0.0018	0.0015	0.0013	0.0011	0.0010	0.0009	0.0008	0.0007
	#5	0.31	0.0069	0.0046	0.0034	0.0028	0.0023	0.0020	0.0017	0.0015	0.0014	0.0013	0.0011
7.625 8" Nom.	#3	0.11	0.0018	0.0012	0.0009	0.0007	0.0006	0.0005	0.0005	0.0004	0.0004	0.0003	0.0003
	#4	0.20	0.0033	0.0022	0.0016	0.0013	0.0011	0.0009	0.0008	0.0007	0.0007	0.0006	0.0005
	#5	0.31	0.0051	0.0034	0.0025	0.0020	0.0017	0.0015	0.0013	0.0011	0.0010	0.0009	0.0008
	#6	0.44	0.0072	0.0048	0.0036	0.0029	0.0024	0.0021	0.0018	0.0016	0.0014	0.0013	0.0012
	#7	0.60	0.0098	0.0066	0.0049	0.0039	0.0033	0.0028	0.0025	0.0022	0.0020	0.0018	0.0016
	#8	0.79	0.0130	0.0086	0.0065	0.0052	0.0043	0.0037	0.0032	0.0029	0.0026	0.0024	0.0022
	#9	1.00	0.0164	0.0109	0.0082	0.0066	0.0055	0.0047	0.0041	0.0036	0.0033	0.0030	0.0027
	#10	1.27	0.0208	0.0139	0.0104	0.0083	0.0069	0.0059	0.0052	0.0046	0.0042	0.0038	0.0035
	#11	1.56	0.0256	0.0170	0.0128	0.0102	0.0085	0.0073	0.0064	0.0057	0.0051	0.0046	0.0043
9.625 10" Nom.	#4	0.20	0.0026	0.0017	0.0013	0.0010	0.0009	0.0007	0.0006	0.0006	0.0005	0.0005	0.0004
	#5	0.31	0.0040	0.0027	0.0020	0.0016	0.0013	0.0012	0.0010	0.0009	0.0008	0.0007	0.0007
	#6	0.44	0.0057	0.0038	0.0029	0.0023	0.0019	0.0016	0.0014	0.0013	0.0011	0.0010	0.0010
	#7	0.60	0.0078	0.0052	0.0039	0.0031	0.0026	0.0022	0.0019	0.0017	0.0016	0.0014	0.0013
	#8	0.79	0.0103	0.0068	0.0051	0.0041	0.0034	0.0029	0.0026	0.0023	0.0021	0.0019	0.0017
	#9	1.00	0.0130	0.0087	0.0065	0.0052	0.0043	0.0037	0.0032	0.0029	0.0026	0.0024	0.0022
	#10	1.27	0.0165	0.0110	0.0082	0.0066	0.0055	0.0047	0.0041	0.0037	0.0033	0.0030	0.0027
	#11	1.56	0.0203	0.0135	0.0101	0.0081	0.0068	0.0058	0.0051	0.0045	0.0041	0.0037	0.0034
11.625 12" Nom.	#4	0.20	0.0022	0.0014	0.0011	0.0009	0.0007	0.0006	0.0005	0.0005	0.0004	0.0004	0.0004
	#5	0.31	0.0033	0.0022	0.0017	0.0013	0.0011	0.0010	0.0008	0.0007	0.0007	0.0006	0.0006
	#6	0.44	0.0047	0.0032	0.0024	0.0019	0.0016	0.0014	0.0012	0.0011	0.0009	0.0009	0.0008
	#7	0.60	0.0065	0.0043	0.0032	0.0026	0.0022	0.0018	0.0016	0.0014	0.0013	0.0012	0.0011
	#8	0.79	0.0085	0.0057	0.0042	0.0034	0.0028	0.0024	0.0021	0.0019	0.0017	0.0015	0.0014
	#9	1.00	0.0108	0.0072	0.0054	0.0043	0.0036	0.0031	0.0027	0.0024	0.0022	0.0020	0.0018
	#10	1.27	0.0137	0.0091	0.0068	0.0055	0.0046	0.0039	0.0034	0.0030	0.0027	0.0025	0.0023
	#11	1.56	0.0168	0.0112	0.0084	0.0067	0.0056	0.0048	0.0042	0.0037	0.0034	0.0030	0.0028
15.625 16" Nom.	#4	0.20	0.0016	0.0011	0.0008	0.0006	0.0005	0.0005	0.0004	0.0004	0.0003	0.0003	0.0003
	#5	0.31	0.0025	0.0017	0.0012	0.0010	0.0008	0.0007	0.0006	0.0006	0.0005	0.0005	0.0004
	#6	0.44	0.0035	0.0023	0.0018	0.0014	0.0012	0.0010	0.0009	0.0008	0.0007	0.0006	0.0006
	#7	0.60	0.0048	0.0032	0.0024	0.0019	0.0016	0.0014	0.0012	0.0011	0.0010	0.0009	0.0008
	#8	0.79	0.0063	0.0042	0.0032	0.0025	0.0021	0.0018	0.0016	0.0014	0.0013	0.0011	0.0011
	#9	1.00	0.0080	0.0053	0.0040	0.0032	0.0027	0.0023	0.0020	0.0018	0.0016	0.0015	0.0013
	#11	1.56	0.0125	0.0083	0.0062	0.0050	0.0042	0.0036	0.0031	0.0028	0.0025	0.0023	0.0021

1. UBC Minimum and ACI/ASCE Minimum = $0.0007bt$ for seismic zones 3 and 4. OSA Minimum = $0.001bt$ for seismic zones 3 and 4.

TABLE C-10   Maximum Area of Steel per CMU Cell

Nom. Thickness (inches)	Actual Thickness (inches)	Cell Area (sq. in.)	5% Recommended Steel Area (sq. in.)	6% Code Allowance Steel Area (sq. in.)[1]
4	$3\frac{5}{8}$	12.6	0.63	0.76
6	$5\frac{5}{8}$	21.0	1.05	1.26
8	$7\frac{5}{8}$	30.0	1.50	1.80
10	$9\frac{5}{8}$	42.0	2.10	2.52
12	$11\frac{5}{8}$	54.0	2.70	3.24

1. Based on UBC Section 2409(e)1.

TABLE C-11   Maximum Number of Reinforcing Bars per Cell

Nominal Thickness (inches)		Area of Steel (sq. in.)[1]	Bar Size and Area Per Bar							
			#4 0.20	#5 0.31	#6 0.44	#7 0.60	#8 0.79	#9 1.00	#10 1.27	#11 1.56
5% Recommended	4	0.63	3	2	1	1	xx	xx	xx	xx
	6	1.05	x	3	2	1	1	1	xx	xx
	8	1.50	x	4	3	2	1	1	1	xx
	10	2.10	x	x	5	3	2	2	1	1
	12	2.70	x	x	x	4	3	2	2	1
6% Code Allowable	4	0.76	3	2	1	1	1	xx	xx	xx
	6	1.26	6	4	2	2	1	1	1	xx
	8	1.80	9	5	4	3	1	1	1	1
	10	2.52	x	x	5	4	3	2	2	1
	12	3.24	x	x	x	5	4	3	2	2

x   Not Recommended.

xx   Exceeds UBC Allowance.

1.   Values based on Table C-10.

TABLE E-1a Flexural Design Coefficients for Half Allowable Stresses: $f'_m = 1500$ psi, $f_y = 60,000$ psi and $n = 25.8$

For use of Table, see Example 5-A

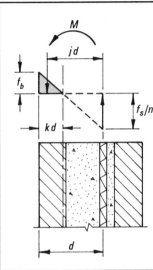

DESIGN DATA

$f'_m = 1500$ psi $\qquad\qquad$ $f_y = 60,000$ psi

$F_b = \frac{1}{2}(0.33 f'_m) = 250$ psi $\qquad$ $F_s = 24,000$ psi

$E_m = 750 f'_m = 1,125,000$ psi $\qquad$ $E_s = 29,000,000$ psi

DESIGN EQUATIONS

$$n = \frac{E_s}{E_m} = 25.8 \qquad K = \frac{M}{F} = \frac{M(\text{ft kips})}{bd^2/12,000} \text{ or } \frac{M(\text{in. lbs})}{bd^2}$$

$$K = \frac{1}{2} jk f_b$$

$$k = \frac{1}{1 + f_s/n f_b} \qquad j = 1 - \frac{k}{3}$$

$$A_s = \frac{M}{f_s jd} \qquad p = \frac{A_s}{bd} = \frac{K}{f_s j}$$

Increase for wind or earthquake

$1.33 F_b = 333$ psi

$1.33 F_s = 32,000$ psi

f_b	f_s	K	p	np	k	j	2/jk	f_b	f_s	K
100	24,000	4.7	0.0002	0.005	0.097	0.968	21.29	133	32,000	6.3
125	24,000	7.1	0.0003	0.008	0.118	0.961	17.58	167	32,000	9.5
150	24,000	9.9	0.0004	0.011	0.139	0.954	15.10	200	32,000	13.2
175	24,000	13.1	0.0006	0.015	0.158	0.947	13.34	233	32,000	17.5
200	24,000	16.7	0.0007	0.019	0.177	0.941	12.01	267	32,000	22.2
225	24,000	20.5	0.0009	0.024	0.195	0.935	10.98	300	32,000	27.3
250	24,000	24.6	0.0011	0.028	0.212	0.929	10.16	333	32,000	32.8
250	23,000	25.4	0.0012	0.031	0.219	0.927	9.85	333	30,667	33.8
250	22,000	26.2	0.0013	0.033	0.227	0.924	9.54	333	29,333	34.9
250	21,000	27.1	0.0014	0.036	0.235	0.922	9.23	333	28,000	36.1
250	20,000	28.0	0.0015	0.039	0.244	0.919	8.93	333	26,667	37.3
250	19,000	29.0	0.0017	0.043	0.253	0.916	8.62	333	25,333	38.7
250	18,000	30.1	0.0018	0.047	0.264	0.912	8.31	333	24,000	40.1
250	17,000	31.2	0.0020	0.052	0.275	0.908	8.01	333	22,667	41.6
250	16,000	32.5	0.0022	0.058	0.287	0.904	7.70	333	21,333	43.3
250	15,000	33.8	0.0025	0.065	0.301	0.900	7.39	333	20,000	45.1
250	14,000	35.3	0.0028	0.073	0.315	0.895	7.09	333	18,667	47.0
250	13,000	36.9	0.0032	0.082	0.332	0.889	6.78	333	17,333	49.2
250	12,000	38.6	0.0036	0.094	0.350	0.883	6.48	333	16,000	51.5
250	11,000	40.5	0.0042	0.108	0.370	0.877	6.17	333	14,667	54.0
250	10,000	42.6	0.0049	0.126	0.392	0.869	5.87	333	13,333	56.8

TABLE E-1b   Flexural Design Coefficients for Full Allowable Stresses: For use of Table, see Example 5-A
f'_m = 1500 psi, f_y = 60,000 psi and n = 25.8

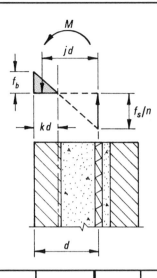

DESIGN DATA

$f'_m = 1500$ psi $\qquad\qquad f_y = 60,000$ psi

$F_b = 0.33 f'_m = 500$ psi $\qquad F_s = 24,000$ psi

$E_m = 750 f'_m = 1,125,000$ psi $\qquad E_s = 29,000,000$ psi

DESIGN EQUATIONS

$n = \dfrac{E_s}{E_m} = 25.8 \qquad K = \dfrac{M}{F} = \dfrac{M(\text{ft kips})}{bd^2/12,000}$ or $\dfrac{M(\text{in. lbs})}{bd^2}$

$K = \frac{1}{2} jk f_b$

$k = \dfrac{1}{1 + f_s/nf_b} \qquad j = 1 - \dfrac{k}{3}$

$A_s = \dfrac{M}{f_s jd} \qquad\qquad p = \dfrac{A_s}{bd} = \dfrac{K}{f_s j}$

Increase for wind or earthquake
$1.33 F_b = 667$ psi
$1.33 F_s = 32,000$ psi

f_b	f_s	K	p	np	k	j	$2/jk$	f_b	f_s	K
125	24,000	7.1	0.0003	0.008	0.118	0.961	17.58	167	32,000	9.5
150	24,000	9.9	0.0004	0.011	0.139	0.954	15.10	200	32,000	13.2
175	24,000	13.1	0.0006	0.015	0.158	0.947	13.34	233	32,000	17.5
200	24,000	16.7	0.0007	0.019	0.177	0.941	12.01	267	32,000	22.2
225	24,000	20.5	0.0009	0.024	0.195	0.935	10.98	300	32,000	27.3
250	24,000	24.6	0.0011	0.028	0.212	0.929	10.16	333	32,000	32.8
275	24,000	29.0	0.0013	0.034	0.228	0.924	9.49	367	32,000	38.6
300	24,000	33.6	0.0015	0.039	0.244	0.919	8.93	400	32,000	44.8
325	24,000	38.4	0.0018	0.045	0.259	0.914	8.45	433	32,000	51.3
350	24,000	43.5	0.0020	0.051	0.273	0.909	8.05	467	32,000	58.0
375	24,000	48.7	0.0022	0.058	0.287	0.904	7.70	500	32,000	64.9
400	24,000	54.1	0.0025	0.065	0.301	0.900	7.39	533	32,000	72.1
425	24,000	59.7	0.0028	0.072	0.314	0.895	7.12	567	32,000	79.6
450	24,000	65.4	0.0031	0.079	0.326	0.891	6.88	600	32,000	87.2
475	24,000	71.2	0.0033	0.086	0.338	0.887	6.67	633	32,000	95.0
500	24,000	77.2	0.0036	0.094	0.350	0.883	6.48	667	32,000	103.0
500	23,000	79.1	0.0039	0.101	0.359	0.880	6.32	667	30,667	105.4
500	22,000	81.0	0.0042	0.108	0.370	0.877	6.17	667	29,333	108.0
500	21,000	83.1	0.0045	0.117	0.381	0.873	6.02	667	28,000	110.8
500	20,000	85.2	0.0049	0.126	0.392	0.869	5.87	667	26,667	113.6
500	19,000	87.5	0.0053	0.137	0.404	0.865	5.72	667	25,333	116.6
500	18,000	89.8	0.0058	0.150	0.417	0.861	5.57	667	24,000	119.8
500	17,000	92.3	0.0063	0.164	0.431	0.856	5.41	667	22,667	123.1
500	16,000	95.0	0.0070	0.180	0.446	0.851	5.26	667	21,333	126.7
500	15,000	97.8	0.0077	0.199	0.462	0.846	5.11	667	20,000	130.4
500	14,000	100.7	0.0086	0.221	0.480	0.840	4.96	667	18,667	134.3
500	13,000	103.8	0.0096	0.247	0.498	0.834	4.81	667	17,333	138.5
500	12,000	107.2	0.0108	0.278	0.518	0.827	4.67	667	16,000	142.9
500	11,000	110.7	0.0123	0.316	0.540	0.820	4.52	667	14,667	147.5
500	10,000	114.4	0.0141	0.363	0.563	0.812	4.37	667	13,333	152.5

WSD

TABLE E-2a   Flexural Design Coefficients for Half Allowable Stresses:
f'_m = 2000 psi, f_y = 60,000 psi and n = 19.3

For use of Table, see Example 5-A

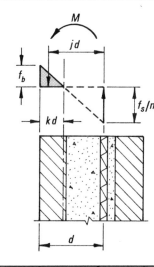

DESIGN DATA

$f'_m = 2000$ psi	$f_y = 60,000$ psi
$F_b = \frac{1}{2}(0.33 f'_m) = 333$ psi	$F_s = 24,000$ psi
$E_m = 750 f'_m = 1,500,000$ psi	$E_s = 29,000,000$ psi

DESIGN EQUATIONS

$$n = \frac{E_s}{E_m} = 19.3 \qquad K = \frac{M}{F} = \frac{M(\text{ft kips})}{bd^2/12,000} \text{ or } \frac{M(\text{in. lbs})}{bd^2}$$

$$K = \frac{1}{2} jk f_b$$

$$k = \frac{1}{1 + f_s/n f_b} \qquad j = 1 - \frac{k}{3}$$

$$A_s = \frac{M}{f_s jd} \qquad p = \frac{A_s}{bd} = \frac{K}{f_s j}$$

Increase for wind or earthquake
$1.33 F_b = 444$ psi
$1.33 F_s = 32,000$ psi

f_b	f_s	K	p	np	k	j	$2/jk$	f_b	f_s	K
150	24,000	7.8	0.0003	0.006	0.108	0.964	19.27	200	32,000	10.4
175	24,000	10.4	0.0004	0.009	0.123	0.959	16.91	233	32,000	13.8
200	24,000	13.2	0.0006	0.011	0.139	0.954	15.13	267	32,000	17.6
225	24,000	16.4	0.0007	0.014	0.153	0.949	13.76	300	32,000	21.8
250	24,000	19.8	0.0009	0.017	0.167	0.944	12.65	333	32,000	26.3
275	24,000	23.4	0.0010	0.020	0.181	0.940	11.75	367	32,000	31.2
300	24,000	27.3	0.0012	0.023	0.194	0.935	11.00	400	32,000	36.4
325	24,000	31.3	0.0014	0.027	0.207	0.931	10.37	433	32,000	41.8
333	24,000	32.7	0.0015	0.028	0.212	0.929	10.19	444	32,000	43.6
333	23,000	33.7	0.0016	0.031	0.218	0.927	9.88	444	30,667	45.0
333	22,000	34.8	0.0017	0.033	0.226	0.925	9.57	444	29,333	46.4
333	21,000	36.0	0.0019	0.036	0.234	0.922	9.26	444	28,000	48.0
333	20,000	37.2	0.0020	0.039	0.243	0.919	8.95	444	26,667	49.6
333	19,000	38.5	0.0022	0.043	0.253	0.916	8.64	444	25,333	51.4
333	18,000	40.0	0.0024	0.047	0.263	0.912	8.33	444	24,000	53.3
333	17,000	41.5	0.0027	0.052	0.274	0.909	8.02	444	22,667	55.3
333	16,000	43.2	0.0030	0.058	0.287	0.904	7.72	444	21,333	57.5
333	15,000	44.9	0.0033	0.064	0.300	0.900	7.41	444	20,000	59.9
333	14,000	46.9	0.0037	0.072	0.315	0.895	7.10	444	18,667	62.5
333	13,000	49.0	0.0042	0.082	0.331	0.890	6.79	444	17,333	65.3
333	12,000	51.3	0.0048	0.093	0.349	0.884	6.49	444	16,000	68.4
333	11,000	53.9	0.0056	0.108	0.369	0.877	6.18	444	14,667	71.8
333	10,000	56.6	0.0065	0.126	0.391	0.870	5.88	444	13,333	75.5

TABLE E-2b   Flexural Design Coefficients for Full Allowable Stresses: For use of Table,
f'_m = 2000 psi, f_y = 60,000 psi and n = 19.3   see Example 5-A

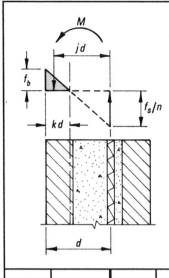

DESIGN DATA

$f'_m = 2000$ psi $f_y = 60,000$ psi

$F_b = 0.33 f'_m = 667$ psi $F_s = 24,000$ psi

$E_m = 750 f'_m = 1,500,000$ psi $E_s = 29,000,000$ psi

DESIGN EQUATIONS

$$n = \frac{E_s}{E_m} = 19.3 \qquad K = \frac{M}{F} = \frac{M(\text{ft kips})}{bd^2/12,000} \text{ or } \frac{M(\text{in. lbs})}{bd^2}$$

$$K = \tfrac{1}{2} jk f_b$$

$$k = \frac{1}{1 + f_s/n f_b} \qquad j = 1 - \frac{k}{3}$$

$$A_s = \frac{M}{f_s jd} \qquad p = \frac{A_s}{bd} = \frac{K}{f_s j}$$

Increase for wind or earthquake
$1.33 F_b = 889$ psi
$1.33 F_s = 32,000$ psi

f_b	f_s	K	p	np	k	j	$2/jk$	f_b	f_s	K
150	24,000	7.8	0.0003	0.006	0.108	0.964	19.27	200	32,000	10.4
200	24,000	13.2	0.0006	0.011	0.139	0.954	15.13	267	32,000	17.6
250	24,000	19.8	0.0009	0.017	0.167	0.944	12.65	333	32,000	26.3
300	24,000	27.3	0.0012	0.023	0.194	0.935	11.00	400	32,000	36.4
350	24,000	35.6	0.0016	0.031	0.220	0.927	9.83	467	32,000	47.5
400	24,000	44.7	0.0020	0.039	0.243	0.919	8.94	533	32,000	59.6
450	24,000	54.5	0.0025	0.048	0.266	0.911	8.26	600	32,000	72.7
500	24,000	64.8	0.0030	0.058	0.287	0.904	7.71	667	32,000	86.5
550	24,000	75.7	0.0035	0.068	0.307	0.898	7.26	733	32,000	100.9
600	24,000	87.0	0.0041	0.079	0.325	0.892	6.89	800	32,000	116.1
650	24,000	98.8	0.0046	0.090	0.343	0.886	6.58	867	32,000	131.7
667	24,000	102.9	0.0049	0.094	0.350	0.883	6.48	889	32,000	137.2
667	23,000	105.4	0.0052	0.100	0.359	0.880	6.33	889	30,667	140.5
667	22,000	108.0	0.0056	0.108	0.369	0.877	6.18	889	29,333	143.9
667	21,000	110.7	0.0060	0.116	0.380	0.873	6.03	889	28,000	147.6
667	20,000	113.6	0.0065	0.126	0.392	0.869	5.87	889	26,667	151.4
667	19,000	116.6	0.0071	0.137	0.404	0.865	5.72	889	25,333	155.4
667	18,000	119.7	0.0077	0.149	0.417	0.861	5.57	889	24,000	159.6
667	17,000	123.1	0.0085	0.163	0.431	0.856	5.42	889	22,667	164.1
667	16,000	126.6	0.0093	0.179	0.446	0.851	5.27	889	21,333	168.8
667	15,000	130.3	0.0103	0.198	0.462	0.846	5.12	889	20,000	173.8
667	14,000	134.2	0.0114	0.220	0.479	0.840	4.97	889	18,667	179.0
667	13,000	138.4	0.0128	0.246	0.498	0.834	4.82	889	17,333	184.5
667	12,000	142.8	0.0144	0.278	0.518	0.827	4.67	889	16,000	190.4
667	11,000	147.5	0.0163	0.316	0.539	0.820	4.52	889	14,667	196.7
667	10,000	152.5	0.0188	0.362	0.563	0.812	4.37	889	13,333	203.3

TABLE E-3a Flexural Design Coefficients for Half Allowable Stresses:
f'_m = 2500 psi, f_y = 60,000 psi and n = 15.5

For use of Table, see Example 5-A

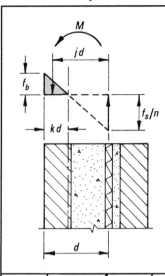

DESIGN DATA

$f'_m = 2500$ psi $f_y = 60,000$ psi

$F_b = \frac{1}{2}(0.33f'_m) = 417$ psi $F_s = 24,000$ psi

$E_m = 750 f'_m = 1,875,000$ psi $E_s = 29,000,000$ psi

DESIGN EQUATIONS

$$n = \frac{E_s}{E_m} = 15.5 \qquad K = \frac{M}{F} = \frac{M(\text{ft kips})}{bd^2/12,000} \text{ or } \frac{M(\text{in. lbs})}{bd^2}$$

$$K = \frac{1}{2} jk f_b$$

$$k = \frac{1}{1 + f_s/n f_b} \qquad j = 1 - \frac{k}{3}$$

$$A_s = \frac{M}{f_s jd} \qquad p = \frac{A_s}{bd} = \frac{K}{f_s j}$$

Increase for wind or earthquake
$1.33 F_b = 556$ psi
$1.33 F_s = 32,000$ psi

f_b	f_s	K	p	np	k	j	2/jk	f_b	f_s	K
200	24,000	11.0	0.0005	0.007	0.114	0.962	18.18	267	32,000	14.7
225	24,000	13.7	0.0006	0.009	0.127	0.958	16.46	300	32,000	18.2
250	24,000	16.6	0.0007	0.011	0.139	0.954	15.09	333	32,000	22.1
275	24,000	19.7	0.0009	0.013	0.151	0.950	13.96	367	32,000	26.3
300	24,000	23.0	0.0010	0.016	0.162	0.946	13.03	400	32,000	30.7
325	24,000	26.6	0.0012	0.018	0.173	0.942	12.24	433	32,000	35.4
350	24,000	30.3	0.0013	0.021	0.184	0.939	11.56	467	32,000	40.4
375	24,000	34.2	0.0015	0.024	0.195	0.935	10.97	500	32,000	45.6
400	24,000	38.2	0.0017	0.027	0.205	0.932	10.46	533	32,000	51.0
417	24,000	41.1	0.0018	0.029	0.212	0.929	10.14	556	32,000	54.8
417	23,000	42.4	0.0020	0.031	0.219	0.927	9.84	556	30,667	56.5
417	22,000	43.8	0.0022	0.033	0.227	0.924	9.53	556	29,333	58.3
417	21,000	45.2	0.0023	0.036	0.235	0.922	9.22	556	28,000	60.3
417	20,000	46.8	0.0025	0.039	0.244	0.919	8.91	556	26,667	62.4
417	19,000	48.4	0.0028	0.043	0.254	0.915	8.61	556	25,333	64.6
417	18,000	50.2	0.0031	0.047	0.264	0.912	8.30	556	24,000	67.0
417	17,000	52.2	0.0034	0.052	0.275	0.908	7.99	556	22,667	69.5
417	16,000	54.2	0.0037	0.058	0.288	0.904	7.69	556	21,333	72.3
417	15,000	56.5	0.0042	0.065	0.301	0.900	7.38	556	20,000	75.3
417	14,000	58.9	0.0047	0.073	0.316	0.895	7.08	556	18,667	78.6
417	13,000	61.6	0.0053	0.083	0.332	0.889	6.77	556	17,333	82.1
417	12,000	64.5	0.0061	0.094	0.350	0.883	6.47	556	16,000	86.0
417	11,000	67.6	0.0070	0.109	0.370	0.877	6.16	556	14,667	90.2
417	10,000	71.1	0.0082	0.127	0.393	0.869	5.86	556	13,333	94.9

TABLE E-3b Flexural Design Coefficients for Full Allowable Stresses:
$f'_m = 2500$ psi, $f_y = 60,000$ psi and $n = 15.5$

For use of Table, see Example 5-A

WSD

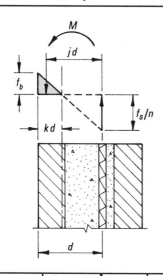

DESIGN DATA

$f'_m = 2500$ psi	$f_y = 60,000$ psi
$F_b = 0.33 f'_m = 833$ psi	$F_s = 24,000$ psi
$E_m = 750 f'_m = 1,875,000$ psi	$E_s = 29,000,000$ psi

DESIGN EQUATIONS

$$n = \frac{E_s}{E_m} = 15.5 \qquad K = \frac{M}{F} = \frac{M(\text{ft kips})}{bd^2/12,000} \text{ or } \frac{M(\text{in. lbs})}{bd^2}$$

$$K = \tfrac{1}{2} jk f_b$$

$$k = \frac{1}{1 + f_s/n f_b} \qquad j = 1 - \frac{k}{3}$$

$$A_s = \frac{M}{f_s jd} \qquad p = \frac{A_s}{bd} = \frac{K}{f_s j}$$

Increase for wind or earthquake
$1.33 F_b = 1111$ psi
$1.33 F_s = 32,000$ psi

f_b	f_s	K	p	np	k	j	2/jk	f_b	f_s	K
200	24,000	11.0	0.0005	0.007	0.114	0.962	18.18	267	32,000	14.7
250	24,000	16.6	0.0007	0.011	0.139	0.954	15.09	333	32,000	22.1
300	24,000	23.0	0.0010	0.016	0.162	0.946	13.03	400	32,000	30.7
350	24,000	30.3	0.0013	0.021	0.184	0.939	11.56	467	32,000	40.4
400	24,000	38.2	0.0017	0.027	0.205	0.932	10.46	533	32,000	51.0
450	24,000	46.9	0.0021	0.033	0.225	0.925	9.60	600	32,000	62.5
500	24,000	56.1	0.0025	0.039	0.244	0.919	8.92	667	32,000	74.7
550	24,000	65.8	0.0030	0.047	0.262	0.913	8.36	733	32,000	87.7
600	24,000	76.0	0.0035	0.054	0.279	0.907	7.90	800	32,000	101.3
650	24,000	86.6	0.0040	0.062	0.296	0.901	7.50	867	32,000	115.5
700	24,000	97.7	0.0045	0.070	0.311	0.896	7.17	933	32,000	130.2
750	24,000	109.1	0.0051	0.079	0.326	0.891	6.88	1,000	32,000	145.4
800	24,000	120.8	0.0057	0.088	0.341	0.886	6.62	1,067	32,000	161.1
833	24,000	128.7	0.0061	0.094	0.350	0.883	6.47	1,111	32,000	171.6
833	23,000	131.8	0.0065	0.101	0.360	0.880	6.32	1,111	30,667	175.7
833	22,000	135.0	0.0070	0.109	0.370	0.877	6.17	1,111	29,333	180.1
833	21,000	138.5	0.0076	0.117	0.381	0.873	6.02	1,111	28,000	184.6
833	20,000	142.0	0.0082	0.127	0.392	0.869	5.86	1,111	26,667	189.4
833	19,000	145.8	0.0089	0.137	0.405	0.865	5.71	1,111	25,333	194.4
833	18,000	149.7	0.0097	0.150	0.418	0.861	5.56	1,111	24,000	199.7
833	17,000	153.9	0.0106	0.164	0.432	0.856	5.41	1,111	22,667	205.2
833	16,000	158.3	0.0116	0.180	0.447	0.851	5.26	1,111	21,333	211.1
833	15,000	163.0	0.0128	0.199	0.463	0.846	5.11	1,111	20,000	217.3
833	14,000	167.9	0.0143	0.221	0.480	0.840	4.96	1,111	18,667	223.8
833	13,000	173.1	0.0160	0.247	0.498	0.834	4.81	1,111	17,333	230.8
833	12,000	178.6	0.0180	0.279	0.518	0.827	4.66	1,111	16,000	238.1
833	11,000	184.4	0.0204	0.317	0.540	0.820	4.52	1,111	14,667	245.9
833	10,000	190.6	0.0235	0.364	0.564	0.812	4.37	1,111	13,333	254.2

TABLE E-4a Flexural Design Coefficients for Half Allowable Stresses: $f'_m = 3000$ psi, $f_y = 60,000$ psi and $n = 12.9$

For use of Table, see Example 5-A

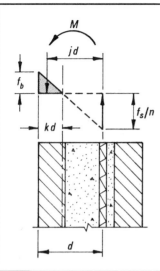

DESIGN DATA

$f'_m = 3000$ psi	$f_y = 60,000$ psi
$F_b = \frac{1}{2}(0.33f'_m) = 500$ psi	$F_s = 24,000$ psi
$E_m = 750 f'_m = 2,250,000$ psi	$E_s = 29,000,000$ psi

DESIGN EQUATIONS

$$n = \frac{E_s}{E_m} = 12.9 \qquad K = \frac{M}{F} = \frac{M(\text{ft kips})}{bd^2/12,000} \text{ or } \frac{M(\text{in. lbs})}{bd^2}$$

$$K = \frac{1}{2}jkf_b$$

$$k = \frac{1}{1 + f_s/nf_b} \qquad j = 1 - \frac{k}{3}$$

$$A_s = \frac{M}{f_s jd} \qquad p = \frac{A_s}{bd} = \frac{K}{f_s j}$$

Increase for wind or earthquake
$1.33F_b = 667$ psi
$1.33F_s = 32,000$ psi

f_b	f_s	K	p	np	k	j	$2/jk$	f_b	f_s	K
200	24,000	9.4	0.0004	0.005	0.097	0.968	21.29	267	32,000	12.5
225	24,000	11.7	0.0005	0.007	0.108	0.964	19.23	300	32,000	15.6
250	24,000	14.2	0.0006	0.008	0.118	0.961	17.58	333	32,000	19.0
275	24,000	16.9	0.0007	0.010	0.129	0.957	16.23	367	32,000	22.6
300	24,000	19.9	0.0009	0.011	0.139	0.954	15.10	400	32,000	26.5
325	24,000	23.0	0.0010	0.013	0.149	0.950	14.15	433	32,000	30.6
350	24,000	26.2	0.0012	0.015	0.158	0.947	13.34	467	32,000	35.0
375	24,000	29.7	0.0013	0.017	0.168	0.944	12.63	500	32,000	39.6
400	24,000	33.3	0.0015	0.019	0.177	0.941	12.01	533	32,000	44.4
425	24,000	37.1	0.0016	0.021	0.186	0.938	11.47	567	32,000	49.4
450	24,000	41.0	0.0018	0.024	0.195	0.935	10.98	600	32,000	54.6
475	24,000	45.0	0.0020	0.026	0.203	0.932	10.55	633	32,000	60.0
500	24,000	49.2	0.0022	0.028	0.212	0.929	10.16	667	32,000	65.6
500	23,000	50.8	0.0024	0.031	0.219	0.927	9.85	667	30,667	67.7
500	22,000	52.4	0.0026	0.033	0.227	0.924	9.54	667	29,333	69.9
500	21,000	54.1	0.0028	0.036	0.235	0.922	9.23	667	28,000	72.2
500	20,000	56.0	0.0030	0.039	0.244	0.919	8.93	667	26,667	74.7
500	19,000	58.0	0.0033	0.043	0.253	0.916	8.62	667	25,333	77.3
500	18,000	60.2	0.0037	0.047	0.264	0.912	8.31	667	24,000	80.2
500	17,000	62.5	0.0040	0.052	0.275	0.908	8.01	667	22,667	83.3
500	16,000	64.9	0.0045	0.058	0.287	0.904	7.70	667	21,333	86.6
500	15,000	67.6	0.0050	0.065	0.301	0.900	7.39	667	20,000	90.2
500	14,000	70.6	0.0056	0.073	0.315	0.895	7.09	667	18,667	94.1
500	13,000	73.7	0.0064	0.082	0.332	0.889	6.78	667	17,333	98.3
500	12,000	77.2	0.0073	0.094	0.350	0.883	6.48	667	16,000	103.0
500	11,000	81.0	0.0084	0.108	0.370	0.877	6.17	667	14,667	108.0
500	10,000	85.2	0.0098	0.126	0.392	0.869	5.87	667	13,333	113.6

TABLE E-4b Flexural Design Coefficients for Full Allowable Stresses:
f'_m = 3000 psi, f_y = 60,000 psi and n = 12.9

For use of Table, see Example 5-A

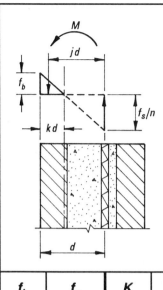

DESIGN DATA

$f'_m = 3000$ psi	$f_y = 60,000$ psi
$F_b = 0.33 f'_m = 1000$ psi	$F_s = 24,000$ psi
$E_m = 750 f'_m = 2,250,000$ psi	$E_s = 29,000,000$ psi

DESIGN EQUATIONS

$$n = \frac{E_s}{E_m} = 12.9 \qquad K = \frac{M}{F} = \frac{M(\text{ft kips})}{bd^2/12,000} \text{ or } \frac{M(\text{in. lbs})}{bd^2}$$

$$K = \tfrac{1}{2} jk f_b$$

$$k = \frac{1}{1 + f_s/n f_b} \qquad j = 1 - \frac{k}{3}$$

$$A_s = \frac{M}{f_s jd} \qquad p = \frac{A_s}{bd} = \frac{K}{f_s j}$$

Increase for wind or earthquake
$1.33 F_b = 1333$ psi
$1.33 F_s = 32,000$ psi

f_b	f_s	K	p	np	k	j	$2/jk$	f_b	f_s	K
250	24,000	14.2	0.0006	0.008	0.118	0.961	17.58	333	32,000	19.0
300	24,000	19.9	0.0009	0.011	0.139	0.954	15.10	400	32,000	26.5
350	24,000	26.2	0.0012	0.015	0.158	0.947	13.34	467	32,000	35.0
400	24,000	33.3	0.0015	0.019	0.177	0.941	12.01	533	32,000	44.4
450	24,000	41.0	0.0018	0.024	0.195	0.935	10.98	600	32,000	54.6
500	24,000	49.2	0.0022	0.028	0.212	0.929	10.16	667	32,000	65.6
550	24,000	58.0	0.0026	0.034	0.228	0.924	9.49	733	32,000	77.3
600	24,000	67.2	0.0030	0.039	0.244	0.919	8.93	800	32,000	89.6
650	24,000	76.9	0.0035	0.045	0.259	0.914	8.45	867	32,000	102.5
700	24,000	87.0	0.0040	0.051	0.273	0.909	8.05	933	32,000	116.0
750	24,000	97.4	0.0045	0.058	0.287	0.904	7.70	1,000	32,000	129.9
800	24,000	108.2	0.0050	0.065	0.301	0.900	7.39	1,067	32,000	144.3
850	24,000	119.3	0.0056	0.072	0.314	0.895	7.12	1,133	32,000	159.1
900	24,000	130.8	0.0061	0.079	0.326	0.891	6.88	1,200	32,000	174.4
950	24,000	142.5	0.0067	0.086	0.338	0.887	6.67	1,267	32,000	190.0
1000	24,000	154.4	0.0073	0.094	0.350	0.883	6.48	1,333	32,000	205.9
1000	23,000	158.1	0.0078	0.101	0.359	0.880	6.32	1,333	30,667	210.9
1000	22,000	162.0	0.0084	0.108	0.370	0.877	6.17	1,333	29,333	216.1
1000	21,000	166.1	0.0091	0.117	0.381	0.873	6.02	1,333	28,000	221.5
1000	20,000	170.4	0.0098	0.126	0.392	0.869	5.87	1,333	26,667	227.2
1000	19,000	174.9	0.0106	0.137	0.404	0.865	5.72	1,333	25,333	233.3
1000	18,000	179.7	0.0116	0.150	0.417	0.861	5.57	1,333	24,000	239.6
1000	17,000	184.7	0.0127	0.164	0.431	0.856	5.41	1,333	22,667	246.3
1000	16,000	190.0	0.0139	0.180	0.446	0.851	5.26	1,333	21,333	253.3
1000	15,000	195.6	0.0154	0.199	0.462	0.846	5.11	1,333	20,000	260.7
1000	14,000	201.4	0.0171	0.221	0.480	0.840	4.96	1,333	18,667	268.6
1000	13,000	207.7	0.0192	0.247	0.498	0.834	4.81	1,333	17,333	276.9
1000	12,000	214.3	0.0216	0.278	0.518	0.827	4.67	1,333	16,000	285.7
1000	11,000	221.3	0.0245	0.316	0.540	0.820	4.52	1,333	14,667	295.1
1000	10,000	228.8	0.0282	0.363	0.563	0.812	4.37	1,333	13,333	305.0

WSD

TABLE E-5 Flexural Design Coefficients for Full Allowable Stresses:
$f'_m = 3500$ psi, $f_y = 60,000$ psi and $n = 11.0$ For use of Table, see Example 5-A

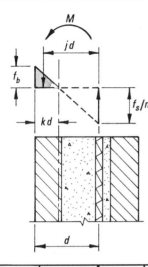

DESIGN DATA

$f'_m = 3500$ psi $f_y = 60,000$ psi

$F_b = 0.33 f'_m = 1167$ psi $F_s = 24,000$ psi

$E_m = 750 f'_m = 2,625,000$ psi $E_s = 29,000,000$ psi

DESIGN EQUATIONS

$$n = \frac{E_s}{E_m} = 11.0 \qquad K = \frac{M}{F} = \frac{M(\text{ft kips})}{bd^2/12,000} \text{ or } \frac{M(\text{in. lbs})}{bd^2}$$

$$K = \tfrac{1}{2} jk f_b$$

$$k = \frac{1}{1 + f_s/n f_b} \qquad j = 1 - \frac{k}{3}$$

$$A_s = \frac{M}{f_s jd} \qquad p = \frac{A_s}{bd} = \frac{K}{f_s j}$$

Increase for wind or earthquake
$1.33 F_b = 1556$ psi
$1.33 F_s = 32,000$ psi

f_b	f_s	K	p	np	k	j	2/jk	f_b	f_s	K
300	24,000	17.4	0.0008	0.008	0.121	0.960	17.24	400	32,000	23.2
350	24,000	23.1	0.0010	0.011	0.138	0.954	15.17	467	32,000	30.8
400	24,000	29.4	0.0013	0.014	0.155	0.948	13.61	533	32,000	39.2
450	24,000	36.3	0.0016	0.018	0.171	0.943	12.40	600	32,000	48.4
500	24,000	43.7	0.0019	0.021	0.186	0.938	11.44	667	32,000	58.3
550	24,000	51.7	0.0023	0.025	0.201	0.933	10.65	733	32,000	68.9
600	24,000	60.1	0.0027	0.030	0.216	0.928	9.99	800	32,000	80.1
650	24,000	68.9	0.0031	0.034	0.230	0.923	9.44	867	32,000	91.9
700	24,000	78.1	0.0035	0.039	0.243	0.919	8.96	933	32,000	104.2
750	24,000	87.8	0.0040	0.044	0.256	0.915	8.55	1,000	32,000	117.0
800	24,000	97.7	0.0045	0.049	0.268	0.911	8.19	1,067	32,000	130.3
850	24,000	108.0	0.0050	0.055	0.280	0.907	7.87	1,133	32,000	144.0
900	24,000	118.6	0.0055	0.060	0.292	0.903	7.59	1,200	32,000	158.2
950	24,000	129.5	0.0060	0.066	0.303	0.899	7.33	1,267	32,000	172.7
1,000	24,000	140.7	0.0065	0.072	0.314	0.895	7.11	1,333	32,000	187.6
1,050	24,000	152.1	0.0071	0.078	0.325	0.892	6.90	1,400	32,000	202.8
1,100	24,000	163.8	0.0077	0.084	0.335	0.888	6.72	1,467	32,000	218.3
1,150	24,000	175.6	0.0083	0.091	0.345	0.885	6.55	1,533	32,000	234.2
1,167	24,000	179.7	0.0085	0.093	0.350	0.883	6.49	1,556	32,000	239.6
1,167	23,000	184.1	0.0091	0.100	0.358	0.881	6.34	1,556	30,667	245.4
1,167	22,000	188.6	0.0098	0.108	0.368	0.877	6.19	1,556	29,333	251.5
1,167	21,000	193.4	0.0105	0.116	0.379	0.874	6.03	1,556	28,000	257.8
1,167	20,000	198.4	0.0114	0.125	0.391	0.870	5.88	1,556	26,667	264.5
1,167	19,000	203.7	0.0124	0.136	0.403	0.866	5.73	1,556	25,333	271.5
1,167	18,000	209.2	0.0135	0.148	0.416	0.861	5.58	1,556	24,000	278.9
1,167	17,000	215.0	0.0148	0.162	0.430	0.857	5.43	1,556	22,667	286.7
1,167	16,000	221.2	0.0162	0.179	0.445	0.852	5.28	1,556	21,333	294.9
1,167	15,000	227.7	0.0179	0.197	0.461	0.846	5.12	1,556	20,000	303.6
1,167	14,000	234.6	0.0199	0.219	0.478	0.841	4.97	1,556	18,667	312.8
1,167	13,000	241.9	0.0223	0.245	0.497	0.834	4.82	1,556	17,333	322.5
1,167	12,000	249.6	0.0251	0.276	0.517	0.828	4.68	1,556	16,000	332.8

TABLE E-6 Flexural Design Coefficients for Full Allowable Stresses: f'_m = 4000 psi, f_y = 60,000 psi and n = 9.7

For use of Table, see Example 5-A

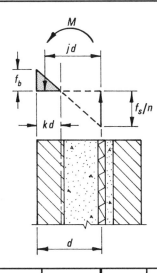

DESIGN DATA

$f'_m = 4000$ psi $\qquad f_y = 60,000$ psi

$F_b = 0.33 f'_m = 1333$ psi $\qquad F_s = 24,000$ psi

$E_{m'} = 750 f'_m = 3,000,000$ psi $\qquad E_s = 29,000,000$ psi

DESIGN EQUATIONS

$$n = \frac{E_s}{E_m} = 9.7 \qquad K = \frac{M}{F} = \frac{M(\text{ft kips})}{bd^2/12,000} \text{ or } \frac{M(\text{in. lbs})}{bd^2}$$

$$K = \tfrac{1}{2} jk f_b$$

$$k = \frac{1}{1 + f_s/n f_b} \qquad j = 1 - \frac{k}{3}$$

$$A_s = \frac{M}{f_s jd} \qquad p = \frac{A_s}{bd} = \frac{K}{f_s j}$$

Increase for wind or earthquake
$1.33F_b = 1777$ psi
$1.33F_s = 32,000$ psi

f_b	f_s	K	p	np	k	j	$2/jk$	f_b	f_s	K
400	24,000	26.5	0.0012	0.011	0.139	0.954	15.09	533	32,000	35.3
450	24,000	32.8	0.0014	0.014	0.154	0.949	13.70	600	32,000	43.8
500	24,000	39.7	0.0018	0.017	0.168	0.944	12.60	667	32,000	52.9
550	24,000	47.0	0.0021	0.020	0.182	0.939	11.71	733	32,000	62.6
600	24,000	54.7	0.0024	0.024	0.195	0.935	10.96	800	32,000	73.0
650	24,000	62.9	0.0028	0.027	0.208	0.931	10.33	867	32,000	83.9
700	24,000	71.5	0.0032	0.031	0.221	0.926	9.79	933	32,000	95.3
750	24,000	80.5	0.0036	0.035	0.233	0.922	9.32	1,000	32,000	107.3
800	24,000	89.8	0.0041	0.040	0.244	0.919	8.91	1,067	32,000	119.7
850	24,000	99.4	0.0045	0.044	0.256	0.915	8.55	1,133	32,000	132.5
900	24,000	109.4	0.0050	0.049	0.267	0.911	8.23	1,200	32,000	145.8
950	24,000	119.6	0.0055	0.053	0.277	0.908	7.94	1,267	32,000	159.5
1,000	24,000	130.1	0.0060	0.058	0.288	0.904	7.69	1,333	32,000	173.5
1,050	24,000	140.9	0.0065	0.063	0.298	0.901	7.45	1,400	32,000	187.8
1,100	24,000	151.9	0.0071	0.068	0.308	0.897	7.24	1,467	32,000	202.5
1,150	24,000	163.2	0.0076	0.074	0.317	0.894	7.05	1,533	32,000	217.5
1,200	24,000	174.6	0.0082	0.079	0.327	0.891	6.87	1,600	32,000	232.8
1,250	24,000	186.3	0.0087	0.085	0.336	0.888	6.71	1,667	32,000	248.4
1,300	24,000	198.2	0.0093	0.090	0.344	0.885	6.56	1,733	32,000	264.2
1,333	24,000	206.1	0.0097	0.094	0.350	0.883	6.47	1,777	32,000	274.8
1,333	23,000	211.1	0.0104	0.101	0.360	0.880	6.32	1,777	30,667	281.4
1,333	22,000	216.3	0.0112	0.109	0.370	0.877	6.16	1,777	29,333	288.4
1,333	21,000	221.7	0.0121	0.117	0.381	0.873	6.01	1,777	28,000	295.6
1,333	20,000	227.5	0.0131	0.127	0.393	0.869	5.86	1,777	26,667	303.3
1,333	19,000	233.5	0.0142	0.138	0.405	0.865	5.71	1,777	25,333	311.3
1,333	18,000	239.8	0.0155	0.150	0.418	0.861	5.56	1,777	24,000	319.7
1,333	17,000	246.5	0.0169	0.164	0.432	0.856	5.41	1,777	22,667	328.6
1,333	16,000	253.5	0.0186	0.181	0.447	0.851	5.26	1,777	21,333	338.0
1,333	15,000	260.9	0.0206	0.200	0.463	0.846	5.11	1,777	20,000	347.9
1,333	14,000	268.8	0.0229	0.222	0.480	0.840	4.96	1,777	18,667	358.4
1,333	13,000	277.1	0.0256	0.248	0.499	0.834	4.81	1,777	17,333	369.5
1,333	12,000	285.9	0.0288	0.279	0.519	0.827	4.66	1,777	16,000	381.2

TABLE E-7   Flexural Design Coefficients for Full Allowable Stresses,
f'_m = 4500 psi, f_y = 60,000 psi and n = 9.7
For use of Table, see Example 5-A

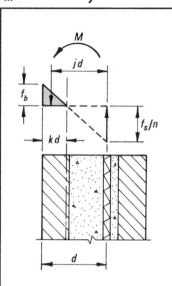

DESIGN DATA

$f'_m = 4500$ psi $\qquad\qquad f_y = 60,000$ psi

$F_b = 0.33 f'_m = 1500$ psi $\qquad F_s = 24,000$ psi

$E_m = 750 f'_m = 3,000,000$ psi $\qquad E_s = 29,000,000$ psi

DESIGN EQUATIONS

$$n = \frac{E_s}{E_m} = 9.7 \qquad K = \frac{M}{F} = \frac{M(\text{ft kips})}{bd^2/12,000} \text{ or } \frac{M(\text{in. lbs})}{bd^2}$$

$$K = \tfrac{1}{2} jk f_b$$

$$k = \frac{1}{1 + f_s/n f_b} \qquad j = 1 - \frac{k}{3}$$

$$A_s = \frac{M}{f_s jd} \qquad p = \frac{A_s}{bd} = \frac{K}{f_s j}$$

Increase for wind or earthquake
$1.33 F_b = 2000$ psi
$1.33 F_s = 32,000$ psi

f_b	f_s	K	p	np	k	j	2/jk	f_b	f_s	K
500	24,000	39.7	0.0018	0.017	0.168	0.944	12.60	667	32,000	52.9
550	24,000	47.0	0.0021	0.020	0.182	0.939	11.71	733	32,000	62.6
600	24,000	54.7	0.0024	0.024	0.195	0.935	10.96	800	32,000	73.0
650	24,000	62.9	0.0028	0.027	0.208	0.931	10.33	867	32,000	83.9
700	24,000	71.5	0.0032	0.031	0.221	0.926	9.79	933	32,000	95.3
750	24,000	80.5	0.0036	0.035	0.233	0.922	9.32	1,000	32,000	107.3
800	24,000	89.8	0.0041	0.040	0.244	0.919	8.91	1,067	32,000	119.7
850	24,000	99.4	0.0045	0.044	0.256	0.915	8.55	1,133	32,000	132.5
900	24,000	109.4	0.0050	0.049	0.267	0.911	8.23	1,200	32,000	145.8
950	24,000	119.6	0.0055	0.053	0.277	0.908	7.94	1,267	32,000	159.5
1,000	24,000	130.1	0.0060	0.058	0.288	0.904	7.69	1,333	32,000	173.5
1,050	24,000	140.9	0.0065	0.063	0.298	0.901	7.45	1,400	32,000	187.8
1,100	24,000	151.9	0.0071	0.068	0.308	0.897	7.24	1,467	32,000	202.5
1,150	24,000	163.2	0.0076	0.074	0.317	0.894	7.05	1,533	32,000	217.5
1,200	24,000	174.6	0.0082	0.079	0.327	0.891	6.87	1,600	32,000	232.8
1,250	24,000	186.3	0.0087	0.085	0.336	0.888	6.71	1,667	32,000	248.4
1,300	24,000	198.2	0.0093	0.090	0.344	0.885	6.56	1,733	32,000	264.2
1,350	24,000	210.2	0.0099	0.096	0.353	0.882	6.42	1,800	32,000	280.3
1,400	24,000	222.5	0.0105	0.102	0.361	0.880	6.29	1,867	32,000	296.6
1,450	24,000	234.9	0.0112	0.108	0.369	0.877	6.17	1,933	32,000	313.2
1,500	24,000	247.5	0.0118	0.114	0.377	0.874	6.06	2,000	32,000	329.9
1,500	23,000	253.1	0.0126	0.123	0.387	0.871	5.93	2,000	30,667	337.4
1,500	22,000	258.9	0.0136	0.132	0.398	0.867	5.79	2,000	29,333	345.3
1,500	21,000	265.1	0.0146	0.142	0.409	0.864	5.66	2,000	28,000	353.4
1,500	20,000	271.5	0.0158	0.153	0.421	0.860	5.52	2,000	26,667	362.0
1,500	19,000	278.2	0.0171	0.166	0.434	0.855	5.39	2,000	25,333	371.0
1,500	18,000	285.3	0.0186	0.181	0.447	0.851	5.26	2,000	24,000	380.4
1,500	17,000	292.7	0.0203	0.197	0.461	0.846	5.12	2,000	22,667	390.3
1,500	16,000	300.5	0.0223	0.217	0.476	0.841	4.99	2,000	21,333	400.7
1,500	15,000	308.7	0.0246	0.239	0.492	0.836	4.86	2,000	20,000	411.6

TABLE E-8   Flexural Design Coefficients for Full Allowable Stresses: For use of Table,
f'_m = 5000 psi, f_y = 60,000 psi and n = 9.7   see Example 5-A

WSD

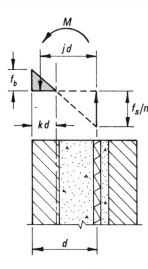

DESIGN DATA

$f'_m = 5000$ psi	$f_y = 60,000$ psi
$F_b = 0.33 f'_m = 1667$ psi	$F_s = 24,000$ psi
$E_m = 750 f'_m = 3,000,000$ psi	$E_s = 29,000,000$ psi

DESIGN EQUATIONS

$$n = \frac{E_s}{E_m} = 9.7 \qquad K = \frac{M}{F} = \frac{M(\text{ft kips})}{bd^2/12,000} \text{ or } \frac{M(\text{in. lbs})}{bd^2}$$

$$K = \tfrac{1}{2} jk f_b$$

$$k = \frac{1}{1 + f_s/n f_b} \qquad j = 1 - \frac{k}{3}$$

$$A_s = \frac{M}{f_s jd} \qquad p = \frac{A_s}{bd} = \frac{K}{f_s j}$$

Increase for wind or earthquake
$1.33 F_b = 2223$ psi
$1.33 F_s = 32,000$ psi

f_b	f_s	K	p	np	k	j	$2/jk$	f_b	f_s	K
600	24,000	54.7	0.0024	0.024	0.195	0.935	10.96	800	32,000	73.0
650	24,000	62.9	0.0028	0.027	0.208	0.931	10.33	867	32,000	83.9
700	24,000	71.5	0.0032	0.031	0.221	0.926	9.79	933	32,000	95.3
750	24,000	80.5	0.0036	0.035	0.233	0.922	9.32	1,000	32,000	107.3
800	24,000	89.8	0.0041	0.040	0.244	0.919	8.91	1,067	32,000	119.7
850	24,000	99.4	0.0045	0.044	0.256	0.915	8.55	1,133	32,000	132.5
900	24,000	109.4	0.0050	0.049	0.267	0.911	8.23	1,200	32,000	145.8
950	24,000	119.6	0.0055	0.053	0.277	0.908	7.94	1,267	32,000	159.5
1,000	24,000	130.1	0.0060	0.058	0.288	0.904	7.69	1,333	32,000	173.5
1,050	24,000	140.9	0.0065	0.063	0.298	0.901	7.45	1,400	32,000	187.8
1,100	24,000	151.9	0.0071	0.068	0.308	0.897	7.24	1,467	32,000	202.5
1,150	24,000	163.2	0.0076	0.074	0.317	0.894	7.05	1,533	32,000	217.5
1,200	24,000	174.6	0.0082	0.079	0.327	0.891	6.87	1,600	32,000	232.8
1,250	24,000	186.3	0.0087	0.085	0.336	0.888	6.71	1,667	32,000	248.4
1,300	24,000	198.2	0.0093	0.090	0.344	0.885	6.56	1,733	32,000	264.2
1,350	24,000	210.2	0.0099	0.096	0.353	0.882	6.42	1,800	32,000	280.3
1,400	24,000	222.5	0.0105	0.102	0.361	0.880	6.29	1,867	32,000	296.6
1,450	24,000	234.9	0.0112	0.108	0.369	0.877	6.17	1,933	32,000	313.2
1,500	24,000	247.5	0.0118	0.114	0.377	0.874	6.06	2,000	32,000	329.9
1,550	24,000	260.2	0.0124	0.121	0.385	0.872	5.96	2,067	32,000	346.9
1,600	24,000	273.0	0.0131	0.127	0.393	0.869	5.86	2,133	32,000	364.1
1,650	24,000	286.0	0.0138	0.133	0.400	0.867	5.77	2,200	32,000	381.4
1,667	24,000	290.5	0.0140	0.136	0.403	0.866	5.74	2,223	32,000	387.3
1,667	23,000	296.7	0.0150	0.145	0.413	0.862	5.62	2,223	30,667	395.6
1,667	22,000	303.2	0.0160	0.156	0.424	0.859	5.50	2,223	29,333	404.3
1,667	21,000	310.0	0.0173	0.167	0.435	0.855	5.38	2,223	28,000	413.4
1,667	20,000	317.1	0.0186	0.181	0.447	0.851	5.26	2,223	26,667	422.8
1,667	19,000	324.5	0.0202	0.196	0.460	0.847	5.14	2,223	25,333	432.6
1,667	18,000	332.2	0.0219	0.213	0.473	0.842	5.02	2,223	24,000	442.9
1,667	17,000	340.3	0.0239	0.232	0.487	0.838	4.90	2,223	22,667	453.7
1,667	16,000	348.8	0.0262	0.254	0.503	0.832	4.78	2,223	21,333	465.0
1,667	15,000	357.6	0.0288	0.280	0.519	0.827	4.66	2,223	20,000	476.8

TABLE E-9a　Flexural Coefficients Based on *np* Values　　　For use of table, see Example 5-F

$$n = \frac{E_s}{E_m} \qquad p = \frac{A_s}{bd} \qquad k = \sqrt{2np + (np)^2} - np \qquad f_b = \frac{M}{bd^2}\left(\frac{2}{jk}\right)$$

$$2/jk = \frac{bd^2 f_b}{M} \qquad npj = \frac{nM}{bd^2 f_s} \qquad j = 1 - \frac{k}{3} \qquad f_s = \frac{M}{bd^2}\left(\frac{1}{pj}\right)$$

$$M_m = \frac{f_b\, jk bd^2}{2} = f_b\, bd^2\left(\frac{1}{2/jk}\right) \qquad\qquad M_s = f_s\, p j bd^2$$

np	2/jk	j	k	npj	np	2/jk	j	k	npj
0.001	46.409	0.985	0.044	0.0010	0.051	8.075	0.909	0.272	0.0464
0.002	33.319	0.980	0.061	0.0020	0.052	8.016	0.908	0.275	0.0472
0.003	27.523	0.975	0.075	0.0029	0.053	7.958	0.908	0.277	0.0481
0.004	24.069	0.971	0.086	0.0039	0.054	7.902	0.907	0.279	0.0490
0.005	21.713	0.968	0.095	0.0048	0.055	7.848	0.906	0.281	0.0498
0.006	19.975	0.965	0.104	0.0058	0.056	7.795	0.906	0.283	0.0507
0.007	18.625	0.963	0.112	0.0067	0.057	7.744	0.905	0.285	0.0516
0.008	17.537	0.960	0.119	0.0077	0.058	7.694	0.904	0.287	0.0524
0.009	16.636	0.958	0.125	0.0086	0.059	7.645	0.903	0.290	0.0533
0.010	15.875	0.956	0.132	0.0096	0.060	7.598	0.903	0.292	0.0542
0.011	15.220	0.954	0.138	0.0105	0.061	7.552	0.902	0.294	0.0550
0.012	14.649	0.952	0.143	0.0114	0.062	7.507	0.901	0.296	0.0559
0.013	14.145	0.950	0.149	0.0124	0.063	7.462	0.901	0.298	0.0568
0.014	13.697	0.949	0.154	0.0133	0.064	7.419	0.900	0.299	0.0576
0.015	13.294	0.947	0.159	0.0142	0.065	7.378	0.900	0.301	0.0585
0.016	12.930	0.945	0.164	0.0151	0.066	7.337	0.899	0.303	0.0593
0.017	12.599	0.944	0.168	0.0160	0.067	7.296	0.898	0.305	0.0602
0.018	12.296	0.942	0.173	0.0170	0.068	7.257	0.898	0.307	0.0610
0.019	12.017	0.941	0.177	0.0179	0.069	7.219	0.897	0.309	0.0619
0.020	11.759	0.940	0.181	0.0188	0.070	7.182	0.896	0.311	0.0628
0.021	11.521	0.938	0.185	0.0197	0.071	7.145	0.896	0.312	0.0636
0.022	11.298	0.937	0.189	0.0206	0.072	7.109	0.895	0.314	0.0645
0.023	11.091	0.936	0.193	0.0215	0.073	7.074	0.895	0.316	0.0653
0.024	10.897	0.935	0.196	0.0224	0.074	7.040	0.894	0.318	0.0662
0.025	10.714	0.933	0.200	0.0233	0.075	7.006	0.894	0.319	0.0670
0.026	10.543	0.932	0.204	0.0242	0.076	6.973	0.893	0.321	0.0679
0.027	10.381	0.931	0.207	0.0251	0.077	6.941	0.892	0.323	0.0687
0.028	10.227	0.930	0.210	0.0260	0.078	6.909	0.892	0.325	0.0696
0.029	10.082	0.929	0.214	0.0269	0.079	6.878	0.891	0.326	0.0704
0.030	9.945	0.928	0.217	0.0278	0.080	6.848	0.891	0.328	0.0713
0.031	9.814	0.927	0.220	0.0287	0.081	6.818	0.890	0.330	0.0721
0.032	9.689	0.926	0.223	0.0296	0.082	6.788	0.890	0.331	0.0729
0.033	9.570	0.925	0.226	0.0305	0.083	6.759	0.889	0.333	0.0738
0.034	9.456	0.924	0.229	0.0314	0.084	6.731	0.889	0.334	0.0746
0.035	9.348	0.923	0.232	0.0323	0.085	6.703	0.888	0.336	0.0755
0.036	9.244	0.922	0.235	0.0332	0.086	6.676	0.887	0.338	0.0763
0.037	9.144	0.921	0.238	0.0341	0.087	6.649	0.887	0.339	0.0772
0.038	9.048	0.920	0.240	0.0350	0.088	6.623	0.886	0.341	0.0780
0.039	8.956	0.919	0.243	0.0358	0.089	6.597	0.886	0.342	0.0788
0.040	8.868	0.918	0.246	0.0367	0.090	6.572	0.885	0.344	0.0797
0.041	8.782	0.917	0.248	0.0376	0.091	6.547	0.885	0.345	0.0805
0.042	8.700	0.916	0.251	0.0385	0.092	6.522	0.884	0.347	0.0814
0.043	8.621	0.916	0.253	0.0394	0.093	6.498	0.884	0.348	0.0822
0.044	8.545	0.915	0.256	0.0402	0.094	6.474	0.883	0.350	0.0830
0.045	8.471	0.914	0.258	0.0411	0.095	6.451	0.883	0.351	0.0839
0.046	8.399	0.913	0.261	0.0420	0.096	6.428	0.882	0.353	0.0847
0.047	8.330	0.912	0.263	0.0429	0.097	6.405	0.882	0.354	0.0856
0.048	8.263	0.911	0.266	0.0438	0.098	6.383	0.882	0.355	0.0864
0.049	8.199	0.911	0.268	0.0446	0.099	6.361	0.881	0.357	0.0872
0.050	8.136	0.910	0.270	0.0455	0.100	6.340	0.881	0.358	0.0881

TABLE E-9b   Flexural Coefficients Based on *np* Values   For use of table, see Example 5-F

$$n = \frac{E_s}{E_m} \qquad p = \frac{A_s}{bd} \qquad k = \sqrt{2np + (np)^2} - np \qquad f_b = \frac{M}{bd^2}\left(\frac{2}{jk}\right)$$

$$2/jk = \frac{bd^2 f_b}{M} \qquad npj = \frac{nM}{bd^2 f_s} \qquad j = 1 - \frac{k}{3} \qquad f_s = \frac{M}{bd^2}\left(\frac{1}{pj}\right)$$

$$M_m = \frac{f_b jkbd^2}{2} = f_b bd^2\left(\frac{1}{2/jk}\right) \qquad\qquad M_s = f_s pjbd^2$$

np	2/jk	j	k	npj	np	2/jk	j	k	npj
0.101	6.318	0.880	0.360	0.0889	0.162	5.431	0.857	0.430	0.1388
0.102	6.297	0.880	0.361	0.0897	0.164	5.411	0.856	0.432	0.1404
0.103	6.277	0.879	0.362	0.0906	0.166	5.392	0.855	0.434	0.1420
0.104	6.257	0.879	0.364	0.0914	0.168	5.372	0.855	0.436	0.1436
0.105	6.237	0.878	0.365	0.0922	0.170	5.353	0.854	0.437	0.1452
0.106	6.217	0.878	0.366	0.0931	0.172	5.335	0.854	0.439	0.1468
0.107	6.197	0.877	0.368	0.0939	0.174	5.316	0.853	0.441	0.1484
0.108	6.178	0.877	0.369	0.0947	0.176	5.298	0.852	0.443	0.1500
0.109	6.159	0.877	0.370	0.0955	0.178	5.281	0.852	0.445	0.1516
0.110	6.141	0.876	0.372	0.0964	0.180	5.263	0.851	0.446	0.1532
0.111	6.122	0.876	0.373	0.0972	0.182	5.246	0.851	0.448	0.1548
0.112	6.104	0.875	0.374	0.0980	0.184	5.230	0.850	0.450	0.1564
0.113	6.086	0.875	0.376	0.0989	0.186	5.213	0.849	0.452	0.1580
0.114	6.069	0.874	0.377	0.0997	0.188	5.197	0.849	0.453	0.1596
0.115	6.051	0.874	0.378	0.1005	0.190	5.181	0.848	0.455	0.1612
0.116	6.034	0.874	0.379	0.1013	0.192	5.165	0.848	0.457	0.1628
0.117	6.017	0.873	0.381	0.1022	0.194	5.150	0.847	0.458	0.1644
0.118	6.001	0.873	0.382	0.1030	0.196	5.135	0.847	0.460	0.1659
0.119	5.984	0.872	0.383	0.1038	0.198	5.120	0.846	0.462	0.1675
0.120	5.968	0.872	0.384	0.1046	0.200	5.105	0.846	0.463	0.1691
0.121	5.952	0.871	0.386	0.1054	0.202	5.091	0.845	0.465	0.1707
0.122	5.936	0.871	0.387	0.1063	0.204	5.076	0.844	0.467	0.1723
0.123	5.920	0.871	0.388	0.1071	0.206	5.062	0.844	0.468	0.1739
0.124	5.905	0.870	0.389	0.1079	0.208	5.049	0.843	0.470	0.1754
0.125	5.890	0.870	0.390	0.1087	0.210	5.035	0.843	0.471	0.1770
0.126	5.874	0.869	0.392	0.1096	0.212	5.022	0.842	0.473	0.1786
0.127	5.860	0.869	0.393	0.1104	0.214	5.008	0.842	0.474	0.1802
0.128	5.845	0.869	0.394	0.1112	0.216	4.995	0.841	0.476	0.1817
0.129	5.830	0.868	0.395	0.1120	0.218	4.983	0.841	0.477	0.1833
0.130	5.816	0.868	0.396	0.1128	0.220	4.970	0.840	0.479	0.1849
0.131	5.802	0.868	0.397	0.1136	0.222	4.957	0.840	0.480	0.1865
0.132	5.788	0.867	0.398	0.1145	0.224	4.945	0.839	0.482	0.1880
0.133	5.774	0.867	0.400	0.1153	0.226	4.933	0.839	0.483	0.1896
0.134	5.760	0.866	0.401	0.1161	0.228	4.921	0.838	0.485	0.1912
0.135	5.747	0.866	0.402	0.1169	0.230	4.909	0.838	0.486	0.1927
0.136	5.733	0.866	0.403	0.1177	0.232	4.898	0.837	0.488	0.1943
0.137	5.720	0.865	0.404	0.1185	0.234	4.886	0.837	0.489	0.1959
0.138	5.707	0.865	0.405	0.1194	0.236	4.875	0.837	0.490	0.1974
0.139	5.694	0.865	0.406	0.1202	0.238	4.864	0.836	0.492	0.1990
0.140	5.681	0.864	0.407	0.1210	0.240	4.853	0.836	0.493	0.2005
0.142	5.656	0.863	0.410	0.1226	0.242	4.842	0.835	0.495	0.2021
0.144	5.631	0.863	0.412	0.1242	0.244	4.831	0.835	0.496	0.2037
0.146	5.607	0.862	0.414	0.1259	0.246	4.821	0.834	0.497	0.2052
0.148	5.584	0.861	0.416	0.1275	0.248	4.810	0.834	0.499	0.2068
0.150	5.560	0.861	0.418	0.1291	0.250	4.800	0.833	0.500	0.2083
0.152	5.538	0.860	0.420	0.1307	0.252	4.790	0.833	0.501	0.2099
0.154	5.516	0.859	0.422	0.1323	0.254	4.780	0.832	0.503	0.2114
0.156	5.494	0.859	0.424	0.1340	0.256	4.770	0.832	0.504	0.2130
0.158	5.473	0.858	0.426	0.1356	0.258	4.760	0.832	0.505	0.2145
0.160	5.452	0.857	0.428	0.1372	0.260	4.750	0.831	0.507	0.2161

WSD

Diagram F-1 *K* vs *p* for Various Masonry and Steel Stresses:
$f'_m = 1500$ psi, $n = 25.8$

For use of Diagram,
see Example 5-Y

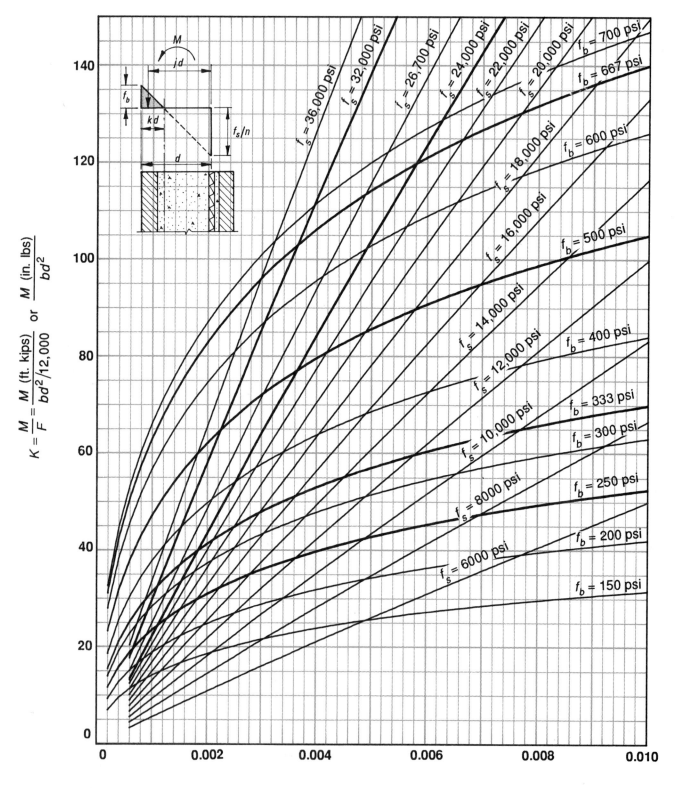

Steel Ratio, $p = \dfrac{A_s}{bd}$

Diagram F-2 *K* vs *p* for Various Masonry and Steel Stresses:
f'_m = 2000 psi, *n* = 19.3

For use of Diagram,
see Example 5-Y

WSD

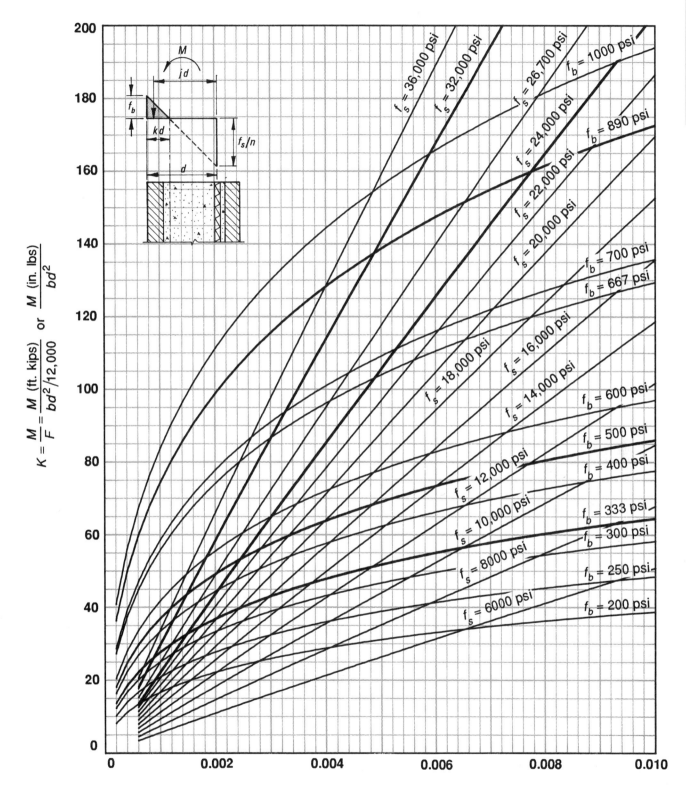

$$K = \frac{M}{F} = \frac{M \text{ (ft. kips)}}{bd^2/12,000} \quad \text{or} \quad \frac{M \text{ (in. lbs)}}{bd^2}$$

Steel Ratio, $p = \dfrac{A_s}{bd}$

Diagram F-3 *K* vs *p* for Various Masonry and Steel Stresses:
f'_m = 2500 psi, *n* = 15.5

For use of Diagram,
see Example 5-Y

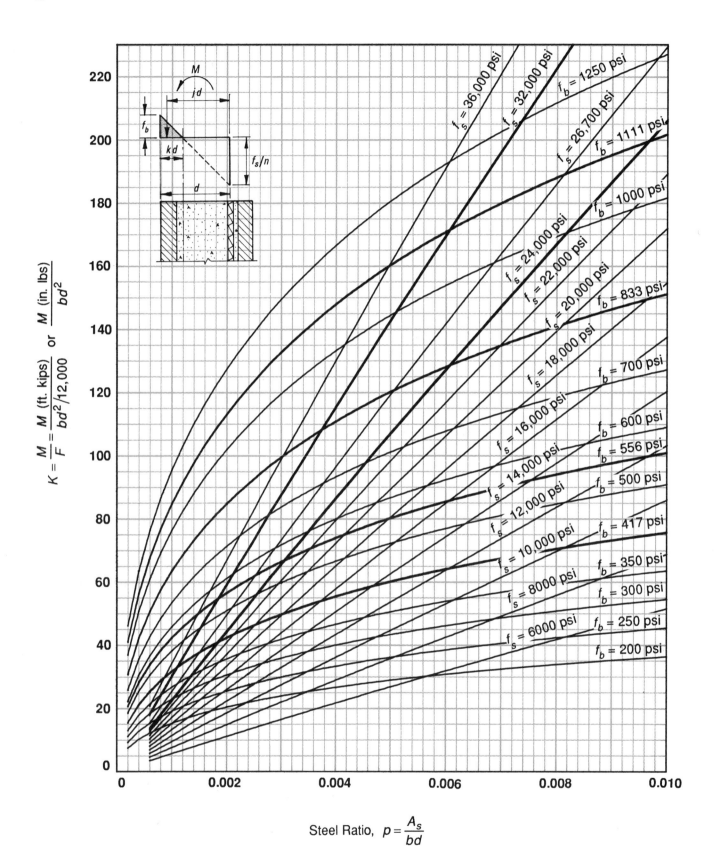

$$K = \frac{M}{F} = \frac{M \text{ (ft. kips)}}{bd^2/12,000} \quad \text{or} \quad \frac{M \text{ (in. lbs)}}{bd^2}$$

Steel Ratio, $p = \dfrac{A_s}{bd}$

Diagram F-4   K vs p for Various Masonry and Steel Stresses:
f'_m = 3000 psi, n = 12.9

For use of Diagram,
see Example 5-Y

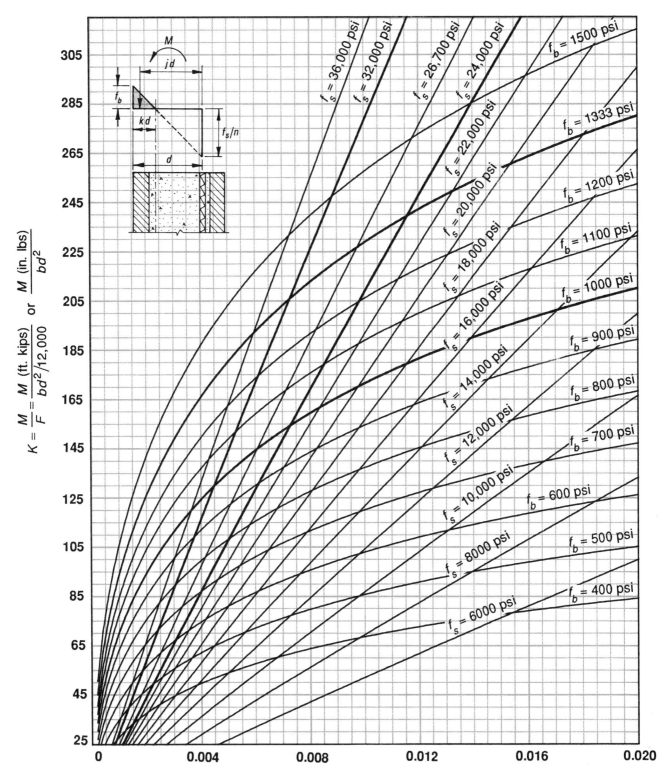

$$K = \frac{M}{F} = \frac{M \text{ (ft. kips)}}{bd^2/12,000} \text{ or } \frac{M \text{ (in. lbs)}}{bd^2}$$

Steel Ratio, $p = \dfrac{A_s}{bd}$

Diagram F-5 *K* vs *p* for Various Masonry and Steel Stresses:
f'_m = 3500 psi, *n* = 11.0

For use of Diagram,
see Example 5-Y

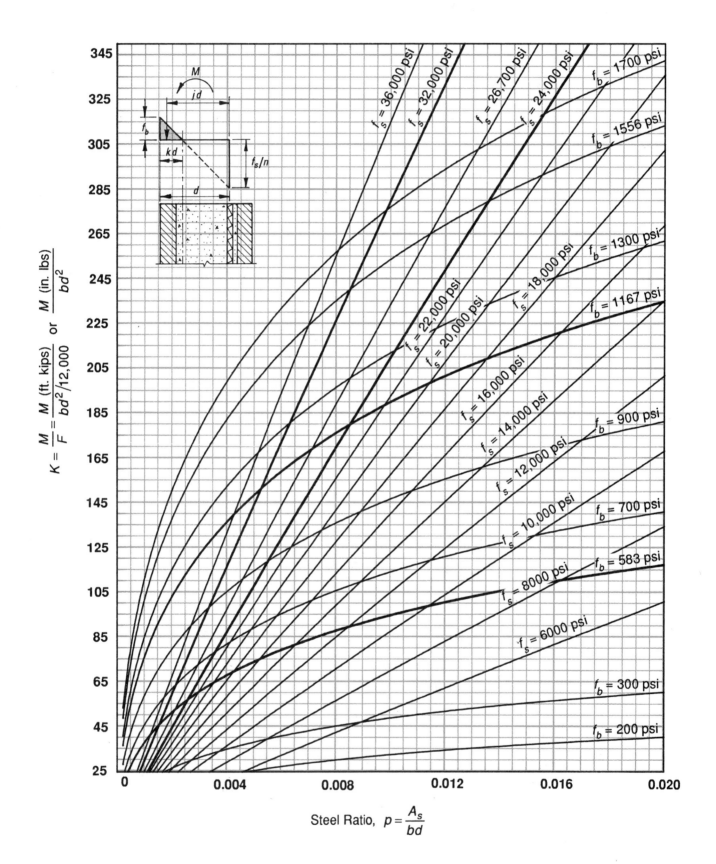

Steel Ratio, $p = \dfrac{A_s}{bd}$

Diagram F-6 *K* vs *p* for Various Masonry and Steel Stresses:
$f'_m = 4000$ psi, $n = 9.7$

For use of Diagram, see Example 5-Y

WSD

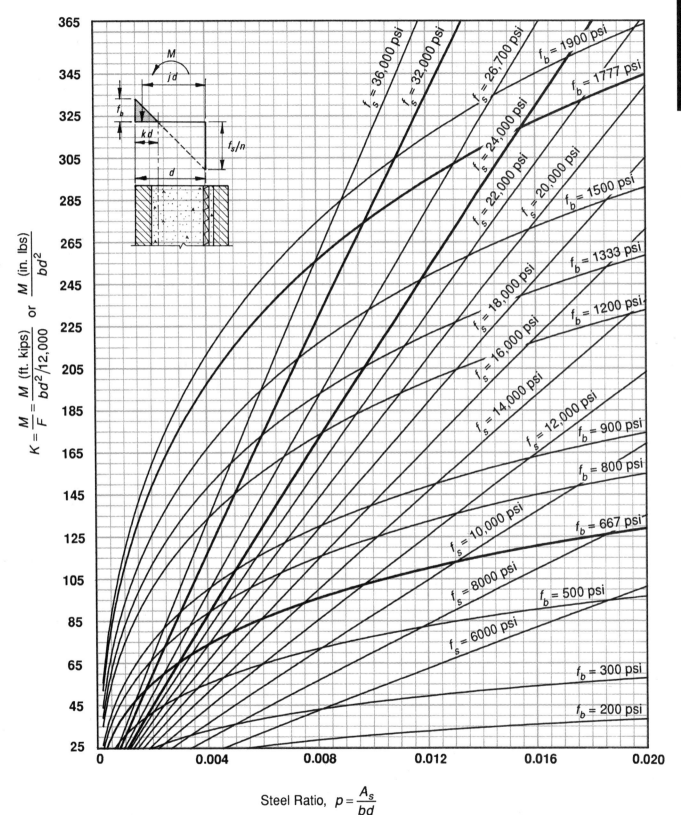

$$K = \frac{M}{F} = \frac{M \text{ (ft. kips)}}{bd^2/12,000} \quad or \quad \frac{M \text{ (in. lbs)}}{bd^2}$$

Steel Ratio, $p = \dfrac{A_s}{bd}$

$f_s = 36,000$ psi
$f_s = 32,000$ psi
$f_s = 26,700$ psi
$f_s = 24,000$ psi
$f_s = 22,000$ psi
$f_s = 20,000$ psi
$f_s = 18,000$ psi
$f_s = 16,000$ psi
$f_s = 14,000$ psi
$f_s = 12,000$ psi
$f_s = 10,000$ psi
$f_s = 8000$ psi
$f_s = 6000$ psi

$f_b = 1900$ psi
$f_b = 1777$ psi
$f_b = 1500$ psi
$f_b = 1333$ psi
$f_b = 1200$ psi
$f_b = 900$ psi
$f_b = 800$ psi
$f_b = 667$ psi
$f_b = 500$ psi
$f_b = 300$ psi
$f_b = 200$ psi

Diagram F-7 *K* vs *p* for Various Masonry and Steel Stresses:
f'_m = 4500 psi, *n* = 9.7

For use of Diagram,
see Example 5-Y

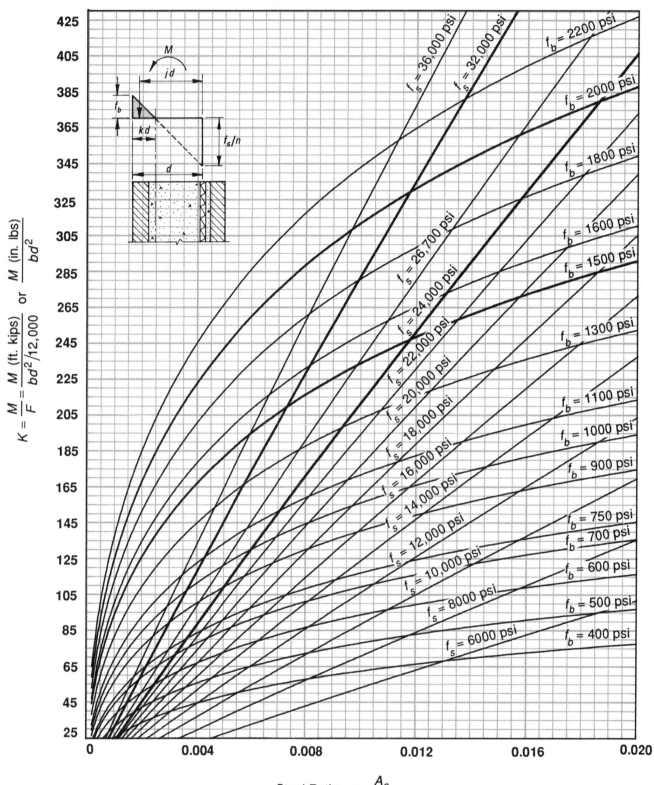

$$K = \frac{M}{F} = \frac{M \text{ (ft. kips)}}{bd^2/12,000} \quad \text{or} \quad \frac{M \text{ (in. lbs)}}{bd^2}$$

Steel Ratio, $p = \dfrac{A_s}{bd}$

Diagram F-8 *K* vs *p* for Various Masonry and Steel Stresses:
f'_m = 5000 psi, *n* = 9.7

For use of Diagram,
see Example 5-Y

WSD

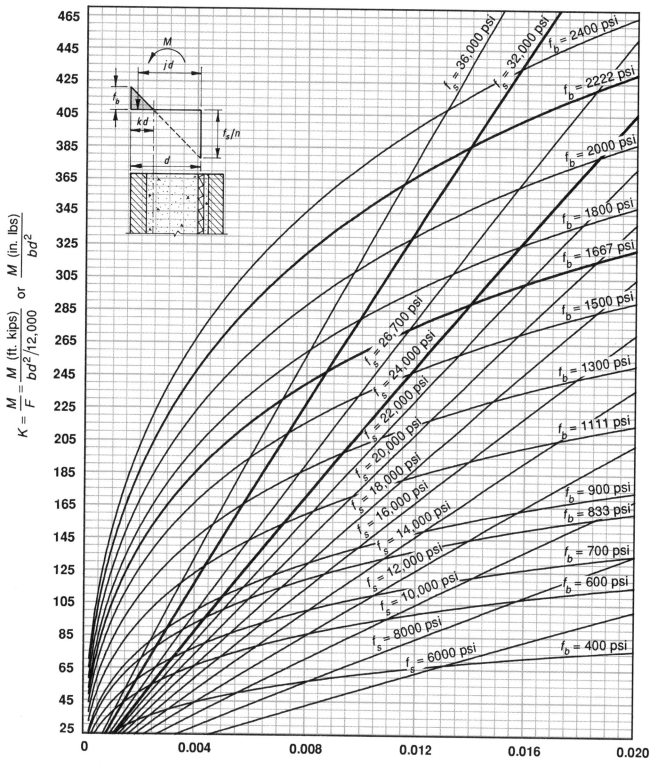

$$K = \frac{M}{F} = \frac{M \text{ (ft. kips)}}{bd^2/12{,}000} \quad \text{or} \quad \frac{M \text{ (in. lbs)}}{bd^2}$$

Steel Ratio, $p = \dfrac{A_s}{bd}$

Diagram F-9 *K* vs *np* for Various Masonry Stresses

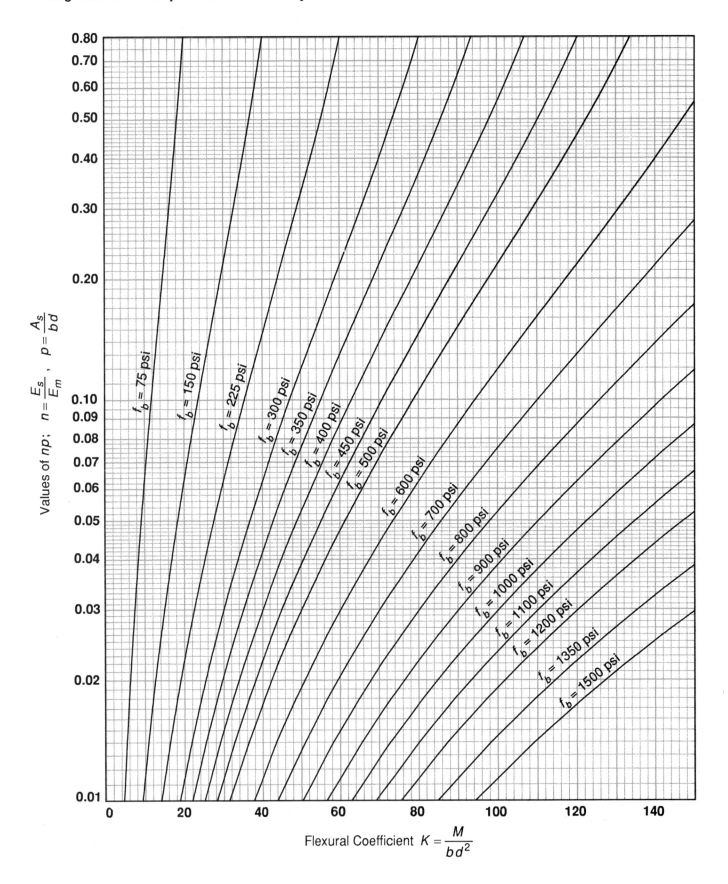

$$n = \frac{E_s}{E_m}, \quad p = \frac{A_s}{bd}$$

Values of *np*

Flexural Coefficient $K = \dfrac{M}{bd^2}$

TABLE H-1 Moment Capacity of Walls and Beams for Balanced Design Conditions: $f'_m = 1500$ psi and $f_y = 60,000$ psi

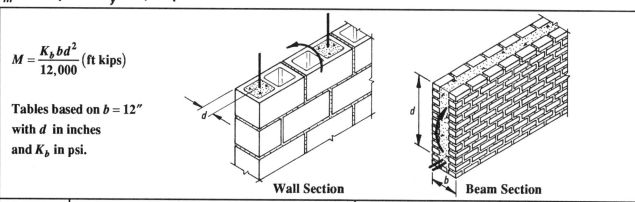

$$M = \frac{K_b bd^2}{12,000} \text{ (ft kips)}$$

Tables based on $b = 12''$ with d in inches and K_b in psi.

Wall Section   **Beam Section**

Effective Depth to Reinforcing Steel, d (inches)	No Special Inspection — Half Allowable Stresses			Special Inspection — Full Allowable Stresses		
	$f_b = 250$ psi $f_s = 24,000$ psi $K_b = 24.6$	$1.33f_b = 333$ psi $1.33f_s = 32,000$ psi $K_b = 32.8$	$p_b = 0.0011$	$f_b = 500$ psi $f_s = 24,000$ psi $K_b = 77.2$	$1.33f_b = 667$ psi $1.33f_s = 32,000$ psi $K_b = 102.9$	$p_b = 0.0036$
	Moment (ft kips)	Moment (ft kips)	A_s (sq. in.)	Moment (ft kips)	Moment (ft kips)	A_s (sq. in.)
2.75	0.19	0.25	0.04	0.58	0.78	0.12
3.00	0.22	0.30	0.04	0.69	0.93	0.13
3.50	0.30	0.40	0.05	0.95	1.26	0.15
4.00	0.39	0.52	0.05	1.24	1.65	0.17
4.50	0.50	0.66	0.06	1.56	2.08	0.19
5.00	0.62	0.82	0.07	1.93	2.57	0.22
5.50	0.74	0.99	0.07	2.34	3.11	0.24
6.00	0.89	1.18	0.08	2.78	3.71	0.26
6.50	1.04	1.39	0.09	3.26	4.35	0.28
7.00	1.21	1.61	0.09	3.78	5.04	0.30
7.50	1.38	1.85	0.10	4.34	5.79	0.32
8.00	1.57	2.10	0.11	4.94	6.59	0.35
9.00	1.99	2.66	0.12	6.25	8.34	0.39
10.00	2.46	3.28	0.13	7.72	10.29	0.43
11.00	2.98	3.97	0.15	9.34	12.45	0.48
12.00	3.54	4.72	0.16	11.12	14.82	0.52
13.00	4.16	5.54	0.17	13.05	17.40	0.56
14.00	4.82	6.43	0.18	15.13	20.17	0.60
15.00	5.54	7.38	0.20	17.37	23.16	0.65
16.00	6.30	8.40	0.21	19.76	26.35	0.69
20.00	9.84	13.12	0.26	30.88	41.17	0.86
24.00	14.17	18.89	0.32	44.47	59.29	1.04
28.00	19.29	25.72	0.37	60.52	80.70	1.21
32.00	25.19	33.59	0.42	79.05	105.40	1.38
36.00	31.88	42.51	0.48	100.05	133.40	1.56
40.00	39.36	52.48	0.53	123.52	164.69	1.73
44.00	47.63	63.50	0.58	149.46	199.28	1.90
48.00	56.68	75.57	0.63	177.87	237.16	2.07

WSD

TABLE H-2   Moment Capacity of Walls and Beams for Balanced Design Conditions:
f'_m = 2000 psi and f_y = 60,000 psi

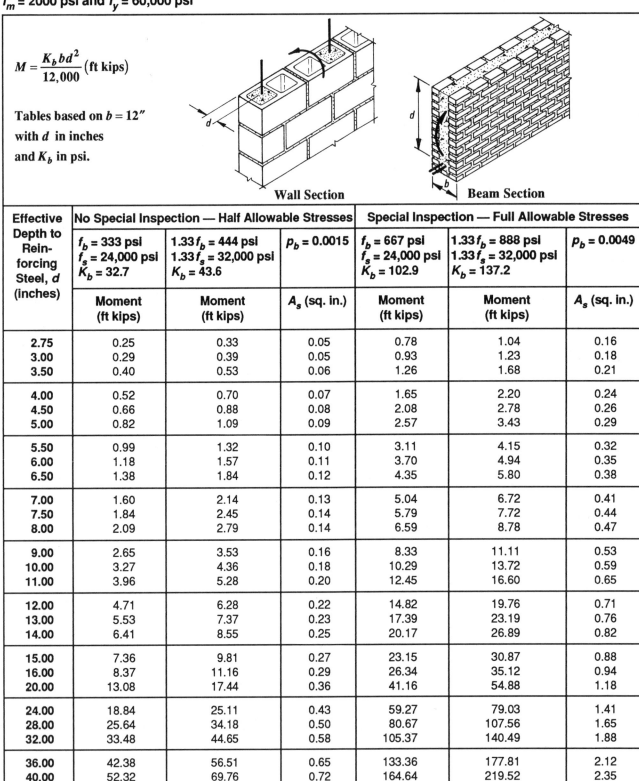

$$M = \frac{K_b b d^2}{12,000} \text{ (ft kips)}$$

Tables based on $b = 12''$
with d in inches
and K_b in psi.

Wall Section   **Beam Section**

Effective Depth to Reinforcing Steel, d (inches)	No Special Inspection — Half Allowable Stresses			Special Inspection — Full Allowable Stresses		
	f_b = 333 psi f_s = 24,000 psi K_b = 32.7	1.33f_b = 444 psi 1.33f_s = 32,000 psi K_b = 43.6	p_b = 0.0015	f_b = 667 psi f_s = 24,000 psi K_b = 102.9	1.33f_b = 888 psi 1.33f_s = 32,000 psi K_b = 137.2	p_b = 0.0049
	Moment (ft kips)	Moment (ft kips)	A_s (sq. in.)	Moment (ft kips)	Moment (ft kips)	A_s (sq. in.)
2.75	0.25	0.33	0.05	0.78	1.04	0.16
3.00	0.29	0.39	0.05	0.93	1.23	0.18
3.50	0.40	0.53	0.06	1.26	1.68	0.21
4.00	0.52	0.70	0.07	1.65	2.20	0.24
4.50	0.66	0.88	0.08	2.08	2.78	0.26
5.00	0.82	1.09	0.09	2.57	3.43	0.29
5.50	0.99	1.32	0.10	3.11	4.15	0.32
6.00	1.18	1.57	0.11	3.70	4.94	0.35
6.50	1.38	1.84	0.12	4.35	5.80	0.38
7.00	1.60	2.14	0.13	5.04	6.72	0.41
7.50	1.84	2.45	0.14	5.79	7.72	0.44
8.00	2.09	2.79	0.14	6.59	8.78	0.47
9.00	2.65	3.53	0.16	8.33	11.11	0.53
10.00	3.27	4.36	0.18	10.29	13.72	0.59
11.00	3.96	5.28	0.20	12.45	16.60	0.65
12.00	4.71	6.28	0.22	14.82	19.76	0.71
13.00	5.53	7.37	0.23	17.39	23.19	0.76
14.00	6.41	8.55	0.25	20.17	26.89	0.82
15.00	7.36	9.81	0.27	23.15	30.87	0.88
16.00	8.37	11.16	0.29	26.34	35.12	0.94
20.00	13.08	17.44	0.36	41.16	54.88	1.18
24.00	18.84	25.11	0.43	59.27	79.03	1.41
28.00	25.64	34.18	0.50	80.67	107.56	1.65
32.00	33.48	44.65	0.58	105.37	140.49	1.88
36.00	42.38	56.51	0.65	133.36	177.81	2.12
40.00	52.32	69.76	0.72	164.64	219.52	2.35
44.00	63.31	84.41	0.79	199.21	265.62	2.59
48.00	75.34	100.45	0.86	237.08	316.11	2.82

TABLE H-3 Moment Capacity of Walls and Beams for Balanced Design Conditions:
f'_m = 2500 psi and f_y = 60,000 psi

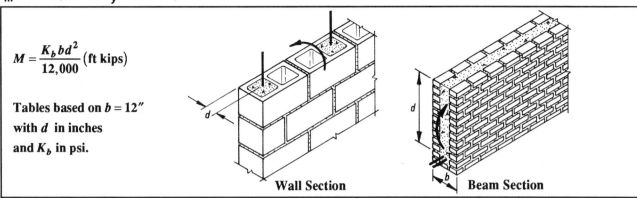

$$M = \frac{K_b b d^2}{12,000} \text{ (ft kips)}$$

Tables based on $b = 12''$
with d in inches
and K_b in psi.

Wall Section   **Beam Section**

Effective Depth to Reinforcing Steel, d (inches)	No Special Inspection — Half Allowable Stresses			Special Inspection — Full Allowable Stresses		
	f_b = 417 psi f_s = 24,000 psi K_b = 41.1	1.33f_b = 556 psi 1.33f_s = 32,000 psi K_b = 54.8	p_b = 0.0018	f_b = 833 psi f_s = 24,000 psi K_b = 128.7	1.33f_b = 1111 psi 1.33f_s = 32,000 psi K_b = 171.6	p_b = 0.0061
	Moment (ft kips)	Moment (ft kips)	A_s (sq. in.)	Moment (ft kips)	Moment (ft kips)	A_s (sq. in.)
2.75	0.31	0.41	0.06	0.97	1.30	0.20
3.00	0.37	0.49	0.06	1.16	1.54	0.22
3.50	0.50	0.67	0.08	1.58	2.10	0.26
4.00	0.66	0.88	0.09	2.06	2.75	0.29
4.50	0.83	1.11	0.10	2.61	3.47	0.33
5.00	1.03	1.37	0.11	3.22	4.29	0.37
5.50	1.24	1.66	0.12	3.89	5.19	0.40
6.00	1.48	1.97	0.13	4.63	6.18	0.44
6.50	1.74	2.32	0.14	5.44	7.25	0.48
7.00	2.01	2.69	0.15	6.31	8.41	0.51
7.50	2.31	3.08	0.16	7.24	9.65	0.55
8.00	2.63	3.51	0.17	8.24	10.98	0.59
9.00	3.33	4.44	0.19	10.42	13.90	0.66
10.00	4.11	5.48	0.22	12.87	17.16	0.73
11.00	4.97	6.63	0.24	15.57	20.76	0.81
12.00	5.92	7.89	0.26	18.53	24.71	0.88
13.00	6.95	9.26	0.28	21.75	29.00	0.95
14.00	8.06	10.74	0.30	25.23	33.63	1.02
15.00	9.25	12.33	0.32	28.96	38.61	1.10
16.00	10.52	14.03	0.35	32.95	43.93	1.17
20.00	16.44	21.92	0.43	51.48	68.64	1.46
24.00	23.67	31.56	0.52	74.13	98.84	1.76
28.00	32.22	42.96	0.60	100.90	134.53	2.05
32.00	42.09	56.12	0.69	131.79	175.72	2.34
36.00	53.27	71.02	0.78	166.80	222.39	2.64
40.00	65.76	87.68	0.86	205.92	274.56	2.93
44.00	79.57	106.09	0.95	249.16	332.22	3.22
48.00	94.69	126.26	1.04	296.52	395.37	3.51

TABLE H-4   Moment Capacity of Walls and Beams for Balanced Design Conditions:
$f_m' = 3000$ psi and $f_y = 60,000$ psi

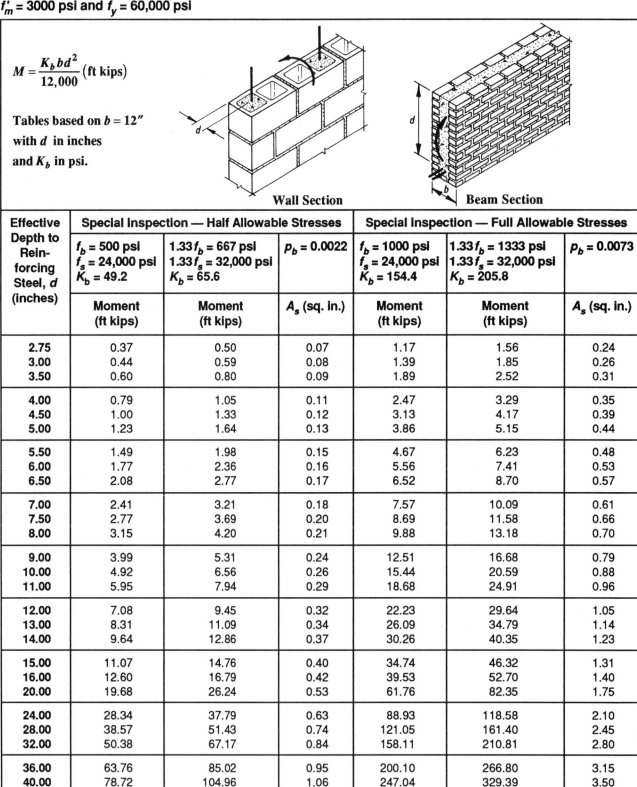

$$M = \frac{K_b b d^2}{12,000} \text{ (ft kips)}$$

Tables based on $b = 12''$
with d in inches
and K_b in psi.

Wall Section

Beam Section

Effective Depth to Reinforcing Steel, d (inches)	Special Inspection — Half Allowable Stresses			Special Inspection — Full Allowable Stresses		
	$f_b = 500$ psi $f_s = 24,000$ psi $K_b = 49.2$	$1.33 f_b = 667$ psi $1.33 f_s = 32,000$ psi $K_b = 65.6$	$p_b = 0.0022$	$f_b = 1000$ psi $f_s = 24,000$ psi $K_b = 154.4$	$1.33 f_b = 1333$ psi $1.33 f_s = 32,000$ psi $K_b = 205.8$	$p_b = 0.0073$
	Moment (ft kips)	Moment (ft kips)	A_s (sq. in.)	Moment (ft kips)	Moment (ft kips)	A_s (sq. in.)
2.75	0.37	0.50	0.07	1.17	1.56	0.24
3.00	0.44	0.59	0.08	1.39	1.85	0.26
3.50	0.60	0.80	0.09	1.89	2.52	0.31
4.00	0.79	1.05	0.11	2.47	3.29	0.35
4.50	1.00	1.33	0.12	3.13	4.17	0.39
5.00	1.23	1.64	0.13	3.86	5.15	0.44
5.50	1.49	1.98	0.15	4.67	6.23	0.48
6.00	1.77	2.36	0.16	5.56	7.41	0.53
6.50	2.08	2.77	0.17	6.52	8.70	0.57
7.00	2.41	3.21	0.18	7.57	10.09	0.61
7.50	2.77	3.69	0.20	8.69	11.58	0.66
8.00	3.15	4.20	0.21	9.88	13.18	0.70
9.00	3.99	5.31	0.24	12.51	16.68	0.79
10.00	4.92	6.56	0.26	15.44	20.59	0.88
11.00	5.95	7.94	0.29	18.68	24.91	0.96
12.00	7.08	9.45	0.32	22.23	29.64	1.05
13.00	8.31	11.09	0.34	26.09	34.79	1.14
14.00	9.64	12.86	0.37	30.26	40.35	1.23
15.00	11.07	14.76	0.40	34.74	46.32	1.31
16.00	12.60	16.79	0.42	39.53	52.70	1.40
20.00	19.68	26.24	0.53	61.76	82.35	1.75
24.00	28.34	37.79	0.63	88.93	118.58	2.10
28.00	38.57	51.43	0.74	121.05	161.40	2.45
32.00	50.38	67.17	0.84	158.11	210.81	2.80
36.00	63.76	85.02	0.95	200.10	266.80	3.15
40.00	78.72	104.96	1.06	247.04	329.39	3.50
44.00	95.25	127.00	1.16	298.92	398.56	3.85
48.00	113.36	151.14	1.27	355.74	474.32	4.20

TABLE H-5 Moment Capacity of Walls and Beams for Balanced Design Conditions: either f'_m = 3500 psi or f'_m = 4000 psi and f_y = 60,000 psi

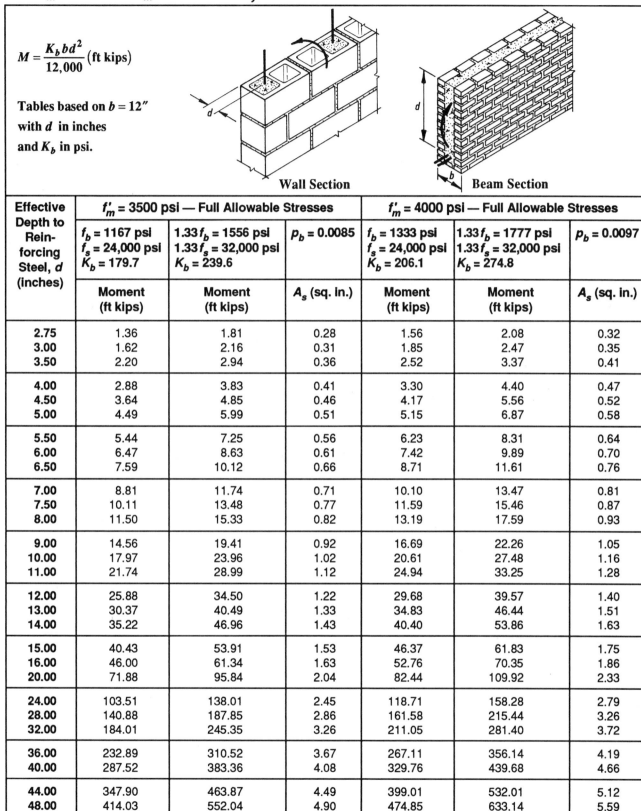

$$M = \frac{K_b bd^2}{12,000} \text{ (ft kips)}$$

Tables based on $b = 12''$
with d in inches
and K_b in psi.

Wall Section   **Beam Section**

Effective Depth to Reinforcing Steel, d (inches)	f'_m = 3500 psi — Full Allowable Stresses			f'_m = 4000 psi — Full Allowable Stresses		
	f_b = 1167 psi f_s = 24,000 psi K_b = 179.7	1.33f_b = 1556 psi 1.33f_s = 32,000 psi K_b = 239.6	p_b = 0.0085	f_b = 1333 psi f_s = 24,000 psi K_b = 206.1	1.33f_b = 1777 psi 1.33f_s = 32,000 psi K_b = 274.8	p_b = 0.0097
	Moment (ft kips)	Moment (ft kips)	A_s (sq. in.)	Moment (ft kips)	Moment (ft kips)	A_s (sq. in.)
2.75	1.36	1.81	0.28	1.56	2.08	0.32
3.00	1.62	2.16	0.31	1.85	2.47	0.35
3.50	2.20	2.94	0.36	2.52	3.37	0.41
4.00	2.88	3.83	0.41	3.30	4.40	0.47
4.50	3.64	4.85	0.46	4.17	5.56	0.52
5.00	4.49	5.99	0.51	5.15	6.87	0.58
5.50	5.44	7.25	0.56	6.23	8.31	0.64
6.00	6.47	8.63	0.61	7.42	9.89	0.70
6.50	7.59	10.12	0.66	8.71	11.61	0.76
7.00	8.81	11.74	0.71	10.10	13.47	0.81
7.50	10.11	13.48	0.77	11.59	15.46	0.87
8.00	11.50	15.33	0.82	13.19	17.59	0.93
9.00	14.56	19.41	0.92	16.69	22.26	1.05
10.00	17.97	23.96	1.02	20.61	27.48	1.16
11.00	21.74	28.99	1.12	24.94	33.25	1.28
12.00	25.88	34.50	1.22	29.68	39.57	1.40
13.00	30.37	40.49	1.33	34.83	46.44	1.51
14.00	35.22	46.96	1.43	40.40	53.86	1.63
15.00	40.43	53.91	1.53	46.37	61.83	1.75
16.00	46.00	61.34	1.63	52.76	70.35	1.86
20.00	71.88	95.84	2.04	82.44	109.92	2.33
24.00	103.51	138.01	2.45	118.71	158.28	2.79
28.00	140.88	187.85	2.86	161.58	215.44	3.26
32.00	184.01	245.35	3.26	211.05	281.40	3.72
36.00	232.89	310.52	3.67	267.11	356.14	4.19
40.00	287.52	383.36	4.08	329.76	439.68	4.66
44.00	347.90	463.87	4.49	399.01	532.01	5.12
48.00	414.03	552.04	4.90	474.85	633.14	5.59

TABLE H-6   Moment Capacity of Walls and Beams for Balanced Design Conditions: either f'_m = 4500 psi or f'_m = 5000 psi and f_y = 60,000 psi

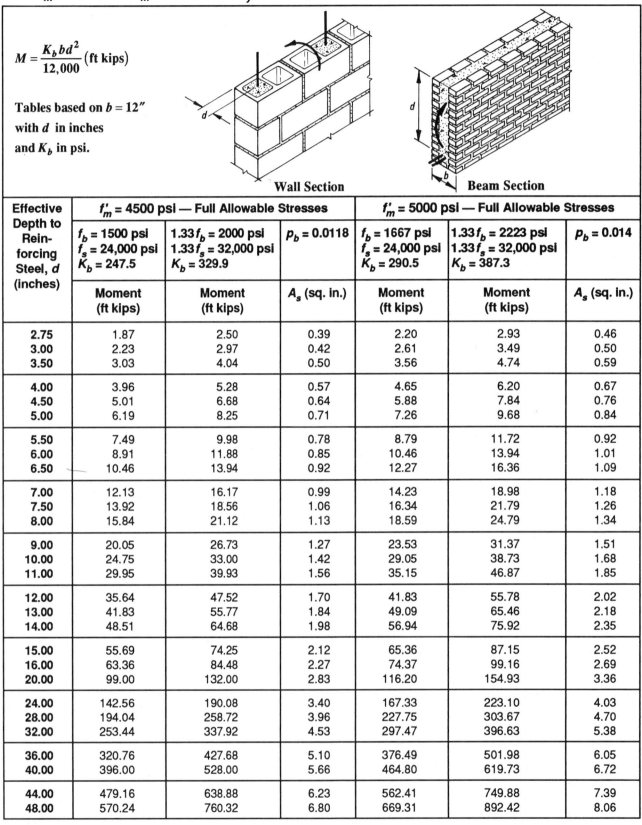

$$M = \frac{K_b bd^2}{12,000} \text{ (ft kips)}$$

Tables based on $b = 12''$ with d in inches and K_b in psi.

Wall Section

Beam Section

Effective Depth to Reinforcing Steel, d (inches)	f'_m = 4500 psi — Full Allowable Stresses			f'_m = 5000 psi — Full Allowable Stresses		
	f_b = 1500 psi f_s = 24,000 psi K_b = 247.5	1.33f_b = 2000 psi 1.33f_s = 32,000 psi K_b = 329.9	p_b = 0.0118	f_b = 1667 psi f_s = 24,000 psi K_b = 290.5	1.33f_b = 2223 psi 1.33f_s = 32,000 psi K_b = 387.3	p_b = 0.014
	Moment (ft kips)	Moment (ft kips)	A_s (sq. in.)	Moment (ft kips)	Moment (ft kips)	A_s (sq. in.)
2.75	1.87	2.50	0.39	2.20	2.93	0.46
3.00	2.23	2.97	0.42	2.61	3.49	0.50
3.50	3.03	4.04	0.50	3.56	4.74	0.59
4.00	3.96	5.28	0.57	4.65	6.20	0.67
4.50	5.01	6.68	0.64	5.88	7.84	0.76
5.00	6.19	8.25	0.71	7.26	9.68	0.84
5.50	7.49	9.98	0.78	8.79	11.72	0.92
6.00	8.91	11.88	0.85	10.46	13.94	1.01
6.50	10.46	13.94	0.92	12.27	16.36	1.09
7.00	12.13	16.17	0.99	14.23	18.98	1.18
7.50	13.92	18.56	1.06	16.34	21.79	1.26
8.00	15.84	21.12	1.13	18.59	24.79	1.34
9.00	20.05	26.73	1.27	23.53	31.37	1.51
10.00	24.75	33.00	1.42	29.05	38.73	1.68
11.00	29.95	39.93	1.56	35.15	46.87	1.85
12.00	35.64	47.52	1.70	41.83	55.78	2.02
13.00	41.83	55.77	1.84	49.09	65.46	2.18
14.00	48.51	64.68	1.98	56.94	75.92	2.35
15.00	55.69	74.25	2.12	65.36	87.15	2.52
16.00	63.36	84.48	2.27	74.37	99.16	2.69
20.00	99.00	132.00	2.83	116.20	154.93	3.36
24.00	142.56	190.08	3.40	167.33	223.10	4.03
28.00	194.04	258.72	3.96	227.75	303.67	4.70
32.00	253.44	337.92	4.53	297.47	396.63	5.38
36.00	320.76	427.68	5.10	376.49	501.98	6.05
40.00	396.00	528.00	5.66	464.80	619.73	6.72
44.00	479.16	638.88	6.23	562.41	749.88	7.39
48.00	570.24	760.32	6.80	669.31	892.42	8.06

TABLE J-1 Moment Capacity (ft k/ft) of Walls with $A_s = 0.0007bt$, $b = 12''$ and $F_s = 24,000$ psi[3]

For use of Table, see Example 5-S

Wall Type	Nominal Thickness	Actual Thickness, t (Inches)	Effective Depth, d (Inches)	$f'_m =$ $f_b =$ $E_m =$ $n =$	1500 500 1,125,000 25.8	2000 667 1,500,000 19.3	2500 833 1,875,000 15.5	3000 1000 2,250,000 12.9	3500 1167 2,625,000 11.0	4000 1333 3,000,000 9.7	5000 1667 3,000,000 9.7	A_s (Sq. in./Ft)
Concrete Masonry Units	6	5.625	2.8[1]	$M_m{}^3$	0.43	0.51	0.58	0.64	0.70	0.76	0.95	0.047
				M_s	0.24	0.25	0.25	0.25	0.25	0.25	0.25	
	8	7.625	3.8[1]	$M_m{}^3$	0.79	0.93	1.06	1.18	1.29	1.39	1.74	0.064
				M_s	0.45	0.45	0.46	0.46	0.46	0.46	0.46	
			5.1[2]	$M_m{}^3$	1.26	1.50	1.70	1.89	2.06	2.22	2.77	
				M_s	0.61	0.62	0.62	0.62	0.63	0.63	0.63	
			5.3[2]	$M_m{}^3$	1.36	1.61	1.83	2.03	2.22	2.39	2.98	
				M_s	0.64	0.65	0.65	0.66	0.66	0.66	0.66	
	10	9.625	4.8[1]	$M_m{}^3$	1.25	1.49	1.70	1.88	2.06	2.22	2.77	0.081
				M_s	0.72	0.72	0.73	0.73	0.74	0.74	0.74	
			7.1[2]	$M_m{}^3$	2.35	2.77	3.15	3.49	3.81	4.10	5.13	
				M_s	1.08	1.09	1.09	1.10	1.10	1.10	1.10	
	12	11.625	5.8[1]	$M_m{}^3$	1.83	2.17	2.47	2.75	3.00	3.24	4.05	0.098
				M_s	1.05	1.06	1.06	1.07	1.07	1.08	1.08	
			9.1[2]	$M_m{}^3$	3.76	4.44	5.04	5.58	6.08	6.55	8.19	
				M_s	1.67	1.68	1.69	1.70	1.71	1.71	1.71	
Concrete Masonry Component (Expandable) Walls	10	10	5.0[1]	$M_m{}^3$	1.35	1.61	1.83	2.04	2.22	2.40	3.00	0.084
				M_s	0.77	0.78	0.79	0.79	0.79	0.80	0.80	
			6.7[2]	$M_m{}^3$	2.19	2.59	2.94	3.26	3.56	3.83	4.79	
				M_s	1.06	1.07	1.07	1.08	1.08	1.08	1.08	
	12	12	6.0[1]	$M_m{}^3$	1.95	2.31	2.64	2.93	3.20	3.45	4.32	0.101
				M_s	1.11	1.13	1.13	1.14	1.14	1.15	1.15	
			8.7[2]	$M_m{}^3$	3.56	4.21	4.78	5.30	5.78	6.22	7.78	
				M_s	1.65	1.66	1.67	1.68	1.68	1.69	1.69	
	16	16	12.7[2]	$M_m{}^3$	7.28	8.60	9.76	10.82	11.79	12.69	15.87	0.134
				M_s	3.21	3.24	3.25	3.27	3.28	3.29	3.29	
Hollow Clay Brick Units	6	5.5	2.7[1]	$M_m{}^3$	0.41	0.49	0.55	0.62	0.67	0.73	0.91	0.046
				M_s	0.24	0.24	0.24	0.24	0.25	0.25	0.25	
	8	7.5	3.7[1]	$M_m{}^3$	0.76	0.90	1.03	1.14	1.25	1.35	1.69	0.063
				M_s	0.44	0.44	0.44	0.45	0.45	0.45	0.45	
			5.2[2]	$M_m{}^3$	1.30	1.54	1.75	1.94	2.12	2.28	2.85	
				M_s	0.62	0.62	0.63	0.63	0.63	0.63	0.63	
Two Wythe Clay Brick Walls	9	9	4.5[1]	$M_m{}^3$	1.10	1.30	1.48	1.65	1.80	1.94	2.43	0.076
				M_s	0.63	0.63	0.64	0.64	0.64	0.65	0.65	
	10	10	5.0[1]	$M_m{}^3$	1.35	1.61	1.83	2.04	2.22	2.40	3.00	0.084
				M_s	0.77	0.78	0.79	0.79	0.79	0.80	0.80	
	12	12	6.0[1]	$M_m{}^3$	1.95	2.31	2.64	2.93	3.20	3.45	4.32	0.101
				M_s	1.11	1.13	1.13	1.14	1.14	1.15	1.15	
			7.5[2]	$M_m{}^3$	2.78	3.30	3.75	4.16	4.54	4.89	6.12	
				M_s	1.40	1.42	1.43	1.43	1.44	1.44	1.44	
	16	16	11.5[2]	$M_m{}^3$	6.18	7.31	8.31	9.21	10.04	10.81	13.52	0.134
				M_s	2.88	2.91	2.93	2.94	2.95	2.96	2.96	

1. Based on $d = t/2$.
2. Based on a 1″ distance between the center of reinforcing and the inside of the face shell or wythe. Therefore $d = (t - 1'' -$ common face shell or wythe thickness).
3. Table values are based on special inspection being provided. For noninspected masonry, the capacity of the masonry, M_m, must be reduced by one half per UBC Section 2406(c)1.

WSD

TABLE J-2 Moment Capacity (ft k/ft) of Walls with $A_s = 0.0013bt$, $b = 12''$ and $F_s = 24{,}000$ psi[3]

For use of Table, see Example 5-S

Wall Type	Nominal Thickness	Actual Thickness, t (Inches)	Effective Depth, d (Inches)		$f'_m =$ 1500 $f_b =$ 500 $E_m =$ 1,125,000 $n =$ 25.8	2000 667 1,500,000 19.3	2500 833 1,875,000 15.5	3000 1000 2,250,000 12.9	3500 1167 2,625,000 11.0	4000 1333 3,000,000 9.7	5000 1667 3,000,000 9.7	A_s (Sq. in./Ft)
Concrete Masonry Units	6	5.625	2.8[1]	M_m[3]	0.54	0.65	0.74	0.83	0.91	0.99	1.23	0.088
				M_s	0.44	0.45	0.45	0.46	0.46	0.46	0.46	
	8	7.625	3.8[1]	M_m[3]	1.00	1.19	1.37	1.53	1.67	1.81	2.26	0.119
				M_s	0.81	0.82	0.83	0.84	0.84	0.85	0.85	
			5.1[2]	M_m[3]	1.62	1.93	2.20	2.45	2.68	2.90	3.62	
				M_s	1.11	1.12	1.13	1.14	1.14	1.15	1.15	
			5.3[2]	M_m[3]	1.74	2.08	2.38	2.64	2.89	3.12	3.91	
				M_s	1.17	1.18	1.19	1.20	1.20	1.21	1.21	
	10	9.625	4.8[1]	M_m[3]	1.59	1.90	2.18	2.43	2.67	2.89	3.61	0.150
				M_s	1.30	1.31	1.33	1.33	1.34	1.35	1.35	
			7.1[2]	M_m[3]	3.01	3.58	4.09	4.55	4.98	5.38	6.72	
				M_s	1.96	1.98	1.99	2.00	2.01	2.02	2.02	
	12	11.625	5.8[1]	M_m[3]	2.31	2.77	3.18	3.55	3.89	4.21	5.26	0.181
				M_s	1.89	1.92	1.93	1.95	1.96	1.97	1.97	
			9.1[2]	M_m[3]	4.82	5.74	6.55	7.28	7.96	8.60	10.74	
				M_s	3.03	3.07	3.09	3.11	3.12	3.13	3.13	
Concrete Masonry Component (Expandable) Walls	10	10	5.0[1]	M_m[3]	1.71	2.05	2.35	2.63	2.88	3.12	3.90	0.156
				M_s	1.40	1.42	1.43	1.44	1.45	1.46	1.46	
			6.7[2]	M_m[3]	2.79	3.33	3.81	4.24	4.64	5.01	6.27	
				M_s	1.92	1.94	1.95	1.97	1.98	1.98	1.98	
	12	12	6.0[1]	M_m[3]	2.47	2.96	3.39	3.79	4.15	4.49	5.61	0.187
				M_s	2.02	2.04	2.06	2.08	2.09	2.10	2.10	
			8.7[2]	M_m[3]	4.56	5.43	6.20	6.90	7.55	8.15	10.19	
				M_s	2.99	3.02	3.05	3.07	3.08	3.09	3.09	
	16	16	12.7[2]	M_m[3]	9.35	11.12	12.69	14.12	15.43	16.66	20.82	0.250
				M_s	5.83	5.89	5.94	5.97	6.00	6.02	6.02	
Hollow Clay Brick Units	6	5.5	2.7[1]	M_m[3]	0.52	0.62	0.71	0.80	0.87	0.94	1.18	0.086
				M_s	0.43	0.44	0.44	0.45	0.45	0.45	0.45	
	8	7.5	3.7[1]	M_m[3]	0.96	1.15	1.32	1.48	1.62	1.75	2.19	0.117
				M_s	0.79	0.80	0.81	0.81	0.82	0.82	0.82	
			5.2[2]	M_m[3]	1.67	1.99	2.27	2.53	2.76	2.99	3.73	
				M_s	1.12	1.13	1.14	1.15	1.15	1.16	1.16	
Two Wythe Clay Brick Walls	9	9	4.5[1]	M_m[3]	1.10	1.30	1.48	1.65	1.80	1.94	2.43	0.076
				M_s	0.63	0.63	0.64	0.64	0.64	0.65	0.65	
	10	10	5.0[1]	M_m[3]	1.35	1.61	1.83	2.04	2.22	2.40	3.00	0.084
				M_s	0.77	0.78	0.79	0.79	0.79	0.80	0.80	
	12	12	6.0[1]	M_m[3]	1.95	2.31	2.64	2.93	3.20	3.45	4.32	0.101
				M_s	1.11	1.13	1.13	1.14	1.14	1.15	1.15	
			7.5[2]	M_m[3]	2.78	3.30	3.75	4.16	4.54	4.89	6.12	
				M_s	1.40	1.42	1.43	1.43	1.44	1.44	1.44	
	16	16	11.5[2]	M_m[3]	6.18	7.31	8.31	9.21	10.04	10.81	13.52	0.134
				M_s	2.88	2.91	2.93	2.94	2.95	2.96	2.96	

1. Based on $d = t/2$.

2. Based on a 1″ distance between the center of reinforcing and the inside of the face shell or wythe. Therefore $d = (t - 1'' -$ common face shell or wythe thickness).

3. Table values are based on special inspection being provided. For noninspected masonry, the capacity of the masonry, M_m, must be reduced by one half per UBC Section 2406(c)1.

TABLE J-3 Moment Capacity (ft k/ft) of Walls with $A_s = 0.001bt$, $b = 12''$ and $F_s = 24{,}000$ psi[3]

For use of Table, see Example 5-S

Wall Type	Nominal Thickness	Actual Thickness, t (Inches)	Effective Depth, d (Inches)	f'_m = f_b = E_m = n =	1500 500 1,125,000 25.8	2000 667 1,500,000 19.3	2500 833 1,875,000 15.5	3000 1000 2,250,000 12.9	3500 1167 2,625,000 11.0	4000 1333 3,000,000 9.7	5000 1667 3,000,000 9.7	A_s (Sq. in./Ft)
Concrete Masonry Units	6	5.625	2.8^1	M_m^3	0.49	0.59	0.67	0.75	0.82	0.88	1.10	0.068
				M_s	0.34	0.35	0.35	0.35	0.36	0.36	0.36	
	8	7.625	3.8^1	M_m^3	0.90	1.08	1.23	1.37	1.50	1.62	2.03	0.092
				M_s	0.63	0.64	0.65	0.65	0.65	0.66	0.66	
			5.1^2	M_m^3	1.46	1.73	1.98	2.20	2.40	2.59	3.24	
				M_s	0.86	0.87	0.88	0.88	0.89	0.89	0.89	
			5.3^2	M_m^3	1.57	1.87	2.13	2.37	2.59	2.79	3.49	
				M_s	0.91	0.92	0.92	0.93	0.93	0.93	0.93	
	10	9.625	4.8^1	M_m^3	1.44	1.72	1.96	2.19	2.39	2.59	3.23	0.116
				M_s	1.01	1.02	1.03	1.04	1.04	1.05	1.05	
			7.1^2	M_m^3	2.71	3.22	3.67	4.08	4.45	4.80	6.00	
				M_s	1.52	1.54	1.55	1.55	1.56	1.57	1.57	
	12	11.625	5.8^1	M_m^3	2.10	2.50	2.86	3.19	3.49	3.77	4.72	0.140
				M_s	1.47	1.49	1.50	1.51	1.52	1.52	1.52	
			9.1^2	M_m^3	4.35	5.15	5.87	6.52	7.11	7.67	9.59	
				M_s	2.36	2.38	2.39	2.41	2.42	2.42	2.42	
Concrete Masonry Component (Expandable) Walls	10	10	5.0^1	M_m^3	1.55	1.85	2.12	2.36	2.59	2.79	3.49	0.120
				M_s	1.09	1.10	1.11	1.12	1.12	1.13	1.13	
			6.7^2	M_m^3	2.52	3.00	3.42	3.80	4.15	4.48	5.60	
				M_s	1.49	1.51	1.52	1.52	1.53	1.54	1.54	
	12	12	6.0^1	M_m^3	2.24	2.67	3.05	3.40	3.72	4.02	5.03	0.144
				M_s	1.57	1.59	1.60	1.61	1.62	1.63	1.63	
			8.7^2	M_m^3	4.11	4.88	5.56	6.18	6.75	7.28	9.10	
				M_s	2.32	2.35	2.36	2.38	2.39	2.39	2.39	
	16	16	12.7^2	M_m^3	8.43	9.99	11.38	12.63	13.79	14.87	18.58	0.192
				M_s	4.53	4.57	4.60	4.63	4.65	4.66	4.66	
Hollow Clay Brick Units	6	5.5	2.7^1	M_m^3	0.47	0.56	0.64	0.71	0.78	0.85	1.06	0.066
				M_s	0.34	0.34	0.34	0.35	0.35	0.35	0.35	
	8	7.5	3.7^1	M_m^3	0.87	1.04	1.19	1.33	1.45	1.57	1.96	0.090
				M_s	0.61	0.62	0.63	0.63	0.63	0.63	0.63	
			5.2^2	M_m^3	1.50	1.79	2.04	2.26	2.47	2.67	3.33	
				M_s	0.87	0.88	0.89	0.89	0.89	0.90	0.90	
Two Wythe Clay Brick Walls	9	9	4.5^1	M_m^3	1.26	1.50	1.72	1.91	2.09	2.26	2.83	0.108
				M_s	0.88	0.89	0.90	0.91	0.91	0.91	0.91	
	10	10	5.0^1	M_m^3	1.55	1.85	2.12	2.36	2.59	2.79	3.49	0.120
				M_s	1.09	1.10	1.11	1.12	1.12	1.13	1.13	
	12	12	6.0^1	M_m^3	2.24	2.67	3.05	3.40	3.72	4.02	5.03	0.144
				M_s	1.57	1.59	1.60	1.61	1.62	1.63	1.63	
			7.5^2	M_m^3	3.21	3.82	4.36	4.85	5.30	5.72	7.15	
				M_s	1.98	2.00	2.02	2.03	2.04	2.04	2.04	
	16	16	11.5^2	M_m^3	7.14	8.48	9.66	10.74	11.73	12.65	15.81	0.192
				M_s	4.07	4.11	4.14	4.16	4.18	4.19	4.19	

1. Based on $d = t/2$.

2. Based on a 1″ distance between the center of reinforcing and the inside of the face shell or wythe. Therefore $d = (t - 1'' -$ common face shell or wythe thickness).

3. Table values are based on special inspection being provided. For noninspected masonry, the capacity of the masonry, M_m, must be reduced by one half per UBC Section 2406(c)1.

DIAGRAM K-1   Spacing of Shear Reinforcing (Inches) Based on the Shear Stress and the Reinforcing Bar Size[4] for Nominal 6″ Wide Sections

For use of Table, see Example 5-K

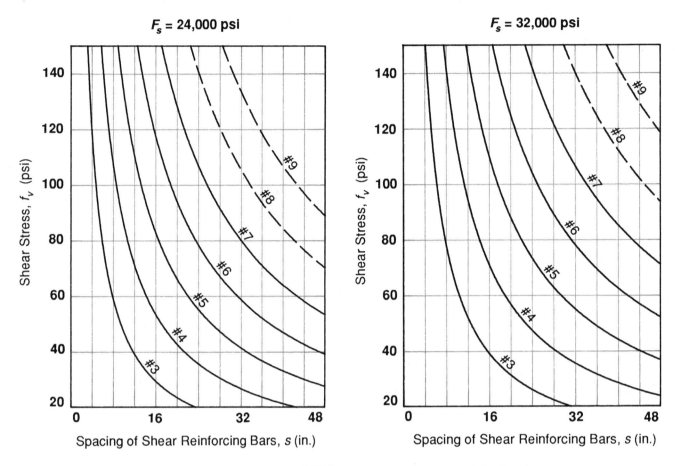

TABLE K-1   Allowable Shear Stress Capacity[1,2,3,4] (psi) for Nominal 6″ Wide Sections — Reinforcing Steel Designed to Carry Entire Shear Force with F_s = 24,000 psi

$$A_v = \frac{sV}{F_s d}$$ (UBC Ch. 24, Eq. 9-10, ACI/ASCE Eq. 7.10)

or rewritten, $$V = \frac{A_v F_s d}{s}$$

Since $F_v = \dfrac{V}{bjd}$; $F_v = \dfrac{A_v F_s}{jbs}$

where: $b = 5.625''$, $F_s = 24,000$ psi and j was conservatively taken as 1.0

Spacing of Shear Reinforcing Bars, s (Inches)	Allowable Shear Stress, F_v (psi)[1,2]						
	Shear Reinforcing Bar Size and Area (Square Inches)						
	#3	#4	#5	#6	#7	#8[4]	#9[4]
	0.11	0.20	0.31	0.44	0.60	0.79	1.00
8	59	107	150[1]	150[1]	150[1]	150[1]	150[1]
12	39	71	110	150[1]	150[1]	150[1]	150[1]
16	29	53	83	117	150[1]	150[1]	150[1]
20	23	43	66	94	128	150[1]	150[1]
24	20	36	55	78	107	140	150[1]
28	17	30	47	67	91	120	150[1]
32	15	27	41	59	80	105	133
36	13	24	37	52	71	94	119
40	12	21	33	47	64	84	107
48	10	18	28	39	53	70	89

1. For flexural members, F_v may not exceed 150 psi nor $3.0(f'_m)^{1/2}$.

2. F_v may be limited to lower values for shear walls. See Tables A-5 and A-6 for specific allowable values for shear walls.

3. Table values may be increased by one-third when considering wind or earthquake forces (UBC Section 2303 (d)).

4. #8 and #9 bars NOT RECOMMENDED for 6″ thick masonry.

DIAGRAM K-2 Spacing of Shear Reinforcing (Inches) Based on the Shear Stress and the Reinforcing Bar Size for Nominal 8″ Wide Sections

For use of Table, see Example 5-K

WSD

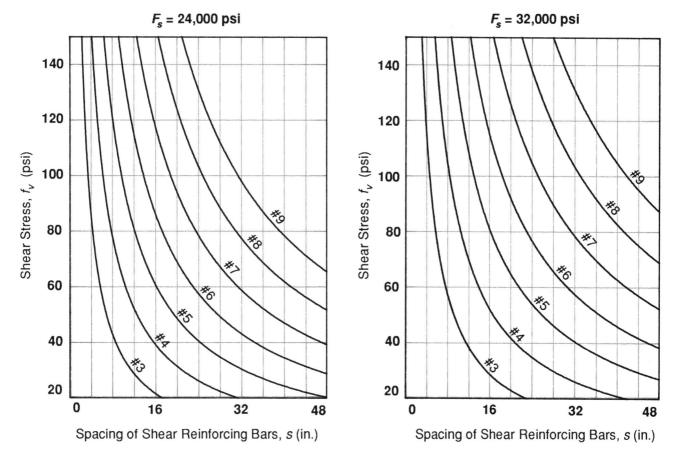

$F_s = 24,000$ psi

$F_s = 32,000$ psi

Spacing of Shear Reinforcing Bars, s (in.)

Spacing of Shear Reinforcing Bars, s (in.)

TABLE K-2 Allowable Shear Stress Capacity[1,2,3] (psi) for Nominal 8″ Wide Sections — Reinforcing Steel Designed to Carry Entire Shear Force with $F_s = 24,000$ psi

$$A_v = \frac{sV}{F_s d} \quad \text{(UBC Ch. 24, Eq. 9-10, ACI/ASCE Eq. 7.10)}$$

or rewritten, $V = \dfrac{A_v F_s d}{s}$

Since $F_v = \dfrac{V}{bjd}$; $F_v = \dfrac{A_v F_s}{jbs}$

where: $b = 7.625″$, $F_s = 24,000$ psi[3] and j was conservatively taken as 1.0

Spacing of Shear Reinforcing Bars, *s* (Inches)	Allowable Shear Stress, F_v (psi)[1,2]						
	Shear Reinforcing Bar Size and Area (Square Inches)						
	#3	#4	#5	#6	#7	#8	#9
	0.11	0.20	0.31	0.44	0.60	0.79	1.00
8	43	79	122	150[1]	150[1]	150[1]	150[1]
12	29	52	81	115	150[1]	150[1]	150[1]
16	22	39	61	87	118	150[1]	150[1]
20	17	31	49	69	94	124	150[1]
24	14	26	41	58	79	104	131
28	12	22	35	49	67	89	112
32	11	20	30	43	59	78	98
36	10	17	27	38	52	69	87
40	9	16	24	35	47	62	79
48	7	13	20	29	39	52	66

1. For flexural members, F_v may not exceed 150 psi nor $3.0(f'_m)^{1/2}$.

2. F_v may be limited to lower values for shear walls. See Tables A-5 and A-6 for specific allowable values for shear walls.

3. Table values may be increased by one-third when considering wind or earthquake forces (UBC Section 2303 (d)).

**DIAGRAM K-3   Spacing of Shear Reinforcing (Inches) Based on the
Shear Stress and the Reinforcing Bar Size for Nominal 10″ Wide Sections**

For use of Table,
see Example 5-K

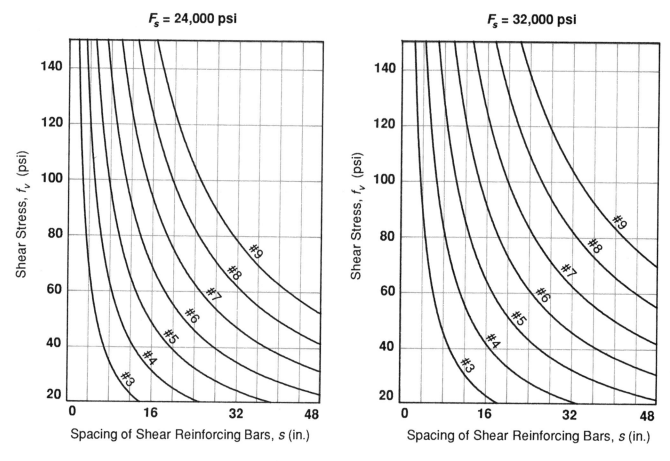

Spacing of Shear Reinforcing Bars, *s* (in.)

**TABLE K-3   Allowable Shear Stress Capacity[1,2,3] (psi) for Nominal 10″ Wide Sections —
Reinforcing Steel Designed to Carry Entire Shear Force with F_s = 24,000 psi**

$$A_v = \frac{sV}{F_s d}$$ (UBC Ch. 24, Eq. 9-10,
ACI / ASCE Eq. 7.10)

or rewritten, $V = \dfrac{A_v F_s d}{s}$

Since $F_v = \dfrac{V}{bjd}$; $F_v = \dfrac{A_v F_s}{jbs}$

where: b = 9.625″, F_s = 24,000 psi and j was
conservatively taken as 1.0

Spacing of Shear Reinforcing Bars (Inches)	Allowable Shear Stress, F_v (psi)[1,2]						
	Shear Reinforcing Bar Size and Area (Square Inches)						
	#3	#4	#5	#6	#7	#8	#9
	0.11	0.20	0.31	0.44	0.60	0.79	1.00
8	34	62	97	137	150[1]	150[1]	150[1]
12	23	42	64	91	125	150[1]	150[1]
16	17	31	48	69	94	123	150[1]
20	14	25	39	55	75	98	125
24	11	21	32	46	62	82	104
28	10	18	28	39	53	70	89
32	9	16	24	34	47	62	78
36	8	14	21	30	42	55	69
40	7	12	19	27	37	49	62
48	6	10	16	23	31	41	52

1. For flexural members, F_v may not exceed 150 psi nor $3.0(f'_m)^{1/2}$.

2. F_v may be limited to lower values for shear walls. See Tables A-5 and A-6 for specific allowable values for shear walls.

3. Table values may be increased by one-third when considering wind or earthquake forces (UBC Section 2303 (d)).

DIAGRAM K-4   Spacing of Shear Reinforcing (Inches) Based on the Shear Stress and the Reinforcing Bar Size for Nominal 12″ Wide Sections

For use of Table, see Example 5-K

WSD

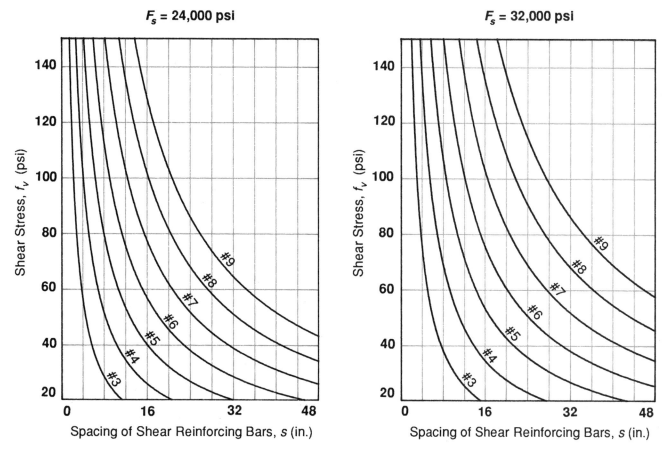

F_s = 24,000 psi

F_s = 32,000 psi

Shear Stress, f_v (psi)

Spacing of Shear Reinforcing Bars, s (in.)

TABLE K-4   Allowable Shear Stress Capacity[1,2,3] (psi) for Nominal 12″ Wide Sections — Reinforcing Steel Designed to Carry Entire Shear Force with F_s = 24,000 psi

$$A_v = \frac{s\,V}{F_s\,d} \quad \text{(UBC Ch. 24, Eq. 9-10,}$$
$$\text{ACI / ASCE Eq. 7.10)}$$

or rewritten, $\quad V = \dfrac{A_v\,F_s\,d}{s}$

Since $F_v = \dfrac{V}{b\,j\,d}$; $\;F_v = \dfrac{A_v\,F_s}{j\,b\,s}$

where: $b = 11.625''$, $F_s = 24,000$ psi and j was conservatively taken as 1.0

Spacing of Shear Reinforcing Bars (Inches)	Allowable Shear Stress, F_v (psi)[1,2]						
	Shear Reinforcing Bar Size and Area (Square Inches)						
	#3	#4	#5	#6	#7	#8	#9
	0.11	0.20	0.31	0.44	0.60	0.79	1.00
8	28	52	80	114	150[1]	150[1]	150[1]
12	19	34	53	76	103	136	150[1]
16	14	26	40	57	77	102	129
20	11	21	32	45	62	82	103
24	9	17	27	38	52	68	86
28	8	15	23	32	44	58	74
32	7	13	20	28	39	51	65
36	6	11	18	25	34	45	57
40	6	10	16	23	31	41	52
48	5	9	13	19	26	34	43

1. For flexural members, F_v may not exceed 150 psi nor $3.0(f'_m)^{1/2}$

2. F_v may be limited to lower values for shear walls. See Tables A-5 and A-6 for specific allowable values for shear walls.

3. Table values may be increased by one-third when considering wind or earthquake forces (UBC Section 2303 (d)).

TABLE N-1a Coefficients p and p' for Tension and Compression Steel in a Flexural Member for Half Allowable Stresses: f'_m = 1500 psi, F_s = 24,000 psi and n = 25.8

For use of Table, see Example 5-I

DESIGN DATA

$f'_m = 1500$ psi	$f_y = 60,000$ psi
$f_b = 250$ psi	$F_s = 24,000$ psi
$E_m = 1,125,000$ psi	$E_s = 29,000,000$ psi
$n = 25.8$	$k = 0.212$
$K_b = 24.6$	$p_b = 0.0011$

DESIGN EQUATIONS

$$K^1 = \frac{M}{F} = \frac{M\,(\text{ft kips})}{bd^2/12{,}000} \ \text{or}\ \frac{M\,(\text{in. lbs})}{bd^2}$$

$$p = p_b + \frac{K - K_b}{f_s\left(1 - d'/d\right)}$$

$$p' = \frac{K - K_b}{(2n-1)\left(\dfrac{k-d'/d}{k}\right)\left(1-\dfrac{d'}{d}\right)f_b}$$

d'/d	Steel Ratio p', p	K_b 24.6	\multicolumn{9}{c}{K^1}
		24.6	40	45	50	55	60	65	70	75	80
0.02	p'	—	0.0014	0.0018	0.0023	0.0027	0.0032	0.0036	0.0040	0.0045	0.0049
	p	0.0011	0.0018	0.0020	0.0022	0.0024	0.0026	0.0028	0.0030	0.0032	0.0035
0.04	p'	—	0.0016	0.0021	0.0026	0.0031	0.0036	0.0041	0.0046	0.0051	0.0056
	p	0.0011	0.0018	0.0020	0.0022	0.0024	0.0026	0.0029	0.0031	0.0033	0.0035
0.06	p'	—	0.0018	0.0024	0.0030	0.0036	0.0042	0.0047	0.0053	0.0059	0.0065
	p	0.0011	0.0018	0.0020	0.0022	0.0024	0.0027	0.0029	0.0031	0.0033	0.0036
0.08	p'	—	0.0021	0.0028	0.0035	0.0042	0.0049	0.0056	0.0063	0.0070	0.0076
	p	0.0011	0.0018	0.0020	0.0023	0.0025	0.0027	0.0029	0.0032	0.0034	0.0036
0.10	p'	—	0.0026	0.0034	0.0042	0.0051	0.0059	0.0067	0.0075	0.0084	0.0092
	p	0.0011	0.0018	0.0020	0.0023	0.0025	0.0027	0.0030	0.0032	0.0034	0.0037
0.12	p'	—	0.0032	0.0042	0.0053	0.0063	0.0073	0.0084	0.0094	0.0104	0.0115
	p	0.0011	0.0018	0.0021	0.0023	0.0025	0.0028	0.0030	0.0032	0.0035	0.0037
0.14[2]	p'	—	0.0042	0.0055	0.0069	0.0082	0.0096	0.0109	0.0123	0.0136	0.0150
	p	0.0011	0.0018	0.0021	0.0023	0.0026	0.0028	0.0031	0.0033	0.0035	0.0038

1. When K is determined where wind or seismic conditions are considered and a $1/3$ increase in stress is permitted, multiply the K obtained by $3/4$ to use this table.

2. For d'/d values greater than 0.14 the effect of the compression steel becomes increasingly negligible.

TABLE N-1b Coefficients p and p' for Tension and Compression Steel in a Flexural Member for Full Allowable Stresses: $f'_m = 1500$ psi, $F_s = 24,000$ psi and $n = 25.8$

For use of Table, see Example 5-I

WSD

DESIGN DATA

$f'_m = 1500$ psi	$f_y = 60,000$ psi
$f_b = 500$ psi	$F_s = 24,000$ psi
$E_m = 1,125,000$ psi	$E_s = 29,000,000$ psi
$n = 25.8$	$k = 0.350$
$K_b = 77.2$	$p_b = 0.0036$

DESIGN EQUATIONS

$$K^1 = \frac{M}{F} = \frac{M\,(\text{ft kips})}{bd^2/12,000} \ \text{or}\ \frac{M\,(\text{in. lbs})}{bd^2}$$

$$p = p_b + \frac{K - K_b}{f_s(1 - d'/d)}$$

$$p' = \frac{K - K_b}{(2n-1)\left(\dfrac{k - d'/d}{k}\right)\left(1 - \dfrac{d'}{d}\right)f_b}$$

d'/d	Steel Ratio p', p	K_b 77.2	90	95	100	105	110	115	120	125	130
0.02	p'	—	0.0005	0.0008	0.0010	0.0012	0.0014	0.0016	0.0018	0.0020	0.0023
	p	0.0036	0.0041	0.0044	0.0046	0.0048	0.0050	0.0052	0.0054	0.0056	0.0058
0.04	p'	—	0.0006	0.0008	0.0011	0.0013	0.0015	0.0018	0.0020	0.0022	0.0025
	p	0.0036	0.0042	0.0044	0.0046	0.0048	0.0050	0.0052	0.0055	0.0057	0.0059
0.06	p'	—	0.0006	0.0009	0.0012	0.0014	0.0017	0.0019	0.0022	0.0024	0.0027
	p	0.0036	0.0042	0.0044	0.0046	0.0048	0.0051	0.0053	0.0055	0.0057	0.0059
0.08	p'	—	0.0007	0.0010	0.0013	0.0015	0.0018	0.0021	0.0024	0.0027	0.0029
	p	0.0036	0.0042	0.0044	0.0046	0.0049	0.0051	0.0053	0.0055	0.0058	0.0060
0.10	p'	—	0.0008	0.0011	0.0014	0.0017	0.0020	0.0023	0.0026	0.0029	0.0032
	p	0.0036	0.0042	0.0044	0.0047	0.0049	0.0051	0.0054	0.0056	0.0058	0.0060
0.12	p'	—	0.0009	0.0012	0.0016	0.0019	0.0022	0.0026	0.0029	0.0033	0.0036
	p	0.0036	0.0042	0.0044	0.0047	0.0049	0.0052	0.0054	0.0056	0.0059	0.0061
0.14	p'	—	0.0010	0.0014	0.0017	0.0021	0.0025	0.0029	0.0033	0.0037	0.0040
	p	0.0036	0.0042	0.0045	0.0047	0.0049	0.0052	0.0054	0.0057	0.0059	0.0062
0.16	p'	—	0.0011	0.0015	0.0020	0.0024	0.0028	0.0033	0.0037	0.0041	0.0046
	p	0.0036	0.0042	0.0045	0.0047	0.0050	0.0052	0.0055	0.0057	0.0060	0.0062
0.18	p'	—	0.0013	0.0018	0.0023	0.0028	0.0033	0.0038	0.0042	0.0047	0.0052
	p	0.0036	0.0043	0.0045	0.0048	0.0050	0.0053	0.0055	0.0058	0.0060	0.0063
0.20	p'	—	0.0015	0.0021	0.0026	0.0032	0.0038	0.0044	0.0049	0.0055	0.0061
	p	0.0036	0.0043	0.0045	0.0048	0.0050	0.0053	0.0056	0.0058	0.0061	0.0064
0.22	p'	—	0.0017	0.0024	0.0031	0.0038	0.0045	0.0052	0.0058	0.0065	0.0072
	p	0.0036	0.0043	0.0046	0.0048	0.0051	0.0054	0.0056	0.0059	0.0062	0.0064
0.24²	p'	—	0.0021	0.0029	0.0038	0.0046	0.0054	0.0063	0.0071	0.0079	0.0087
	p	0.0036	0.0043	0.0046	0.0049	0.0051	0.0054	0.0057	0.0059	0.0062	0.0065

1. When K is determined where wind or seismic conditions are considered and a $1/3$ increase in stress is permitted, multiply the K obtained by $3/4$ to use this table.

2. For d'/d values greater than 0.24 the effect of the compression steel becomes increasingly negligible.

TABLE N-2a   Coefficients p and p' for Tension and Compression Steel in a Flexural Member for Half Allowable Stresses:
$f'_m = 2000$ psi, $F_s = 24,000$ psi and $n = 19.3$

For use of Table, see Example 5-I

DESIGN DATA

$f'_m = 2000$ psi	$f_y = 60,000$ psi
$f_b = 333$ psi	$F_s = 24,000$ psi
$E_m = 1,500,000$ psi	$E_s = 29,000,000$ psi
$n = 19.3$	$k = 0.212$
$K_b = 32.7$	$p_b = 0.0015$

DESIGN EQUATIONS

$$K^1 = \frac{M}{F} = \frac{M \text{ (ft kips)}}{bd^2/12,000} \text{ or } \frac{M \text{ (in. lbs)}}{bd^2}$$

$$p = p_b + \frac{K - K_b}{f_s(1 - d'/d)}$$

$$p' = \frac{K - K_b}{(2n-1)\left(\frac{k - d'/d}{k}\right)\left(1 - \frac{d'}{d}\right)f_b}$$

d'/d	Steel Ratio p', p	K_b	K^1								
		32.7	40	45	50	55	60	65	70	75	80
0.02	p'	—	0.0007	0.0011	0.0016	0.0020	0.0025	0.0029	0.0034	0.0038	0.0043
	p	0.0015	0.0018	0.0020	0.0022	0.0024	0.0027	0.0029	0.0031	0.0033	0.0035
0.04	p'	—	0.0007	0.0013	0.0018	0.0023	0.0028	0.0033	0.0038	0.0043	0.0049
	p	0.0015	0.0018	0.0020	0.0023	0.0025	0.0027	0.0029	0.0031	0.0033	0.0036
0.06	p'	—	0.0009	0.0015	0.0021	0.0026	0.0032	0.0038	0.0044	0.0050	0.0056
	p	0.0015	0.0018	0.0020	0.0023	0.0025	0.0027	0.0029	0.0032	0.0034	0.0036
0.08	p'	—	0.0010	0.0017	0.0024	0.0031	0.0038	0.0045	0.0052	0.0059	0.0066
	p	0.0015	0.0018	0.0021	0.0023	0.0025	0.0027	0.0030	0.0032	0.0034	0.0036
0.10	p'	—	0.0012	0.0021	0.0029	0.0038	0.0046	0.0054	0.0063	0.0071	0.0080
	p	0.0015	0.0018	0.0021	0.0023	0.0025	0.0028	0.0030	0.0032	0.0035	0.0037
0.12	p'	—	0.0015	0.0026	0.0036	0.0047	0.0057	0.0068	0.0078	0.0089	0.0100
	p	0.0015	0.0018	0.0021	0.0023	0.0026	0.0028	0.0030	0.0033	0.0035	0.0037
0.14	p'	—	0.0020	0.0034	0.0048	0.0062	0.0075	0.0089	0.0103	0.0117	0.0131
	p	0.0015	0.0019	0.0021	0.0023	0.0026	0.0028	0.0031	0.0033	0.0035	0.0038
0.16	p'	—	0.0029	0.0048	0.0068	0.0088	0.0107	0.0127	0.0147	0.0166	0.0186
	p	0.0015	0.0019	0.0021	0.0024	0.0026	0.0029	0.0031	0.0034	0.0036	0.0038
0.18[2]	p'	—	0.0048	0.0082	0.0115	0.0148	0.0181	0.0214	0.0247	0.0280	0.0314
	p	0.0015	0.0019	0.0021	0.0024	0.0026	0.0029	0.0031	0.0034	0.0036	0.0039

1. When K is determined where wind or seismic conditions are considered and a $^1/_3$ increase in stress is permitted, multiply the K obtained by $^3/_4$ to use this table.

2. For d'/d values greater than 0.18 the effect of the compression steel becomes increasingly negligible.

TABLE N-2b   Coefficients p and p' for Tension and Compression Steel in a Flexural Member for Full Allowable Stresses: $f'_m = 2000$ psi, $F_s = 24,000$ psi and $n = 19.3$

For use of Table, see Example 5-I

WSD

DESIGN DATA

$f'_m = 2000$ psi	$f_y = 60,000$ psi
$f_b = 667$ psi	$F_s = 24,000$ psi
$E_m = 1,500,000$ psi	$E_s = 29,000,000$ psi
$n = 19.3$	$k = 0.350$
$K_b = 102.9$	$p_b = 0.0049$

DESIGN EQUATIONS

$$K^1 = \frac{M}{F} = \frac{M\,(\text{ft kips})}{bd^2/12,000} \text{ or } \frac{M\,(\text{in. lbs})}{bd^2}$$

$$p = p_b + \frac{K - K_b}{f_s(1 - d'/d)}$$

$$p' = \frac{K - K_b}{(2n-1)\left(\dfrac{k - d'/d}{k}\right)\left(1 - \dfrac{d'}{d}\right)f_b}$$

d'/d	Steel Ratio p', p	K_b 102.9	120	125	130	135	140	145	150	155	160
0.02	p'	—	0.0007	0.0010	0.0012	0.0014	0.0016	0.0018	0.0020	0.0022	0.0025
	p	0.0049	0.0056	0.0058	0.0061	0.0063	0.0065	0.0067	0.0069	0.0071	0.0073
0.04	p'	—	0.0008	0.0010	0.0013	0.0015	0.0017	0.0020	0.0022	0.0024	0.0027
	p	0.0049	0.0056	0.0059	0.0061	0.0063	0.0065	0.0067	0.0069	0.0072	0.0074
0.06	p'	—	0.0009	0.0011	0.0014	0.0016	0.0019	0.0022	0.0024	0.0027	0.0029
	p	0.0049	0.0057	0.0059	0.0061	0.0063	0.0065	0.0068	0.0070	0.0072	0.0074
0.08	p'	—	0.0010	0.0012	0.0015	0.0018	0.0021	0.0024	0.0026	0.0029	0.0032
	p	0.0049	0.0057	0.0059	0.0061	0.0064	0.0066	0.0068	0.0070	0.0073	0.0075
0.10	p'	—	0.0011	0.0014	0.0017	0.0020	0.0023	0.0026	0.0029	0.0032	0.0035
	p	0.0049	0.0057	0.0059	0.0062	0.0064	0.0066	0.0068	0.0071	0.0073	0.0075
0.12	p'	—	0.0012	0.0015	0.0019	0.0022	0.0026	0.0029	0.0033	0.0036	0.0039
	p	0.0049	0.0057	0.0059	0.0062	0.0064	0.0067	0.0069	0.0071	0.0074	0.0076
0.14	p'	—	0.0013	0.0017	0.0021	0.0025	0.0029	0.0033	0.0036	0.0040	0.0044
	p	0.0049	0.0057	0.0060	0.0062	0.0065	0.0067	0.0069	0.0072	0.0074	0.0077
0.16	p'	—	0.0015	0.0019	0.0024	0.0028	0.0033	0.0037	0.0041	0.0046	0.0050
	p	0.0049	0.0057	0.0060	0.0062	0.0065	0.0067	0.0070	0.0072	0.0075	0.0077
0.18	p'	—	0.0017	0.0022	0.0027	0.0032	0.0037	0.0042	0.0047	0.0052	0.0057
	p	0.0049	0.0058	0.0060	0.0063	0.0065	0.0068	0.0070	0.0073	0.0075	0.0078
0.20	p'	—	0.0020	0.0026	0.0032	0.0037	0.0043	0.0049	0.0055	0.0061	0.0067
	p	0.0049	0.0058	0.0061	0.0063	0.0066	0.0068	0.0071	0.0074	0.0076	0.0079
0.22	p'	—	0.0024	0.0031	0.0037	0.0044	0.0051	0.0058	0.0065	0.0072	0.0079
	p	0.0049	0.0058	0.0061	0.0063	0.0066	0.0069	0.0071	0.0074	0.0077	0.0080
0.24[2]	p'		0.0029	0.0037	0.0046	0.0054	0.0062	0.0071	0.0079	0.0088	0.0096
	p	0.0049	0.0058	0.0061	0.0064	0.0067	0.0069	0.0072	0.0075	0.0078	0.0080

1. When K is determined where wind or seismic conditions are considered and a $1/3$ increase in stress is permitted, multiply the K obtained by $3/4$ to use this table.

2. For d'/d values greater than 0.24 the effect of the compression steel becomes increasingly negligible.

TABLE N-3a   Coefficients *p* and *p'* for Tension and Compression Steel in a Flexural Member for Half Allowable Stresses:
f'_m = 2500 psi, F_s = 24,000 psi and *n* = 15.5

For use of Table, see Example 5-I

DESIGN DATA

$f'_m = 2500$ psi	$f_y = 60,000$ psi
$f_b = 417$ psi	$F_s = 24,000$ psi
$E_m = 1,875,000$ psi	$E_s = 29,000,000$ psi
$n = 15.5$	$k = 0.212$
$K_b = 41.1$	$p_b = 0.0018$

DESIGN EQUATIONS

$$K^1 = \frac{M}{F} = \frac{M\,(\text{ft kips})}{bd^2/12,000} \text{ or } \frac{M\,(\text{in. lbs})}{bd^2}$$

$$p = p_b + \frac{K - K_b}{f_s\,(1 - d'/d)}$$

$$p' = \frac{K - K_b}{(2n-1)\left(\dfrac{k - d'/d}{k}\right)\left(1 - \dfrac{d'}{d}\right)f_b}$$

d'/d	Steel Ratio p', p	K_b 41.1	K^1 60	65	70	75	80	85	90	95	100
0.02	p'	—	0.0017	0.0022	0.0026	0.0031	0.0035	0.0040	0.0044	0.0049	0.0053
	p	0.0018	0.0026	0.0028	0.0030	0.0032	0.0035	0.0037	0.0039	0.0041	0.0043
0.04	p'	—	0.0019	0.0025	0.0030	0.0035	0.0040	0.0045	0.0050	0.0055	0.0060
	p	0.0018	0.0026	0.0028	0.0031	0.0033	0.0035	0.0037	0.0039	0.0041	0.0044
0.06	p'	—	0.0022	0.0028	0.0034	0.0040	0.0046	0.0052	0.0058	0.0064	0.0070
	p	0.0018	0.0026	0.0029	0.0031	0.0033	0.0035	0.0037	0.0040	0.0042	0.0044
0.08	p'	—	0.0026	0.0033	0.0040	0.0047	0.0054	0.0061	0.0068	0.0075	0.0082
	p	0.0018	0.0027	0.0029	0.0031	0.0033	0.0036	0.0038	0.0040	0.0042	0.0045
0.10	p'	—	0.0032	0.0040	0.0049	0.0057	0.0065	0.0074	0.0082	0.0091	0.0099
	p	0.0018	0.0027	0.0029	0.0031	0.0034	0.0036	0.0038	0.0041	0.0043	0.0045
0.12	p'	—	0.0040	0.0050	0.0060	0.0071	0.0081	0.0092	0.0102	0.0113	0.0123
	p	0.0018	0.0027	0.0029	0.0032	0.0034	0.0036	0.0039	0.0041	0.0044	0.0046
0.14	p'	—	0.0052	0.0065	0.0079	0.0093	0.0106	0.0120	0.0134	0.0148	0.0161
	p	0.0018	0.0027	0.0030	0.0032	0.0034	0.0037	0.0039	0.0042	0.0044	0.0047
0.16	p'	—	0.0073	0.0093	0.0112	0.0132	0.0151	0.0170	0.0190	0.0209	0.0229
	p	0.0018	0.0027	0.0030	0.0032	0.0035	0.0037	0.0040	0.0042	0.0045	0.0047
0.18[2]	p'	—	0.0122	0.0154	0.0187	0.0219	0.0251	0.0284	0.0316	0.0348	0.0380
	p	0.0018	0.0028	0.0030	0.0033	0.0035	0.0038	0.0040	0.0043	0.0045	0.0048

1. When *K* is determined where wind or seismic conditions are considered and a $^1/_3$ increase in stress is permitted, multiply the *K* obtained by $^3/_4$ to use this table.

2. For *d'/d* values greater than 0.18 the effect of the compression steel becomes increasingly negligible.

TABLE N-3b Coefficients p and p' for Tension and Compression Steel in a Flexural Member for Full Allowable Stresses: $f'_m = 2500$ psi, $F_s = 24,000$ psi and $n = 15.5$

For use of Table, see Example 5-I

WSD

DESIGN DATA

$f'_m = 2500$ psi	$f_y = 60,000$ psi
$f_b = 833$ psi	$F_s = 24,000$ psi
$E_m = 1,875,000$ psi	$E_s = 29,000,000$ psi
$n = 15.5$	$k = 0.350$
$K_b = 128.7$	$p_b = 0.0061$

DESIGN EQUATIONS

$$K^1 = \frac{M}{F} = \frac{M\,(\text{ft kips})}{bd^2/12,000} \text{ or } \frac{M\,(\text{in. lbs})}{bd^2}$$

$$p = p_b + \frac{K - K_b}{f_s(1 - d'/d)}$$

$$p' = \frac{K - K_b}{(2n-1)\left(\dfrac{k - d'/d}{k}\right)\left(1 - \dfrac{d'}{d}\right)f_b}$$

d'/d	Steel Ratio p', p	K_b 128.7	140	145	150	155	160	165	170	175	180
0.02	p'	—	0.0005	0.0007	0.0009	0.0011	0.0014	0.0016	0.0018	0.0020	0.0022
	p	0.0061	0.0066	0.0068	0.0070	0.0072	0.0074	0.0076	0.0079	0.0081	0.0083
0.04	p'	—	0.0005	0.0008	0.0010	0.0012	0.0015	0.0017	0.0019	0.0022	0.0024
	p	0.0061	0.0066	0.0068	0.0070	0.0072	0.0075	0.0077	0.0079	0.0081	0.0083
0.06	p'	—	0.0006	0.0008	0.0011	0.0014	0.0016	0.0019	0.0021	0.0024	0.0026
	p	0.0061	0.0066	0.0068	0.0070	0.0073	0.0075	0.0077	0.0079	0.0082	0.0084
0.08	p'	—	0.0006	0.0009	0.0012	0.0015	0.0018	0.0020	0.0023	0.0026	0.0029
	p	0.0061	0.0066	0.0068	0.0071	0.0073	0.0075	0.0077	0.0080	0.0082	0.0084
0.10	p'	—	0.0007	0.0010	0.0013	0.0016	0.0019	0.0023	0.0026	0.0029	0.0032
	p	0.0061	0.0066	0.0069	0.0071	0.0073	0.0075	0.0078	0.0080	0.0082	0.0085
0.12	p'	—	0.0008	0.0011	0.0015	0.0018	0.0022	0.0025	0.0029	0.0032	0.0035
	p	0.0061	0.0066	0.0069	0.0071	0.0073	0.0076	0.0078	0.0081	0.0083	0.0085
0.14	p'	—	0.0009	0.0013	0.0017	0.0020	0.0024	0.0028	0.0032	0.0036	0.0040
	p	0.0061	0.0066	0.0069	0.0071	0.0074	0.0076	0.0079	0.0081	0.0083	0.0086
0.16	p'	—	0.0010	0.0014	0.0019	0.0023	0.0027	0.0032	0.0036	0.0041	0.0045
	p	0.0061	0.0067	0.0069	0.0072	0.0074	0.0077	0.0079	0.0081	0.0084	0.0086
0.18	p'	—	0.0011	0.0016	0.0021	0.0026	0.0031	0.0036	0.0041	0.0047	0.0052
	p	0.0061	0.0067	0.0069	0.0072	0.0074	0.0077	0.0079	0.0082	0.0085	0.0087
0.20	p'	—	0.0013	0.0019	0.0025	0.0031	0.0037	0.0042	0.0048	0.0054	0.0060
	p	0.0061	0.0067	0.0069	0.0072	0.0075	0.0077	0.0080	0.0083	0.0085	0.0088
0.22	p'	—	0.0016	0.0023	0.0029	0.0036	0.0043	0.0050	0.0057	0.0064	0.0071
	p	0.0061	0.0067	0.0070	0.0072	0.0075	0.0078	0.0080	0.0083	0.0086	0.0088
0.24[2]	p'	—	0.0019	0.0027	0.0036	0.0044	0.0052	0.0061	0.0069	0.0078	0.0086
	p	0.0061	0.0067	0.0070	0.0073	0.0075	0.0078	0.0081	0.0084	0.0086	0.0089

1. When K is determined where wind or seismic conditions are considered and a $1/3$ increase in stress is permitted, multiply the K obtained by $3/4$ to use this table.

2. For d'/d values greater than 0.24 the effect of the compression steel becomes increasingly negligible.

TABLE N-4a   Coefficients p and p' for Tension and Compression Steel in a Flexural Member for Half Allowable Stresses: $f'_m = 3000$ psi, $F_s = 24,000$ psi and $n = 12.9$

For use of Table, see Example 5-I

DESIGN DATA

$f'_m = 3000$ psi	$f_y = 60,000$ psi
$f_b = 500$ psi	$F_s = 24,000$ psi
$E_m = 2,250,000$ psi	$E_s = 29,000,000$ psi
$n = 12.9$	$k = 0.212$
$K_b = 49.2$	$p_b = 0.0022$

DESIGN EQUATIONS

$$K^1 = \frac{M}{F} = \frac{M\,(\text{ft kips})}{bd^2/12,000} \text{ or } \frac{M\,(\text{in. lbs})}{bd^2}$$

$$p = p_b + \frac{K - K_b}{f_s(1 - d'/d)}$$

$$p' = \frac{K - K_b}{(2n-1)\left(\dfrac{k - d'/d}{k}\right)\left(1 - \dfrac{d'}{d}\right)f_b}$$

d'/d	Steel Ratio p', p	K_b 49.2	K^1 60	65	70	75	80	85	90	95	100
0.02	p'	—	0.0010	0.0014	0.0019	0.0023	0.0028	0.0033	0.0037	0.0042	0.0046
	p	0.0022	0.0027	0.0029	0.0031	0.0033	0.0035	0.0037	0.0039	0.0041	0.0044
0.04	p'	—	0.0011	0.0016	0.0022	0.0027	0.0032	0.0037	0.0042	0.0047	0.0053
	p	0.0022	0.0027	0.0029	0.0031	0.0033	0.0035	0.0038	0.0040	0.0042	0.0044
0.06	p'	—	0.0013	0.0019	0.0025	0.0031	0.0037	0.0043	0.0049	0.0055	0.0061
	p	0.0022	0.0027	0.0029	0.0031	0.0033	0.0036	0.0038	0.0040	0.0042	0.0045
0.08	p'	—	0.0015	0.0022	0.0029	0.0036	0.0043	0.0050	0.0057	0.0064	0.0072
	p	0.0022	0.0027	0.0029	0.0031	0.0034	0.0036	0.0038	0.0040	0.0043	0.0045
0.10	p'	—	0.0018	0.0027	0.0035	0.0044	0.0052	0.0061	0.0069	0.0078	0.0086
	p	0.0022	0.0027	0.0029	0.0032	0.0034	0.0036	0.0039	0.0041	0.0043	0.0046
0.12	p'		0.0023	0.0033	0.0044	0.0054	0.0065	0.0076	0.0086	0.0097	0.0107
	p	0.0022	0.0027	0.0029	0.0032	0.0034	0.0037	0.0039	0.0041	0.0044	0.0046
0.14	p'	—	0.0030	0.0044	0.0057	0.0071	0.0085	0.0099	0.0113	0.0126	0.0140
	p	0.0022	0.0027	0.0030	0.0032	0.0035	0.0037	0.0039	0.0042	0.0044	0.0047
0.16	p'	—	0.0042	0.0062	0.0081	0.0101	0.0121	0.0140	0.0160	0.0179	0.0199
	p	0.0022	0.0027	0.0030	0.0032	0.0035	0.0037	0.0040	0.0042	0.0045	0.0047
0.18^2	p'	—	0.0070	0.0103	0.0136	0.0168	0.0201	0.0233	0.0266	0.0298	0.0331
	p	0.0022	0.0027	0.0030	0.0033	0.0035	0.0038	0.0040	0.0043	0.0045	0.0048

1. When K is determined where wind or seismic conditions are considered and a $1/3$ increase in stress is permitted, multiply the K obtained by $3/4$ to use this table.

2. For d'/d values greater than 0.18 the effect of the compression steel becomes increasingly negligible.

TABLE N-4b Coefficients *p* and *p'* for Tension and Compression Steel in a Flexural Member for Full Allowable Stresses: f'_m = 3000 psi, F_s = 24,000 psi and *n* = 12.9

For use of Table, see Example 5-I

DESIGN DATA

$f'_m = 3000$ psi	$f_y = 60,000$ psi
$f_b = 1000$ psi	$F_s = 24,000$ psi
$E_m = 2,250,000$ psi	$E_s = 29,000,000$ psi
$n = 12.9$	$k = 0.350$
$K_b = 154.4$	$p_b = 0.0073$

DESIGN EQUATIONS

$$K^1 = \frac{M}{F} = \frac{M(\text{ft kips})}{bd^2/12,000} \text{ or } \frac{M(\text{in. lbs})}{bd^2}$$

$$p = p_b + \frac{K - K_b}{f_s(1 - d'/d)}$$

$$p' = \frac{K - K_b}{(2n-1)\left(\dfrac{k - d'/d}{k}\right)\left(1 - \dfrac{d'}{d}\right)f_b}$$

d'/d	Steel Ratio p', p	K_b 154.4	170	175	180	185	190	195	200	205	210
0.02	p'	—	0.0007	0.0009	0.0011	0.0013	0.0016	0.0018	0.0020	0.0022	0.0024
	p	0.0073	0.0080	0.0082	0.0084	0.0086	0.0088	0.0090	0.0092	0.0095	0.0097
0.04	p'	—	0.0007	0.0010	0.0012	0.0015	0.0017	0.0019	0.0022	0.0024	0.0026
	p	0.0073	0.0080	0.0082	0.0084	0.0086	0.0088	0.0091	0.0093	0.0095	0.0097
0.06	p'	—	0.0008	0.0011	0.0013	0.0016	0.0018	0.0021	0.0024	0.0026	0.0029
	p	0.0073	0.0080	0.0082	0.0084	0.0087	0.0089	0.0091	0.0093	0.0095	0.0098
0.08	p'	—	0.0009	0.0012	0.0015	0.0017	0.0020	0.0023	0.0026	0.0029	0.0032
	p	0.0073	0.0080	0.0082	0.0085	0.0087	0.0089	0.0091	0.0094	0.0096	0.0098
0.10	p'	—	0.0010	0.0013	0.0016	0.0019	0.0022	0.0025	0.0029	0.0032	0.0035
	p	0.0073	0.0080	0.0083	0.0085	0.0087	0.0089	0.0092	0.0094	0.0096	0.0099
0.12	p'	—	0.0011	0.0014	0.0018	0.0021	0.0025	0.0028	0.0032	0.0035	0.0039
	p	0.0073	0.0080	0.0083	0.0085	0.0087	0.0090	0.0092	0.0095	0.0097	0.0099
0.14	p'	—	0.0012	0.0016	0.0020	0.0024	0.0028	0.0032	0.0036	0.0040	0.0043
	p	0.0073	0.0081	0.0083	0.0085	0.0088	0.0090	0.0093	0.0095	0.0098	0.0100
0.16	p'	—	0.0014	0.0018	0.0023	0.0027	0.0031	0.0036	0.0040	0.0045	0.0049
	p	0.0073	0.0081	0.0083	0.0086	0.0088	0.0091	0.0093	0.0096	0.0098	0.0101
0.18	p'	—	0.0016	0.0021	0.0026	0.0031	0.0036	0.0041	0.0046	0.0051	0.0056
	p	0.0073	0.0081	0.0083	0.0086	0.0089	0.0091	0.0094	0.0096	0.0099	0.0101
0.20	p'	—	0.0018	0.0024	0.0030	0.0036	0.0042	0.0048	0.0054	0.0060	0.0065
	p	0.0073	0.0081	0.0084	0.0086	0.0089	0.0092	0.0094	0.0097	0.0099	0.0102
0.22	p'	—	0.0022	0.0029	0.0036	0.0043	0.0050	0.0057	0.0063	0.0070	0.0077
	p	0.0073	0.0081	0.0084	0.0087	0.0089	0.0092	0.0095	0.0097	0.0100	0.0103
0.24[2]	p'	—	0.0026	0.0035	0.0043	0.0052	0.0060	0.0069	0.0077	0.0085	0.0094
	p	0.0073	0.0082	0.0084	0.0087	0.0090	0.0093	0.0095	0.0098	0.0101	0.0103

1. When *K* is determined where wind or seismic conditions are considered and a $1/3$ increase in stress is permitted, multiply the *K* obtained by $3/4$ to use this table.

2. For *d'*/*d* values greater than 0.24 the effect of the compression steel becomes increasingly negligible.

TABLE N-5   Coefficients _p_ and _p'_ for Tension and Compression Steel in a Flexural Member for Full Allowable Stresses: f'_m = 3500 psi, F_s = 24,000 psi and _n_ = 11.0

For use of Table, see Example 5-I

DESIGN DATA

$f'_m = 3500$ psi	$f_y = 60,000$ psi
$f_b = 1167$ psi	$F_s = 24,000$ psi
$E_m = 2,625,000$ psi	$E_s = 29,000,000$ psi
$n = 11.0$	$k = 0.350$
$K_b = 179.7$	$p_b = 0.0085$

DESIGN EQUATIONS

$$K^1 = \frac{M}{F} = \frac{M\,(\text{ft kips})}{bd^2/12,000} \text{ or } \frac{M\,(\text{in. lbs})}{bd^2}$$

$$p = p_b + \frac{K - K_b}{f_s(1 - d'/d)}$$

$$p' = \frac{K - K_b}{(2n-1)\left(\dfrac{k - d'/d}{k}\right)\left(1 - \dfrac{d'}{d}\right)f_b}$$

d'/d	Steel Ratio p', p	K_b	K^1								
		179.7	195	200	205	210	215	220	225	230	235
0.02	p'	—	0.0007	0.0009	0.0011	0.0013	0.0016	0.0018	0.0020	0.0022	0.0024
	p	0.0085	0.0092	0.0094	0.0096	0.0098	0.0100	0.0102	0.0104	0.0106	0.0109
0.04	p'	—	0.0007	0.0010	0.0012	0.0015	0.0017	0.0019	0.0022	0.0024	0.0027
	p	0.0085	0.0092	0.0094	0.0096	0.0098	0.0100	0.0102	0.0105	0.0107	0.0109
0.06	p'	—	0.0008	0.0011	0.0013	0.0016	0.0019	0.0021	0.0024	0.0026	0.0029
	p	0.0085	0.0092	0.0094	0.0096	0.0098	0.0101	0.0103	0.0105	0.0107	0.0110
0.08	p'	—	0.0009	0.0012	0.0015	0.0017	0.0020	0.0023	0.0026	0.0029	0.0032
	p	0.0085	0.0092	0.0094	0.0096	0.0099	0.0101	0.0103	0.0106	0.0108	0.0110
0.10	p'	—	0.0010	0.0013	0.0016	0.0019	0.0022	0.0026	0.0029	0.0032	0.0035
	p	0.0085	0.0092	0.0094	0.0097	0.0099	0.0101	0.0104	0.0106	0.0108	0.0111
0.12	p'	—	0.0011	0.0014	0.0018	0.0021	0.0025	0.0029	0.0032	0.0036	0.0039
	p	0.0085	0.0092	0.0095	0.0097	0.0099	0.0102	0.0104	0.0106	0.0109	0.0111
0.14	p'	—	0.0012	0.0016	0.0020	0.0024	0.0028	0.0032	0.0036	0.0040	0.0044
	p	0.0085	0.0092	0.0095	0.0097	0.0100	0.0102	0.0105	0.0107	0.0109	0.0112
0.16	p'	—	0.0014	0.0018	0.0023	0.0027	0.0032	0.0036	0.0041	0.0045	0.0050
	p	0.0085	0.0093	0.0095	0.0098	0.0100	0.0103	0.0105	0.0107	0.0110	0.0112
0.18	p'	—	0.0016	0.0021	0.0026	0.0031	0.0036	0.0042	0.0047	0.0052	0.0057
	p	0.0085	0.0093	0.0095	0.0098	0.0100	0.0103	0.0105	0.0108	0.0111	0.0113
0.20	p'	—	0.0018	0.0024	0.0030	0.0036	0.0042	0.0048	0.0054	0.0060	0.0066
	p	0.0085	0.0093	0.0096	0.0098	0.0101	0.0103	0.0106	0.0109	0.0111	0.0114
0.22	p'	—	0.0022	0.0029	0.0036	0.0043	0.0050	0.0057	0.0064	0.0072	0.0079
	p	0.0085	0.0093	0.0096	0.0099	0.0101	0.0104	0.0107	0.0109	0.0112	0.0115
0.24[2]	p'	—	0.0026	0.0035	0.0044	0.0052	0.0061	0.0070	0.0078	0.0087	0.0096
	p	0.0085	0.0093	0.0096	0.0099	0.0102	0.0104	0.0107	0.0110	0.0113	0.0115

1. When _K_ is determined where wind or seismic conditions are considered and a $^1/_3$ increase in stress is permitted, multiply the _K_ obtained by $^3/_4$ to use this table.

2. For _d'/d_ values greater than 0.24 the effect of the compression steel becomes increasingly negligible.

TABLE N-6 Coefficients p and p' for Tension and Compression Steel in a Flexural Member for Full Allowable Stresses: $f'_m = 4000$ psi, $F_s = 24,000$ psi and $n = 9.7$

For use of Table, see Example 5-I

WSD

DESIGN DATA

$f'_m = 4000$ psi	$f_y = 60,000$ psi
$f_b = 1333$ psi	$F_s = 24,000$ psi
$E_m = 3,000,000$ psi	$E_s = 29,000,000$ psi
$n = 9.7$	$k = 0.350$
$K_b = 206.1$	$p_b = 0.0097$

DESIGN EQUATIONS

$$K^1 = \frac{M}{F} = \frac{M(\text{ft kips})}{bd^2/12,000} \text{ or } \frac{M(\text{in. lbs})}{bd^2}$$

$$p = p_b + \frac{K - K_b}{f_s(1 - d'/d)}$$

$$p' = \frac{K - K_b}{(2n-1)\left(\dfrac{k - d'/d}{k}\right)\left(1 - \dfrac{d'}{d}\right)f_b}$$

d'/d	Steel Ratio p', p	K_b 206.1	K^1 225	230	235	240	245	250	255	260	265
0.02	p'	—	0.0008	0.0011	0.0013	0.0015	0.0017	0.0019	0.0022	0.0024	0.0026
	p	0.0097	0.0105	0.0107	0.0109	0.0111	0.0114	0.0116	0.0118	0.0120	0.0122
0.04	p'	—	0.0009	0.0011	0.0014	0.0016	0.0019	0.0021	0.0023	0.0026	0.0028
	p	0.0097	0.0105	0.0107	0.0110	0.0112	0.0114	0.0116	0.0118	0.0120	0.0123
0.06	p'	—	0.0010	0.0013	0.0015	0.0018	0.0020	0.0023	0.0026	0.0028	0.0031
	p	0.0097	0.0105	0.0108	0.0110	0.0112	0.0114	0.0116	0.0119	0.0121	0.0123
0.08	p'	—	0.0011	0.0014	0.0017	0.0019	0.0022	0.0025	0.0028	0.0031	0.0034
	p	0.0097	0.0106	0.0108	0.0110	0.0112	0.0115	0.0117	0.0119	0.0121	0.0124
0.10	p'	—	0.0012	0.0015	0.0018	0.0021	0.0025	0.0028	0.0031	0.0034	0.0037
	p	0.0097	0.0106	0.0108	0.0110	0.0113	0.0115	0.0117	0.0120	0.0122	0.0124
0.12	p'	—	0.0013	0.0017	0.0020	0.0024	0.0027	0.0031	0.0034	0.0038	0.0042
	p	0.0097	0.0106	0.0108	0.0111	0.0113	0.0115	0.0118	0.0120	0.0123	0.0125
0.14	p'	—	0.0015	0.0019	0.0023	0.0027	0.0031	0.0035	0.0039	0.0043	0.0047
	p	0.0097	0.0106	0.0109	0.0111	0.0113	0.0116	0.0118	0.0121	0.0123	0.0126
0.16	p'	—	0.0017	0.0021	0.0026	0.0030	0.0035	0.0039	0.0044	0.0048	0.0053
	p	0.0097	0.0106	0.0109	0.0111	0.0114	0.0116	0.0119	0.0121	0.0124	0.0126
0.18	p'	—	0.0019	0.0024	0.0030	0.0035	0.0040	0.0045	0.0050	0.0055	0.0060
	p	0.0097	0.0107	0.0109	0.0112	0.0114	0.0117	0.0119	0.0122	0.0124	0.0127
0.20	p'	—	0.0022	0.0028	0.0034	0.0040	0.0046	0.0052	0.0058	0.0064	0.0070
	p	0.0097	0.0107	0.0109	0.0112	0.0115	0.0117	0.0120	0.0122	0.0125	0.0128
0.22	p'	—	0.0027	0.0034	0.0041	0.0048	0.0055	0.0062	0.0069	0.0076	0.0083
	p	0.0097	0.0107	0.0110	0.0112	0.0115	0.0118	0.0120	0.0123	0.0126	0.0128
0.24[2]	p'	—	0.0032	0.0041	0.0049	0.0058	0.0066	0.0075	0.0083	0.0092	0.0101
	p	0.0097	0.0107	0.0110	0.0113	0.0116	0.0118	0.0121	0.0124	0.0127	0.0129

1. When K is determined where wind or seismic conditions are considered and a $1/3$ increase in stress is permitted, multiply the K obtained by $3/4$ to use this table.

2. For d'/d values greater than 0.24 the effect of the compression steel becomes increasingly negligible.

TABLE N-7 Coefficients _p_ and _p'_ for Tension and Compression Steel in a Flexural Member for Full Allowable Stresses: _f'ₘ_ = 4500 psi, _Fₛ_ = 24,000 psi and _n_ = 9.7

For use of Table, see Example 5-I

DESIGN DATA

$f'_m = 4500$ psi $f_y = 60,000$ psi
$f_b = 1500$ psi $F_s = 24,000$ psi
$E_m = 3,000,000$ psi $E_s = 29,000,000$ psi
$n = 9.7$ $k = 0.377$
$K_b = 247.5$ $p_b = 0.0118$

DESIGN EQUATIONS

$$K^1 = \frac{M}{F} = \frac{M\,(\text{ft kips})}{bd^2/12,000} \text{ or } \frac{M\,(\text{in. lbs})}{bd^2}$$

$$p = p_b + \frac{K - K_b}{f_s(1 - d'/d)}$$

$$p' = \frac{K - K_b}{(2n-1)\left(\dfrac{k - d'/d}{k}\right)\left(1 - \dfrac{d'}{d}\right)f_b}$$

d'/d	Steel Ratio p', p	K_b 247.5	275	280	285	290	295	300	305	310	315
0.02	p'	—	0.0011	0.0013	0.0015	0.0017	0.0019	0.0020	0.0022	0.0024	0.0026
	p	0.0118	0.0130	0.0132	0.0134	0.0136	0.0138	0.0140	0.0142	0.0145	0.0147
0.04	p'	—	0.0012	0.0014	0.0016	0.0018	0.0020	0.0022	0.0024	0.0026	0.0028
	p	0.0118	0.0130	0.0132	0.0134	0.0136	0.0139	0.0141	0.0143	0.0145	0.0147
0.06	p'	—	0.0013	0.0015	0.0017	0.0019	0.0022	0.0024	0.0026	0.0029	0.0031
	p	0.0118	0.0130	0.0132	0.0135	0.0137	0.0139	0.0141	0.0143	0.0146	0.0148
0.08	p'	—	0.0014	0.0016	0.0019	0.0021	0.0024	0.0026	0.0029	0.0031	0.0034
	p	0.0118	0.0130	0.0133	0.0135	0.0137	0.0140	0.0142	0.0144	0.0146	0.0149
0.10	p'	—	0.0015	0.0018	0.0021	0.0023	0.0026	0.0029	0.0032	0.0034	0.0037
	p	0.0118	0.0131	0.0133	0.0135	0.0138	0.0140	0.0142	0.0145	0.0147	0.0149
0.12	p'	—	0.0017	0.0020	0.0023	0.0026	0.0029	0.0032	0.0035	0.0038	0.0041
	p	0.0118	0.0131	0.0133	0.0136	0.0138	0.0140	0.0143	0.0145	0.0148	0.0150
0.14	p'	—	0.0018	0.0022	0.0025	0.0028	0.0032	0.0035	0.0039	0.0042	0.0045
	p	0.0118	0.0131	0.0134	0.0136	0.0139	0.0141	0.0143	0.0146	0.0148	0.0151
0.16	p'	—	0.0021	0.0024	0.0028	0.0032	0.0036	0.0039	0.0043	0.0047	0.0051
	p	0.0118	0.0132	0.0134	0.0137	0.0139	0.0142	0.0144	0.0147	0.0149	0.0151
0.18	p'	—	0.0023	0.0027	0.0032	0.0036	0.0040	0.0044	0.0049	0.0053	0.0057
	p	0.0118	0.0132	0.0135	0.0137	0.0140	0.0142	0.0145	0.0147	0.0150	0.0152
0.20	p'	—	0.0027	0.0031	0.0036	0.0041	0.0046	0.0051	0.0055	0.0060	0.0065
	p	0.0118	0.0132	0.0135	0.0138	0.0140	0.0143	0.0145	0.0148	0.0151	0.0153
0.22	p'	—	0.0031	0.0036	0.0042	0.0047	0.0053	0.0059	0.0064	0.0070	0.0075
	p	0.0118	0.0133	0.0135	0.0138	0.0141	0.0143	0.0146	0.0149	0.0151	0.0154
0.24²	p'	—	0.0036	0.0043	0.0049	0.0056	0.0062	0.0069	0.0075	0.0082	0.0089
	p	0.0118	0.0133	0.0136	0.0139	0.0141	0.0144	0.0147	0.0150	0.0152	0.0155

1. When _K_ is determined where wind or seismic conditions are considered and a ¹/₃ increase in stress is permitted, multiply the _K_ obtained by ³/₄ to use this table.

2. For _d'/d_ values greater than 0.24 the effect of the compression steel becomes increasingly negligible.

TABLE N-8 Coefficients *p* and *p'* for Tension and Compression Steel in a Flexural Member for Full Allowable Stresses: f'_m = 5000 psi, F_s = 24,000 psi and *n* = 9.7

For use of Table, see Example 5-I

For use of Table, see Example 5-I

DESIGN DATA

$f'_m = 5000$ psi	$f_y = 60,000$ psi
$f_b = 1667$ psi	$F_s = 24,000$ psi
$E_m = 3,000,000$ psi	$E_s = 29,000,000$ psi
$n = 9.7$	$k = 0.403$
$K_b = 290.5$	$p_b = 0.014$

DESIGN EQUATIONS

$$K^1 = \frac{M}{F} = \frac{M \, (\text{ft kips})}{bd^2/12{,}000} \text{ or } \frac{M \, (\text{in. lbs})}{bd^2}$$

$$p = p_b + \frac{K - K_b}{f_s(1 - d'/d)}$$

$$p' = \frac{K - K_b}{(2n-1)\left(\dfrac{k - d'/d}{k}\right)\left(1 - \dfrac{d'}{d}\right)f_b}$$

d'/d	Steel Ratio p', p	K_b 290.5	305	315	320	325	330	335	340	345	350
0.02	p'	—	0.0005	0.0007	0.0009	0.0010	0.0012	0.0014	0.0016	0.0017	0.0019
	p	0.0140	0.0146	0.0148	0.0150	0.0153	0.0155	0.0157	0.0159	0.0161	0.0163
0.04	p'	—	0.0005	0.0007	0.0009	0.0011	0.0013	0.0015	0.0017	0.0019	0.0021
	p	0.0140	0.0146	0.0148	0.0151	0.0153	0.0155	0.0157	0.0159	0.0161	0.0164
0.06	p'	—	0.0006	0.0008	0.0010	0.0012	0.0014	0.0016	0.0018	0.0020	0.0022
	p	0.0140	0.0146	0.0149	0.0151	0.0153	0.0155	0.0158	0.0160	0.0162	0.0164
0.08	p'	—	0.0006	0.0009	0.0011	0.0013	0.0015	0.0017	0.0020	0.0022	0.0024
	p	0.0140	0.0147	0.0149	0.0151	0.0153	0.0156	0.0158	0.0160	0.0162	0.0165
0.10	p'	—	0.0007	0.0009	0.0012	0.0014	0.0017	0.0019	0.0021	0.0024	0.0026
	p	0.0140	0.0147	0.0149	0.0151	0.0154	0.0156	0.0158	0.0161	0.0163	0.0165
0.12	p'	—	0.0008	0.0010	0.0013	0.0016	0.0018	0.0021	0.0023	0.0026	0.0029
	p	0.0140	0.0147	0.0149	0.0152	0.0154	0.0156	0.0159	0.0161	0.0163	0.0166
0.14	p'	—	0.0008	0.0011	0.0014	0.0017	0.0020	0.0023	0.0026	0.0029	0.0032
	p	0.0140	0.0147	0.0149	0.0152	0.0154	0.0157	0.0159	0.0162	0.0164	0.0166
0.16	p'	—	0.0009	0.0013	0.0016	0.0019	0.0022	0.0025	0.0029	0.0032	0.0035
	p	0.0140	0.0147	0.0150	0.0152	0.0155	0.0157	0.0160	0.0162	0.0165	0.0167
0.18	p'	—	0.0010	0.0014	0.0018	0.0021	0.0025	0.0028	0.0032	0.0036	0.0039
	p	0.0140	0.0147	0.0150	0.0152	0.0155	0.0158	0.0160	0.0163	0.0165	0.0168
0.20	p'	—	0.0012	0.0016	0.0020	0.0024	0.0028	0.0032	0.0036	0.0040	0.0044
	p	0.0140	0.0148	0.0150	0.0153	0.0155	0.0158	0.0161	0.0163	0.0166	0.0168
0.22	p'	—	0.0013	0.0018	0.0023	0.0027	0.0032	0.0036	0.0041	0.0046	0.0050
	p	0.0140	0.0148	0.0150	0.0153	0.0156	0.0158	0.0161	0.0164	0.0166	0.0169
0.24[2]	p'	—	0.0015	0.0021	0.0026	0.0031	0.0037	0.0042	0.0047	0.0052	0.0058
	p	0.0140	0.0148	0.0151	0.0153	0.0156	0.0159	0.0162	0.0164	0.0167	0.0170

1. When *K* is determined where wind or seismic conditions are considered and a $^1/_3$ increase in stress is permitted, multiply the *K* obtained by $^3/_4$ to use this table.

2. For *d'/d* values greater than 0.24 the effect of the compression steel becomes increasingly negligible.

WSD

Diagram P-1a   Steel Ratio p & p′ vs K, Half Allowable, Stresses:
f'_m = 1500 psi, F_s = 24,000 psi and n = 25.8

For use of Diagram,
see Example 5-I

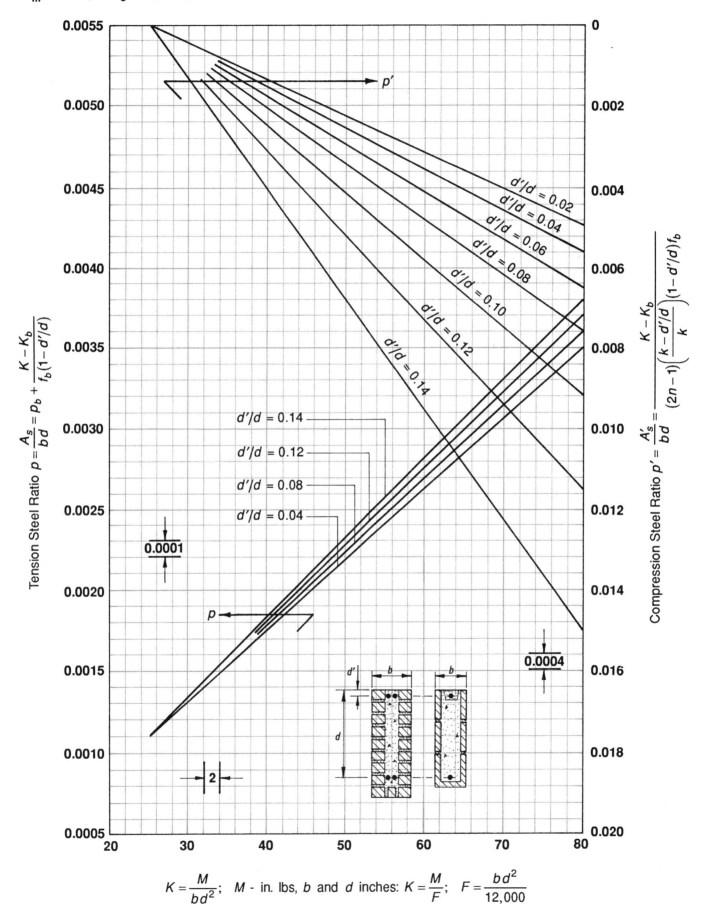

$$K = \frac{M}{bd^2}; \quad M \text{ - in. lbs, } b \text{ and } d \text{ inches: } K = \frac{M}{F}; \quad F = \frac{bd^2}{12,000}$$

WSD

Diagram P-1b   Steel Ratio p & p' vs K, Full Allowable, Stresses:
$f'_m = 1500$ psi, $F_s = 24{,}000$ psi and $n = 25.8$

For use of Diagram,
see Example 5-I

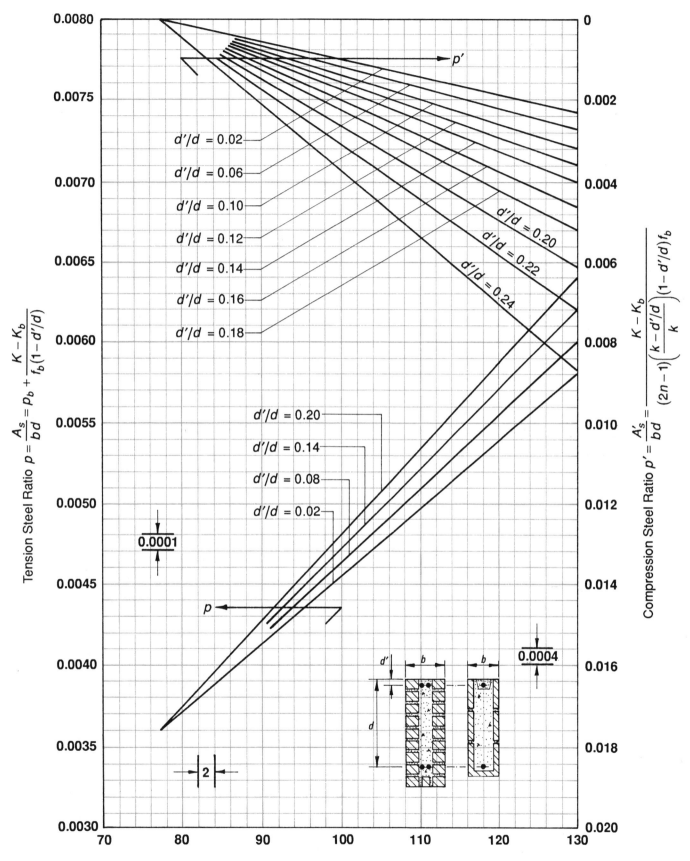

Tension Steel Ratio $p = \dfrac{A_s}{bd} = p_b + \dfrac{K - K_b}{f_b(1 - d'/d)}$

Compression Steel Ratio $p' = \dfrac{A'_s}{bd} = \dfrac{K - K_b}{(2n-1)\left(\dfrac{k - d'/d}{k}\right)(1 - d'/d)f_b}$

$d'/d = 0.02$
$d'/d = 0.06$
$d'/d = 0.10$
$d'/d = 0.12$
$d'/d = 0.14$
$d'/d = 0.16$
$d'/d = 0.18$
$d'/d = 0.20$
$d'/d = 0.22$
$d'/d = 0.24$

$d'/d = 0.20$
$d'/d = 0.14$
$d'/d = 0.08$
$d'/d = 0.02$

$K = \dfrac{M}{bd^2}$; M - in. lbs, b and d inches: $K = \dfrac{M}{F}$; $F = \dfrac{bd^2}{12{,}000}$

Diagram P-2a   Steel Ratio p & p' vs K, Half Allowable, Stresses:
$f'_m = 2000$ psi, $F_s = 24,000$ psi and $n = 19.3$

For use of Diagram,
see Example 5-I

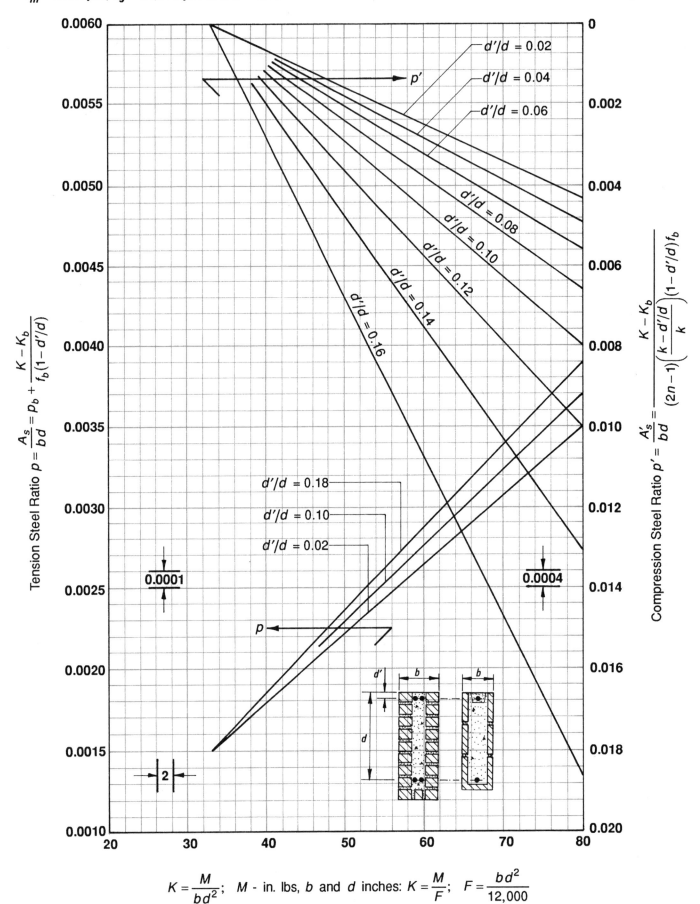

Tension Steel Ratio $p = \dfrac{A_s}{bd} = p_b + \dfrac{K - K_b}{f_b(1 - d'/d)}$

Compression Steel Ratio $p' = \dfrac{A'_s}{bd} = \dfrac{K - K_b}{(2n-1)\left(\dfrac{k - d'/d}{k}\right)(1 - d'/d)f_b}$

$K = \dfrac{M}{bd^2}$; M - in. lbs, b and d inches: $K = \dfrac{M}{F}$; $F = \dfrac{bd^2}{12,000}$

Diagram P-2b   Steel Ratio p & p' vs K, Full Allowable, Stresses: f'_m = 2000 psi, F_s = 24,000 psi and n = 19.3

For use of Diagram, see Example 5-I

WSD

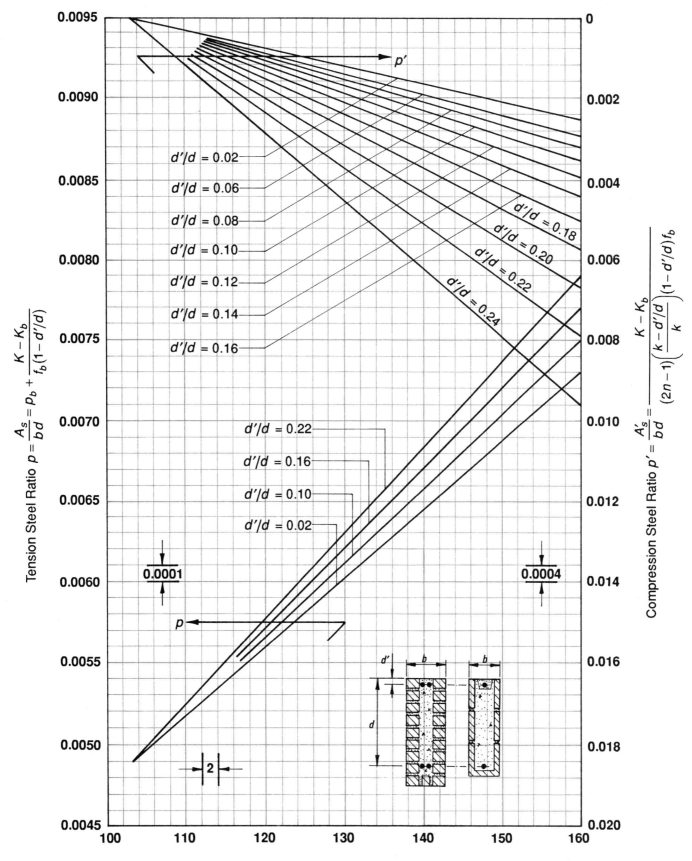

Tension Steel Ratio $p = \dfrac{A_s}{bd} = p_b + \dfrac{K - K_b}{f_b(1 - d'/d)}$

Compression Steel Ratio $p' = \dfrac{A'_s}{bd} = \dfrac{K - K_b}{(2n - 1)\left(\dfrac{k - d'/d}{k}\right)(1 - d'/d)f_b}$

d'/d = 0.02
d'/d = 0.06
d'/d = 0.08
d'/d = 0.10
d'/d = 0.12
d'/d = 0.14
d'/d = 0.16

d'/d = 0.18
d'/d = 0.20
d'/d = 0.22
d'/d = 0.24

d'/d = 0.22
d'/d = 0.16
d'/d = 0.10
d'/d = 0.02

0.0001

0.0004

p'

p

2

$K = \dfrac{M}{bd^2}$; M - in. lbs, b and d inches: $K = \dfrac{M}{F}$; $F = \dfrac{bd^2}{12,000}$

Diagram P-3a   Steel Ratio p & p' vs K, Half Allowable, Stresses:
f'_m = 2500 psi, F_s = 24,000 psi and n = 15.5

For use of Diagram,
see Example 5-I

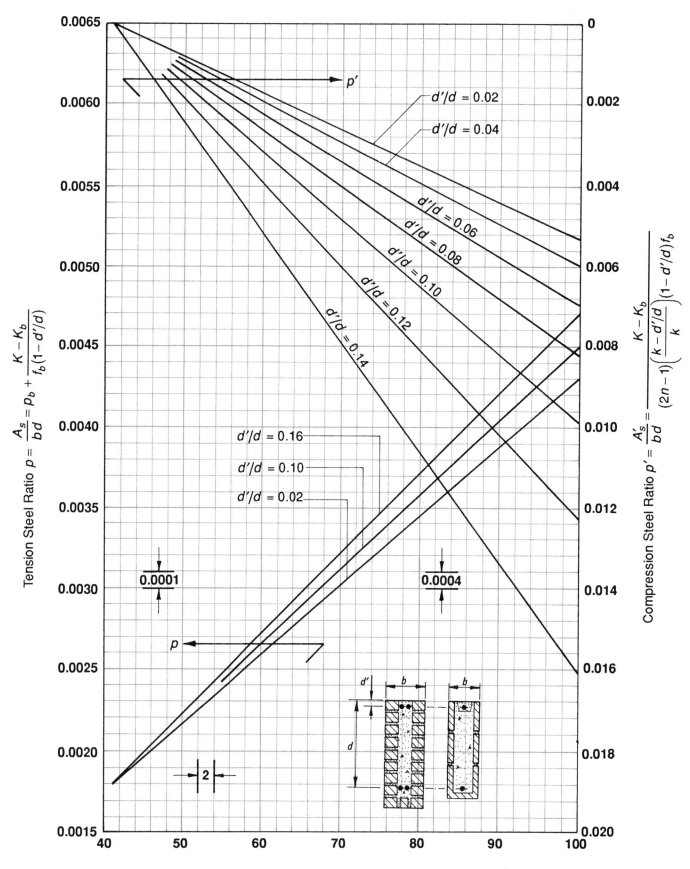

$$K = \frac{M}{bd^2}; \quad M \text{ - in. lbs, } b \text{ and } d \text{ inches: } K = \frac{M}{F}; \quad F = \frac{bd^2}{12,000}$$

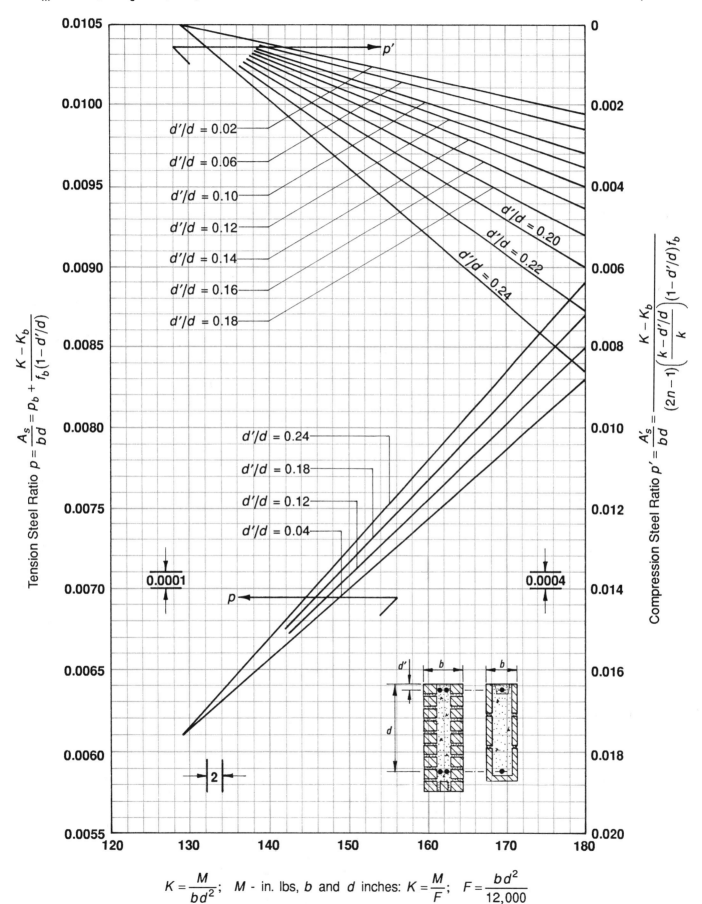

Diagram P-3b   Steel Ratio p & p' vs K, Full Allowable, Stresses:
f'_m = 2500 psi, F_s = 24,000 psi and n = 15.5

For use of Diagram,
see Example 5-I

WSD

Tension Steel Ratio $p = \dfrac{A_s}{bd} = p_b + \dfrac{K - K_b}{f_b(1 - d'/d)}$

Compression Steel Ratio $p' = \dfrac{A'_s}{bd} = \dfrac{\dfrac{K - K_b}{(2n-1)}\left(\dfrac{k - d'/d}{k}\right)(1 - d'/d)f_b}$

d'/d = 0.02
d'/d = 0.06
d'/d = 0.10
d'/d = 0.12
d'/d = 0.14
d'/d = 0.16
d'/d = 0.18

d'/d = 0.20
d'/d = 0.22
d'/d = 0.24

p'

d'/d = 0.24
d'/d = 0.18
d'/d = 0.12
d'/d = 0.04

0.0001

0.0004

p

$K = \dfrac{M}{bd^2}$;   M - in. lbs, b and d inches: $K = \dfrac{M}{F}$; $F = \dfrac{bd^2}{12,000}$

Diagram P-4a   Steel Ratio *p* & *p'* vs *K*, Half Allowable, Stresses:
f'_m = 3000 psi, F_s = 24,000 psi and *n* = 12.9

For use of Diagram, see Example 5-I

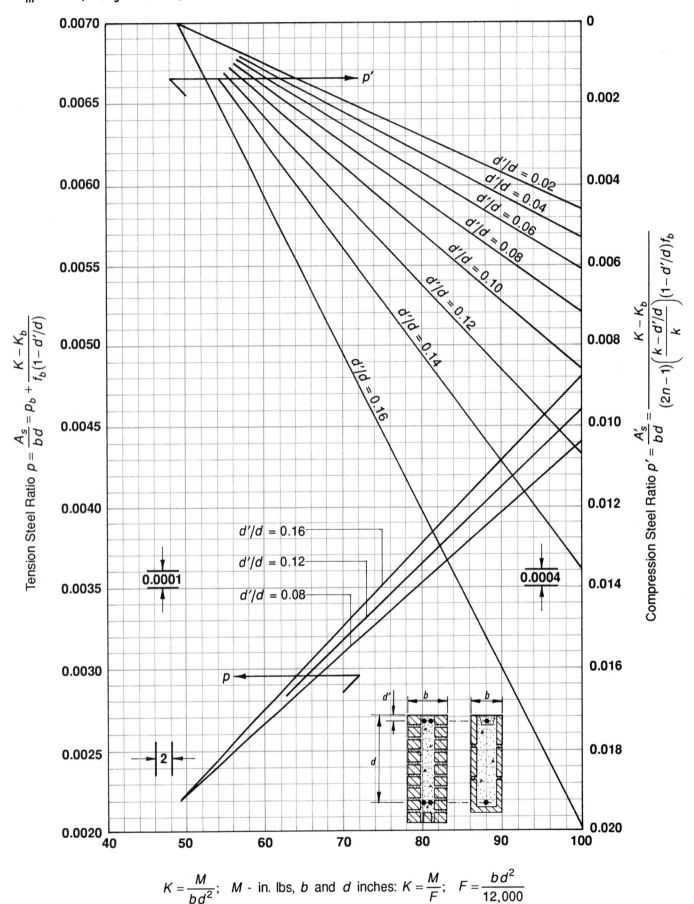

$$K = \frac{M}{bd^2}; \quad M \text{ - in. lbs, } b \text{ and } d \text{ inches: } K = \frac{M}{F}; \quad F = \frac{bd^2}{12,000}$$

Diagram P-4b   Steel Ratio p & p' vs K, Full Allowable, Stresses:
f'_m = 3000 psi, F_s = 24,000 psi and n = 12.9

For use of Diagram,
see Example 5-I

WSD

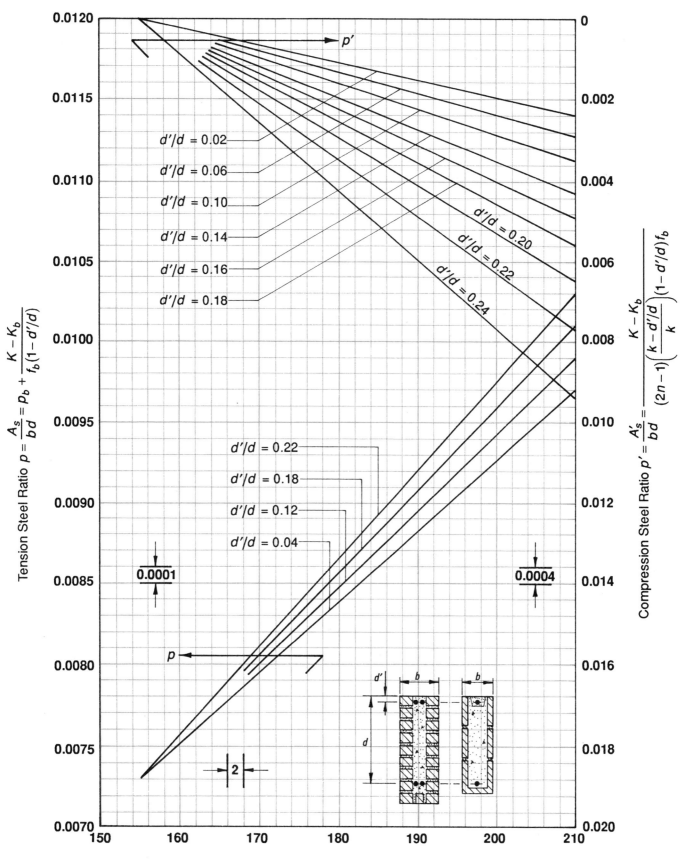

Tension Steel Ratio $p = \dfrac{A_s}{bd} = p_b + \dfrac{K - K_b}{f_b(1 - d'/d)}$

Compression Steel Ratio $p' = \dfrac{A'_s}{bd} = \dfrac{K - K_b}{(2n-1)\left(\dfrac{k - d'/d}{k}\right)(1 - d'/d)f_b}$

$d'/d = 0.02$
$d'/d = 0.06$
$d'/d = 0.10$
$d'/d = 0.14$
$d'/d = 0.16$
$d'/d = 0.18$
$d'/d = 0.20$
$d'/d = 0.22$
$d'/d = 0.24$

$d'/d = 0.22$
$d'/d = 0.18$
$d'/d = 0.12$
$d'/d = 0.04$

0.0001

0.0004

p'

p

2

$$K = \frac{M}{bd^2}; \quad M \text{ - in. lbs, } b \text{ and } d \text{ inches: } K = \frac{M}{F}; \quad F = \frac{bd^2}{12,000}$$

Diagram P-5　Steel Ratio p & p' vs K, Full Allowable, Stresses:
f'_m = 3500 psi, F_s = 24,000 psi and n = 11.0

For use of Diagram,
see Example 5-I

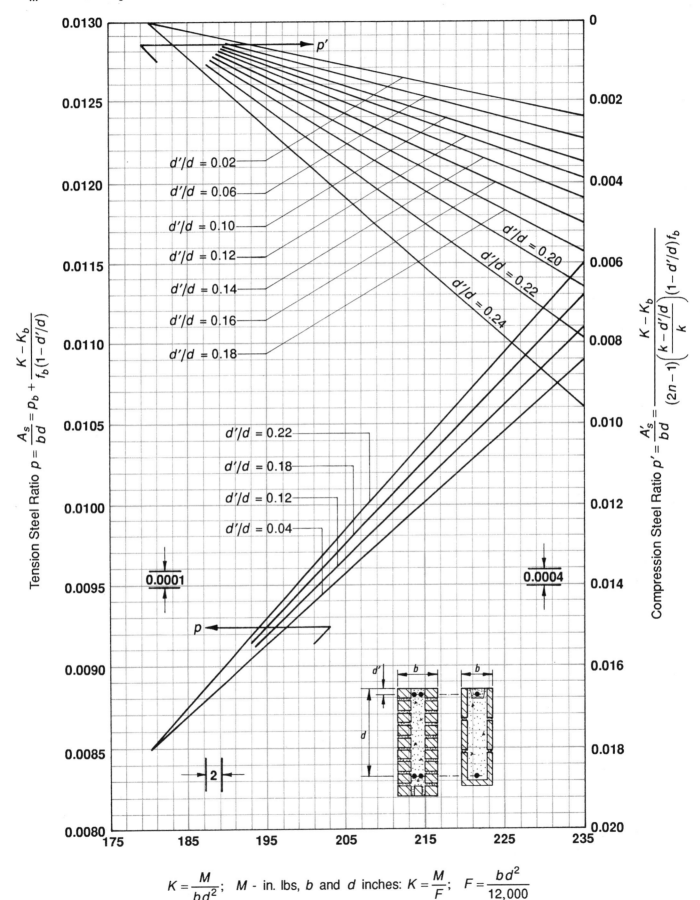

$$K = \frac{M}{bd^2}; \quad M \text{ - in. lbs, } b \text{ and } d \text{ inches: } K = \frac{M}{F}; \quad F = \frac{bd^2}{12,000}$$

Diagram P-6 Steel Ratio p & p' vs K, Full Allowable, Stresses:
$f'_m = 4000$ psi, $F_s = 24{,}000$ psi and $n = 9.7$

For use of Diagram,
see Example 5-I

WSD

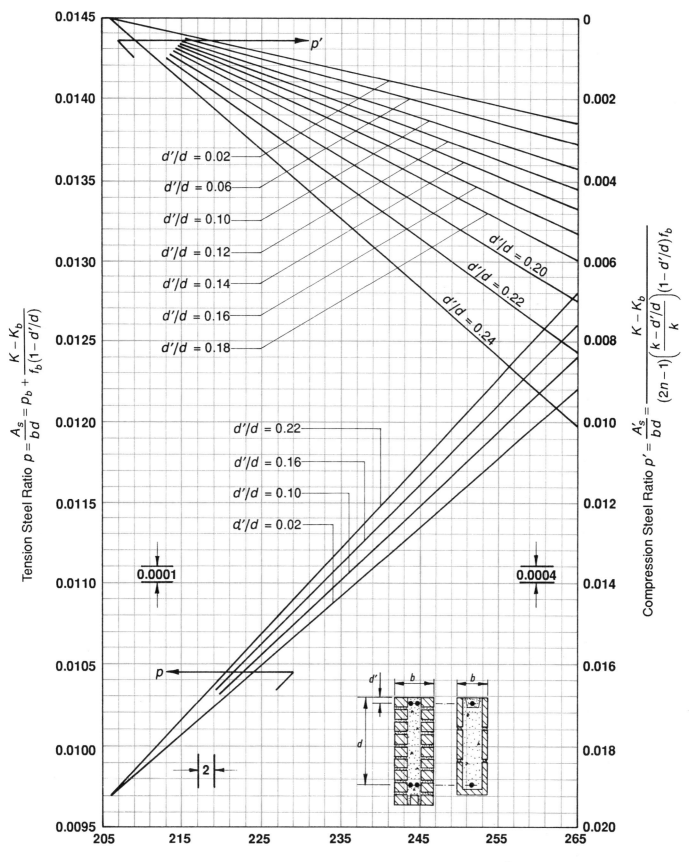

$$K = \frac{M}{bd^2}; \quad M\text{ - in. lbs, } b \text{ and } d \text{ inches: } K = \frac{M}{F}; \quad F = \frac{bd^2}{12{,}000}$$

Diagram P-7   Steel Ratio *p* & *p'* vs *K*, Full Allowable, Stresses:
f'_m = 4500 psi, F_s = 24,000 psi and n = 9.7

For use of Diagram,
see Example 5-I

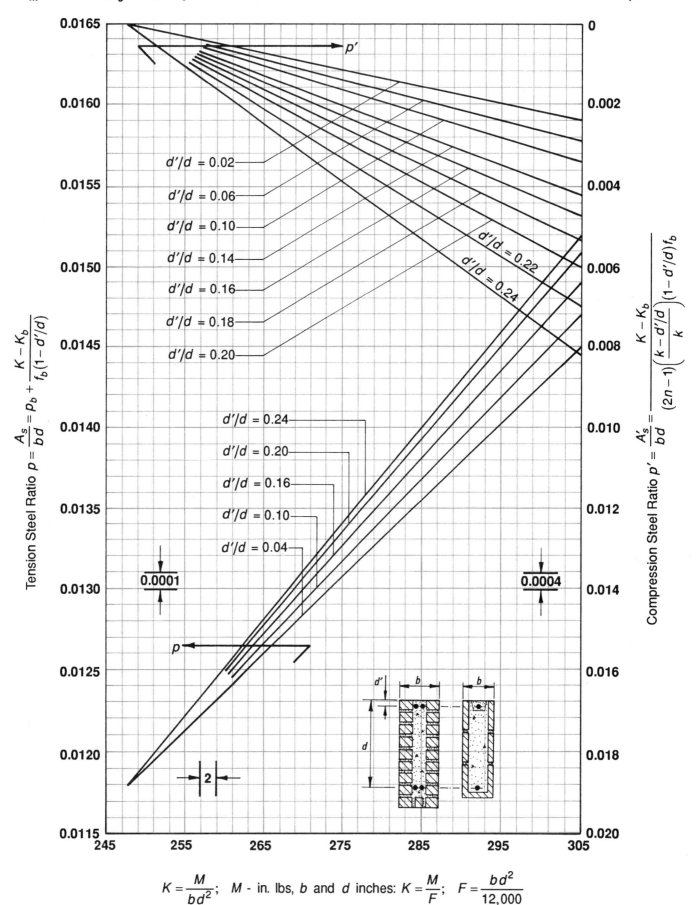

$$K = \frac{M}{bd^2}; \quad M \text{ - in. lbs, } b \text{ and } d \text{ inches: } K = \frac{M}{F}; \quad F = \frac{bd^2}{12,000}$$

Diagram P-8 Steel Ratio p & p' vs K, Full Allowable, Stresses: f'_m = 5000 psi, F_s = 24,000 psi and n = 9.7

For use of Diagram, see Example 5-I

WSD

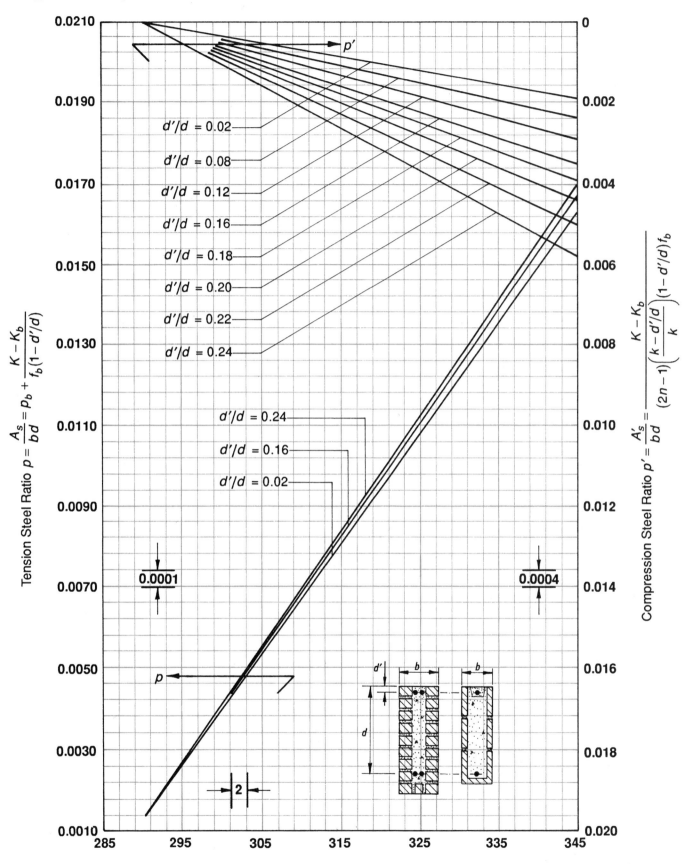

Tension Steel Ratio $p = \dfrac{A_s}{bd} = p_b + \dfrac{K - K_b}{f_b(1 - d'/d)}$

$d'/d = 0.02$
$d'/d = 0.08$
$d'/d = 0.12$
$d'/d = 0.16$
$d'/d = 0.18$
$d'/d = 0.20$
$d'/d = 0.22$
$d'/d = 0.24$

$d'/d = 0.24$
$d'/d = 0.16$
$d'/d = 0.02$

p'
p

0.0001
0.0004
2

Compression Steel Ratio $p' = \dfrac{A'_s}{bd} = \dfrac{K - K_b}{(2n - 1)\left(\dfrac{k - d'/d}{k}\right)(1 - d'/d)f_b}$

$$K = \frac{M}{bd^2}; \quad M\text{ - in. lbs, } b \text{ and } d \text{ inches: } K = \frac{M}{F}; \quad F = \frac{bd^2}{12,000}$$

TABLE Q-1 Stress Reduction Coefficients, R^1, Based on h'/t Ratio   For use of Table, see Example 5-Y

h'/t	R^1	\multicolumn{12}{Actual Wall Thickness, t (inches)}	h'/t
		15.625	14	12	11.625	10	9.625	9	8.5	7.625	5.625	4	3.5	
		\multicolumn{12}{Effective Height of Wall, h' (feet)}	
5	0.998	6.51	5.83	5.00	4.84	4.17	4.01	3.75	3.54	3.18	2.34	1.67	1.46	5
6	0.997	7.81	7.00	6.00	5.81	5.00	4.81	4.50	4.25	3.81	2.81	2.00	1.75	6
7	0.995	9.11	8.17	7.00	6.78	5.83	5.61	5.25	4.96	4.45	3.28	2.33	2.04	7
8	0.993	10.42	9.33	8.00	7.75	6.67	6.42	6.00	5.67	5.08	3.75	2.67	2.33	8
9	0.990	11.72	10.50	9.00	8.72	7.50	7.22	6.75	6.38	5.72	4.22	3.00	2.63	9
10	0.987	13.92	11.67	10.00	9.69	8.33	8.02	7.50	7.08	6.35	4.69	3.33	2.92	10
11	0.982	14.32	12.83	11.00	10.66	9.17	8.82	8.25	7.79	6.99	5.16	3.67	3.21	11
12	0.977	15.63	14.00	12.00	11.63	10.00	9.63	9.00	8.50	7.63	5.63	4.00	3.50	12
13	0.970	16.93	15.17	13.00	12.59	10.83	10.43	9.75	9.21	8.26	6.09	4.33	3.79	13
14	0.963	18.23	16.33	14.00	13.56	11.67	11.23	10.50	9.92	8.90	6.56	4.67	4.08	14
15	0.954	19.53	17.50	15.00	14.53	12.50	12.03	11.25	10.63	9.53	7.03	5.00	4.38	15
16	0.945	20.83	18.67	16.00	15.50	13.33	12.83	12.00	11.33	10.17	7.50	5.33	4.67	16
17	0.934	22.14	19.83	17.00	16.47	14.17	13.64	12.75	12.04	10.80	7.97	5.67	4.96	17
18	0.921	23.44	21.00	18.00	17.44	15.00	14.44	13.50	12.75	11.44	8.44	6.00	5.25	18
19	0.907	24.74	22.17	19.00	18.41	15.83	15.24	14.25	13.46	12.07	8.91	6.33	5.54	19
20	0.892	26.04	23.33	20.00	19.38	16.67	16.04	15.00	14.17	12.71	9.38	6.67	5.83	20
21	0.875	27.34	24.50	21.00	20.34	17.50	16.84	15.75	14.88	13.34	9.84	7.00	6.13	21
22	0.856	28.65	25.67	22.00	21.31	18.33	17.65	16.50	15.58	13.98	10.31	7.33	6.42	22
23	0.836	29.95	26.83	23.00	22.28	19.17	18.45	17.25	16.29	14.61	10.78	7.67	6.71	23
24	0.813	31.25	28.00	24.00	23.25	20.00	19.25	18.00	17.00	15.25	11.25	8.00	7.00	24
25	0.789	32.55	29.17	25.00	24.22	20.83	20.05	18.75	17.71	15.89	11.72	8.33	7.29	25
26	0.763	33.85	30.33	26.00	25.19	21.67	20.85	19.50	18.42	16.52	12.19	8.67	7.58	26
27	0.734	35.16	31.50	27.00	26.16	22.50	21.66	20.25	19.13	17.16	12.66	9.00	7.88	27
28	0.704	36.46	32.67	28.00	27.13	23.33	22.46	21.00	19.83	17.79	13.13	9.33	8.17	28
29	0.671	37.76	33.83	29.00	28.09	24.17	23.26	21.75	20.54	18.43	13.59	9.67	8.46	29
30	0.636	39.06	35.00	30.00	29.06	25.00	24.06	22.50	21.25	19.06	14.06	10.00	8.75	30
31	0.598	40.36	36.17	31.00	30.03	25.83	24.86	23.25	21.96	19.70	14.53	10.33	9.04	31
32	0.558	41.67	37.33	32.00	31.00	26.67	25.67	24.00	22.67	20.33	15.00	10.67	9.33	32
33	0.515	42.97	38.50	33.00	31.97	27.50	26.47	24.75	23.38	20.97	15.47	11.00	9.63	33
34	0.469	44.27	39.67	34.00	32.94	28.33	27.27	25.50	24.08	21.60	15.94	11.33	9.92	34
35	0.421	45.57	40.83	35.00	33.91	29.17	28.07	26.25	24.79	22.24	16.41	11.67	10.21	35
36	0.370	46.88	42.00	36.00	34.88	30.00	28.88	27.00	25.50	22.88	16.88	12.00	10.50	36
37	0.316	48.18	43.17	37.00	35.84	30.83	29.68	27.75	26.21	23.51	17.34	12.33	10.79	37
38	0.259	49.48	44.33	38.00	36.81	31.67	30.48	28.50	26.92	24.15	17.81	12.67	11.08	38
39	0.199	50.78	45.50	39.00	37.78	32.50	31.28	29.25	27.63	24.78	18.28	13.00	11.38	39
40	0.136	52.08	46.67	40.00	38.75	33.33	32.08	30.00	28.33	25.42	18.75	13.33	11.67	40
41	0.070	53.39	47.83	41.00	39.72	34.17	32.89	30.75	29.04	26.05	19.22	13.67	11.96	41
42	0.000	54.69	49.00	42.00	40.69	35.00	33.69	31.50	29.75	26.69	19.69	14.00	12.25	42

1. $R = 1 - (h'/42t)^3$

TABLE Q-2   Allowable Axial Wall Compressive Stresses, $F_a = 0.2 f'_m(R)^{1,2}$, (psi)   For use of Table, see Example 5-Y

WSD

h'/t	R^1	Specified Strength of Masonry, f'_m (psi)												h'/t
		No Special Inspection				Special Inspection								
		1500	2000	2500	3000	1500	2000	2500	3000	3500	4000	4500	5000	
		Allowable Axial Wall Stress, F_a (psi)												
5	0.998	150	200	250	299	299	399	499	599	699	799	898	998	5
6	0.997	150	199	249	299	299	399	499	598	698	798	897	997	6
7	0.995	149	199	249	299	299	398	498	597	697	796	896	995	7
8	0.993	149	199	248	298	298	397	497	596	695	794	894	993	8
9	0.990	149	198	248	297	297	396	495	594	693	792	891	990	9
10	0.987	148	197	247	296	296	395	493	592	691	789	888	987	10
11	0.982	147	196	246	295	295	393	491	589	687	786	884	982	11
12	0.977	147	195	244	293	293	391	488	586	684	781	879	977	12
13	0.970	146	194	243	291	291	388	485	582	679	776	873	970	13
14	0.963	144	193	241	289	289	385	481	578	674	770	867	963	14
15	0.954	143	191	239	286	286	382	477	573	668	764	859	954	15
16	0.945	142	189	236	283	283	378	472	567	661	756	850	945	16
17	0.934	140	187	233	280	280	373	467	560	654	747	840	934	17
18	0.921	138	184	230	276	276	369	461	553	645	737	829	921	18
19	0.907	136	181	227	272	272	363	454	544	635	726	817	907	19
20	0.892	134	178	223	268	268	357	446	535	624	714	803	892	20
21	0.875	131	175	219	263	263	350	438	525	613	700	788	875	21
22	0.856	128	171	214	257	257	343	428	514	599	685	771	856	22
23	0.836	125	167	209	251	251	334	418	501	585	669	752	836	23
24	0.813	122	163	203	244	244	325	407	488	569	651	732	813	24
25	0.789	118	158	197	237	237	316	395	473	552	631	710	789	25
26	0.763	114	153	191	229	229	305	381	458	534	610	686	763	26
27	0.734	110	147	184	220	220	294	367	441	514	587	661	734	27
28	0.704	106	141	176	211	211	281	352	422	493	563	633	704	28
29	0.671	101	134	168	201	201	268	335	402	470	537	604	671	29
30	0.636	95	127	159	191	191	254	318	381	445	508	572	636	30
31	0.598	90	120	149	179	179	239	299	359	419	478	538	598	31
32	0.558	84	112	139	167	167	223	279	335	390	446	502	558	32
33	0.515	77	103	129	154	154	206	257	309	360	412	463	515	33
34	0.469	70	94	117	141	141	188	235	282	329	376	423	469	34
35	0.421	63	84	105	126	126	169	211	253	295	337	379	421	35
36	0.370	56	74	93	111	111	148	185	222	259	296	333	370	36
37	0.316	47	63	79	95	95	127	158	190	221	253	285	316	37
38	0.259	39	52	65	78	78	104	130	156	182	207	233	259	38
39	0.199	30	40	50	60	60	80	100	120	140	159	179	199	39
40	0.136	20	27	34	41	41	54	68	82	95	109	123	136	40
41	0.070	10	14	17	21	21	28	35	42	49	56	63	70	41
42	0.000	0	0	0	0	0	0	0	0	0	0	0	0	42

1. $R = 1 - (h'/42t)^3$
2. Based on UBC Chapter 24, Eq. 6-3.

TABLE S-1a Tied Masonry Column Capacity for Columns Constructed with 3/8″ Head Joints[1,2]

For use of Table, see Example 5-T

$$P_a\ (\text{kips}) = \left(0.20 f'_m A_e + 0.65 F_{sc}\, p A_e\right)\left[1 - \left(\frac{h'}{42t}\right)^3\right] \div 1000$$

per UBC Section 2406 Equation 6 - 4

$$P_a = \frac{\left(P_{masonry} + P_{steel}\right)}{1000}\left[1 - \left(\frac{h'}{42t}\right)^3\right]$$

Note for $f_y = 60{,}000$ psi, $F_{sc} = 24{,}000$ psi per UBC Chapter 24, Eq. 6 - 19

Nominal Column Size (inches)	Effective Area A_e (sq. in.)	$P_{masonry}$[2] (kips) = 0.20 f'ₘ Aₑ ÷ 1000 — f'_m (psi)								P_{steel} (kips) = 0.65 F_{sc} p Aₑ ÷ 1000	
		1500 (psi)	2000 (psi)	2500 (psi)	3000 (psi)	3500 (psi)	4000 (psi)	4500 (psi)	5000 (psi)	Min.[3]	Max.[4]
8 x 8[1]	58.1	8.7	11.6	14.5	17.4	20.3	23.3	26.2	29.1	2.3	18.1
8 x 10	73.4	11.0	14.7	18.3	22.0	25.7	29.4	33.0	36.7	2.9	22.9
8 x 12	88.6	13.3	17.7	22.2	26.6	31.0	35.5	39.9	44.3	3.5	27.7
8 x 14	103.9	15.6	20.8	26.0	31.2	36.4	41.6	46.8	51.9	4.1	32.4
8 x 16	119.1	17.9	23.8	29.8	35.7	41.7	47.7	53.6	59.6	4.6	37.2
8 x 18	134.4	20.2	26.9	33.6	40.3	47.0	53.8	60.5	67.2	5.2	41.9
8 x 20	149.6	22.4	29.9	37.4	44.9	52.4	59.9	67.3	74.8	5.8	46.7
8 x 22	164.9	24.7	33.0	41.2	49.5	57.7	66.0	74.2	82.4	6.4	51.4
8 x 24	180.1	27.0	36.0	45.0	54.0	63.0	72.1	81.1	90.1	7.0	56.2
10 x 10[1]	92.6	13.9	18.5	23.2	27.8	32.4	37.1	41.7	46.3	3.6	28.9
10 x 12	111.9	16.8	22.4	28.0	33.6	39.2	44.8	50.4	55.9	4.4	34.9
10 x 14	131.1	19.7	26.2	32.8	39.3	45.9	52.5	59.0	65.6	5.1	40.9
10 x 16	150.4	22.6	30.1	37.6	45.1	52.6	60.2	67.7	75.2	5.9	46.9
10 x 18	169.6	25.4	33.9	42.4	50.9	59.4	67.9	76.3	84.8	6.6	52.9
10 x 20	188.9	28.3	37.8	47.2	56.7	66.1	75.6	85.0	94.4	7.4	58.9
10 x 22	208.1	31.2	41.6	52.0	62.4	72.8	83.3	93.7	104.1	8.1	64.9
10 x 24	227.4	34.1	45.5	56.8	68.2	79.6	91.0	102.3	113.7	8.9	70.9
10 x 26	246.6	37.0	49.3	61.7	74.0	86.3	98.7	111.0	123.3	9.6	77.0
10 x 28	265.9	39.9	53.2	66.5	79.8	93.1	106.4	119.7	132.9	10.4	83.0
10 x 30	285.1	42.8	57.0	71.3	85.5	99.8	114.1	128.3	142.6	11.1	89.0
12 x 12	135.1	40.5	54.1	67.6	81.1	47.3	108.1	60.8	135.1	10.5	84.3
12 x 14	158.4	47.5	63.4	79.2	95.0	55.4	126.7	71.3	158.4	12.4	98.8
12 x 16	181.6	54.5	72.7	90.8	109.0	63.6	145.3	81.7	181.6	14.2	113.3
12 x 18	204.9	61.5	82.0	102.4	122.9	71.7	163.9	92.2	204.9	16.0	127.9
12 x 20	228.1	68.4	91.3	114.1	136.9	79.8	182.5	102.7	228.1	17.8	142.4
12 x 22	251.4	75.4	100.6	125.7	150.8	88.0	201.1	113.1	251.4	19.6	156.9
12 x 24	274.6	82.4	109.9	137.3	164.8	96.1	219.7	123.6	274.6	21.4	171.4
12 x 26	297.9	89.4	119.2	148.9	178.7	104.3	238.3	134.1	297.9	23.2	185.9
12 x 28	321.1	96.3	128.5	160.6	192.7	112.4	256.9	144.5	321.1	25.0	200.4
12 x 30	344.4	103.3	137.8	172.2	206.6	120.5	275.5	155.0	344.4	26.9	214.9
12 x 32	367.6	110.3	147.1	183.8	220.6	128.7	294.1	165.4	367.6	28.7	229.4
12 x 34	390.9	117.3	156.4	195.4	234.5	136.8	312.7	175.9	390.9	30.5	243.9
12 x 36	414.1	124.2	165.7	207.1	248.5	144.9	331.3	186.4	414.1	32.3	258.4
16 x 16	244.1	73.2	97.7	122.1	146.5	85.4	195.3	109.9	244.1	19.0	152.3
16 x 20	306.6	92.0	122.7	153.3	184.0	107.3	245.3	138.0	306.6	23.9	191.3
16 x 24	369.1	110.7	147.7	184.6	221.5	129.2	295.3	166.1	369.1	28.8	230.3
16 x 28	431.6	129.5	172.7	215.8	259.0	151.1	345.3	194.2	431.6	33.7	269.3
16 x 32	494.1	148.2	197.7	247.1	296.5	172.9	395.3	222.4	494.1	38.5	308.3
16 x 36	556.6	167.0	222.7	278.3	334.0	194.8	445.3	250.5	556.6	43.4	347.3
16 x 40	619.1	185.7	247.7	309.6	371.5	216.7	495.3	278.6	619.1	48.3	386.3
16 x 44	681.6	204.5	272.7	340.8	409.0	238.6	545.3	306.7	681.6	53.2	425.3
16 x 48	744.1	223.2	297.7	372.1	446.5	260.4	595.3	334.9	744.1	58.0	464.3

1. Columns in Seismic Zone Nos. 3 and 4 with a minimum dimension less than 12″ must be designed for one half the allowable stresses per UBC Section 2407h4E(ii). These tabular values reflect this code requirement.

2. Table values are based on special inspection being pro-vided. For noninspected masonry, the capacity of the masonry, p_m, must be reduced by one half per UBC Section 2406(c)1.

3. Based on $p_{min} = 0.5\%$.

4. Based on $p_{max} = 4.0\%$.

TABLE S-1b Tied Masonry Column Capacity for Columns Constructed with ³⁄₈″ Head Joints[1,2]

For use of Table, see Example 5-T

WSD

$$P_a \text{ (kips)} = \left(0.20 f'_m A_e + 0.65 F_{sc} p A_e\right)\left[1 - \left(\frac{h'}{42\,t}\right)^3\right] \div 1000$$

per UBC Section 2406 Equation 6 - 4

$$P_a = \frac{\left(P_{masonry} + P_{steel}\right)}{1000}\left[1 - \left(\frac{h'}{42\,t}\right)^3\right]$$

Note for $f_y = 60,000$ psi, $F_{sc} = 24,000$ psi per UBC Chapter 24, Eq. 6 - 19

Nominal Column Size (inches)	Effective Area A_e (sq. in.)	$P_{masonry}$[2] (kips) = $0.20 f'_m A_e \div 1000$								P_{steel} (kips) = $0.65 F_{sc} p A_e \div 1000$	
		f'_m (psi)									
		1500 (psi)	2000 (psi)	2500 (psi)	3000 (psi)	3500 (psi)	4000 (psi)	4500 (psi)	5000 (psi)	Min.[3]	Max.[4]
20 x 20	385.1	115.5	154.1	192.6	231.1	134.8	308.1	173.3	385.1	30.0	240.3
20 x 24	463.6	139.1	185.5	231.8	278.2	162.3	370.9	208.6	463.6	36.2	289.3
20 x 28	542.1	162.6	216.9	271.1	325.3	189.7	433.7	244.0	542.1	42.3	338.3
20 x 32	620.6	186.2	248.3	310.3	372.4	217.2	496.5	279.3	620.6	48.4	387.3
20 x 36	699.1	209.7	279.7	349.6	419.5	244.7	559.3	314.6	699.1	54.5	436.3
20 x 40	777.6	233.3	311.1	388.8	466.6	272.2	622.1	349.9	777.6	60.7	485.2
20 x 44	856.1	256.8	342.5	428.1	513.7	299.6	684.9	385.3	856.1	66.8	534.2
20 x 48	934.6	280.4	373.9	467.3	560.8	327.1	747.7	420.6	934.6	72.9	583.2
20 x 52	1013.1	303.9	405.3	506.6	607.9	354.6	810.5	455.9	1013.1	79.0	632.2
20 x 56	1091.6	327.5	436.7	545.8	655.0	382.1	873.3	491.2	1091.6	85.1	681.2
20 x 60	1170.1	351.0	468.1	585.1	702.1	409.5	936.1	526.6	1170.1	91.3	730.2
24 x 24	558.1	167.4	223.3	279.1	334.9	195.3	446.5	251.2	558.1	43.5	348.3
24 x 28	652.6	195.8	261.1	326.3	391.6	228.4	522.1	293.7	652.6	50.9	407.2
24 x 32	747.1	224.1	298.9	373.6	448.3	261.5	597.7	336.2	747.1	58.3	466.2
24 x 36	841.6	252.5	336.7	420.8	505.0	294.6	673.3	378.7	841.6	65.6	525.2
24 x 40	936.1	280.8	374.5	468.1	561.7	327.6	748.9	421.3	936.1	73.0	584.2
24 x 44	1030.6	309.2	412.3	515.3	618.4	360.7	824.5	463.8	1030.6	80.4	643.1
24 x 48	1125.1	337.5	450.1	562.6	675.1	393.8	900.1	506.3	1125.1	87.8	702.1
24 x 52	1219.6	365.9	487.9	609.8	731.8	426.9	975.7	548.8	1219.6	95.1	761.1
24 x 56	1314.1	394.2	525.7	657.1	788.5	459.9	1051.3	591.4	1314.1	102.5	820.0
24 x 60	1408.6	422.6	563.5	704.3	845.2	493.0	1126.9	633.9	1408.6	109.9	879.0
28 x 28	763.1	228.9	305.3	381.6	457.9	267.1	610.5	343.4	763.1	59.5	476.2
28 x 32	873.6	262.1	349.5	436.8	524.2	305.8	698.9	393.1	873.6	68.1	545.2
28 x 36	984.1	295.2	393.7	492.1	590.5	344.4	787.3	442.9	984.1	76.8	614.1
28 x 40	1094.6	328.4	437.9	547.3	656.8	383.1	875.7	492.6	1094.6	85.4	683.1
28 x 44	1205.1	361.5	482.1	602.6	723.1	421.8	964.1	542.3	1205.1	94.0	752.0
28 x 48	1315.6	394.7	526.3	657.8	789.4	460.5	1052.5	592.0	1315.6	102.6	821.0
28 x 52	1426.1	427.8	570.5	713.1	855.7	499.1	1140.9	641.8	1426.1	111.2	889.9
28 x 56	1536.6	461.0	614.7	768.3	922.0	537.8	1229.3	691.5	1536.6	119.9	958.9
28 x 60	1647.1	494.1	658.9	823.6	988.3	576.5	1317.7	741.2	1647.1	128.5	1027.8
32 x 32	1000.1	300.0	400.1	500.1	600.1	350.0	800.1	450.1	1000.1	78.0	624.1
32 x 36	1126.6	338.0	450.7	563.3	676.0	394.3	901.3	507.0	1126.6	87.9	703.0
32 x 40	1253.1	375.9	501.3	626.6	751.9	438.6	1002.5	563.9	1253.1	97.7	782.0
32 x 44	1379.6	413.9	551.9	689.8	827.8	482.9	1103.7	620.8	1379.6	107.6	860.9
32 x 48	1506.1	451.8	602.5	753.1	903.7	527.1	1204.9	677.8	1506.1	117.5	939.8
32 x 52	1632.6	489.8	653.1	816.3	979.6	571.4	1306.1	734.7	1632.6	127.3	1018.8
32 x 56	1759.1	527.7	703.7	879.6	1055.5	615.7	1407.3	791.6	1759.1	137.2	1097.7
32 x 60	1885.6	565.7	754.3	942.8	1131.4	660.0	1508.5	848.5	1885.6	147.1	1176.6

1. Columns in Seismic Zone Nos. 3 and 4 with a minimum dimension less than 12″ must be designed for one half the allowable stresses per UBC Section 2407h4E(ii). These tabular values reflect this code requirement.

2. Table values are based on special inspection being provided. For noninspected masonry, the capacity of the masonry, p_m, must be reduced by one half per UBC Section 2406(c)1.

3. Based on $p_{min} = 0.5\%$.

4. Based on $p_{max} = 4.0\%$.

TABLE S-2a Tied Masonry Column Capacity for Columns Constructed with ¹/₂″ Head Joints[1,2]

For use of Table, see Example 5-T

$$P_a \text{ (kips)} = \left(0.20 f'_m A_e + 0.65 F_{sc} p A_e\right)\left[1 - \left(\frac{h'}{42t}\right)^3\right] \div 1000$$

per UBC Section 2406 Equation 6 - 4

$$P_a = \frac{\left(P_{masonry} + P_{steel}\right)}{1000}\left[1 - \left(\frac{h'}{42t}\right)^3\right]$$

Note for $f_y = 60,000$ psi, $F_{sc} = 24,000$ psi per UBC Chapter 24, Eq. 6 - 19

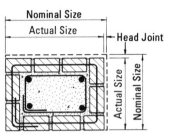

Nominal Column Size (inches)	Effective Area A_e (sq. in.)	$P_{masonry}$[2] (kips) = 0.20 f'_m A_e ÷ 1000 f'_m (psi)								P_{steel} (kips) = 0.65F_{sc} p A_e ÷ 1000	
		1500 (psi)	2000 (psi)	2500 (psi)	3000 (psi)	3500 (psi)	4000 (psi)	4500 (psi)	5000 (psi)	Min.[3]	Max.[4]
8 x 8[1]	56.3	8.4	11.3	14.1	16.9	19.7	22.5	25.3	28.1	2.2	17.6
8 x 10	71.3	10.7	14.3	17.8	21.4	24.9	28.5	32.1	35.6	2.8	22.2
8 x 12	86.3	12.9	17.3	21.6	25.9	30.2	34.5	38.8	43.1	3.4	26.9
8 x 14	101.3	15.2	20.3	25.3	30.4	35.4	40.5	45.6	50.6	3.9	31.6
8 x 16	116.3	17.4	23.3	29.1	34.9	40.7	46.5	52.3	58.1	4.5	36.3
8 x 18	131.3	19.7	26.3	32.8	39.4	45.9	52.5	59.1	65.6	5.1	41.0
8 x 20	146.3	21.9	29.3	36.6	43.9	51.2	58.5	65.8	73.1	5.7	45.6
8 x 22	161.3	24.2	32.3	40.3	48.4	56.4	64.5	72.6	80.6	6.3	50.3
8 x 24	176.3	26.4	35.3	44.1	52.9	61.7	70.5	79.3	88.1	6.9	55.0
10 x 10[1]	90.3	13.5	18.1	22.6	27.1	31.6	36.1	40.6	45.1	3.5	28.2
10 x 12	109.3	16.4	21.9	27.3	32.8	38.2	43.7	49.2	54.6	4.3	34.1
10 x 14	128.3	19.2	25.7	32.1	38.5	44.9	51.3	57.7	64.1	5.0	40.0
10 x 16	147.3	22.1	29.5	36.8	44.2	51.5	58.9	66.3	73.6	5.7	45.9
10 x 18	166.3	24.9	33.3	41.6	49.9	58.2	66.5	74.8	83.1	6.5	51.9
10 x 20	185.3	27.8	37.1	46.3	55.6	64.8	74.1	83.4	92.6	7.2	57.8
10 x 22	204.3	30.6	40.9	51.1	61.3	71.5	81.7	91.9	102.1	8.0	63.7
10 x 24	223.3	33.5	44.7	55.8	67.0	78.1	89.3	100.5	111.6	8.7	69.7
10 x 26	242.3	36.3	48.5	60.6	72.7	84.8	96.9	109.0	121.1	9.4	75.6
10 x 28	261.3	39.2	52.3	65.3	78.4	91.4	104.5	117.6	130.6	10.2	81.5
10 x 30	280.3	42.0	56.1	70.1	84.1	98.1	112.1	126.1	140.1	10.9	87.4
12 x 12	132.3	39.7	52.9	66.1	79.4	92.6	105.8	59.5	66.1	10.3	82.5
12 x 14	155.3	46.6	62.1	77.6	93.2	108.7	124.2	69.9	77.6	12.1	96.9
12 x 16	178.3	53.5	71.3	89.1	107.0	124.8	142.6	80.2	89.1	13.9	111.2
12 x 18	201.3	60.4	80.5	100.6	120.8	140.9	161.0	90.6	100.6	15.7	125.6
12 x 20	224.3	67.3	89.7	112.1	134.6	157.0	179.4	100.9	112.1	17.5	139.9
12 x 22	247.3	74.2	98.9	123.6	148.4	173.1	197.8	111.3	123.6	19.3	154.3
12 x 24	270.3	81.1	108.1	135.1	162.2	189.2	216.2	121.6	135.1	21.1	168.6
12 x 26	293.3	88.0	117.3	146.6	176.0	205.3	234.6	132.0	146.6	22.9	183.0
12 x 28	316.3	94.9	126.5	158.1	189.8	221.4	253.0	142.3	158.1	24.7	197.3
12 x 30	339.3	101.8	135.7	169.6	203.6	237.5	271.4	152.7	169.6	26.5	211.7
12 x 32	362.3	108.7	144.9	181.1	217.4	253.6	289.8	163.0	181.1	28.3	226.0
12 x 34	385.3	115.6	154.1	192.6	231.2	269.7	308.2	173.4	192.6	30.0	240.4
12 x 36	408.3	122.5	163.3	204.1	245.0	285.8	326.6	183.7	204.1	31.8	254.7
16 x 16	240.3	72.1	96.1	120.1	144.2	168.2	192.2	108.1	120.1	18.7	149.9
16 x 20	302.3	90.7	120.9	151.1	181.4	211.6	241.8	136.0	151.1	23.6	188.6
16 x 24	364.3	109.3	145.7	182.1	218.6	255.0	291.4	163.9	182.1	28.4	227.3
16 x 28	426.3	127.9	170.5	213.1	255.8	298.4	341.0	191.8	213.1	33.2	266.0
16 x 32	488.3	146.5	195.3	244.1	293.0	341.8	390.6	219.7	244.1	38.1	304.7
16 x 36	550.3	165.1	220.1	275.1	330.2	385.2	440.2	247.6	275.1	42.9	343.4
16 x 40	612.3	183.7	244.9	306.1	367.4	428.6	489.8	275.5	306.1	47.8	382.0
16 x 44	674.3	202.3	269.7	337.1	404.6	472.0	539.4	303.4	337.1	52.6	420.7
16 x 48	736.3	220.9	294.5	368.1	441.8	515.4	589.0	331.3	368.1	57.4	459.4

1. Columns in Seismic Zone Nos. 3 and 4 with a minimum dimension less than 12″ must be designed for one half the allowable stresses per UBC Section 2407h4E(ii). These tabular values reflect this code requirement.

2. Table values are based on special inspection being pro-vided. For noninspected masonry, the capacity of the masonry, p_m, must be reduced by one half per UBC Section 2406(c)1.

3. Based on $p_{min} = 0.5\%$.

4. Based on $p_{max} = 4.0\%$.

TABLE S-2b Tied Masonry Column Capacity for Columns Constructed with $\frac{1}{2}''$ Head Joints[1,2]

For use of Table, see Example 5-T

$$P_a \text{ (kips)} = \left(0.20 f'_m A_e + 0.65 F_{sc}\, p A_e\right)\left[1 - \left(\frac{h'}{42t}\right)^3\right] \div 1000$$

per UBC Section 2406 Equation 6 - 4

$$P_a = \frac{\left(P_{masonry} + P_{steel}\right)}{1000}\left[1 - \left(\frac{h'}{42t}\right)^3\right]$$

Note for $f_y = 60{,}000$ psi, $F_{sc} = 24{,}000$ psi per UBC Chapter 24, Eq. 6 - 19

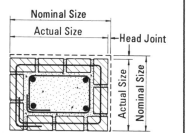

Nominal Column Size (inches)	Effective Area A_e (sq. in.)	$P_{masonry}$[2] (kips) = $0.20 f'_m A_e \div 1000$								P_{steel} (kips) = $0.65 F_{sc}\, p A_e \div 1000$	
		f'_m (psi)									
		1500 (psi)	2000 (psi)	2500 (psi)	3000 (psi)	3500 (psi)	4000 (psi)	4500 (psi)	5000 (psi)	Min.[3]	Max.[4]
20 x 20	380.3	114.1	152.1	190.1	228.2	266.2	304.2	171.1	190.1	29.7	237.3
20 x 24	458.3	137.5	183.3	229.1	275.0	320.8	366.6	206.2	229.1	35.7	285.9
20 x 28	536.3	160.9	214.5	268.1	321.8	375.4	429.0	241.3	268.1	41.8	334.6
20 x 32	614.3	184.3	245.7	307.1	368.6	430.0	491.4	276.4	307.1	47.9	383.3
20 x 36	692.3	207.7	276.9	346.1	415.4	484.6	553.8	311.5	346.1	54.0	432.0
20 x 40	770.3	231.1	308.1	385.1	462.2	539.2	616.2	346.6	385.1	60.1	480.6
20 x 44	848.3	254.5	339.3	424.1	509.0	593.8	678.6	381.7	424.1	66.2	529.3
20 x 48	926.3	277.9	370.5	463.1	555.8	648.4	741.0	416.8	463.1	72.2	578.0
20 x 52	1004.3	301.3	401.7	502.1	602.6	703.0	803.4	451.9	502.1	78.3	626.7
20 x 56	1082.3	324.7	432.9	541.1	649.4	757.6	865.8	487.0	541.1	84.4	675.3
20 x 60	1160.3	348.1	464.1	580.1	696.2	812.2	928.2	522.1	580.1	90.5	724.0
24 x 24	552.3	165.7	220.9	276.1	331.4	386.6	441.8	248.5	276.1	43.1	344.6
24 x 28	646.3	193.9	258.5	323.1	387.8	452.4	517.0	290.8	323.1	50.4	403.3
24 x 32	740.3	222.1	296.1	370.1	444.2	518.2	592.2	333.1	370.1	57.7	461.9
24 x 36	834.3	250.3	333.7	417.1	500.6	584.0	667.4	375.4	417.1	65.1	520.6
24 x 40	928.3	278.5	371.3	464.1	557.0	649.8	742.6	417.7	464.1	72.4	579.2
24 x 44	1022.3	306.7	408.9	511.1	613.4	715.6	817.8	460.0	511.1	79.7	637.9
24 x 48	1116.3	334.9	446.5	558.1	669.8	781.4	893.0	502.3	558.1	87.1	696.5
24 x 52	1210.3	363.1	484.1	605.1	726.2	847.2	968.2	544.6	605.1	94.4	755.2
24 x 56	1304.3	391.3	521.7	652.1	782.6	913.0	1043.4	586.9	652.1	101.7	813.9
24 x 60	1398.3	419.5	559.3	699.1	839.0	978.8	1118.6	629.2	699.1	109.1	872.5
28 x 28	756.3	226.9	302.5	378.1	453.8	529.4	605.0	340.3	378.1	59.0	471.9
28 x 32	866.3	259.9	346.5	433.1	519.8	606.4	693.0	389.8	433.1	67.6	540.5
28 x 36	976.3	292.9	390.5	488.1	585.8	683.4	781.0	439.3	488.1	76.1	609.2
28 x 40	1086.3	325.9	434.5	543.1	651.8	760.4	869.0	488.8	543.1	84.7	677.8
28 x 44	1196.3	358.9	478.5	598.1	717.8	837.4	957.0	538.3	598.1	93.3	746.5
28 x 48	1306.3	391.9	522.5	653.1	783.8	914.4	1045.0	587.8	653.1	101.9	815.1
28 x 52	1416.3	424.9	566.5	708.1	849.8	991.4	1133.0	637.3	708.1	110.5	883.7
28 x 56	1526.3	457.9	610.5	763.1	915.8	1068.4	1221.0	686.8	763.1	119.0	952.4
28 x 60	1636.3	490.9	654.5	818.1	981.8	1145.4	1309.0	736.3	818.1	127.6	1021.0
32 x 32	992.3	297.7	396.9	496.1	595.4	694.6	793.8	446.5	496.1	77.4	619.2
32 x 36	1118.3	335.5	447.3	559.1	671.0	782.8	894.6	503.2	559.1	87.2	697.8
32 x 40	1244.3	373.3	497.7	622.1	746.6	871.0	995.4	559.9	622.1	97.1	776.4
32 x 44	1370.3	411.1	548.1	685.1	822.2	959.2	1096.2	616.6	685.1	106.9	855.0
32 x 48	1496.3	448.9	598.5	748.1	897.8	1047.4	1197.0	673.3	748.1	116.7	933.7
32 x 52	1622.3	486.7	648.9	811.1	973.4	1135.6	1297.8	730.0	811.1	126.5	1012.3
32 x 56	1748.3	524.5	699.3	874.1	1049.0	1223.8	1398.6	786.7	874.1	136.4	1090.9
32 x 60	1874.3	562.3	749.7	937.1	1124.6	1312.0	1499.4	843.4	937.1	146.2	1169.5

1. Columns in Seismic Zone Nos. 3 and 4 with a minimum dimension less than 12″ must be designed for one half the allowable stresses per UBC Section 2407h4E(ii). These tabular values reflect this code requirement.

2. Table values are based on special inspection being pro-vided. For noninspected masonry, the capacity of the masonry, p_m, must be reduced by one half per UBC Section 2406(c)1.

3. Based on $p_{min} = 0.5\%$.

4. Based on $p_{max} = 4.0\%$.

TABLE S-3a Tied Masonry Column Capacity for Columns Constructed So That the Nominal Column Dimension Equals the Actual Column Dimension[1,2]

For use of Table, see Example 5-T

$$P_a \text{ (kips)} = \left(0.20 f'_m A_e + 0.65 F_{sc}\, p A_e\right)\left[1 - \left(\frac{h'}{42\,t}\right)^3\right] \div 1000$$

per UBC Section 2406 Equation 6 - 4

$$P_a = \frac{\left(P_{masonry} + P_{steel}\right)}{1000}\left[1 - \left(\frac{h'}{42\,t}\right)^3\right]$$

Note for $f_y = 60{,}000$ psi, $F_{sc} = 24{,}000$ psi per UBC Chapter 24, Eq. 6 - 19

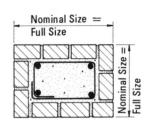

Nominal Size = Full Size

Nominal Size = Full Size

Nominal Column Size (inches)	Effective Area A_e (sq. in.)	$P_{masonry}$[2] (kips) = $0.20 f'_m A_e \div 1000$								P_{steel} (kips) = $0.65 F_{sc}\, p A_e \div 1000$	
		f'_m (psi)									
		1500 (psi)	2000 (psi)	2500 (psi)	3000 (psi)	3500 (psi)	4000 (psi)	4500 (psi)	5000 (psi)	Min.[3]	Max.[4]
8 x 8[1]	64.0	9.6	12.8	16.0	19.2	22.4	25.6	28.8	32.0	2.5	20.0
8 x 10	80.0	12.0	16.0	20.0	24.0	28.0	32.0	36.0	40.0	3.1	25.0
8 x 12	96.0	14.4	19.2	24.0	28.8	33.6	38.4	43.2	48.0	3.7	30.0
8 x 14	112.0	16.8	22.4	28.0	33.6	39.2	44.8	50.4	56.0	4.4	34.9
8 x 16	128.0	19.2	25.6	32.0	38.4	44.8	51.2	57.6	64.0	5.0	39.9
8 x 18	144.0	21.6	28.8	36.0	43.2	50.4	57.6	64.8	72.0	5.6	44.9
8 x 20	160.0	24.0	32.0	40.0	48.0	56.0	64.0	72.0	80.0	6.2	49.9
8 x 22	176.0	26.4	35.2	44.0	52.8	61.6	70.4	79.2	88.0	6.9	54.9
8 x 24	192.0	28.8	38.4	48.0	57.6	67.2	76.8	86.4	96.0	7.5	59.9
10 x 10[1]	100.0	15.0	20.0	25.0	30.0	35.0	40.0	45.0	50.0	3.9	31.2
10 x 12	120.0	18.0	24.0	30.0	36.0	42.0	48.0	54.0	60.0	4.7	37.4
10 x 14	140.0	21.0	28.0	35.0	42.0	49.0	56.0	63.0	70.0	5.5	43.7
10 x 16	160.0	24.0	32.0	40.0	48.0	56.0	64.0	72.0	80.0	6.2	49.9
10 x 18	180.0	27.0	36.0	45.0	54.0	63.0	72.0	81.0	90.0	7.0	56.2
10 x 20	200.0	30.0	40.0	50.0	60.0	70.0	80.0	90.0	100.0	7.8	62.4
10 x 22	220.0	33.0	44.0	55.0	66.0	77.0	88.0	99.0	110.0	8.6	68.6
10 x 24	240.0	36.0	48.0	60.0	72.0	84.0	96.0	108.0	120.0	9.4	74.9
10 x 26	260.0	39.0	52.0	65.0	78.0	91.0	104.0	117.0	130.0	10.1	81.1
10 x 28	280.0	42.0	56.0	70.0	84.0	98.0	112.0	126.0	140.0	10.9	87.4
10 x 30	300.0	45.0	60.0	75.0	90.0	105.0	120.0	135.0	150.0	11.7	93.6
12 x 12	144.0	43.2	57.6	72.0	86.4	100.8	115.2	64.8	72.0	11.2	89.9
12 x 14	168.0	50.4	67.2	84.0	100.8	117.6	134.4	75.6	84.0	13.1	104.8
12 x 16	192.0	57.6	76.8	96.0	115.2	134.4	153.6	86.4	96.0	15.0	119.8
12 x 18	216.0	64.8	86.4	108.0	129.6	151.2	172.8	97.2	108.0	16.8	134.8
12 x 20	240.0	72.0	96.0	120.0	144.0	168.0	192.0	108.0	120.0	18.7	149.8
12 x 22	264.0	79.2	105.6	132.0	158.4	184.8	211.2	118.8	132.0	20.6	164.7
12 x 24	288.0	86.4	115.2	144.0	172.8	201.6	230.4	129.6	144.0	22.5	179.7
12 x 26	312.0	93.6	124.8	156.0	187.2	218.4	249.6	140.4	156.0	24.3	194.7
12 x 28	336.0	100.8	134.4	168.0	201.6	235.2	268.8	151.2	168.0	26.2	209.7
12 x 30	360.0	108.0	144.0	180.0	216.0	252.0	288.0	162.0	180.0	28.1	224.6
12 x 32	384.0	115.2	153.6	192.0	230.4	268.8	307.2	172.8	192.0	30.0	239.6
12 x 34	408.0	122.4	163.2	204.0	244.8	285.6	326.4	183.6	204.0	31.8	254.6
12 x 36	432.0	129.6	172.8	216.0	259.2	302.4	345.6	194.4	216.0	33.7	269.6
16 x 16	256.0	76.8	102.4	128.0	153.6	179.2	204.8	115.2	128.0	20.0	159.7
16 x 20	320.0	96.0	128.0	160.0	192.0	224.0	256.0	144.0	160.0	25.0	199.7
16 x 24	384.0	115.2	153.6	192.0	230.4	268.8	307.2	172.8	192.0	30.0	239.6
16 x 28	448.0	134.4	179.2	224.0	268.8	313.6	358.4	201.6	224.0	34.9	279.6
16 x 32	512.0	153.6	204.8	256.0	307.2	358.4	409.6	230.4	256.0	39.9	319.5
16 x 36	576.0	172.8	230.4	288.0	345.6	403.2	460.8	259.2	288.0	44.9	359.4
16 x 40	640.0	192.0	256.0	320.0	384.0	448.0	512.0	288.0	320.0	49.9	399.4
16 x 44	704.0	211.2	281.6	352.0	422.4	492.8	563.2	316.8	352.0	54.9	439.3
16 x 48	768.0	230.4	307.2	384.0	460.8	537.6	614.4	345.6	384.0	59.9	479.2

1. Columns in Seismic Zone Nos. 3 and 4 with a minimum dimension less than 12″ must be designed for one half the allowable stresses per UBC Section 2407h4E(ii). These tabular values reflect this code requirement.

2. Table values are based on special inspection being pro-
vided. For noninspected masonry, the capacity of the masonry, p_m, must be reduced by one half per UBC Section 2406(c)1.

3. Based on $p_{min} = 0.5\%$.

4. Based on $p_{max} = 4.0\%$.

TABLE S-3b   Tied Masonry Column Capacity for Columns Constructed So That the Nominal Column Dimension Equals the Actual Column Dimension[1,2]

For use of Table, see Example 5-T

$$P_a \text{ (kips)} = \left(0.20 f'_m A_e + 0.65 F_{sc}\, p A_e\right)\left[1 - \left(\frac{h'}{42\,t}\right)^3\right] \div 1000$$

per UBC Section 2406 Equation 6 - 4

$$P_a = \frac{\left(P_{masonry} + P_{steel}\right)}{1000}\left[1 - \left(\frac{h'}{42\,t}\right)^3\right]$$

Note for $f_y = 60{,}000$ psi, $F_{sc} = 24{,}000$ psi per UBC Chapter 24, Eq. 6 - 19

Nominal Column Size (inches)	Effective Area A_e (sq. in.)	$P_{masonry}^{(2)}$ (kips) $= 0.20 f'_m A_e \div 1000$ — f'_m (psi)								P_{steel} (kips) $= 0.65 F_{sc}\, p A_e \div 1000$	
		1500 (psi)	2000 (psi)	2500 (psi)	3000 (psi)	3500 (psi)	4000 (psi)	4500 (psi)	5000 (psi)	Min.[3]	Max.[4]
20 x 20	400.0	120.0	160.0	200.0	240.0	280.0	320.0	180.0	200.0	31.2	249.6
20 x 24	480.0	144.0	192.0	240.0	288.0	336.0	384.0	216.0	240.0	37.4	299.5
20 x 28	560.0	168.0	224.0	280.0	336.0	392.0	448.0	252.0	280.0	43.7	349.4
20 x 32	640.0	192.0	256.0	320.0	384.0	448.0	512.0	288.0	320.0	49.9	399.4
20 x 36	720.0	216.0	288.0	360.0	432.0	504.0	576.0	324.0	360.0	56.2	449.3
20 x 40	800.0	240.0	320.0	400.0	480.0	560.0	640.0	360.0	400.0	62.4	499.2
20 x 44	880.0	264.0	352.0	440.0	528.0	616.0	704.0	396.0	440.0	68.6	549.1
20 x 48	960.0	288.0	384.0	480.0	576.0	672.0	768.0	432.0	480.0	74.9	599.0
20 x 52	1040.0	312.0	416.0	520.0	624.0	728.0	832.0	468.0	520.0	81.1	649.0
20 x 56	1120.0	336.0	448.0	560.0	672.0	784.0	896.0	504.0	560.0	87.4	698.9
20 x 60	1200.0	360.0	480.0	600.0	720.0	840.0	960.0	540.0	600.0	93.6	748.8
24 x 24	576.0	172.8	230.4	288.0	345.6	403.2	460.8	259.2	288.0	44.9	359.4
24 x 28	672.0	201.6	268.8	336.0	403.2	470.4	537.6	302.4	336.0	52.4	419.3
24 x 32	768.0	230.4	307.2	384.0	460.8	537.6	614.4	345.6	384.0	59.9	479.2
24 x 36	864.0	259.2	345.6	432.0	518.4	604.8	691.2	388.8	432.0	67.4	539.1
24 x 40	960.0	288.0	384.0	480.0	576.0	672.0	768.0	432.0	480.0	74.9	599.0
24 x 44	1056.0	316.8	422.4	528.0	633.6	739.2	844.8	475.2	528.0	82.4	658.9
24 x 48	1152.0	345.6	460.8	576.0	691.2	806.4	921.6	518.4	576.0	89.9	718.8
24 x 52	1248.0	374.4	499.2	624.0	748.8	873.6	998.4	561.6	624.0	97.3	778.8
24 x 56	1344.0	403.2	537.6	672.0	806.4	940.8	1075.2	604.8	672.0	104.8	838.7
24 x 60	1440.0	432.0	576.0	720.0	864.0	1008.0	1152.0	648.0	720.0	112.3	898.6
28 x 28	784.0	235.2	313.6	392.0	470.4	548.8	627.2	352.8	392.0	61.2	489.2
28 x 32	896.0	268.8	358.4	448.0	537.6	627.2	716.8	403.2	448.0	69.9	559.1
28 x 36	1008.0	302.4	403.2	504.0	604.8	705.6	806.4	453.6	504.0	78.6	629.0
28 x 40	1120.0	336.0	448.0	560.0	672.0	784.0	896.0	504.0	560.0	87.4	698.9
28 x 44	1232.0	369.6	492.8	616.0	739.2	862.4	985.6	554.4	616.0	96.1	768.8
28 x 48	1344.0	403.2	537.6	672.0	806.4	940.8	1075.2	604.8	672.0	104.8	838.7
28 x 52	1456.0	436.8	582.4	728.0	873.6	1019.2	1164.8	655.2	728.0	113.6	908.5
28 x 56	1568.0	470.4	627.2	784.0	940.8	1097.6	1254.4	705.6	784.0	122.3	978.4
28 x 60	1680.0	504.0	672.0	840.0	1008.0	1176.0	1344.0	756.0	840.0	131.0	1048.3
32 x 32	1024.0	307.2	409.6	512.0	614.4	716.8	819.2	460.8	512.0	79.9	639.0
32 x 36	1152.0	345.6	460.8	576.0	691.2	806.4	921.6	518.4	576.0	89.9	718.8
32 x 40	1280.0	384.0	512.0	640.0	768.0	896.0	1024.0	576.0	640.0	99.8	798.7
32 x 44	1408.0	422.4	563.2	704.0	844.8	985.6	1126.4	633.6	704.0	109.8	878.6
32 x 48	1536.0	460.8	614.4	768.0	921.6	1075.2	1228.8	691.2	768.0	119.8	958.5
32 x 52	1664.0	499.2	665.6	832.0	998.4	1164.8	1331.2	748.8	832.0	129.8	1038.3
32 x 56	1792.0	537.6	716.8	896.0	1075.2	1254.4	1433.6	806.4	896.0	139.8	1118.2
32 x 60	1920.0	576.0	768.0	960.0	1152.0	1344.0	1536.0	864.0	960.0	149.8	1198.1

1. Columns in Seismic Zone Nos. 3 and 4 with a minimum dimension less than 12″ must be designed for one half the allowable stresses per UBC Section 2407h4E(ii). These tabular values reflect this code requirement.

2. Table values are based on special inspection being pro-vided. For noninspected masonry, the capacity of the masonry, p_m, must be reduced by one half per UBC Section 2406(c)1.

3. Based on $p_{min} = 0.5\%$.

4. Based on $p_{max} = 4.0\%$.

TABLE S-4   Capacity of Reinforcing Steel in Tied Masonry Columns (kips)

Capacity of Reinforcing Steel, P_s (kips) $= 0.65 F_{sc} A_s / 1000$ (From UBC Chapter 24, Eq. 6-4)

where: $F_{sc} = 0.4 F_y$ or for grade 60 reinforcement, $F_{sc} = 24,000$ psi
$A_s = p_g A_g$

Excerpts from the 1991 Uniform Building Code

UBC Section 2409(b)5.

5. Reinforcement for columns. A. Vertical Reinforcement. The area of vertical reinforcement shall not be less than $0.005 \, A_e$ and not more than $0.04 \, A_e$. At least four No. 3 bars shall be provided.

UBC Section 2409(e)1.

(e) Reinforcing Requirements and Details. 1. Maximum reinforcing size. The maximum size of reinforcing shall be No. 11.

Bar Size	Number of Bars						
	4	6	8	10	12	14	16
#3	6.9	10.3	13.7	17.2	20.6	24.0	27.5
#4	12.5	18.7	25.0	31.2	37.4	43.7	49.9
#5	19.3	29.0	38.7	48.4	58.0	67.7	77.4
#6	27.5	41.2	54.9	68.6	82.4	96.1	109.8
#7	37.4	56.2	74.9	93.6	112.3	131.0	149.8
#8	49.3	73.9	98.6	123.2	147.9	172.5	197.2
#9	62.4	93.6	124.8	156.0	187.2	218.4	249.6
#10	79.2	118.9	158.5	198.1	237.7	277.4	317.0
#11	97.3	146.0	194.7	243.4	292.0	340.7	389.4

TABLE S-5   Maximum Spacing of Column Ties (inches)[1,2]

Tie Bar Size	Longitudinal Bar Size								
	#3	#4	#5	#6	#7	#8	#9	#10	#11
1/4″	6	8	10	12	12	—	—	—	—
#3	6	8	10	12	14	16	18	18	18
#4	—	—	10	12	14	16	18	18	18
#5	—	—	—	12	14	16	18	18	18

1. This table shows the maximum spacing of ties permitted by UBC Sec. 2409(B)5B based on the tie diameter, the longitudinal bar diameter and a maximum spacing of 18″. The spacing determined from this table may not exceed the least column dimension.

2. Per UBC Sec. 2409(b)5B, 1/4″ diameter ties may be used for only #7 and smaller longitudinal bars.

TABLE T-1a Coefficients for Deflection and Rigidity of Walls or Piers for Distribution of Horizontal Forces

For use of Table, see Example 4-F

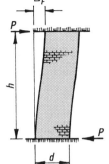

Fixed Wall or Pier

Δ_F = Deflection of wall or pier fixed top and bottom.

$$\Delta_F = \frac{P}{E_m t}\left[\left(\frac{h}{d}\right)^3 + 3\left(\frac{h}{d}\right)\right]$$

$$\Delta_F = 0.1\left(\frac{h}{d}\right)^3 + 0.3\left(\frac{h}{d}\right)$$

$R_F = \dfrac{1}{\Delta_F}$ Rigidity of fixed wall or pier.

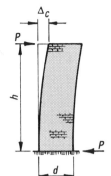

Cantilever Wall or Pier

Δ_C = Deflection of cantilever wall or pier.

$$\Delta_C = \frac{P}{E_m t}\left[4\left(\frac{h}{d}\right)^3 + 3\left(\frac{h}{d}\right)\right]$$

$$\Delta_C = 0.4\left(\frac{h}{d}\right)^3 + 0.3\left(\frac{h}{d}\right)$$

$R_C = \dfrac{1}{\Delta_C}$ Rigidity of cantilever wall or pier.

$$P = 100{,}000 \text{ pounds; } t = 1''; \ E_m = 1{,}000{,}000 \text{ psi}$$

h/d	Δ_F	Δ_C	R_F	R_C	h/d	Δ_F	Δ_C	R_F	R_C	h/d	Δ_F	Δ_C	R_F	R_C
0.10	0.030	0.030	33.223	32.895	0.45	0.144	0.171	6.939	5.833	0.80	0.291	0.445	3.434	2.248
0.11	0.033	0.034	30.181	29.822	0.46	0.148	0.177	6.769	5.652	0.81	0.296	0.456	3.377	2.195
0.12	0.036	0.037	27.645	27.254	0.47	0.151	0.183	6.606	5.479	0.82	0.301	0.467	3.321	2.143
0.13	0.039	0.040	25.497	25.076	0.48	0.155	0.188	6.449	5.312	0.83	0.306	0.478	3.266	2.093
0.14	0.042	0.043	23.655	23.203	0.49	0.159	0.194	6.299	5.153	0.84	0.311	0.489	3.213	2.045
0.15	0.045	0.046	22.057	21.575	0.50	0.163	0.200	6.154	5.000	0.85	0.316	0.501	3.160	1.997
0.16	0.048	0.050	20.657	20.146	0.51	0.166	0.206	6.014	4.853	0.86	0.322	0.512	3.109	1.952
0.17	0.051	0.053	19.421	18.880	0.52	0.170	0.212	5.880	4.712	0.87	0.327	0.524	3.060	1.907
0.18	0.055	0.056	18.321	17.752	0.53	0.174	0.219	5.751	4.576	0.88	0.332	0.537	3.011	1.864
0.19	0.058	0.060	17.335	16.738	0.54	0.178	0.225	5.626	4.445	0.89	0.337	0.549	2.963	1.822
0.20	0.061	0.063	16.447	15.823	0.55	0.182	0.232	5.505	4.319	0.90	0.343	0.562	2.916	1.781
0.21	0.064	0.067	15.643	14.992	0.56	0.186	0.238	5.389	4.197	0.91	0.348	0.574	2.871	1.741
0.22	0.067	0.070	14.911	14.233	0.57	0.190	0.245	5.277	4.080	0.92	0.354	0.587	2.826	1.702
0.23	0.070	0.074	14.242	13.538	0.58	0.194	0.252	5.168	3.968	0.93	0.359	0.601	2.782	1.665
0.24	0.073	0.078	13.627	12.898	0.59	0.198	0.259	5.062	3.859	0.94	0.365	0.614	2.739	1.628
0.25	0.077	0.081	13.061	12.308	0.60	0.202	0.266	4.960	3.754	0.95	0.371	0.628	2.697	1.592
0.26	0.080	0.085	12.538	11.760	0.61	0.206	0.274	4.861	3.652	0.96	0.376	0.642	2.656	1.558
0.27	0.083	0.089	12.053	11.252	0.62	0.210	0.281	4.766	3.555	0.97	0.382	0.656	2.616	1.524
0.28	0.086	0.093	11.602	10.778	0.63	0.214	0.289	4.673	3.460	0.98	0.388	0.670	2.577	1.491
0.29	0.089	0.097	11.181	10.335	0.64	0.218	0.297	4.583	3.369	0.99	0.394	0.685	2.538	1.460
0.30	0.093	0.101	10.787	9.921	0.65	0.222	0.305	4.495	3.280	1.00	0.400	0.700	2.500	1.429
0.31	0.096	0.105	10.419	9.531	0.66	0.227	0.313	4.410	3.195	1.01	0.406	0.715	2.463	1.398
0.32	0.099	0.109	10.073	9.165	0.67	0.231	0.321	4.328	3.112	1.02	0.412	0.730	2.426	1.369
0.33	0.103	0.113	9.747	8.820	0.68	0.235	0.330	4.247	3.032	1.03	0.418	0.746	2.391	1.340
0.34	0.106	0.118	9.440	8.495	0.69	0.240	0.338	4.169	2.955	1.04	0.424	0.762	2.356	1.312
0.35	0.109	0.122	9.150	8.187	0.70	0.244	0.347	4.093	2.880	1.05	0.431	0.778	2.321	1.285
0.36	0.113	0.127	8.876	7.895	0.71	0.249	0.356	4.019	2.808	1.06	0.437	0.794	2.288	1.259
0.37	0.116	0.131	8.616	7.618	0.72	0.253	0.365	3.948	2.737	1.07	0.444	0.811	2.255	1.233
0.38	0.119	0.136	8.369	7.356	0.73	0.258	0.375	3.877	2.669	1.08	0.450	0.828	2.222	1.208
0.39	0.123	0.141	8.135	7.106	0.74	0.263	0.384	3.809	2.604	1.09	0.457	0.845	2.191	1.183
0.40	0.126	0.146	7.911	6.868	0.75	0.267	0.394	3.743	2.540	1.10	0.463	0.862	2.159	1.160
0.41	0.130	0.151	7.699	6.641	0.76	0.272	0.404	3.678	2.478	1.11	0.470	0.880	2.129	1.136
0.42	0.133	0.156	7.496	6.425	0.77	0.277	0.414	3.615	2.418	1.12	0.476	0.898	2.099	1.114
0.43	0.137	0.161	7.302	6.219	0.78	0.281	0.424	3.553	2.359	1.13	0.483	0.916	2.069	1.092
0.44	0.141	0.166	7.117	6.021	0.79	0.286	0.434	3.493	2.303	1.14	0.490	0.935	2.040	1.070

WSD

TABLE T-1b Coefficients for Deflection and Rigidity of Walls or Piers for Distribution of Horizontal Forces

For use of Table, see Example 4-F

Fixed Wall or Pier

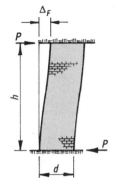

Δ_F = Deflection of wall or pier fixed top and bottom.

$$\Delta_F = \frac{P}{E_m t}\left[\left(\frac{h}{d}\right)^3 + 3\left(\frac{h}{d}\right)\right]$$

$$\Delta_F = 0.1\left(\frac{h}{d}\right)^3 + 0.3\left(\frac{h}{d}\right)$$

$$R_F = \frac{1}{\Delta_F} \quad \text{Rigidity of fixed wall or pier.}$$

Cantilever Wall or Pier

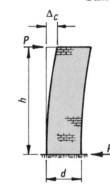

Δ_C = Deflection of cantilever wall or pier.

$$\Delta_C = \frac{P}{E_m t}\left[4\left(\frac{h}{d}\right)^3 + 3\left(\frac{h}{d}\right)\right]$$

$$\Delta_C = 0.4\left(\frac{h}{d}\right)^3 + 0.3\left(\frac{h}{d}\right)$$

$$R_C = \frac{1}{\Delta_C} \quad \text{Rigidity of cantilever wall or pier.}$$

$$P = 100,000 \text{ pounds}; \ t = 1''; \ E_m = 1,000,000 \text{ psi}$$

h/d	Δ_F	Δ_C	R_F	R_C	h/d	Δ_F	Δ_C	R_F	R_C	h/d	Δ_F	Δ_C	R_F	R_C
1.15	0.497	0.953	2.012	1.049	1.50	0.788	1.800	1.270	0.556	1.85	1.188	3.088	0.842	0.324
1.16	0.504	0.972	1.984	1.028	1.51	0.797	1.830	1.254	0.546	1.86	1.201	3.132	0.832	0.319
1.17	0.511	0.992	1.956	1.008	1.52	0.807	1.861	1.239	0.537	1.87	1.215	3.177	0.823	0.315
1.18	0.518	1.011	1.929	0.989	1.53	0.817	1.892	1.224	0.529	1.88	1.228	3.222	0.814	0.310
1.19	0.526	1.031	1.903	0.970	1.54	0.827	1.923	1.209	0.520	1.89	1.242	3.268	0.805	0.306
1.20	0.533	1.051	1.877	0.951	1.55	0.837	1.955	1.194	0.512	1.90	1.256	3.314	0.796	0.302
1.21	0.540	1.072	1.851	0.933	1.56	0.848	1.987	1.180	0.503	1.91	1.270	3.360	0.788	0.298
1.22	0.548	1.092	1.826	0.915	1.57	0.858	2.019	1.166	0.495	1.92	1.284	3.407	0.779	0.293
1.23	0.555	1.113	1.802	0.898	1.58	0.868	2.052	1.152	0.487	1.93	1.298	3.455	0.770	0.289
1.24	0.563	1.135	1.777	0.881	1.59	0.879	2.085	1.138	0.480	1.94	1.312	3.503	0.762	0.286
1.25	0.570	1.156	1.753	0.865	1.60	0.890	2.118	1.124	0.472	1.95	1.326	3.551	0.754	0.282
1.26	0.578	1.178	1.730	0.849	1.61	0.900	2.152	1.111	0.465	1.96	1.341	3.600	0.746	0.278
1.27	0.586	1.200	1.707	0.833	1.62	0.911	2.187	1.098	0.457	1.97	1.356	3.649	0.738	0.274
1.28	0.594	1.223	1.684	0.818	1.63	0.922	2.221	1.085	0.450	1.98	1.370	3.699	0.730	0.270
1.29	0.602	1.246	1.662	0.803	1.64	0.933	2.256	1.072	0.443	1.99	1.385	3.749	0.722	0.267
1.30	0.610	1.269	1.640	0.788	1.65	0.944	2.292	1.059	0.436	2.00	1.400	3.800	0.714	0.263
1.31	0.618	1.292	1.619	0.774	1.66	0.955	2.328	1.047	0.430	2.01	1.415	3.851	0.707	0.260
1.32	0.626	1.316	1.597	0.760	1.67	0.967	2.364	1.034	0.423	2.02	1.430	3.903	0.699	0.256
1.33	0.634	1.340	1.577	0.746	1.68	0.978	2.401	1.022	0.417	2.03	1.446	3.955	0.692	0.253
1.34	0.643	1.364	1.556	0.733	1.69	0.990	2.438	1.010	0.410	2.04	1.461	4.008	0.684	0.250
1.35	0.651	1.389	1.536	0.720	1.70	1.001	2.475	0.999	0.404	2.05	1.477	4.061	0.677	0.246
1.36	0.660	1.414	1.516	0.707	1.71	1.013	2.513	0.987	0.398	2.06	1.492	4.115	0.670	0.243
1.37	0.668	1.440	1.497	0.695	1.72	1.025	2.551	0.976	0.392	2.07	1.508	4.169	0.663	0.240
1.38	0.677	1.465	1.478	0.682	1.73	1.037	2.590	0.965	0.386	2.08	1.524	4.224	0.656	0.237
1.39	0.686	1.491	1.459	0.671	1.74	1.049	2.629	0.953	0.380	2.09	1.540	4.279	0.649	0.234
1.40	0.694	1.518	1.440	0.659	1.75	1.061	2.669	0.943	0.375	2.10	1.556	4.334	0.643	0.231
1.41	0.703	1.544	1.422	0.648	1.76	1.073	2.709	0.932	0.369	2.11	1.572	4.391	0.636	0.228
1.42	0.712	1.571	1.404	0.636	1.77	1.086	2.749	0.921	0.364	2.12	1.589	4.447	0.629	0.225
1.43	0.721	1.599	1.386	0.626	1.78	1.098	2.790	0.911	0.358	2.13	1.605	4.504	0.623	0.222
1.44	0.731	1.626	1.369	0.615	1.79	1.111	2.831	0.900	0.353	2.14	1.622	4.562	0.617	0.219
1.45	0.740	1.654	1.352	0.604	1.80	1.123	2.873	0.890	0.348	2.15	1.639	4.620	0.610	0.216
1.46	0.749	1.683	1.335	0.594	1.81	1.136	2.915	0.880	0.343	2.16	1.656	4.679	0.604	0.214
1.47	0.759	1.712	1.318	0.584	1.82	1.149	2.957	0.870	0.338	2.17	1.673	4.738	0.598	0.211
1.48	0.768	1.741	1.302	0.574	1.83	1.162	3.000	0.861	0.333	2.18	1.690	4.798	0.592	0.208
1.49	0.778	1.770	1.286	0.565	1.84	1.175	3.044	0.851	0.329	2.19	1.707	4.858	0.586	0.206

TABLE T-1c   Coefficients for Deflection and Rigidity of Walls or Piers
for Distribution of Horizontal Forces

For use of Table,
see Example 4-F

WSD

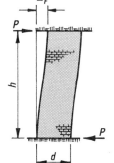

Fixed Wall or Pier

Δ_F = Deflection of wall or pier fixed top and bottom.

$$\Delta_F = \frac{P}{E_m t}\left[\left(\frac{h}{d}\right)^3 + 3\left(\frac{h}{d}\right)\right]$$

$$\Delta_F = 0.1\left(\frac{h}{d}\right)^3 + 0.3\left(\frac{h}{d}\right)$$

$R_F = \dfrac{1}{\Delta_F}$ **Rigidity of fixed wall or pier.**

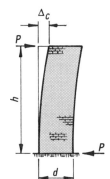

Cantilever Wall or Pier

Δ_C = Deflection of cantilever wall or pier.

$$\Delta_C = \frac{P}{E_m t}\left[4\left(\frac{h}{d}\right)^3 + 3\left(\frac{h}{d}\right)\right]$$

$$\Delta_C = 0.4\left(\frac{h}{d}\right)^3 + 0.3\left(\frac{h}{d}\right)$$

$R_C = \dfrac{1}{\Delta_C}$ **Rigidity of cantilever wall or pier.**

$$P = 100{,}000 \text{ pounds}; \ t = 1''; \ E_m = 1{,}000{,}000 \text{ psi}$$

h/d	Δ_F	Δ_C	R_F	R_C	h/d	Δ_F	Δ_C	R_F	R_C	h/d	Δ_F	Δ_C	R_F	R_C
2.20	1.725	4.919	0.580	0.203	2.55	2.423	7.398	0.413	0.135	2.90	3.309	10.626	0.302	0.094
2.21	1.742	4.981	0.574	0.201	2.56	2.446	7.479	0.409	0.134	2.91	3.337	10.730	0.300	0.093
2.22	1.760	5.042	0.568	0.198	2.57	2.468	7.561	0.405	0.132	2.92	3.366	10.835	0.297	0.092
2.23	1.778	5.105	0.562	0.196	2.58	2.491	7.643	0.401	0.131	2.93	3.394	10.941	0.295	0.091
2.24	1.796	5.168	0.557	0.194	2.59	2.514	7.727	0.398	0.129	2.94	3.423	11.047	0.292	0.091
2.25	1.814	5.231	0.551	0.191	2.60	2.538	7.810	0.394	0.128	2.95	3.452	11.154	0.290	0.090
2.26	1.832	5.295	0.546	0.189	2.61	2.561	7.895	0.390	0.127	2.96	3.481	11.262	0.287	0.089
2.27	1.851	5.360	0.540	0.187	2.62	2.584	7.980	0.387	0.125	2.97	3.511	11.370	0.285	0.088
2.28	1.869	5.425	0.535	0.184	2.63	2.608	8.066	0.383	0.124	2.98	3.540	11.479	0.282	0.087
2.29	1.888	5.491	0.530	0.182	2.64	2.632	8.152	0.380	0.123	2.99	3.570	11.589	0.280	0.086
2.30	1.907	5.557	0.524	0.180	2.65	2.656	8.239	0.377	0.121	3.00	3.600	11.700	0.278	0.085
2.31	1.926	5.624	0.519	0.178	2.66	2.680	8.326	0.373	0.120	3.01	3.630	11.811	0.275	0.085
2.32	1.945	5.691	0.514	0.176	2.67	2.704	8.415	0.370	0.119	3.02	3.660	11.923	0.273	0.084
2.33	1.964	5.759	0.509	0.174	2.68	2.729	8.504	0.366	0.118	3.03	3.691	12.036	0.271	0.083
2.34	1.983	5.827	0.504	0.172	2.69	2.754	8.593	0.363	0.116	3.04	3.721	12.150	0.269	0.082
2.35	2.003	5.896	0.499	0.170	2.70	2.778	8.683	0.360	0.115	3.05	3.752	12.264	0.267	0.082
2.36	2.022	5.966	0.494	0.168	2.71	2.803	8.774	0.357	0.114	3.06	3.783	12.379	0.264	0.081
2.37	2.042	6.036	0.490	0.166	2.72	2.828	8.865	0.354	0.113	3.07	3.814	12.495	0.262	0.080
2.38	2.062	6.107	0.485	0.164	2.73	2.854	8.958	0.350	0.112	3.08	3.846	12.611	0.260	0.079
2.39	2.082	6.178	0.480	0.162	2.74	2.879	9.050	0.347	0.110	3.09	3.877	12.728	0.258	0.079
2.40	2.102	6.250	0.476	0.160	2.75	2.905	9.144	0.344	0.109	3.10	3.909	12.846	0.256	0.078
2.41	2.123	6.322	0.471	0.158	2.76	2.930	9.238	0.341	0.108	3.11	3.941	12.965	0.254	0.077
2.42	2.143	6.395	0.467	0.156	2.77	2.956	9.333	0.338	0.107	3.12	3.973	13.085	0.252	0.076
2.43	2.164	6.469	0.462	0.155	2.78	2.982	9.428	0.335	0.106	3.13	4.005	13.205	0.250	0.076
2.44	2.185	6.543	0.458	0.153	2.79	3.009	9.524	0.332	0.105	3.14	4.038	13.326	0.248	0.075
2.45	2.206	6.617	0.453	0.151	2.80	3.035	9.621	0.329	0.104	3.15	4.071	13.447	0.246	0.074
2.46	2.227	6.693	0.449	0.149	2.81	3.062	9.718	0.327	0.103	3.16	4.103	13.570	0.244	0.074
2.47	2.248	6.769	0.445	0.148	2.82	3.089	9.816	0.324	0.102	3.17	4.137	13.693	0.242	0.073
2.48	2.269	6.845	0.441	0.146	2.83	3.116	9.915	0.321	0.101	3.18	4.170	13.817	0.240	0.072
2.49	2.291	6.922	0.437	0.144	2.84	3.143	10.015	0.318	0.100	3.19	4.203	13.942	0.238	0.072
2.50	2.312	7.000	0.432	0.143	2.85	3.170	10.115	0.315	0.099	3.20	4.237	14.067	0.236	0.071
2.51	2.334	7.078	0.428	0.141	2.86	3.197	10.215	0.313	0.098	3.21	4.271	14.193	0.234	0.070
2.52	2.356	7.157	0.424	0.140	2.87	3.225	10.317	0.310	0.097	3.22	4.305	14.320	0.232	0.070
2.53	2.378	7.237	0.420	0.138	2.88	3.253	10.419	0.307	0.096	3.23	4.339	14.448	0.230	0.069
2.54	2.401	7.317	0.417	0.137	2.89	3.281	10.522	0.305	0.095	3.24	4.373	14.577	0.229	0.069

TABLE T-1d Coefficients for Deflection and Rigidity of Walls or Piers for Distribution of Horizontal Forces

For use of Table, see Example 4-F

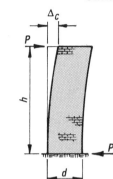

Fixed Wall or Pier	Cantilever Wall or Pier
Δ_F = Deflection of wall or pier fixed top and bottom.	Δ_C = Deflection of cantilever wall or pier.
$\Delta_F = \dfrac{P}{E_m t}\left[\left(\dfrac{h}{d}\right)^3 + 3\left(\dfrac{h}{d}\right)\right]$	$\Delta_C = \dfrac{P}{E_m t}\left[4\left(\dfrac{h}{d}\right)^3 + 3\left(\dfrac{h}{d}\right)\right]$
$\Delta_F = 0.1\left(\dfrac{h}{d}\right)^3 + 0.3\left(\dfrac{h}{d}\right)$	$\Delta_C = 0.4\left(\dfrac{h}{d}\right)^3 + 0.3\left(\dfrac{h}{d}\right)$
$R_F = \dfrac{1}{\Delta_F}$ Rigidity of fixed wall or pier.	$R_C = \dfrac{1}{\Delta_C}$ Rigidity of cantilever wall or pier.

$$P = 100{,}000 \text{ pounds};\ t = 1'';\ E_m = 1{,}000{,}000 \text{ psi}$$

h/d	Δ_F	Δ_C	R_F	R_C	h/d	Δ_F	Δ_C	R_F	R_C	h/d	Δ_F	Δ_C	R_F	R_C
3.25	4.408	14.706	0.227	0.068	3.60	5.746	19.742	0.174	0.051	3.95	7.348	25.837	0.136	0.039
3.26	4.443	14.836	0.225	0.067	3.61	5.788	19.901	0.173	0.050	3.96	7.398	26.028	0.135	0.038
3.27	4.478	14.967	0.223	0.067	3.62	5.830	20.061	0.172	0.050	3.97	7.448	26.219	0.134	0.038
3.28	4.513	15.099	0.222	0.066	3.63	5.872	20.222	0.170	0.049	3.98	7.498	26.412	0.133	0.038
3.29	4.548	15.232	0.220	0.066	3.64	5.915	20.383	0.169	0.049	3.99	7.549	26.605	0.132	0.038
3.30	4.584	15.365	0.218	0.065	3.65	5.958	20.546	0.168	0.049	4.00	7.600	26.800	0.132	0.037
3.31	4.619	15.499	0.216	0.065	3.66	6.001	20.709	0.167	0.048	4.01	7.651	26.995	0.131	0.037
3.32	4.655	15.634	0.215	0.064	3.67	6.044	20.873	0.165	0.048	4.02	7.702	27.192	0.130	0.037
3.33	4.692	15.769	0.213	0.063	3.68	6.088	21.038	0.164	0.048	4.03	7.754	27.389	0.129	0.037
3.34	4.728	15.906	0.212	0.063	3.69	6.131	21.204	0.163	0.047	4.04	7.806	27.588	0.128	0.036
3.35	4.765	16.043	0.210	0.062	3.70	6.175	21.371	0.162	0.047	4.05	7.858	27.787	0.127	0.036
3.36	4.801	16.181	0.208	0.062	3.71	6.219	21.539	0.161	0.046	4.06	7.910	27.987	0.126	0.036
3.37	4.838	16.320	0.207	0.061	3.72	6.264	21.708	0.160	0.046	4.07	7.963	28.189	0.126	0.035
3.38	4.875	16.460	0.205	0.061	3.73	6.309	21.877	0.159	0.046	4.08	8.016	28.391	0.125	0.035
3.39	4.913	16.600	0.204	0.060	3.74	6.353	22.047	0.157	0.045	4.09	8.069	28.594	0.124	0.035
3.40	4.950	16.742	0.202	0.060	3.75	6.398	22.219	0.156	0.045	4.10	8.122	28.798	0.123	0.035
3.41	4.988	16.884	0.200	0.059	3.76	6.444	22.391	0.155	0.045	4.11	8.176	29.004	0.122	0.034
3.42	5.026	17.027	0.199	0.059	3.77	6.489	22.564	0.154	0.044	4.12	8.229	29.210	0.122	0.034
3.43	5.064	17.170	0.197	0.058	3.78	6.535	22.738	0.153	0.044	4.13	8.283	29.417	0.121	0.034
3.44	5.103	17.315	0.196	0.058	3.79	6.581	22.913	0.152	0.044	4.14	8.338	29.625	0.120	0.034
3.45	5.141	17.460	0.195	0.057	3.80	6.627	23.089	0.151	0.043	4.15	8.392	29.834	0.119	0.034
3.46	5.180	17.607	0.193	0.057	3.81	6.674	23.266	0.150	0.043	4.16	8.447	30.045	0.118	0.033
3.47	5.219	17.754	0.192	0.056	3.82	6.720	23.443	0.149	0.043	4.17	8.502	30.256	0.118	0.033
3.48	5.258	17.902	0.190	0.056	3.83	6.767	23.622	0.148	0.042	4.18	8.557	30.468	0.117	0.033
3.49	5.298	18.050	0.189	0.055	3.84	6.814	23.801	0.147	0.042	4.19	8.613	30.681	0.116	0.033
3.50	5.337	18.200	0.187	0.055	3.85	6.862	23.982	0.146	0.042	4.20	8.669	30.895	0.115	0.032
3.51	5.377	18.350	0.186	0.054	3.86	6.909	24.163	0.145	0.041	4.21	8.725	31.110	0.115	0.032
3.52	5.417	18.502	0.185	0.054	3.87	6.957	24.345	0.144	0.041	4.22	8.781	31.327	0.114	0.032
3.53	5.458	18.654	0.183	0.054	3.88	7.005	24.528	0.143	0.041	4.23	8.838	31.544	0.113	0.032
3.54	5.498	18.807	0.182	0.053	3.89	7.053	24.713	0.142	0.040	4.24	8.895	31.762	0.112	0.031
3.55	5.539	18.961	0.181	0.053	3.90	7.102	24.898	0.141	0.040	4.25	8.952	31.981	0.112	0.031
3.56	5.580	19.115	0.179	0.052	3.91	7.151	25.084	0.140	0.040	4.26	9.009	32.202	0.111	0.031
3.57	5.621	19.271	0.178	0.052	3.92	7.200	25.271	0.139	0.040	4.27	9.066	32.423	0.110	0.031
3.58	5.662	19.427	0.177	0.051	3.93	7.249	25.458	0.138	0.039	4.28	9.124	32.645	0.110	0.031
3.59	5.704	19.584	0.175	0.051	3.94	7.298	25.647	0.137	0.039	4.29	9.182	32.868	0.109	0.030

TABLE T-1e   Coefficients for Deflection and Rigidity of Walls or Piers for Distribution of Horizontal Forces

For use of Table, see Example 4-F

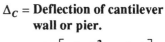

WSD

Fixed Wall or Pier

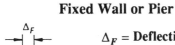

Δ_F = Deflection of wall or pier fixed top and bottom.

$$\Delta_F = \frac{P}{E_m t}\left[\left(\frac{h}{d}\right)^3 + 3\left(\frac{h}{d}\right)\right]$$

$$\Delta_F = 0.1\left(\frac{h}{d}\right)^3 + 0.3\left(\frac{h}{d}\right)$$

$$R_F = \frac{1}{\Delta_F}\quad\text{Rigidity of fixed wall or pier.}$$

Cantilever Wall or Pier

Δ_C = Deflection of cantilever wall or pier.

$$\Delta_C = \frac{P}{E_m t}\left[4\left(\frac{h}{d}\right)^3 + 3\left(\frac{h}{d}\right)\right]$$

$$\Delta_C = 0.4\left(\frac{h}{d}\right)^3 + 0.3\left(\frac{h}{d}\right)$$

$$R_C = \frac{1}{\Delta_C}\quad\text{Rigidity of cantilever wall or pier.}$$

$$P = 100,000 \text{ pounds}; \ t = 1''; \ E_m = 1,000,000 \text{ psi}$$

h/d	Δ_F	Δ_C	R_F	R_C	h/d	Δ_F	Δ_C	R_F	R_C	h/d	Δ_F	Δ_C	R_F	R_C
4.30	9.241	33.093	0.108	0.030	4.65	11.449	41.613	0.087	0.024	5.00	14.000	51.500	0.071	0.019
4.31	9.299	33.318	0.108	0.030	4.66	11.517	41.876	0.087	0.024	5.01	14.078	51.804	0.071	0.019
4.32	9.358	33.545	0.107	0.030	4.67	11.586	42.140	0.086	0.024	5.02	14.157	52.108	0.071	0.019
4.33	9.417	33.772	0.106	0.030	4.68	11.654	42.405	0.086	0.024	5.03	14.235	52.414	0.070	0.019
4.34	9.477	34.001	0.106	0.029	4.69	11.723	42.672	0.085	0.023	5.04	14.314	52.722	0.070	0.019
4.35	9.536	34.230	0.105	0.029	4.70	11.792	42.939	0.085	0.023	5.05	14.394	53.030	0.069	0.019
4.36	9.596	34.461	0.104	0.029	4.71	11.862	43.208	0.084	0.023	5.06	14.473	53.340	0.069	0.019
4.37	9.656	34.692	0.104	0.029	4.72	11.931	43.478	0.084	0.023	5.07	14.553	53.651	0.069	0.019
4.38	9.717	34.925	0.103	0.029	4.73	12.001	43.749	0.083	0.023	5.08	14.634	53.963	0.068	0.019
4.39	9.777	35.159	0.102	0.028	4.74	12.072	44.021	0.083	0.023	5.09	14.714	54.276	0.068	0.018
4.40	9.838	35.394	0.102	0.028	4.75	12.142	44.294	0.082	0.023	5.10	14.795	54.590	0.068	0.018
4.41	9.900	35.629	0.101	0.028	4.76	12.213	44.568	0.082	0.022	5.11	14.876	54.906	0.067	0.018
4.42	9.961	35.866	0.100	0.028	4.77	12.284	44.844	0.081	0.022	5.12	14.958	55.223	0.067	0.018
4.43	10.023	36.104	0.100	0.028	4.78	12.356	45.120	0.081	0.022	5.13	15.040	55.541	0.066	0.018
4.44	10.085	36.343	0.099	0.028	4.79	12.427	45.398	0.080	0.022	5.14	15.122	55.861	0.066	0.018
4.45	10.147	36.583	0.099	0.027	4.80	12.499	45.677	0.080	0.022	5.15	15.204	56.181	0.066	0.018
4.46	10.210	36.825	0.098	0.027	4.81	12.571	45.957	0.080	0.022	5.16	15.287	56.503	0.065	0.018
4.47	10.272	37.067	0.097	0.027	4.82	12.644	46.238	0.079	0.022	5.17	15.370	56.826	0.065	0.018
4.48	10.336	37.310	0.097	0.027	4.83	12.717	46.520	0.079	0.021	5.18	15.453	57.151	0.065	0.017
4.49	10.399	37.555	0.096	0.027	4.84	12.790	46.804	0.078	0.021	5.19	15.537	57.476	0.064	0.017
4.50	10.462	37.800	0.096	0.026	4.85	12.863	47.089	0.078	0.021	5.20	15.621	57.803	0.064	0.017
4.51	10.526	38.047	0.095	0.026	4.86	12.937	47.375	0.077	0.021	5.21	15.705	58.131	0.064	0.017
4.52	10.591	38.294	0.094	0.026	4.87	13.011	47.662	0.077	0.021	5.22	15.790	58.461	0.063	0.017
4.53	10.655	38.543	0.094	0.026	4.88	13.085	47.950	0.076	0.021	5.23	15.875	58.791	0.063	0.017
4.54	10.720	38.793	0.093	0.026	4.89	13.160	48.239	0.076	0.021	5.24	15.960	59.123	0.063	0.017
4.55	10.785	39.044	0.093	0.026	4.90	13.235	48.530	0.076	0.021	5.25	16.045	59.456	0.062	0.017
4.56	10.850	39.296	0.092	0.025	4.91	13.310	48.821	0.075	0.020	5.26	16.131	59.791	0.062	0.017
4.57	10.915	39.549	0.092	0.025	4.92	13.386	49.114	0.075	0.020	5.27	16.217	60.126	0.062	0.017
4.58	10.981	39.803	0.091	0.025	4.93	13.461	49.408	0.074	0.020	5.28	16.304	60.463	0.061	0.017
4.59	11.047	40.058	0.091	0.025	4.94	13.537	49.704	0.074	0.020	5.29	16.391	60.801	0.061	0.016
4.60	11.114	40.314	0.090	0.025	4.95	13.614	50.000	0.073	0.020	5.30	16.478	61.141	0.061	0.016
4.61	11.180	40.572	0.089	0.025	4.96	13.690	50.298	0.073	0.020	5.31	16.565	61.482	0.060	0.016
4.62	11.247	40.830	0.089	0.024	4.97	13.767	50.596	0.073	0.020	5.32	16.653	61.824	0.060	0.016
4.63	11.314	41.090	0.088	0.024	4.98	13.845	50.896	0.072	0.020	5.33	16.741	62.167	0.060	0.016
4.64	11.382	41.351	0.088	0.024	4.99	13.922	51.198	0.072	0.020	5.34	16.829	62.511	0.059	0.016

**TABLE T-1f Coefficients for Deflection and Rigidity of Walls or Piers
for Distribution of Horizontal Forces**

For use of Table,
see Example 4-F

Fixed Wall or Pier

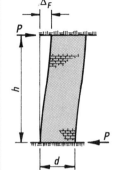

Δ_F = Deflection of wall or pier fixed top and bottom.

$$\Delta_F = \frac{P}{E_m t}\left[\left(\frac{h}{d}\right)^3 + 3\left(\frac{h}{d}\right)\right]$$

$$\Delta_F = 0.1\left(\frac{h}{d}\right)^3 + 0.3\left(\frac{h}{d}\right)$$

$R_F = \dfrac{1}{\Delta_F}$ Rigidity of fixed wall or pier.

Cantilever Wall or Pier

Δ_C = Deflection of cantilever wall or pier.

$$\Delta_C = \frac{P}{E_m t}\left[4\left(\frac{h}{d}\right)^3 + 3\left(\frac{h}{d}\right)\right]$$

$$\Delta_C = 0.4\left(\frac{h}{d}\right)^3 + 0.3\left(\frac{h}{d}\right)$$

$R_C = \dfrac{1}{\Delta_C}$ Rigidity of cantilever wall or pier.

$P = 100{,}000$ pounds; $t = 1''$; $E_m = 1{,}000{,}000$ psi

h/d	Δ_F	Δ_C	R_F	R_C	h/d	Δ_F	Δ_C	R_F	R_C	h/d	Δ_F	Δ_C	R_F	R_C
5.35	16.918	62.857	0.059	0.016	5.70	20.229	75.787	0.049	0.013	6.05	23.960	90.393	0.042	0.011
5.36	17.007	63.204	0.059	0.016	5.71	20.330	76.181	0.049	0.013	6.06	24.073	90.836	0.042	0.011
5.37	17.096	63.553	0.058	0.016	5.72	20.431	76.576	0.049	0.013	6.07	24.186	91.280	0.041	0.011
5.38	17.186	63.902	0.058	0.016	5.73	20.532	76.972	0.049	0.013	6.08	24.300	91.726	0.041	0.011
5.39	17.276	64.253	0.058	0.016	5.74	20.634	77.370	0.048	0.013	6.09	24.414	92.174	0.041	0.011
5.40	17.366	64.606	0.058	0.015	5.75	20.736	77.769	0.048	0.013	6.10	24.528	92.622	0.041	0.011
5.41	17.457	64.959	0.057	0.015	5.76	20.838	78.169	0.048	0.013	6.11	24.643	93.073	0.041	0.011
5.42	17.548	65.314	0.057	0.015	5.77	20.941	78.571	0.048	0.013	6.12	24.758	93.524	0.040	0.011
5.43	17.639	65.670	0.057	0.015	5.78	21.044	78.974	0.048	0.013	6.13	24.874	93.978	0.040	0.011
5.44	17.731	66.028	0.056	0.015	5.79	21.147	79.379	0.047	0.013	6.14	24.990	94.432	0.040	0.011
5.45	17.823	66.386	0.056	0.015	5.80	21.251	79.785	0.047	0.013	6.15	25.106	94.888	0.040	0.011
5.46	17.915	66.747	0.056	0.015	5.81	21.355	80.192	0.047	0.012	6.16	25.222	95.346	0.040	0.010
5.47	18.008	67.108	0.056	0.015	5.82	21.460	80.601	0.047	0.012	6.17	25.340	95.805	0.039	0.010
5.48	18.101	67.471	0.055	0.015	5.83	21.565	81.011	0.046	0.012	6.18	25.457	96.266	0.039	0.010
5.49	18.194	67.835	0.055	0.015	5.84	21.670	81.423	0.046	0.012	6.19	25.575	96.728	0.039	0.010
5.50	18.287	68.200	0.055	0.015	5.85	21.775	81.836	0.046	0.012	6.20	25.693	97.191	0.039	0.010
5.51	18.381	68.567	0.054	0.015	5.86	21.881	82.250	0.046	0.012	6.21	25.811	97.656	0.039	0.010
5.52	18.476	68.935	0.054	0.015	5.87	21.987	82.666	0.045	0.012	6.22	25.930	98.123	0.039	0.010
5.53	18.570	69.304	0.054	0.014	5.88	22.094	83.083	0.045	0.012	6.23	26.049	98.591	0.038	0.010
5.54	18.665	69.675	0.054	0.014	5.89	22.201	83.502	0.045	0.012	6.24	26.169	99.060	0.038	0.010
5.55	18.760	70.047	0.053	0.014	5.90	22.308	83.922	0.045	0.012	6.25	26.289	99.531	0.038	0.010
5.56	18.856	70.420	0.053	0.014	5.91	22.416	84.343	0.045	0.012	6.26	26.409	100.004	0.038	0.010
5.57	18.952	70.794	0.053	0.014	5.92	22.523	84.766	0.044	0.012	6.27	26.530	100.478	0.038	0.010
5.58	19.048	71.170	0.052	0.014	5.93	22.632	85.190	0.044	0.012	6.28	26.651	100.953	0.038	0.010
5.59	19.145	71.548	0.052	0.014	5.94	22.740	85.616	0.044	0.012	6.29	26.773	101.430	0.037	0.010
5.60	19.242	71.926	0.052	0.014	5.95	22.849	86.043	0.044	0.012	6.30	26.895	101.909	0.037	0.010
5.61	19.339	72.306	0.052	0.014	5.96	22.959	86.471	0.044	0.012	6.31	27.017	102.389	0.037	0.010
5.62	19.436	72.688	0.051	0.014	5.97	23.069	86.901	0.043	0.012	6.32	27.140	102.870	0.037	0.010
5.63	19.534	73.070	0.051	0.014	5.98	23.179	87.333	0.043	0.011	6.33	27.263	103.353	0.037	0.010
5.64	19.633	73.454	0.051	0.014	5.99	23.289	87.766	0.043	0.011	6.34	27.386	103.838	0.037	0.010
5.65	19.731	73.840	0.051	0.014	6.00	23.400	88.200	0.043	0.011	6.35	27.510	104.324	0.036	0.010
5.66	19.830	74.227	0.050	0.013	6.01	23.511	88.636	0.043	0.011	6.36	27.634	104.812	0.036	0.010
5.67	19.929	74.615	0.050	0.013	6.02	23.623	89.073	0.042	0.011	6.37	27.758	105.301	0.036	0.009
5.68	20.029	75.004	0.050	0.013	6.03	23.735	89.511	0.042	0.011	6.38	27.883	105.792	0.036	0.009
5.69	20.129	75.395	0.050	0.013	6.04	23.847	89.952	0.042	0.011	6.39	28.009	106.284	0.036	0.009

TABLE T-1g Coefficients for Defection and Rigidity of Walls or Piers for Distribution of Horizontal Forces

For use of Table, see Example 4-F

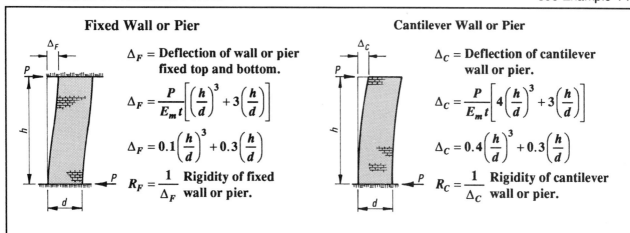

Fixed Wall or Pier

Δ_F = Deflection of wall or pier fixed top and bottom.

$$\Delta_F = \frac{P}{E_m t}\left[\left(\frac{h}{d}\right)^3 + 3\left(\frac{h}{d}\right)\right]$$

$$\Delta_F = 0.1\left(\frac{h}{d}\right)^3 + 0.3\left(\frac{h}{d}\right)$$

$R_F = \dfrac{1}{\Delta_F}$ Rigidity of fixed wall or pier.

Cantilever Wall or Pier

Δ_C = Deflection of cantilever wall or pier.

$$\Delta_C = \frac{P}{E_m t}\left[4\left(\frac{h}{d}\right)^3 + 3\left(\frac{h}{d}\right)\right]$$

$$\Delta_C = 0.4\left(\frac{h}{d}\right)^3 + 0.3\left(\frac{h}{d}\right)$$

$R_C = \dfrac{1}{\Delta_C}$ Rigidity of cantilever wall or pier.

$$P = 100,000 \text{ pounds}; \ t = 1''; \ E_m = 1,000,000 \text{ psi}$$

h/d	Δ_F	Δ_C	R_F	R_C	h/d	Δ_F	Δ_C	R_F	R_C	h/d	Δ_F	Δ_C	R_F	R_C
6.40	28.134	106.778	0.036	0.009	6.75	32.780	125.044	0.031	0.008	7.10	37.921	145.294	0.026	0.007
6.41	28.260	107.273	0.035	0.009	6.76	32.920	125.594	0.030	0.008	7.11	38.076	145.903	0.026	0.007
6.42	28.387	107.770	0.035	0.009	6.77	33.060	126.146	0.030	0.008	7.12	38.230	146.514	0.026	0.007
6.43	28.514	108.268	0.035	0.009	6.78	33.201	126.700	0.030	0.008	7.13	38.386	147.126	0.026	0.007
6.44	28.641	108.768	0.035	0.009	6.79	33.342	127.256	0.030	0.008	7.14	38.541	147.740	0.026	0.007
6.45	28.769	109.269	0.035	0.009	6.80	33.483	127.813	0.030	0.008	7.15	38.698	148.355	0.026	0.007
6.46	28.897	109.772	0.035	0.009	6.81	33.625	128.371	0.030	0.008	7.16	38.854	148.973	0.026	0.007
6.47	29.025	110.277	0.034	0.009	6.82	33.767	128.932	0.030	0.008	7.17	39.011	149.592	0.026	0.007
6.48	29.154	110.783	0.034	0.009	6.83	33.910	129.494	0.029	0.008	7.18	39.169	150.212	0.026	0.007
6.49	29.283	111.291	0.034	0.009	6.84	34.053	130.057	0.029	0.008	7.19	39.326	150.835	0.025	0.007
6.50	29.412	111.800	0.034	0.009	6.85	34.197	130.623	0.029	0.008	7.20	39.485	151.459	0.025	0.007
6.51	29.542	112.311	0.034	0.009	6.86	34.341	131.190	0.029	0.008	7.21	39.644	152.085	0.025	0.007
6.52	29.673	112.823	0.034	0.009	6.87	34.485	131.758	0.029	0.008	7.22	39.803	152.713	0.025	0.007
6.53	29.804	113.337	0.034	0.009	6.88	34.630	132.328	0.029	0.008	7.23	39.962	153.342	0.025	0.007
6.54	29.935	113.853	0.033	0.009	6.89	34.775	132.900	0.029	0.008	7.24	40.122	153.973	0.025	0.006
6.55	30.066	114.370	0.033	0.009	6.90	34.921	133.474	0.029	0.007	7.25	40.283	154.606	0.025	0.006
6.56	30.198	114.888	0.033	0.009	6.91	35.067	134.049	0.029	0.007	7.26	40.444	155.241	0.025	0.006
6.57	30.330	115.408	0.033	0.009	6.92	35.213	134.626	0.028	0.007	7.27	40.605	155.877	0.025	0.006
6.58	30.463	115.930	0.033	0.009	6.93	35.360	135.204	0.028	0.007	7.28	40.767	156.515	0.025	0.006
6.59	30.596	116.453	0.033	0.009	6.94	35.508	135.784	0.028	0.007	7.29	40.929	157.155	0.024	0.006
6.60	30.730	116.978	0.033	0.009	6.95	35.655	136.366	0.028	0.007	7.30	41.092	157.797	0.024	0.006
6.61	30.863	117.505	0.032	0.009	6.96	35.803	136.949	0.028	0.007	7.31	41.255	158.440	0.024	0.006
6.62	30.998	118.033	0.032	0.008	6.97	35.952	137.535	0.028	0.007	7.32	41.418	159.085	0.024	0.006
6.63	31.132	118.563	0.032	0.008	6.98	36.101	138.121	0.028	0.007	7.33	41.582	159.732	0.024	0.006
6.64	31.267	119.094	0.032	0.008	6.99	36.250	138.710	0.028	0.007	7.34	41.747	160.381	0.024	0.006
6.65	31.403	119.627	0.032	0.008	7.00	36.400	139.300	0.027	0.007	7.35	41.912	161.031	0.024	0.006
6.66	31.539	120.161	0.032	0.008	7.01	36.550	139.892	0.027	0.007	7.36	42.077	161.683	0.024	0.006
6.67	31.675	120.697	0.032	0.008	7.02	36.701	140.485	0.027	0.007	7.37	42.243	162.337	0.024	0.006
6.68	31.812	121.235	0.031	0.008	7.03	36.852	141.081	0.027	0.007	7.38	42.409	162.993	0.024	0.006
6.69	31.949	121.774	0.031	0.008	7.04	37.003	141.677	0.027	0.007	7.39	42.575	163.650	0.023	0.006
6.70	32.086	122.315	0.031	0.008	7.05	37.155	142.276	0.027	0.007	7.40	42.742	164.310	0.023	0.006
6.71	32.224	122.858	0.031	0.008	7.06	37.308	142.876	0.027	0.007	7.41	42.910	164.971	0.023	0.006
6.72	32.362	123.402	0.031	0.008	7.07	37.460	143.478	0.027	0.007	7.42	43.078	165.633	0.023	0.006
6.73	32.501	123.947	0.031	0.008	7.08	37.613	144.082	0.027	0.007	7.43	43.246	166.298	0.023	0.006
6.74	32.640	124.495	0.031	0.008	7.09	37.767	144.687	0.026	0.007	7.44	43.415	166.964	0.023	0.006

TABLE U-1   Conversion of Measurement Systems

English Measurement to SI (Metric) Measurement

Unit	Exact Conversion	Approximate Conversion
Length		
1 mile	1.609344 kilometres	1.6 km or 1 1/2 km
1 yard	0.9144 metres	0.9 m or 1 metre
1 foot	0.3048 metres	0.3 m or 1/3 metre
1 inch	25.40 millimetres	25 mm or 1/40 metre
Speed		
1 mile per hour	1.609344 kilometres per hour	1.6 km/h or 1 1/2 km/h
1 foot per second	0.3048 metres per second	0.3 m/s or 1/3 m/s
Area		
1 acre	4046.856 square metres	4000 m²
1 square foot	0.0929 square metres	1/10 m² or 1000 cm²
1 square inch	645.2 square millimetres	6 cm² or 650 mm²
Weight or Mass		
1 ounce (Avdp)	28.35 grams	30 grams
1 pound	0.45359 kilograms or 453 grams	1/2 kg or 500 g
1 kip	453.59 kilograms	500 kg or 0.5 Mg
1 Ton (short)	907.18 kilograms	1 Megagram
Volume		
1 cubic yard	0.7646 cubic metres or	3/4 m³ or
	764.56 litres	750 litres
1 cubic foot	0.0283 cubic metres or	1/35 m³ or
	28.217 litres	30 litres
1 cubic inch	16.387 cubic centimetres	16 cm³ or 16,000 mm³
1 gallon	3785.4 cubic centimetres or	4000 cm³ or
	3.785 litres	4 litres
1 quart	946.35 cubic centimetres or	1000 cm³ or
	0.94635 litres	1 litre
Density		
1 pound/cubic foot	16.018 kilograms/cubic metre (grams/litre)	16 kg/m³
1 pound/gallon	119.83 kilograms/cubic metre (grams/litre)	120 kg/m³
Force		
1 pound force	4.448 newtons	4 1/2 N
1 kip force	4.448 kilo newtons	4500 N or 4 1/2 kN
1 pound force/lin ft	14.59 newtons/metre	14 1/2 N/m
1 kip force/lin ft	14.59 kilo newtons/metre	14 1/2 kN/m
Pressure		
1 pounds/square inch	6894.8 pascals	7000 Pa or 7 kPa
1 kips/square inch	6.895 mega pascals	7 MPa
1 pound force/sq. ft	47.9 pascals	48 Pa
1 kip force/sq. ft.	47.9 kilo pascals	48 kPa
Moment		
1 foot pound force	1.356 newton metre	1.36 Nm
1 foot kip force	1.356 kilo newton metre	1.36 kNm
1 foot pound force/foot	4448 newton metre/metre	4.45 Nm/m
Energy		
1 BTU	1054.35 joule or 1.054 kj	1 kj
Temperature		
°Fahrenheit	$[(°F-32)5/9]$ °Celcius	

SI (Metric) Measurement to English Measurement

Unit	Exact Conversion	Approximate Conversion
Length		
1 kilometre	0.6214 miles	5/8 miles or 0.6 miles
1 metre	3.2808 feet or 39 3/8 inches	3 ft 3 in. or 3 ft+
1 centimetre	0.3937 inches	0.4 inches or 3/8 inches
1 millimetre	0.0394 inches	1/32 inch
Speed		
1 kilometre per hour	0.6214 mile per hour	5/8 mph or 0.6 mph
1 metre per second	3.2808 feet per second	3 ft/s or 1 yd/s
Area		
1 square kilometre	0.3861 square miles or 247.1 acres	1/3 mi² or 250 acres
1 square metre	1.196 square yds or 10.764 square ft	1/2 yd² or 10 ft²
1 square centimetre	0.155 square inches	1/6 square inches
Weight or Mass		
1 gram	0.03527 ounces (Avdp)	1/30 ounce
1 kilogram	2.205 pounds	2 1/4 pounds or 2 pounds
1 megagram	2.205 kips or 2205 pounds	2 1/4 kips or 2000 pounds
1 gigagram	.1102 Tons or 2,205,000 lbs	1000 Tons or 2 million lbs
Volume		
1 cubic metre	35.315 cubic feet or	35 ft³ or 265 gal
	264.17 gallons	
1 litre	0.0353 cubic feet or	
	0.2642 gallons or 61.024 in³	1/4 gal or 1 quart or 60 in³
1 cubic centimetre	0.061 cubic inches	1/16 in³
Density		
1 gram/cu centimetre	8.345 lbs/gal or 62.428 lbs/cu ft	8 1/2 lbs/gal or 62 lbs/ft³
1 kg/cubic metre	0.0081345 lbs/gal or 0.062428 lbs/cu ft	1/8 oz/gal or 1/16 lb/ft³
Force		
1 newton	0.2248 pound force	1/4 pound force
1 kilo newton	224.8 pound force	225 pound force
Pressure		
1 pascal	0.000145 pounds/square inch	1/7 psi
1 kilo pascals	0.145 pounds/square inch	1/7 psi
1 mega pascals	145 pounds/square inch	150 psi
Moment		
1 newton metre	0.737 foot pound force	3/4 ft lb
1 kilo newton metre	0.737 foot kip force	3/4 ft k
1 newton metre/metre	0.225 foot pound force/foot.	1/4 ft lb/ft
Energy		
1000 joules	0.94845 BTU	1 BTU
Temperature		
°Celcius	$[(1.8 °C) + 32]$ °Fahrenheit	

TABLE U-2 SI Prefixes for Magnitude

Amount	Exponential Factor	Prefix	Symbol
1 000 000 000 000	10^{12}	tera	T
1 000 000 000	10^9	giga	G
1 000 000	10^6	mega	M
1 000	10^3	kilo	k
100	10^2	hecta	h
10	10^1	deka	da
0.1	10^{-1}	deci	d
0.01	10^{-2}	centi	c
0.001	10^{-3}	milli	m
0.000 001	10^{-6}	micro	μ
0.000 000 001	10^{-9}	nano	n

TABLE U-3 SI Properties of Reinforcing Bars

Bar Size	Weight (kg·m)	Diameter (mm)	Area (mm^2)	Perimeter (mm)
#2	0.25	6.4	32	20
#3	0.56	9.6	71	30
#4	0.99	12.7	129	40
#5	1.55	15.9	200	50
#6	2.34	19.1	284	60
#7	3.04	22.3	387	70
#8	3.97	25.4	510	80
#9	5.06	28.6	645	90
#10	6.40	33.3	819	100
#11	7.91	35.8	1006	113

WSD

TABLE U-4 Masonry and Steel Stresses[1]—Psi to MPa and kg/cm^2

psi	MPa	kg/cm^2	psi	MPa	kg/cm^2	psi	MPa	kg/cm^2
100	0.69	7.03	1000	6.90	70.30	4500	31.03	316.35
150	1.03	10.55	1100	7.58	77.33	5000	34.48	351.50
200	1.38	14.06	1167	8.05	82.04	5300	36.54	372.59
250	1.72	17.58	1200	8.27	84.36	10000	68.95	703.00
300	2.07	21.09	1333	9.19	93.71	16000	110.32	1124.80
333	2.30	23.41	1500	10.34	105.45	18000	124.11	1265.40
400	2.76	28.12	1667	11.49	117.19	20000	137.90	1406.00
500	3.45	35.15	2000	13.79	140.60	24000	165.48	1687.20
600	4.14	42.18	2400	16.55	168.72	26700	184.10	1877.01
667	4.60	46.89	2500	17.24	175.75	30000	206.85	2109.00
750	5.17	52.73	2700	18.62	189.81	32000	220.64	2249.60
800	5.52	56.24	3000	20.69	210.90	40000	275.80	2812.00
833	5.74	58.56	3500	24.13	246.05	50000	344.75	3515.00
900	6.21	63.27	4000	27.58	281.20	60000	413.70	4218.00

Note: Modulus of reinforcing steel = 29,000,000 psi = 199 955 MPa = 2 038 700 kg/cm^2.

1. Values in this table are based on the relations, 1 psi = 0.006 895 MPa = 0.0703 kg/cm^2.

TABLE U-5 Length Equivalents[1]—Inches to Millimetres

Inches	0	1	2	3	4	5	6	7	8	9
	Millimetres									
0	0.0	25.4	50.8	76.2	101.6	127.0	152.4	177.8	203.2	228.6
10	254.0	279.4	304.8	330.2	355.6	381.0	406.4	431.8	457.2	482.6
20	508.0	533.4	558.8	584.2	609.6	635.0	660.4	685.8	711.2	736.6
30	762.0	787.4	812.8	838.2	863.6	889.0	914.4	939.8	965.2	990.6
40	1016.0	1041.4	1066.8	1092.2	1117.6	1143.0	1168.4	1193.8	1219.2	1244.6
50	1270.0	1295.4	1320.8	1346.2	1371.6	1397.0	1422.4	1447.8	1473.2	1498.6
60	1524.0	1549.4	1574.8	1600.2	1625.6	1651.0	1676.4	1701.8	1727.2	1752.6
70	1778.0	1803.4	1828.8	1854.2	1879.6	1905.0	1930.4	1955.8	1981.2	2006.6
80	2032.0	2057.4	2082.8	2108.2	2133.6	2159.0	2184.4	2209.8	2235.2	2260.6
90	2286.0	2311.4	2336.8	2362.2	2387.6	2413.0	2438.4	2463.8	2489.2	2514.6
100	2540.0	2565.4	2590.8	2616.2	2641.6	2667.0	2692.4	2717.8	2743.2	2768.6

1. All values in this table are based on the relation, 1 in. = 25.4 mm. By manipulation of the decimal point any decimal value or multiple of an inch may be converted to its equivalent in millimetres.

TABLE U-6   Length Equivalents[1] — Feet to Metres

Feet	0	1	2	3	4	5	6	7	8	9
	Metres									
0	0.00	0.30	0.61	0.91	1.22	1.52	1.83	2.13	2.44	2.74
10	3.05	3.35	3.66	3.96	4.27	4.57	4.88	5.18	5.49	5.79
20	6.10	6.40	6.71	7.01	7.32	7.62	7.92	8.23	8.53	8.84
30	9.14	9.45	9.75	10.06	10.36	10.67	10.97	11.28	11.58	11.89
40	12.19	12.50	12.80	13.11	13.41	13.72	14.02	14.33	14.63	14.94
50	15.24	15.54	15.85	16.15	16.46	16.76	17.07	17.37	17.68	17.98
60	18.29	18.59	18.90	19.20	19.51	19.81	20.12	20.42	20.73	21.03
70	21.34	21.64	21.95	22.25	22.56	22.86	23.16	23.47	23.77	24.08
80	24.38	24.69	24.99	25.30	25.60	25.91	26.21	26.52	26.82	27.13
90	27.43	27.74	28.04	28.35	28.65	28.96	29.26	29.57	29.87	30.18
100	30.48	30.78	31.09	31.39	31.70	32.00	32.31	32.61	32.92	33.22

1. All values in this table are based on the relation, 1 ft = 0.3048 m. By manipulation of the decimal point any decimal value or multiple of a foot may be converted to its equivalent in metres.

TABLE U-7   Force Equivalents — Pounds Force to Newtons[1]

Pounds Force	0	1	2	3	4	5	6	7	8	9
	Newtons, N									
0	0.0	4.5	8.9	13.3	17.8	22.2	26.7	31.1	35.6	40.0
10	44.5	48.9	53.4	57.8	62.3	66.7	71.2	75.6	80.1	84.5
20	89.0	93.4	97.9	102.3	106.8	111.2	115.7	120.1	124.5	129.0
30	133.4	137.9	142.3	146.8	151.2	155.7	160.1	164.6	169.0	173.5
40	177.9	182.4	186.8	191.3	195.7	200.2	204.6	209.1	213.5	218.0
50	222.4	226.9	231.3	235.7	240.2	244.6	249.1	253.5	258.0	262.4
60	266.9	271.3	275.8	280.2	284.7	289.1	293.6	298.0	302.5	306.9
70	311.4	315.8	320.3	324.7	329.2	333.6	338.1	342.5	346.9	351.4
80	355.8	360.3	364.7	369.2	373.6	378.1	382.5	387.0	391.4	395.9
90	400.3	404.8	409.2	413.7	418.1	422.6	427.0	431.5	435.9	440.4
100	444.8	449.3	453.7	458.1	462.6	467.0	471.5	475.9	480.4	484.8

1. All values in this table are based on the relation, 1 lbf = 4.448 N. By manipulation of the decimal point any decimal value or multiple of a pounds force may be converted to its equivalent in Newtons.

TABLE U-8   Pressure and Stress Equivalents[1] — Psi to Kilopascals

Ft lb/ft	0	1	2	3	4	5	6	7	8	9
	Kilopascals									
0	0.00	6.89	13.79	20.68	27.58	34.47	41.37	48.26	55.16	62.05
10	68.95	75.84	82.74	89.63	96.53	103.42	110.32	117.21	124.11	131.00
20	137.90	144.79	151.69	158.58	165.48	172.37	179.26	186.16	193.05	199.95
30	206.84	213.74	220.63	227.53	234.42	241.32	248.21	255.11	262.00	268.90
40	275.79	282.69	289.58	296.48	303.37	310.27	317.16	324.06	330.95	337.85
50	344.74	351.63	358.53	365.42	372.32	379.21	386.11	393.00	399.90	406.79
60	413.69	420.58	427.48	434.37	441.27	448.16	455.06	461.95	468.85	475.74
70	482.64	489.53	496.43	503.32	510.22	517.11	524.00	530.90	537.79	544.69
80	551.58	558.48	565.37	572.27	579.16	586.06	592.95	599.85	606.74	613.64
90	620.53	627.43	634.32	641.22	648.11	655.01	661.90	668.80	675.69	682.59
100	689.48	696.37	703.27	710.16	717.06	723.95	730.85	737.74	744.64	751.53

1. All values in this table are based on the relation, 1 psi = 6.895 kPa. By manipulation of the decimal point any decimal value or multiple of psi may be converted to its equivalent in kilopascals.

WSD

TABLE U-9 Pressure and Stress Equivalents[1] — Pounds per Square Foot to Pascals

Psf	0	1	2	3	4	5	6	7	8	9
	Pascals									
0	0.0	47.9	95.8	143.7	191.6	239.5	287.4	335.3	383.2	431.1
10	479.0	526.9	574.8	622.7	670.6	718.5	766.4	814.3	862.2	910.1
20	958.0	1005.9	1053.8	1101.7	1149.6	1197.5	1245.4	1293.3	1341.2	1389.1
30	1437.0	1484.9	1532.8	1580.7	1628.6	1676.5	1724.4	1772.3	1820.2	1868.1
40	1916.0	1963.9	2011.8	2059.7	2107.6	2155.5	2203.4	2251.3	2299.2	2347.1
50	2395.0	2442.9	2490.8	2538.7	2586.6	2634.5	2682.4	2730.3	2778.2	2826.1
60	2874.0	2921.9	2969.8	3017.7	3065.6	3113.5	3161.4	3209.3	3257.2	3305.1
70	3353.0	3400.9	3448.8	3496.7	3544.6	3592.5	3640.4	3688.3	3736.2	3784.1
80	3832.0	3879.9	3927.8	3975.7	4023.6	4071.5	4119.4	4167.3	4215.2	4263.1
90	4311.0	4358.9	4406.8	4454.7	4502.6	4550.5	4598.4	4646.3	4694.2	4742.1
100	4790.0	4837.9	4885.8	4933.7	4981.6	5029.5	5077.4	5125.3	5173.2	5221.1

1. All values in this table are based on the relation, 1 psf = 47.90 Pa. By manipulation of the decimal point any decimal value or multiple of psf may be converted to its equivalent in pascals.

TABLE U-10 Moment Equivalents[1] — Foot Pounds Force to Newton Metres

Ft lb	0	1	2	3	4	5	6	7	8	9
	N•m									
0	0.00	1.36	2.72	4.08	5.44	6.80	8.15	9.51	10.87	12.23
10	13.59	14.95	16.31	17.67	19.03	20.39	21.74	23.10	24.46	25.82
20	27.18	28.54	29.90	31.26	32.62	33.98	35.33	36.69	38.05	39.41
30	40.77	42.13	43.49	44.85	46.21	47.57	48.92	50.28	51.64	53.00
40	54.36	55.72	57.08	58.44	59.80	61.16	62.51	63.87	65.23	66.59
50	67.95	69.31	70.67	72.03	73.39	74.75	76.10	77.46	78.82	80.18
60	81.54	82.90	84.26	85.62	86.98	88.34	89.69	91.05	92.41	93.77
70	95.13	96.49	97.85	99.21	100.57	101.93	103.28	104.64	106.00	107.36
80	108.72	110.08	111.44	112.80	114.16	115.52	116.87	118.23	119.59	120.95
90	122.31	123.67	125.03	126.39	127.75	129.11	130.46	131.82	133.18	134.54
100	135.90	137.26	138.62	139.98	141.34	142.70	144.05	145.41	146.77	148.13

1. All values in this table are based on the relation, 1 ft lb = 1.359 N•m. By manipulation of the decimal point any decimal value or multiple of foot pounds may be converted to its equivalent in Newton metres.

TABLE U-11 Moment per Unit Length Equivalents — Foot Pounds Force per Foot to Newton Metres Per Metre

Ft lb/ft	0	1	2	3	4	5	6	7	8	9
	N•m/m									
0	0.00	4.45	8.90	13.34	17.79	22.24	26.69	31.14	35.58	40.03
10	44.48	48.93	53.38	57.82	62.27	66.72	71.17	75.62	80.06	84.51
20	88.96	93.41	97.86	102.30	106.75	111.20	115.65	120.10	124.54	128.99
30	133.44	137.89	142.34	146.78	151.23	155.68	160.13	164.58	169.02	173.47
40	177.92	182.37	186.82	191.26	195.71	200.16	204.61	209.06	213.50	217.95
50	222.40	226.85	231.30	235.74	240.19	244.64	249.09	253.54	257.98	262.43
60	266.88	271.33	275.78	280.22	284.67	289.12	293.57	298.02	302.46	306.91
70	311.36	315.81	320.26	324.70	329.15	333.60	338.05	342.50	346.94	351.39
80	355.84	360.29	364.74	369.18	373.63	378.08	382.53	386.98	391.42	395.87
90	400.32	404.77	409.22	413.66	418.11	422.56	427.01	431.46	435.90	440.35
100	444.80	449.25	453.70	458.14	462.59	467.04	471.49	475.94	480.38	484.83

1. All values in this table are based on the relation, 1 ft lb/ft = 4.448 N•m/m. By manipulation of the decimal point any decimal value or multiple of foot pounds per foot may be converted to its equivalent in newton metres per metre.

TABLE U-12 Pressure and Stress Equivalents — Pounds per Square Inch to Kilograms per Square Centimetre

Psi	0	1	2	3	4	5	6	7	8	9
	kg/cm²									
0	0.00	0.07	0.14	0.21	0.28	0.35	0.42	0.49	0.56	0.63
10	0.71	0.78	0.85	0.92	0.99	1.06	1.13	1.20	1.27	1.34
20	1.41	1.48	1.55	1.62	1.69	1.76	1.83	1.90	1.97	2.05
30	2.12	2.19	2.26	2.33	2.40	2.47	2.54	2.61	2.68	2.75
40	2.82	2.89	2.96	3.03	3.10	3.17	3.24	3.31	3.39	3.46
50	3.53	3.60	3.67	3.74	3.81	3.88	3.95	4.02	4.09	4.16
60	4.23	4.30	4.37	4.44	4.51	4.58	4.65	4.73	4.80	4.87
70	4.94	5.01	5.08	5.15	5.22	5.29	5.36	5.43	5.50	5.57
80	5.64	5.71	5.78	5.85	5.92	5.99	6.07	6.14	6.21	6.28
90	6.35	6.42	6.49	6.56	6.63	6.70	6.77	6.84	6.91	6.98
100	7.05	7.12	7.19	7.26	7.33	7.41	7.48	7.55	7.62	7.69

1. All values in this table are based on the relation, 1 psi = 0.0705 kg/cm². By manipulation of the decimal point any decimal value or multiple of pounds per square inch may be converted to its equivalent in kilograms per square centimetre.

TABLE U-13 Moment Equivalents[1] — Foot Kips to Kilogram Metres

Foot Kips	0	1	2	3	4	5	6	7	8	9
	kg·m									
0	0.00	1.39	2.77	4.16	5.55	6.94	8.32	9.71	11.10	12.48
10	13.87	15.26	16.64	18.03	19.42	20.81	22.19	23.58	24.97	26.35
20	27.74	29.13	30.51	31.90	33.29	34.68	36.06	37.45	38.84	40.22
30	41.61	43.00	44.38	45.77	47.16	48.55	49.93	51.32	52.71	54.09
40	55.48	56.87	58.25	59.64	61.03	62.42	63.80	65.19	66.58	67.96
50	69.35	70.74	72.12	73.51	74.90	76.29	77.67	79.06	80.45	81.83
60	83.22	84.61	85.99	87.38	88.77	90.16	91.54	92.93	94.32	95.70
70	97.09	98.48	99.86	101.25	102.64	104.03	105.41	106.80	108.19	109.57
80	110.96	112.35	113.73	115.12	116.51	117.90	119.28	120.67	122.06	123.44
90	124.83	126.22	127.60	128.99	130.38	131.77	133.15	134.54	135.93	137.31
100	138.70	140.09	141.47	142.86	144.25	145.64	147.02	148.41	149.80	151.18

1. All values in this table are based on the relation, 1 ft k = 1.387 kg·m. By manipulation of the decimal value or multiple of foot kips may be converted to its equivalent in kilogram metres.

TABLE U-14 Pounds per Linear Foot Equivalents to Kilograms per Metre

Lb/Ft	0	1	2	3	4	5	6	7	8	9
	kg·m									
0	0.00	1.49	2.99	4.48	5.97	7.46	8.96	10.45	11.94	13.44
10	14.93	16.42	17.91	19.41	20.90	22.39	23.88	25.38	26.87	28.36
20	29.86	31.35	32.84	34.33	35.83	37.32	38.81	40.31	41.80	43.29
30	44.78	46.28	47.77	49.26	50.75	52.25	53.74	55.23	56.73	58.22
40	59.71	61.20	62.70	64.19	65.68	67.18	68.67	70.16	71.65	73.15
50	74.64	76.13	77.62	79.12	80.61	82.10	83.60	85.09	86.58	88.07
60	89.57	91.06	92.55	94.05	95.54	97.03	98.52	100.02	101.51	103.00
70	104.49	105.99	107.48	108.97	110.47	111.96	113.45	114.94	116.44	117.93
80	119.42	120.92	122.41	123.90	125.39	126.89	128.38	129.87	131.36	132.86
90	134.35	135.84	137.34	138.83	140.32	141.81	143.31	144.80	146.29	147.79
100	149.28	150.77	152.26	153.76	155.25	156.74	158.23	159.73	161.22	162.71

1. All values in this table are based on the relation, 1 lb/ft = 1.49 kg·m. By manipulation of the decimal point any decimal value or multiple of pounds per foot may be converted to its equivalent in kilograms per metre.

STRENGTH DESIGN TABLES and DIAGRAMS

Based on the Uniform Building Code Requirements

STR. DES. TABLE A-1   Coefficients for Flexural Strength Design: For use of Table,
f'_m = 1500 psi and f_y = 60,000 psi[1,2]   see Example 6-C

$$f'_m = 1500 \text{ psi} \qquad f_y = 60,000 \text{ psi}$$

$$K_u = \phi f'_m q(1 - 0.59q) = \frac{M_u}{bd^2} \qquad q = p\left(\frac{f_y}{f'_m}\right)$$

$$a_u = \frac{K_u}{12,000\,p} = \frac{\phi f_y(1 - 0.59q)}{12,000} \qquad p = q\left(\frac{f'_m}{f_y}\right) = \frac{A_s}{bd}$$

$$A_s = \frac{M_u}{a_u d} \qquad\qquad \phi = 0.8$$

$$\frac{a}{d} = 0.85\frac{c}{d} \qquad \frac{c}{d} = 1.39q \qquad j = 1 - \frac{a}{2}$$

q	K_u	a_u	p	c/d	a/d	j
0.030	35.4	3.93	0.0008	0.041	0.035	0.983
0.040	46.9	3.91	0.0010	0.055	0.047	0.977
0.050	58.2	3.88	0.0013	0.068	0.058	0.971
0.060	69.5	3.86	0.0015	0.082	0.070	0.965
0.070	80.5	3.83	0.0018	0.096	0.081	0.959
0.080	91.5	3.81	0.0020	0.109	0.093	0.953
0.090	102.3	3.79	0.0023	0.123	0.105	0.948
0.100	112.9	3.76	0.0025	0.137	0.116	0.942
0.110	123.4	3.74	0.0028	0.150	0.128	0.936
0.120	133.8	3.72	0.0030	0.164	0.140	0.930
0.130	144.0	3.69	0.0033	0.178	0.151	0.924
0.140	154.1	3.67	0.0035	0.192	0.163	0.919
0.150	164.1	3.65	0.0038	0.205	0.174	0.913
0.160	173.9	3.62	0.0040	0.219	0.186	0.907
0.170	183.5	3.60	0.0043	0.233	0.198	0.901
0.180	193.1	3.58	0.0045	0.246	0.209	0.895
0.190	202.4	3.55	0.0048	0.260	0.221	0.890
0.200	211.7	3.53	0.0050	0.274	0.233	0.884
0.210	220.8	3.50	0.0053	0.287	0.244	0.878
0.220	229.7	3.48	0.0055	0.301	0.256	0.872
0.230	238.5	3.46	0.0058	0.315	0.267	0.866
0.240	247.2	3.43	0.0060	0.328	0.279	0.860
0.250	255.8	3.41	0.0063	0.342	0.291	0.855
0.260	264.1	3.39	0.0065	0.356	0.302	0.849
0.270	272.4	3.36	0.0068	0.369	0.314	0.843
0.280	280.5	3.34	0.0070	0.383	0.326	0.837
0.290	288.5	3.32	0.0073	0.397	0.337	0.831
0.300	296.3	3.29	0.0075	0.410	0.349	0.826
0.310	304.0	3.27	0.0078	0.424	0.360	0.820
0.320	311.5	3.24	0.0080	0.438	0.372	0.814

1. Values below the solid bold line exceed the limit of $0.5p_b$ ($0.5p_b$ = 0.0053) for slender walls (UBC Sec. 2411(b)).

2. Maximum steel ratios shown are based on the limit of $0.75p_b$ for reinforced concrete (UBC Sec. 2610(d)3).

STR. DES. TABLE A-2 Coefficients for Flexural Strength Design: For use of Table,
f'_m = 2000 psi and f_y = 60,000 psi[1,2] see Example 6-C

$f'_m = 2000$ psi $\qquad f_y = 60,000$ psi

$$K_u = \phi f'_m q(1 - 0.59q) = \frac{M_u}{b d^2} \qquad q = p\left(\frac{f_y}{f'_m}\right)$$

$$a_u = \frac{K_u}{12,000\,p} = \frac{\phi f_y(1 - 0.59q)}{12,000} \qquad p = q\left(\frac{f'_m}{f_y}\right) = \frac{A_s}{b\,d}$$

$$A_s = \frac{M_u}{a_u d} \qquad \phi = 0.8$$

$$\frac{a}{d} = 0.85\frac{c}{d} \qquad \frac{c}{d} = 1.39q \qquad j = 1 - \frac{a}{2}$$

q	K_u	a_u	p	c/d	a/d	j
0.030	47.2	3.93	0.0010	0.041	0.035	0.983
0.040	62.5	3.91	0.0013	0.055	0.047	0.977
0.050	77.6	3.88	0.0017	0.068	0.058	0.971
0.060	92.6	3.86	0.0020	0.082	0.070	0.965
0.070	107.4	3.83	0.0023	0.096	0.081	0.959
0.080	122.0	3.81	0.0027	0.109	0.093	0.953
0.090	136.4	3.79	0.0030	0.123	0.105	0.948
0.100	150.6	3.76	0.0033	0.137	0.116	0.942
0.110	164.6	3.74	0.0037	0.150	0.128	0.936
0.120	178.4	3.72	0.0040	0.164	0.140	0.930
0.130	192.0	3.69	0.0043	0.178	0.151	0.924
0.140	205.5	3.67	0.0047	0.192	0.163	0.919
0.150	218.8	3.65	0.0050	0.205	0.174	0.913
0.160	231.8	3.62	0.0053	0.219	0.186	0.907
0.170	244.7	3.60	0.0057	0.233	0.198	0.901
0.180	257.4	3.58	0.0060	0.246	0.209	0.895
0.190	269.9	3.55	0.0063	0.260	0.221	0.890
0.200	282.2	3.53	0.0067	0.274	0.233	0.884
0.210	294.4	3.50	0.0070	0.287	0.244	0.878
0.220	306.3	3.48	0.0073	0.301	0.256	0.872
0.230	318.1	3.46	0.0077	0.315	0.267	0.866
0.240	329.6	3.43	0.0080	0.328	0.279	0.860
0.250	341.0	3.41	0.0083	0.342	0.291	0.855
0.260	352.2	3.39	0.0087	0.356	0.302	0.849
0.270	363.2	3.36	0.0090	0.369	0.314	0.843
0.280	374.0	3.34	0.0093	0.383	0.326	0.837
0.290	384.6	3.32	0.0097	0.397	0.337	0.831
0.300	395.0	3.29	0.0100	0.410	0.349	0.826
0.310	405.3	3.27	0.0103	0.424	0.360	0.820
0.320	415.3	3.24	0.0107	0.438	0.372	0.814

1. Values below the solid bold line exceed the limit of $0.5p_b$ ($0.5p_b = 0.0071$) for slender walls (UBC Sec. 2411(b)).

2. Maximum steel ratios shown are based on the limit of $0.75p_b$ for reinforced concrete (UBC Sec. 2610(d)3).

STR. DES. TABLE A-3 Coefficients for Flexural Strength Design: For use of Table,
f'_m = 2500 psi and f_y = 60,000 psi[1,2]   see Example 6-C

$$f'_m = 2500 \text{ psi} \qquad f_y = 60,000 \text{ psi}$$

$$K_u = \phi f'_m q (1 - 0.59q) = \frac{M_u}{b\,d^2} \qquad q = p\left(\frac{f_y}{f'_m}\right)$$

$$a_u = \frac{K_u}{12,000\,p} = \frac{\phi f_y (1 - 0.59q)}{12,000} \qquad p = q\left(\frac{f'_m}{f_y}\right) = \frac{A_s}{b\,d}$$

$$A_s = \frac{M_u}{a_u\,d} \qquad\qquad \phi = 0.8$$

$$\frac{a}{d} = 0.85\frac{c}{d} \qquad \frac{c}{d} = 1.39q \qquad j = 1 - \frac{a}{2}$$

q	K_u	a_u	p	c/d	a/d	j
0.030	58.9	3.93	0.0013	0.041	0.035	0.983
0.040	78.1	3.91	0.0017	0.055	0.047	0.977
0.050	97.1	3.88	0.0021	0.068	0.058	0.971
0.060	115.8	3.86	0.0025	0.082	0.070	0.965
0.070	134.2	3.83	0.0029	0.096	0.081	0.959
0.080	152.4	3.81	0.0033	0.109	0.093	0.953
0.090	170.4	3.79	0.0038	0.123	0.105	0.948
0.100	188.2	3.76	0.0042	0.137	0.116	0.942
0.110	205.7	3.74	0.0046	0.150	0.128	0.936
0.120	223.0	3.72	0.0050	0.164	0.140	0.930
0.130	240.1	3.69	0.0054	0.178	0.151	0.924
0.140	256.9	3.67	0.0058	0.192	0.163	0.919
0.150	273.5	3.65	0.0063	0.205	0.174	0.913
0.160	289.8	3.62	0.0067	0.219	0.186	0.907
0.170	305.9	3.60	0.0071	0.233	0.198	0.901
0.180	321.8	3.58	0.0075	0.246	0.209	0.895
0.190	337.4	3.55	0.0079	0.260	0.221	0.890
0.200	352.8	3.53	0.0083	0.274	0.233	0.884
0.210	368.0	3.50	0.0088	0.287	0.244	0.878
0.220	382.9	3.48	0.0092	0.301	0.256	0.872
0.230	397.6	3.46	0.0096	0.315	0.267	0.866
0.240	412.0	3.43	0.0100	0.328	0.279	0.860
0.250	426.3	3.41	0.0104	0.342	0.291	0.855
0.260	440.2	3.39	0.0108	0.356	0.302	0.849
0.270	454.0	3.36	0.0113	0.369	0.314	0.843
0.280	467.5	3.34	0.0117	0.383	0.326	0.837
0.290	480.8	3.32	0.0121	0.397	0.337	0.831
0.300	493.8	3.29	0.0125	0.410	0.349	0.826
0.310	506.6	3.27	0.0129	0.424	0.360	0.820
0.320	519.2	3.24	0.0133	0.438	0.372	0.814

1. Values below the solid bold line exceed the limit of $0.5p_b$ ($0.5p_b$ = 0.0089) for slender walls (UBC Sec. 2411(b)).

2. Maximum steel ratios shown are based on the limit of $0.75p_b$ for reinforced concrete (UBC Sec. 2610(d)3).

STR. DES. TABLE A-4 Coefficients for Flexural Strength Design: For use of Table,
f'_m = 3000 psi and f_y = 60,000 psi[1,2] see Example 6-C

$f'_m = 3000$ psi $f_y = 60,000$ psi

$$K_u = \phi f'_m q(1 - 0.59q) = \frac{M_u}{b\,d^2} \qquad q = p\left(\frac{f_y}{f'_m}\right)$$

$$a_u = \frac{K_u}{12,000\,p} = \frac{\phi f_y(1 - 0.59q)}{12,000} \qquad p = q\left(\frac{f'_m}{f_y}\right) = \frac{A_s}{b\,d}$$

$$A_s = \frac{M_u}{a_u\,d} \qquad \phi = 0.8$$

$$\frac{a}{d} = 0.85\frac{c}{d} \qquad \frac{c}{d} = 1.39q \qquad j = 1 - \frac{a}{2}$$

q	K_u	a_u	p	c/d	a/d	j
0.020	47.4	3.95	0.0010	0.027	0.023	0.988
0.030	70.7	3.93	0.0015	0.041	0.035	0.983
0.040	93.7	3.91	0.0020	0.055	0.047	0.977
0.050	116.5	3.88	0.0025	0.068	0.058	0.971
0.060	138.9	3.86	0.0030	0.082	0.070	0.965
0.070	161.1	3.83	0.0035	0.096	0.081	0.959
0.080	182.9	3.81	0.0040	0.109	0.093	0.953
0.090	204.5	3.79	0.0045	0.123	0.105	0.948
0.100	225.8	3.76	0.0050	0.137	0.116	0.942
0.110	246.9	3.74	0.0055	0.150	0.128	0.936
0.120	267.6	3.72	0.0060	0.164	0.140	0.930
0.130	288.1	3.69	0.0065	0.178	0.151	0.924
0.140	308.2	3.67	0.0070	0.192	0.163	0.919
0.150	328.1	3.65	0.0075	0.205	0.174	0.913
0.160	347.8	3.62	0.0080	0.219	0.186	0.907
0.170	367.1	3.60	0.0085	0.233	0.198	0.901
0.180	386.1	3.58	0.0090	0.246	0.209	0.895
0.190	404.9	3.55	0.0095	0.260	0.221	0.890
0.200	423.4	3.53	0.0100	0.274	0.233	0.884
0.210	441.6	3.50	0.0105	0.287	0.244	0.878
0.220	459.5	3.48	0.0110	0.301	0.256	0.872
0.230	477.1	3.46	0.0115	0.315	0.267	0.866
0.240	494.4	3.43	0.0120	0.328	0.279	0.860
0.250	511.5	3.41	0.0125	0.342	0.291	0.855
0.260	528.3	3.39	0.0130	0.356	0.302	0.849
0.270	544.8	3.36	0.0135	0.369	0.314	0.843
0.280	561.0	3.34	0.0140	0.383	0.326	0.837
0.290	576.9	3.32	0.0145	0.397	0.337	0.831
0.300	592.6	3.29	0.0150	0.410	0.349	0.826
0.310	607.9	3.27	0.0155	0.424	0.360	0.820
0.320	623.0	3.24	0.0160	0.438	0.372	0.814

1. Values below the solid bold line exceed the limit of $0.5p_b$ ($0.5p_b = 0.0107$) for slender walls (UBC Sec. 2411(b)).

2. Maximum steel ratios shown are based on the limit of $0.75p_b$ for reinforced concrete (UBC Sec. 2610(d)3).

STR. DES. TABLE A-5 Coefficients for Flexural Strength Design:
$f'_m = 3500$ psi and $f_y = 60,000$ psi[1,2]

For use of Table, see Example 6-C

$f'_m = 3500$ psi $f_y = 60,000$ psi

$$K_u = \phi f'_m q(1 - 0.59q) = \frac{M_u}{bd^2}$$

$$q = p\left(\frac{f_y}{f_m}\right)$$

$$a_u = \frac{K_u}{12,000\,p} = \frac{\phi f_y(1 - 0.59q)}{12,000}$$

$$p = q\left(\frac{f'_m}{f_y}\right) = \frac{A_s}{bd}$$

$$A_s = \frac{M_u}{a_u d}$$

$$\phi = 0.8$$

$$\frac{a}{d} = 0.85\frac{c}{d} \qquad \frac{c}{d} = 1.39q \qquad j = 1 - \frac{a}{2}$$

q	K_u	a_u	p	c/d	a/d	j
0.020	55.3	3.95	0.0012	0.027	0.023	0.988
0.030	82.5	3.93	0.0018	0.041	0.035	0.983
0.040	109.4	3.91	0.0023	0.055	0.047	0.977
0.050	135.9	3.88	0.0029	0.068	0.058	0.971
0.060	162.1	3.86	0.0035	0.082	0.070	0.965
0.070	187.9	3.83	0.0041	0.096	0.081	0.959
0.080	213.4	3.81	0.0047	0.109	0.093	0.953
0.090	238.6	3.79	0.0053	0.123	0.105	0.948
0.100	263.5	3.76	0.0058	0.137	0.116	0.942
0.110	288.0	3.74	0.0064	0.150	0.128	0.936
0.120	312.2	3.72	0.0070	0.164	0.140	0.930
0.130	336.1	3.69	0.0076	0.178	0.151	0.924
0.140	359.6	3.67	0.0082	0.192	0.163	0.919
0.150	382.8	3.65	0.0088	0.205	0.174	0.913
0.160	405.7	3.62	0.0093	0.219	0.186	0.907
0.170	428.3	3.60	0.0099	0.233	0.198	0.901
0.180	450.5	3.58	0.0105	0.246	0.209	0.895
0.190	472.4	3.55	0.0111	0.260	0.221	0.890
0.200	493.9	3.53	0.0117	0.274	0.233	0.884
0.210	515.1	3.50	0.0123	0.287	0.244	0.878
0.220	536.0	3.48	0.0128	0.301	0.256	0.872
0.230	556.6	3.46	0.0134	0.315	0.267	0.866
0.240	576.8	3.43	0.0140	0.328	0.279	0.860
0.250	596.8	3.41	0.0146	0.342	0.291	0.855
0.260	616.3	3.39	0.0152	0.356	0.302	0.849
0.270	635.6	3.36	0.0158	0.369	0.314	0.843
0.280	654.5	3.34	0.0163	0.383	0.326	0.837
0.290	673.1	3.32	0.0169	0.397	0.337	0.831
0.300	691.3	3.29	0.0175	0.410	0.349	0.826
0.310	709.2	3.27	0.0181	0.424	0.360	0.820
0.320	726.8	3.24	0.0187	0.438	0.372	0.814

1. Values below the solid bold line exceed the limit of $0.5p_b$ ($0.5p_b = 0.0125$) for slender walls (UBC Sec. 2411(b)).

2. Maximum steel ratios shown are based on the limit of $0.75p_b$ for reinforced concrete (UBC Sec. 2610(d)3).

STR. DES. TABLE A-6 Coefficients for Flexural Strength Design: For use of Table,
f'_m = 4000 psi and f_y = 60,000 psi[1,2]   see Example 6-C

$$f'_m = 4000 \text{ psi} \qquad f_y = 60,000 \text{ psi}$$

$$K_u = \phi f'_m q(1 - 0.59q) = \frac{M_u}{bd^2} \qquad q = p\left(\frac{f_y}{f'_m}\right)$$

$$a_u = \frac{K_u}{12,000\,p} = \frac{\phi f_y(1 - 0.59q)}{12,000} \qquad p = q\left(\frac{f'_m}{f_y}\right) = \frac{A_s}{bd}$$

$$A_s = \frac{M_u}{a_u d} \qquad \phi = 0.8$$

$$\frac{a}{d} = 0.85\frac{c}{d} \qquad \frac{c}{d} = 1.39q \qquad j = 1 - \frac{a}{2}$$

q	K_u	a_u	p	c/d	a/d	j
0.010	31.8	3.98	0.0007	0.014	0.012	0.994
0.020	63.2	3.95	0.0013	0.027	0.023	0.988
0.030	94.3	3.93	0.0020	0.041	0.035	0.983
0.040	125.0	3.91	0.0027	0.055	0.047	0.977
0.050	155.3	3.88	0.0033	0.068	0.058	0.971
0.060	185.2	3.86	0.0040	0.082	0.070	0.965
0.070	214.7	3.83	0.0047	0.096	0.081	0.959
0.080	243.9	3.81	0.0053	0.109	0.093	0.953
0.090	272.7	3.79	0.0060	0.123	0.105	0.948
0.100	301.1	3.76	0.0067	0.137	0.116	0.942
0.110	329.2	3.74	0.0073	0.150	0.128	0.936
0.120	356.8	3.72	0.0080	0.164	0.140	0.930
0.130	384.1	3.69	0.0087	0.178	0.151	0.924
0.140	411.0	3.67	0.0093	0.192	0.163	0.919
0.150	437.5	3.65	0.0100	0.205	0.174	0.913
0.160	463.7	3.62	0.0107	0.219	0.186	0.907
0.170	489.4	3.60	0.0113	0.233	0.198	0.901
0.180	514.8	3.58	0.0120	0.246	0.209	0.895
0.190	539.8	3.55	0.0127	0.260	0.221	0.890
0.200	564.5	3.53	0.0133	0.274	0.233	0.884
0.210	588.7	**3.50**	**0.0140**	0.287	0.244	0.878
0.220	612.6	3.48	0.0147	0.301	0.256	0.872
0.230	636.1	3.46	0.0153	0.315	0.267	0.866
0.240	659.3	3.43	0.0160	0.328	0.279	0.860
0.250	682.0	3.41	0.0167	0.342	0.291	0.855
0.260	704.4	3.39	0.0173	0.356	0.302	0.849
0.270	726.4	3.36	0.0180	0.369	0.314	0.843
0.280	748.0	3.34	0.0187	0.383	0.326	0.837
0.290	769.2	3.32	0.0193	0.397	0.337	0.831
0.300	790.1	3.29	0.0200	0.410	0.349	0.826
0.310	810.6	3.27	0.0207	0.424	0.360	0.820
0.320	830.7	3.24	0.0213	0.438	0.372	0.814

1. Values below the solid bold line exceed the limit of $0.5p_b$ ($0.5p_b = 0.0143$) for slender walls (UBC Sec. 2411(b)).

2. Maximum steel ratios shown are based on the limit of $0.75p_b$ for reinforced concrete (UBC Sec. 2610(d)3).

STR. DES. TABLE A-7 Coefficients for Flexural Strength Design: For use of Table, f'_m = 4500 psi and f_y = 60,000 psi[1,2] see Example 6-C

$$f'_m = 4500 \text{ psi} \qquad f_y = 60,000 \text{ psi}$$

$$K_u = \phi f'_m q(1 - 0.59q) = \frac{M_u}{b\,d^2} \qquad q = p\left(\frac{f_y}{f'_m}\right)$$

$$a_u = \frac{K_u}{12,000\,p} = \frac{\phi f_y(1 - 0.59q)}{12,000} \qquad p = q\left(\frac{f'_m}{f_y}\right) = \frac{A_s}{b\,d}$$

$$A_s = \frac{M_u}{a_u\,d} \qquad\qquad \phi = 0.8$$

$$\frac{a}{d} = 0.85\frac{c}{d} \qquad \frac{c}{d} = 1.39q \qquad j = 1 - \frac{a}{2}$$

q	K_u	a_u	p	c/d	a/d	j
0.010	35.8	3.98	0.0008	0.014	0.012	0.994
0.020	71.2	3.95	0.0015	0.027	0.023	0.988
0.030	106.1	3.93	0.0023	0.041	0.035	0.983
0.040	140.6	3.91	0.0030	0.055	0.047	0.977
0.050	174.7	3.88	0.0038	0.068	0.058	0.971
0.060	208.4	3.86	0.0045	0.082	0.070	0.965
0.070	241.6	3.83	0.0053	0.096	0.081	0.959
0.080	274.4	3.81	0.0060	0.109	0.093	0.953
0.090	306.8	3.79	0.0068	0.123	0.105	0.948
0.100	338.8	3.76	0.0075	0.137	0.116	0.942
0.110	370.3	3.74	0.0083	0.150	0.128	0.936
0.120	401.4	3.72	0.0090	0.164	0.140	0.930
0.130	432.1	3.69	0.0098	0.178	0.151	0.924
0.140	462.4	3.67	0.0105	0.192	0.163	0.919
0.150	492.2	3.65	0.0113	0.205	0.174	0.913
0.160	521.6	3.62	0.0120	0.219	0.186	0.907
0.170	550.6	3.60	0.0128	0.233	0.198	0.901
0.180	579.2	3.58	0.0135	0.246	0.209	0.895
0.190	607.3	3.55	0.0143	0.260	0.221	0.890
0.200	635.0	3.53	0.0150	0.274	0.233	0.884
0.210	662.3	3.50	0.0158	0.287	0.244	0.878
0.220	689.2	3.48	0.0165	0.301	0.256	0.872
0.230	715.6	3.46	0.0173	0.315	0.267	0.866
0.240	741.7	3.43	0.0180	0.328	0.279	0.860
0.250	767.3	3.41	0.0188	0.342	0.291	0.855
0.260	792.4	3.39	0.0195	0.356	0.302	0.849
0.270	817.2	3.36	0.0203	0.369	0.314	0.843
0.280	841.5	3.34	0.0210	0.383	0.326	0.837
0.290	865.4	3.32	0.0218	0.397	0.337	0.831
0.300	888.8	3.29	0.0225	0.410	0.349	0.826
0.310	911.9	3.27	0.0233	0.424	0.360	0.820
0.320	934.5	3.24	0.0240	0.438	0.372	0.814

1. Values below the solid bold line exceed the limit of $0.5p_b$ ($0.5p_b$ = 0.0156) for slender walls (UBC Sec. 2411(b)).

2. Maximum steel ratios shown are based on the limit of $0.75p_b$ for reinforced concrete (UBC Sec. 2610(d)3).

STR. DES. TABLE A-8 Coefficients for Flexural Strength Design: For use of Table,
f'_m = 5000 psi and f_y = 60,000 psi[1,2] see Example 6-C

$f'_m = 5000$ psi $\qquad f_y = 60,000$ psi

$$K_u = \phi f'_m q (1 - 0.59q) = \frac{M_u}{bd^2} \qquad q = p\left(\frac{f_y}{f'_m}\right)$$

$$a_u = \frac{K_u}{12,000\,p} = \frac{\phi f_y (1 - 0.59q)}{12,000} \qquad p = q\left(\frac{f'_m}{f_y}\right) = \frac{A_s}{bd}$$

$$A_s = \frac{M_u}{a_u d} \qquad\qquad \phi = 0.8$$

$$\frac{a}{d} = 0.85\frac{c}{d} \qquad \frac{c}{d} = 1.39q \qquad j = 1 - \frac{a}{2}$$

q	K_u	a_u	p	c/d	a/d	j
0.010	39.8	3.98	0.0008	0.014	0.012	0.994
0.020	79.1	3.95	0.0017	0.027	0.023	0.988
0.030	117.9	3.93	0.0025	0.041	0.035	0.983
0.040	156.2	3.91	0.0033	0.055	0.047	0.977
0.050	194.1	3.88	0.0042	0.068	0.058	0.971
0.060	231.5	3.86	0.0050	0.082	0.070	0.965
0.070	268.4	3.83	0.0058	0.096	0.081	0.959
0.080	304.9	3.81	0.0067	0.109	0.093	0.953
0.090	340.9	3.79	0.0075	0.123	0.105	0.948
0.100	376.4	3.76	0.0083	0.137	0.116	0.942
0.110	411.4	3.74	0.0092	0.150	0.128	0.936
0.120	446.0	3.72	0.0100	0.164	0.140	0.930
0.130	480.1	3.69	0.0108	0.178	0.151	0.924
0.140	513.7	3.67	0.0117	0.192	0.163	0.919
0.150	546.9	3.65	0.0125	0.205	0.174	0.913
0.160	579.6	3.62	0.0133	0.219	0.186	0.907
0.170	611.8	3.60	0.0142	0.233	0.198	0.901
0.180	643.5	3.58	0.0150	0.246	0.209	0.895
0.190	674.8	3.55	0.0158	0.260	0.221	0.890
0.200	705.6	3.53	0.0167	0.274	0.233	0.884
0.210	735.9	3.50	0.0175	0.287	0.244	0.878
0.220	765.8	3.48	0.0183	0.301	0.256	0.872
0.230	795.2	3.46	0.0192	0.315	0.267	0.866
0.240	824.1	3.43	0.0200	0.328	0.279	0.860
0.250	852.5	3.41	0.0208	0.342	0.291	0.855
0.260	880.5	3.39	0.0217	0.356	0.302	0.849
0.270	908.0	3.36	0.0225	0.369	0.314	0.843
0.280	935.0	3.34	0.0233	0.383	0.326	0.837
0.290	961.5	3.32	0.0242	0.397	0.337	0.831
0.300	987.6	3.29	0.0250	0.410	0.349	0.826
0.310	1013.2	3.27	0.0258	0.424	0.360	0.820
0.320	1038.3	3.24	0.0267	0.438	0.372	0.814

1. Values below the solid bold line exceed the limit of $0.5p_b$ ($0.5p_b = 0.0168$) for slender walls (UBC Sec. 2411(b)).

2. Maximum steel ratios shown are based on the limit of $0.75p_b$ for reinforced concrete (UBC Sec. 2610(d)3).

STR. DES. TABLE B-1 Design Coefficient, q, for the Determination of the Reinforcing Ratio, p, Based on the Factored Moment, M_u

For use of Table, see Example 6-F

$$\frac{M_u}{f'_m b d^2} = q(1 - 0.59q) \qquad q = \frac{p f_y}{f'_m} \qquad p = \frac{q f'_m}{f_y} \qquad \text{Nominal moment, } M_n = \phi M_u$$

$$\frac{M_n}{f'_m b d^2} = A_s f_y \left(d - \frac{a}{2} \right) f'_m b d^2 = q(1 - 0.59q), \text{ where } q = \frac{p f_y}{f'_m} \text{ and } a = \frac{A_s f_y}{0.85 f'_m b}$$

Design: Using the factored moment M_u, enter the table with $\dfrac{M_u}{\phi f'_m b d^2}$; find a and compute the steel percentage p from $p = \dfrac{q f'_m}{f'_y}$.

Investigation: Enter the table with q from $q = \dfrac{p f_y}{f'_m}$; find the value of $\dfrac{M_n}{f'_m b d^2}$ and solve for the nominal moment strength, M_n.

q	$M_u/f'_m b d^2$									
	0	0.001	0.002	0.003	0.004	0.005	0.006	0.007	0.008	0.009
0	0.0000	0.0010	0.0020	0.0030	0.0040	0.0050	0.0060	0.0070	0.0080	0.0090
0.01	0.0099	0.0109	0.0119	0.0129	0.0139	0.0149	0.0158	0.0168	0.0178	0.0188
0.02	0.0198	0.0207	0.0217	0.0227	0.0237	0.0246	0.0256	0.0266	0.0275	0.0285
0.03	0.0295	0.0304	0.0314	0.0324	0.0333	0.0343	0.0352	0.0362	0.0371	0.0381
0.04	0.0391	0.0400	0.0410	0.0419	0.0429	0.0438	0.0448	0.0457	0.0466	0.0476
0.05	0.0485	0.0495	0.0504	0.0513	0.0523	0.0532	0.0541	0.0551	0.0560	0.0569
0.06	0.0579	0.0588	0.0597	0.0607	0.0616	0.0625	0.0634	0.0644	0.0653	0.0662
0.07	0.0671	0.0680	0.0689	0.0699	0.0708	0.0717	0.0726	0.0735	0.0744	0.0753
0.08	0.0762	0.0771	0.0780	0.0789	0.0798	0.0807	0.0816	0.0825	0.0834	0.0843
0.09	0.0852	0.0861	0.0870	0.0879	0.0888	0.0897	0.0906	0.0914	0.0923	0.0932
0.10	0.0941	0.0950	0.0959	0.0967	0.0976	0.0985	0.0994	0.1002	0.1011	0.1020
0.11	0.1029	0.1037	0.1046	0.1055	0.1063	0.1072	0.1081	0.1089	0.1098	0.1106
0.12	0.1115	0.1124	0.1132	0.1141	0.1149	0.1158	0.1166	0.1175	0.1183	0.1192
0.13	0.1200	0.1209	0.1217	0.1226	0.1234	0.1242	0.1251	0.1259	0.1268	0.1276
0.14	0.1284	0.1293	0.1301	0.1309	0.1318	0.1326	0.1334	0.1343	0.1351	0.1359
0.15	0.1367	0.1375	0.1384	0.1392	0.1400	0.1408	0.1416	0.1425	0.1433	0.1441
0.16	0.1449	0.1457	0.1465	0.1473	0.1481	0.1489	0.1497	0.1505	0.1513	0.1521
0.17	0.1529	0.1537	0.1545	0.1553	0.1561	0.1569	0.1577	0.1585	0.1593	0.1601
0.18	0.1609	0.1617	0.1625	0.1632	0.1640	0.1648	0.1656	0.1664	0.1671	0.1679
0.19	0.1687	0.1695	0.1703	0.1710	0.1718	0.1726	0.1733	0.1741	0.1749	0.1756
0.20	0.1764	0.1772	0.1779	0.1787	0.1794	0.1802	0.1810	0.1817	0.1825	0.1832
0.21	0.1840	0.1847	0.1855	0.1862	0.1870	0.1877	0.1885	0.1892	0.1900	0.1907
0.22	0.1914	0.1922	0.1929	0.1937	0.1944	0.1951	0.1959	0.1966	0.1973	0.1981
0.23	0.1988	0.1995	0.2002	0.2010	0.2017	0.2024	0.2031	0.2039	0.2046	0.2053
0.24	0.2060	0.2067	0.2074	0.2082	0.2089	0.2096	0.2103	0.2110	0.2117	0.2124
0.25	0.2131	0.2138	0.2145	0.2152	0.2159	0.2166	0.2173	0.2180	0.2187	0.2194
0.26	0.2201	0.2208	0.2215	0.2222	0.2229	0.2236	0.2243	0.2249	0.2256	0.2263
0.27	0.2270	0.2277	0.2283	0.2290	0.2297	0.2304	0.2311	0.2317	0.2324	0.2331
0.28	0.2337	0.2344	0.2351	0.2357	0.2364	0.2371	0.2377	0.2384	0.2391	0.2397
0.29	0.2404	0.2410	0.2417	0.2423	0.2430	0.2437	0.2443	0.2450	0.2456	0.2463
0.30	0.2469	0.2475	0.2482	0.2488	0.2495	0.2501	0.2508	0.2514	0.2520	0.2527
0.31	0.2533	0.2539	0.2546	0.2552	0.2558	0.2565	0.2571	0.2577	0.2583	0.2590
0.32	0.2596	0.2602	0.2608	0.2614	0.2621	0.2627	0.2633	0.2639	0.2645	0.2651
0.33	0.2657	0.2664	0.2670	0.2676	0.2682	0.2688	0.2694	0.2700	0.2706	0.2712
0.34	0.2718	0.2724	0.2730	0.2736	0.2742	0.2748	0.2754	0.2760	0.2765	0.2771
0.35	0.2777	0.2783	0.2789	0.2795	0.2801	0.2806	0.2812	0.2818	0.2824	0.2830
0.36	0.2835	0.2841	0.2847	0.2853	0.2858	0.2864	0.2870	0.2875	0.2881	0.2887
0.37	0.2892	0.2898	0.2904	0.2909	0.2915	0.2920	0.2926	0.2931	0.2937	0.2943
0.38	0.2948	0.2954	0.2959	0.2965	0.2970	0.2975	0.2981	0.2986	0.2992	0.2997
0.39	0.3003	0.3008	0.3013	0.3019	0.3024	0.3029	0.3035	0.3040	0.3045	0.3051

STR. DES. TABLE C-1   Moment Capacity of Walls and Beams:
$f'_m = 1500$ psi and $f_y = 60,000$ psi

$$M_u = \phi\left(d^2\right)\left(f'_m\right)q\left(1 - 0.59q\right) \text{ ft kips}$$

$f'_m = 1500$ psi $f_y = 60,000$ psi

$$q = p\left(\frac{f_y}{f'_m}\right)$$ $b = 12$ inches
                                                        $= 1$ foot

$\phi = 0.8$

Moment Capacity in ft kips

Wall Section   **Beam Section**

$p^{1,2}$	d (inches)											
	2.8	**3.8**	**4.8**	**5.3**	**7.3**	**9.0**	**12.0**	**18.0**	**22.0**	**26.0**	**30.0**	**36.0**
0.0010	0.37	0.68	1.08	1.32	2.50	3.80	6.75	15.18	22.68	31.68	42.18	60.74
0.0015	0.54	1.00	1.60	1.95	3.70	5.63	10.00	22.50	33.61	46.95	62.51	90.01
0.0020	0.72	1.32	2.11	2.57	4.87	7.41	13.17	29.64	44.27	61.83	82.32	118.54
0.0025	0.89	1.63	2.60	3.17	6.02	9.15	16.26	36.59	54.65	76.33	101.63	146.34
0.0030	1.05	1.93	3.08	3.76	7.13	10.84	19.27	43.35	64.76	90.45	120.42	173.41
0.0035	1.21	2.23	3.55	4.33	8.21	12.48	22.19	49.94	74.60	104.19	138.71	199.74
0.0040	1.36	2.51	4.01	4.88	9.27	14.08	25.04	56.34	84.16	117.54	156.49	225.34
0.0045	1.51	2.79	4.45	5.42	10.29	15.64	27.80	62.55	93.44	130.51	173.75	250.21
0.0050	1.66	3.06	4.88	5.95	11.28	17.15	30.48	68.58	102.45	143.10	190.51	274.34
0.0055	1.80	3.32	5.29	6.45	12.24	18.61	33.08	74.43	111.19	155.30	206.76	297.73
0.0060	1.94	3.57	5.70	6.94	13.17	20.02	35.60	80.10	119.65	167.12	222.50	320.40
0.0065	2.07	3.81	6.09	7.42	14.08	21.40	38.04	85.58	127.84	178.56	237.73	342.32
0.0070	2.20	4.05	6.46	7.88	14.95	22.72	40.39	90.88	135.76	189.61	252.44	363.52
0.0075	2.32	4.28	6.83	8.32	15.79	24.00	42.66	95.99	143.40	200.29	266.65	383.98
0.0080	2.44	4.50	7.18	8.75	16.60	25.23	44.86	100.93	150.77	210.57	280.35	403.71

1. UBC Section 2411(b) limits p to $0.5p_b = 0.0053$ for slender wall design.

2. UBC Section 2610(d)3 limits p to $0.75p_b = 0.0080$ for reinforced concrete design.

STR. DES.

STR. DES. TABLE C-2 Moment Capacity of Walls and Beams:
f'_m = 2000 psi and f_y = 60,000 psi

$$M_u = \phi\left(d^2\right)\left(f'_m\right)q\left(1-0.59q\right) \text{ ft kips}$$

$f'_m = 2000 \text{ psi} \qquad f_y = 60,000 \text{ psi}$

$q = p\left(\dfrac{f_y}{f'_m}\right) \qquad b = 12 \text{ inches}$

$\qquad\qquad\qquad\qquad = 1 \text{ foot}$

$\phi = 0.8$

Moment Capacity in ft kips

Wall Section   **Beam Section**

$p^{1,2}$	d (inches)											
	2.8	**3.8**	**4.8**	**5.3**	**7.3**	**9.0**	**12.0**	**18.0**	**22.0**	**26.0**	**30.0**	**36.0**
0.0010	0.37	0.68	1.09	1.32	2.51	3.82	6.79	15.28	22.82	31.87	42.44	61.11
0.0015	0.55	1.01	1.61	1.97	3.74	5.68	10.09	22.71	33.92	47.38	63.08	90.83
0.0020	0.73	1.34	2.13	2.60	4.93	7.50	13.33	30.00	44.82	62.60	83.34	120.01
0.0025	0.90	1.66	2.64	3.22	6.11	9.29	16.52	37.16	55.51	77.53	103.22	148.64
0.0030	1.07	1.97	3.14	3.83	7.27	11.04	19.63	44.18	66.00	92.18	122.72	176.71
0.0035	1.24	2.28	3.63	4.43	8.40	12.76	22.69	51.06	76.27	106.53	141.83	204.24
0.0040	1.40	2.58	4.11	5.01	9.51	14.45	25.69	57.80	86.35	120.60	160.57	231.21
0.0045	1.56	2.87	4.58	5.58	10.59	16.10	28.63	64.41	96.22	134.39	178.92	257.64
0.0050	1.72	3.16	5.04	6.14	11.66	17.72	31.50	70.88	105.88	147.88	196.88	283.51
0.0055	1.87	3.44	5.49	6.69	12.70	19.30	34.32	77.21	115.34	161.09	214.47	308.84
0.0060	2.02	3.72	5.93	7.23	13.72	20.85	37.07	83.40	124.59	174.01	231.67	333.61
0.0065	2.16	3.99	6.36	7.76	14.71	22.36	39.76	89.46	133.63	186.65	248.49	357.83
0.0070	2.31	4.25	6.78	8.27	15.69	23.84	42.39	95.38	142.47	198.99	264.93	381.50
0.0075	2.45	4.51	7.19	8.77	16.64	25.29	44.96	101.16	151.11	211.05	280.99	404.62
0.0080	2.58	4.76	7.59	9.26	17.57	26.70	47.47	106.80	159.54	222.83	296.66	427.19
0.0085	2.72	5.01	7.99	9.74	18.47	28.08	49.91	112.30	167.76	234.31	311.95	449.21
0.0090	2.85	5.24	8.37	10.20	19.35	29.42	52.30	117.67	175.78	245.51	326.86	470.68
0.0095	2.97	5.48	8.74	10.66	20.21	30.73	54.62	122.90	183.59	256.42	341.39	491.60
0.0100	3.10	5.70	9.10	11.10	21.05	32.00	56.89	127.99	191.20	267.05	355.54	511.97
0.0105	3.22	5.93	9.45	11.53	21.87	33.24	59.09	132.95	198.60	277.38	369.30	531.79
0.0110	3.33	6.14	9.80	11.94	22.66	34.44	61.23	137.76	205.80	287.43	382.68	551.06

1. UBC Section 2411(b) limits p to $0.5p_b = 0.0071$ for slender wall design.

2. UBC Section 2610(d)3 limits p to $0.75p_b = 0.0107$ for reinforced concrete design.

STR. DES. TABLE C-3 Moment Capacity of Walls and Beams:
f'_m = 2500 psi and f_y = 60,000 psi

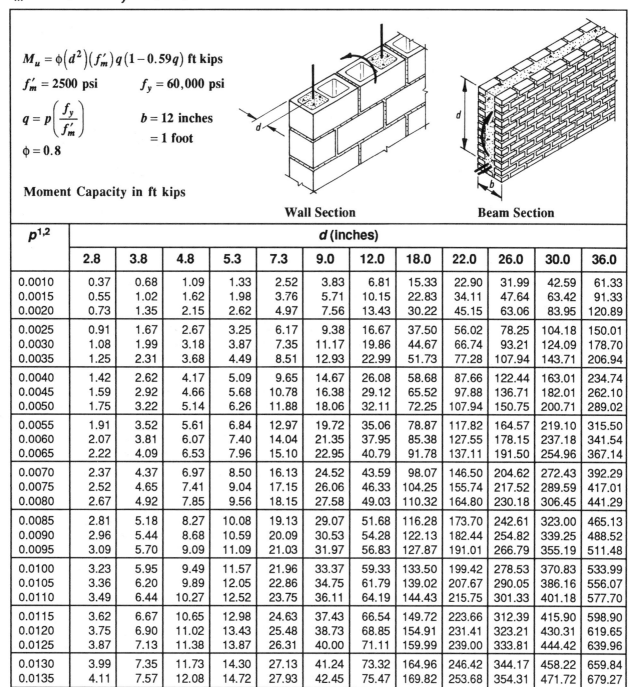

$$M_u = \phi\left(d^2\right)\left(f'_m\right)q(1-0.59q) \text{ ft kips}$$

$f'_m = 2500 \text{ psi}$ $f_y = 60,000 \text{ psi}$

$q = p\left(\dfrac{f_y}{f'_m}\right)$ $b = 12 \text{ inches}$ $= 1 \text{ foot}$

$\phi = 0.8$

Moment Capacity in ft kips

Wall Section **Beam Section**

$p^{1,2}$	d (inches)											
	2.8	**3.8**	**4.8**	**5.3**	**7.3**	**9.0**	**12.0**	**18.0**	**22.0**	**26.0**	**30.0**	**36.0**
0.0010	0.37	0.68	1.09	1.33	2.52	3.83	6.81	15.33	22.90	31.99	42.59	61.33
0.0015	0.55	1.02	1.62	1.98	3.76	5.71	10.15	22.83	34.11	47.64	63.42	91.33
0.0020	0.73	1.35	2.15	2.62	4.97	7.56	13.43	30.22	45.15	63.06	83.95	120.89
0.0025	0.91	1.67	2.67	3.25	6.17	9.38	16.67	37.50	56.02	78.25	104.18	150.01
0.0030	1.08	1.99	3.18	3.87	7.35	11.17	19.86	44.67	66.74	93.21	124.09	178.70
0.0035	1.25	2.31	3.68	4.49	8.51	12.93	22.99	51.73	77.28	107.94	143.71	206.94
0.0040	1.42	2.62	4.17	5.09	9.65	14.67	26.08	58.68	87.66	122.44	163.01	234.74
0.0045	1.59	2.92	4.66	5.68	10.78	16.38	29.12	65.52	97.88	136.71	182.01	262.10
0.0050	1.75	3.22	5.14	6.26	11.88	18.06	32.11	72.25	107.94	150.75	200.71	289.02
0.0055	1.91	3.52	5.61	6.84	12.97	19.72	35.06	78.87	117.82	164.57	219.10	315.50
0.0060	2.07	3.81	6.07	7.40	14.04	21.35	37.95	85.38	127.55	178.15	237.18	341.54
0.0065	2.22	4.09	6.53	7.96	15.10	22.95	40.79	91.78	137.11	191.50	254.96	367.14
0.0070	2.37	4.37	6.97	8.50	16.13	24.52	43.59	98.07	146.50	204.62	272.43	392.29
0.0075	2.52	4.65	7.41	9.04	17.15	26.06	46.33	104.25	155.74	217.52	289.59	417.01
0.0080	2.67	4.92	7.85	9.56	18.15	27.58	49.03	110.32	164.80	230.18	306.45	441.29
0.0085	2.81	5.18	8.27	10.08	19.13	29.07	51.68	116.28	173.70	242.61	323.00	465.13
0.0090	2.96	5.44	8.68	10.59	20.09	30.53	54.28	122.13	182.44	254.82	339.25	488.52
0.0095	3.09	5.70	9.09	11.09	21.03	31.97	56.83	127.87	191.01	266.79	355.19	511.48
0.0100	3.23	5.95	9.49	11.57	21.96	33.37	59.33	133.50	199.42	278.53	370.83	533.99
0.0105	3.36	6.20	9.89	12.05	22.86	34.75	61.79	139.02	207.67	290.05	386.16	556.07
0.0110	3.49	6.44	10.27	12.52	23.75	36.11	64.19	144.43	215.75	301.33	401.18	577.70
0.0115	3.62	6.67	10.65	12.98	24.63	37.43	66.54	149.72	223.66	312.39	415.90	598.90
0.0120	3.75	6.90	11.02	13.43	25.48	38.73	68.85	154.91	231.41	323.21	430.31	619.65
0.0125	3.87	7.13	11.38	13.87	26.31	40.00	71.11	159.99	239.00	333.81	444.42	639.96
0.0130	3.99	7.35	11.73	14.30	27.13	41.24	73.32	164.96	246.42	344.17	458.22	659.84
0.0135	4.11	7.57	12.08	14.72	27.93	42.45	75.47	169.82	253.68	354.31	471.72	679.27

1. UBC Section 2411(b) limits p to $0.5p_b = 0.0089$ for slender wall design.

2. UBC Section 2610(d)3 limits p to $0.75p_b = 0.0134$ for reinforced concrete design.

STR. DES. TABLE C-4 Moment Capacity of Walls and Beams:
f'_m = 3000 psi and f_y = 60,000 psi

$$M_u = \phi\left(d^2\right)\left(f'_m\right)q\left(1-0.59q\right) \text{ ft kips}$$

$f'_m = 3000$ psi $\qquad f_y = 60,000$ psi

$q = p\left(\dfrac{f_y}{f'_m}\right)$ $\qquad b = 12$ inches

$\qquad\qquad\qquad\qquad = 1$ foot

$\phi = 0.8$

Moment Capacity in ft kips

Wall Section **Beam Section**

$p^{1,2}$	d (inches)											
	2.8	**3.8**	**4.8**	**5.3**	**7.3**	**9.0**	**12.0**	**18.0**	**22.0**	**26.0**	**30.0**	**36.0**
0.0010	0.37	0.68	1.09	1.33	2.53	3.84	6.83	15.37	22.96	32.07	42.69	61.47
0.0015	0.55	1.02	1.63	1.99	3.77	5.73	10.18	22.92	34.23	47.81	63.65	91.66
0.0020	0.73	1.35	2.16	2.63	5.00	7.59	13.50	30.37	45.37	63.36	84.36	121.48
0.0025	0.91	1.68	2.68	3.27	6.21	9.43	16.77	37.73	56.37	78.73	104.81	150.93
0.0030	1.09	2.01	3.20	3.90	7.40	11.25	20.00	45.00	67.23	93.90	125.01	180.02
0.0035	1.26	2.33	3.71	4.52	8.58	13.05	23.19	52.18	77.95	108.88	144.96	208.74
0.0040	1.43	2.64	4.21	5.14	9.75	14.82	26.34	59.27	88.54	123.67	164.64	237.09
0.0045	1.60	2.95	4.71	5.75	10.90	16.57	29.45	66.27	98.99	138.26	184.08	265.07
0.0050	1.77	3.26	5.20	6.34	12.04	18.29	32.52	73.17	109.31	152.67	203.26	292.69
0.0055	1.94	3.56	5.69	6.93	13.16	20.00	35.55	79.98	119.48	166.88	222.18	319.94
0.0060	2.10	3.86	6.17	7.52	14.26	21.68	38.54	86.71	129.52	180.90	240.85	346.82
0.0065	2.26	4.16	6.64	8.09	15.35	23.33	41.48	93.33	139.43	194.74	259.26	373.34
0.0070	2.42	4.45	7.10	8.66	16.43	24.97	44.39	99.87	149.19	208.37	277.42	399.49
0.0075	2.57	4.74	7.56	9.22	17.49	26.58	47.25	106.32	158.82	221.82	295.33	425.27
0.0080	2.73	5.02	8.01	9.77	18.53	28.17	50.08	112.67	168.31	235.08	312.98	450.68
0.0085	2.88	5.30	8.46	10.31	19.56	29.73	52.86	118.93	177.67	248.14	330.37	475.73
0.0090	3.03	5.58	8.90	10.85	20.58	31.28	55.60	125.10	186.88	261.02	347.51	500.41
0.0095	3.17	5.85	9.33	11.37	21.58	32.80	58.30	131.18	195.96	273.70	364.39	524.73
0.0100	3.32	6.11	9.75	11.89	22.56	34.29	60.96	137.17	204.91	286.19	381.02	548.67
0.0105	3.46	6.38	10.17	12.40	23.53	35.77	63.58	143.06	213.71	298.49	397.40	572.25
0.0110	3.60	6.63	10.59	12.91	24.48	37.22	66.16	148.87	222.38	310.60	413.52	595.47
0.0115	3.74	6.89	10.99	13.40	25.42	38.64	68.70	154.58	230.91	322.52	429.38	618.31
0.0120	3.88	7.14	11.39	13.89	26.35	40.05	71.20	160.20	239.31	334.24	444.99	640.79
0.0125	4.01	7.39	11.78	14.37	27.26	41.43	73.66	165.73	247.57	345.77	460.35	662.90
0.0130	4.14	7.63	12.17	14.84	28.15	42.79	76.07	171.16	255.69	357.12	475.45	684.65
0.0135	4.27	7.87	12.55	15.30	29.03	44.13	78.45	176.51	263.67	368.27	490.30	706.03
0.0140	4.40	8.10	12.93	15.76	29.89	45.44	80.78	181.76	271.52	379.23	504.89	727.04
0.0145	4.52	8.33	13.29	16.21	30.74	46.73	83.08	186.92	279.23	389.99	519.22	747.68
0.0150	4.65	8.56	13.65	16.65	31.58	48.00	85.33	191.99	286.80	400.57	533.30	767.96
0.0155	4.77	8.78	14.01	17.08	32.40	49.24	87.54	196.97	294.23	410.96	547.13	787.87
0.0160	4.88	9.00	14.35	17.50	33.20	50.46	89.71	201.85	301.53	421.15	560.70	807.41

1. UBC Section 2411(b) limits p to $0.5p_b = 0.0107$ for slender wall design.

2. UBC Section 2610(d)3 limits p to $0.75p_b = 0.0161$ for reinforced concrete design.

STR. DES. TABLE C-5 Moment Capacity of Walls and Beams:
f'_m = 3500 psi and f_y = 60,000 psi

$$M_u = \phi(d^2)(f'_m)q(1-0.59q) \text{ ft kips}$$

$f'_m = 3500 \text{ psi} \qquad f_y = 60,000 \text{ psi}$

$$q = p\left(\frac{f_y}{f'_m}\right) \qquad b = 12 \text{ inches} = 1 \text{ foot}$$

$\phi = 0.8$

Moment Capacity in ft kips

Wall Section   **Beam Section**

$p^{1,2}$	\multicolumn{12}{c}{d (inches)}
	2.8	3.8	4.8	5.3	7.3	9.0	12.0	18.0	22.0	26.0	30.0	36.0
0.0010	0.37	0.69	1.09	1.33	2.53	3.85	6.84	15.39	23.00	32.12	42.76	61.58
0.0015	0.56	1.02	1.63	1.99	3.78	5.74	10.21	22.97	34.32	47.93	63.82	91.90
0.0020	0.74	1.36	2.17	2.64	5.01	7.62	13.54	30.47	45.52	63.58	84.65	121.90
0.0025	0.92	1.69	2.69	3.29	6.23	9.47	16.84	37.90	56.61	79.07	105.27	151.59
0.0030	1.09	2.02	3.22	3.92	7.44	11.31	20.11	45.24	67.58	94.39	125.67	180.96
0.0035	1.27	2.34	3.73	4.55	8.64	13.13	23.34	52.51	78.43	109.55	145.85	210.02
0.0040	1.44	2.66	4.24	5.18	9.82	14.92	26.53	59.69	89.17	124.54	165.81	238.76
0.0045	1.62	2.98	4.75	5.79	10.99	16.70	29.69	66.80	99.79	139.37	185.55	267.19
0.0050	1.79	3.29	5.25	6.40	12.14	18.46	32.81	73.83	110.29	154.04	205.08	295.31
0.0055	1.95	3.60	5.74	7.00	13.29	20.19	35.90	80.78	120.67	168.54	224.38	323.11
0.0060	2.12	3.91	6.23	7.60	14.42	21.91	38.96	87.65	130.93	182.87	243.47	350.60
0.0065	2.29	4.21	6.72	8.19	15.53	23.61	41.97	94.44	141.08	197.05	262.34	377.77
0.0070	2.45	4.51	7.19	8.77	16.64	25.29	44.96	101.16	151.11	211.05	280.99	404.63
0.0075	2.61	4.80	7.67	9.35	17.73	26.95	47.91	107.79	161.02	224.90	299.42	431.17
0.0080	2.77	5.10	8.13	9.91	18.81	28.59	50.82	114.35	170.82	238.58	317.64	457.40
0.0085	2.92	5.39	8.59	10.48	19.87	30.21	53.70	120.83	180.50	252.10	335.63	483.31
0.0090	3.08	5.67	9.05	11.03	20.93	31.81	56.55	127.23	190.06	265.45	353.41	508.91
0.0095	3.23	5.95	9.50	11.58	21.97	33.39	59.35	133.55	199.50	278.64	370.97	534.19
0.0100	3.38	6.23	9.94	12.12	22.99	34.95	62.13	139.79	208.82	291.66	388.31	559.16
0.0105	3.53	6.50	10.38	12.65	24.01	36.49	64.87	145.95	218.03	304.52	405.43	583.82
0.0110	3.68	6.78	10.81	13.18	25.01	38.01	67.57	152.04	227.12	317.22	422.33	608.16
0.0115	3.82	7.04	11.24	13.70	25.99	39.51	70.24	158.05	236.09	329.75	439.02	632.18
0.0120	3.97	7.31	11.66	14.22	26.97	40.99	72.88	163.97	244.95	342.12	455.48	655.89
0.0125	4.11	7.57	12.08	14.72	27.93	42.46	75.48	169.82	253.69	354.32	471.73	679.29
0.0130	4.25	7.83	12.49	15.22	28.88	43.90	78.04	175.59	262.31	366.36	487.76	702.37
0.0135	4.39	8.08	12.89	15.72	29.82	45.32	80.57	181.28	270.81	378.24	503.57	725.14
0.0140	4.52	8.33	13.29	16.20	30.74	46.72	83.07	186.90	279.19	389.95	519.16	747.59
0.0145	4.66	8.58	13.68	16.68	31.65	48.11	85.53	192.43	287.46	401.49	534.53	769.73
0.0150	4.79	8.82	14.07	17.16	32.55	49.47	87.95	197.89	295.61	412.88	549.69	791.55
0.0155	4.92	9.06	14.45	17.62	33.43	50.82	90.34	203.27	303.64	424.10	564.63	813.06
0.0160	5.05	9.30	14.83	18.08	34.30	52.14	92.70	208.56	311.56	435.15	579.34	834.26
0.0165	5.17	9.53	15.20	18.53	35.16	53.45	95.02	213.78	319.36	446.04	593.84	855.14
0.0170	5.30	9.76	15.57	18.98	36.01	54.73	97.30	218.93	327.04	456.77	608.13	875.70
0.0175	5.42	9.98	15.93	19.42	36.84	56.00	99.55	223.99	334.60	467.33	622.19	895.95
0.0180	5.54	10.20	16.28	19.85	37.66	57.24	101.77	228.97	342.04	477.73	636.03	915.89
0.0185	5.66	10.42	16.63	20.28	38.47	58.47	103.95	233.88	349.37	487.97	649.66	935.51

1. UBC Section 2411(b) limits p to $0.5p_b = 0.0125$ for slender wall design.
2. UBC Section 2610(d)3 limits p to $0.75p_b = 0.0188$ for reinforced concrete design.

STR. DES. TABLE C-6 Moment Capacity of Walls and Beams:
f'_m = 4000 psi and f_y = 60,000 psi

$M_u = \phi\left(d^2\right)\left(f'_m\right)q\left(1-0.59q\right)$ ft kips

$f'_m = 4000$ psi $f_y = 60,000$ psi

$q = p\left(\dfrac{f_y}{f'_m}\right)$ $b = 12$ inches
                                    $= 1$ foot

$\phi = 0.8$

Moment Capacity in ft kips

Wall Section   **Beam Section**

$p^{1,2}$	d (inches)											
	2.8	**3.8**	**4.8**	**5.3**	**7.3**	**9.0**	**12.0**	**18.0**	**22.0**	**26.0**	**30.0**	**36.0**
0.0010	0.37	0.69	1.10	1.34	2.54	3.85	6.85	15.41	23.03	32.16	42.82	61.66
0.0015	0.56	1.03	1.64	2.00	3.79	5.75	10.23	23.02	34.39	48.03	63.94	92.07
0.0020	0.74	1.36	2.17	2.65	5.03	7.64	13.58	30.55	45.64	63.75	84.87	122.21
0.0025	0.92	1.69	2.70	3.30	6.25	9.50	16.90	38.02	56.79	79.33	105.61	152.08
0.0030	1.10	2.02	3.23	3.94	7.47	11.35	20.19	45.42	67.85	94.76	126.16	181.67
0.0035	1.28	2.35	3.75	4.57	8.68	13.19	23.44	52.75	78.79	110.05	146.52	210.98
0.0040	1.45	2.67	4.27	5.20	9.87	15.00	26.67	60.01	89.64	125.20	166.68	240.02
0.0045	1.63	2.99	4.78	5.83	11.05	16.80	29.87	67.20	100.38	140.20	186.66	268.79
0.0050	1.80	3.31	5.28	6.44	12.22	18.58	33.03	74.32	111.02	155.06	206.44	297.28
0.0055	1.97	3.63	5.79	7.05	13.38	20.34	36.17	81.37	121.56	169.78	226.03	325.49
0.0060	2.14	3.94	6.28	7.66	14.53	22.09	39.27	88.36	131.99	184.35	245.44	353.43
0.0065	2.31	4.25	6.77	8.26	15.67	23.82	42.34	95.27	142.32	198.78	264.65	381.09
0.0070	2.47	4.55	7.26	8.85	16.80	25.53	45.39	102.12	152.55	213.06	283.67	408.48
0.0075	2.64	4.85	7.74	9.44	17.91	27.22	48.40	108.90	162.67	227.21	302.49	435.59
0.0080	2.80	5.15	8.22	10.02	19.01	28.90	51.38	115.61	172.70	241.21	321.13	462.43
0.0085	2.96	5.45	8.69	10.60	20.11	30.56	54.33	122.25	182.62	255.06	339.58	488.99
0.0090	3.12	5.74	9.16	11.17	21.19	32.20	57.25	128.82	192.43	268.77	357.83	515.28
0.0095	3.27	6.03	9.62	11.73	22.26	33.83	60.14	135.32	202.15	282.34	375.90	541.29
0.0100	3.43	6.32	10.08	12.29	23.32	35.44	63.00	141.76	211.76	295.76	393.77	567.03
0.0105	3.58	6.60	10.53	12.84	24.36	37.03	65.83	148.12	221.27	309.04	411.45	592.49
0.0110	3.74	6.88	10.98	13.39	25.40	38.60	68.63	154.42	230.67	322.18	428.94	617.67
0.0115	3.89	7.16	11.42	13.93	26.42	40.16	71.40	160.65	239.98	335.17	446.24	642.58
0.0120	4.04	7.43	11.86	14.46	27.44	41.70	74.14	166.80	249.18	348.02	463.35	667.22
0.0125	4.18	7.71	12.29	14.99	28.44	43.22	76.84	172.89	258.27	360.73	480.26	691.58
0.0130	4.33	7.97	12.72	15.51	29.43	44.73	79.52	178.92	267.27	373.29	496.99	715.66
0.0135	4.47	8.24	13.15	16.03	30.41	46.22	82.16	184.87	276.16	385.71	513.52	739.47
0.0140	4.62	8.50	13.56	16.54	31.37	47.69	84.78	190.75	284.95	397.99	529.87	763.01
0.0145	4.76	8.76	13.98	17.04	32.33	49.14	87.36	196.57	293.64	410.12	546.02	786.26
0.0150	4.90	9.02	14.39	17.54	33.28	50.58	89.92	202.31	302.22	422.11	561.98	809.25
0.0155	5.03	9.27	14.79	18.03	34.21	52.00	92.44	207.99	310.70	433.95	577.75	831.96
0.0160	5.17	9.52	15.19	18.52	35.13	53.40	94.93	213.60	319.08	445.65	593.33	854.39
0.0165	5.30	9.77	15.58	19.00	36.04	54.78	97.39	219.14	327.35	457.21	608.71	876.55
0.0170	5.43	10.01	15.97	19.47	36.94	56.15	99.83	224.61	335.52	468.63	623.91	898.43
0.0175	5.57	10.25	16.36	19.94	37.83	57.50	102.23	230.01	343.59	479.90	638.91	920.04
0.0180	5.69	10.49	16.74	20.40	38.71	58.84	104.60	235.34	351.56	491.02	653.73	941.37
0.0185	5.82	10.72	17.11	20.86	39.57	60.15	106.94	240.61	359.42	502.01	668.35	962.43

1. UBC Section 2411(b) limits p to $0.5p_b = 0.0143$ for slender wall design.

2. UBC Section 2610(d)3 limits p to $0.75p_b = 0.0215$ for reinforced concrete design.

STR. DES. TABLE C-7   Moment Capacity of Walls and Beams:
f'_m = 4500 psi and f_y = 60,000 psi

$$M_u = \phi\left(d^2\right)\left(f'_m\right)q\left(1-0.59q\right) \text{ ft kips}$$

$f'_m = 4500$ psi $f_y = 60,000$ psi

$$q = p\left(\frac{f_y}{f'_m}\right)$$ $b = 12$ inches
                                = 1 foot

$\phi = 0.8$

Moment Capacity in ft kips

Wall Section   Beam Section

$p^{1,2}$	\multicolumn{12}{c}{d (inches)}
	2.8	3.8	4.8	5.3	7.3	9.0	12.0	18.0	22.0	26.0	30.0	36.0
0.0010	0.37	0.69	1.10	1.34	2.54	3.86	6.86	15.43	23.05	32.19	42.86	61.72
0.0015	0.56	1.03	1.64	2.00	3.79	5.76	10.25	23.05	34.44	48.10	64.04	92.21
0.0020	0.74	1.36	2.18	2.65	5.04	7.65	13.61	30.61	45.73	63.87	85.04	122.46
0.0025	0.92	1.70	2.71	3.30	6.27	9.53	16.94	38.12	56.94	79.52	105.88	152.46
0.0030	1.10	2.03	3.24	3.95	7.49	11.39	20.25	45.55	68.05	95.05	126.54	182.22
0.0035	1.28	2.36	3.76	4.59	8.71	13.23	23.53	52.93	79.07	110.44	147.04	211.73
0.0040	1.46	2.69	4.28	5.22	9.91	15.06	26.78	60.25	90.00	125.71	167.36	241.00
0.0045	1.63	3.01	4.80	5.85	11.10	16.88	30.00	67.51	100.84	140.85	187.52	270.03
0.0050	1.81	3.33	5.31	6.48	12.29	18.68	33.20	74.70	111.59	155.86	207.50	298.81
0.0055	1.98	3.65	5.82	7.09	13.46	20.46	36.37	81.84	122.25	170.74	227.32	327.34
0.0060	2.15	3.96	6.32	7.71	14.62	22.23	39.51	88.91	132.81	185.50	246.97	355.63
0.0065	2.32	4.27	6.82	8.32	15.78	23.98	42.63	95.92	143.29	200.13	266.44	383.68
0.0070	2.49	4.58	7.32	8.92	16.92	25.72	45.72	102.87	153.67	214.63	285.75	411.48
0.0075	2.66	4.89	7.81	9.52	18.05	27.44	48.78	109.76	163.96	229.00	304.88	439.03
0.0080	2.82	5.20	8.29	10.11	19.18	29.15	51.82	116.59	174.16	243.25	323.85	466.34
0.0085	2.98	5.50	8.77	10.69	20.29	30.84	54.82	123.35	184.27	257.37	342.65	493.41
0.0090	3.15	5.80	9.25	11.28	21.39	32.51	57.80	130.06	194.28	271.36	361.27	520.23
0.0095	3.31	6.09	9.72	11.85	22.48	34.18	60.76	136.70	204.21	285.22	379.73	546.81
0.0100	3.47	6.39	10.19	12.42	23.57	35.82	63.68	143.29	214.04	298.95	398.02	573.14
0.0105	3.62	6.68	10.65	12.99	24.64	37.45	66.58	149.81	223.79	312.56	416.13	599.23
0.0110	3.78	6.96	11.11	13.55	25.70	39.07	69.45	156.27	233.44	326.04	434.08	625.07
0.0115	3.94	7.25	11.57	14.10	26.75	40.67	72.30	162.67	243.00	339.39	451.86	650.67
0.0120	4.09	7.53	12.02	14.65	27.80	42.25	75.11	169.01	252.47	352.62	469.46	676.03
0.0125	4.24	7.81	12.46	15.20	28.83	43.82	77.90	175.28	261.84	365.72	486.90	701.14
0.0130	4.39	8.09	12.91	15.74	29.85	45.38	80.67	181.50	271.13	378.69	504.17	726.00
0.0135	4.54	8.36	13.34	16.27	30.86	46.91	83.40	187.66	280.32	391.53	521.26	750.62
0.0140	4.69	8.63	13.78	16.80	31.87	48.44	86.11	193.75	289.43	404.24	538.19	775.00
0.0145	4.83	8.90	14.21	17.32	32.86	49.95	88.79	199.78	298.44	416.83	554.95	799.13
0.0150	4.98	9.17	14.63	17.84	33.84	51.44	91.45	205.75	307.36	429.29	571.54	823.01
0.0155	5.12	9.43	15.05	18.35	34.81	52.92	94.07	211.66	316.19	441.62	587.95	846.65
0.0160	5.26	9.69	15.47	18.86	35.78	54.38	96.67	217.51	324.93	453.82	604.20	870.05
0.0165	5.40	9.95	15.88	19.36	36.73	55.83	99.24	223.30	333.57	465.90	620.28	893.20
0.0170	5.54	10.21	16.29	19.86	37.67	57.26	101.79	229.03	342.13	477.85	636.19	916.11
0.0175	5.68	10.46	16.69	20.35	38.60	58.67	104.31	234.69	350.59	489.67	651.92	938.77
0.0180	5.81	10.71	17.09	20.83	39.52	60.07	106.80	240.30	358.96	501.36	667.49	961.19
0.0185	5.95	10.96	17.48	21.31	40.43	61.46	109.26	245.84	367.24	512.93	682.89	983.36

1. UBC Section 2411(b) limits p to $0.5p_b = 0.0156$ for slender wall design.

2. UBC Section 2610(d)3 limits p to $0.75p_b = 0.0233$ for reinforced concrete design.

STR. DES.

STR. DES. TABLE C-8 Moment Capacity of Walls and Beams:
f'_m = 5000 psi and f_y = 60,000 psi

$$M_u = \phi\left(d^2\right)\left(f'_m\right)q\left(1-0.59q\right) \text{ ft kips}$$

$$f'_m = 5000 \text{ psi} \qquad f_y = 60,000 \text{ psi}$$

$$q = p\left(\frac{f_y}{f'_m}\right) \qquad b = 12 \text{ inches}$$
$$\qquad\qquad\qquad\qquad = 1 \text{ foot}$$
$$\phi = 0.8$$

Moment Capacity in ft kips

Wall Section   **Beam Section**

$p^{1,2}$	d (inches)											
	2.8	**3.8**	**4.8**	**5.3**	**7.3**	**9.0**	**12.0**	**18.0**	**22.0**	**26.0**	**30.0**	**36.0**
0.0010	0.37	0.69	1.10	1.34	2.54	3.86	6.86	15.44	23.07	32.22	42.89	61.77
0.0015	0.56	1.03	1.64	2.00	3.80	5.77	10.26	23.08	34.48	48.16	64.11	92.32
0.0020	0.74	1.37	2.18	2.66	5.04	7.67	13.63	30.66	45.81	63.98	85.18	122.65
0.0025	0.92	1.70	2.72	3.31	6.28	9.55	16.97	38.19	57.05	79.68	106.09	152.77
0.0030	1.10	2.04	3.25	3.96	7.51	11.42	20.30	45.67	68.22	95.28	126.85	182.66
0.0035	1.28	2.37	3.77	4.60	8.73	13.27	23.59	53.08	79.30	110.75	147.45	212.33
0.0040	1.46	2.69	4.30	5.24	9.94	15.11	26.87	60.45	90.30	126.12	167.91	241.79
0.0045	1.64	3.02	4.82	5.87	11.14	16.94	30.11	67.75	101.21	141.36	188.21	271.02
0.0050	1.81	3.34	5.33	6.50	12.34	18.75	33.34	75.01	112.05	156.50	208.35	300.03
0.0055	1.99	3.66	5.85	7.13	13.52	20.55	36.54	82.21	122.80	171.51	228.35	328.82
0.0060	2.16	3.98	6.35	7.75	14.70	22.34	39.71	89.35	133.47	186.42	248.19	357.39
0.0065	2.33	4.30	6.86	8.36	15.86	24.11	42.86	96.44	144.06	201.21	267.88	385.74
0.0070	2.50	4.61	7.36	8.97	17.02	25.87	45.99	103.47	154.56	215.88	287.41	413.87
0.0075	2.67	4.92	7.85	9.58	18.17	27.61	49.09	110.45	164.99	230.44	306.80	441.79
0.0080	2.84	5.23	8.35	10.18	19.30	29.34	52.16	117.37	175.33	244.88	326.03	469.48
0.0085	3.01	5.54	8.83	10.77	20.43	31.06	55.22	124.24	185.59	259.21	345.10	496.95
0.0090	3.17	5.84	9.32	11.36	21.55	32.76	58.24	131.05	195.76	273.42	364.03	524.20
0.0095	3.33	6.14	9.80	11.95	22.67	34.45	61.25	137.81	205.86	287.52	382.80	551.23
0.0100	3.50	6.44	10.28	12.53	23.77	36.13	64.23	144.51	215.87	301.51	401.41	578.04
0.0105	3.66	6.74	10.75	13.10	24.86	37.79	67.18	151.16	225.80	315.38	419.88	604.63
0.0110	3.82	7.03	11.22	13.68	25.95	39.44	70.11	157.75	235.65	329.13	438.19	631.00
0.0115	3.98	7.32	11.68	14.24	27.02	41.07	73.02	164.29	245.42	342.77	456.35	657.14
0.0120	4.13	7.61	12.14	14.81	28.09	42.69	75.90	170.77	255.10	356.29	474.36	683.07
0.0125	4.29	7.90	12.60	15.36	29.14	44.30	78.75	177.20	264.70	369.70	492.21	708.78
0.0130	4.44	8.18	13.05	15.91	30.19	45.89	81.59	183.57	274.22	383.00	509.91	734.27
0.0135	4.59	8.46	13.50	16.46	31.23	47.47	84.39	189.88	283.66	396.18	527.46	759.54
0.0140	4.75	8.74	13.95	17.01	32.26	49.04	87.18	196.15	293.01	409.24	544.85	784.59
0.0145	4.90	9.02	14.39	17.54	33.28	50.59	89.94	202.35	302.28	422.19	562.09	809.42
0.0150	5.05	9.29	14.83	18.08	34.29	52.13	92.67	208.51	311.47	435.03	579.18	834.02
0.0155	5.19	9.56	15.26	18.61	35.30	53.65	95.38	214.60	320.58	447.75	596.12	858.41
0.0160	5.34	9.83	15.69	19.13	36.29	55.16	98.06	220.64	329.60	460.36	612.90	882.58
0.0165	5.48	10.10	16.12	19.65	37.28	56.66	100.72	226.63	338.55	472.85	629.53	906.52
0.0170	5.63	10.36	16.54	20.16	38.25	58.14	103.36	232.56	347.41	485.22	646.01	930.25
0.0175	5.77	10.63	16.96	20.67	39.22	59.61	105.97	238.44	356.19	497.48	662.33	953.76
0.0180	5.91	10.89	17.37	21.18	40.17	61.07	108.56	244.26	364.88	509.63	678.50	977.04
0.0185	6.05	11.14	17.78	21.68	41.12	62.51	111.12	250.03	373.50	521.66	694.52	1000.11

1. UBC Section 2411(b) limits p to $0.5p_b = 0.0168$ for slender wall design.

2. UBC Section 2610(d)3 limits p to $0.75p_b = 0.0252$ for reinforced concrete design.

STR. DES. TABLE D-1   Modulus of Rupture, f_r, of Masonry (psi)[1]

f'_m (psi)	Two-wythe Brick Masonry Fully Grouted $f_r = 2.0\sqrt{f'_m}$ (125 psi maximum)	Hollow-unit Masonry Partially Grouted $f_r = 2.5\sqrt{f'_m}$ (125 psi maximum)	Hollow-unit Masonry Fully Grouted $f_r = 4.0\sqrt{f'_m}$ (235 psi maximum)
1500	77	97	155
2000	89	112	179
2500	100	125	200
3000	110	125	219
3500	118	125	235
4000 +	125	125	235

1. Based on UBC Section 2411(b)4.

STR. DES. TABLE D-2   Strength Reduction Factor, $\phi^{1,4}$ for Axial Loads for Combined Axial and Flexural Loading

A^2	0.10	0.09	0.08	0.07	0.06	0.05	0.04	0.03	0.02	0.01	0
B^3	0.250	0.225	0.200	0.175	0.150	0.125	0.100	0.075	0.050	0.025	0
ϕ	0.65	0.67	0.69	0.72	0.74	0.76	0.78	0.81	0.83	0.85	0.85

1. Based on UBC Sec. 2412(c)2.A.

   Strength reduction factor ϕ shall be as follows: A. Axial load and axial load with flexure $\phi = 0.65$. For members in which f_y does not exceed 60,000 psi, with symmetrical reinforcement, ϕ may be increased linearly to 0.85 as ϕP_n decreases from $0.10 f'_m A_e$ or $0.25 P_b$ to zero.

2. A is used to designate the multiplier preceeding the term $f'_m A_e$ (see footnote 1).

3. B is used to designate the multiplier preceeding the term P_b (see footnote 1).

4. Designs should be based on the minimum ϕ value determined from the limiting A or B value.

STR. DES. Diagram E-1   Maximum Nominal Shear Stress Provided by the Masonry, v_m, psi

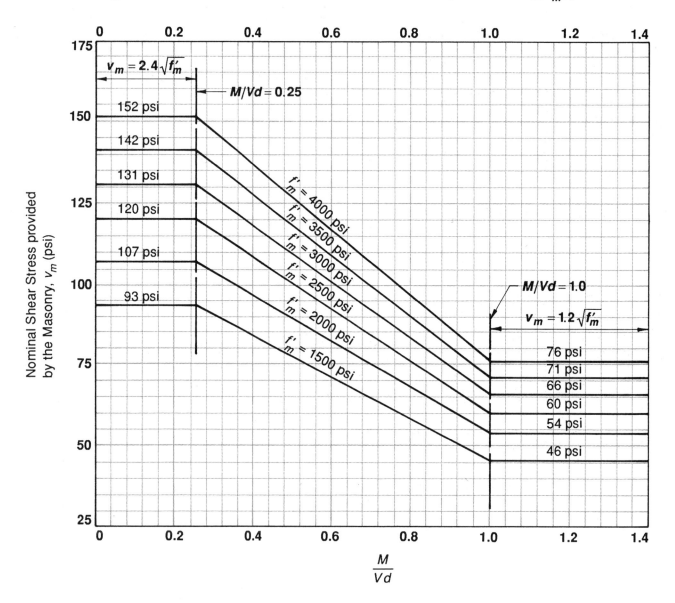

$$\frac{M}{Vd}$$

STR. DES. TABLE E-1   Maximum Nominal Shear Stress Provided by the Masonry, v_m, psi

M/Vd	0.00	0.10	0.20	0.25	0.30	0.40	0.50	0.60	0.70	0.80	0.90	1.00
C_d^1	2.40	2.40	2.40	2.40	2.32	2.16	2.00	1.84	1.68	1.52	1.36	1.20
f_m' (psi)	Maximum Nominal Shear Stress Provided by the Masonry, v_m, psi[2]											
4000	152	152	152	152	147	137	126	116	106	96	86	76
3500	142	142	142	142	137	128	118	109	99	90	80	71
3000	131	131	131	131	127	118	110	101	92	83	74	66
2500	120	120	120	120	116	108	100	92	84	76	68	60
2000	107	107	107	107	104	97	89	82	75	68	61	54
1500	93	93	93	93	90	84	77	71	65	59	53	46

1. Based on UBC Table No. 24-O.

2. Value equal to $C_d \times \sqrt{f_m'}$. To be used in UBC Chapter 24, Equation 12 - 14.

STR. DES. Diagram E-2   Maximum Nominal Shear Stress of Masonry and Reinforcing Steel, v_n, psi

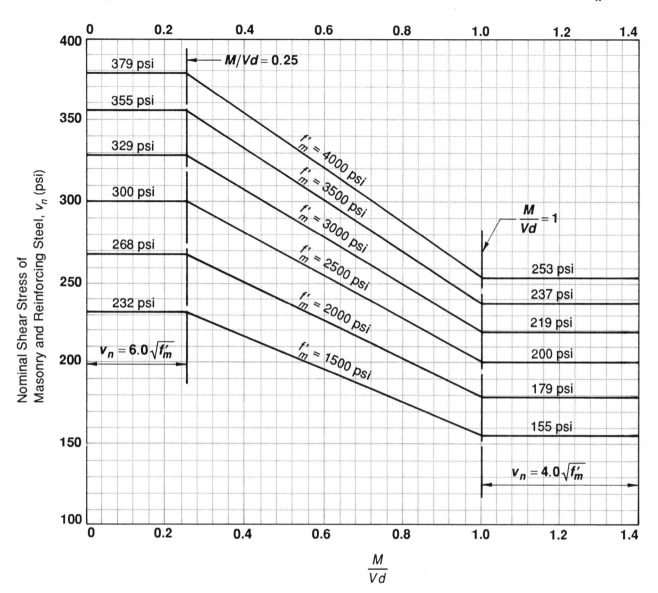

STR. DES. TABLE E-2   Maximum Nominal Shear Stress of Masonry and Reinforcing Steel, v_n, psi[1,2]

M/Vd	0.00	0.10	0.20	0.25	0.30	0.40	0.50	0.60	0.70	0.80	0.90	1.00
Coefficient, $\dfrac{V_n}{A_e\sqrt{f'_m}}$	6.00	6.00	6.00	6.00	5.87	5.60	5.33	5.07	4.80	4.53	4.27	4.00
f'_m (psi)												
4000	379	379	379	379	371	354	337	320	304	287	270	253
3500	355	355	355	355	347	331	316	300	284	268	252	237
3000	329	329	329	329	321	307	292	278	263	248	234	219
2500	300	300	300	300	293	280	267	253	240	227	213	200
2000	268	268	268	268	262	250	239	227	215	203	191	179
1500	232	232	232	232	227	217	207	196	186	176	165	155

The row under f'_m (psi) header spans: **Maximum Nominal Shear Stress Values, v_n, psi[2]**

1. Based on UBC Table No. 24-N.

2. Value equal to Coefficient times $\sqrt{f'_m}$. Maximum Nominal Shear Strength Value, V_n, can thus be determined by multiplying values by A_e.

WORKING
STRESS
DESIGN
TABLES

**Based on the
ACI/ASCE
Masonry Building
Code and Specifications**

The following A/A Tables are based specifically on the ACI/ASCE Masonry Building Code and Specifications. Additionally the tables listed below may also be used in designing structures by the ACI code:

Diagram and Table A-5 and A-6

Tables A-7a, b and c

Table A-8

Tables B-1 to B-6

Tables C-1 to C-11

Tables E-9a and b

Diagrams and Tables K-1 to K-4

Tables T-1a to T-1g

Tables U-1 to U-14

A/A TABLE A-1 CLAY Masonry f'_m, E_m, n and E_v Values Based on the Clay Masonry Unit Strength and the Mortar Type

Net Area Compressive Strength of Clay Masonry Units (psi)	Net Area Compressive Strength of Clay Masonry Assemblage[1], f'_m (psi)	Modulus of Elasticity[2], E_m (psi)	Modular Ratio, $n = E_s/E_m$ where $E_s = 29{,}000{,}000$ psi	Modulus of Rigidity[3], $E_v = 0.4E_m$ (psi)
Type N Mortar				
5,500	1,500	1,500,000	19.3	600,000
8,000	2,000	2,000,000	14.5	800,000
10,500	2,500	2,500,000	11.6	1,000,000
13,000	3,000	2,800,000	10.4	1,120,000
Type S Mortar				
4,400	1,500	1,500,000	19.3	600,000
6,400	2,000	2,000,000	14.5	800,000
8,400	2,500	2,500,000	11.6	1,000,000
10,400	3,000	2,920,000	9.9	1,168,000
12,400	3,500	3,000,000	9.7	1,200,000
14,400	4,000	3,000,000	9.7	1,200,000
Type M Mortar				
4,400	1,500	1,720,000	16.9	688,000
6,400	2,000	2,320,000	12.5	928,000
8,400	2,500	2,840,000	10.2	1,136,000
10,400	3,000	3,000,000	9.7	1,200,000
12,400	3,500	3,000,000	9.7	1,200,000
14,400	4,000	3,000,000	9.7	1,200,000

1. Based on ACI/ASCE Standard Table 5.5.1.2.

2. Based on ACI/ASCE Specification Table 1.6.2.1.

3. Based on ACI/ASCE Section 5.5.1.2(b).

A/A TABLE A-2 Correction Factors for CLAY Masonry Prism Strength[1]

Prism Height to Thickness Ratio	2.0	2.5	3.0	3.5	4.0	4.5	5.0
Correction Factor	0.82	0.85	0.88	0.91	0.94	0.97	1.0

1. Based on ACI/ASCE Specifications Table 1.6.3.3(b).

A/A TABLE B-1a   Flexural Design Coefficients for *CLAY* Masonry
Constructed with Type *S* or *N* Mortar: f'_m = 1500 psi, f_y = 60,000 psi and n = 19.3

For use of Table, see Example 5-B

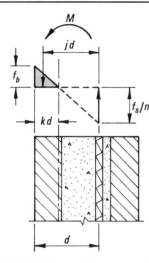

DESIGN DATA

$f'_m = 1500$ psi $\qquad\qquad f_y = 60,000$ psi

$F_b = 0.33 f'_m = 500$ psi $\qquad F_s = 24,000$ psi

$E_m = 1,500,000$ psi $\qquad E_s = 29,000,000$ psi

DESIGN EQUATIONS

$$n = \frac{E_s}{E_m} = 19.3 \qquad K = \frac{M}{F} = \frac{M(\text{ft kips})}{bd^2/12,000} \text{ or } \frac{M(\text{in. lbs})}{bd^2}$$

$$K = \tfrac{1}{2} jk f_b$$

$$k = \frac{1}{1 + f_s/n f_b} \qquad j = 1 - \frac{k}{3}$$

$$A_s = \frac{M}{f_s j d} \qquad p = \frac{A_s}{bd} = \frac{K}{f_s j}$$

Increase for wind or earthquake
$1.33 F_b = 667$ psi
$1.33 F_s = 32,000$ psi

f_b	f_s	K	p	np	k	j	$2/jk$	f_b	f_s	K
125	24,000	5.5	0.0002	0.005	0.091	0.970	22.58	167	32,000	7.4
150	24,000	7.8	0.0003	0.006	0.108	0.964	19.27	200	32,000	10.4
175	24,000	10.4	0.0004	0.009	0.123	0.959	16.91	233	32,000	13.8
200	24,000	13.2	0.0006	0.011	0.139	0.954	15.13	267	32,000	17.6
225	24,000	16.4	0.0007	0.014	0.153	0.949	13.76	300	32,000	21.8
250	24,000	19.8	0.0009	0.017	0.167	0.944	12.65	333	32,000	26.3
275	24,000	23.4	0.0010	0.020	0.181	0.940	11.75	367	32,000	31.2
300	24,000	27.3	0.0012	0.023	0.194	0.935	11.00	400	32,000	36.4
325	24,000	31.3	0.0014	0.027	0.207	0.931	10.37	433	32,000	41.8
350	24,000	35.6	0.0016	0.031	0.220	0.927	9.83	467	32,000	47.5
375	24,000	40.1	0.0018	0.035	0.232	0.923	9.35	500	32,000	53.4
400	24,000	44.7	0.0020	0.039	0.243	0.919	8.94	533	32,000	59.6
425	24,000	49.5	0.0023	0.044	0.255	0.915	8.58	567	32,000	66.0
450	24,000	54.5	0.0025	0.048	0.266	0.911	8.26	600	32,000	72.7
475	24,000	59.6	0.0027	0.053	0.276	0.908	7.97	633	32,000	79.5
500	24,000	64.8	0.0030	0.058	0.287	0.904	7.71	667	32,000	86.5
500	23,000	66.6	0.0032	0.062	0.296	0.901	7.51	667	30,667	88.8
500	22,000	68.5	0.0035	0.067	0.305	0.898	7.30	667	29,333	91.3
500	21,000	70.5	0.0037	0.072	0.315	0.895	7.10	667	28,000	93.9
500	20,000	72.5	0.0041	0.079	0.325	0.892	6.89	667	26,667	96.7
500	19,000	74.8	0.0044	0.086	0.337	0.888	6.69	667	25,333	99.7
500	18,000	77.1	0.0048	0.094	0.349	0.884	6.49	667	24,000	102.8
500	17,000	79.6	0.0053	0.103	0.362	0.879	6.28	667	22,667	106.1
500	16,000	82.3	0.0059	0.113	0.376	0.875	6.08	667	21,333	109.7
500	15,000	85.1	0.0065	0.126	0.391	0.870	5.88	667	20,000	113.5
500	14,000	88.1	0.0073	0.141	0.408	0.864	5.67	667	18,667	117.5
500	13,000	91.4	0.0082	0.158	0.426	0.858	5.47	667	17,333	121.8
500	12,000	94.9	0.0093	0.179	0.446	0.851	5.27	667	16,000	126.5
500	11,000	98.6	0.0106	0.205	0.467	0.844	5.07	667	14,667	131.5
500	10,000	102.7	0.0123	0.237	0.491	0.836	4.87	667	13,333	136.9

A/A TABLE B-1b   Flexural Design Coefficients for *CLAY* Masonry
Constructed with Type *M* Mortar: f'_m = 1500 psi, f_y = 60,000 psi and *n* = 16.9
For use of Table, see Example 5-B

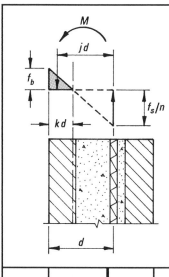

DESIGN DATA

$f'_m = 1500$ psi $f_y = 60,000$ psi

$F_b = 0.33 f'_m = 500$ psi $F_s = 24,000$ psi

$E_m = 1,720,000$ psi $E_s = 29,000,000$ psi

DESIGN EQUATIONS

$$n = \frac{E_s}{E_m} = 16.9 \qquad K = \frac{M}{F} = \frac{M(\text{ft kips})}{bd^2/12,000} \text{ or } \frac{M(\text{in. lbs})}{bd^2}$$

$$K = \frac{1}{2} jk f_b$$

$$k = \frac{1}{1 + f_s/n f_b} \qquad j = 1 - \frac{k}{3}$$

$$A_s = \frac{M}{f_s j d} \qquad p = \frac{A_s}{bd} = \frac{K}{f_s j}$$

Increase for wind or earthquake
$1.33 F_b = 667$ psi
$1.33 F_s = 32,000$ psi

f_b	f_s	K	p	np	k	j	2/jk	f_b	f_s	K
125	24,000	4.9	0.0002	0.004	0.081	0.973	25.46	167	32,000	6.5
150	24,000	6.9	0.0003	0.005	0.095	0.968	21.67	200	32,000	9.2
175	24,000	9.2	0.0004	0.007	0.109	0.964	18.96	233	32,000	12.3
200	24,000	11.8	0.0005	0.009	0.123	0.959	16.93	267	32,000	15.8
225	24,000	14.7	0.0006	0.011	0.136	0.955	15.35	300	32,000	19.5
250	24,000	17.7	0.0008	0.013	0.149	0.950	14.09	333	32,000	23.7
275	24,000	21.1	0.0009	0.016	0.162	0.946	13.06	367	32,000	28.1
300	24,000	24.6	0.0011	0.018	0.174	0.942	12.20	400	32,000	32.8
325	24,000	28.3	0.0013	0.021	0.186	0.938	11.47	433	32,000	37.8
350	24,000	32.3	0.0014	0.024	0.197	0.934	10.85	467	32,000	43.0
375	24,000	36.4	0.0016	0.027	0.209	0.930	10.31	500	32,000	48.5
400	24,000	40.7	0.0018	0.031	0.219	0.927	9.84	533	32,000	54.2
425	24,000	45.1	0.0020	0.034	0.230	0.923	9.42	567	32,000	60.2
450	24,000	49.7	0.0023	0.038	0.240	0.920	9.05	600	32,000	66.3
475	24,000	54.5	0.0025	0.042	0.250	0.917	8.72	633	32,000	72.6
500	24,000	59.4	0.0027	0.046	0.260	0.913	8.42	667	32,000	79.1
500	23,000	61.1	0.0029	0.049	0.268	0.911	8.19	667	30,667	81.4
500	22,000	62.9	0.0031	0.053	0.277	0.908	7.95	667	29,333	83.8
500	21,000	64.8	0.0034	0.057	0.286	0.905	7.72	667	28,000	86.4
500	20,000	66.8	0.0037	0.062	0.297	0.901	7.48	667	26,667	89.1
500	19,000	69.0	0.0040	0.068	0.307	0.898	7.25	667	25,333	91.9
500	18,000	71.3	0.0044	0.075	0.319	0.894	7.02	667	24,000	95.0
500	17,000	73.7	0.0049	0.082	0.332	0.889	6.78	667	22,667	98.3
500	16,000	76.3	0.0054	0.091	0.345	0.885	6.55	667	21,333	101.8
500	15,000	79.2	0.0060	0.101	0.360	0.880	6.32	667	20,000	105.5
500	14,000	82.2	0.0067	0.113	0.376	0.875	6.08	667	18,667	109.6
500	13,000	85.4	0.0076	0.128	0.393	0.869	5.85	667	17,333	113.9
500	12,000	89.0	0.0086	0.145	0.413	0.862	5.62	667	16,000	118.6
500	11,000	92.8	0.0099	0.166	0.434	0.855	5.39	667	14,667	123.7
500	10,000	96.9	0.0114	0.193	0.457	0.848	5.16	667	13,333	129.2

ACI/ASCE

A/A TABLE B-2a　Flexural Design Coefficients for *CLAY* Masonry
Constructed with Type *S* or *N* Mortar: f'_m = 2000 psi, f_y = 60,000 psi and n = 14.5　　For use of Table, see Example 5-B

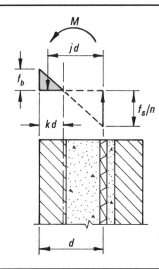

DESIGN DATA

$f'_m = 2000$ psi　　　　　　$f_y = 60,000$ psi

$F_b = 0.33 f'_m = 667$ psi　　$F_s = 24,000$ psi

$E_m = 2,000,000$ psi　　　　$E_s = 29,000,000$ psi

DESIGN EQUATIONS

$$n = \frac{E_s}{E_m} = 14.5 \qquad K = \frac{M}{F} = \frac{M(\text{ft kips})}{bd^2/12,000} \text{ or } \frac{M(\text{in. lbs})}{bd^2}$$

$$K = \tfrac{1}{2} jkf_b$$

$$k = \frac{1}{1 + f_s/n f_b} \qquad j = 1 - \frac{k}{3}$$

$$A_s = \frac{M}{f_s j d} \qquad p = \frac{A_s}{bd} = \frac{K}{f_s j}$$

			Increase for wind or earthquake		
			$1.33 F_b = 889$ psi		
			$1.33 F_s = 32,000$ psi		

f_b	f_s	K	p	np	k	j	$2/jk$	f_b	f_s	K
150	24,000	6.1	0.0003	0.004	0.083	0.972	24.75	200	32,000	8.1
200	24,000	10.4	0.0004	0.007	0.108	0.964	19.24	267	32,000	13.9
250	24,000	15.7	0.0007	0.010	0.131	0.956	15.94	333	32,000	20.9
300	24,000	21.8	0.0010	0.014	0.153	0.949	13.74	400	32,000	29.1
350	24,000	28.8	0.0013	0.018	0.175	0.942	12.17	467	32,000	38.4
400	24,000	36.4	0.0016	0.024	0.195	0.935	10.99	533	32,000	48.5
450	24,000	44.7	0.0020	0.029	0.214	0.929	10.07	600	32,000	59.6
500	24,000	53.5	0.0024	0.035	0.232	0.923	9.34	667	32,000	71.4
550	24,000	62.9	0.0029	0.041	0.249	0.917	8.75	733	32,000	83.8
600	24,000	72.7	0.0033	0.048	0.266	0.911	8.25	800	32,000	97.0
650	24,000	83.0	0.0038	0.055	0.282	0.906	7.83	867	32,000	110.7
667	24,000	86.6	0.0040	0.058	0.287	0.904	7.70	889	32,000	115.5
667	23,000	89.0	0.0043	0.062	0.296	0.901	7.50	889	30,667	118.6
667	22,000	91.5	0.0046	0.067	0.305	0.898	7.29	889	29,333	122.0
667	21,000	94.1	0.0050	0.073	0.315	0.895	7.09	889	28,000	125.5
667	20,000	96.9	0.0054	0.079	0.326	0.891	6.88	889	26,667	129.2
667	19,000	99.8	0.0059	0.086	0.337	0.888	6.68	889	25,333	133.1
667	18,000	103.0	0.0065	0.094	0.350	0.883	6.48	889	24,000	137.3
667	17,000	106.3	0.0071	0.103	0.363	0.879	6.27	889	22,667	141.8
667	16,000	109.9	0.0079	0.114	0.377	0.874	6.07	889	21,333	146.5
667	15,000	113.7	0.0087	0.126	0.392	0.869	5.87	889	20,000	151.5
667	14,000	117.7	0.0097	0.141	0.409	0.864	5.67	889	18,667	156.9
667	13,000	122.0	0.0109	0.159	0.427	0.858	5.47	889	17,333	162.7
667	12,000	126.7	0.0124	0.180	0.446	0.851	5.26	889	16,000	168.9
667	11,000	131.7	0.0142	0.206	0.468	0.844	5.06	889	14,667	175.6
667	10,000	137.1	0.0164	0.238	0.492	0.836	4.87	889	13,333	182.8

A/A TABLE B-2b   Flexural Design Coefficients for *CLAY* Masonry — For use of Table,
Constructed with Type *M* Mortar: f'_m = 2000 psi, f_y = 60,000 psi and n = 12.5 — see Example 5-B

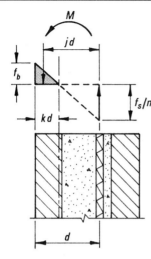

DESIGN DATA

$f'_m = 2000$ psi	$f_y = 60,000$ psi
$F_b = 0.33 f'_m = 667$ psi	$F_s = 24,000$ psi
$E_m = 2,320,000$ psi	$E_s = 29,000,000$ psi

DESIGN EQUATIONS

$$n = \frac{E_s}{E_m} = 12.5 \qquad K = \frac{M}{F} = \frac{M(\text{ft kips})}{bd^2/12,000} \text{ or } \frac{M(\text{in. lbs})}{bd^2}$$

$$K = \tfrac{1}{2} j k f_b$$

$$k = \frac{1}{1 + f_s/n f_b} \qquad j = 1 - \frac{k}{3}$$

$$A_s = \frac{M}{f_s j d} \qquad p = \frac{A_s}{bd} = \frac{K}{f_s j}$$

Increase for wind or earthquake
$1.33 F_b = 889$ psi
$1.33 F_s = 32,000$ psi

f_b	f_s	K	p	np	k	j	$2/jk$	f_b	f_s	K
200	24,000	9.1	0.0004	0.005	0.094	0.969	21.89	267	32,000	12.2
250	24,000	13.8	0.0006	0.008	0.115	0.962	18.05	333	32,000	18.5
300	24,000	19.4	0.0008	0.011	0.135	0.955	15.50	400	32,000	25.8
350	24,000	25.6	0.0011	0.014	0.154	0.949	13.67	467	32,000	34.1
400	24,000	32.5	0.0014	0.018	0.172	0.943	12.31	533	32,000	43.3
450	24,000	40.0	0.0018	0.022	0.190	0.937	11.25	600	32,000	53.4
500	24,000	48.1	0.0022	0.027	0.207	0.931	10.40	667	32,000	64.1
550	24,000	56.7	0.0026	0.032	0.223	0.926	9.70	733	32,000	75.6
600	24,000	65.8	0.0030	0.037	0.238	0.921	9.12	800	32,000	87.7
650	24,000	75.3	0.0034	0.043	0.253	0.916	8.64	867	32,000	100.4
667	24,000	78.6	0.0036	0.045	0.258	0.914	8.49	889	32,000	104.8
667	23,000	80.9	0.0039	0.048	0.266	0.911	8.25	889	30,667	107.8
667	22,000	83.3	0.0042	0.052	0.275	0.908	8.01	889	29,333	111.0
667	21,000	85.8	0.0045	0.056	0.284	0.905	7.77	889	28,000	114.4
667	20,000	88.5	0.0049	0.061	0.294	0.902	7.54	889	26,667	118.0
667	19,000	91.4	0.0054	0.067	0.305	0.898	7.30	889	25,333	121.8
667	18,000	94.4	0.0059	0.073	0.317	0.894	7.06	889	24,000	125.9
667	17,000	97.7	0.0065	0.081	0.329	0.890	6.83	889	22,667	130.3
667	16,000	101.2	0.0071	0.089	0.343	0.886	6.59	889	21,333	134.9
667	15,000	105.0	0.0079	0.099	0.357	0.881	6.35	889	20,000	139.9
667	14,000	109.0	0.0089	0.111	0.373	0.876	6.12	889	18,667	145.3
667	13,000	113.3	0.0100	0.125	0.391	0.870	5.88	889	17,333	151.1
667	12,000	118.0	0.0114	0.142	0.410	0.863	5.65	889	16,000	157.4
667	11,000	123.1	0.0131	0.163	0.431	0.856	5.42	889	14,667	164.2
667	10,000	128.7	0.0152	0.190	0.455	0.848	5.18	889	13,333	171.5

ACI/ASCE

A/A TABLE B-3a Flexural Design Coefficients for *CLAY* Masonry
Constructed with Type *S* or *N* Mortar: f'_m = 2500 psi, f_y = 60,000 psi and n = 11.6

For use of Table, see Example 5-B

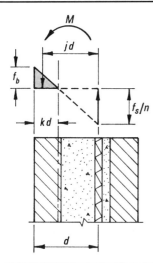

DESIGN DATA

$f'_m = 2500$ psi $\qquad f_y = 60,000$ psi

$F_b = 0.33 f'_m = 833$ psi $\qquad F_s = 24,000$ psi

$E_m = 2,500,000$ psi $\qquad E_s = 29,000,000$ psi

DESIGN EQUATIONS

$$n = \frac{E_s}{E_m} = 11.6 \qquad K = \frac{M}{F} = \frac{M(\text{ft kips})}{bd^2/12,000} \text{ or } \frac{M(\text{in. lbs})}{bd^2}$$

$$K = \tfrac{1}{2} jk f_b$$

$$k = \frac{1}{1 + f_s/n f_b} \qquad j = 1 - \frac{k}{3}$$

$$A_s = \frac{M}{f_s\,jd} \qquad p = \frac{A_s}{bd} = \frac{K}{f_s\,j}$$

Increase for wind or earthquake

$1.33 F_b = 1111$ psi

$1.33 F_s = 32,000$ psi

f_b	f_s	K	p	np	k	j	$2/jk$	f_b	f_s	K
200	24,000	8.6	0.0004	0.004	0.088	0.971	23.38	267	32,000	11.4
250	24,000	13.0	0.0006	0.007	0.108	0.964	19.24	333	32,000	17.3
300	24,000	18.2	0.0008	0.009	0.127	0.958	16.49	400	32,000	24.3
350	24,000	24.1	0.0011	0.012	0.145	0.952	14.52	467	32,000	32.1
400	24,000	30.7	0.0014	0.016	0.162	0.946	13.05	533	32,000	40.9
450	24,000	37.8	0.0017	0.019	0.179	0.940	11.90	600	32,000	50.4
500	24,000	45.5	0.0020	0.024	0.195	0.935	10.99	667	32,000	60.7
550	24,000	53.7	0.0024	0.028	0.210	0.930	10.24	733	32,000	71.6
600	24,000	62.4	0.0028	0.033	0.225	0.925	9.62	800	32,000	83.2
650	24,000	71.5	0.0032	0.038	0.239	0.920	9.09	867	32,000	95.3
700	24,000	81.0	0.0037	0.043	0.253	0.916	8.64	933	32,000	108.0
750	24,000	90.9	0.0042	0.048	0.266	0.911	8.25	1,000	32,000	121.2
800	24,000	101.2	0.0046	0.054	0.279	0.907	7.91	1,067	32,000	134.9
833	24,000	108.1	0.0050	0.058	0.287	0.904	7.70	1,111	32,000	144.2
833	23,000	111.1	0.0054	0.062	0.296	0.901	7.50	1,111	30,667	148.1
833	22,000	114.2	0.0058	0.067	0.305	0.898	7.30	1,111	29,333	152.2
833	21,000	117.5	0.0063	0.073	0.315	0.895	7.09	1,111	28,000	156.6
833	20,000	120.9	0.0068	0.079	0.326	0.891	6.89	1,111	26,667	161.3
833	19,000	124.6	0.0074	0.086	0.337	0.888	6.68	1,111	25,333	166.2
833	18,000	128.5	0.0081	0.094	0.349	0.884	6.48	1,111	24,000	171.4
833	17,000	132.7	0.0089	0.103	0.362	0.879	6.28	1,111	22,667	176.9
833	16,000	137.1	0.0098	0.114	0.377	0.874	6.07	1,111	21,333	182.9
833	15,000	141.9	0.0109	0.126	0.392	0.869	5.87	1,111	20,000	189.2
833	14,000	146.9	0.0121	0.141	0.408	0.864	5.67	1,111	18,667	195.9
833	13,000	152.3	0.0137	0.158	0.426	0.858	5.47	1,111	17,333	203.1
833	12,000	158.2	0.0155	0.180	0.446	0.851	5.27	1,111	16,000	210.9
833	11,000	164.4	0.0177	0.205	0.468	0.844	5.07	1,111	14,667	219.2
833	10,000	171.2	0.0205	0.237	0.491	0.836	4.87	1,111	13,333	228.2

A/A TABLE B-3b Flexural Design Coefficients for *CLAY* Masonry
Constructed with Type *M* Mortar: f'_m = 2500 psi, f_y = 60,000 psi and n = 10.2

For use of Table,
see Example 5-B

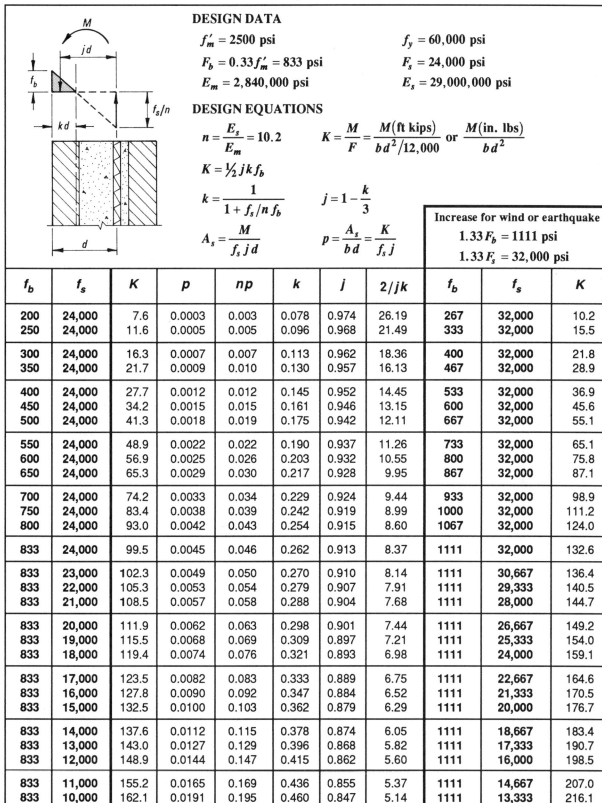

DESIGN DATA

$f'_m = 2500$ psi

$F_b = 0.33 f'_m = 833$ psi

$E_m = 2,840,000$ psi

$f_y = 60,000$ psi

$F_s = 24,000$ psi

$E_s = 29,000,000$ psi

DESIGN EQUATIONS

$$n = \frac{E_s}{E_m} = 10.2 \qquad K = \frac{M}{F} = \frac{M(\text{ft kips})}{bd^2/12,000} \text{ or } \frac{M(\text{in. lbs})}{bd^2}$$

$$K = \tfrac{1}{2} jk f_b$$

$$k = \frac{1}{1 + f_s/n f_b} \qquad j = 1 - \frac{k}{3}$$

$$A_s = \frac{M}{f_s j d} \qquad p = \frac{A_s}{bd} = \frac{K}{f_s j}$$

Increase for wind or earthquake
$1.33 F_b = 1111$ psi
$1.33 F_s = 32,000$ psi

f_b	f_s	K	p	np	k	j	$2/jk$	f_b	f_s	K
200	24,000	7.6	0.0003	0.003	0.078	0.974	26.19	267	32,000	10.2
250	24,000	11.6	0.0005	0.005	0.096	0.968	21.49	333	32,000	15.5
300	24,000	16.3	0.0007	0.007	0.113	0.962	18.36	400	32,000	21.8
350	24,000	21.7	0.0009	0.010	0.130	0.957	16.13	467	32,000	28.9
400	24,000	27.7	0.0012	0.012	0.145	0.952	14.45	533	32,000	36.9
450	24,000	34.2	0.0015	0.015	0.161	0.946	13.15	600	32,000	45.6
500	24,000	41.3	0.0018	0.019	0.175	0.942	12.11	667	32,000	55.1
550	24,000	48.9	0.0022	0.022	0.190	0.937	11.26	733	32,000	65.1
600	24,000	56.9	0.0025	0.026	0.203	0.932	10.55	800	32,000	75.8
650	24,000	65.3	0.0029	0.030	0.217	0.928	9.95	867	32,000	87.1
700	24,000	74.2	0.0033	0.034	0.229	0.924	9.44	933	32,000	98.9
750	24,000	83.4	0.0038	0.039	0.242	0.919	8.99	1000	32,000	111.2
800	24,000	93.0	0.0042	0.043	0.254	0.915	8.60	1067	32,000	124.0
833	24,000	99.5	0.0045	0.046	0.262	0.913	8.37	1111	32,000	132.6
833	23,000	102.3	0.0049	0.050	0.270	0.910	8.14	1111	30,667	136.4
833	22,000	105.3	0.0053	0.054	0.279	0.907	7.91	1111	29,333	140.5
833	21,000	108.5	0.0057	0.058	0.288	0.904	7.68	1111	28,000	144.7
833	20,000	111.9	0.0062	0.063	0.298	0.901	7.44	1111	26,667	149.2
833	19,000	115.5	0.0068	0.069	0.309	0.897	7.21	1111	25,333	154.0
833	18,000	119.4	0.0074	0.076	0.321	0.893	6.98	1111	24,000	159.1
833	17,000	123.5	0.0082	0.083	0.333	0.889	6.75	1111	22,667	164.6
833	16,000	127.8	0.0090	0.092	0.347	0.884	6.52	1111	21,333	170.5
833	15,000	132.5	0.0100	0.103	0.362	0.879	6.29	1111	20,000	176.7
833	14,000	137.6	0.0112	0.115	0.378	0.874	6.05	1111	18,667	183.4
833	13,000	143.0	0.0127	0.129	0.396	0.868	5.82	1111	17,333	190.7
833	12,000	148.9	0.0144	0.147	0.415	0.862	5.60	1111	16,000	198.5
833	11,000	155.2	0.0165	0.169	0.436	0.855	5.37	1111	14,667	207.0
833	10,000	162.1	0.0191	0.195	0.460	0.847	5.14	1111	13,333	216.1
833	9,000	169.6	0.0225	0.230	0.486	0.838	4.91	1111	12,000	226.1

ACI/ASCE

A/A TABLE B-4a Flexural Design Coefficients for *CLAY* Masonry Constructed with Type *N* Mortar: f'_m = 3000 psi, f_y = 60,000 psi and n = 10.4

For use of Table, see Example 5-B

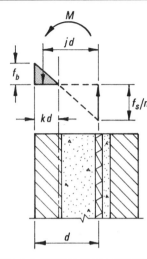

DESIGN DATA

$f'_m = 3000$ psi $\qquad\qquad f_y = 60,000$ psi

$F_b = 0.33 f'_m = 1000$ psi $\qquad F_s = 24,000$ psi

$E_m = 2,840,000$ psi $\qquad\quad E_s = 29,000,000$ psi

DESIGN EQUATIONS

$$n = \frac{E_s}{E_m} = 10.4 \qquad K = \frac{M}{F} = \frac{M(\text{ft kips})}{bd^2/12,000} \text{ or } \frac{M(\text{in. lbs})}{bd^2}$$

$$K = \tfrac{1}{2} jk f_b$$

$$k = \frac{1}{1 + f_s/n f_b} \qquad j = 1 - \frac{k}{3}$$

$$A_s = \frac{M}{f_s j d} \qquad p = \frac{A_s}{bd} = \frac{K}{f_s j}$$

Increase for wind or earthquake
$1.33 F_b = 1333$ psi
$1.33 F_s = 32,000$ psi

f_b	f_s	K	p	np	k	j	$2/jk$	f_b	f_s	K
250	24,000	11.8	0.0005	0.005	0.098	0.967	21.15	333	32,000	15.8
300	24,000	16.6	0.0007	0.007	0.115	0.962	18.08	400	32,000	22.1
350	24,000	22.0	0.0010	0.010	0.132	0.956	15.88	467	32,000	29.4
400	24,000	28.1	0.0012	0.013	0.148	0.951	14.24	533	32,000	37.5
450	24,000	34.7	0.0015	0.016	0.163	0.946	12.96	600	32,000	46.3
500	24,000	41.9	0.0019	0.019	0.178	0.941	11.94	667	32,000	55.8
550	24,000	49.5	0.0022	0.023	0.192	0.936	11.10	733	32,000	66.0
600	24,000	57.6	0.0026	0.027	0.206	0.931	10.41	800	32,000	76.9
650	24,000	66.2	0.0030	0.031	0.220	0.927	9.82	867	32,000	88.3
700	24,000	75.1	0.0034	0.035	0.233	0.922	9.32	933	32,000	100.2
750	24,000	84.5	0.0038	0.040	0.245	0.918	8.88	1000	32,000	112.6
800	24,000	94.1	0.0043	0.045	0.257	0.914	8.50	1067	32,000	125.5
850	24,000	104.1	0.0048	0.050	0.269	0.910	8.16	1133	32,000	138.9
900	24,000	114.5	0.0053	0.055	0.281	0.906	7.86	1200	32,000	152.6
950	24,000	125.1	0.0058	0.060	0.292	0.903	7.60	1267	32,000	166.7
1000	24,000	135.9	0.0063	0.066	0.302	0.899	7.36	1333	32,000	181.2
1000	23,000	139.5	0.0068	0.070	0.311	0.896	7.17	1333	30,667	186.0
1000	22,000	143.3	0.0073	0.076	0.321	0.893	6.98	1333	29,333	191.1
1000	21,000	147.3	0.0079	0.082	0.331	0.890	6.79	1333	28,000	196.4
1000	20,000	151.5	0.0086	0.089	0.342	0.886	6.60	1333	26,667	202.1
1000	19,000	156.0	0.0093	0.097	0.354	0.882	6.41	1333	25,333	208.0
1000	18,000	160.7	0.0102	0.106	0.366	0.878	6.22	1333	24,000	214.3
1000	17,000	165.8	0.0112	0.116	0.380	0.873	6.03	1333	22,667	221.0
1000	16,000	171.1	0.0123	0.128	0.394	0.869	5.84	1333	21,333	228.1
1000	15,000	176.8	0.0136	0.142	0.409	0.864	5.66	1333	20,000	235.7
1000	14,000	182.8	0.0152	0.158	0.426	0.858	5.47	1333	18,667	243.8
1000	13,000	189.3	0.0171	0.178	0.444	0.852	5.28	1333	17,333	252.4
1000	12,000	196.2	0.0193	0.201	0.464	0.845	5.10	1333	16,000	261.6
1000	11,000	203.6	0.0221	0.230	0.486	0.838	4.91	1333	14,667	271.5
1000	10,000	211.6	0.0255	0.265	0.510	0.830	4.73	1333	13,333	282.1

A/A TABLE B-4b   Flexural Design Coefficients for *CLAY* Masonry Constructed with Type *S* Mortar: f'_m = 3000 psi, f_y = 60,000 psi and *n* = 9.9   For use of Table, see Example 5-B

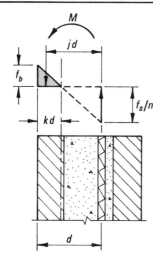

DESIGN DATA

$f'_m = 3000$ psi	$f_y = 60,000$ psi
$F_b = 0.33 f'_m = 1000$ psi	$F_s = 24,000$ psi
$E_m = 2,920,000$ psi	$E_s = 29,000,000$ psi

DESIGN EQUATIONS

$$n = \frac{E_s}{E_m} = 9.9 \qquad K = \frac{M}{F} = \frac{M(\text{ft kips})}{bd^2/12,000} \text{ or } \frac{M(\text{in. lbs})}{bd^2}$$

$$K = \frac{1}{2} jk f_b$$

$$k = \frac{1}{1 + f_s/n f_b} \qquad j = 1 - \frac{k}{3}$$

$$A_s = \frac{M}{f_s j d} \qquad p = \frac{A_s}{bd} = \frac{K}{f_s j}$$

Increase for wind or earthquake
$1.33 F_b = 1333$ psi
$1.33 F_s = 32,000$ psi

f_b	f_s	K	p	np	k	j	2/jk	f_b	f_s	K
250	24,000	11.3	0.0005	0.005	0.093	0.969	22.08	333	32,000	15.1
300	24,000	15.9	0.0007	0.007	0.110	0.963	18.85	400	32,000	21.2
350	24,000	21.1	0.0009	0.009	0.126	0.958	16.55	467	32,000	28.2
400	24,000	27.0	0.0012	0.012	0.142	0.953	14.82	533	32,000	36.0
450	24,000	33.4	0.0015	0.015	0.157	0.948	13.48	600	32,000	44.5
500	24,000	40.3	0.0018	0.018	0.171	0.943	12.40	667	32,000	53.7
550	24,000	47.7	0.0021	0.021	0.185	0.938	11.53	733	32,000	63.6
600	24,000	55.6	0.0025	0.025	0.198	0.934	10.79	800	32,000	74.1
650	24,000	63.9	0.0029	0.028	0.211	0.930	10.18	867	32,000	85.2
700	24,000	72.6	0.0033	0.032	0.224	0.925	9.65	933	32,000	96.7
750	24,000	81.6	0.0037	0.037	0.236	0.921	9.19	1000	32,000	108.8
800	24,000	91.0	0.0041	0.041	0.248	0.917	8.79	1067	32,000	121.4
850	24,000	100.8	0.0046	0.046	0.260	0.913	8.43	1133	32,000	134.4
900	24,000	110.8	0.0051	0.050	0.271	0.910	8.12	1200	32,000	147.8
950	24,000	121.2	0.0056	0.055	0.282	0.906	7.84	1267	32,000	161.6
1000	24,000	131.8	0.0061	0.060	0.292	0.903	7.59	1333	32,000	175.7
1000	23,000	135.4	0.0065	0.065	0.301	0.900	7.39	1333	30,667	180.5
1000	22,000	139.1	0.0071	0.070	0.310	0.897	7.19	1333	29,333	185.5
1000	21,000	143.1	0.0076	0.076	0.320	0.893	6.99	1333	28,000	190.8
1000	20,000	147.3	0.0083	0.082	0.331	0.890	6.79	1333	26,667	196.4
1000	19,000	151.7	0.0090	0.089	0.343	0.886	6.59	1333	25,333	202.3
1000	18,000	156.4	0.0099	0.098	0.355	0.882	6.39	1333	24,000	208.6
1000	17,000	161.4	0.0108	0.107	0.368	0.877	6.19	1333	22,667	215.3
1000	16,000	166.8	0.0119	0.118	0.382	0.873	6.00	1333	21,333	222.4
1000	15,000	172.4	0.0133	0.131	0.398	0.867	5.80	1333	20,000	229.9
1000	14,000	178.5	0.0148	0.146	0.414	0.862	5.60	1333	18,667	238.0
1000	13,000	185.0	0.0166	0.165	0.432	0.856	5.41	1333	17,333	246.7
1000	12,000	192.0	0.0188	0.186	0.452	0.849	5.21	1333	16,000	256.0
1000	11,000	199.4	0.0215	0.213	0.474	0.842	5.01	1333	14,667	265.9
1000	10,000	207.5	0.0249	0.246	0.497	0.834	4.82	1333	13,333	276.7

ACI/ASCE

A/A TABLE B-4c   Flexural Design Coefficients for *CLAY* Masonry
Constructed with Type *M* Mortar: f'_m = 3000 psi, f_y = 60,000 psi and n = 9.7

For use of Table, see Example 5-B

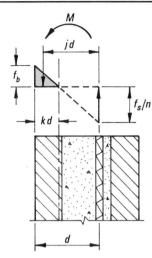

DESIGN DATA

$f'_m = 3000$ psi $\qquad f_y = 60,000$ psi

$F_b = 0.33 f'_m = 1000$ psi $\qquad F_s = 24,000$ psi

$E_m = 3,000,000$ psi $\qquad E_s = 29,000,000$ psi

DESIGN EQUATIONS

$$n = \frac{E_s}{E_m} = 9.7 \qquad K = \frac{M}{F} = \frac{M(\text{ft kips})}{bd^2/12,000} \text{ or } \frac{M(\text{in. lbs})}{bd^2}$$

$$K = \tfrac{1}{2} jk f_b$$

$$k = \frac{1}{1 + f_s/n f_b} \qquad j = 1 - \frac{k}{3}$$

$$A_s = \frac{M}{f_s j d} \qquad p = \frac{A_s}{bd} = \frac{K}{f_s j}$$

Increase for wind or earthquake
$1.33 F_b = 1333$ psi
$1.33 F_s = 32,000$ psi

f_b	f_s	K	p	np	k	j	2/jk	f_b	f_s	K
250	24,000	11.1	0.0005	0.005	0.091	0.970	22.55	333	32,000	14.8
300	24,000	15.6	0.0007	0.007	0.108	0.964	19.24	400	32,000	20.8
350	24,000	20.7	0.0009	0.009	0.124	0.959	16.88	467	32,000	27.6
400	24,000	26.5	0.0012	0.011	0.139	0.954	15.11	533	32,000	35.3
450	24,000	32.8	0.0014	0.014	0.153	0.949	13.74	600	32,000	43.7
500	24,000	39.6	0.0017	0.017	0.168	0.944	12.64	667	32,000	52.8
550	24,000	46.9	0.0021	0.020	0.181	0.940	11.74	733	32,000	62.5
600	24,000	54.6	0.0024	0.024	0.195	0.935	10.99	800	32,000	72.8
650	24,000	62.8	0.0028	0.027	0.207	0.931	10.36	867	32,000	83.7
700	24,000	71.3	0.0032	0.031	0.220	0.927	9.81	933	32,000	95.1
750	24,000	80.3	0.0036	0.035	0.232	0.923	9.34	1000	32,000	107.0
800	24,000	89.6	0.0041	0.039	0.244	0.919	8.93	1067	32,000	119.4
850	24,000	99.2	0.0045	0.044	0.255	0.915	8.57	1133	32,000	132.2
900	24,000	109.1	0.0050	0.048	0.266	0.911	8.25	1200	32,000	145.5
950	24,000	119.3	0.0055	0.053	0.277	0.908	7.96	1267	32,000	159.1
1000	24,000	129.8	0.0060	0.058	0.288	0.904	7.70	1333	32,000	173.1
1000	23,000	133.4	0.0064	0.062	0.296	0.901	7.50	1333	30,667	177.8
1000	22,000	137.1	0.0069	0.067	0.305	0.898	7.29	1333	29,333	182.8
1000	21,000	141.0	0.0075	0.073	0.315	0.895	7.09	1333	28,000	188.1
1000	20,000	145.2	0.0081	0.079	0.326	0.891	6.89	1333	26,667	193.6
1000	19,000	149.7	0.0089	0.086	0.337	0.888	6.68	1333	25,333	199.5
1000	18,000	154.4	0.0097	0.094	0.349	0.884	6.48	1333	24,000	205.8
1000	17,000	159.3	0.0107	0.103	0.363	0.879	6.28	1333	22,667	212.5
1000	16,000	164.7	0.0118	0.114	0.377	0.874	6.07	1333	21,333	219.6
1000	15,000	170.3	0.0131	0.126	0.392	0.869	5.87	1333	20,000	227.1
1000	14,000	176.4	0.0146	0.141	0.408	0.864	5.67	1333	18,667	235.2
1000	13,000	182.9	0.0164	0.159	0.426	0.858	5.47	1333	17,333	243.9
1000	12,000	189.9	0.0186	0.180	0.446	0.851	5.27	1333	16,000	253.2
1000	11,000	197.4	0.0213	0.206	0.468	0.844	5.07	1333	14,667	263.2
1000	10,000	205.5	0.0246	0.238	0.492	0.836	4.87	1333	13,333	274.0

A/A TABLE B-5 Flexural Design Coefficients for *CLAY* Masonry
Constructed with Type *S* Mortar: f'_m = 3500 psi, f_y = 60,000 psi and n = 9.7

For use of Table, see Example 5-B

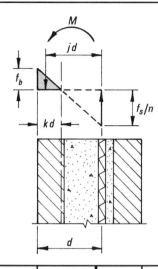

DESIGN DATA

$f'_m = 3500$ psi $\qquad f_y = 60,000$ psi

$F_b = 0.33 f'_m = 1167$ psi $\qquad F_s = 24,000$ psi

$E_m = 3,000,000$ psi $\qquad E_s = 29,000,000$ psi

DESIGN EQUATIONS

$$n = \frac{E_s}{E_m} = 9.7 \qquad K = \frac{M}{F} = \frac{M(\text{ft kips})}{bd^2/12,000} \text{ or } \frac{M(\text{in. lbs})}{bd^2}$$

$$K = \tfrac{1}{2} jkf_b$$

$$k = \frac{1}{1 + f_s/n f_b} \qquad j = 1 - \frac{k}{3}$$

$$A_s = \frac{M}{f_s j d} \qquad p = \frac{A_s}{bd} = \frac{K}{f_s j}$$

Increase for wind or earthquake
$1.33 F_b = 1556$ psi
$1.33 F_s = 32,000$ psi

f_b	f_s	K	p	np	k	j	$2/jk$	f_b	f_s	K
300	24,000	15.6	0.0007	0.007	0.108	0.964	19.19	400	32,000	20.8
350	24,000	20.8	0.0009	0.009	0.124	0.959	16.83	467	32,000	27.7
400	24,000	26.5	0.0012	0.011	0.139	0.954	15.07	533	32,000	35.4
450	24,000	32.8	0.0014	0.014	0.154	0.949	13.70	600	32,000	43.8
500	24,000	39.7	0.0018	0.017	0.168	0.944	12.60	667	32,000	52.9
550	24,000	47.0	0.0021	0.020	0.182	0.939	11.71	733	32,000	62.6
600	24,000	54.7	0.0024	0.024	0.195	0.935	10.96	800	32,000	73.0
650	24,000	62.9	0.0028	0.027	0.208	0.931	10.33	867	32,000	83.9
700	24,000	71.5	0.0032	0.031	0.221	0.926	9.79	933	32,000	95.3
750	24,000	80.5	0.0036	0.035	0.233	0.922	9.32	1,000	32,000	107.3
800	24,000	89.8	0.0041	0.040	0.244	0.919	8.91	1,067	32,000	119.7
850	24,000	99.4	0.0045	0.044	0.256	0.915	8.55	1,133	32,000	132.5
900	24,000	109.4	0.0050	0.049	0.267	0.911	8.23	1,200	32,000	145.8
950	24,000	119.6	0.0055	0.053	0.277	0.908	7.94	1,267	32,000	159.5
1,000	24,000	130.1	0.0060	0.058	0.288	0.904	7.69	1,333	32,000	173.5
1,050	24,000	140.9	0.0065	0.063	0.298	0.901	7.45	1,400	32,000	187.8
1,100	24,000	151.9	0.0071	0.068	0.308	0.897	7.24	1,467	32,000	202.5
1,150	24,000	163.2	0.0076	0.074	0.317	0.894	7.05	1,533	32,000	217.5
1,167	24,000	167.0	0.0078	0.076	0.320	0.893	6.99	1,556	32,000	222.7
1,167	23,000	171.3	0.0084	0.081	0.330	0.890	6.81	1,556	30,667	228.4
1,167	22,000	175.8	0.0090	0.087	0.340	0.887	6.64	1,556	29,333	234.4
1,167	21,000	180.5	0.0097	0.094	0.350	0.883	6.47	1,556	28,000	240.7
1,167	20,000	185.5	0.0105	0.102	0.361	0.880	6.29	1,556	26,667	247.3
1,167	19,000	190.7	0.0115	0.111	0.373	0.876	6.12	1,556	25,333	254.3
1,167	18,000	196.3	0.0125	0.121	0.386	0.871	5.95	1,556	24,000	261.7
1,167	17,000	202.2	0.0137	0.133	0.400	0.867	5.77	1,556	22,667	269.5
1,167	16,000	208.4	0.0151	0.147	0.414	0.862	5.60	1,556	21,333	277.8
1,167	15,000	215.0	0.0167	0.162	0.430	0.857	5.43	1,556	20,000	286.6
1,167	14,000	222.0	0.0186	0.181	0.447	0.851	5.26	1,556	18,667	296.0
1,167	13,000	229.5	0.0209	0.203	0.465	0.845	5.09	1,556	17,333	305.9
1,167	12,000	237.4	0.0236	0.229	0.485	0.838	4.92	1,556	16,000	316.5

ACI/ASCE

A/A TABLE B-6   Flexural Design Coefficients for *CLAY* Masonry
Constructed with Type *S* Mortar: f'_m = 4000 psi, f_y = 60,000 psi and n = 9.7

For use of Table,
see Example 5-B

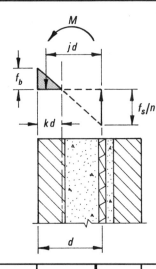

DESIGN DATA

$f'_m = 4000$ psi $f_y = 60,000$ psi

$F_b = 0.33 f'_m = 1333$ psi $F_s = 24,000$ psi

$E_m = 3,000,000$ psi $E_s = 29,000,000$ psi

DESIGN EQUATIONS

$$n = \frac{E_s}{E_m} = 9.7 \qquad K = \frac{M}{F} = \frac{M(\text{ft kips})}{bd^2/12,000} \text{ or } \frac{M(\text{in. lbs})}{bd^2}$$

$$K = \tfrac{1}{2} jkf_b$$

$$k = \frac{1}{1 + f_s/n f_b} \qquad j = 1 - \frac{k}{3}$$

$$A_s = \frac{M}{f_s j d} \qquad p = \frac{A_s}{bd} = \frac{K}{f_s j}$$

Increase for wind or earthquake
$1.33 F_b = 1777$ psi
$1.33 F_s = 32,000$ psi

f_b	f_s	K	p	np	k	j	$2/jk$	f_b	f_s	K
400	24,000	26.5	0.0012	0.011	0.139	0.954	15.07	533	32,000	35.4
450	24,000	32.8	0.0014	0.014	0.154	0.949	13.70	600	32,000	43.8
500	24,000	39.7	0.0018	0.017	0.168	0.944	12.60	667	32,000	52.9
550	24,000	47.0	0.0021	0.020	0.182	0.939	11.71	733	32,000	62.6
600	24,000	54.7	0.0024	0.024	0.195	0.935	10.96	800	32,000	73.0
650	24,000	62.9	0.0028	0.027	0.208	0.931	10.33	867	32,000	83.9
700	24,000	71.5	0.0032	0.031	0.221	0.926	9.79	933	32,000	95.3
750	24,000	80.5	0.0036	0.035	0.233	0.922	9.32	1,000	32,000	107.3
800	24,000	89.8	0.0041	0.040	0.244	0.919	8.91	1,067	32,000	119.7
850	24,000	99.4	0.0045	0.044	0.256	0.915	8.55	1,133	32,000	132.5
900	24,000	109.4	0.0050	0.049	0.267	0.911	8.23	1,200	32,000	145.8
950	24,000	119.6	0.0055	0.053	0.277	0.908	7.94	1,267	32,000	159.5
1,000	24,000	130.1	0.0060	0.058	0.288	0.904	7.69	1,333	32,000	173.5
1,050	24,000	140.9	0.0065	0.063	0.298	0.901	7.45	1,400	32,000	187.8
1,100	24,000	151.9	0.0071	0.068	0.308	0.897	7.24	1,467	32,000	202.5
1,150	24,000	163.2	0.0076	0.074	0.317	0.894	7.05	1,533	32,000	217.5
1,200	24,000	174.6	0.0082	0.079	0.327	0.891	6.87	1,600	32,000	232.8
1,250	24,000	186.3	0.0087	0.085	0.336	0.888	6.71	1,667	32,000	248.4
1,300	24,000	198.2	0.0093	0.090	0.344	0.885	6.56	1,733	32,000	264.2
1,333	24,000	206.1	0.0097	0.094	0.350	0.883	6.47	1,777	32,000	274.8
1,333	23,000	211.1	0.0104	0.101	0.360	0.880	6.32	1,777	30,667	281.4
1,333	22,000	216.3	0.0112	0.109	0.370	0.877	6.16	1,777	29,333	288.4
1,333	21,000	221.7	0.0121	0.117	0.381	0.873	6.01	1,777	28,000	295.6
1,333	20,000	227.5	0.0131	0.127	0.393	0.869	5.86	1,777	26,667	303.3
1,333	19,000	233.5	0.0142	0.138	0.405	0.865	5.71	1,777	25,333	311.3
1,333	18,000	239.8	0.0155	0.150	0.418	0.861	5.56	1,777	24,000	319.7
1,333	17,000	246.5	0.0169	0.164	0.432	0.856	5.41	1,777	22,667	328.6
1,333	16,000	253.5	0.0186	0.181	0.447	0.851	5.26	1,777	21,333	338.0
1,333	15,000	260.9	0.0206	0.200	0.463	0.846	5.11	1,777	20,000	347.9
1,333	14,000	268.8	0.0229	0.222	0.480	0.840	4.96	1,777	18,667	358.4
1,333	13,000	277.1	0.0256	0.248	0.499	0.834	4.81	1,777	17,333	369.5
1,333	12,000	285.9	0.0288	0.279	0.519	0.827	4.66	1,777	16,000	381.2

A/A TABLE C-1 *CONCRETE* Masonry f'_m, E_m, n and E_v Values Based on the Concrete Masonry Unit Strength and the Mortar Type

Type N Mortar				
Net Area Compressive Strength of Concrete Masonry Units (psi)	Net Area Compressive Strength of Concrete Masonry Assemblage[1], f'_m (psi)	Modulus of Elasticity[2], E_m (psi)	Modular Ratio, $n = E_s/E_m$ where $E_s =$ 29,000,000 psi	Modulus of Rigidity[3], $E_v = 0.4 E_m$ (psi)
2,150	1,500	1,920,000	15.1	768,000
3,050	2,000	2,320,000	12.5	928,000
4,050	2,500	2,610,000	11.1	1,044,000
5,250	3,000	2,800,000	10.4	1,120,000
Type M or S Mortar				
Net Area Compressive Strength of Concrete Masonry Units (psi)	Net Area Compressive Strength of Concrete Masonry Assemblage[1], f'_m (psi)	Modulus of Elasticity[2], E_m (psi)	Modular Ratio, $n = E_s/E_m$ where $E_s =$ 29,000,000 psi	Modulus of Rigidity[3], $E_v = 0.4 E_m$ (psi)
1,900	1,500	2,080,000	13.9	832,000
2,800	2,000	2,460,000	11.8	984,000
3,750	2,500	2,800,000	10.4	1,120,000
4,800	3,000	3,140,000	9.2	1,256,000

1. Based on ACI/ASCE Standard Table 5.5.1.3.

2. Based on ACI/ASCE Specification Table 1.6.2.2.

3. Based on ACI/ASCE Section 5.5.1.3(b).

ACI/ASCE

A/A TABLE C-2 Correction Factors for *CONCRETE* Masonry Prism Strength[1]

Prism Height to Thickness Ratio	1.33	2.0	3.0	4.0	5.0
Correction Factor	0.75	1.00	1.07	1.15	1.22

1. Based on ACI/ASCE Specifications Table 1.6.3.3(c).

A/A TABLE D-1a   Flexural Design Coefficients for *CONCRETE* Masonry Constructed with Type *N* Mortar: f'_m = 1500 psi, f_y = 60,000 psi and n = 15.1

For use of Table, see Example 5-B

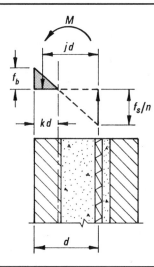

DESIGN DATA

$f'_m = 1500$ psi $f_y = 60,000$ psi

$F_b = 0.33 f'_m = 500$ psi $F_s = 24,000$ psi

$E_m = 1,920,000$ psi $E_s = 29,000,000$ psi

DESIGN EQUATIONS

$$n = \frac{E_s}{E_m} = 15.1 \qquad K = \frac{M}{F} = \frac{M(\text{ft kips})}{bd^2/12,000} \text{ or } \frac{M(\text{in. lbs})}{bd^2}$$

$$K = \tfrac{1}{2} jk f_b$$

$$k = \frac{1}{1 + f_s/n f_b} \qquad j = 1 - \frac{k}{3}$$

$$A_s = \frac{M}{f_s j d} \qquad p = \frac{A_s}{bd} = \frac{K}{f_s j}$$

Increase for wind or earthquake

$1.33 F_b = 667$ psi

$1.33 F_s = 32,000$ psi

f_b	f_s	K	p	np	k	j	$2/jk$	f_b	f_s	K
125	24,000	4.4	0.0002	0.003	0.073	0.976	28.11	167	32,000	5.9
150	24,000	6.3	0.0003	0.004	0.086	0.971	23.88	200	32,000	8.4
175	24,000	8.4	0.0004	0.005	0.099	0.967	20.85	233	32,000	11.2
200	24,000	10.8	0.0005	0.007	0.112	0.963	18.59	267	32,000	14.3
225	24,000	13.4	0.0006	0.009	0.124	0.959	16.82	300	32,000	17.8
250	24,000	16.2	0.0007	0.011	0.136	0.955	15.41	333	32,000	21.6
275	24,000	19.3	0.0008	0.013	0.148	0.951	14.26	367	32,000	25.7
300	24,000	22.6	0.0010	0.015	0.159	0.947	13.30	400	32,000	30.1
325	24,000	26.0	0.0011	0.017	0.170	0.943	12.49	433	32,000	34.7
350	24,000	29.7	0.0013	0.020	0.180	0.940	11.79	467	32,000	39.6
375	24,000	33.5	0.0015	0.023	0.191	0.936	11.19	500	32,000	44.7
400	24,000	37.5	0.0017	0.025	0.201	0.933	10.66	533	32,000	50.0
425	24,000	41.7	0.0019	0.028	0.211	0.930	10.20	567	32,000	55.6
450	24,000	46.0	0.0021	0.031	0.221	0.926	9.78	600	32,000	61.3
475	24,000	50.5	0.0023	0.034	0.230	0.923	9.41	633	32,000	67.3
500	24,000	55.1	0.0025	0.038	0.239	0.920	9.08	667	32,000	73.4
500	23,000	56.7	0.0027	0.041	0.247	0.918	8.82	667	30,667	75.6
500	22,000	58.4	0.0029	0.044	0.255	0.915	8.56	667	29,333	77.9
500	21,000	60.3	0.0031	0.048	0.264	0.912	8.29	667	28,000	80.4
500	20,000	62.3	0.0034	0.052	0.274	0.909	8.03	667	26,667	83.0
500	19,000	64.4	0.0037	0.056	0.284	0.905	7.77	667	25,333	85.8
500	18,000	66.6	0.0041	0.062	0.295	0.902	7.51	667	24,000	88.8
500	17,000	69.0	0.0045	0.068	0.308	0.897	7.25	667	22,667	92.0
500	16,000	71.6	0.0050	0.076	0.321	0.893	6.98	667	21,333	95.4
500	15,000	74.4	0.0056	0.084	0.335	0.888	6.72	667	20,000	99.1
500	14,000	77.4	0.0063	0.094	0.350	0.883	6.46	667	18,667	103.1
500	13,000	80.6	0.0071	0.107	0.367	0.878	6.20	667	17,333	107.5
500	12,000	84.1	0.0080	0.121	0.386	0.871	5.94	667	16,000	112.2
500	11,000	87.9	0.0093	0.140	0.407	0.864	5.69	667	14,667	117.3
500	10,000	92.1	0.0108	0.162	0.430	0.857	5.43	667	13,333	122.8

A/A TABLE D-1b   Flexural Design Coefficients for _CONCRETE_ Masonry Constructed with Type _M_ or _S_ Mortar: f'_m = 1500 psi, f_y = 60,000 psi and _n_ = 13.9

For use of Table, see Example 5-B

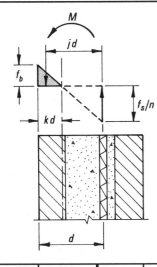

DESIGN DATA

$f'_m = 1500$ psi $f_y = 60,000$ psi

$F_b = 0.33 f'_m = 500$ psi $F_s = 24,000$ psi

$E_m = 2,080,000$ psi $E_s = 29,000,000$ psi

DESIGN EQUATIONS

$$n = \frac{E_s}{E_m} = 13.9 \qquad K = \frac{M}{F} = \frac{M(\text{ft kips})}{bd^2/12,000} \text{ or } \frac{M(\text{in. lbs})}{bd^2}$$

$$K = \tfrac{1}{2} jk f_b$$

$$k = \frac{1}{1 + f_s/n f_b} \qquad j = 1 - \frac{k}{3}$$

$$A_s = \frac{M}{f_s j d} \qquad p = \frac{A_s}{bd} = \frac{K}{f_s j}$$

Increase for wind or earthquake
$1.33 F_b = 667$ psi
$1.33 F_s = 32,000$ psi

f_b	f_s	K	p	np	k	j	2/jk	f_b	f_s	K
125	24,000	4.1	0.0002	0.002	0.068	0.977	30.31	167	32,000	5.5
150	24,000	5.8	0.0002	0.003	0.080	0.973	25.71	200	32,000	7.8
175	24,000	7.8	0.0003	0.005	0.092	0.969	22.42	233	32,000	10.4
200	24,000	10.0	0.0004	0.006	0.104	0.965	19.96	267	32,000	13.4
225	24,000	12.5	0.0005	0.008	0.115	0.962	18.04	300	32,000	16.6
250	24,000	15.1	0.0007	0.009	0.126	0.958	16.51	333	32,000	20.2
275	24,000	18.0	0.0008	0.011	0.137	0.954	15.26	367	32,000	24.0
300	24,000	21.1	0.0009	0.013	0.148	0.951	14.21	400	32,000	28.1
325	24,000	24.4	0.0011	0.015	0.158	0.947	13.33	433	32,000	32.5
350	24,000	27.8	0.0012	0.017	0.169	0.944	12.57	467	32,000	37.1
375	24,000	31.5	0.0014	0.019	0.178	0.941	11.92	500	32,000	42.0
400	24,000	35.3	0.0016	0.022	0.188	0.937	11.34	533	32,000	47.0
425	24,000	39.2	0.0017	0.024	0.198	0.934	10.84	567	32,000	52.3
450	24,000	43.3	0.0019	0.027	0.207	0.931	10.39	600	32,000	57.7
475	24,000	47.6	0.0021	0.030	0.216	0.928	9.99	633	32,000	63.4
500	24,000	51.9	0.0023	0.033	0.225	0.925	9.63	667	32,000	69.2
500	23,000	53.5	0.0025	0.035	0.232	0.923	9.34	667	30,667	71.4
500	22,000	55.2	0.0027	0.038	0.240	0.920	9.06	667	29,333	73.6
500	21,000	57.0	0.0030	0.041	0.249	0.917	8.77	667	28,000	76.0
500	20,000	58.9	0.0032	0.045	0.258	0.914	8.48	667	26,667	78.6
500	19,000	61.0	0.0035	0.049	0.268	0.911	8.20	667	25,333	81.3
500	18,000	63.2	0.0039	0.054	0.279	0.907	7.91	667	24,000	84.2
500	17,000	65.5	0.0043	0.059	0.290	0.903	7.63	667	22,667	87.4
500	16,000	68.1	0.0047	0.066	0.303	0.899	7.35	667	21,333	90.8
500	15,000	70.8	0.0053	0.073	0.317	0.894	7.06	667	20,000	94.4
500	14,000	73.8	0.0059	0.082	0.332	0.889	6.78	667	18,667	98.4
500	13,000	77.0	0.0067	0.093	0.348	0.884	6.50	667	17,333	102.6
500	12,000	80.5	0.0076	0.106	0.367	0.878	6.21	667	16,000	107.3
500	11,000	84.3	0.0088	0.122	0.387	0.871	5.93	667	14,667	112.4
500	10,000	88.5	0.0103	0.142	0.410	0.863	5.65	667	13,333	118.0

ACI/ASCE

A/A TABLE D-2a Flexural Design Coefficients for *CONCRETE* Masonry
Constructed with Type *N* Mortar: f'_m = 2000 psi, f_y = 60,000 psi and n = 12.5

For use of Table, see Example 5-B

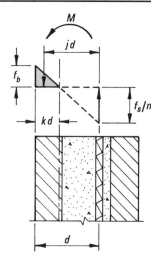

DESIGN DATA

$f'_m = 2000$ psi	$f_y = 60,000$ psi
$F_b = 0.33 f'_m = 667$ psi	$F_s = 24,000$ psi
$E_m = 2,320,000$ psi	$E_s = 29,000,000$ psi

DESIGN EQUATIONS

$$n = \frac{E_s}{E_m} = 12.5 \qquad K = \frac{M}{F} = \frac{M(\text{ft kips})}{bd^2/12,000} \text{ or } \frac{M(\text{in. lbs})}{bd^2}$$

$$K = \tfrac{1}{2} jk f_b$$

$$k = \frac{1}{1 + f_s/n f_b} \qquad j = 1 - \frac{k}{3}$$

$$A_s = \frac{M}{f_s j d} \qquad p = \frac{A_s}{bd} = \frac{K}{f_s j}$$

Increase for wind or earthquake
$1.33 F_b = 889$ psi
$1.33 F_s = 32,000$ psi

f_b	f_s	K	p	np	k	j	$2/jk$	f_b	f_s	K
150	24,000	5.3	0.0002	0.003	0.072	0.976	28.28	200	32,000	7.1
200	24,000	9.1	0.0004	0.005	0.094	0.969	21.89	267	32,000	12.2
250	24,000	13.8	0.0006	0.008	0.115	0.962	18.05	333	32,000	18.5
300	24,000	19.4	0.0008	0.011	0.135	0.955	15.50	400	32,000	25.8
350	24,000	25.6	0.0011	0.014	0.154	0.949	13.67	467	32,000	34.1
400	24,000	32.5	0.0014	0.018	0.172	0.943	12.31	533	32,000	43.3
450	24,000	40.0	0.0018	0.022	0.190	0.937	11.25	600	32,000	53.4
500	24,000	48.1	0.0022	0.027	0.207	0.931	10.40	667	32,000	64.1
550	24,000	56.7	0.0026	0.032	0.223	0.926	9.70	733	32,000	75.6
600	24,000	65.8	0.0030	0.037	0.238	0.921	9.12	800	32,000	87.7
650	24,000	75.3	0.0034	0.043	0.253	0.916	8.64	867	32,000	100.4
667	24,000	78.6	0.0036	0.045	0.258	0.914	8.49	889	32,000	104.8
667	23,000	80.9	0.0039	0.048	0.266	0.911	8.25	889	30,667	107.8
667	22,000	83.3	0.0042	0.052	0.275	0.908	8.01	889	29,333	111.0
667	21,000	85.8	0.0045	0.056	0.284	0.905	7.77	889	28,000	114.4
667	20,000	88.5	0.0049	0.061	0.294	0.902	7.54	889	26,667	118.0
667	19,000	91.4	0.0054	0.067	0.305	0.898	7.30	889	25,333	121.8
667	18,000	94.4	0.0059	0.073	0.317	0.894	7.06	889	24,000	125.9
667	17,000	97.7	0.0065	0.081	0.329	0.890	6.83	889	22,667	130.3
667	16,000	101.2	0.0071	0.089	0.343	0.886	6.59	889	21,333	134.9
667	15,000	105.0	0.0079	0.099	0.357	0.881	6.35	889	20,000	139.9
667	14,000	109.0	0.0089	0.111	0.373	0.876	6.12	889	18,667	145.3
667	13,000	113.3	0.0100	0.125	0.391	0.870	5.88	889	17,333	151.1
667	12,000	118.0	0.0114	0.142	0.410	0.863	5.65	889	16,000	157.4
667	11,000	123.1	0.0131	0.163	0.431	0.856	5.42	889	14,667	164.2
667	10,000	128.7	0.0152	0.190	0.455	0.848	5.18	889	13,333	171.5

A/A TABLE D-2b Flexural Design Coefficients for *CONCRETE* Masonry
Constructed with Type *M* or *S* Mortar: f'_m = 2000 psi, f_y = 60,000 psi and n = 11.8
For use of Table, see Example 5-B

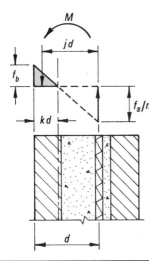

DESIGN DATA

$f'_m = 2000$ psi $f_y = 60,000$ psi

$F_b = 0.33 f'_m = 667$ psi $F_s = 24,000$ psi

$E_m = 2,460,000$ psi $E_s = 29,000,000$ psi

DESIGN EQUATIONS

$$n = \frac{E_s}{E_m} = 11.8 \qquad K = \frac{M}{F} = \frac{M(\text{ft kips})}{bd^2/12,000} \text{ or } \frac{M(\text{in. lbs})}{bd^2}$$

$$K = \tfrac{1}{2} jk f_b$$

$$k = \frac{1}{1 + f_s/n f_b} \qquad j = 1 - \frac{k}{3}$$

$$A_s = \frac{M}{f_s j d} \qquad p = \frac{A_s}{bd} = \frac{K}{f_s j}$$

Increase for wind or earthquake
$1.33 F_b = 889$ psi
$1.33 F_s = 32,000$ psi

f_b	f_s	K	p	np	k	j	$2/jk$	f_b	f_s	K
150	24,000	5.0	0.0002	0.003	0.069	0.977	29.80	200	32,000	6.7
200	24,000	8.7	0.0004	0.004	0.090	0.970	23.03	267	32,000	11.6
250	24,000	13.2	0.0006	0.007	0.109	0.964	18.96	333	32,000	17.6
300	24,000	18.5	0.0008	0.009	0.129	0.957	16.26	400	32,000	24.6
350	24,000	24.4	0.0011	0.013	0.147	0.951	14.32	467	32,000	32.6
400	24,000	31.1	0.0014	0.016	0.164	0.945	12.87	533	32,000	41.4
450	24,000	38.3	0.0017	0.020	0.181	0.940	11.75	600	32,000	51.1
500	24,000	46.1	0.0021	0.024	0.197	0.934	10.85	667	32,000	61.4
550	24,000	54.4	0.0024	0.029	0.213	0.929	10.11	733	32,000	72.5
600	24,000	63.2	0.0028	0.034	0.228	0.924	9.50	800	32,000	84.2
650	24,000	72.4	0.0033	0.039	0.242	0.919	8.98	867	32,000	96.5
667	24,000	75.6	0.0034	0.040	0.247	0.918	8.83	889	32,000	100.8
667	23,000	77.8	0.0037	0.044	0.255	0.915	8.57	889	30,667	103.7
667	22,000	80.2	0.0040	0.047	0.263	0.912	8.32	889	29,333	106.9
667	21,000	82.7	0.0043	0.051	0.273	0.909	8.07	889	28,000	110.2
667	20,000	85.3	0.0047	0.056	0.282	0.906	7.82	889	26,667	113.8
667	19,000	88.1	0.0051	0.061	0.293	0.902	7.57	889	25,333	117.5
667	18,000	91.2	0.0056	0.067	0.304	0.899	7.32	889	24,000	121.6
667	17,000	94.4	0.0062	0.073	0.316	0.895	7.07	889	22,667	125.9
667	16,000	97.9	0.0069	0.081	0.330	0.890	6.81	889	21,333	130.5
667	15,000	101.6	0.0077	0.090	0.344	0.885	6.56	889	20,000	135.5
667	14,000	105.6	0.0086	0.101	0.360	0.880	6.32	889	18,667	140.8
667	13,000	110.0	0.0097	0.114	0.377	0.874	6.07	889	17,333	146.6
667	12,000	114.7	0.0110	0.130	0.396	0.868	5.82	889	16,000	152.9
667	11,000	119.8	0.0126	0.149	0.417	0.861	5.57	889	14,667	159.7
667	10,000	125.3	0.0147	0.173	0.440	0.853	5.32	889	13,333	167.1

ACI/ASCE

A/A TABLE D-3a   Flexural Design Coefficients for *CONCRETE* Masonry Constructed with Type *N* Mortar: f'_m = 2500 psi, f_y = 60,000 psi and *n* = 11.1

For use of Table, see Example 5-B

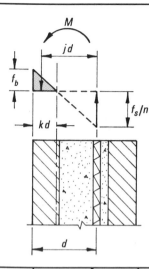

DESIGN DATA

$f'_m = 2500$ psi	$f_y = 60{,}000$ psi
$F_b = 0.33f'_m = 833$ psi	$F_s = 24{,}000$ psi
$E_m = 2{,}610{,}000$ psi	$E_s = 29{,}000{,}000$ psi

DESIGN EQUATIONS

$$n = \frac{E_s}{E_m} = 11.1 \qquad K = \frac{M}{F} = \frac{M(\text{ft kips})}{bd^2/12{,}000} \text{ or } \frac{M(\text{in. lbs})}{bd^2}$$

$$K = \tfrac{1}{2}jkf_b$$

$$k = \frac{1}{1 + f_s/nf_b} \qquad j = 1 - \frac{k}{3}$$

$$A_s = \frac{M}{f_s\,jd} \qquad p = \frac{A_s}{bd} = \frac{K}{f_s\,j}$$

Increase for wind or earthquake
$1.33 F_b = 1111$ psi
$1.33 F_s = 32{,}000$ psi

f_b	f_s	K	p	np	k	j	$2/jk$	f_b	f_s	K
200	24,000	8.2	0.0004	0.004	0.085	0.972	24.31	267	32,000	11.0
250	24,000	12.5	0.0005	0.006	0.104	0.965	19.99	333	32,000	16.7
300	24,000	17.5	0.0008	0.008	0.122	0.959	17.11	400	32,000	23.4
350	24,000	23.2	0.0010	0.011	0.139	0.954	15.05	467	32,000	31.0
400	24,000	29.6	0.0013	0.014	0.156	0.948	13.51	533	32,000	39.5
450	24,000	36.5	0.0016	0.018	0.172	0.943	12.32	600	32,000	48.7
500	24,000	44.0	0.0020	0.022	0.188	0.937	11.36	667	32,000	58.7
550	24,000	52.0	0.0023	0.026	0.203	0.932	10.58	733	32,000	69.3
600	24,000	60.4	0.0027	0.030	0.217	0.928	9.93	800	32,000	80.6
650	24,000	69.3	0.0031	0.035	0.231	0.923	9.38	867	32,000	92.4
700	24,000	78.6	0.0036	0.040	0.245	0.918	8.90	933	32,000	104.8
750	24,000	88.3	0.0040	0.045	0.258	0.914	8.50	1,000	32,000	117.7
800	24,000	98.3	0.0045	0.050	0.270	0.910	8.14	1,067	32,000	131.1
833	24,000	105.1	0.0048	0.054	0.278	0.907	7.93	1,111	32,000	140.1
833	23,000	108.0	0.0052	0.058	0.287	0.904	7.71	1,111	30,667	144.0
833	22,000	111.1	0.0056	0.062	0.296	0.901	7.50	1,111	29,333	148.1
833	21,000	114.3	0.0061	0.067	0.306	0.898	7.28	1,111	28,000	152.5
833	20,000	117.8	0.0066	0.073	0.316	0.895	7.07	1,111	26,667	157.1
833	19,000	121.5	0.0072	0.080	0.327	0.891	6.86	1,111	25,333	162.0
833	18,000	125.4	0.0079	0.087	0.339	0.887	6.65	1,111	24,000	167.1
833	17,000	129.5	0.0086	0.096	0.352	0.883	6.43	1,111	22,667	172.7
833	16,000	133.9	0.0095	0.106	0.366	0.878	6.22	1,111	21,333	178.6
833	15,000	138.6	0.0106	0.118	0.381	0.873	6.01	1,111	20,000	184.9
833	14,000	143.7	0.0118	0.131	0.398	0.867	5.80	1,111	18,667	191.6
833	13,000	149.1	0.0133	0.148	0.416	0.861	5.59	1,111	17,333	198.8
833	12,000	155.0	0.0151	0.168	0.435	0.855	5.38	1,111	16,000	206.6
833	11,000	161.3	0.0173	0.192	0.457	0.848	5.17	1,111	14,667	215.0
833	10,000	168.1	0.0200	0.222	0.480	0.840	4.96	1,111	13,333	224.1

A/A TABLE D-3b Flexural Design Coefficients for *CONCRETE* Masonry
Constructed with Type *M* or *S* Mortar: f'_m = 2500 psi, f_y = 60,000 psi and n = 10.4

For use of Table,
see Example 5-B

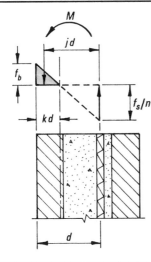

DESIGN DATA

$f'_m = 2500$ psi	$f_y = 60,000$ psi
$F_b = 0.33 f'_m = 833$ psi	$F_s = 24,000$ psi
$E_m = 2,800,000$ psi	$E_s = 29,000,000$ psi

DESIGN EQUATIONS

$$n = \frac{E_s}{E_m} = 10.4 \qquad K = \frac{M}{F} = \frac{M(\text{ft kips})}{bd^2/12,000} \text{ or } \frac{M(\text{in. lbs})}{bd^2}$$

$$K = \tfrac{1}{2} jk f_b$$

$$k = \frac{1}{1 + f_s/n f_b} \qquad j = 1 - \frac{k}{3}$$

$$A_s = \frac{M}{f_s j d} \qquad p = \frac{A_s}{bd} = \frac{K}{f_s j}$$

Increase for wind or earthquake
$1.33 F_b = 1111$ psi
$1.33 F_s = 32,000$ psi

f_b	f_s	K	p	np	k	j	2/jk	f_b	f_s	K
200	24,000	7.8	0.0003	0.003	0.080	0.973	25.76	267	32,000	10.4
250	24,000	11.8	0.0005	0.005	0.098	0.967	21.15	333	32,000	15.8
300	24,000	16.6	0.0007	0.007	0.115	0.962	18.08	400	32,000	22.1
350	24,000	22.0	0.0010	0.010	0.132	0.956	15.88	467	32,000	29.4
400	24,000	28.1	0.0012	0.013	0.148	0.951	14.24	533	32,000	37.5
450	24,000	34.7	0.0015	0.016	0.163	0.946	12.96	600	32,000	46.3
500	24,000	41.9	0.0019	0.019	0.178	0.941	11.94	667	32,000	55.8
550	24,000	49.5	0.0022	0.023	0.192	0.936	11.10	733	32,000	66.0
600	24,000	57.6	0.0026	0.027	0.206	0.931	10.41	800	32,000	76.9
650	24,000	66.2	0.0030	0.031	0.220	0.927	9.82	867	32,000	88.3
700	24,000	75.1	0.0034	0.035	0.233	0.922	9.32	933	32,000	100.2
750	24,000	84.5	0.0038	0.040	0.245	0.918	8.88	1,000	32,000	112.6
800	24,000	94.1	0.0043	0.045	0.257	0.914	8.50	1,067	32,000	125.5
833	24,000	100.7	0.0046	0.048	0.265	0.912	8.27	1,111	32,000	134.3
833	23,000	103.6	0.0050	0.052	0.274	0.909	8.04	1,111	30,667	138.1
833	22,000	106.6	0.0053	0.056	0.283	0.906	7.81	1,111	29,333	142.1
833	21,000	109.8	0.0058	0.060	0.292	0.903	7.59	1,111	28,000	146.4
833	20,000	113.2	0.0063	0.065	0.302	0.899	7.36	1,111	26,667	150.9
833	19,000	116.8	0.0069	0.071	0.313	0.896	7.13	1,111	25,333	155.8
833	18,000	120.7	0.0075	0.078	0.325	0.892	6.90	1,111	24,000	160.9
833	17,000	124.8	0.0083	0.086	0.338	0.887	6.68	1,111	22,667	166.4
833	16,000	129.2	0.0091	0.095	0.351	0.883	6.45	1,111	21,333	172.2
833	15,000	133.9	0.0102	0.106	0.366	0.878	6.22	1,111	20,000	178.5
833	14,000	138.9	0.0114	0.118	0.382	0.873	6.00	1,111	18,667	185.2
833	13,000	144.4	0.0128	0.133	0.400	0.867	5.77	1,111	17,333	192.5
833	12,000	150.2	0.0146	0.151	0.419	0.860	5.55	1,111	16,000	200.3
833	11,000	156.6	0.0167	0.173	0.441	0.853	5.32	1,111	14,667	208.7
833	10,000	163.4	0.0193	0.201	0.464	0.845	5.10	1,111	13,333	217.9

ACI/ASCE

A/A TABLE D-4a Flexural Design Coefficients for *CONCRETE* Masonry Constructed with Type *N* Mortar: f'_m = 3000 psi, f_y = 60,000 psi and *n* = 10.4

For use of Table, see Example 5-B

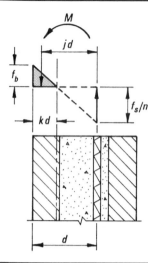

DESIGN DATA

$f'_m = 3000$ psi	$f_y = 60,000$ psi
$F_b = 0.33 f'_m = 1000$ psi	$F_s = 24,000$ psi
$E_m = 2,800,000$ psi	$E_s = 29,000,000$ psi

DESIGN EQUATIONS

$$n = \frac{E_s}{E_m} = 10.4 \qquad K = \frac{M}{F} = \frac{M(\text{ft kips})}{bd^2/12,000} \text{ or } \frac{M(\text{in. lbs})}{bd^2}$$

$$K = \tfrac{1}{2} jk f_b$$

$$k = \frac{1}{1 + f_s/n f_b} \qquad j = 1 - \frac{k}{3}$$

$$A_s = \frac{M}{f_s j d} \qquad p = \frac{A_s}{bd} = \frac{K}{f_s j}$$

Increase for wind or earthquake
$1.33 F_b = 1333$ psi
$1.33 F_s = 32,000$ psi

f_b	f_s	K	p	np	k	j	2/jk	f_b	f_s	K
250	24,000	11.8	0.0005	0.005	0.098	0.967	21.15	333	32,000	15.8
300	24,000	16.6	0.0007	0.007	0.115	0.962	18.08	400	32,000	22.1
350	24,000	22.0	0.0010	0.010	0.132	0.956	15.88	467	32,000	29.4
400	24,000	28.1	0.0012	0.013	0.148	0.951	14.24	533	32,000	37.5
450	24,000	34.7	0.0015	0.016	0.163	0.946	12.96	600	32,000	46.3
500	24,000	41.9	0.0019	0.019	0.178	0.941	11.94	667	32,000	55.8
550	24,000	49.5	0.0022	0.023	0.192	0.936	11.10	733	32,000	66.0
600	24,000	57.6	0.0026	0.027	0.206	0.931	10.41	800	32,000	76.9
650	24,000	66.2	0.0030	0.031	0.220	0.927	9.82	867	32,000	88.3
700	24,000	75.1	0.0034	0.035	0.233	0.922	9.32	933	32,000	100.2
750	24,000	84.5	0.0038	0.040	0.245	0.918	8.88	1000	32,000	112.6
800	24,000	94.1	0.0043	0.045	0.257	0.914	8.50	1067	32,000	125.5
850	24,000	104.1	0.0048	0.050	0.269	0.910	8.16	1133	32,000	138.9
900	24,000	114.5	0.0053	0.055	0.281	0.906	7.86	1200	32,000	152.6
950	24,000	125.1	0.0058	0.060	0.292	0.903	7.60	1267	32,000	166.7
1000	24,000	135.9	0.0063	0.066	0.302	0.899	7.36	1333	32,000	181.2
1000	23,000	139.5	0.0068	0.070	0.311	0.896	7.17	1333	30,667	186.0
1000	22,000	143.3	0.0073	0.076	0.321	0.893	6.98	1333	29,333	191.1
1000	21,000	147.3	0.0079	0.082	0.331	0.890	6.79	1333	28,000	196.4
1000	20,000	151.5	0.0086	0.089	0.342	0.886	6.60	1333	26,667	202.1
1000	19,000	156.0	0.0093	0.097	0.354	0.882	6.41	1333	25,333	208.0
1000	18,000	160.7	0.0102	0.106	0.366	0.878	6.22	1333	24,000	214.3
1000	17,000	165.8	0.0112	0.116	0.380	0.873	6.03	1333	22,667	221.0
1000	16,000	171.1	0.0123	0.128	0.394	0.869	5.84	1333	21,333	228.1
1000	15,000	176.8	0.0136	0.142	0.409	0.864	5.66	1333	20,000	235.7
1000	14,000	182.8	0.0152	0.158	0.426	0.858	5.47	1333	18,667	243.8
1000	13,000	189.3	0.0171	0.178	0.444	0.852	5.28	1333	17,333	252.4
1000	12,000	196.2	0.0193	0.201	0.464	0.845	5.10	1333	16,000	261.6
1000	11,000	203.6	0.0221	0.230	0.486	0.838	4.91	1333	14,667	271.5
1000	10,000	211.6	0.0255	0.265	0.510	0.830	4.73	1333	13,333	282.1

A/A TABLE D-4b Flexural Design Coefficients for *CONCRETE* Masonry Constructed with Type *M* or *S* Mortar: f'_m = 3000 psi, f_y = 60,000 psi and n = 9.2

For use of Table, see Example 5-B

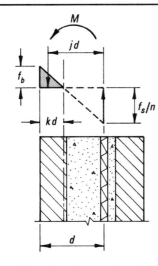

DESIGN DATA

$f'_m = 3000$ psi $f_y = 60,000$ psi

$F_b = 0.33 f'_m = 1000$ psi $F_s = 24,000$ psi

$E_m = 3,140,000$ psi $E_s = 29,000,000$ psi

DESIGN EQUATIONS

$$n = \frac{E_s}{E_m} = 9.2 \qquad K = \frac{M}{F} = \frac{M(\text{ft kips})}{bd^2/12,000} \text{ or } \frac{M(\text{in. lbs})}{bd^2}$$

$$K = \tfrac{1}{2} jk f_b$$

$$k = \frac{1}{1 + f_s/n f_b} \qquad j = 1 - \frac{k}{3}$$

$$A_s = \frac{M}{f_s jd} \qquad p = \frac{A_s}{bd} = \frac{K}{f_s j}$$

Increase for wind or earthquake
$1.33 F_b = 1333$ psi
$1.33 F_s = 32,000$ psi

f_b	f_s	K	p	np	k	j	$2/jk$	f_b	f_s	K
250	24,000	10.6	0.0005	0.004	0.087	0.971	23.56	333	32,000	14.2
300	24,000	14.9	0.0006	0.006	0.103	0.966	20.08	400	32,000	19.9
350	24,000	19.9	0.0009	0.008	0.118	0.961	17.60	467	32,000	26.5
400	24,000	25.4	0.0011	0.010	0.133	0.956	15.74	533	32,000	33.9
450	24,000	31.5	0.0014	0.013	0.147	0.951	14.30	600	32,000	42.0
500	24,000	38.1	0.0017	0.015	0.161	0.946	13.14	667	32,000	50.7
550	24,000	45.1	0.0020	0.018	0.174	0.942	12.19	733	32,000	60.1
600	24,000	52.6	0.0023	0.022	0.187	0.938	11.41	800	32,000	70.1
650	24,000	60.5	0.0027	0.025	0.199	0.934	10.74	867	32,000	80.7
700	24,000	68.8	0.0031	0.028	0.212	0.929	10.17	933	32,000	91.8
750	24,000	77.5	0.0035	0.032	0.223	0.926	9.68	1000	32,000	103.3
800	24,000	86.5	0.0039	0.036	0.235	0.922	9.24	1067	32,000	115.4
850	24,000	95.9	0.0044	0.040	0.246	0.918	8.86	1133	32,000	127.9
900	24,000	105.6	0.0048	0.044	0.257	0.914	8.53	1200	32,000	140.7
950	24,000	115.5	0.0053	0.049	0.267	0.911	8.22	1267	32,000	154.0
1000	24,000	125.8	0.0058	0.053	0.277	0.908	7.95	1333	32,000	167.7
1000	23,000	129.3	0.0062	0.057	0.286	0.905	7.74	1333	30,667	172.3
1000	22,000	132.9	0.0067	0.062	0.295	0.902	7.52	1333	29,333	177.3
1000	21,000	136.9	0.0073	0.067	0.305	0.898	7.31	1333	28,000	182.5
1000	20,000	141.0	0.0079	0.072	0.315	0.895	7.09	1333	26,667	188.0
1000	19,000	145.4	0.0086	0.079	0.326	0.891	6.88	1333	25,333	193.8
1000	18,000	150.1	0.0094	0.086	0.338	0.887	6.66	1333	24,000	200.1
1000	17,000	155.0	0.0103	0.095	0.351	0.883	6.45	1333	22,667	206.7
1000	16,000	160.3	0.0114	0.105	0.365	0.878	6.24	1333	21,333	213.8
1000	15,000	166.0	0.0127	0.117	0.380	0.873	6.02	1333	20,000	221.3
1000	14,000	172.1	0.0142	0.130	0.397	0.868	5.81	1333	18,667	229.4
1000	13,000	178.6	0.0159	0.147	0.414	0.862	5.60	1333	17,333	238.1
1000	12,000	185.6	0.0181	0.166	0.434	0.855	5.39	1333	16,000	247.5
1000	11,000	193.2	0.0207	0.190	0.455	0.848	5.18	1333	14,667	257.5
1000	10,000	201.3	0.0240	0.220	0.479	0.840	4.97	1333	13,333	268.4

ACI/ASCE

A/A TABLE E-1   Radius of Gyration[1], *r*, for Hollow Concrete Masonry; Unit Length = 16 inches[2,3]

Grout Spacing (inches)	Nominal Thickness (inches)				
	4	6	8	10	12
Solid Grouted	1.1	1.6	2.2	2.8	3.4
16	1.2	1.8	2.4	3.0	3.7
24	1.2	1.9	2.5	3.2	3.8
32	1.2	1.9	2.6	3.2	3.9
40	1.3	1.9	2.6	3.3	4.0
48	1.3	2.0	2.7	3.3	4.0
56	1.3	2.0	2.7	3.4	4.1
64	1.3	2.0	2.7	3.4	4.1
72	1.3	2.0	2.7	3.4	4.1
No Grout	1.4	2.1	2.8	3.6	4.3

A/A TABLE E-2   Radius of Gyration[1], *r*, for Hollow Clay Masonry; Unit Length = 16 inches[2,3]

Grout Spacing (inches)	Nominal Thickness (inches)				
	4	6	8	10	12
Solid Grouted	1.1	1.6	2.2	2.8	3.4
16	1.2	1.8	2.4	3.0	3.7
24	1.2	1.8	2.5	3.1	3.8
32	1.2	1.9	2.6	3.2	3.8
40	1.3	1.9	2.6	3.2	3.9
48	1.3	1.9	2.6	3.3	3.9
56	1.3	1.9	2.6	3.3	4.0
64	1.3	1.9	2.6	3.3	4.0
72	1.3	2.0	2.7	3.3	4.0
No Grout	1.4	2.0	2.7	3.4	4.1

A/A TABLE E-3   Radius of Gyration[1], *r*, for Hollow Clay Masonry; Unit Length = 12 inches[2,3]

Grout Spacing (inches)	Nominal Thickness (inches)				
	4	6	8	10	12
Solid Grouted	1.1	1.7	2.3	2.9	3.5
12	1.2	1.8	2.5	3.1	3.7
18	1.2	1.9	2.5	3.2	3.8
24	1.2	1.9	2.6	3.2	3.9
30	1.3	1.9	2.6	3.3	3.9
36	1.3	1.9	2.6	3.3	4.0
42	1.3	1.9	2.6	3.3	4.0
48	1.3	2.0	2.7	3.3	4.0
54	1.3	2.0	2.7	3.3	4.0
60	1.3	2.0	2.7	3.3	4.0
66	1.3	2.0	2.7	3.3	4.0
72	1.3	2.0	2.7	3.4	4.0
No Grout	1.3	2.0	2.7	3.4	4.1

1. For single wythe masonry or for an individual wythe of a cavity wall.

$$r = \sqrt{\frac{I}{A}}$$

2. Based on minimum thickness of face shells and webs. In accordance with UBC Standard No. 24-4. Face shells mortar bedding except webs adjacent to grouted cell. Two cell units.

3. Courtesy of the Concrete Masonry Association of California and Nevada.

A/A TABLE E-4   Maximum Allowable Axial Stress on Walls and Columns, F_a

$$F_a = (\tfrac{1}{4})f'_m\left[1-\left(\frac{h}{140r}\right)^2\right] \quad \text{for } \frac{h}{r} \leq 99 \quad \text{(ACI/ASCE Eq. 7-1)}$$

$$F_a = (\tfrac{1}{4})f'_m\left(\frac{70r}{h}\right)^2 \quad \text{for } \frac{h}{r} > 99 \quad \text{(ACI/ASCE Eq. 7-2)}$$

$\dfrac{h}{r}$	f'_m (psi)								$\dfrac{h}{r}$
	1500	2000	2500	3000	3500	4000	4500	5000	
20	367	490	612	735	857	980	1102	1224	20
24	364	485	607	728	849	971	1092	1213	24
28	360	480	600	720	840	960	1080	1200	28
32	355	474	592	711	829	948	1066	1185	32
36	350	467	584	700	817	934	1051	1167	36
40	344	459	574	689	804	918	1033	1148	40
44	338	451	563	676	789	901	1014	1127	44
48	331	441	552	662	772	882	993	1103	48
52	323	431	539	647	754	862	970	1078	52
56	315	420	525	630	735	840	945	1050	56
60	306	408	510	612	714	816	918	1020	60
64	297	396	494	593	692	791	890	989	64
68	287	382	478	573	669	764	860	955	68
72	276	368	460	552	644	736	827	919	72
76	264	353	441	529	617	705	793	882	76
80	253	337	421	505	589	673	758	842	80
84	240	320	400	480	560	640	720	800	84
88	227	302	378	454	529	605	681	756	88
92	213	284	355	426	497	568	639	710	92
96	199	265	331	397	464	530	596	662	96
99	187	250	312	375	437	500	562	625	99
100	184	245	306	368	429	490	551	613	100
104	170	227	283	340	396	453	510	566	104
108	158	210	263	315	368	420	473	525	108
112	146	195	244	293	342	391	439	488	112
116	137	182	228	273	319	364	410	455	116
120	128	170	213	255	298	340	383	425	120
124	120	159	199	239	279	319	359	398	124
128	112	150	187	224	262	299	336	374	128
132	105	141	176	211	246	281	316	352	132
136	99	132	166	199	232	265	298	331	136
140	94	125	156	188	219	250	281	313	140
144	89	118	148	177	207	236	266	295	144
148	84	112	140	168	196	224	252	280	148
152	80	106	133	159	186	212	239	265	152
156	76	101	126	151	176	201	227	252	156
160	72	96	120	144	167	191	215	239	160
164	68	91	114	137	159	182	205	228	164
168	65	87	109	130	152	174	195	217	168
172	62	83	104	124	145	166	186	207	172
176	59	79	99	119	138	158	178	198	176
180	57	76	95	113	132	151	170	189	180
184	54	72	90	109	127	145	163	181	184
188	52	69	87	104	121	139	156	173	188
192	50	66	83	100	116	133	150	166	192
196	48	64	80	96	112	128	143	159	196
200	46	61	77	92	107	123	138	153	200

ACI/ASCE

References

General

American Concrete Institute & American Society of Civil Engineers. (1988). *Building Code Requirements for Concrete Masonry Structures (ACI 530-88/ASCE 5-88) and Specifications for Masonry Structures (ACI 530.1-88/ASCE 6-88)*. Detroit: ACI/ASCE.

American Society for Testing and Materials. (1991). *1991 Annual Book of ASTM Standards*. Vol. 04.05. Philadelphia: ASTM.

Amrhein, et al. (1979). *Masonry Design Manual*, 3rd Edition. Los Angeles: the Masonry Advancement Committee.

Brick Institute of America. (Feb. 1990). "Building Code Requirements for Masonry Structures ACI 530/ASCE 5 and Specifications for Masonry Structures ACI 530/ASCE 6." *Technical Notes on Brick Construction*. No. 3. Reston: BIA.

International Conference of Building Officials. (1991). *1991 Uniform Building Code*. Whittier: ICBO.

International Conference of Building Officials. (1991). *Uniform Building Code Standards, 1991*. Whittier: ICBO.

Section 1   Materials

Amrhein, J.E. (June 1977). "Grout...The Third Ingredient." *Masonry Industry Magazine*, pp. 9-14.

Amrhein, J.E. (1991). *Reinforced Grouted Brick Masonry*, 13th Edition. Los Angeles: Masonry Institute of America.

Brick Institute of America. (1989). *Principles of Brick Masonry*. Reston: BIA.

Brick Institute of America. (March 1986). "Manufacturing, Classification and Selection of Brick, Manufacturing - Part 1." *Technical Notes on Brick Construction*. No. 9. Revised.

Brick Institute of America. (Jan. 1989). "Manufacturing, Classification and Selection of Brick, Selection - Part 3." *Technical Notes on Brick Construction*. No. 9B.

Brick Institute of America. (June 1989). "Manufacturing, Classification and Selection of Brick, Classification - Part 2." *Technical Notes on Brick Construction*. No. 9A.

Building News, Inc. (1981). *Concrete Masonry Design Manual*, 4th Edition. Los Angeles: Building News, Inc.

National Concrete Masonry Association. (1980). "ACI 531 Building Code Requirements For Concrete Masonry." *NCMA TEK Notes*. No. 113. Herndon.

National Concrete Masonry Association. (1990). "Building Code Requirements For Masonry Structures." *NCMA TEK Notes*. No. 113A. Herndon. Revised.

National Concrete Masonry Association. (1979). "Laboratory and Field testing of Mortar and Grout." *NCMA TEK Notes*. No. 107. Herndon.

Panarese, W.; Kosmatka, S.H.; Randall, F. (1991). *Concrete Masonry Handbook*. Skokie: Portland Cement Association.

Senbu, O.; Abe, M.; Matsushima, Y.; Baba, A.; Sugiyama, M. (1991). "Effect of Admixtures on Compactibility and Properties of Grout." *9th International Brick/Block Masonry Conference*, Vol.1. Berlin: pp. 109.

Section 2   Masonry Assemblies

Bexten, Karen A.; Tadros,Maher K.; Horton,Richard T. (1989). "Compression Strength of Masonry." *5th Canadian Masonry Symposium*, Vol. 2. Vancouver, BC: University of British Columbia. pp. 629.

Brown, R. (1975). *Prediction of Brick Masonry Prism Strength from Reduced Constraint Brick Tests*, ASTM STP589.

California Concrete Masonry Technical Committee. (1975). *Recommended Testing Procedures for Concrete Masonry Units, Prisms. Grout and Mortar*, Los Angeles: CCMTC.

Fishburn, C.C. (1961). *Effect of Mortar Properties on Strength of Masonry*, Washington DC: National Bureau of Standards Monograph 36.

Hamid, A.A.; Drysdale, R.G.; Heidebrecht, A.C. (Aug. 1978). "Effect of Grouting on the Strength Characteristics of Concrete Block Masonry." *Proceedings of the North American Masonry Conference*. Boulder: The Masonry Society. paper #11.

Holm, T.A. (Aug. 1978). "Structural Properties of Block Concrete." *Proceedings of the North American Masonry Conference*. Boulder: The Masonry Society. paper #5.

Kingsley, G.R.; Atkinson, R.H.; Noland, J.R.; Hart, G.C. (1989). "The Effect of Height on Stress-Strain Measurements on Grouted Masonry Prisms." *5th Canadian Masonry Symposium*, Vol. 2. Vancouver, BC: University of British Columbia, pp. 587.

National Concrete Masonry Association. (Oct. 1969). "Compressive Strength of Concrete Masonry." *NCMA TEK Notes*. No. 15. Herndon.

National Concrete Masonry Association. (1975). "Concrete Masonry Prism Strength." *NCMA TEK Notes*. No. 70. Herndon.

National Concrete Masonry Association. (1979). "Testing Concrete Masonry Assemblages." *NCMA TEK Notes*. No. 108. Herndon.

Page, A.W.; Kleeman, P.W. (1991). "The Influence of Capping Material and Platen Restraint of the Failure of Hollow Masonry Units and Prisms." *9th International Brick/Block Masonry Conference*, Vol. 1. Berlin: pp. 662.

Redmond, T.; Allen, M. (1975). *Compressive Strength of Composite Brick and Concrete Masonry Walls*, ASTM STP589.

Sahlin, Sven. (1971). *Structural Masonry*. Englewood: Prentis-Hall, Inc.

U.S. Department of Commerce. (Sept. 1977). "Earthquake Resistant Masonry Construction." *National Workshop*. National Bureau of Standards Building Science Series 106.

Yao, Chichao; Nathan, N.D. (1989). "Axial Capacity of Grouted Concrete Masonry." *5th Canadian Masonry Symposium*, Vol. 1. Vancouver, BC: University of British Columbia, pp. 45.

Section 3   Loads

American National Standards Institute. (1982). *Minimum Design Loads for Buildings and Other Structures*. New York: ANSI.

American Society of Civil Engineers. (1970). *Lateral Stresses in the Ground and Design of Earth Retaining Structures*. New York: Speciality Proceedings Conference.

Applied Technology Council. (1979). *Seismic Design Guidelines for Highway Bridges*. ATC-306, Palo Alto: ATC.

Blume; Corning; Newmark. (1961). *Design of Multistory Reinforced Concrete Buildings for Earthquake Motions*. Skokie: Portland Cement Association.

Dimarogons, P.D. (Dec. 1983). "Distribution of Lateral Earthquake Pressure on a Retaining Wall." *Soils and Foundations* (Japanese Society of Soil Mechanics). Vol. 23, No. 4.

Los Angeles City. (1990). *Los Angeles City 1990 Building Code*, Los Angeles.

Mononobe, N. (1929). "Earthquake-Proof Construction of Masonry Dams," *Proceedings, World Engineering Conference*, Vol. 9.

Okbe, S. (1926). "General Theory of Earth Pressure," *Journal, Japanese Society of Civil Engineers*, Vol. 12.

Seed, Bolton H. & Whitman, Robert V. (1970). *Design of Earth Retaining Structures for Dynamic Loads*. New York: ASCE.

Structural Engineers Association of California. (1988). *Recommended Lateral Force Requirements and Commentary*. Sacramento: SEAOC.

Terry, Phillip. (March-April, 1991). "Reviewing the Seismic Provisions of the 1988 Uniform Building Code." *Building Standards*.

Terzaghi, & Peck. (1948). *Soil Mechanics in Engineering Practice*. New York: John Wiley & Sons, Inc.

Virdee, A. S. (1966). *Soil Pressures on Structures Due to Backfill Under Seismic Conditions*. Sacramento: Department of Water Resources, State of California.

Section 4   Distribution of Loads

Ambrose, James, & Vergun. (1987). *Design for Lateral Forces*. New York: John Wiley & Sons.

Blakeley, R.W.G. (June, 1979). "Recommendations for the Design and Construction of Base Isolated Structures." *Bulletin of the New Zealand National Socitey for Earthquake Engineering*. Vol. 12, No. 2.

Buckle, Ian. (June, 1988). *Basic Principles, Real World Case Studies, Evaluation of SEAOC Provisions*. Los Angeles: SEAOC Seminar Notes.

Mayes, R.L.; Weissberg, S.M.; Jones, L.R.; & Van Volkinburg. (Spring, 1991). *Seismic Isolation: Enhancing the Earthquake Resistance of Masonry*. Herndon: Council for Masonry Research Report. Vol. 4, No. 1.

Priestley, M.; Crosbie, R.; Carr, A. (June, 1977). "Seismic Forces in Base-Isolated Masonry Structures." *Bulletin of the New Zealand National Society for Earthquake Engineering*, Vol. 10, No. 2.

Structural Engineers Association of California Notes. (June, 1988). *Design and Construction of Base Isolated Buildings*. Los Angeles: SEAOC Seminar.

Section 5   Structural Design, WSD

American Plywood Association. (1987). *APA Design/Construction Guide, Diaphragms*. Tacoma: APA.

Blume, J.A. (1968). *Shear in Grouted Brick Masonry Wall Elements*. San Francisco: Western States Clay Products Association.

Fried, A.N. (1991). "The Position of the Neutral Axis in Masonry Joints." *9th International Brick/Block Masonry Conference,* Vol. 1. Berlin: pp. 188.

Hamid, A.A.; Ghanem,G.M. (1991). "Partially Reinforced Concrete Masonry." *9th International Brick/Block Masonry Conference,* Vol. 1. Berlin: pp. 368.

Holmes, I.L. (1969). *Masonry Building in High Intensity Seismic Zones; Designing Engineering and Constructing with Masonry Products.* Houston: Gulf Publishing Co.

Hosny, A.H.; Essawy, A.S.; Abou-Elenain, A.; Higazy, E.M. (1991). "Behavior of Reinforced Block Masonry Walls Under Out-of-Plane Bending." *9th International Brick/Block Masonry Conference,* Vol. 1. Berlin: pp. 387.

Leet, Kenneth. (1982). *Reinforced Concrete Design.* New York: McGraw-Hill Book Co.

Luttrell, Larry. (1987). *Diaphragm Design Manual.* Canton: Steel Deck Institute.

Mayes, R.L.; Clough, R.W. (1975). *A Literature Survey-Compressive, Tensile, Bond, and Shear Strength of Masonry, EERC* 75-21. Berkeley: University of California.

Mayes, R.L.; Clough, R.W. (1975). *State-of-the-Art in Seismic Shear Strength of Masonry; An Evaluation and Review, EERC* 75-21. Berkeley: University of California.

McGinley, W.M.; Borchlet, J.G. (1989) "Friction Between Brick and Its Support." *5th Canadian Masonry Symposium,* Vol 2. Vancouver, BC: University of British Columbia, pp. 713.

McGinley, W.M.; Borchelt, J.G. (1991). "Influence of Materials on the Friction Developed at the Base of Clay Brick Walls." *9th International Brick/Block Masonry Conference,* Vol. 1. Berlin: pp. 292.

Pfeffermann, I.O.; Van de Loock, I.G. (1991). "20 Years Experience with Bed Joint Reinforced Masonry in Belgium and Europe." *9th International Brick/Block Masonry Conference,* Vol. 1. Berlin: pp. 427.

Schneider, R.R., Dickey, W.L. (1987). *Reinforced Masonry Design,* 2nd Edition. Inglewood Cliffs: Prentiss Hall, Inc.

Soric, Z.; Tulin, L.G. (August 1987). "Bond & Splices in Reinforced Masonry." *U.S. - Japan Coordinated Program for Masonry Building Research,* Report No. 6.2-2. Boulder: University of Colorado.

Section 6   Strength Design

ACI/SEAOSC Task Force Committee on Slender Walls. (1982). *Test Report on Slender Walls.* Los Angeles: SEASC and the Southern California Chapter of the American Concrete Institute.

Adham, S. & Amrhein, J.E. (1991). "Dynamic and Testing of Tall Slender Reinforced Masonry Walls." *9th International Brick/Block Masonry Conference,* Vol. 1. Berlin: pp. 465.

Amrhein, J.E. & Lee. (1986). *Design of Reinforced Masonry Tall Slender Walls,* 2nd Edition. San Francisco: Western States Clay Products Association.

Amrhein, J.E. & Lee. (1988). *Slender Wall Design for Los Angeles, Estimating Curves.* Los Angeles: Masonry Institute of America.

Amrhein, J.E. & Lee. (1985) *Tall Slender Walls, Estimating Curves.* Los Angeles: Masonry Institute of America.

Atkinson, R.H.; Noland, J.L.; Hart, G.C. (1991). "Properties of Masonry Materials for Limit States Design." *9th International Brick/Block Masonry Conference,* Vol. 2. Berlin: pp. 678.

Curtin, W.G.; Shaw, G.; Beek, J.K. (1988). *Design of Reinforced and Prestressed Masonry.* London: Thomas Telford LTD.

Ferguson, P.M. (1973). *Reinforced Concrete Fundamentals,* 3rd Edition. New York: John Wiley and Sons.

Fling, R.S. (1987). *Practical Design of Reinforced Concrete.* New York: John Wiley and Sons.

Hart, G.C. (July-Dec. 1989). "Limit State Design Criteria for Minimum Flexural Steel in Concrete Masonry Beams," *The Masonry Society Journal,* Vol 8, No. 2. pp. 7-18.

Hart, G.C.; Noland, J.L. (1991). "Expected Value Limit State Design Criteria for Structural Masonry." *9th International Brick/Block Masonry Conference,* Vol. 2. Berlin: pp. 752.

Hart, G.C.; Englekirk, R.E.; Sabol, T.A. (July-Dec. 1986). "Limit State Design Criteria for One to Four Story Reinforced Concrete Masonry Buildings," *The Masonry Society Journal,* Vol. 5, No. 2. pp. 21-24.

Hart, G.C.; Noland, J.; Kingsley, G.; Englekirk, R.; Sajjad, N. (July-Dec.1988). "The Use of Confinement Steel to Increase Ductility in Reinforced Concrete Masonry Sheer Walls." *The Masonry Journal,* Vol. 7, No. 2. pp. 19-42.

Heeringa, R.L., McLean, D.L. (July-Dec. 1989). "Ultimate Strength Flexural Behavior of Concrete Masonry Walls," *The Masonry Society Journal*, Vol. 8, No. 2. pp. 19-30.

Hegemier, G.A. (1975). *Mechanics of Reinforced Concrete Masonry, A Literature Survey*, AMES-NSF TR 75-5. San Diego: University of California.

Hogan, Mark. (April,1991). "Limit States Design Provisions." *The Concrete Specifier*.

Leet, Kenneth. (1982). *Reinforced Concrete Design*. New York: McGraw-Hill Book Co.

Mayes, R.L.; Omote, Y.; Clough, R.W. (1976). *Cyclic Shear Testing of Masonry Piers*, Vol. 1, *Test Results*. EERC-76-8. Berkeley: University of California.

Paulay, T. (September 1972). "Some Aspects of Shear Wall Design." *Bulletin of New Zealand Society for Earthquake Engineering*, Vol. 5, No. 3.

Priestley, M.J.N. (July-Dec. 1986). "Flexural Strength of Rectangular Unconfined Masonry Shear Walls with Distributed Reinforcement." *The Masonry Society Journal*, Vol. 5, No. 2. pp. 1-15.

Priestly, M.J.N. (1987). *Strength Design of Masonry*. Los Angeles: Fourth North American Masonry Conference.

Selna, L.G. & Asher, J.W. (1986). "Multistory Slender Masonry Walls; Analysis, Design and Construction." Redondo Beach: Higgins Brick Co.

Shing, P.B.; Schuller, M.; Hoskere, V.S.; Carter, E. (Nov.-Dec. 1990). "Flexural and Shear Response of Reinforced Masonry Walls." *ACI Journal*: Paper No. 87-S66.

Structural Engineers Association of Southern California Seismology Committee of the SEAOC Strength Design. (1991). *Masonry Moment Resisting Wall Frames*. San Francisco: SEAOC.

Wang, C.K. & Salmon, C.J. (1985). *Reinforced Concrete Design*. New York: Harper & Rowe.

Section 7   Reinforcing Steel

Amrhein, J.E. (1991). *Reinforcing Steel in Masonry*. Los Angeles: Masonry Institute of America.

Beall, Christine. (1987). *Masonry Design and Detailing*. New York: McGraw-Hill Book Co.

Concrete Reinforcing Steel Institute. (1991). *CRSI Handbook*. Schaumburg: Concrete Reinforcing Steel Institute.

Newman, Morton.(1976). *Standard Cantilever Retaining Walls*. New York: McGraw-Hill Book Co.

Newman, Morton. (1968). *Standard Structural Details for Building Construction*. New York: McGraw-Hill Book Co.

Virdee, Ajit. (Oct. 1988). *Fundamentals of Reinforced Masonry Design*. Citrus Heights: Concrete Masonry Association of California and Nevada.

Section 8   Building Details

Beall, Christine. (1987). Masonry Design and Detailing. New York: McGraw-Hill Book Co.

Curtin, W.G.; Shaw, G.; Beck, J.K.; Parkinson, J.I. (1984). *Structural Masonry Detailing*. London: Granada Publishing.

Elmiger, A. (1976). *Architectural and Engineering Concrete Masonry Details for Building Construction*. McLean: National Concrete Masonry Association.

Newman, Morton. (1968). *Standard Structural Details For Building Construction*. New York: McGraw-Hill Book Co.

Section 9   Special Topics

Beall, Christine. (1987). *Masonry Design and Detailing*. New York: McGraw-Hill Book Co.

Brick Institute of America. (Feb. 1985). "Water Resistance of Brick Masonry, Design and Detailing — Part 1 of 2." *Technical Notes on Brick Construction*, No. 7. Revised.

Brick Institute of America. (March 1985). "Water Resistance of Brick Masonry Materials — Part 2 of 3." *Technical Notes on Brick Construction*, No. 7A. Revised.

Brick Institute of America. (Dec. 1985). "Painting Brick Masonry." *Technical Notes on Brick Construction*, No. 6. Revised.

Brick Institute of America. (June 1986). "Fire Resistance Requirements Relating to Brick Bearing Wall Buildings." *Technical Notes on Brick Construction*, No. 16A. Revised.

Brick Institute of America. (Jan. 1987). "Moisture Resistance of Brick Masonry Maintenance." *Technical Notes on Brick Construction*, No. 7F. Reissued.

Brick Institute of America. (Feb. 1987). "Colorless Coatings for Brick Masonry." *Technical Notes on Brick Construction*, No. 7E. Reissued.

Brick Institute of America. (May 1987). "Brick Masonry Cavity Walls." *Technical Notes on Brick Construction*, No. 21. Reissued.

Brick Institute of America. (May 1987). "Fire Resistance." *Technical Notes on Brick Construction*, No. 16. Revised, Reissued.

Brick Institute of America. (Sept. 1988). "Differential Movement - Expansion Joints, Part 2 of 3." *Technical Notes on Brick Construction*, No. 18A. Reissued.

Brick Institute of America. (Jan. 1991). "Movement — Volume Changes and Effect of Movement, Part 1." *Technical Notes on Brick Construction*, No. 18. Revised.

Brick Institute of America. (June 1991). "Calculated Fire Resistance." *Technical Notes on Brick Construction*, No. 16B.

Concrete Masonry Association of California and Nevada. (1986). *Waterproofing Concrete Masonry*, Citrus Heights: CMACN.

Concrete Masonry Association of California and Nevada. *Fire Resistive Construction Using Concrete Masonry*, Citrus Heights: CMACN.

Lauersdorf, Lyn R. (May 1988). "Stopping Rainwater Penetration." *The Magazine of Masonry Construction*, pp. 74-77.

Masonry Advancement Committee. *Guidelines for Clear Waterproofing Masonry Walls*, Los Angeles: MAC.

National Concrete Masonry Association. (1970). "Concrete Masonry Basements." *NCMA TEK Notes* No. 1(2A). Herndon.

National Concrete Masonry Association. (1972). "Concrete Masonry Foundation Walls." *NCMA TEK Notes*, No. 43. Herndon.

National Concrete Masonry Association. (1972). "Control of Wall Movement With Concrete Masonry." *NCMA TEK Notes*, No. 3(5B). Herndon.

National Concrete Masonry Association. (1972). "Maintenance of Concrete Masonry Walls." *NCMA TEK Notes*, No. 44(17A). Herndon.

National Concrete Masonry Association. (1973). "Design of Concrete Masonry for Crack Control." *NCMA TEK Notes*, No. 53. Herndon.

National Concrete Masonry Association. (1973). "Waterproof Coatings for Concrete Masonry." *NCMA TEK Notes*, No. 55. Herndon.

National Concrete Masonry Association. (1977). "Building Weathertight Concrete Masonry Walls." *NCMA TEK Notes*, No. 85 (33C). Herndon.

National Concrete Masonry Association. (1981). "Decorative Waterproofing of Concrete Masonry Walls." *NCMA TEK Notes*, No. 10A. Herndon.

National Concrete Masonry Association. (1981). "Waterproofing Concrete Masonry Basements and Earth-Sheltered Structures." *NCMA TEK Notes*, No. 121(33B). Herndon.

National Concrete Masonry Association. (1982). "Concrete Basement Walls." *NCMA TEK Notes*, No. 56A(2A). Herndon.

National Concrete Masonry Association. (1983). "Design for Dry Concrete Masonry Walls." *NCMA TEK Notes*, No. 13A(33A). Herndon.

National Concrete Masonry Association. (1985). "Increasing the Fire Resistance of Concrete Masonry." *NCMA TEK Notes*, No. 80A(10B). Herndon.

National Concrete Masonry Association. (1987). "Fire Safety With Concrete Masonry." *NCMA TEK Notes*, No. 35C. Herndon.

National Concrete Masonry Association. (1991). "Balanced Design Fire Protection for Multi-family Housing." *NCMA TEK Notes*, No. 17B. Herndon.

National Concrete Masonry Association. (1991). "Fire Resistance Rating of Concrete Masonry Assemblies." *NCMA TEK Notes*, No. 6A. Herndon.

Panarese, W.C.; Kosmatka, S.H.; Randall, Jr, F.A. (1991). *Concrete Masonry Handbook*, Skokie: Portland Cement Association.

Suprenant, Bruce. (March 1989). "Painting Concrete Masonry." *The Magazine of Masonry Construction*, pp. 100-103.

Suprenant, Bruce. (August 1989). "Repelling Water from the Inside." *The Magazine of Masonry Construction*, pp. 358-360.

Suprenant, Bruce. (April 1990). "Choosing a Water Repellent." *The Magazine of Masonry Construction*, pp. 5-11.

Section 13   Retaining Walls

Bowles, Joseph E. (1977). *Foundation Analysis & Design*. New York: McGraw-Hill Book Co.

Das, Braja M. (1984). *Principles of Foundation Engineering*. Boston: PWS Engineering.

Newman, Morton. (1976). *Standard Cantilever Retaining Walls*. New York: McGraw-Hill Book Co.

Supplemental References

Section 1   Materials

Matthys, J. (1990). "Concrete Masonry Flexural Bond Strength Prisms vs Wall Tests." *5th North American Masonry Conference*, Vol. 2. Urbana-Champaign: University of Illinois, pp. 677.

NCMA Engineered Concrete Masonry Design Committee. (August,1988). *Research Investigation of the Properties of Masonry Grout in Concrete Masonry*. National Concrete Masonry Association.

Qui-Gu, Hu. (1987). "Quality Requirements & Control of Masonry Materials." *4th North American Masonry Conference*. Los Angeles: University of California, pp. 12.

Section 2   Strength of Masonry

Assis, G.; Hamid, A.; Harris, H.G. (1990). "Compressive Behavior of Block Masonry Prisms Under Strain Gradient." *5th North American Masonry Conference*, Vol. 1. Urbana-Champaign: University of Illinois, pp. 615.

Atkinson, R.H. (September 15, 1991). *Development of a Database for Compressive Stress-Strain Behavior of Masonry*. Boulder: Atkinson-Noland & Associates, Inc. Final Report.

Atkinson, R.H. (November 1990). *Evaluation of Strength and Modulus Tables for Grouted and Ungrouted Hollow Unit Masonry*. Boulder: Atkinson-Noland & Associates, Inc.

Baba, A. & Senbu, O. (1986). "Influencing Factors on Prism Strength of Grouted Masonry and Fracture Mechanism Under Uniaxial Loading." *4th Canadian Masonry Symposium*, Vol. 2. Fredericton, NB: University of New Brunswick, pp. 1081.

Baba, A. & Senbu, O. (1986). "Mechanical Properties of Masonry Components." *4th Canadian Masonry Symposium, Vol. 2*. Fredericton, NB: University of New Brunswick, pp. 1066.

Becica, I.J. & Harris, H.G. (1982). "Ultimate Strength Behavior of Hollow Concrete Masonry Prisms Under Axial Load and Bending." *2nd North American Masonry Conference*. College Park: University of Maryland. paper #3.

Colville, J. & Wolde-Tinsae, A. (1990). "Compressive Strength of Hollow Concrete Masonry." *5th North American Masonry Conference*, Vol. 2. Urbana-Champaign: University of Illinois, pp. 663.

Drysdale, R. & Hamid, A.A. (1982). "Influence of the Unit Strength of Block Masonry." *2nd North American Masonry Conference*. College Park: University of Maryland. paper #2.

Ghosh, S. & Neis, V. (1990). "A Photoelastic Examination of Stress-Strain Behavior of Grouted Concrete Block Prisms." *5th North American Masonry Conference*, Vol. 2. Urbana-Champaign: University of Illinois, pp. 627.

Hamid, A.A.; Assis, G.F.; Harris, H.G. (1987). "Compression Behavior of Grouted Concrete Block Masonry — Some Preliminary Results." *4th North American Masonry Conference*, Vol. 2. Los Angeles: University of California. paper #43.

Hamid, A.A.; Ziab, G.; ElNawawy, O. (1987). "Modulus of Elasticity of Concrete Block Masonry." *4th North American Masonry Conference*, Vol. 1. Los Angeles: University of California. paper #7.

Hendry, A.W. (1987). "Testing Methods in Masonry Engineering." *4th North American Masonry Conference*, Vol. 2. Los Angeles: University of California. paper #49.

Khalaf, F.; Handry, A.; Fairbairn, D. (1990). "Concrete Block Masonry Prisms Compressed Normal & Parallel to Bed Face." *5th North American Masonry Conference*, Vol.1. Urbana-Champaign: University of Illinois, pp. 595.

Lenczner, D.; Foster, D. (1979). "Strength and Deformation of Brickwork Prisms in Three Directions." *5th International Brick Masonry Conference, Session 2*. Washington DC: B.I.A. paper #4.

Maurenbrecher, A.H.P. (1983). "Compressive Strength of Eccentrically Loaded Masonry Prisms." *3rd Canadian Masonry Symposium*. Edmonton: University of Alberta, pp. 10.

Maurenbrecher, A.H.P. (1986). "Compressive Strength of Hollow Concrete Blockwork." *4th Canadian Masonry Symposium*, Vol. 2. Fredericton, NB: University of New Brunswick, pp. 1000.

Maurenbrecker, A.H.P. (1980). "The Effect of Test Procedures on the Compressive Strength of Masonry Prisms." *2nd Canadian Masonry Symposium*. Ottawa: Carleton University, pp. 119.

McAskill, N. & Morgan, D.R. (1983). "Inspection and Testing of Reinforced Masonry." *3rd Canadian Masonry Symposium*. Edmonton: University of Alberta, pp. 26.

Miller, D.; Noland, J.; Feng, C. (1979). "Factors Influencing the Compressive Strength of Hollow Clay Unit Prisms." *5th International Brick Masonry Conference, Session 2.* Washington DC: B.I.A., paper #15.

Schubert, P. (1979). "Modulus of Elasticity of Masonry." *5th International Brick Masonry Conference, Session 2.* Washington DC: B.I.A., paper #17.

Yao, Chichao. (1986). "Joint Effect on Fully Bedded Plain Concrete Masonry." *5th Canadian Masonry Symposium, Vol. 1.* Vancouver, BC: University of British Columbia, pp. 55.

Section 4   Distribution of Loads

Merryman, K.M.; Leiva, G.; Antrobus, N.; Klingner, R.E. (May, 1990). "In-Plane Seismic Resistance of Two-Story Concrete Masonry Coupled Shear Walls." *U.S.-Japan Coordinated Program for Masonry Building Research,* Report No. 3.1(c)-1. Austin: The University of Texas.

Shing, P.B.; Noland, J.L.; Klamerus, E.W.; Schuller, M.P. (January, 1991). "Response of Single-Story Reinforced Masonry Shear Walls to In-Plane Lateral Loads." *U.S. - Japan Coordinated Program for Masonry Building Research.* Report No. 3.1(a)-2. Boulder: University of Colorado.

Section 5   Structural Design WSD

Borchelt, G. (1990). "Friction at Supports of Clay Brick Walls." *5th North American Masonry Conference,* Vol. 3. Urbana-Champaign: University of Illinois, pp. 1053.

Grimm, C. (1990). "Masonry Flexural Strength vs Course Height." *5th North American Masonry Conference,* Vol. 2. Urbana-Champaign: University of Illinois, pp. 673.

Limin, H. & Priestly, M.J.N. (May, 1988). "Seismic Behavior of Flanged Masonry Shear Walls." *Structural Systems Research Project.* Report No. SSRP-88/01. La Jolla: University of California, San Diego.

Modena, C. & Cecchinato, P. (1987). "Researches on the Interaction Mechanisms Between Steel Bars & Hollow Clay Unit Masonry." *4th North American Masonry Conference.* Los Angeles: University of California, pp. 16.

Pfeir, I.M. (1987). "Analytical Investigations of Masonry Walls Subjected to Axial Compressive Forces & Biaxial Bending Moments." *4th North American Masonry Conference.* Los Angeles: University of California, pp. 13.

Scrivener, J.C. (July, 1986). "Bond Reinforcement in Grouted Hollow-Unit Masonry: A State-of-the-Art." *U.S. - Japan Coordinated Program for Masonry Building Research.* Report No. 6.2-1. Boulder: Atkinson-Noland & Associates, Inc.

Soric, Z.; Tulin, L.G. (1987). "Comparison Between Predicted & Observed Responses for Bond Stress & Relative Displacement in Reinforced Concrete Masonry." *4th North American Masonry Conference.* Los Angeles: University of California, pp. 44.

Tawresey, J.G. (1987). "Walls with Axial Load Combined with Bending Moment — Interaction Equations for Masonry." *4th North American Masonry Conference.* Los Angeles: University of California, pp. 34.

Section 6   Strength Design

Agbabian, M.S.; Adham, S.A.; Masri, S.F.; Avanessian, V.; Traina, I. (July 1989). "Out-of-Plane Dynamic Testing of Concrete Masonry Walls." Volume 1: Final Report; Volume 2: Test Results. *U.S.-Japan Coordinated Program for Masonry Building Research,* Report No. 3.2(b1). Los Angeles: The University of Southern California.

Asher, J. & Selna, L. (1990). "Multistory Slender Wall Design." *5th North American Masonry Conference,* Vol. 3. Urbana-Champaign: University of Illinois, pp. 915.

Atkinson, R.H. (June, 1991). "An Assessment of Current Material Test Standards for Masonry Limit States Design Methods." *U.S. - Japan Coordinated Program for Masonry Building Research,* Report No. 1.3-1. Boulder: Atkinson-Noland & Associates, Inc.

Essawy, A.S.; Drysdale, R.G. (1987). "Evaluation of Available Design Methods for Masonry Walls Subject to Out-of-Plane Loading." *4th North American Masonry Conference.* Los Angeles: University of California, pp. 32.

Hart, G.C.; Bashartchah, M.A.; Zorapapel, G.T. (1987). "Limit State Design Criteria for Minimum Flexural Steel." *4th North American Masonry Conference.* Los Angeles: University of California, pp. 31.

Hart, G.C.; Noland, J.L.; Kingsley, G.R.; Englekirk, R.E. (1987). "Confinement Steel in Reinforced Block Masonry Walls." *4th North American Masonry Conference.* Los Angeles: University of California, pp. 52.

Hart, G.C. (1987). "Technology Transfer, Limit State Design & the Critical Need for a New Direction in Masonry Code." *4th North American Masonry Conference.* Los Angeles: University of California, pp. 41.

Heeringa, R.; McLean, D. (1990). "Ultimate Strength Behavior of Reinforced Concrete Block Walls." *5th North American Masonry Conference,* Vol. 3. Urbana-Champaign: University of Illinois, pp. 1041.

Masonry Society, The. (March, 1991). *Limit States Design of Masonry.* The Masonry Society.

Matsumura, A. (1987). "Shear Strength of Reinforced Hollow Unit Masonry Walls." *4th North American Masonry Conference.* Los Angeles: University of California, pp. 6.

Nakaki, D.K.; Hart, G.C. (1987). "A Proposed Seismic Design Approach for Masonry Shear Walls Incorporating Foundation Uplift." *4th North American Masonry Conference.* Los Angeles: University of California, pp. 25.

Porter, M.L.; Wolde-Tinsae, A.M.; Ahmed, M.H. (1987). "Strength Design Method for Brick Composite Walls." *4th North American Masonry Conference.* Los Angeles: University of California, pp. 37.

Sveinsson, B.I.; Blondet, M.; Mayes, R.L. (December 1988). "The Transverse Response of Clay Masonry Walls Subjected to Strong Motion Earthquakes." *U.S. - Japan Coordinated Program for Masonry Building Research,* Report No. 3.2 (b2)-10. Berkeley: Computech Engineering Services, Inc.

Sveinsson, B.I.; Kelley, T.E.; Mayes, R.L.; Jones, L.R. (1987). "Out-of-Plane Response of Masonry Walls to Seismic Loads." *4th North American Masonry Conference.* Los Angeles: University of California, pp. 46.

Section 7   Steel in Masonry

Snell, L.M.; Rutledge, R.B. (1987). "Methodology for Accurately Determining the Location of Reinforcement within Masonry." *4th North American Masonry Conference.* Los Angeles: University of California, pp. 11.

Section 9   Special Topics

Schaffler, M.; Chin, I.; Slaton, D. (1990). "Moisture Expansion of Fired Bricks." *5th North American Masonry Conference,* Vol. 2. Urbana-Champaign: University of Illinois, pp. 549.

INDEX

Z